Proceedings of the Tenth International Conference on Raman Spectroscopy

31 August–5 September 1986

Eugene, Oregon

U.S.A.

Raman 86
EUGENE

Editors
W. L. Peticolas and B. Hudson

University Printing Department
University of Oregon
Eugene, Oregon

Proceedings of the Tenth International Conference on Raman Spectroscopy

ISBN 0-87114-176-0

Copies may be obtained for $45.00 by contacting:

Secretary, ICORS–X
Chemical Physics Institute
University of Oregon
Eugene OR 97403

Published by the University Printing Department, University of Oregon

Designed by Julie Schaum, University Publications

ORGANIZING COMMITTEE

Durig, J. R., University of South Carolina
Harvey, A., National Science Foundation
Hudson, B., University of Oregon (Secretary)
Lascombe, J., University of Bordeaux
Loehr, T., Oregon Graduate Center
Nibler, J., Oregon State University (Treasurer)
Peticolas, W. L., University of Oregon (Chair)
Rabolt, J., IBM Research Laboratory
Tsuboi, M., University of Tokyo

INTERNATIONAL STEERING COMMITTEE

CHAIR: Lascombe, J. (France)
VICE-CHAIR: Tsuboi, M.(Japan)

MEMBERS:
Burstein, E. (U.S.A.)
Cardona, M. (F.R.G.)
Carey, P. R. (Canada)
Colombo, L. (Yugoslavia)
Delhaye, M. (France)
Durig, J. R. (U.S.A.)
Hester, R. E. (U.K.)
Kiefer, W. (Austria)
Koningstein, A. (Canada)
Long, D. A. (U.K.)
Lutz, M. (France)
Peticolas, W.L. (U.S.A.)
Schmid, E. D. (F.R.G.)
Schrotter, H. W. (F.R.G.)
Spiro, T. G. (U.S.A.)
Tasumi, M. (Japan)
Zerbi, G. (Italy)

PROGRAM ADVISORY COMMITTEE

CHAIR: Peticolas, W. L.
VICE-CHAIR: Griffiths, J. E.

MEMBERS:
Adar, F.
Albrecht, A. C.
Atkinson, G.
Bloembergen, N.
Burstein, E.
Carreira, L. A.
Chang, R.
Cummins, H. Z.
Durig, J.
El Sayed, M. A.
Fateley, W.
Harvey, A.
Hochstrasser, R. M.
Hudson, B.
Krimm, S.
Laane, J.
Levin, I.
Loehr, T.
Mathies, R.
Murray, C.
Nibler, J.
Owyoung, A.
Page, J. B.
Perry, C. H.
Rabolt, J.
Ramdas, A. K.
Richmond, G.
Rosasco, G.
Rousseau, D.
Spiro, T.
Thomas, G. J., Jr
Walrafen, G. E.
Weber, A.
Worlock, J.

FUNDED BY:
Allied Corporation
Bomem, Inc.
Coherent, Inc.
IBM Corporation
Instruments S.A., Inc.
National Institutes of Health
National Science Foundation
NSF USA-India Science Program
Newport Corporation
Office of Naval Research
The Procter and Gamble Company
Shell Oil Company
Spex Industries, Inc.
Standard Oil Company (Ohio)

SPONSORED BY:
Chemical Physics Institute,
University of Oregon
The Coblentz Society
International Union of Pure and
Applied Chemistry
International Union of Pure and
Applied Physics
Oregon Scientific and Scholarly
Conferences, Inc.

TABLE OF CONTENTS

PLENARY LECTURES

"A Quarter Century of Stimulated Raman Effect"
Bloembergen, N. — P-1

"Generation of Coherent and Tunable Radiation by Scattering Processes"
Stoicheff, B.P. — P-2

"Light-Induced Spectral Redistribution of Spontaneous Emission"
Melinger, J.S.; Albrecht, A.C. — P-3

CHAPTER 1 - PROTEINS & PROTEIN COMPONENTS

"Protein Structure and Dynamics Via Ultraviolet Resonance Raman Spectroscopy"
Spiro, T.G. — 1-1

"UV Resonance Raman Enhancement: From Small Molecules to Proteins"
Asher, S.A. — 1-2

"Primary Donor Bacteriochlorophylls in Photosynthetic Reaction Centers: Ground-State Interactions from Resonance Raman Spectroscopy"
Lutz, M.; Robert, B.; Zhou, Q. — 1-4

Heme Proteins

"Resonance Raman Spectra of Catalytic Intermediates of Heme Enzymes"
Kitagawa, T.; Hashimoto, S.; Ogura, T. — 1-6

"Spectral Broadening Mechanisms in Biomolecules: Cytochrome C vs Myoglobin"
Champion, P.M.; Schomacker, K.T.; Srajer, V. — 1-8

"Time-Resolved Resonance Raman Spectroscopy of Peroxidase and Catalase Intermediates"
Terner, J.; Sitter, A.J.; Reczek, C.M. — 1-9

"Resonance Raman Scattering Studies of Photodissociated MbCO"
Rousseau, D.L.; Sassaroli, M.; Dasgupta, S. — 1-11

"The Cobalt-Nitrosyl Stretching Vibration as a Sensitive Resonance Raman Probe for Distal Histidine-Nitrosyl Interaction in Monomeric Hemoglobins"
Yu, N-T.; Thompson, H.M.; Mizukami, H.; Gersonde, K. 1-13

"Axial Coordination in Nickel Porphyrins and Nickel-Reconstituted Heme Proteins Investigated by Raman-Difference and Transient-Raman Spectroscopy"
Shellnutt, J.A.; Alston, K.; Findsen, E.W.; Ondrias, M.R.; Rifkind, J.M. 1-15

"Correlations Among ν(Fe-CO), ν(C-O), and ν(Fe-C-O) Vibrations in the CO Complexes of Various Heme Proteins"
Uno, T.; Nishimura, Y.; Tsuboi, M.; Kita, K.; Anraku, Y.; Makino, R.; Iizuka, T.; Ishimura, Y. 1-17

"Correlation of ν(Fe-CO) with Ligand Dissociation Rates for Monomeric *Glycera* Hemoglobins"
Armstrong, R.S.; Cooke, R.M.; Reedy, B.J.; Walsh, M.J.; Wright, P.E. 1-19

"Resonance Raman Study on Oxygenation and Deoxygenation of Hemoglobins Hb-A and Hb-F"
Li, H-L; Chen, S-C. 1-21

"Resonance Raman Detection of Bound Dioxygen in Cytochrome $P450_{cam}$"
Champion, P.M.; Bangcharoenpaurpong, O.; Rizos, A.K.; Jollie, D.; Sligar, S.G. 1-23

"Resonance Raman Detection of Manganese-Nitrogen Triple Bond. Formation of Nitridomanganese(V) Protoporphyrin IX in Manganese Myoglobin"
Tsubaki, M.; Hiroshi, H.; Hotta, T.; Hiwatashi, A.; Ichikawa, Y.; Yu, N-T. 1-24

"Resonance Raman Spectroscopy of Nitrosyl-Manganese Hemoglobin and Myoglobin"
Parthasarathi, N.; Spiro, T.G. 1-26

"Resonance Raman Spectra of Oxo-Ferryl Porphyrin Model Complexes"
Hashimoto, S.; Tatsuno,Y.; Kitagawa, T. 1-28

"Resonance Raman Spectroscopic Characterization of Dihydroporphyrins, the Prosthetic Groups of Hemes D and Chlorophylls"
Loehr, T.M.; Andersson, L.A.; Chang, C.K.; Smith, K.M. 1-30

"Anomalous Resonant Enhancement of the Quasi-Elastic Light Scattering of Cytochrome-c"
Reinisch, L.; Schomacker, K.T.; Champion, P.M. 1-32

"The Effects of Dehydration upon the Local Heme Environment in Equilibrium and Photolyzed Carbonmonoxy Hemoglobin"
Findsen, E.W.; Simons, P.; Ondrias, M.R. 1-34

"Resonance Raman Spectra: Role of Disorder Mode In Tunneling of Cytochrome c"
Bhattacharyya, B.D. 1-36

Other Chromophoric Proteins

"Determination of Retinal Chromophore Structure in Rhodopsins with Resonance Raman Spectroscopy"
Mathies, R.A.; Smith, S.O.; Palings, I.; Lugtenburg, J. 1-38

"Kinetic Studies on Bacteriorhodopsin by RR Spectroscopy"
Diller, R.; Stockburger, M. 1-40

"Picosecond Conformational Intermediates in the Bacteriorhodopsin Photocycle"
Brack, T.L.; Blanchard, D.; Siemankowski, L.; Atkinson, G.H. 1-42

"Raman Studies on Pigments Relating to Photosynthesis"
Tasumi, M.; Hayashi, H.; Fujiwara, M.; Okamoto, H.; Iwata, K.; Noguchi, T. 1-44

"Resonance Raman Studies on Lipoxygenase-Inhibitor Complexes"
Gustafson, W.G.; McFarland, J.T. 1-46

"Determination of Flavin Oxidation State by Resonance Raman Spectroscopy"
Lively, C.R.; Gustafson, W.G.; McFarland, J.T. 1-47

"Raman and Resonance Raman Studies of Some Dehydrogenases"
Hall, W.H.; Price, R.A.; Jagodzinski, P.W. 1-48

"Resonance Raman Spectroscopy of Iron-Sulfur Proteins"
Czernuszewicz, R.S.; Han, S.; Macor, K.A.; Johnson, M.K.; Adams, M.W.; Spiro, T.G. 1-50

"Ultra-violet (239 nm) Resonance Raman Spectroscopy of An Enzyme-Substrate Intermediate of Papain"
Peticolas, W.L.; Bajdor, K.; Wharton, C.W.; Hester, R.E. 1-52

"Using the Furylacrylolyl Chromophore and UV Resonance Raman and Classical Raman Difference Spectroscopy to Probe Active Site Environments in Proteins"
Wilson, J.; Peticolas, W.L. 1-54

Non-resonance Studies of Proteins

"The Influence of pH and Ca^{2+} Concentration on the Tyrosine Doublet Intensity Ratio in the Bovine α-Lactalbumin Raman Spectrum"
Lafaut, J.P.; Van Dael, H. 1-56

"Amide I Region of Resolution-Enhanced Raman Spectra of Proteins"
Byler, D.M.; Susi, H. 1-58

"Protein Conformation by Laser Raman Spectroscopy: Quantitative Determination of the Secondary Structure from the Amide I and III Bands by the Reference Intensity Profiles Method"
Alix, A.J.P.; Berjot, M.; ElJoundi, A.; Marx, J. 1-60

"Polarized Raman Spectra of Proteins Using the Four Channel Divided Cell Technique"
Bussian, B.M.; Eysel, H.H. 1-62

"Changes in the Lysozyme Raman Spectra After Iodination: Relationship with Enzymatic Activity"
Jacquot, J.; Marx, J.; Lambert, B.; Berjot, M.; Alix, A.J.P. 1-64

"The Effect of Water on the Secondary Structure of the Protein Bovine Serum Albumin"
Bolton, B.A.; Scherer, J.R. 1-66

Protein Components

"Raman Spectroscopic Studies of Structure in Polypeptides"
Krimm, S. 1-68

"Hydrogen Bonding in Amides and Proteins. A Low Frequency Study (10-400 cm^{-1})"
Nielsen, O.F.; Bigio, I.; Berquier, J.-M.; Olsen, I. 1-70

"Vibrations of Tryptophan and Tyrosine Side Chains in Proteins"
Harada, I.; Miura, T.; Watanabe, N.; Takeuchi, H. 1-72

"Experimental Investigation of the Lattice Modes of α-Glycine"
Ramnarine, R.; Sherman, W. 1-74

"Raman Spectra and Conformational Analysis of N-Acyl-Homooligoglycines"
Taga, K.; Okuoka, T.; Debnath, R.; Okabayashi, H. 1-76

"The Use of Raman Spectroscopy to Solve the Conformational Features of Small Peptides"
Lagant, P.; Vergoten, G.; Thomas, G.A.; Peticolas, W.L. 1-78

"Raman Spectroscopic Studies of Solid Bovine Somatotropin and Related Analogs: Estimation of Secondary Structure, Tyrosine Environment, and Disulfide Conformation"
Thamann, T.J. 1-80

"UV Resonance Raman Studies of Aromatic Amino Acids and Proteins"
Ludwig, M.; Asher, S.A. 1-82

"Selective Enhancement of Proline Raman Signals with Ultraviolet Excitation"
Mayne, L.; Hudson, B. 1-84

"Ultraviolet Resonance Raman Study of Arginine"
Hildebrandt, P.; Spiro, T.G. 1-86

"Laser Raman Spectra of β-Alanine Silver Nitrate and β-Alanine Copper Nitrate"
Bhoopathy, T.J.; Mohan, S.; Payami, F. 1-87

"Raman Spectroscopy Study of pH and Pressure Induced Conformational Changes in Chymotrypsin"
Heremans, K.; Heremans, L. 1-89

CHAPTER 2 - NUCLEIC ACIDS AND VIRUSES

"Nucleic Acid Structures in Viruses from Correlations Established by Raman Spectroscopy of DNA and RNA Crystals"
Thomas, G.J., Jr. 2-1

"A Raman Spectroscopic Study on Sequence Dependent Modulation of DNA Conformations"
Nishimura, Y. 2-3

"DNA Structures and DNA-Protein Interactions in UV Resonance Raman Spectroscopy"
Turpin, P.Y.; Chinsky, L.; Laigle, A.; Otto, K. 2-5

"Raman Spectra of the Model B-DNA Oligomer, d(CGCGAATTCGFCG)$_2$ and of the DNA in Living Salmon Sperm Show That Both Have Very Similar B-Type Conformations"
Kubasek, W.L.; Wang, Y.; Thomas, G.A.; Patapoff, T.W.; Peticolas, W.L. 2-7

"Ethanol-Induced Conformational Changes in PolydA.PolydT, Polyd(A-T), PolydG.PolydC and Polyd(G-C) Monitored by Raman Spectroscopy and Interpreted by Base Stacking Energies"
Wang, Y.; Peticolas, W.L. 2-9

"Ultra-Violet Resonance Raman Spectroscopy of Complexes of cis- and trans-Dichlorodiammineplatinum with Guanosine as a Function of pH"
Bajdor, K.; Wang, Y.; Peticolas, W.L. 2-11

"Laser Raman Spectroscopy of Complexes of Nucleotides with Acetamide"
Molina, M.; Carmona, P. 2-13

"Raman Spectroscopic Study of the Binding of Copper(II) to Nucleotides"
Carmona, P.; Huong, P.V.; Gredilla, E. 2-15

"Softening of the Lowest Optic Modes of DNA"
Lee, S.A.; Lindsay, S.M.; Powell, J.W.; Tao, N.J.; DeMarco, C.; Rupprecht, A. 2-17

"Tautomeric Equilibria and Vibrational Spectroscopy of Matrix-Isolated Guanines"
Szczesniak, M.; Szczepaniak, K.; Hernandez, L.; Person, W.B. 2-19

"Resonance Raman Studies of Interaction of Adriamycin with DNA and Membranes"
Dutta, P.K.; Hutt, J.; Lee, B.S. 2-20

"Normal Coordinate Analysis of the Nucleic Acid Vibrational Modes"
Letellier, R.; Ghomi, M.; Taillandier, E. 2-21

CHAPTER 3 - LIPIDS/BILAYERS AND RELATED SYSTEMS

"Raman Spectroscopic Study of Boundary Lipid in Apolipoprotein A-1/Dimyristoylphosphatidylcholine Recombinants"
Bush, S.F.; Levin, I.W. 3-1

"Interactions of Ferricytochrome *c* with Model Phospholipid Bilayers"
Vincent, J.S.; Levin, I.W. 3-3

"Raman Spectra of Lipid Membranes and Head Group Hydration"
Scherer, J.R.; Kint, S.; Bolton, B.A. 3-5

"Raman and Calorimetric Studies of Phase Transitions in Several Lipid Bilayers"
Rich, N.H.; Stevenson, C.; Keough, K.M.W. 3-7

"Vibrational Spectroscopic Studies of the Structure of Crystalline and Membrane Bound Gramicidin A"
Naik, V.M.; Krimm, S. 3-9

"Melittin-Phosphatidylcholine Complexes: A Model to Study Lipid-Protein Interactions"
Lafleur, M.; Dasseux, J.-L. ; Pezolet, M. 3-11

"Polarized UV Resonance Raman and Absorption Spectra of Monolayers Adsorbed at the Aqueous Solution-Carbon Tetrachloride Interface"
Takenaka, T.; Umemura, J.; Isono, N. 3-13

"Raman Spectroscopic Estimation of Solubilization Equilibria in Aqueous Polymer/Surfactant Solutions"
Shih, L.B.; Luccas, M.H.; Warner, R.J. 3-15

"Dynamics of Alkylammonium Chains in $C_nH_{2n+1}NH_3Cl$ Compounds (n = 8,10). A Spectroscopic Study"
Schenk, K.J.; Rey-Lafon, M. 3-17

"Raman Spectroscopic Studies of Chain Interactions and Mobility in Liquid Crystalline Phases of Cationic Surfactants"
Pacynko,W.F.; Tiddy,G.J.T.; Yarwood, J. 3-19

"Raman Band Shape Analysis and Molecular Inertial Motion in Anisotropic Aligned Fluids"
Kirov, N.; Dozov, I.; Fontana, M.; Petroff, B. 3-21

CHAPTER 4 - OTHER BIOLOGICAL AND BIOMEDICAL APPLICATIONS

"Raman Investigations of Interactions Between Glutathione and Antimicrotubular Agents"
Turrell, S.; Bouchaour, T.; Lesieur,I. 4-1

"Metallo-Proteins and Metallo-Nucleotides"
Huong, P.V.; Carmona, P.; Lorosch, J. 4-3

"Rapid Headspace Analysis in Sealed Drug Vials By Multichannel Raman Spectroscopy"
Powell, L.P.; Campion, A. 4-5

"Laser Raman Spectroscopic Studies of Galactose Cataract in Rats"
Cai, M.-Z.; Chen, Y.-L. 4-6

CHAPTER 5 - SURFACES AND SURFACE ENHANCED RAMAN SCATTERING

"Does SERS Work Without Electromagnetic Field Enhancement?"
Pettenkofer, C.; Otto, A. 5-1

"Theory of Surface Enhanced Raman Scattering"
Zeyher, R. 5-3

"Raman Spectroscopy of Molecles Adsorbed on Solid Surfaces"
Campion, A. 5-5

"Developments in the Use of Raman Scattering to Investigate the Excitation of Adsorbate-Substrate Complexes"
Brotman, A.D. 5-7

"Surface Enhanced Spectroscopy - Calculation of Local Fields"
Gersten, J.I.; Nitzan, A. 5-9

"Polarization Resolved Resonance Scattering by Crystal Violet Adsorbed on Smooth Metal Surfaces: Evidence for a Charge-Transfer Excitation"
Jiang, J.D.; Burstein, E.; Kobayashi, H. 5-10

"Surface Enhanced Resonance Raman Scattering (SERS) from Tri-(p-dimethylamino)-phenyl Carbonium Ion (Crystal Violet)"
Liang, N.T.; Tse, W.S.; Chou, Y.C. 5-12

"The Role of the Cavity Model in SERS"
Seki, H. 5-14

"Coverage Dependence of Surface Enhanced Resonance Raman Scattering and Surface Fluorescence"
Gerolymatou, A.; Pettinger, B. 5-16

"Quantum-Mechanical Model for Surface Enhanced Raman Scattering: An Additional Energy Transfer Induced Gain"
Pettinger, B. 5-17

"Electromagnetic Short Range Enhancement of Surface-Enhanced Raman Scattering"
Lopez-Rios, T.; Gao, Y. 5-18

"Surface-Enhanced Raman Spectroscopy of Guanine Derivatives"
Koglin, E.; Sequaris, J.-M. 5-20

"Surface Enhanced Raman Scattering of Macromolecules"
Garrell, R.L.; Beer, K.D.; Tadayyoni, M.-A.; Ahern, A.M. 5-22

"Adsorption of Aniline in Acidic Solutions Studied by SERS"
Shindo, H.; Nishihara, C. 5-24

"Surface Enhanced Resonance Raman Spectra of Transition Metal α-Diimine Complexes"
Dines, T.J.; Peacock, R.D. 5-26

"Spin and Redox Equilibria of Cytochrome c Adsorbed at the Ag/Electrolyte Interface Probed by Surface Enhanced Resonance Raman Spectroscopy (SERS)"
Hildebrandt, P.; Stockburger, M. 5-28

"SERS as a Probe of Surface Geometry Effects on the Polymerization of Acrylic Acid on Silver"
Suh, J.S.; Michaelian, K.H. 5-30

"Charge-Transfer Band and SERS Mechanism for Pyridine-Ag System"
Yamada, H.; Nagata, H.; Toba, K. 5-32

"Surface Characterization of Ion-Implanted Polyethylene Films by the Use of SERS Effect"
Ishida, H.; Katagiri, G.; Fukuda, H.; Tomita, S.; Yoshida, K.; Ishitani, A. 5-34

"Structures of Adsorbed 2,2'-Bipyridine and Pyridine on Ag Colloids Studied by SERS Spectroscopy"
Kim, M.; Itoh, K. 5-36

"Surface Enhanced Resonance Raman of Haem Proteins on Colloid Silver"
de Groot, J.; Hester, R.E. 5-38

"Coordination Chemistry of Thiosemicarbizide on Copper and Silver Surfaces - A SERS Study"
Leahey, J.; Lee, Y.G.; Loo, B.H. 5-39

"Surface Enhanced Raman Scattering (SERS) of Benzoic Acid Adsorbed on Colloidal Silver Particles"
Pagannone, M.; Fornari, B.; Mattei, G. 5-41

"Evidence for a Photon-Driven Charge-Transfer Mechanism in the Surface Enhanced Raman Scattering from DABCO at a Silver Electrode"
Guzonas, D.; Irish, D.E.; Atkinson, G.F. 5-43

"In-Situ Identification During the Electrochemical Polarization of Ni and NiMo Alloy in Sulfuric Solutions"
Delichere, P.; Hugot-Le Goff, A.; Yu, N. 5-45

"Raman Spectroscopic Studies of $Cu(HCO_2)_2$ Surface Complexes"
Jeziorowski, H.; Moser, B. 5-46

"Surface-Enhanced Raman Spectra from Thin Films of CTAB on Nonmetallic Surfaces"
Parry, D.B.; Dendramis, A.L. 5-48

"Characterization of the Silver Aggregates Responsible for the Enhancement of Raman Spectra on Silver Halide Crystals and Coated Electrodes"
Guilment, J.; Regis, A.; Corset, J.; Riveccie, M. 5-50

"Voltage-Induced Energy Level Shifting of Adsorbed Molecles in SERS"
Gu, R.-A.; Leng, Y.-Z.; Zhu, H.-R.; Ling, D.-H. 5-52

"Temperature and Electrode Potential Dependency of the Co-Adsorption of Pyridine and Methylpridine on a Silver Electrode"
Saito, H.; Ikezawa, Y.; Toda, G. 5-54

"Simple, Reproducible Chemical Procedure for Preparing SERS-Active Silver Films on Glass Substrates"
Cotton, T.M.; Ni, F.; Feng, Q. 5-56

"Enhanced Raman Scattering from Thin Chromia Film Substrates"
Graves, P.R. 5-58

"The Effects of Electrochemical Processes on the Surface Structural Symmetry of Single Crystal Electrodes as Studied by Optical Second Harmonic Generation"
Shannon, V.L.; Richmond, G.L. 5-60

"Interfacial Studies of Silver-Aqueous Electrolytes by Optical Second Harmonic and Differential Capacitance"
Rojhantalab, H.M.; Richmond, G.L. 5-61

"Optical Studies of Electrochemical Interfaces: A Comparative Analysis by Second Harmonic Generation and Photoacoustic Methods"
Richmond, G.L.; Chu, P. 5-64

"Static and Time-Resolved Nonlinear Optical Studies of Semiconductor Interfacial Properties"
Robinson, J.M.; Richmond, G.L. 5-66

"Tentative Investigation on Mechanism of SERS - Experimental Study of Electrode Surface"
Zhu, H.; Leng, Y.; Gu, R.; Ling, D. 5-68

CHAPTER 6 - THIN FILMS

"Raman Spectroscopy of Surfaces and Layered Structures With Optical Waveguides and Surface Plasmons"
Swalen, J.D. 6-1

"Raman Spectrometry of Interfaces and Ultra-Thin Films"
Dupeyrat, R.; Harrand, M.; Masson, M. 6-3

"Unenhanced Raman Spectroscopy of Thin Organic Films on Surfaces"
Schlotter, N.E. 6-5

"Molecular Orientation in Thin Films by Raman Spectroscopy Using Integrated Optical Techniques"
Rabolt, J.F. 6-7

"Characterization of Order in Langmuir-Blodgett Mono- and Multilayers by Multichannel Raman Spectroscopy Using a CCD Detector"
Dierker, S.B. 6-8

"Davydov Splitting in the Raman Spectra of Thin Solid Films"
Aroca, R.; Jennings, C.; Hor, A.-M.; Loutfy, R.O. 6-10

"High-Temperature In-Situ Raman Spectroscopy of Surface Films in Aqueous Environments"
Melendres, C.A.; McMahon, J.J.; Ruther, W. 6-12

"Study of Thin Titanium Oxide Films by Interference-Enhanced Raman Spectroscopy"
Hugot-Le Goff, A. 6-14

"Electrochrome Behavior of WO_3 Films Studied by Raman Spectroscopy"
Delichere, P.; Falaras, P.; Hugot-Le Goff, A. 6-15

"In-Situ Raman Spectroscopy for the Study of Iron Passivity"
Hugot-Le Goff, A.; Pallotta, C. 6-16

CHAPTER 7 - SEMICONDUCTORS

"Raman Scattering by Magnetic Excitations in Diluted Magnetic Semi-Conductors"
Rodriguez, S.; Ramdas, A.K. 7-1

"Raman Scattering Studies of Optically Excited GaAs"
Nather, H. 7-3

"Sub-Picosecond Raman Studies of Carrier Relaxation in GaAs"
Tsang, J.C.; Kash, J.A. 7-5

"Time Resolved CARS Measurements of Coherent Excitations of Semiconductors"
Bron, W.E. 7-7

"Study of Dynamics of Photoexcited Electron-Hole Plasmas in GaAs by Picosecond Time-Resolved Raman Scattering"
Huang, Y.; Yu, P.Y. 7-9

"Theory of Time-Resolved Raman Spectroscopy of Non-Equilibrium Carriers and Phonons in Semiconductors"
Jha, S.S. 7-11

"Green's Function Theory for the Disorder Induced Raman Spectra in CdHgTe System"
Talwar, D.N.; Vandevyver, M. 7-13

"High Pressure Raman Scattering Study of Phase Transition in $Cd_{1-x}Mn_xTe$"
Arora, A.K.; Bartholomew, D.U.; Peterson, D.L.; Ramdas, A.K. 7-15

"Interference Between Allowed and Forbidden Scattering in Zincblende-Type Semiconductors"
Cardona, M.; Kauschke, W.; Mestres, N. 7-17

"Laser-Induced Crystallization and Medium Range Order in Amorphous $GeSe_2$"
Murase, K.; Inoue, K.; Kawamoto, K. 7-19

"Thin Film Reactions at the Aluminum-Gallium Antimonide Interface"
Schwartz, G.P.; Susnow, R.G.; Gualtieri, G.J.; Sunder, W.A. 7-21

"Raman Study of the Phonon Spectra in $Hg_{0.83}Cd_{0.17}Te$ from 10K to 250K"
Bansal, M.L.; Roy, A.P. 7-23

CHAPTER 8 - CONFINED RESONANCES

"Raman Scattering in Optical Fibers"
Stolen, R.H. 8-1

"Raman - Mie - Scattering from Levitated Dielectric Microspheres"
Kiefer, W.; Thurn, R. 8-3

"Enhanced Raman Scattering by Dielectric Spheres"
Inoue, M. 8-5

"Stimulated Raman Scattering Spectra from Single Water Droplets"
Eickmans, J.H.; Chang, R.K. 8-7

"Spectra of Stimulated Raman Effect for Anharmonic Coupling of the Polariton and Stokes Waves on a Slab"
Schulz, V.; Merten, L. 8-9

CHAPTER 9 - SUPERLATTICE STRUCTURES

"Raman Scattering From Phonons in Semiconductor Superlattices"
Klein, M.V. 9-1

"Resonance Raman Studies of Folded Acoustic Phonons in GaAs-AlAs Superlattices"
Gant, T.A.; Klein M.V.; Henderson, T.; Morkoc, H. 9-2

"Raman Investigation of Confined Optic Phonons and Their Annealing Effects in GaAs-AlAs Superlattices"
Zhang, S.-L.; Levi, D.H.; Gant, T.A.; Klein, M.V.; Klem, J.; Morkoc, H. 9-4

"Optical Phonons in GaAs-AlGaAs Superlattices"
Sood, A.K.; Cardona, M. 9-5

"Resonant Raman Spectroscopy of Excitons and Phonons in Quantum Wells"
Zucker, J.E. 9-7

"Raman Scattering from Amorphous Superlattices"
Ley, L.; Santos, P. 9-8

"Brillouin Scattering in Metallic Superlattices"
Grimsditch, M. 9-9

"Inelastic Light Scattering by Free Carriers in Semiconductor Superlattices"
Pinczuk, A. 9-11

"Coherent Anti-Stokes Raman Scattering in Narrow Gap Semiconductor Superlattices"
Pascher, H. 9-13

CHAPTER 10 - LAYERED STRUCTURES

"Resonant Raman Scattering at the Interband Absorption Edge in Graphite-H_2SO_4 Intercalation Compounds"
Eklund, P.C.; Mahan, G.D.; Spolar, J.G.; Zhang, J.M.; Arakawa, E.T.; Hoffman, D.M. 10-1

"Ion Size and Substrate Rigidity in Mixed Layered Solids"
Solin, S.A.; Lee, S.; Leonard, W.; Pinnavaia, T.J. 10-3

"Long-Wavelength Phonons in Metallic Layered Transition Metal Dichalcogenides"
Irwin, J.C.; McMullan, W.G. 10-5

"Intra- and Interlayer Raman Modes in Layered Silicates"
Wada, N. 10-7

"Raman Scattering by Amplitude Modes and Soft Modes in Ionized Transition-Metal Tetrachalcogenides"
Sekine, T. 10-9

"Structural Phase Transitions in Perovskite-Type Layer Compounds"
Couzi, M. 10-11

"Study of Charge-Density-Wave Phase Transitions in Layered Transition-Metal Dichalcogenides 4Hb-TaS_2 and 4Hb-$TaSe_2$ by Raman Scattering"
Nakashizu, T.; Sekine, T.; Uchinokura, K.; Matsuura, E. 10-13

CHAPTER 11 - SOLID STATE: INORGANIC SYSTEMS

"Raman Scattering From Localized Electronic Excitations in Metallic Rare Earth Compounds"
Zirngiebl, E.; Blumenroder, S.; Pofahl, G.; Guntherodt, G. 11-1

"Magnetic Ordering Induced Raman Scattering in MPS_3 (M = Fe, Ni) Layered Compounds"
Balkanski, M.; Jouanne, M.; Scagliotti, M.; Benedek, G. 11-3

"Soft-Mode Spectroscopy in Ferroelectrics by Hyper-Raman Scattering"
Vogt, H. 11-5

"Direct Measurement of Wave Vector Dependent Polariton Energy Velocity and Dephasing Time in NH_4Cl"
Vallee, F.; Gale, G.M.; Flytzanis, C. 11-7

"Raman Spectra of Crystalline Germanium and Silicon Tetrabromides"
Tse, W.S.; Yur, G. 11-9

"Raman Spectra of High-Temperature Phase Changes in Scandium Fluorides"
Hafsi, S.; Turrell, S.; Dhamlincourt, M.-C.; Barbier, P.; Champarnaud, J. 11-11

"Multichannel Raman Studies of Binary Alkaline Borides"
Yahia, Z.; Turrell, S.; Dhamlincourt, M.-C.; Turrell, G. 11-12

"The Intensities and High-Pressure Behaviour of Raman Bands in Rhombohedral AXF_6 Crystals (A = K,Cs,X = Sb,As)"
Heyns, A.M. 11-13

"Raman Scattering in $PbMo_6S_8$, $SnMo_6S_8$, and $Cu_{1.8}Mo_6S_8$"
Demers, R.T.; Holmgren, D.J.; Klein, M.V.; Ginsberg, D.M. 11-15

"Raman Scattering Investigation of Low-Temperature Phase Transitions of $LiTlC_4H_4O_6(H_2O)$ and $LiKSO_4$"
Haller, K.E. 11-17

"Raman Spectra of Solid Complexes Formed of Trihalides of Group V Elements with Trihalides of Group III Elements"
Shamir, J.; Rafaeloff, R. 11-19

"Raman Scattering Study of the Phase Transitions in $TlAlF_4$"
Bulou, A.; Rousseau, M.; Nouet, J. 11-21

"Comparison of Long Wavelength Optical Phonons in $ThSe_3$ and USe_3"
Nouvel, G.; Noel, H.; Zwick, A.; Renucci, M.A. 11-23

"Brillouin Scattering Studies of Superionic $Ba_{1-x} La_x F_{2+x}$ (x = 0.393)"
Ngoepe, P.E.; Comins, J.D. 11-25

"High Temperature Brillouin Scattering Studies in Superionic Conductors: Cubic Zirconia and Lanthanum Fluoride"
Ngoepe, P.E.; Comins, J.D. 11-27

"Hydrogen in $Fe_{90}Zr_{10}$ Magnetic Glass - A Raman Study"
Mathur, M.S.; Marzouk, H.A.; McKee; J.S.C. 11-29

"Two-Magnon and Phonon Raman Scattering in $Cd_{1-x}Mn_xF_2$ Mixed Crystals"
Ciepielewski, P.; Gregora, I.; Vorlicek, V.; Kalinski, Z. 11-30

"Coherent Antistokes Raman Scattering by Oblique Polaritons in the Lithium Iodate Crystal"
Petrosyan, K.B.; Pokhsraryan, K.M. 11-32

"Low Temperature Behaviour of the Lithium Modes in Monoclinic Li_2SO_4"
Cazzanelli, E.; Fontana, A.; Rocca, F. 11-34

"Raman Spectra and Directional Dispersion in $LiNbO_3$ and $LiTaO_3$"
Lan, G.; Wang, H.; Yang, X.; Li, B. 11-36

"Resonance Raman Scattering in $CdGa_2(S_xSe_{1-x})_4$"
Lottici, P.P.; Parisini, A.; Razzetti, C. 11-38

"Dynamic Processes in the Pyroelectric Single Crystals of Uranyl Nitrate Hexahydrate Through Raman Spectroscopy"
Bist, H.D.; Little, T.S.; Durig, J.R. 11-40

"A Variable Temperature Raman Study of the Vibrational Mode in NH_4I"
Xu, Y.; Sherman, W.F.; Wilkinson, G.R. 11-42

"Theory of Raman Scattering in Alkali Halides"
Mahan, G.D.; Subbaswamy, K.R. 11-44

"The 2-Dimensional Raman Map of the Stress Induced Monoclinic Phase in PSZ"
Mizuno, J.; Kudo, S.; Hasegawa, H. 11-46

"Raman Microprobe Study of Phase Transformations in Ceria Stabilized Zirconia"
Begun, G.M.; Becher,P.F.; Funkenbusch, E.F. 11-48

"Evaluation of Transformation Zone Size of Y-TZP by Raman Microprobe"
Katagiri, G.; Teramura, M.; Ishida, H.; Masaki, T.; Ishitani, A. 11-50

"Distribution and Characterisation of High Temperature Air Corrosion Products on Iron-Chromium Alloys by Raman Microscopy"
Gardiner, D.J.; Littleton, C.J.; Thomas, K.M.; Strafford, K.N. 11-52

"Conformational Order in Lead(II) Carboxylates From Skeletal Optical and Longitudinal Acoustic Raman Modes"
Amorim da Costa, A.M. 11-53

"Photoelastic Constants of Aragonite From Raman Intensity Analysis of Calcite"
Rao, N.R.; Wahegoankar, L.D. 11-55

"Raman Spectrum of NaCN"
Sharma, T.P.; Sharma, R.M. 11-57

"Raman Spectrum of KCN"
Sharma, R.M.; Sharma, T.P. 11-59

"Raman and IR Studies and Phase Transitions in Substituted Ammonium Nitrates and Perchlorates"
Mylrajan, M.; Srinivasan, T.K.K. 11-61

"Bromate Ion Dynamics in Orthorhombic, Monoclinic and Cubic Metal Bromates"
Srinivasan, T.K.K.; Devanathan, T. 11-62

"Low Frequency Modes in Incommensurate Phase of Quartz"
Abe, K.; Shionoya, T.; Shigenari, T. 11-63

"Raman Microprobe (RMP) Study of Diamond Anvil Cell Synthetized Germanate Phases"
Boyer, H.; Guyot, F. 11-65

"Raman Microprobe Spectra of Naturally Shocked Quartz"
Boyer, H.; Velde, B. 11-66

"Stokes and Anti-Stokes Fluorescence of Er^{3+} in the Raman Spectra of Erbium Oxide and Erbium Glasses"
Etz, E.S.; Travis, J.C. 11-67

"Raman Spectra of KDP Melts and $(KPO_3)_n$ Glass"
Kim, J.-J.; Choi, B.-K.; Koo, J.-Y.; Lee, K.-S. 11-69

"High Resolution Raman Scattering in Glass Forming Liquids"
Samios, D.; Dorfmuller, T. 11-71

"Oxygen-Oxygen Vibrations in Vitreous Silica"
Galeener, F.L. 11-73

"Raman Spectroscopy for Industrial Glass Materials and Optical Fibers"
Fujiki, F.; Higuchi, S.; Tanaka, S. 11-75

"OH Stretching Band Shape Analysis and Proton Dynamics in OH...O Hydrogen Bonded Crystals at Low Temperature"
Soulard, L.; Fillaux, F. 11-77

"Extended Defect Structures in Bulk As_2S_3 Glass"
Malyj, M.; Griffiths, J.E. 11-79

"Molecular Origin of Glass Forming Tendency in Chalcogenides"
Boolchand, P.; Enzweiler, R.N.; Lemon, G.H.; Ruffolo, D.; Hernandez, J.G. 11-81

"Raman Study of Low Temperature Cured Metaphosphate Coacervate-Glass"
Palavit, G.; Vast, P.; Bera, A.M.; Vergoten, G.; Lagant, P. 11-83

"Raman Scattering from a New Class of Disordered Solids"
Farrow, L.A.; Sandroff, C.J. 11-85

"About the Sharp Increasing of Low Frequency Raman Scattering Spectral Intensity Near the Phase Transition Points in Crystals"
Gorelik, V.S. 11-87

"Raman Scattering of Light in Amorphous and Crystalline Films of Silicon and Carbon"
Gorelik, V.S.; Sagitov, S.I.; Phaizullov, T.P. 11-88

CHAPTER 12 - SOLID STATE: ORGANIC CRYSTALS AND POLYMERS

"Time-Resolved Studies of Phonon Dynamics in Molecular Crystals"
Hill, J.R.; Chronister, E.L., Postlewaite, J.C.; Dlott, D.D. 12-1

"Raman Spectroscopic Investigations on Molecular Crystals: Pressure and Temperature Dependence of External Phonons in Naphthalene-D_8 and Anthracene-D_{10}"
Hafner, W.; Kiefer, W. 12-3

"Polarized Raman and Infrared Study of 5-Aminotetrazole Monohydrate"
Bigotto, A. 12-5

"Raman Spectra of N-H Stretching Modes in Urea Crystal"
Khilji, M.Y.; Sherman, W.F.; Wilkinson, G.R. 12-7

"Raman Study of the Lattice Vibrations of Methanol"
Anderson, A.; Andrews, B.; Meiering, E.; Torrie, B.H. 12-9

"Raman and Far Infrared Spectra of Crystalline Methylene Iodide"
Torrie, B.H.; Anderson, A.; Andrews, B.; Laurin, D.G.; White, J.K.; Zung, W.W.E. 12-11

"Methyl Dynamics and Molecular Mechanism Determining Phase Transition in 4-Methyl-Pyridine"
Le Calve, N.; Pasquier, B.; Soulard, L.; Fillaux, F. 12-13

"Raman Linewidths in Alkane-urea Clathrates"
Strauss, H.L.; Snyder, R.G. 12-15

Polymers

"Raman Spectroscopic Study of the Conformational Variety of n-Alkyloligo(Oxyethylene)S in the Solid State"
Matsuura, H.; Fukuhara, K. 12-17

"Raman and Brillouin Scattering of Ammonium Hydrogen Oxylate, Semi Hydrate $(NH_4)(HC_2O_4)$, 1/2 H_2O and Phase Diagram"
Godet, J.L.; Berger, J.; Krauzman, M.; Toupry, N. 12-19

"Raman Spectroscopic Study of Crystallization Kinetics in Polyester Fibers"
Bulkin, B.J.; DeBlase, F.; Kim, J.; Lewin, M. 12-21

"Raman Mode Attributions and Ferro-Paraelectric Transition in the 70/30 Vinylidene Fluoride-Trifluoroethylene Copolymer"
Abenoza, M.; Armengaud, A. 12-22

Polyacetylene & Other Conducting Polymers

"Picosecond Raman Transients in Polyacetylene"
Weidman, D.L.; Fitchen, D.B. 12-24

"Fully Oriented Trans-$(CH)_x$ Systems: Analysis of Polarized Resonant Raman Scattering"
Lefrant, S.; Mulazzi, E.; Faulques, E.; Brivio, G.P. 12-26

"Disperson of the Depolarization Ratio in the Resonance Raman Spectra of Soluble Polyacetylenes"
Fuso, S.; Cataliotti, R.S.; Dellepiane, G.; Destri, S.; Paliani, G.; Tubino, R. 12-28

"Raman Characterization of an Implanted Conducting Polymer"
Bidarian, A.; Mathur, M.S.; Kadaba,P.K.; Bradley, E.B. 12-30

"Raman Spectra of Polypyrrole and its 2,5-^{13}C-Substituted and C-Deuterated Derivatives at Their Neutral and Doped States"
Furukawa, Y.; Tazawa, S.; Fujii, Y.; Harada, I. 12-31

CHAPTER 13 - SPECIFIC MOLECULAR SYSTEMS: INORGANIC AND ORGANOMETALLIC

"Time-Resolved Resonance Raman Spectroscopy of the $\delta\delta^*$ Excited Electronic State of $Re_2Cl_8^{2-}$"
Dallinger, R.F. 13-1

"Simulation of Resonance Raman Spectra of Molybdenum and Rhenium Complexes"
Jean, J.M.; Belhadj, M.; Friesner, R.A. 13-3

"Resonance Raman Study of the Linear-Chain Mixed-Valence Complex $[Pd(-chxn)_2Br]Br_2$"
Jensen, P.W.; Toftlund, H. 13-4

"Vibrational Spectrum of $MMoO_4$ (M = Mn or Co)"
Matsuura, I.; Hashiba, H.; Kawahara, H.; Kanesaka, I. 13-6

"Tetranuclear Platinum Complexes Studied by Resonance Raman Spectroscopy"
Stein, P.; Mahtani, H.K. 13-8

"Raman and Multinuclear NMR Spectra of Peroxo Complexes of Mo, W and V in Aqueous Solution"
Campbell, N.J.; Dengel, A.C.; Griffith, W.P. 13-10

"Absolute Raman Intensities and EOPs of ClO_n^- Anions in Aqueous Solutions"
Eysel, H.H. 13-12

"Recent Applications of Microraman Spectroscopy in Inorganic Chemistry"
Harvey, P.D.; Butler, I.S. 13-13

"The Pre-Resonance Raman Effect of the Solvated Ions $I^-.SO_2$ and $SCN^-.SO_2$"
Santos, P.S.; Santos, J.G.N.; Faria, D.L.A. 13-14

"Raman Spectra of Halomercuriomethanes $CH_2(HgX)_2$ and $CH(HgX)_3$ (X=Cl, Br and I) in Various States"
Breitinger, D.K.; Kress, W. 13-15

"Resonance Raman Spectra of Metallo-Phthalocyanins"
Thoai, P.V. 13-17

"Out-of-Plane Deformation Modes of Ni(II)-Octaethylporphyrin in Resonance Raman Spectra"
Li, X.-Y.; Czernuszewicz, R.S.; Kincaid, J.R.; Spiro, T.G. 13-19

"Resonance Raman Spectra of Cobalt Phthalocyanine and Lutetium Diphthalocyanine in the Solid State"
Bartholomew, C.R.; McConnell, A.A.; Smith, W.E. 13-21

"Resonance Raman Spectra of Co(II),Ni(II),Cu(II),Zn(II) Protoporphyrin IX Dimethyl Ester Complexes"
He, L.T.; Jiang, T.L. 13-23

"Ligand Deuteration as a Means of Differentiating Between Metal-Metal and Ligand-Sensitive Vibrations of Dimetal-Tetraacetates"
Clark, R.J.H; Flint, C.D.; Hempleman, A.J. 13-25

"The Influence of Selected Metals on the Aromatic System of Salicylic Acid"
Baranska, H.; Lewandowski, W. 13-27

"Raman Spectroscopic Study of Disproportionation of Iodine in Basic Solutions"
Wren, J.C.; Sunder, S.; Paquette, J. 13-29

"In Situ Micro-Raman Study of Discharge Products of Li/$SOCl_2$ Cells"
Dhamelincourt, M.C.; Wallart, F.; Delhaye, M. 13-31

CHAPTER 14 - SPECIFIC MOLECULAR SYSTEMS: ORGANIC

"Conformational Analysis of Some Organic Molecules By Gas Phase Raman Spectroscopy"
Durig, J.R.; Sullivan, J.F.; Little, T.S.; Geyer, T.J.; Godbey, S.E. 14-1

"Determination of Vibrational Potential Energy Surfaces from Raman and Infrared Spectra"
Laane, J. 14-3

"Raman Spectra and Conformations of Dibenzo- and Dicyclohexyl-18-Crown-6"
Takeuchi, H.; Arai, T.; Harada, I. 14-5

"Comparative Study of Phenylacetylene, Diphenylacetylene and Diphenyldiacetylene. Assignment of the Acetylenic Bridge Vibrations"
Colombo, L.; Baranovic, G.; Durig, J.R. 14-7

"On the Validity of Using Raman Intensity Measurements for Determining Conformer Populations as a Function of Pressure and Temperature"
Gardiner, D.J.; Walker, N.A. 14-9

"Dephasing and the Vibrational Frequency Correlation Function in Liquid Crystals"
Gburski, Z. 14-10

"Conformational Analysis of Organophosphorus Compounds by Gas Phase Raman Spectroscopy"
Hizer, T.J.; Johnson, R.D.; Groner, P.; Durig, J.R. 14-12

"Conformational Stability and Low Frequency Vibrational Assignment of 1,1,1-Trifluoroisopropanol, Hexafluoroisopropanol, 1,1,1-Trifluoroisopropylmethylether, and Hexafluoroisopropylmethylether"
Larsen, R.A.; Cox, F.O.; Durig, J.R. 14-13

"Vibrational Spectra and Normal Coordinate Analysis of Nitro Substituted Benzenes"
Tian, Y., Zhao, W.; Liang, Y. 14-14

"Raman Spectroscopic Study on Molecular Conformations of Methanesulfonate Esters and Deuterated Derivatives"
Matsuura, H.; Tamaoki, H. 14-15

"Raman Spectra and Structures of 1,5-Dithiacyclooctane and its Monocation Radical and Dication"
Itoh, K.; Serita, M.; Tamaoki, M.; Shiratori, Y. 14-18

"Raman and Infrared Spectra of 2-Amino-4H-Pyran Derivatives"
Bellanato, J.; Martin, N.; Seoane, C. 14-20

"Intensity Parameters for the Raman Trace Scattering of n-Alkanes"
Gough, K.M.; Murphy, W.F. 14-22

"The Raman Spectrum of 1,4-Naphtoquinone Interpreted by Using a Transferable Valence Force Field"
Mannfors, B.; Palmo,K. 14-24

"Rotational Isomerism of O,O'-Diakyl Esters of Phosphorodithioic Acid"
Waters, D.N.; Paddy, J.L. 14-26

"Raman and Infrared Spectroscopic Studies of Novel Poly-aza Chains and Related Metal Complexes"
Bariz, R.; Bremard, C.; Merlin, J.C. 14-28

"Prediction of the Vibrational Frequencies of Trans-Stilbene and Tolane by Using a Transferable Valence Force Field"
Palmo, K.; Mannfors, B. 14-30

"A Priori Calculations of the Vibrational Frequencies of Methylated Polyenes"
Abo Aly, M.M.; Favrot, J. 14-32

"Increase in the C=N Stretching Frequency Upon Complexation of Trans-retinylidene-N-Butylamine with General Lewis Acids"
Lopez-Garriga, J.J.; Babcock, G.T. 14-34

"Excited State Structure and Dynamics of Isoprene from Absolute Resonance Raman Intensities"
Trulson, M.O.; Mathies, R.A. 14-36

"Excitation Profiles of Resonance Raman Scattering and Luminescence from β-Carotene: The Non-Motional Narrowing Effect"
Kinoshita, S.; Watanabe, J.; Kushida, T. 14-38

"Low-Frequency Raman Spectra of Disordered Chain Molecules"
Snyder, R.G.; Hallmark, V.M.; Strauss, H.L. 14-40

"Raman Spectroscopic Investigation of Branched Nonanes"
Adil, A.; Turrell, G.; Vergoten, G.; Snyder, R.G. 14-42

"Vibrations of Kinks and Folds: Understanding the Phases of the Cycloalkanes"
Shannon, V.L.; Strauss, H.L.; Snyder, R.G. 14-44

"Oxygen Sensitive Background in the Raman Spectra of Woody Tissue"
Agarwal, U.P.; Atalla, R.H. 14-46

"Microraman Study of the Structural Order of the Carbon Network in Pitches"
Dhamelincourt, P.; Cottinet, D.; Couderc, P. 14-48

"Normal Modes of Pentachlorobenzene"
Yadav, R.A.; Shanker, R. 14-49

"Fundamental Frequencies of O-Ethyl Phenol"
Shanker, R.; Yadav, R.A.; Singh, I.S.; Singh, O.N. 14-51

"Vibrational Spectra of Cyclopropyl, 1,-Cyano Carboxamide (CCC) and Cyclopropyl 1,1, Carboxamido Acetate (CCA)"
Gopinath, C.R.; Sathyanarayana, D.N.; Natarajan, A. 14-53

"Laser Raman Spectra of Some Halogeno Cyclo Phosphazenes"
Natarajan, A.; Gopinath, C.R.; Rao, C.R.; Nagaraj, B.S. Kumaraswamy, K.C. 14-55

CHAPTER 15 - GAS PHASE STUDIES

"CARS Diagnostics of Discharges"
Taran, J.P. 15-1

"Highly Time-Resolved CARS Spectroscopy of Photofragments and Reaction Products"
Valentini, J.J. 15-3

"Double-Resonance Techniques for High-Resolution Raman-Spectroscopy"
Esherick, P.; Owyoung, A. 15-4

"Studies of the Self-Broadening and Foreign Gas-Broadening Coefficient Dependence on Temperature for Diatomic Molecules and Monatomic Gases"
Edwards, H.G.M.; Long, D.A.; Sherwood, G. 15-6

"A Temperature Study of the Pressure-Broadening Coefficient Dependence on Rotational Quantum Number, J, for Nitrogen"
Edwards, H.G.M.; Long, D.A.; Sherwood, G. 15-8

"Dual Broadband Approaches to Vibrational and Pure Rotational CARS Spectra"
Eckbreth, A.C.; Anderson, T.J. 15-10

"CARS Measurements of Temperature Profiles and Pressure in a Tungsten Lamp"
Trebino, R.; Farrow, R.L. 15-12

"Raman Spectroscopy Especially of High Vibrational States"
Knippers, W.; van Helvoort, K.; Fantoni, R.; Reuss, J.; Stolte, S. 15-14

"High Resolution Stimulated Raman Spectroscopy: Studies of Molecular Structures and Collisional Effects on Line Shapes"
Lavorel, B. 15-16

"High Resolution Stimulated Raman Spectroscopy of $^{13}CD_4$"
Millot, G.; Lavorel, B.; Chaux, R.; Saint-Loup, R.; Pierre, G.; Berger, H. 15-18

"High-Resolution Nonlinear Rotation-Vibrational Raman Spectroscopy of Gases"
Schrotter, H.W. 15-19

"Time-Resolved CARS Spectroscopy: Application to Photophysics and Photochemistry"
Weisman, R.B. 15-21

"CARS Thermometry at Ultra High Temperatures"
Greenhalgh, D.A.; Dring, I.; Devonshire, R.; Boysan, F. 15-23

"Application of High-Resolution Coherent Raman Spectroscopy to Combustion Measurements"
Rahn, L.A.; Palmer, R.E. 15-25

"High-Resolution CARS for Combustion Diagnostics"
Farrow, R.L.; Lucht, R.P. 15-27

"CARS Nitrogen Thermometry at High Pressure for I.C. Engine Applications"
Williams, D.R.; Baker, C.A.; Greenhalgh, D.A. 15-29

"Resonance CARS Spectroscopy of the OH Radical in a Microwave Discharge"
Attal-Tretout, B.; Houssin, D.; Taran, J.P. 15-31

"CARS Time Resolved Measurements of H_2 Rovibrational Population in a H^- Ion Source"
Lefebvre, M.; Pealat, M.; Taran, J.P.; Hillion, F.; Bacal, M. 15-33

"High Resolution CARS Spectrometer with Injection Locked Pulsed Dye Lasers: Application to the Q Branch of Oxygen"
Boquillon, J.P.; Ouazzany, Y.; Chaux, R. 15-35

"Generation of Single-Pulse Rotational CARS Spectra Using a Multiple Three-Color Scheme"
Alden, M.; Bengtsson, P.-E.; Edner, H. 15-37

"Low Frequency CARS Spectroscopy"
Triggs, N.E.; Bozler, B.J.; Pubanz, G.A.; Yang, J.J.; Nibler, J.W. 15-39

"Free Jet Spectroscopy by Coherent Raman Methods"
Pubanz, G.A.; Maroncelli, M.; Nibler, J.W. 15-41

"Nonlinear Raman Spectroscopy with High-Resolution cw-Lasers"
Rosasco, G.J.; Hurst, W.S. 15-43

"Vibrational Spacings in Molecular Hydrogen"
Veirs, D.K.; Rosenblatt, G.M. 15-45

"Vibrational Raman Cross Sections of the Overtone Transitions in H_2 and N_2"
Huo, W.M.; Herring, G.C.; Bischel, W.K. 15-47

"A New Feature in the Raman Q Branch of D_2"
May, A.D.; Rosasco, G.J.; Hurst, W.S.; Smyth, K.C.; Petway, L.B. 15-49

CHAPTER 16 - COLLISION INDUCED PHENOMENA, MOLECULAR INTERACTIONS, AND LIQUIDS

"Induced Raman Spectroscopy: A General View"
Barocchi, F. 16-1

"Induced Spectra of Spherical Molecules at Low Density"
Tabisz, G.C.; Meinander, N. 16-2

"Light Scattering from Fluid CO_2"
De Santis, A.; Frattini, R.; Sampoli, M.; Vallauri, R. 16-4

"Rotational Raman Scattering from Liquid and Gaseous H_2S"
De Lorenzi, A.; De Santis, A.; Frattini, R.; Sampoli, M. 16-5

"Induced Effects in Molecular Liquids and Local Field Problems"
Ladanyi, B.M. 16-6

"Induced Light Scattering in Crystals and Disordered Solids"
Signorelli, G.; Mazzacurati, V. 16-7

"The Effects of Dephasing Collisions on Raman Polarization"
Hochstrasser, R.M.; Myers, A.B. 16-9

"Collision Induced Raman Scattering in Hydrogen at Low Density: Temperature Dependence"
Bafile, U.; Ulivi, L.; Zoppi, M.; Barocchi, F. 16-11

"The Relation Between the Band Shape, the Strength of H-Bond and the Solution Viscosity"
Abramczyk, H.; Samios, D.; Marcinek, A.; Dorfmuller, Th. 16-13

"A Raman Scattering Study of the Liquid State Dynamics of a High Torque Molecular Liquid"
Moses, D.G.; Baglin, F.G. 16-15

"Rayleigh Scattering of Dense O_2"
Zerda, T.W. 16-17

"High-Resolution Raman and Time Resolved CARS Studies of the Fermi Diad in CO_2"
Ranson, P.; Ouillon, R. 16-19

"Computer Simulation Studies of Raman Scattering From Dense Fluids"
Ladanyi, B.M.; Geiger, L. 16-21

"Spectrum of Collision-Induced Scattering: Role of Multipolar Polarizabilities"
Barreau, A., Berrue, J., Chave, A.; Dumon, B.; Thibeau, M. 16-22

"Strong Collision Model for the Calculation of Collisionally Narrowed Isotropic Raman Spectra"
Sala, J.P.; Bonamy, J.; Robert, D.; Lavorel, B.; Millot, G.; Berger, H. 16-24

"Depolarized Scattering from Hydrogen at Low Temperature"
Le Duff, Y.; Chave, A.; Dumon, B.; Idrissi, A.; Thibeau, M. 16-25

"Collision-Induced Polarized and Depolarized Raman Scattering by Anisotropic Molecules (H_2,D_2)"
Brown, M.S.; Wang, S.-K.; Frommhold, L. 16-26

"Intermolecular Potentials and High Pressure Raman Studies"
Chandrasekharan, V.; Uzan, E. 16-27

"Vibrational and Reorientational Relaxation of Liquid Carbon Disulfide at Pressures up to 10 kbar"
Ikawa, S.-I.; Whalley, E. 16-29

"Motional Narrowing in the Raman Spectrum of Iodine Interacting with Benzene or its Methylated Derivatives"
Paatero, M.; Eskola, S.; Stenman, F. 16-31

"Raman Intensity and Molecular Interaction"
Schmid, E.D.; Brodbek, E.; Moschallski, M.; Hnida, U.E. 16-33

"Resonance Raman Spectra of Hexamethylbenzene-TCNQ Electron Donor-Acceptor Complex"
Merriam, M.J.; McHale, J.L. 16-35

"Depolarized Rayleigh Wing and Isotropic/Anisotropic Raman Splitting in Dimethylformamide/CCl_4 Solutions"
Rodriguez, R.; McHale, J.L. 16-36

"Isotopic Dilution Studies of Liquid Benzene by Raman Difference Spectroscopy"
Meinander, N.; Laane, J. 16-37

"A Molecular Dynamics Study of the OH Stretch Vibration in Liquid Water"
Bansil, R.; Berger, T. 16-39

"Rotational Dynamics of Liquid Water: Isotropic Induced Light Scattering Spectra"
Briganti, G.; Ricci, M.A.; De Santis, A.; Ruocco, G. 16-41

"Very Low Frequency Depolarized Raman Scattering in Liquid and Supercooled Water"
Signorelli, G.; Mazzacurati, V.; Benassi, P. 16-43

"On the Reorientation of α-Substituted Methylbenzene Molecules in Liquid Phase"
Ribeiro-Claro, P.J.A.; Amorim da Costa, A.M.; Teixeira-Dias, J.J.C. 16-45

"Raman Spectral Study of Ion-Pairing and Hydration of Alkaline and Alkaline Earth Perchlorates"
Jeremic, M.; Jovanovic, A.; Lalic, M. 16-46

"Raman Active Vibrations of Cation Complexes in Concentrated Aqueous Solutions"
Alves Marques, M.; Sousa Oliveira, M.A.; Resina Rodrigues, J. 16-48

"Raman Spectral Study of Solutions of $Mg(ClO_4)_2$ in Some Organic Solvents"
Antic-Jovanovic, A.; Jeremic, M.; Lalic, M. 16-49

CHAPTER 17 - RESONANCE RAMAN SCATTERING: THEORY AND EXPERIMENT

"Theory of Raman Excitation Profiles of Multimode System with Dushinsky Rotation and Jahn-Teller Effect"
Hizhnyakov, V. 17-1

"Theory of Temperature-Dependent Resonance Raman Scattering by Multimode Systems--A Many Body Approach"
Page, J.B. 17-3

"Time-Dependent Approaches to Spectroscopy Involving Non-Adiabatically Coupled Excited Electronic States"
Coalson, R.D. 17-5

"Incorporating Protein and Solvent Environments in Resonance Raman Calculations"
Warshel, A. 17-6

"Femtosecond Rotation-Vibration Dynamics as Revealed by Resonance Rotational Raman Studies of Ammonia"
Ziegler, L.D. 17-8

"Ultraviolet Resonance Raman Spectra of NH_3 and ND_3"
Hess, W.; Ziegler, L.D.; Hudson, B. 17-10

"The Inverse Transform in Resonance Raman Scattering"
Cable, J.R.; Albrecht, A.C. 17-12

"Consistent Non-Perturbative Theory of Temperature-Dependent Resonance Raman Scattering for <u>Simultaneous</u> Mode Mixing and Non-Condon Coupling"
Lu, H.M.; Page, J.B. 17-14

"Vibronic Coupling in the First Excited State of Anthracene"
Bree, A. 17-16

"Time Characteristics of Secondary Emission From Resonantly Excited I_2"
Overlaet, W.; Kimura, M.; Kinoshita, S.; Kushida, T. 17-18

"Calculation and Measurement of Polarization Dispersion Curves for the Resonance Raman Scattering of I_2 in Solution"
Kobayashi, T.; Sension, R.J.; Strauss, H.L. 17-20

"Comparison of Theory and Experiment for the Raman Excitation Profile of I_2 in Solution"
Sension, R.J.; Strauss, H.L. 17-22

"Ultraviolet Resonance Raman Scattering Studies of Methylene Cyclopropane: The Propensity of the V State to Form a Trimethylenemethane Structure from the Vertically Excited Structure"
Chadwick, R.R.; Parker, D.W.; Baldwin, J.E.; Hudson, B. 17-23

"Ultraviolet Resonance Raman Spectroscopy of Butadiene: Detection of the Low Energy 2^1A_g State"
Chadwick, R.R.; Hudson, B. 17-24

"Ultraviolet Reonance Raman Studies of Benzene Derivatives"
Li, S.; Hudson, B. 17-26

"Resonance Raman Spectroscopy of Chloroethylenes"
Kelly, P.B.; Hudson, B. 17-28

"Resonance Raman Spectroscopy of Methylethylenes"
Kelly, P.B.; Hudson, B. 17-29

"Preresonance Raman Scattering in Quinoxaline and Some Vitamin A-Type Molecules"
Mallick, P.K. 17-30

"Raman Excitation Profiles of 1.8-Dihydroxyanthraquinone at 8 K"
Smulevich, G.; Marzocchi, M.P. 17-32

"Raman Excitation Profiles of Polyatomic Molecules in Condensed Phases - A Stochastic Theory"
Mukamel, S.; Sue, J.; Yan, Y.; Deng, Z. 17-34

"Interferences in an Raman Excitation Profile of Aggregated Chlorophyll"
Binnie, N.; Haley, L.V.; Mattioli, T.A.; Thibodeau, D.L.; Wang, W.; Koningstein, J.A. 17-36

CHAPTER 18 - TIME RESOLVED RAMAN STUDIES

"Picosecond Time-Resolved Resonance Raman Spectroscopy"
Atkinson, G.H.; Rumbles, G.; Brack, T.L.; Blanchard, D. 18-1

"Picosecond Resonant Light Scattering in Silver Halides"
Stolz, H.; von der Osten, W. 18-3

"Transient Intermediates in Photoreactions of Derivatized Anthraquinone and Duroquinone"
Hester, R.E.; Moore, J.N.; McCubbin, I.; Phillips, D. 18-5

"The Ground and Excited State Raman Spectra of , '-^{13}C-trans-Stilbene"
Gustafson, T.L.; Chernoff, D.A.; Hazle, M.A.; Palmer, J.F.; Roberts, D.M. 18-6

"Time-Resolved Resonance Raman Spectrum of Chrysene in the S_1 and T_1 States"
Koshihara, S.-Y.; Kobayashi, T. 18-8

"Time-Resolved Resonance Raman Study of the Photoionization of N,N,N',N'-Tetramethyl-p-phenylenediamine in Solution"
Takahashi, H.; Isaka, H.; Ohzeki, T.; Suzuki, S. 18-10

"Characterization of the Photodynamic Behavior of Nickel Porphyrins in Noncoordinating Solvents Using Transient Raman Spectroscopy"
Findsen, E.W.; Shelnutt, J.A.; Ondrias, M.R. 18-12

"A Transient Resonance Raman Study on the Lowest Excited Triplet-State of β-carotene Isomers"
Hashimoto, H.; Koyama, Y. 18-14

"The Structure of the Lowest Excited Triplet States of Hexatrienes, Studies by Time-Resolved Resonance Raman Spectroscopy"
Langkilde, F.W.; Jensen, N.-H.; Wilbrandt, R.; Brouwer, A.M.; Jacobs, H.J.C. 18-16

"$O_2(^1\Delta_g)$Photofragment CARS Spectroscopy and the Dynamics of the UV Photolysis of Ozone"
Gerrity, D.P.; Phillips, D.L.; Nieh, J.-C.; Tabor, K.T.; Valentini, J.J. 18-18

"Mechanism of Time-Resolved Ionization Detected Stimulated Raman Scattering From Molecules"
Nomura, Y.; Fujimura, Y.; Kono, H.; Lin, S.H. 18-20

"Time-Resolved Resonance Raman Study of the Pulse Radiolytic Oxidation of Aqueous Trihydroxybenzenes"
Qin, L.; Tripathi, G.N.R.; Schuler, R.H. 18-22

CHAPTER 19 - CARS AND OTHER NON-LINEAR OPTICAL STUDIES

"Time-Resolved Coherent Raman Scattering of Liquids and Gases on the Sub-Picosecond to Nanosecond Time Scale"
Laubereau, A. 19-1

"CARS Studies of the Lowest Excited State of α,ω-Diphenylpolyenes"
Maeda, S. 19-3

"The Application of Nondegenerate CARS to Investigate Solutions"
Apanasevich, P.A.; Kvach, V.V.; Kozich, V.P.; Orlovich, V.A. 19-5

"Finding Real and Imaginary Susceptibilities Going Through Hilbert Transforms, Freight-CARS, A New Method for the Analysis of CARS Lineshapes"
Bostick, J.M.; Carreira, L.A. 19-7

"CSRS with Incoherent Light for Vibrational Dephasing Time Measurements"
Hattori, T.; Kobayashi, T. 19-9

"Higher Order Susceptibility in Non-Linear Optical Mixing. A Comparative Analysis"
Weiszmann, A.N.; Chencinski, N.; Levine, A.M.; Schreiber, W.M. 19-11

"Coherent Anti-Stokes Raman Scattering of Nematic Liquid Crystal"
Aslanyan, L.S.; Badalyan, N.N.; Petrosyan, A.A.; Chilingaryan, Yu.S. 19-13

CHAPTER 20 - NEW RAMAN TECHNIQUES

"Impulsive Stimulated Raman Scattering: An Entirely Time-Domain Vibrational Spectroscopy"
Nelson, K.A.; Farrar, M.R.; Williams, L.R. 20-1

"Instrumental Advances in Raman Optical Activity"
Diem, M. 20-3

"Raman Optical Ativity in CARS"
Schneider, F.W.; Brakel, R. 20-4

"Raman Optical Activity Spectra of (+)-(3R)-Methylcyclohexanone Obtained with a Soleil-Babinet Compensator"
Zimba, C.G.; Freedman, T.B.; Spencer, K.M.; Hu, X.M.; Nafie, L.A. 20-6

"Raman Optical Activity of Small Molecules"
Diem, M.; Calienni, J.; Davies, M.; Gunnia, U.; Lee, O. 20-8

"Influence of Electric Quadrupole vs Magnetic Dipole in Raman Optical Activity"
Escribano, J.R.; Barron, L.D. 20-9

"Techniques for Far UV Resonance Raman Spectroscopy"
Kelly, P.B.; Ruggiero, A.; Li, S.; Harhay, G.; Strahan, G.D.; Hudson, B. 20-11

"A New Method of Studying Enzyme-Substrate Intermediates Using Enzyme Microcrystals and an Intracavity Laser Raman Microscope Technique"
Patapoff, T.W.; Thomas, G.A.; Peticolas, W.L. 20-13

"Mixing Techniques for Time-Resolved Raman Spectroscopy"
Kincaid, J.R.; Paeng, K.-J. 20-15

"Study of Mixing Processes in Liquids by Time Resolved Multichannel Raman Spectroscopy"
Lorriaux-Rubbens, A.; Castaignos, N.; Bridoux,. M.; Dalhaye, M.; Wallart, F. 20-17

"Fluorescence Free Raman Spectroscopy"
Chase, B. 20-19

"Fluorescence Free FT-Raman Spectroscopy in the Near Infrared"
Buijs, H. 20-21

"Development of a New Rayleigh Line Rejection Filter: The Basis for a New Generation of Raman Instrumentation"
Asher, S.A.; Flaugh, P.; Hagopian, S.; Rundquist, P.; Washinger, G.R. 20-22

"Measurements of Very Small Frequency Shifts in Raman Spectroscopy: Difference and Fourier Transform Methods"
Savoie, R.; Levesque, D.; Langlais, M. 20-24

"High-Pressure Microraman Spectroscopy"
Benham, V.; Butler, I.S.; Lord, G.; Gilson, D.F.R. 20-26

"Isotropic and Anisotropic Samples in Polarized Raman Microspectroscopy"
Bremard, C.; Laureyns, J.; Turrell, G. 20-27

"Microprobes with Multichannel Detection and Laser Excitation for Raman and Fluorescence Microspectrometry"
Dhamelincourt, P,; Delhaye, M.; Da Silva, E.; Cornard, J.P.; Lenain, B. 20-29

"Pulsed Multichannel Raman Spectroscopic Investigation of the Spatial Distribution of Molecular Species in an Engine"
Desenne, D.; Crunelle-Cras, M.; Grase, F.; Bridoux, M.; Sawerysyn, J.P.; Sochet, L.R. 20-30

"Remote and In Situ Raman Spectroscopy Using Optical Fibers"
Dao, N.Q.; Jouan, M.; Plaza, P. 20-32

"The ASTM E 13.08 Committee on Raman Spectroscopy and the IUPAC Commission 1.5 on Molecular Spectroscopy - Overview of Activities"
Bulkin, B.J.; Powell, L.; Grasselli, J.G. 20-34

"Raman Spectra of Liquefied Noble Gas Solutions"
Davidson, G.; Davies, C.L. 20-35

AUTHORS INDEX A-1

PREFACE

The first International Conference on Raman Spectroscopy (ICORS-I) was held in Ottawa, Canada, 17 years ago. Who could have forseen then that in 1986 Raman spectroscopy would still be such an exciting and vigorous field, still attracting outstanding young investigators? We are pleased that many of the invited speakers in the ICORS-X program are relative newcomers. We are also pleased that so many of the established investigators in Raman spectroscopy have agreed to participate in this conference. On behalf of the Organizing Committee and the Program Advisory Committee, we thank all of the authors who have generously given the time and effort to provide an abstract. The collected abstracts published in this book will serve as an index to the rapidly developing field of modern Raman spectrosopic techniques and theory. There is no question that this will help advance the field by making available, in a single volume, a means of cross-referencing the work of many investigators. Those who have been doing research in Raman spectrosopy for a long time are well aware of the usefulness of these books, which have come out every two years since 1976. Those who are new to the field will find these volumes an excellent source for references to the considerable amount of work that has been the foundation of this active science

This year a new format for the conference is being tried. With the exception of the opening plenary session, we have scheduled either two parallel lecture sessions (whose subject matter, we hope, does not overlap) or a single poster session. The lecture sessions are divided into brief 30-minute reviews and 15-minute research reports. Since the time alloted the speakers is very short, it seems unlikely that any of them will be able to go into as much depth as those intimately concerned with the work would like. Thus we have encouraged all of the participants to give a poster presentation. There, specialists can convene and have a reasonably open-ended discussion of the details and meaning of the work presented. The poster sessions are in some respects an extension of the abstracts.

The text has been organized with subdivisions roughly corresponding to the sessions of the meeting but arranged in an order that groups related topics together. Talks and poster presentations on similar subjects are grouped together. In general, the primary criterion for classification of papers in the book was the nature of the material being studied, rather than the technique being used, unless the technique itself was the primary subject of the investigation. We hope that the proximity of related papers will help the reader locate topics of interest.

We thank those organizations that have provided money or services to help make this conference a success. We are very aware that Eugene is a bit out of the way and hence difficult to get to. To all of the participants who made the effort to come—many using their own personal or grant funds—we express our appreciation. We hope the experience of participating in this conference, together with this remarkable collection of abstracts, will make the effort well worthwhile.

Warner L. Peticolas
Bruce S. Hudson

Eugene, Oregon

PLENARY LECTURES

A QUARTER CENTURY OF STIMULATED RAMAN EFFECT.

N. Bloembergen

Division of Applied Sciences, Harvard University, Cambridge, MA 02138, USA

A historical review of the stimulated Raman effect starts with its experimental discovery and theoretical interpretation in 1962. The exponential gain at the Stokes frequency presents an instability for the propagation of intense light beams. Coherent anti-Stokes Raman scattering has led to many applications, including analysis of combustion processes. Frequency up- and down-conversion by stimulated Raman processes is also widely used. The transient behavior with picosecond or femtosecond light pulses has led to time-resolved coherent Raman spectroscopy and the study of the time dependence of vibrational coherences. Another recent manifestation is the existence of collision-induced Raman coherences between (nearly) degenerate Zeeman levels in atoms. A look ahead into the next year completes the first quarter century of stimulated Raman spectroscopy.

GENERATION OF COHERENT AND TUNABLE RADIATION BY SCATTERING PROCESSES.

B. P. Stoicheff.

Department of Physics, University of Toronto, Toronto, Ontario, M5S 1A7, Canada.

Since the first observation of stimulated Raman scattering, it was clear that this process could be used to generate coherent radiation at prescribed wavelengths by selection of the primary laser and Raman medium. In the intervening years, Raman Stokes and anti-Stokes radiation has been produced at specific wavelengths in the infrared, visible and reaching into the vacuum ultraviolet region, and has been applied to a variety of spectroscopic studies.

Frequency-mixing processes based on the third-order nonlinear polarizability of atomic and molecular systems provide yet another means of generating coherent radiation. Such radiation is monochromatic, directional, and also readily tunable. To date, continuous tunability over a broad wavelength range from 200 to 86 nm ($\sim$70,000 cm^{-1}) has been achieved, and over limited regions to wavelengths as short as 50 nm. This coherent and tunable radiation has proven to be important for fluorescence and absorption spectroscopy, and promises to become a useful source for resonance Raman scattering.

LIGHT-INDUCED SPECTRAL REDISTRIBUTION OF SPONTANEOUS EMISSION.

J. S. Melinger and A. C. Albrecht

Department of Chemistry, Cornell University, Ithaca, New York 14853

When a molecular system interacting with a thermal bath is driven by a resonant, or nearly resonant, c.w. coherent light source, the resulting spontaneous emission often contains both Raman-like (RL), and fluorescence-like (FL) components.[1] The FL component appears as a result of dephasing processes which are associated with system/bath interactions (pure-dephasing). The phrase, "pure-dephasing induced redistribution of emission" characterizes such an appearance of FL features in the emission. In this communication we focus on another class of dephasing processes associated with the quality of the incident light that, also, can induce the appearance of FL features in the spontaneous emission spectrum.[2] To achieve this "light-induced redistribution of emission," the incident light need only possess a spectral density of a width, Γ_L, that exceeds the damping parameters of the system. This spectral density may originate from phase incoherence of the incident light and/or from the temporal profile of its amplitude.

Within the perturbative density matrix approach to optical spectroscopies, both off-diagonal, and diagonal components contribute to the spontaneous emission lineshape. In the steady state (no pure-dephasing), the off-diagonal component (the source of RL behavior), and the diagonal component (the source of FL behavior) interfere in such a way as to yield a purely RL emission.[1] Pulsed light excitation and/or incoherent light excitation, however, weakens the off-diagonal component with respect to the diagonal component, and, as a consequence, induces the appearance of the FL component in the emission spectrum.

Figure 1 displays model emission spectra from a three-level system for three different experimental situations: 1) frequency resolved emission from c.w. excitation, 2) frequency resolved emission from pulsed excitation, and 3) time and frequency resolved emission (transform limited) from pulsed excitation. In all examples, the detuning of the incident light from resonance is 20 (dimensionless units). Figure (1A) shows the purely RL emission (centered at Δ_D=-20) that arises when the excitation is c.w. coherent. The density matrix approach in this case reduces formally to the Kramers-Heisenberg (KH) equation which, alone, describes only RL behavior.[1] When the incident radiation becomes pulsed (or c.w. incoherent) we can show that the emission lineshape, $I(\omega_L,\omega_R)$, formally reduces to a convolution of the KH expression with the spectral density of the incident light:[2]

$$I(\omega_L,\omega_R) = \int_{-\infty}^{\infty} KH(\omega_L-\omega,\omega_R)\, f(\omega)d\omega. \qquad (1)$$

In Eq. (1), ω_L and ω_R represent frequencies of incident and scattered photons, respectively, and $f(\omega)$ is the spectral density of the light source. The most conspicuous result depicted by Fig. (1B) is how, even in the absence of pure dephasing, pulsed or incoherent excitation, alone, induces both RL and FL components in the emission. The RL component reflects the width of the incident spectral density, whereas the width of the FL component (centered at $\Delta_D=0$) remains unaffected by the broad excitation.[2]

The transient emission spectrum arises whenever the experimental apparatus allows for simultaneous time and frequency resolution.[3,4] Figure (1C) shows the transient emission spectrum resulting from pulsed excitation in the absence of pure-dephasing. At early times the emission shows mainly RL character, though, as time evolves, the emission becomes increasingly FL. Even in this experimental domain, pulsed excitation, alone, causes a strong redistribution of the emission towards fluorescence. This redistribution becomes stronger as the exciting pulses are made shorter in time.[4]

We gratefully acknowledge support from the Materials Science Center of Cornell University and NSF Grant No. CHE-8311401.

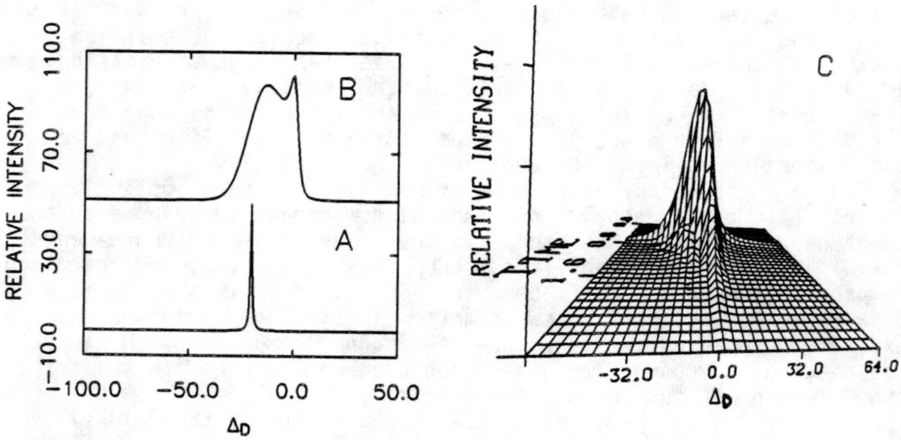

Fig. 1. A) Frequency (Δ_D) resolved emission from c.w. coherent excitation - only RL emission occurs, centered at $\Delta_D=-20$. B) Frequency resolved emission from pulsed excitation, $f(\omega)$ is Gaussian with $\Gamma_L=16$. Both RL and FL components appear. C) Time and frequency resolved emission from pulsed excitation (same as in B). See Refs. [2,4] for a full description of modeling constants.

References

1. D. Lee and A. C. Albrecht, in Advances in Infrared and Raman Spectroscopy, edited by R. J. Clark and R. E. Hester (Wiely-Heyden, New York, 1985), Vol. 12.
2. J. S. Melinger and A. C. Albrecht, (in preparation).
3. J. H. Eberly and K. Wodkiewicz, J. Opt. Soc. Am. **67,** 1252 (1977).
4. J. S. Melinger and A. C. Albrecht, J. Chem. Phys. **84,** 1247 (1986).

chapter 1

PROTEINS

PROTEIN STRUCTURE AND DYNAMICS VIA ULTRAVIOLET RESONANCE RAMAN SPECTROSCOPY. Thomas G. Spiro, Department of Chemistry, Princeton University, Princeton, New Jersey 08544.

Raman spectroscopy provides structural information about complex molecules via their vibrational frequencies. The problem of low sensitivity can be overcome by exploiting the resonance Raman effect, in which tuning the laser wavelength to an electronic transition enhances Raman bands associated with vibrations of the chromophore. Prosthetic groups such as heme, flavin, retinal and metal ion centers can be studied selectively with visible and near-UV laser excitation. The advent of high power pulsed lasers, whose wavelength is easily shifted via non-linear optical techniques permits the acquisition of Raman spectra with excitation throughout the ultraviolet region. UV excitation provides selective enhancement of vibrational modes associated with the amide backbone of proteins, and with aromatic residues: phenylalanine, tyrosine and tryptophan. Environmental effects for tyrosine and tryptophan rings can be seen as distinctive changes in band frequency and/or intensity. Striking hypochromic effects are seen for the amide II mode of α-helical polypeptides, making this band a reliable indicator for α-helical content. The pulsed character of the excitation also makes possible time-resolved Raman studies of protein dynamics. Applications to hemoglobin will be described.

UV RESONANCE RAMAN ENHANCEMENT: FROM SMALL MOLECULES TO PROTEINS

Sanford A. ASHER

Department of Chemistry, University of Pittsburgh, Pittsburgh, Pa. 15260

Our recent extension of Raman spectroscopy into the UV spectral region[1] from 200 to 400 nm has demonstrated the utility of resonance Raman enhancement studies as an important physical technique to explore excited state structure. The selectivity of resonance excitation in the UV permits studies of aromatic chromaphores in complex environments. Analytically and biophysically important examples include studies of polycyclic aromatic hydrocarbons in coal liquids[2,3] and aromatic amino acids in proteins.[4] UV Raman measurements of condensed phase samples avoid the fluorescence interferences that often fatally degrade the signal-to-noise ratios of spectra obtained with visible wavelength excitation.[5]

We will discuss Raman cross section studies of a series of small molecules which are useful as internal standards for Raman spectral studies. We have developed a new technique[6] to measure the dispersion of the absolute Raman cross section between 210 and 800 nm and have measured the cross sections of molecules such as CH_3CN, H_2O, acetone, SO_4, ClO_4, and NO_3. The preresonance dispersion of the cross section of these species (except for NO_3) appears to be dominated by states in the far, far UV. In contrast, the preresonance Raman dispersion of NO_3, acetamide, and N-methyl acetamide is dominated by the 180-200 nm transitions.[7] These observations question the existence of discrete valence electronic transitions in the vacuum UV in condensed phase samples.

We examined resonance enhancement in both gas phase and condensed[8] phase benzene. For excitation between 215 and 250 nm the symmetric ring breathing 992 cm^{-1} vibration derives enhancement mainly from the B_{1u}/E_{1u} states with little evidence of enhancement from the B_{2u} transition. In contrast, in the gas phase the smaller homogeneous linewidth of the vibronic components of the B_{2u} transition results in enhancement with exact resonance.

Both resonance Raman and single vibrational level fluorescence (SVL) are observed for excitation in the vibronic components of the gas phase benzene B_{2u} transition. These phenomena have been differentiated by the fact that Raman intensity must add at the amplitude level to any preresonance enhancement, while SVL adds at the intensity level because it occurs from a state which has dephased by a T_2 process from the virtual intermediate state.

We will also discuss the selection of vibrational modes enhanced by the UV transitions of substituted benzene derivatives including the aromatic amino acids. We will relate the observed cross sections to those observed for the aromatic amino acids as side chains in proteins. The utility of the use of Raman cross sections as probes of protein structure will be discussed.

We will also examine Raman excitation profiles of polycyclic aromatic hydrocarbons such as pyrene and relate the excitation profiles to the underlying vibronic structure.

References

1. S. A. Asher, C. Johnson, and J. Murtaugh, Rev. Sci. Instr., 54, 1657-1662 (1983).

2. C. R. Johnson and S. A. Asher, Anal. Chem., 56, 2258-2261 (1984).

3. S. A. Asher, Anal. Chem., 56, 720-724 (1984).

4. C. R. Johnson M. Ludwig, S. E. O'Donnell and S. A. Asher, J. Am. Chem. Soc., 106, 5008-5010 (1984).

5. S. A. Asher and C. R. Johnson, Science, 225, 311-313 (1984).

6. J. R. Dudik, C. R. Johnson, and S. A. Asher, J. Chem. Phys., 82, 1732-1740, (1985).

7. J. M. Dudik, C. R. Johnson and S. A. Asher, J. Phys. Chem., 89, 3805-3814 (1985).

8. S. A. Asher and C. R. Johnson, J. Phys. Chem., 89, 1375-1379 (1985).

PRIMARY DONOR BACTERIOCHLOROPHYLLS IN PHOTOSYNTHETIC REACTION CENTERS : GROUND-STATE INTERACTIONS FROM RESONANCE RAMAN SPECTROSCOPY

Marc LUTZ, Bruno ROBERT and Qing ZHOU,

Département de Biologie, CEN-Saclay 91191 Gif-sur-Yvette cedex, France

In reaction centers of purple and green photosynthetic bacteria, the photoinduced primary charge separation appears to generally involve two bacteriochlorophyll (BChl) molecules, which assume strong mutual excitonic coupling. Understanding the mechanism of the primary charge separation requires knowledge about the actual strcture of this molecular assembly, and in particular about the ground-state interactions which stabilize it in its host protein.

Resonance Raman (RR) spectroscopy of the BChl molecules can provide such information (1).

However, resonance at the Soret bands of the BChls, which best enhances scattering of key modes such as stretching of the conjugated carbonyls, does not permit selective observations amid the six bacteriochlorin-type pigments present in the reaction center, namely the two primary donor BChls, the two "accessory" BChls and the two bacteriopheophytins.

We thus developed difference methods permitting selective observations of RR spectra of the primary donor in the neutral, ground state (2). Three distinct methods, applied to reaction centers from Rhodobacter sphaeroides, wild type, yielded almost identical RR spectra (3). This permitted to exclude any artefactual origin for these spectra.

These difference spectra, which expectedly arise from two distinct BChl molecules, indicate that the magnesium atoms of both these molecules each bind a single axial ligand. These ligands most likely are side chains of histidines 173 (L subunit) and 200 (M subunit) of the protein (3,4).

Difference RR spectra of the primary donor also show that the conjugated acetyl carbonyl group of one BChl is free from intermolecular bonding, and that the same group in the second BChl is strongly bound. The binding site of this latter group is most probably not the magnesium of the other BChl molecule, but rather a proteic site. The keto carbonyls of both BChls are each interacting with chemically distinct, likely proteic sites.

The topological model of the primary donor of Rhodobacter sphaeroides which can be proposed from these observations (Fig. A) hence does not involve any direct bonding, in the ground state, between the two BChls, at variance with the earlier so called "special-pair" model (5).

The precise relative positioning of the two BChls shown, in particular, by X-ray crystallography (6), essentially results from bonding of each of the BChls with the protein, as well as from $\pi - \pi$ interaction between their conjugated macrocycles.

The preliminary model recently proposed for the primary donor of Rhodopseudomonas viridis, on the basis of X-ray data (6) also differs from the present model. Some of these differences may arise from an interspecific variability of the structure of the bacterial primary donor.

In order to test this variability, we studied the primary donor of Rhodospirillum rubrum, using the same methods. The difference RR spectra obtained for this species are extremely close to those observed for Rhodobacter sphaeroides, and indicate the same basic structures for both primary donors (Fig. B). However, we observed that, in the case of Rsp rubrum, the keto carbonyl of one of the two BChls is free from any intermolecular binding. This indicates that the interaction that the corresponding group assumes in the primary donor of Rhb. sphaeroides is non essential in determining its structure.

REFERENCES

1) Lutz, M. (1984) Adv. Infrared Raman Spectrosc. 11, 211-300
2) Lutz, M. and Robert, B. (1985) in Antennas and Reaction Centers of Photosynthetic Bacteria (Michel-Beyerle, M.E., Ed.) pp 138-146 Springer-Verlag, Berlin
3) Robert, B. and Lutz, M. (1986) Biochemistry, in press
4) Deisenhofer, J., Epp, O., Miki, K., Huber, R., and Michel, H. (1985) Nature 318, 618-624
5) Katz, J.J, Oettmeier, W. and Norris, J.R. (1976) Phil. Trans. R. Soc. London B273, 227-253
6) Deisenhofer, J., Epp, O., Miki, K., Huber, R. and Michel, H. (1984) J. Mol. Biol. 180, 385-398
7) Zhou, Q. (1985) Diplôme d'Etudes Approfondies, Université P. and M. Curie, Paris.

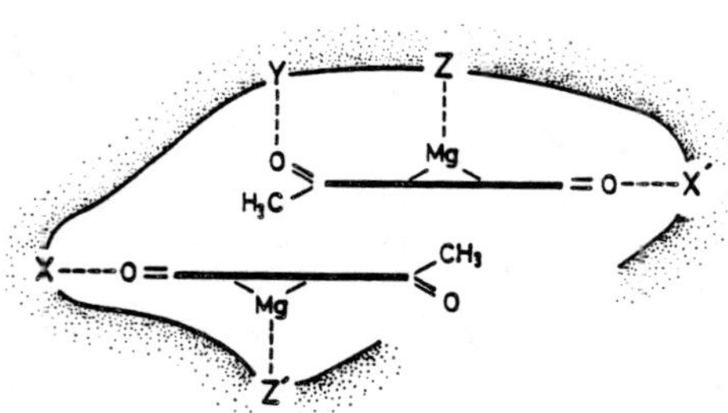

Fig. A - Rhodobacter sphaeroides

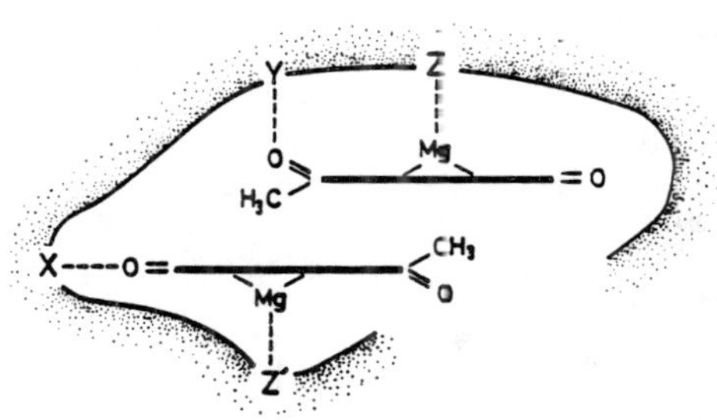

Fig. B - Rhodospirillum rubrum

RESONANCE RAMAN SPECTRA OF CATALYTIC INTERMEDIATES OF HEME ENZYMES

T. Kitagawa, S. Hashimoto and T. Ogura

Institute for Molecular Science, Okazaki National Research Institutes, Myodaiji, Okazaki, 444 Japan

A reaction of an iron-porphyrin with oxygen is specific to each class of heme enzymes, but the iron-porphyrin taken out from the protein does not catalyze the specific reaction. To know how the protein bestows the specific reactivity on the heme iron, we have investigated resonance Raman spectra (RRS) of various heme enzymes. In this paper we report RRS of the reaction intermediates of horseradish peroxidase (HRP) and cytochrome c oxidase (CcO), discussing a role of the protein.

Since the iron-oxygen bond of the intermediates can be photolabile, time-resolved RR measurements with pulse lasers are not always appropriate due to their high photon-density. Therefore we adopted a combination of a CW laser and a mixed flow apparatus along with an OMA II detection system attached to Spex 1404 double- or Jasco CT-50 single-monochromator. This method allows the observation of transient RRS with a time resolution of ∿25 μs in a favorable case. We also used a variable-speed spinning cell for an intermediate with longer life time.

The catalytic cycle of HRP involves the oxidation of the native ferric enzyme by H_2O_2 and successive two-stage reductions by substrates via two intermediates called compound I and compound II. Mössbauer study revealed the presence of the Fe^{IV} ion in compounds I and II. Since the reaction from compound II to the native state is inhibited at alkaline pH with the midpoint pH at 8.6 and its rate constant differs between the H_2O and D_2O solutions, Fe^{IV}-OH structure had been proposed for compound II from biochemical side, while recent ENDOR and NMR studies suggested the Fe^{IV}=O structure.

Our observation of the Fe^{IV}=O stretching ($\nu_{Fe=O}$) Raman line on compound II [1] and the independent study by Terner et al. [2] provided conclusive evidence for the Fe^{IV}=O structure. In addition to it we point out here a new and very unexpected feature for compound II; the oxygen atom is exchanged between the Fe^{IV}=O heme and bulk water and the exchange reaction is catalyzed by a proton of the pK_a=8.6 residue which is hydrogen-bonded to the oxygen atom of the Fe^{IV}=O heme. The hydrogen-bonded Fe^{IV}=O bond gave the $\nu_{Fe=O}$ line at lower frequency than nonhydrogen-bonded one, and when the former ceased to exist at alkaline pH due to deprotonation, the exchange reaction no more took place. In this way, the hydrogen bonded proton from the protein was found to play an essential role in conversion of the $Fe^{IV}=O$ oxygen into water and thus in the catalytic activity of HRP. This has never been noticed for HRP until the present study. The oxygen exchange was also observed for a yeast cytochrome *c* peroxidase [3] and is presumably a characteristic feature of a peroxidase. Compound I has a short life time and is very photolabile. We observed the RRS of compound I using the mixed flow apparatus.

CcO is a terminal enzyme of the respiratory chain and catalyzes reduction of dioxygen. This enzyme is conveniently grouped into two functional units; Cyt *a* and Cyt a_3. Cyt *a* receives electrons from the respiratory chain and transfers them to Cyt a_3, while Cyt a_3 provides the catalytic site. The RRS of intermediates for the electron transfer were reported previously and the rate of the electron transfer from Cyt *a* to Cyt a_3 was discussed [4]. Studies on the catalysis of CcO are currently focused on identification of the oxygenated intermediates. Babcock et al. [5] applied the pulse-laser transient Raman technique to CcO, offering evidence for the existence of a photolabile $Fe-O_2$ type intermediate. We have approached this problem from a different angle [6].

The CO-bound CcO and O_2 saturated buffer was mixed using a jet mixer and was flowed through the cell. Since photodissociation of CO initiates the enzymic reaction in this experiment, the spectrum should arise from intermediates which are produced while the photodissociated molecules stay in the laser beam (0.1 mmϕ). When the CO-bound form was measured without flowing, the RRS of the photodissociated form was obtained. When only the CO-bound enzyme was flowed and excited with a minimal laser power (10 mW) at 416 nm, the RRS of the CO-bound form was observed. Upon raising the laser power (60 mW), the RRS of the mixture of the photodissociated and CO-bound forms was observed, but under the same instrumental conditions as this except that the flowing sample is the mixed solution of the CO-bound enzyme with the O_2-saturated buffer, a RRS distinct from others was obtained. The S/N ratio was much improved compared with that of the CO-bound form, suggesting the increased resonance effect and accordingly detection of an intermediate having the Soret maximumm near 416 nm. Details of the RRS of oxygenated intermediates of CcO will be discussed in a talk.

References

1. S. Hashimoto, T. Tatsuno, & T. Kitagawa, Proc. Japan Acad. 60, Ser B. 345 (1984).
2. J. Terner A. J. Sitter, C. M. Reczek, Biochim. Biophys. Acta 828, 73 (1985).
3. S. Hashimoto, J. Teraoka, T. Inubushi, T. Yonetani, & T. Kitagawa, J. Biol. Chem. in press.
4. T Ogura, S. Yoshikawa, & T. Kitagawa, Biochemistry 24, 7746 (1985).
5. G. T. Babcock, J. M. Jean, C. N. Johnston, G. Palmer, & W. H. Woodruff, J. Am. Chem. Soc. 106 8305 (1984).
6. T. Ogura, S. Yoshikawa, & T. Kitagawa, Biochim. Biophys. Acta 832 220 (1985).

SPECTRAL BROADENING MECHANISMS IN BIOMOLECULES: CYTOCHROME C VS MYOGLOBIN

P.M. Champion, K.T. Schomacker and V. Srajer

Department of Physics, Northeastern University, Boston, MA 02115 USA

We have recently completed a detailed study of the optical lineshape of the intense π-π^* "Soret" transition of cytochrome c as a function of temperature. Use of the electron-nuclear coupling strengths found from the transform analysis of the resonance Raman excitation profiles allows us to simulate the measured lineshape changes using a multimode time correlator model that is exact at all temperatures. Figure 1 displays the fit to the room temperature absorption lineshape and exposes the presence of weak underlying transitions in the amplified ($\times$10) residuals. These transitions increase their intensity by an order of magnitude as the temperature is lowered to 14K. The thermal properties of the higher moments of the Soret band lineshape are not compatible with significant coupling to a low frequency bath; hence, large homogeneous linewidths (Γ = 370 cm^{-1}) are needed to fit the data. Inhomogeneous broadening is found to be negligible in the cytochrome c system; this is compatible with insignificant coupling to the surrounding "bath" and with the "tightly" bound, six-coordinate nature of the central iron atom.

The five coordinate deoxy myoglobin (Mb) system stands in marked contrast to cytochrome c. In this case, inhomogeneous broadening is significant and non-Gaussian. It is proposed that disorder in the iron-porphyrin displacement and orientation is responsible for the onset of the broadening. A quantitative theory to lowest order in coordinate space is presented. The fits to the Soret band lineshape, using input parameters obtained from x-ray crystallography yield a disorder parameter σ=0.27Å and a quadratic mapping constant b=2.2kK/Å^2. This latter constant predicts that a ca.1nm red shift in the Soret band maximum will be accompanied by a 0.15Å motion of the iron atom towards the heme plane. These results are in remarkable agreement with a variety of optical and EXAFS studies of the T→R transition in deoxy hemoglobin. The six coordinate Mb-CO spectra show no evidence of significant inhomogeneous broadening and can be fit using the same methods that are successful for cytochrome c.

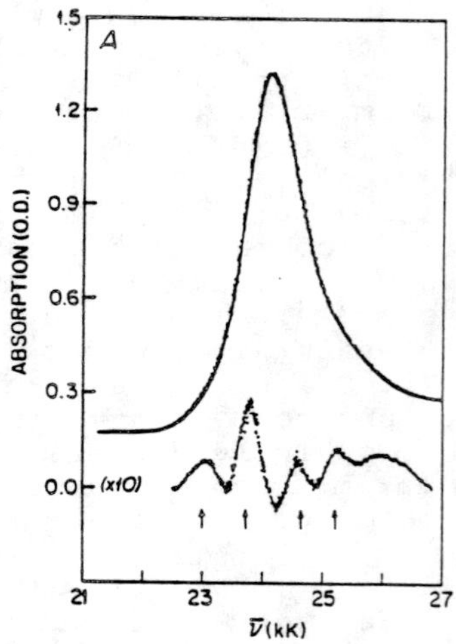

Figure: Theoretical fit to the Soret band lineshape. Random noise is artificially introduced into the abcissa of the data. The four arrows locate weak absorption features that increase dramatically at low temperature.

TIME-RESOLVED RESONANCE RAMAN SPECTROSCOPY OF PEROXIDASE AND CATALASE INTERMEDIATES.

James Terner, Andrew J. Sitter and Catherine M. Reczek

Department of Chemistry, Virginia Commonwealth University
Richmond, VA 23284, U.S.A.

Time-resolved resonance Raman spectroscopy has been shown, over the last few years, to be of great utility towards the elucidation of structural changes that occur during rapid chemical reactions [1-5]. Our most recent interests have been concerned with the structures of enzymatic intermediates of peroxidase and catalase, via rapid mixing techniques.

Peroxidase and catalase are heme enzymes which catalyze reactions of peroxides. Their enzymatic cycles involve oxidation of the heme to oxidation states above Fe(III). Though there are a number of variations, the intermediate states of these enzymes are most commonly known as compounds I and II. Upon reaction with peroxide, the resting enzyme, which contains an Fe(III) heme, is converted to compound I, which is two oxidation equivalents above the resting enzyme. A one electron reduction of compound I results in compound II. Compound I usually contains an Fe(IV) porphyrin π-radical cation. Compound II is an Fe(IV) heme.

The configuration of the peroxide derived oxygen in compound II has been described by several different proposals including Fe(IV)=O, Fe(IV)-OH, or Fe(IV)-OOH. A predominant formalism has been Fe(IV)=O, by analogy to the oxo-groups which stabilize the higher oxidation states of transition metals. Identification by this laboratory, of high iron-oxygen stretching frequencies based on isotopically induced frequency shifts, for horseradish peroxidase compound II [6] and the peroxidase model, ferryl myoglobin [7], are indicative of an iron-oxygen double bond, Fe(IV)=O.

We have found that the Fe(IV)=O vibrational band of horseradish peroxidase compound II can exist as either of two frequencies depending on pH [8,9]. At acidic and neutral pH, the Fe(IV)=O stretching frequency assumes the lower of the two frequencies, 775 cm^{-1}, confirmed by an ^{18}O-induced shift to 743 cm^{-1}. At alkaline pH the Fe(IV)=O stretching frequency assumes the higher of the two values, 788 cm^{-1}, which is confirmed by an ^{18}O-induced frequency shift to 756 cm^{-1}. We have found the pK of this transition to be approximately 8.5 for isoenzymes B and C, in agreement with the previously reported pK of the heme-linked ionization of compound II of isoenzyme C. For isoenzymes A-1 and A-2, we found the pK of the transition to be approximately 7, in agreement with the reported pK of 6.9 for compound II of isoenzyme A-2.

Kinetic rate constants for the reactions of compound II have been shown to increase with decreasing pH, and to be influenced by D_2O. Our resonance Raman data show a sensitivity of the resonance Raman frequencies to deuteration below the pK of the compound II heme-linked ionization, but not above, suggesting the involvement of hydrogen bonding of the oxyferryl group to a distal amino acid group.

Distal histidine groups are believed to be present in isoenzyme C and the A isoenzymes, which are likely responsible for the observed ionizations. Though the pK values of compound II for isoenzymes A-1 and C are 6.9 and 8.5 respectively, the variation in pK of histidine is well known to be dependent on differences in neighboring charged groups.

Myoglobin in the Fe(III) heme state (metmyoglobin) can be made to react with hydrogen peroxide to form a compound known as ferryl myoglobin, which contains an Fe(IV) heme. This compound is similar in structure to horseradish peroxidase compound II, however its peroxidative activity is much lower. Resonance Raman data shows that the heme structures of horseradish peroxidase compound II [8,10] and ferryl myoglobin [7] are quite similar, though there are variances in the Fe(IV)=O stretching frequencies. It has been suggested that the interaction of a distal base with the sixth ligand is weak in myoglobin, but very strong in peroxidases. We observe no shifting of the Fe(IV)=O stretching frequency of ferryl myoglobin as pH is varied from pH 6 to pH 12, contrary to our observations for horseradish peroxidase compound II. Additionally, the ferryl myoglobin Fe(IV)=O resonance Raman frequency is high (797 cm^{-1}), showing no detectable sensitivity to deuterium substitution, suggesting a lack of hydrogen bonding to the oxo-group by a distal amino acid group.

Thus, the Fe(IV)=O stretching frequency of horseradish peroxidase can assume one of two possible values in the pH region from pH 5 to 11. Similar pH dependent frequency shifts were not observed for ferryl myoglobin, above pH 6. Based on the observation of deuteration effects, the shifts were interpreted to be due to hydrogen bonding of the oxygen of the compound II Fe(IV)=O group to a distal amino acid group which is likely to be a histidine. An alkaline deprotonation of the distal amino acid group apparently disrupts hydrogen bonding, resulting in the raised Fe(IV)=O stretching frequency, and is proposed to account for the lowering of compound II reactivity. The transition points of the frequency shifts corresponded to increases in the kinetic rate constants measured for horseradish peroxidase compound II that are observed below the pK of the transition. Thus, hydrogen bonding to the oxygen of the oxyferryl groups in peroxidase intermediates appears to play an essential role in peroxidase activity.

This work is supported by the Jeffress Memorial Trust, the National Institutes of Health (Grant BBCA 1 R01 GM34443-01A1), and a Virginia Commonwealth University Biomedical Grant-In-Aid. James Terner is an Alfred P. Sloan Fellow (1985 - 87).

References

1. J. Terner and M.A. El-Sayed (1985) Accts. Chem. Res. 18, 331.
2. J. Terner and M.A. El-Sayed, Proceedings of the Sixth International Conference on Raman Spectroscopy, Bangalore, India (1978).
3. J. Terner, J.D. Stong, T.G. Spiro, M. Nagumo, M.F. Nicol, and M.A. El-Sayed, Proceedings of the Seventh International Conference on Raman Spectroscopy, Ottawa, Canada (1980).
4. T.G. Spiro and J. Terner, Proceedings of the Eighth International Conference on Raman Spectroscopy, Bordeaux, France (1982).
5. J. Terner and J. Topich, Proceedings of the Ninth International Conference on Raman Spectroscopy, Tokyo, Japan (1984).
6. J. Terner, A.J. Sitter, and C.M. Reczek (1985) Biochim. Biophys. Acta 828, 73-80.
7. A.J. Sitter, C.M. Reczek, and J. Terner (1985) Biochim. Biophys. Acta 828, 229-235.
8. A.J. Sitter, C.M. Reczek, and J. Terner (1985) J. Biol. Chem. 260, 7515-7522.
9. A.J. Sitter, C.M. Reczek, and J. Terner (1986) J. Biol. Chem., in press
10. J. Terner, and D.E. Reed, (1984) Biochim. Biophys. Acta. 789, 80-86.

RESONANCE RAMAN SCATTERING STUDIES OF PHOTODISSOCIATED MbCO.

Denis L. Rousseau, Massimo Sassaroli and Siddharth Dasgupta
AT&T Bell Laboratories, Murray Hill, NJ 07974

To gain a complete understanding of the biological function of heme proteins, the molecular basis of ligand binding and the mechanisms which control ligand affinity must be examined. An approach to this understanding is to study the properties of the ligand-free photoproducts which may be generated by photodissociation of the carbon monoxide adducts. Examination of the photoproducts may be made by either time resolved techniques or cryogenic techniques. Both of these techniques have been applied to carbon monoxide-bound hemoglobin and have advanced significantly our understanding of the protein (1,2). In the studies of HbCO it was shown that the resonance Raman differences between the photoproduct and the deoxy preparation are the same at cryogenic temperatures as they are at room temperature using time resolved techniques (3).

Although myoglobin is the simplest of the heme proteins, many puzzling questions have arisen concerning the properties of its photoproduct. First, although both the deoxy and the carbon monoxide-bound forms of myoglobin are structurally similar to the corresponding forms of hemoglobin, in time resolved studies with 10 nsec pulses, no differences between the resonance Raman spectra of the photoproduct and the deoxy species were detected in myoglobin (4). In contrast, large differences were detected under the same conditions in the case of hemoglobin (1). This implies a very rapid relaxation of the myoglobin photoproduct. Secondly, it has been proposed (5) that at very low temperatures, the Fe-C distance of the photodissociated CO remains the same as that of the bound CO (within 0.05 Å). This must be reconciled with the observations that the properties of the heme are very similar to those in deoxy myoglobin and the properties of the CO are very similar to those of the isolated gaseous molecule (6). Finally, large differences between the resonance Raman spectra of the photoproduct and the deoxy species are observed at low temperatures (7) but corresponding differences are absent in room temperature spectra with 10 nsec pulses (1).

We have addressed these questions by doing a series of Raman experiments on low temperature photodissociated myoglobin and molecular energy minimization calculations of the relaxation processes following photodissociation. In the low temperature resonance Raman experiments we find that the iron-histidine stretching mode of the photoproduct is very weak at the lowest temperature ($\approx$ 4K) and its frequency is several wavenumbers higher than that of deoxy myoglobin. Upon warming toward 77K, the intensity of the mode increases but its frequency is unchanged. From studies of the structure of the distal pocket of the heme and the energy minimization calculations we have been able to determine a pathway for CO movement away from the iron atom.

From our results we are able to draw the following conclusions:

1. At 4K a photoproduct is generated in which the heme core is expanded (compared to deoxy myoglobin) and the iron-histidine bond is perpendicular to the heme plane (8). We attribute the core expansion to the absence of full out-of-plane relaxation of the iron atom in the photoproduct.
2. As the temperature is increased toward 77K the core contracts and the iron atom takes on its full out-of-plane position.
3. At higher temperatures (>77K), histidine tilting occurs and the heme relaxes to its deoxy conformation.
4. Structural analysis shows that there is a clear channel in which the CO can escape from the heme subsequent to photodissociation.
5. Energy minimization places the CO about 4 Å from the iron atom subsequent to simulated low temperature photodissociation.
6. The rapid relaxation of the heme at room temperature can be accounted for by the small shift in the F-helix between deoxy myoglobin and CO-bound myoglobin.

We thank Biodesign, Inc. of Pasadena, CA for the use of their Biograf molecular modeling and simulation system on which we carried out the structural analysis and energy minimumization calculations.

References

1. Friedman, J. M. (1985) Science **228**, 1273-1280.
2. Rousseau, D. L. and Ondrias, M. R. (1985) Biophys. J. 7, 537-545.
3. Ondrias, M. R., Friedman, J. M. and Rousseau, D. L. (1983) Science **220**, 615-617.
4. Findsen, E. W., Scott, T. W., Chance, M. R., Friedman, J. M. and Ondrias, M. R. (1985) J. Am. Chem. Soc. **107**, 3355-3357.
5. Chance, B., Fischetti, R. and Powers, L. (1983) Biochemistry **22**, 3820-3829.
6. Fiamingo, F. G. and Alben, J. O. (1985) Biochemistry **24**, 7964-7970.
7. Sassaroli, M., Dasgupta, S. and Rousseau, D. L., submitted.
8. Rousseau, D. L. and Argade, P. V. (1986) Proc. Nat. Acad. Sci. (USA) **83**, 1310-1314.

THE COBALT-NITROSYL STRETCHING VIBRATION AS A SENSITIVE RESONANCE RAMAN PROBE FOR DISTAL HISTIDINE-NITROSYL INTERACTION IN MONOMERIC HEMOGLOBINS.

Nai-Teng Yu, Helen Mackin Thompson, Hiroshi Mizukami+ and Klaus Gersonde*

School of Chemistry, Georgia Institute of Technology, Atlanta, Georgia 30332, USA; +Department of Biological Sciences, Wayne State University, Detroit, Michigan 48202, USA; *Abteilung Physiologische Chemie, Rheinisch-Westfalische Technische Hochschule (RWTH), D-5100 Aachen, Germany

Cobalt-substituted hemes are capable of binding nitric oxide in the presence of nitrogen bases (1). The cobalt-nitrosyl moiety is isoelectronic with the iron-dioxygen moiety. It is demonstrated in this paper that cobalt-substituted hemoglobins also bind nitric oxide. Therefore, the nitrosylated cobalt hemoglobins are of particular interest for investigating the interactions of the exogenous ligand with the protein side groups on the distal side of the heme. Since the $Fe-O_2$ stretching vibration has been first observed by Brunner in oxy Hb A (2) and later by other authors in many hemoglobins (3-5), it is of particular interest to find the corresponding Co-NO stretching frequency. The assignment of the Co-NO stretching mode is easier than that of the $\nu(Fe-O_2)$ as various isotopically labelled nitric oxides ($^{14}N^{16}O$, $^{15}N^{16}O$, $^{14}N^{18}O$) are available. By comparison of the stretching vibrations of the Co-NO and $Fe-O_2$ bonds we can determine the similarities or differences between the metal-ligand bonding interactions in these isoelectronic systems.

Four cobalt-substituted monomeric hemoglobins were studied via resonance Raman spectroscopy: Sperm whale myoglobin, leghemoglobin from soybean, insect hemoglobin from *Chironomus thummi thummi* (CTT III), and elephant myoglobin. These hemoglobins contain the same protoheme-IX as prosthetic groups but they are different with regard to their heme pockets. Thus, we expect that the porphyrin-protein or ligand-protein interactions are different in these hemoglobins. Sperm whale myoglobin has a distal histidine, whereas in elephant myoglobin this histidine is replaced by glutamine. In CTT III the "distal" histidine is turned to the surface of the molecule, thus incapable of interacting with the bound ligand. Leghemoglobin is characterized by a wide heme pocket providing a distal histidine. For exclusion of protein interactions we also studied the nitrosyl complex of cobalt "picket fence" porphyrins which have coordinated an axial base.

The Co-NO stretching vibration has been assigned in the resonance Raman spectra of various cobalt-substituted monomeric hemoglobins by employing isotope-labeling of nitrosyl ($^{14}N^{16}O$, $^{15}N^{16}O$, $^{14}N^{18}O$). Monomeric hemoglobins with a distal histidine (sperm whale myoglobin and leghemoglobin) exhibit this vibration at 573-575 cm^{-1}, whereas hemoglobins without distal histidine (elephant myoglobin and insect hemoglobin CTT III) show this vibration in the range of 553-558 cm^{-1}. The Fe-NO stretching vibration which occurs in the range of 554-556 cm^{-1}, does not reflect the distal histidine-ligand interaction. Therefore, the Co-NO moiety which is isoelectronic with the $Fe-O_2$ moiety is a good monitor for distal effects on the exogenous ligand of hemoglobins, especially due to the fact that in hemoglobins with distal histidine the $Fe-O_2$ stretching vibration (567-572 cm^{-1}) is similar to the Co-NO stretching vibration.

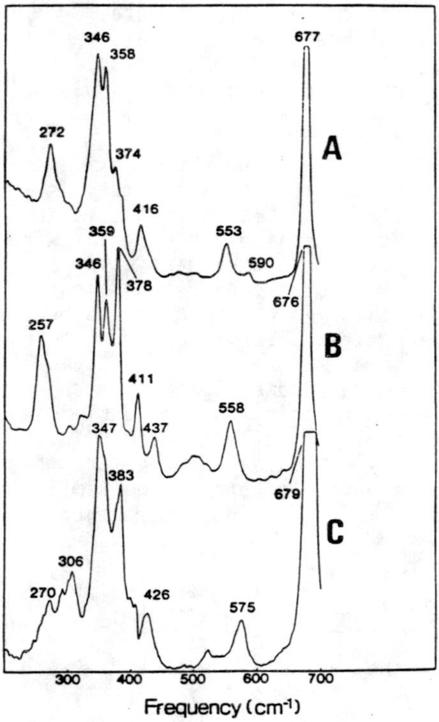

Fig. 1 Comparison of resonance Raman spectra of nitrosyl cobalt hemoglobins. A, insect Hb CTT III; B, elephant myoglobin; C, leghemoglobin. Excitation wavelength, 406.7 nm.

References

1. Hori, H., Ikeda-Saito, M., Leigh, J. S. and Yonetani, T. (1982) Biochemistry 21, 1431.
2. Brunner, H. (1974) Naturwissenschaften 61, 129.
3. Debois, A., Lutz, M. and Banerjee, R. (1979) Biochemistry 18, 1510.
4. Tsubaki, M., Nagai, K. and Kitagawa, T. (1980) Biochemistry 19, 379.
5. Waters, M. A. and Spiro, T. G. (1982) Biochemistry 21, 6989.

AXIAL COORDINATION IN NICKEL PORPHYRINS AND NICKEL-RECONSTITUTED HEME PROTEINS INVESTIGATED BY RAMAN-DIFFERENCE AND TRANSIENT-RAMAN SPECTROSCOPY.*

J. A. Shelnutt[a], K. Alston[b], E. W. Findsen[c], M. R. Ondrias[c], J. M. Rifkind[d]

[a]Sandia National Laboratories, Albuquerque, NM 87185
[b]Benedict College, Columbia, SC 29204
[c]University of New Mexico, Albuquerque, NM 87131
[d]National Institute of Health, Baltimore, MD 21224

Nickel porphyrins are of interest because of their occurrence in coal, shale, and petroleum deposits and because of their key role in biological conversion of CO_2 to methane.[1,2] The enzyme methylreductase is the enzyme that catalyses the final step, and possibly other steps, in the reduction of CO_2 to methane. The active site of methylreductase contains a nickel-sirochlorin derivative called F_{430}.[2] Understanding the involvement of F_{430} in methane production is of importance to the natural gas industry for utilizing abundant inorganic resources for production of gaseous fuels. Recently, we have used resonance Raman spectroscopy to investigate axial ligand processes at the metal in Ni-porphyrin complexes and Ni-reconstituted heme proteins.[3] These studies provide useful models for axial ligation and for Ni-porphyrin-protein interactions that may be important in the function of the methylreductase enzyme. In addition, the Ni-hemoglobin studies provide information about proximal histidine interaction with the metal site and changes in the interaction associated with quaternary structure. Also, for the first time, we have used transient Raman spectroscopic techniques to examine the dynamical aspects of axial ligation at Ni porphyrins in coordinating solvents[4,5] and in the Ni-reconstituted heme proteins.[6,7] The results of transient Raman investigations of the Ni porphyrins, which show photoinitiated axial ligand release and uptake, have important implications for fuel generating photocycles based on these processes.

Nickel-porphyrin complexes and Ni(II)-reconstituted hemoglobin (^{Ni}Hb), α chains (^{Ni}α), and myoglobin (^{Ni}Mb) were investigated using cw and transient resonance Raman spectroscopy. The state of axial coordination at the metal in these materials can be determined using the characteristic frequencies of the Raman marker lines arising from the porphyrin moiety. The marker frequencies distinguish between 4- and 6-coordination of Ni-porphyrins in non-coordinating and coordinating solvents and give evidence of a novel 5-coordinate species in the reconstituted proteins. The existence of only one axial ligand in the proteins is supported by identification of the axial ligand-Ni stretching vibration in ^{Ni}Hb and ^{Ni}Mb by isotopic substitution. The frequency of the Ni-ligand mode (236 cm^{-1} in ^{Ni}Hb) is consistent with histidine as the ligand.

The effects of structural changes in the protein environment of the Ni porphyrin are observed. T->R changes in the protein conformation result in structural changes in the Ni-histidine bond as evidenced by a

~5-cm^{-1} increase in frequency of ν(Ni-his) for ^{Ni}Mb (R-like structure) with respect to ^{Ni}Hb (T-structure). The increase in ν(Ni-his) is 7 cm^{-1} for ^{Ni}α. The changes at the metal also cause changes in the electronic structure of the porphyrin ring. The Raman lines that shift for the ^{Ni}α - ^{Ni}Hb comparison include the oxidation-state marker line ν_4 (+.4 cm^{-1}, 5-coordinate site), and the core-size marker lines ν_3 (-.6 cm^{-1}, 4-coordinate; +.1 cm^{-1}, 5-coordinate), ν_{10} (-.6, 4-coordinate; -1.1 cm^{-1}, 5-coordinate), and ν_2 (-.4, 4-coordinate; -1.7 cm^{-1}, 5-coordinate). For Fe hemoglobin, R/T shifts in ν_4 anti-correlate with shifts in ν(Fe-his). In contrast, for the Ni-reconstituted proteins ν(Ni-his) and ν_4 shift in the same direction. A correlation instead of an anti-correlation is also observed for ν(Fe-his) and ν_4 of peroxidases.[8]

Transient Raman investigations of the photoinitiated ligand release and uptake processes in excited states of the Ni porphyrins and Ni-reconstituted proteins indicate that the protein matrix effectively controls ligation dynamics. Laser excitation of ~10-ns duration induces the release of axial ligands in coordinating solvents, but not for the histidine ligand of Ni-reconstituted hemoglobin and myoglobin. Rapid geminate recombination of the proximal histidine accounts for the apparent lack of ligand release in the proteins. Ligand uptake is observed in ^{Ni}Hb when the 4-coordinate sites are excited. The Raman spectrum of the transiently ligated species is identical with that of the 5-coordinate sites. Binding of a single ligand within the laser pulse indicates the 4-coordinate Ni-porphyrin sites are in the normal heme pocket.

Photoinitiated axial ligation and de-ligation processes in Ni porphyrins can be understood in terms of the electronic states of the porphyrin and metal. Excitation of the 4-coordinate, $(d^2_{xy}, d^2_{xz}, d_{yz}{}^2)d^2_{z^2}$ ground state results in a d-d transition to the $d^1_{z^2}$, $d^1_{x^2-y^2}$ state. The vacancy in the d_{z^2} orbital allows binding of σ-donor ligands. Alternatively, excitation of the ligand-bound $d^1_{z^2}$, $d^1_{x^2-y^2}$ ground state results in the $d^2_{z^2}$ state and ligand loss. The transient unligated $d^1_{z^2}$, $d^1_{x^2-y^2}$ excited state is observed in non-coordinating solvents.

*This work performed at Sandia National Laboratories and supported by the U. S. Department of Energy Contract DE-AC04-76DP-00789 and the Gas Research Institute Contract 5082-260-0767. Work at UNM supported by NIH Grant GM33330, PRF, and Research Corp.

1. Jaenchen, R.; Diekert, G.; Thauer, R. K. FEBS Lett. 1981, 130, 133.
2. Pfaltz, A.; Jaun, B.; Fassler, A.; Eschenmoser, A.; Jaechen, R.; Gilles, H.; Diekert, G.; Thauer, R.K. Helv. Chim. Acta 1982, 65, 825.
3. Shelnutt, J. A.; Alston, K.; Ho, J-Y.; Yu, N-T.; Yamamoto, T.; Rifkind, J. M. Biochemistry 1986, 25, 620.
4. Findsen, E. W.; Shelnutt, J. A.; Friedman, J. M.; Ondrias, M. R. Chem. Phys. Lett. 1986, in press.
5. Findsen, E. W.; Shelnutt, J. A.; Friedman, J. M.; Ondrdias, M. R. Proc. Conf. Excited States Dynamics Porphyrin, Straub, K. D., ed. 1986, in press.
6. Findsen, E. W.; Shelnutt, J. A.; Alston, K.; Ondrias, M. R. J. Am. Chem. Soc. 1986, in press.
7. Shelnutt, J. A.; Alston, K.; Findsen, E. W.; Ondrias, M. R.; Rifkind, J. M. Proc. Conf. Excited States Dynamics Porphyrins, Straub, K. D. ed., 1986, in press.
8. Shelnutt, J. A.; Alden, R. G.; Ondrias, M. R. J. Biol. Chem. 1986, 261, 1720.

CORRELATIONS AMONG ν(Fe-CO), ν(C-O), AND δ(Fe-C-O) VIBRATIONS IN THE CO COMPLEXES OF VARIOUS HEME PROTEINS

T. Uno#1, Y. Nishimura, M. Tsuboi#2, K, Kita*1, Y. Anraku*2, R. Makino*3, T. Iizuka*3, and Y. Ishimura*3

Faculty of Pharmaceutical Sciences, University of Tokyo, Bunkyo-ku, Tokyo, *1 Department of Parasitology, School of Medicene, Juntendo University, Bunkyo-ku, Tokyo, *2 Department of Biology, Faculty of Science, University of Tokyo, Bunkyo-ku, Tokyo, and *3 Department of Biochemistry, School of Medicine, Keio University, Shinjuku-ku, Tokyo

Recently vibrational frequencies of the Fe-C-O linkage have been detected in the CO adducts of heme proteins [1-5] and iron-porphyrin complexes [6]. The ν(Fe-CO) stretching mode is enhanced within the Soret excitations, and is readily assignable by the CO isotope substitution experiment. It has been indicated that the vibrational frequencies of ν(Fe-CO) and ν(C-O) stretching modes are correlated, i.e., the higher the ν(Fe-CO) frequency, the lower the ν(C-O) stretching frequency [6]. To make sure this correlation , we have examined ν(Fe-CO), ν(C-O), and δ(Fe-C-O) vibrations of various CO complexes of cytochrome P450cam [2], cytochrome b562-o complex [3], horseradish peroxidase and its deuteroheme derivative [4], and tryptophan 2,3-dioxygenase.

Fig. 1 shows that the ν(Fe-CO) stretching frequency is linearly-correlated to the δ(Fe-C-O) bending frequency independent of the 5th ligand character. However, linear-correlations between the squares of the ν(Fe-CO) and ν(C-O) stretching frequencies are found to be classified into three groups depending on the 5th ligand; P-450's, which have cysteinyl ligand at the 5th site form the first group, and the other heme proteins, which have histidyl residue as the proximal ligand form the second group. On the other hand CO adduct of TPP(THF) (THF; tetrahydrofuran) forms the third group.

It is a striking feature that cytochrome aa3, which includes heme a as the prosthetic group, does not belong to the second group (see aa3 in Fig.1) even though histidine residue is at the 5th coordination site [5]. This anomaly should not be attributed to the effect of the substitution of the heme periphery since the frequencies of the Fe-C-O linkage observed in the CO adduct of deuteroheme-substituted horseradish peroxidase (deutero HRP) well belong to the second group [4]. It is of considerable interest to see that cytochrome b562-o complex, a terminal oxidase of *E. coli* aerobic respiratory chain, showed two Raman bands

Fig.1
Correlations between ν(Fe-CO) <u>vs</u> δ(Fe-C-O) (upper), <u>and</u> ν(Fe-CO) <u>vs</u> ν(C-O) (lower).

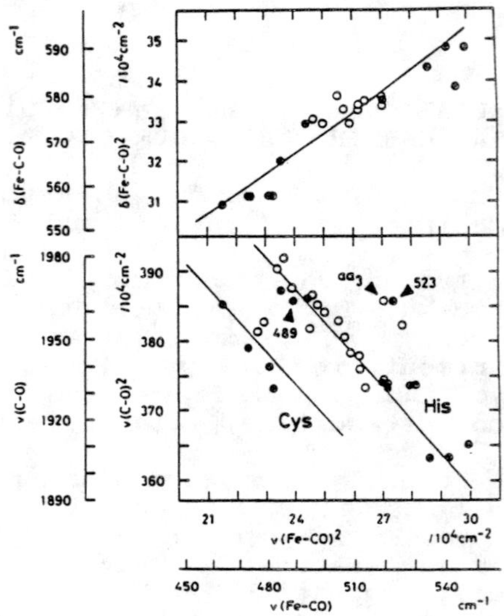

assingnable to ν(Fe-CO) stretching vibration at 489 and 523 cm^{-1} as shown in Fig.1. Upon excitation of strong energy densiy of laser photon, the line at 489 cm-1 disappeared because of the photodissociation of CO molecule, while the latter Raman line at 523 cm-1 has been found to be highly photo-resistant, which might be assigned to the copper bound form [3]. The anomalous CO binding site commonly found in oxidases, which could involve the copper atom, should be essential to the oxidase to express its biological function, i.e., the 4 electron reduction of molecular oxygen.

<u>REFERENCES</u>

1. Tsubaki, M., Srivastava, R. B., and Yu, N.-T. (1982) Biochemistry <u>21</u>, 1132-1140
2. Uno, T., Nishimura, Y., Makino, R., Iizuka, T., Ishimura, Y., and Tsuboi, M. (1985) J. Biol. Chem. <u>260</u>, 2023- 2026
3. Uno, T., Nishimura, Y., Tsuboi, M., Kita, K., and Anraku, Y. (1985) J. Biol. Chem. <u>260</u>, 6755-6760
4. Uno, T., Nishimura, Y., Makino, R., Iizuka, T., Ishimura, Y., and Tsuboi, M. submitted for publication
5. Argade, P.V., Ching, Y.C., and Rousseau, D.L. (1984) Science 225, 329-331
6. Yu, N.-T., Ward., B., Kerr, E. and Chang, C. K. (1983) Biochemistry, <u>22</u>, 4534-4540

#1 Present address: Department of Pharmaceutical Science, Nagoya City University, Mizuho-ku, Nagoya, Japan
#2 Present address: Faculty of Science and Engineering, Meisei University, Hodokubo, Hino, Tokyo

CORRELATION OF ν(Fe-CO) WITH LIGAND DISSOCIATION RATES FOR MONOMERIC *GLYCERA* HEMOGLOBINS.

R.S. Armstrong, R.M. Cooke, B.J. Reedy, M.J. Walsh and P.E. Wright*

School of Chemistry, University of Sydney, Sydney, N.S.W. 2006, Australia
*Department of Molecular Biology, Research Institute of Scripps Clinic, La Jolla, California 92037, U.S.A.

The monomeric hemoglobin fraction from the marine annelid *Glycera dibranchiata* can be separated reproducibly into three major and two minor components [1]. Recent NMR studies [1] have shown that each of these is a discrete hemoglobin with a unique heme electronic environment. The *Glycera* monomer components are of considerable interest in investigations of structure-function relationships in oxygen-binding heme proteins, since they possess an extremely low affinity for oxygen and exhibit the highest O_2 and CO off-rates yet recorded for a hemoglobin or myoglobin [2]. There are significant differences between the kinetics of each component, despite similar heme environments.

Used as a direct probe of metal-ligand binding in the heme, resonance Raman (RR) spectra have been obtained for the first time, and at high resolution, for the CO and deoxy forms of the three distinct major monomer components. Designated Gb(A), Gb(B) and Gb(C), they correspond respectively to the *Glycera* hemoglobins III, II and I obtained by Parkhurst *et al* [2]. Raman bands have been assigned to the ν(Fe–CO), δ(Fe-C-O) and ν(C-O) vibrations in GbCO, and to the $\nu(Fe\text{-}N_\varepsilon)$ mode in deoxy Gb, using isotopic substitution of the CO ligand and laser irradiation at 406.7 nm, 413.1 nm and 457.9 nm (Table 1). (Study of GbO_2 has so far been precluded by the high sensitivity of the proteins to oxygen).

TABLE 1

Adduct	ν(Fe-CO)	δ(Fe-C-O)	ν(C-O)	I_δ/I_ν	$\nu(Fe\text{-}N_\varepsilon)$ (deoxy)
Gb(A)CO	501	573	1968	0.20	221
$Gb(A)^{13}CO$	497	559	1923		
$Gb(A)^{13}C^{18}O$	490	557	1877		
Gb(B)CO	494	572	1968	0.18	222
$Gb(B)^{13}CO$	492	559	1922		
$Gb(B)^{13}C^{18}O$	483	557	1877		
Gb(C)CO	496		1967		
Mixed polymeric GbCO[3]	498,520	580	1963	0.10,0.20	214
Mixed monomeric GbCO[3]	496	575	1970	0.15	221

Figure 1. illustrates an inverse relationship between k_{off} and ν(Fe-CO) for the *Glycera* monomers. Grouping of the other hemeproteins (leghemoglobin, myoglobin and human hemoglobin) can be explained in part by their having in common, a 'distal' histidine capable of hydrogen bonding, as distinct from the leucine group of the *Glycera* hemoglobins.

Isotopic and intensity data allow the calculation and comparison of Fe-C-O bond angles in the monomers Gb(A) and Gb(B), using the method of Yu et al [4]. Differences in Fe-C-O bond geometry can be attributed to the effects of differing heme pocket environments as detected by NMR spectroscopy [1]. In this regard it is noteworthy, that despite differences of up to $7 cm^{-1}$ in ν(Fe-CO) values between monomer components for a given isotope of CO, no differences are found in ν(CO) greater than $1 cm^{-1}$ between the three proteins.

The values of ν($Fe-N_{\varepsilon}$) for deoxy Gb(A) and Gb(B) are almost identical, suggesting very similar protein influence upon the heme through the iron-proximal histidine bond in the two proteins.

FIGURE 1

Plot of rate coefficient for $Xb + CO \underset{k_{off}}{\overset{k_{on}}{\rightleftharpoons}} XbCO$ vs. ν(Fe - CO)

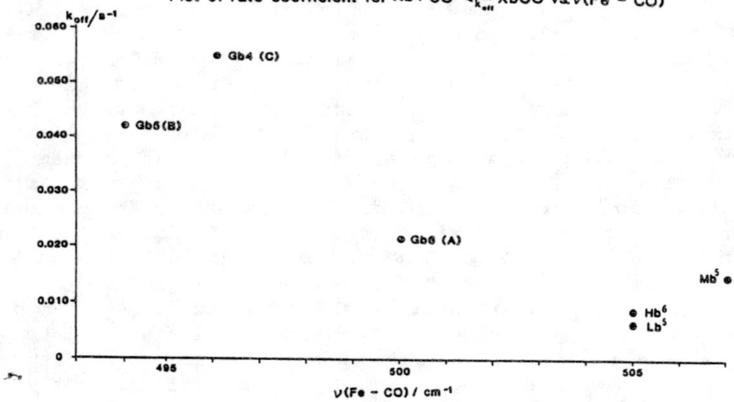

REFERENCES

1. Cooke, R.M. and Wright, P.E. (1985), *Biophys. Acta, 832,* 357-364.
2. Parkhurst, L.J., Sima, P. and Goss, D.J. (1980), *Biochemistry, 19,* 2688-2692.
3. Carson, S.D., Constantinidis, I., Satterlee, J.D. and Ondrias, M.R. (1985), *J. Biol. Chem., 260,* 8741-8745.
4. Yu, N.-T., Kerr, E.A. Ward, B., and Chang, C.K. (1983), *Biochemistry, 22,* 4534-4540.
5. Values of ν(Fe-CO) for LbCO and MbCO from Armstrong, R.S., Irwin, M.J. and Wright, P.E. (1982), *J. Am. Chem. Soc., 104,* 626-627.
6. Value of ν(Fe-CO) for HbCO from Rousseau, D.L., Tan, S.L., Ondrias, M.R., Ogawa, S. and Noble, R.W. (1984), *Biochemistry, 23,* 2857-2865.

RESONANCE RAMAN STUDY ON OXYGENATION AND DEOXYGENATION OF HEMOGLOBINS Hb-A AND Hb-F

Li Huan-Ling and Chen Shun-Cun*

Physics Department, Zhongshan University, Guangzhou, PR of China
* The First Affiliated Hospital of Guangzhou Medical College

Characteristic Raman bands $\nu(Fe-O_2)=572cm^{-1}$ and $\nu(=C-N)=1387cm^{-1}$, $1392cm^{-1}$ of ferroprotoporphyrin can be used as the probe to study the oxygenation and deoxygenation of hemoglobin components[1,2,3]. Otherwise, one can use Oxyhemoglobin Dissociation Curve (ODC) to interpret how the blood to combine with oxygen in lungs and to release oxygen to cell in tissue. In order to have a better understanding of some physiological and pathological phenomena of fetus, newborn baby and the "Hereditary persistence of fetal hemoglobin syndrome" patients, we studied the oxygenation and deoxygenation of hemoglobins Hb-A and Hb-F in different partial pressure of oxygen P_{O_2} by laser Raman measuring and blood-gas analysis. Some interesting results of our works is presented here as follows.

The intensity of Raman band $I_{572cm^{-1}}$ versus partial pressure of oxygen P_{O_2} is shown in Fig. 1. In where curve (a) is experimental results of Hb-A, curve (b) is that of Hb-F which isolated from cord blood and curve (c) is that of Hb-F which isolated from the blood of a "Hereditary persistence of fetal hemoglobin syndrome" patient.

Moreover, the determination of ODC in normal adult's blood (in which the content of Hb-A is 97%) and cord blood (Hb-F is 80%) by used the IL813 and 282 pH/Blood-Gas Analyzer as shown in Fig. 2. The solid curve (a) is that a well-known ODC of normal adult's blood determinated by J.W.Severinghaus[4], in which the data points are measured in our laboratory. The broken curve (b) is the ODC of cord blood measured in our lab., it was never seen in published up-to-date.

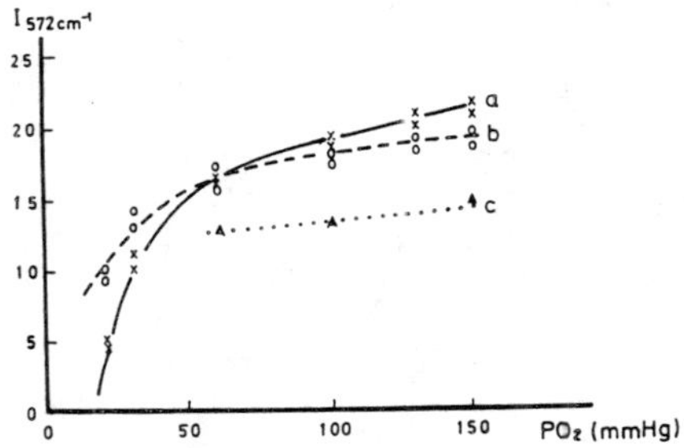

Fig. 1 I-PO_2 Curve of Hb-A (a)、Hb-F (b) and Hb-F↑ (c)

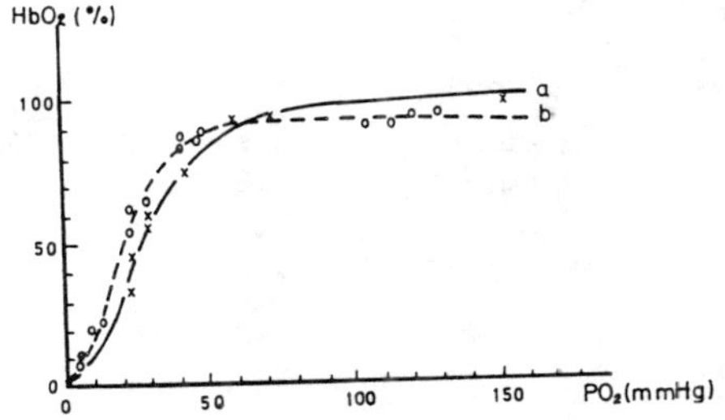

Fig. 2 Oxyhemoglobin Dissociation Curve of Adult's Blood (a) and Cord Blood (b)

Curves (a) and (b) respectively in Fig. 1 and Fig. 2 are similar in shape and tendency, and also intersect at $P_{O_2} \approx 60$mmHg and $HbO_2\% \approx 90\%$. In the range from 0 to 60mmHg P_{O_2}, the $HbO_2\%$ content of cord blood or Hb-F was higher than that of normal adult's blood or Hb-A, it means that the ability of Hb-F to combine with O_2 is better than that of Hb -A. While $P_{O_2} > 60$mmHg, the $HbO_2\%$ content of cord blood or Hb-F is lower than that of adult's blood or Hb-A. In this case, the oxygenation ability of Hb-F here is worse than that of Hb-A. Curve (c) is lower than curves (a) and (b) as shown in Fig. 1, indicated the oxygenation ability of Hb-F which isolated from a patient's blood is worse than that of Hb-F from cord blood.

Some physiological and pathological phenomena could be interpreted on the oxygenation and deoxygenation of hemoglobin from these three curves as follows:

1. Fetus living in maternal body has lack of O_2, i.e. $P_{O_2} \approx 60$mmHg, in this case, Hb-F in cord blood has a better ability of oxygenation and deoxygenation. Thus, fetus is capable of getting oxygen from and discharging waste to maternal body through the medium of cord blood and grows well.

2. Hb-F content in the blood of newborn baby (not older than two months age) always has high value. Once the baby was born, he must respire by lungs. In this time, P_{O_2} in blood is up to 90mmHg, the corresponding content of $HbO_2\%$ in Hb-F is lower than that in Hb-A, thus the ability of oxygenation of Hb-F is lower than that of Hb-A. But the newborn baby can still adapt to this conditions, since they are less in activity. Following that the baby activity increases with months, Hb-F content in blood decreases and Hb-A increases. It gives a better ability of oxygenation of Hb-A to adapt the more need of O_2 for baby. After six months, the content of Hb-F decreases from 80% to less than 2%, while Hb-A content increases to more than 90% in baby's blood. This is a significance mark of well development of a baby.

3. There are a few of childern and adults named "Hereditary persistence of fetal hemoglobin syndrome" patient, they maintain a high Hb-F content over 10~20% in their blood. Since the ability of oxygenation of Hb-F here is worse, they are easily tired with sporting and also have higher red cell count in their blood than the normal person. Obviously, this symptom could pathologically interpreted by our experimental results as see in Fig. 1 (c).

Reference

1. H.Brunner and H.Sussner, J. Mol. Biol. 70, 153 (1972)
2. Li Huan-Ling etal., "Chinese Journal of Laser", 11(4), 244 (1984)
3. Li Huan-Ling etal., IXth ICORS (Aug.27-Sept.1, 1984, Tokyo Japan)
4. J.W.Severinghaus, J. Appl. Physiol. 21, 1108 (1966)

RESONANCE RAMAN DETECTION OF BOUND DIOXYGEN IN CYTOCHROME $P450_{cam}$.

P.M. Champion,[a] O. Bangcharoenpaurpong,[a] A.K. Rizos,[a] D. Jollie[b] and S.G. Sligar[b].

(a) Dept. of Physics, Northeastern Univ., Boston, MA 02115 U.S.A.
(b) Dept. of Biochemistry, Univ. of Illinois, Urbana, IL 61801 U.S.A.

We have used resonance Raman spectroscopy and isotopic labelling techniques to unambiguously assign the dioxygen stretching frequency ($\bar{\nu}_{O-O}$) in the substrate bound oxygenated complex of cytochrome $P450_{cam}$. The frequency found for $\bar{\nu}_{O-O}$ in the $P450_{cam}$ system (1140cm^{-1}) is in remarkable agreement with recent studies of thiolate heme model compounds. The general features of the oxy $P450_{cam}$ Raman spectra are tabulated and comparisons are made with the oxy complexes of hemoglobin, myoglobin and various model compounds. Most of the results are qualitatively explained by consideration of electron donation into the $\pi_g^*(O_2)/d\pi(M)$ orbitals of the oxygenated complex (M = Fe, Co). It is also noted that the effect of the "extra" electron in the nitrogen base Co(II) oxy complexes, in some ways, parallels the effect of the lone pair electrons of thiolate in the oxy $P450_{cam}$ complex. This is evidenced by the enhanced resonance Raman activity of $\bar{\nu}_{O-O}$ in both the Co(II) and P450 systems as well as by the similarity of the $\bar{\nu}_{O-O}$ frequencies.

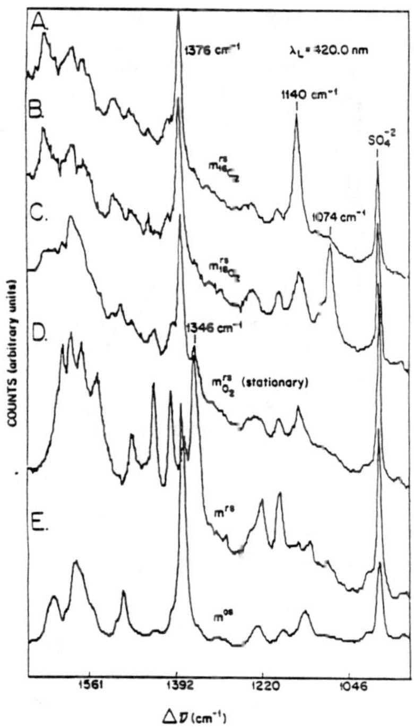

<u>Figure</u>: Resonance Raman spectra of oxygenated $P450_{cam}$ (A-C) along with spectra of the reduced (D) and oxidized (E) complexes for comparison. The sample temperature is <u>ca.</u> -60°C and spectra A and B are taken with a spinning cell and $^{16}O_2$ and $^{18}O_2$, respectively. In spectrum C, the spinning cell has been stopped for <u>ca.</u> 5 min. and the photoproduct spectrum indicates that most of the $m_{O_2}^{rs}$ complex has disappeared.

RESONANCE RAMAN DETECTION OF MANGANESE-NITROGEN TRIPLE BOND. FORMATION OF NITRIDOMANGANESE(V) PROTOPORPHYRIN IX IN MANGANESE MYOGLOBIN.

Motonari Tsubaki, Hiroshi Hori,[*] Takao Hotta,[*] Atsuo Hiwatashi, Yoshiyuki Ichikawa, and Nai-Teng Yu[#]

Department of Biochemistry, Kagawa Medical School, Miki-cho, Kita-gun, Kagawa 761-07, Japan; [*] Department of Biophysical Engineering, Faculty of Engineering Science, Osaka University, Toyonaka, Osaka 560, Japan; [#] School of Chemistry, Georgia Institute of Technology, Atlanta, Georgia 30332, USA

The activation of molecular oxygen via an iron-oxene intermediate with subsequent oxygen transfer to substrate has been suggested in the cytochrome P-450 reaction mechanism. Recently, metalloporphyrins have been shown to catalyze the transfer of a functionalized nitrogen atom from a tosylimide analogue of iodosobenzene. It is further reported that cytochrome P-450 catalyzed transfer and incorporation of a functionalized nitrogen atom into a C-H bond(1). To explain the chemical steps involved in cytochrome P-450-catalyzed substrate processing, the identification of the higher oxidation state of the heme iron reaction center is highly necessary. Model systems employing artificial metalloporphyrins and a variety of oxidants have shown the existence of stable higher oxidation states. By hypochlorite oxidation of Mn(OEP) (or Mn(TPP)) in the presence of ammonia, surprisingly stable nitridomanganese(V) porphyrins, MnN(OEP) (or MnN(TPP)) were obtained(2). This study was conducted to reveal the nature of the higher oxidation state of manganese porphyrin by resonance Raman spectroscopy.

Manganese protoporphyrin IX-substituted myoglobin (Mn(III)Mb) was prepared according to Yonetani and Asakura(3). To the alkaline Mn(III)Mb solution (pH ~10) containing ammonium ion, hypochlorite solution was added aerobically. This treatment caused the formation of a new complex, which was characterized by split α band (peaks located at 585 and 574 nm), β band (544 nm), and a single Soret band (430 nm). This complex was examined by resonance Raman spectroscopy employing 441.6 nm line from He-Cd laser (Kimmon) and a conventional Raman spectrophotometer (JASCO, R-800D). As shown in Figure 1, two Raman lines appeared at 1010 and 2006 cm^{-1} in addition to porphyrin ring modes (line a), the former is very close to the reported value of 1050 cm^{-1} for Mn(V)-N stretching frequency of $MnN(OEPMe_2)$ identified by IR spectroscopy. This manganese porphyrin complex is surprisingly stable. We could extract this complex from myoglobin by using 2-butanone. The extracted manganese porphyrin complex was identical with that prepared from Mn(III) protoporhyrin IX in alkaline aqueous solution in the same manner (i.e. oxidation by hypochlorite in the presence of ammonium ion). The resonance Raman spectra of this manganese porphyrin complex in alkaline aqueous solution gave a sharp Raman line at 1046 cm^{-1},

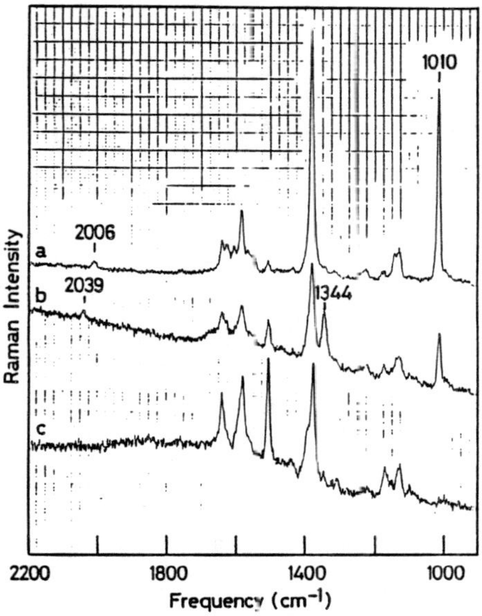

Figure 1. Resonance Raman spectra of Mn(V)≡NPPIX-containing Mb (a), Mn(III)Mb·N_3^- (b), and Mn(III)Mb (c). Excitation wavelength, 441.6 nm. During the measurement of resonance Raman spectra of Mn(III)Mb·N_3^-, a sharp line at 1010 cm^{-1} became evident. The 2039 cm^{-1} line is from manganese porphyrin-bound azide whereas the 1344 cm^{-1} line is from free azide ion (azide concentration, 0.25M).

confirming that this complex is exactly the nitridomanganese(V) protoporphyrin IX (Mn(V)≡NPPIX). It is very interesting to note that incorporation of manganese porphyrin complex into apomyoglobin caused the decrease of Mn(V)-N stretching frequency by as large as 35 cm^{-1}.

We have previously reported that, during the course of resonance Raman examination of Mn(III)Mb·N_3^-, two new additional Raman lines at 1010 and 2006 cm^{-1} appeared in addition to bound azide modes (650 cm^{-1} for bending, 2039 cm^{-1} for antisymmetric stretch) with excitation wavelengths at 413.1 and 406.7 nm(4). These two new additional Raman lines showed shifts to 983 and 1954 cm^{-1} upon $^{15}N_3$ isotope substitution. The identities of the frequencies of two additional Raman lines with those of Mn(V)≡NPPIX-containing myoglobin suggest the occurence of photo-induced decomposition of bound azide to generate nitridomanganese(V). Binding of azide ion to manganese(III) porphyrin is essential to the generation of nitridomanganese(V). When a diatomic harmonic oscillator model is used, the predicted frequency shift for the 1010 cm^{-1} Mn(V)-N stretching frequency is 27.2 cm^{-1} upon $^{14}N \rightarrow {}^{15}N$ isotopic substitution, almost identical with the observed one. The 2006 cm^{-1} mode is assignable to the first overtone of the Mn(V)-N stretching frequency on the basis of frequency shift upon isotopic substitution.

References

1. Svastis, E.W., Dawson, J.H., Breslow, R., and Gellman, S.H. (1985) J. Am. Chem. Soc. 107, 6427-6428.
2. Buchler, J.W., Dreher, C., Lay, K.-L., Lee, Y.J.A., and Scheidt, W.R. (1983) Inorg. Chem. 22, 888-891.
3. Yonetani, T. and Asakura, T. (1969) J. Biol. Chem. 244, 4580-4588.
4. Yu, N.-T. and Tsubaki, M. (1980) Biochemistry 19, 4647-4653.

RESONANCE RAMAN SPECTROSCOPY OF NITROSYL-MANGANESE HEMOGLOBIN AND MYOGLOBIN.

Niraja Parthasarathi and Thomas G. Spiro

Department of Chemistry, Princeton University, Princeton, NJ 08544

Manganese-substituted hemoglobin has been shown (1) to bind nitric oxide cooperatively, just as native hemoglobin bins carbon monoxide. The Mn^{II}-NO and Fe^{II}-CO systems are iso-electronic and a comparison of the resonance Raman spectra of the two is of interest. We present a resonance Raman characterization of nitrosyl manganese hemoglobin (NOMnHb), nitrosyl manganese myoglobin (NOMnMb), and a model compound, nitrosyl manganese(II) piperidine protoporphyrin dimethyl ester, (NO)Mn^{II}(Pip)(PPDME). [Figs. 1, 2 & 3]. All spectra were obtained with 413.1nm excitation. Isotope substitution studies show that the Mn^{II}-NO stretch is around 627cm^{-1} and the Mn^{II}-N-O bend is around 570cm^{-1}. These frequencies are markedly different from those for Fe^{II}-CO (Table 1) (2,3), implying a significant difference in electronic charge distribution between the two complexes. The N-O stretch was observed as a weak band at 1713cm^{-1} for (NO)MnMb, 1721cm^{-1} for (NO)MnHb and 1735cm^{-1} for (NO)Mn^{II}(Pip)(PPDME). A comparison with the C-O stretch in the carboxy-iron(II) complexes indicates a greater metal (d_π)-to-ligand (π^*) back-donation in the Mn^{II}-NO system. This results in a stronger Mn^{II}-NO bond (than Fe^{II}-CO) and a weaker N-O bond (as compared to C-O).

Figure 1

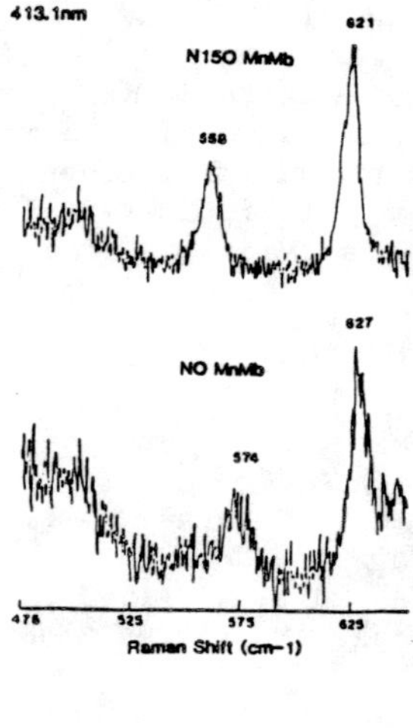

Figure 2

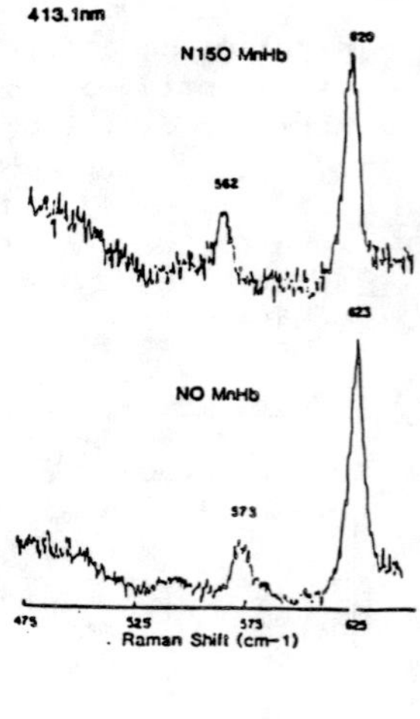

Figure 3

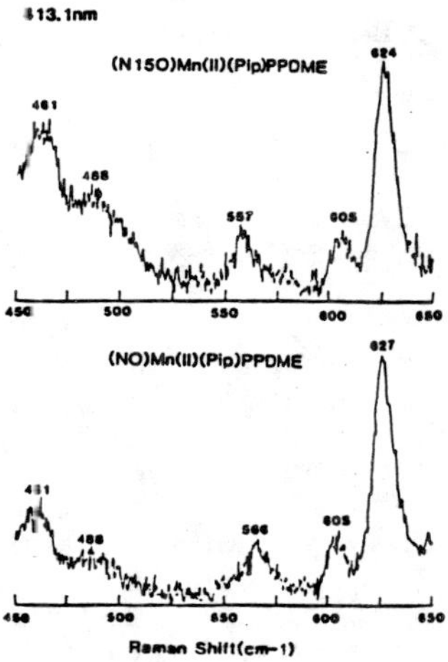

TABLE I

Complex	Fe-CO stretch (cm^{-1})	Fe-C-O bend (cm^{-1})	C-O stretch (cm^{-1})
(CO)FeMb (2)	512	577	1944
(CO)FeHb (2)	507	578	1951
(CO)Fe(ImH) PPDME (3)	495	-	1960

References:

(1) Gibson, Q.H., and Hoffman, B.M., J. Biol. Chem. 254 (11) 4691-4697 (1979).
(2) Tsubaki, M., Srivastava, R.B., and Yu, N.-T., Biochemistry 21, 1132-1140 (1982).
(3) Evangelista-Kirkup, et al. (1985) submitted for publication.

RESONANCE RAMAN SPECTRA OF OXO-FERRYL PORPHYRIN MODEL COMPLEXES

S. Hashimoto, Y. Tatsuno* and T. Kitagawa

Institute for Molecular Science, Myodaiji, Okazaki, 444 Japan
*Faculty of Engineering Science, Osaka University, Toyonaka, Osaka, 560 Japan

Extensive effort has been made to detect the Fe(IV)=O stretching Raman line on highly oxidized porphyrin by several research groups. Nakamoto et. al.(1) observed for the first time the Fe(IV)=O stretching Raman line at 852 cm^{-1}. Our group(2) and Terner et. al.(3) confirmed the Fe(IV)=O bond on horseradish peroxidase compound II and its frequency was determined to be 787 cm^{-1}. The frequency difference is about 60 cm^{-1}. This may be attributed to the difference between two compounds, that is, the coordination of the axial base to ferryl complex lowers the Fe(IV)=O bond order. Recently, Balch et. al.(4) characterized the five and six coordinated ferryl model complex at low temperature with absorption and NMR spectra; (tetramesitylporphyrin)iron(II)[TMPFe(II)] is oxygenated at -70°C to produce peroxo complex[TMPFe-O-O-FeTMP] and it decomposes into five coordinated ferryl complex [TMPFe(IV)=O] and ferric hydroxy complex[TMPFe(III)OH] at -30°C. The six coordinated ferryl complex is obtained by adding a base to tetraphenylporphyrin[TPP] peroxo complex. We have tried to measure the resonance Raman spectra of these five- and six-coordinated ferryl complexes and discuss the reaction mechanism of oxidation of ferrous porphyrin.

The resonance Raman spectra of TMPFe-O-O-FeTMP is shown in Fig. 1(A). The Raman line at 843 cm^{-1} and 646 cm^{-1} are not observed in the spectrum of TMPFe(II). The Raman line at 843 cm^{-1} is shifted to 810 cm^{-1} with $^{18}O_2$(B). When $^{16}O_2$ is used, no Raman line is observed around 810 cm^{-1}(C). For (B) and (C), toluene-d_8 is used to avoid the interference of solvent Raman line. The 843 cm^{-1} line is shifted to 847 cm^{-1} with ^{54}Fe. This Raman line may possibly be assigned to the O-O stretching mode of peroxo complex. But the frequency and the amount of isotope shift are very close to those of Fe(IV)=O stretching Raman line of the five coordinated ferryl complex. Therefore, the 843 cm^{-1} line is assigned to the Fe(IV)=O stretching mode of five coordinated ferryl complex. Spectrum (E) is obtained when the peroxo complex is warmed to room temperature and recooled to -70°C. Intensity of the Fe(IV)=O stretching Raman line decreased upon the warming but the Raman line at 648 cm^{-1} still remained.

Fig. 1(F) shows the spectrum of ferric hydroxy complex [TMPFe(III)OH] synthesized separately. Since the spectrum (E) and (F) are very similar, it is likely that the peroxo complex decomposes into TMPFe(III)OH upon warming. As the 644 cm^{-1} line is also observed on the spectrum (F), the Raman line is assigned to TMPFe(III)OH. Raman lines of TMPFe(IV)=O and TMPFe(III)OH are observed at the stage of peroxo complex, formation of which was confirmed with absorption spectrum. Therefore, the O-O bond of peroxo complex is cleaved by irradiation of laser light to form TMPFe(IV)=O. In addition, the intensity of Fe(IV)=O stretching Raman line decreased when the excitation laser power increased up to 20 mW. Therefore, TMPFe(IV)=O is further decomposed into TMPFe(III)OH with laser light. The same experiment was made on TPPFe(II), but the Raman line assignable to Fe(IV)=O stretching mode was not observed. Since the electronic structure is similar between TPP and TMP, the O-O bond of TPP peroxo complex must be cleaved by the laser light. Since TPPFe(IV)=O is unstable in toluene solution, it results in TPPFe-O-FeTPP. The steric hinderance to prevent porphyrin from dimerization is important to stabilize the Fe(IV)=O bond in solution.

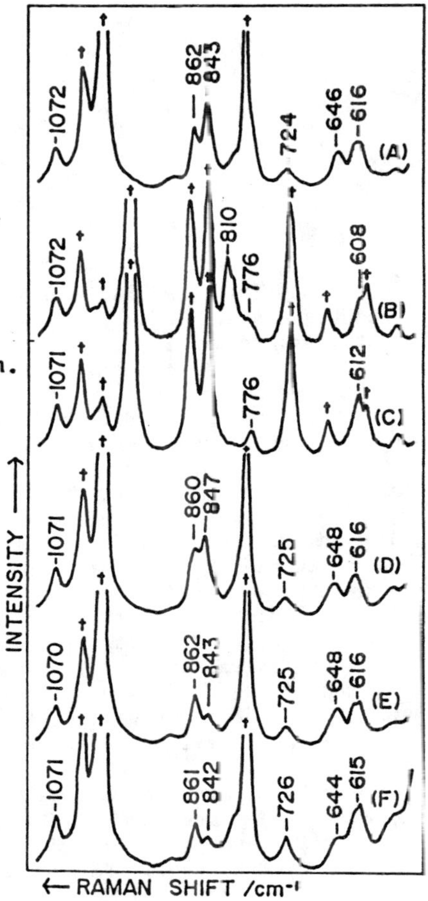

Fig. 1.
The resonance Raman spectra of TMPFe-O-O-FeTMP at -70°C. See text.

REFERENCE

(1) Bajor, K. and Nakamoto, K.; J. Am. Chem. Soc., 105, 3045-3046 (1984)

(2) Hashimoto, S., Tatsuno, Y. and Kitagawa, T.; Proc. Jpn. Acad. Ser B 60, 345-348 (1984)

(3) Terner, J., Sitter, A. J. and Reczek, C. M.; Biochim. Biophys. Acta 828, 73-80 (1985)

(4) Balch, A. L., Chan, Y., Cheng, R., La Mar, G. N., Latos-Grazynski, L. and Renner, M. W.; J. Am. Chem. Soc., 106, 7779-7785 (1984)

RESONANCE RAMAN SPECTROSCOPIC CHARACTERIZATION OF DIHYDROPORPHYRINS, THE PROSTHETIC GROUPS OF HEMES D AND CHLOROPHYLLS.

Thomas M. Loehr,[a] Laura A. Andersson,[a] Chi K. Chang,[b] and Kevin M. Smith[c]

[a]Department of Chemical, Biological, and Environmental Sciences, Oregon Graduate Center, Beaverton, OR 97006, [b]Department of Chemistry, Michigan State University, East Lansing, MI 48824, and [c]Department of Chemistry, University of California, Davis, CA 95616

Heme proteins and heme enzymes, containing iron-porphyrins as prosthetic groups, are ubiquitous biological molecules whose fundamental structural, chemical, and spectroscopic properties are well known from years of extensive investigation. In contrast, the properties of metallo-chlorins (metallated dihydroporphyrins) are considerably less well understood despite the fact that numerous examples exist in nature, including the bacterial iron hemes <u>d</u> and plant magnesium chlorophylls. Although heme prosthetic groups can generally be extracted from their proteins with full retention of the chemical integrity of the heme moiety, thus facilitating study of their properties, the dihydroporphyrins are often not amenable to extraction, suffering degradation or reverting to the oxidized porphyrin analog. It is, therefore, desirable to develop a structural probe for metallochlorins that is non-destructive and can be applied to intact biological systems for the characterization of prosthetic groups of unknown structure.

We have investigated the resonance Raman (rR) and infrared (ir) spectra of a number of metallochlorin complexes as models for the biological macrocyclic groups. Our heme <u>d</u> models included the metallo-chlorins Fe(III)-deuterochlorin IX and Fe(III)-photoprotoporphyrin IX [1], Ni(II)-tetramethylchlorin and Cu(II)-tetraphenylchlorin [2], and Cu(II)-<u>vic</u>-dihydroxy-2,4-dimethyldeuterochlorin IX and Cu(II)-hydroxy-gamma-<u>spiro</u>-lactone-2,4-dimethyldeuterochlorin IX [3]. We have also utilized the rR spectra of the synthetic chlorin-containing protein, sulfmyoglobin, as a spectroscopic model for these reduced hemes [4]. The Fe(III) complexes of methylpheophorbide <u>a</u>, mesopheophorbide <u>a</u>, and mesopyrophorbide were used as structural models for chlorophyll [5].

In these studies, the vibrational spectra of the model complexes were compared with those of analogous metalloporphyrins. The results have established a set of definitive criteria diagnostic of a metallochlorin that reflect the loss in symmetry of the macrocycle upon reduction. For a comparison of a chlorin with the corresponding porphyrin, these criteria include: (a) an increase in the total number of rR and ir bands; (b) largely identical frequencies between the rR and ir spectra of the chlorins; (c) the predominance of polarized rR bands with both Soret and visible excitation; (d) the distinct presence of one or more new, polarized rR bands in the vicinity of the oxidation-state marker band, ν_4; (e) a decreased frequency of the dominant band in the ν_4 region; (f) strong enhancement of the C_m-H bending modes as well as low-frequency modes with red excitation. Several of these features are demonstrated by the rR spectrum of Ni(II)-tetramethylporphyrin vs. Ni(II)-tetramethylchlorin shown in the Figure.

These spectroscopic properties appear to be generally characteristic of the dihydroporphyrin macrocycle. They have been consistently observed, regardless of the identity of the central metal ion, the pattern of pyrrole ring or meso-position substituents, or the chemical nature of the substitue In addition, they hold whether the macrocycle exists as a free molecule [1-3,5], or as the prosthetic group of a protein [4]. Furthermore, we have observed that the rR spectra are surprisingly sensitive to the nature of the substituents on the reduced pyrrole ring of the metallochlorin, even if they are no longer in conjugation with the macrocycle. This sensitivity extends to structural variations on the isocyclic ring of the chlorophyll derivatives. Work is in progress that extends these studies to complexes of tetrahydroporphyrin ring systems as models for the prosthet groups of bacteriochlorins, isobacteriochlorins, and sirohemes.

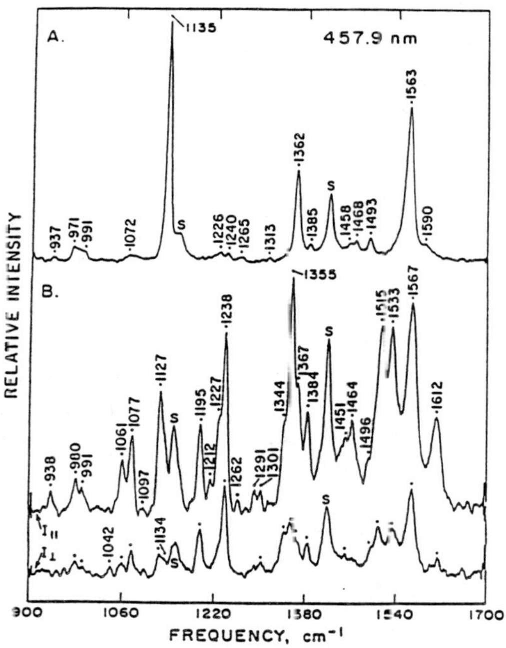

FIGURE
Resonance Raman spectra of CH_2Cl_2 solutions of

A. Ni(II)-TMP

B. Ni(II)-TMC upper (∥); lower (⊥).

REFERENCES

[1] L.A. Andersson, T.M. Loehr, C.K. Chang, & A.G. Mauk, J. Am. Chem. Soc., 107, 182 (1985).

[2] L.A. Andersson, T.M. Loehr, C. Sotiriou, W. Wu, & C.K. Chang, J. Am. Chem. Soc., in press.

[3] L.A. Andersson, C. Sotiriou, C.K. Chang, & T.M. Loehr, submitted.

[4] L.A. Andersson, T.M. Loehr, A.R. Lim, & A.G. Mauk, J. Biol. Chem., 259, 15340 (1984).

[5] L.A. Andersson, K.M. Smith, & T.M. Loehr, submitted.

ANOMALOUS RESONANT ENHANCEMENT OF THE QUASI-ELASTIC LIGHT SCATTERING OF CYTOCHROME-C.

L. Reinisch, K.T. Schomacker and P.M. Champion
Department of Physics, Northeastern University, Boston, MA 02115

We have measured anomalously large resonace enhanced quasi-elastic light scattering (QELS) from reduced cytochrome-c (cyt-c), a heme containing protein where a $\pi \rightarrow \pi*$ electronic excitation (the Soret band) of the porphyrin ring affords resonant excitation in the region near 24000 cm^{-1} . Resonant enhancement of scattering is often used in Raman experiments and has recently been observed in QELS.[1] Additionally, Rayleigh scattering cross sections of porphyrins and other chromophores have been reported by Anglister and Steinberg.[2] Their measured cross sections for scattering are typically a factor of 2 - 4 larger than those calculated from dispersion theory. We also measure scattering cross sections that are nearly three times larger than predicted from the dispersion relation.

Scattering experiments are performed at 90° using buffered solutions of reduced cyt-c with 1 M sulfate added as an internal standard. Both $\parallel$ and $\perp$ QELS and the 982 cm^{-1} Raman scattering of the sulfate are measured as function of protein concentration with incident light frequencies throughout the Soret region.

In figure 1 we plot the ratio of $\perp$ Rayleigh scattering to the $\parallel$ sulfate scattering (corrected for reabsorption and system response, η) versus the protein concentration, C_p. We write the scattered intensity ratio as

$$\eta \, \frac{I_{Ray}}{I_{SO_4}}(\tilde{\nu}) = C_p \, \frac{\sigma_p(\tilde{\nu})}{\sigma_{SO_4} \, C_{SO_4}} + \frac{\sigma_s \, C_s}{\sigma_{SO_4} \, C_{SO_4}} \qquad (1)$$

where I_{Ray} is the integrated $\perp$ (or $\parallel$) Rayleigh scattering, and I_{SO_4} is the integrated $\parallel$ Raman scattering of sulfate. The subscripts p, SO_4 and s represent the protein, the sulfate and the solvent, respectively. σ_p and σ_s are the Rayleigh scattering cross sections $\perp$ (or $\parallel$) to the incident polarization vector, and σ_{SO_4} is the $\parallel$ Raman scattering cross section of the sulfate. The correction for reabsorption involves the difference in the extinction coefficient between $\tilde{\nu}$ and $\tilde{\nu}-982$, as well as the depth of the laser from the front surface of the sample, ℓ. This depth is obtained from the intensity ratio of the Stokes to the anti-Stokes scattering from sulfate.[3] Incorrect determination of ℓ by 0.1 mm would result in σ_p changing by 30% or less. The slope of the least squares fit to the data in figure 1 is used to find $\sigma_p(\tilde{\nu})$ according to Eq. (1). The y-intercept in figure 1 contains the ratio of the solvent to sulfate cross sections. This is constant (within error) as expected. The straight line fits to the data not only permit us to separate the protein scattering from the solvent scattering, but they also show that there are no non-linear effects (such as thermal blooming). We have also varied laser intensity by more than a factor of 100 and obtain the same value for $\sigma_p(\tilde{\nu})$.

We observe resonant enhancement in both the $\perp$ and the $\parallel$ scattering, with the expected large background in the $\parallel$ data due to the small depolarization ratio, ρ_ℓ, of the nearly spherically symmetric protein material. The data attributed to the scattering from the resonant heme group are consistant with a depolarization ratio of $\frac{1}{8}$ as one would expect for D_{4h} symmetry. However, to avoid the issue of the exact magnitude of the protein background to be extracted from the ∥ data, we use only the ⊥ scattering and $\rho_\ell = \frac{1}{8}$ to find the total cross sections.

Shown in figure 2 are the total Raman and the QELS cross sections for cyt-c as a function of the laser frequency. The points are the measured data and the curves are derived from calculations consistent with the observed absorption band ($\Gamma = 370\ cm^{-1}$).[4] S^0 represents the predicted zero order (Rayleigh) scattering and S^1 the sum of all scattering first order in electron–nuclear coupling (*i.e.*, the Raman fundamentals). Overtones and difference/combination frequencies are given in the higher orders (S^2, S^3, *etc.*). The theory fits the Raman data quite well, but falls nearly a factor of 3 below the QELS data. The excess scattering of Anglister and Steinberg has been tentatively explained as due to interference from the UV resonances in the scattering molecule.[5] We have extensively explored this possibility and find it to be an unlikely explanation of our data. We are currently investigating local field effects and ensemble averaging approximations as possible alternative explanations of the anomalous scattering.

[1]D. R. Bauer, B. Hudson and R. Pecora, J. Chem. Phys. **63**, 588 (1975).
[2]J. Anglister and I. Z. Steinberg, Chem. Phys. **75**, 443 (1983).
[3]K.T. Schomacker, O. Bangcharoenpaurpong and P. M. Champion, J. Chem. Phys. **80**, 4701 (1984).
[4]K. T. Schomacker and P. M. Champion, J. Chem. Phys. (in press).
[5]S. G. Stanton, R. Pecora and B. Hudson, J. Chem. Phys. **75**, 5615 (1981).

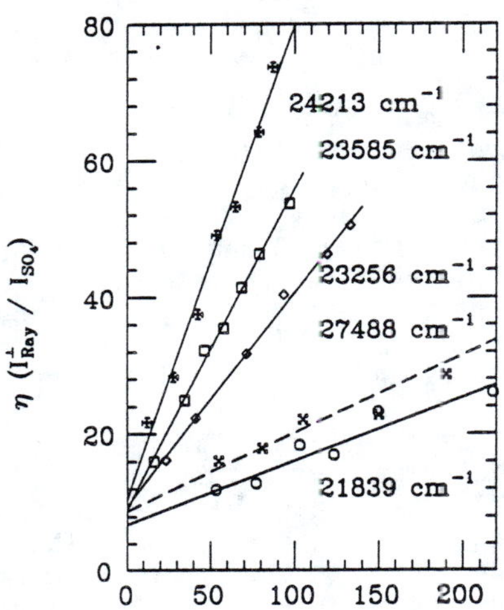

Figure 1 C_p [μM]

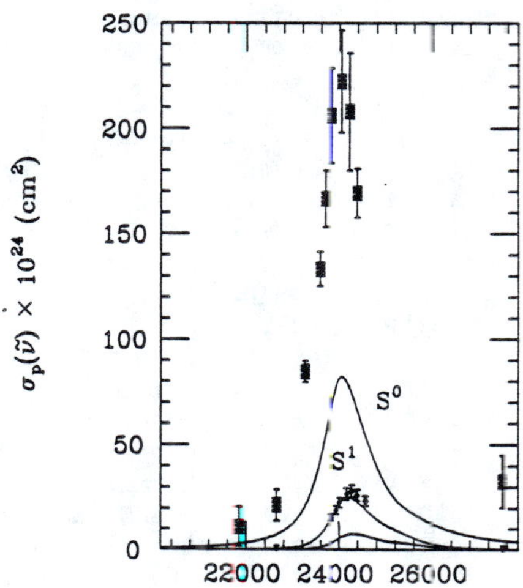

Figure 2 $\tilde{\nu}$ (cm^{-1})

THE EFFECTS OF DEHYDRATION UPON THE LOCAL HEME ENVIRONMENT IN EQUILIBRIUM AND PHOTOLYZED CARBONMONOXY HEMOGLOBIN.

E.W. Findsen, P. Simons, and M.R. Ondrias *

Department of Chemistry, University of New Mexico
Albuquerque, New Mexico 87131

In recent years various spectroscopic and theoretical studies have revealed that proteins are dynamic systems (1-3). The ligand binding heme proteins hemoglobin (Hb) and myoglobin (Mb) have been shown to exhibit both conformational and kinetic behavior at the heme active site which can be directly related to their ligand binding properties. In particular, these proteins possess internal motions that are strongly influenced by globin-solvent interactions. Changes in solvent viscosity and pH or dehydration of the solvent sphere surrounding the globin produce alterations in both the steady-state ligand conformation at the heme site and ligand binding kinetics (4-5). Steady-state liganded Mb displays a number of different heme-CO conformers which are interconvertible and can be spectroscopically identified by their distinct C-O bond strengths (vibrational frequencies) (4). The relative stability of these conformers has been found to be quite solvent dependent. Pulsinelli and coworkers (5) have demonstrated using reflectance infrared spectroscopy that removal of the globin hydration sphere profoundly affects the bound C-O frequency in both Hb and Mb.

We have employed resonance Raman spectroscopy to examine both the steady-state and kinetic behavior of the heme active site of Hb in various stages of dehydration. Resonance Raman scattering has been shown to be a highly specific probe of the heme protein active site (6). The molecular specificity of this technique allows for the direct assessment of the local heme structure. Specifically, the Fe-C bond strength and the C-O vibrational frequency of bound CO can be determined. Moreover, by employing time-resolved techniques, the response of the proximal heme pocket immediately subsequent to ligand photolysis can be monitored (2).

Resonance Raman spectra were collected using a nitrogen/dye laser system for an excitation source. Spectra of the dehydrated carbonmonoxy Hb were obtained with a defocused laser beam at 406 nm. Laser pulses of 440 nm were used to generate the photolyzed transient data. These pulses were tightly focused through a cylindrical lens. Each pulse served to both photolyze the HbCO and generate a transient Raman spectrum. Thus, the transient spectra represent the average heme species generated within the laser pulse width (~10 nsec) subsequent to photolysis. All data was obtained in a backscattering geometry. The samples were prepared on one face of an optical cuvette which was placed in a configuration such that the laser beam first penetrated the window opposite the sample before striking the sample itself. Thus, all spectra were gathered from the side of the dehydrated film that was exposed to the internal CO atmosphere of the cell and not the quartz cuvette face. This eliminated any potential interfering effects from the glass itself or from anomalous glass-protein interactions.

Spectra of carbonmonoxy HbA in various stages of dehydration (from solution down to .1 gram of water per gram of protein) have been obtained. They reveal that the heme skeletal mode frequencies are insensitive to protein dehydration. The C=O stretching mode is weakly present in these spectra. Its frequency was markedly affected by dehydration. Two conformers, one at ~1970 cm^{-1} and another at ~1950 cm^{-1}, were evident in the dehydrated samples while only the 1950 cm^{-1} mode is present in solution species. This corroborates previous IR studies (5). In contrast, the Fe-C mode (at ~507 cm^{-1}) was largely insensitive to protein dehydration, exhibiting a broadening to lower frequency only at the lowest hydration level (.1 g H_2O/g protein). The spectra of photolytic transients of dehydrated HbA are quite similar to analogous solution spectra indicating that the initial (10 nsec) proximal hemepocket geometry subsequent to photolysis is not greatly dependent upon extent of hydration.

We conclude that the solvent dependence of the C-O stretching mode must derive primarily from distal pocket effects and not from changes in either heme geometry or the degree of proximal heme pocket strain in the dehydrated protein.

1) Debrunner, P.G. and Frauenfelder, H (1982) Ann. Rev. of Phys. Chem., 33, 283.
2) Friedman, J.M., Ondrias, M.R. and Rousseau, D.L. (1982) Ann. Rev. of Phys. Chem., 33, 471.
3) Case, D.A. and Karplus, M. (1979) J. Mol. Biol., 132, 343.
4) Caughey, W.S. and Shimada, H. (1982) J. Biol. Chem., 257, 11893.
5) Brown, W.D., Sutcliffe, J.W. and Pulsinelli, P.D. (1983) Biochemistry, 22, 2914.
6) Spiro, T.G. (1983) in "Iron Porphyrins" part 2 (Lever, A.B.P. and Gray, H.B. eds) pg.89, Addison-Wesley.

This work supported by the NIH (GM33330).

RESONANCE RAMAN SPECTRA : ROLE OF DISORDER MODE IN TUNNELING OF CYTOCHROME c

B.D.Bhattacharyya

Department of Physics, St.Xavier's College, Calcutta 700 016 and Institute of Theoretical Biophysics, Baidyabati 712 222, Hooghly, India.

In addition to the Xray (Takano and Dickerson,1980), EPR (Schejter and Eaton,1984) and tunneling (Hopfield,1974 Chang and Austin 1982) studies, an extensive research on Resonance Raman (RR) spectra has been made in recent years (Spiro and Strekas,1974; Spaulding Chang et al 1975; Spiro et al 1979; Adar 1978, and Schomacker et al 1984)which revealed a clear understanding of the coordination environment of the active site of ferricytochrome c (cytc). Cyt c is a low-spin heme protein of the mitochondrial electron transport chain with two redox levels (cyt c^{3+}/cyt c^{2+}) and the transition metal ion (Fe^{3+}/Fe^{2+}) covalently bound to the heme prothetic group that is responsible for the intense Soret (406.7 nm) and vissible (530.9 nm) bands observed in the absorption spectra of cyt c. These bands arising from allowed $\pi-\pi^*$ porphyrin electronic transitions via the asymmetric (A_{2g}, B_{1g}, B_{2g}) and breathing (A_{1g}) modes in D_{4h} nomenclature.

Schomacker et al (1984) have calculated the Stokes shift parameter (ΔM) from RR excitation profiles and found that this value Δ_M is significantly smaller than that obtained in the tunneling mechanism. They suggested this discrepancy due to some unobserved bath of low frequency modes coupled to the Soret transition. But the value of Δ_M so obtained in the tunneling process could not correlate the EPR results.

The present paper is an extension of our work (Bhattacharyya 1976) based on the very recent spin-glass model of proteins (Stein 1985; Frauenfelder and Young 1985, Bhattacharyya 1986) where the concept of a configurational mode,or a tunneling state is attached to a localized disorder mode (Anderson, Halperin and Varma, 1972). Coupling this disorder mode to the depolarized Stokes fundamental of B2g symmetry

$$b\hbar\omega \begin{bmatrix} -q & 0 & 0 \\ 0 & q & 0 \\ 0 & 0 & 0 \end{bmatrix} + \frac{1}{2}\alpha q^2 + \text{disorder mode}$$

we obtain strain shift potential energy surfaces (if). With the usual adiabatic approximation the vibronic wave function s are

$$\Psi_i = \phi_i (r,x)\, N_n H_n [\alpha (q - \Delta q^i_{disorder})] \exp[-\tfrac{1}{2}\alpha^2 (q - \Delta q^i)^2]$$

where ϕ_i as given by

$$\phi_i = a_i|-,1/2\rangle + b_i|+,1/2\rangle + c_i|0,-1/2\rangle$$

are the ground and excited electronic wavefunctions. We obtain the tunneling rate expression

$$K_{if}(T) \simeq A \exp[-(\Delta E - \tfrac{1}{2}\Delta_M^2 \hbar\omega_M)^2 / 2\Delta_M^2 \hbar^2 \omega_M^2 (\bar{n}_M + \tfrac{1}{2})]$$

where $2\Delta_M$ is the Stokes shift parameter. If we estimate Δ_M from EPR results we find it is much smaller than that obtained from tunneling process. Our calculated values of $2\Delta_M \simeq$ 0.14 ev - which is slightly higher than the value obtained by Schomacker (1984).

Warshel et al (1981) found that the reorganization energy of cyt c is much smaller than the corresponding value of reorganization energy for electron transfer in water.

Our disorder mode model is extended to obtain an expression for solvent reorganization energy. Details of theoretical calculations will be reported at the conference.

References:

1. T.Takano and R.E.Dickerson, Proc.Natl.Acad.Sc. USA,77 6371 (1980)
2. A.Schejter and W.A.Eaton, Biochemistry 23, 1081 (1984)
3. J.J.Hopfield, Proc.Natl.Acad.Sc. USA 71, 3640 (1974)
4. A.Chang and R.Austin, J.Chem.Phys. 77, 5272 (1982)
5. T.G.Spiro and T.C.Strekas, J.Am.Chem.Soc. 96,338 (1974)
6. L.D.Spaulding,R.C.Chang,N.T.Yu and R.H.Felton,ibid 97, 2517 (1975)
7. T.G.Spiro,J.D.Stong and P.Stein, ibid 101, 2648 (1979)
8. F.Adar, J.Phys.Chem. 82, 230 (1978)
9. K.T.Schomacker,O.Bangcharoenpaurpong and P.M.Chanpion J.Chem. Phys. 80, 4701 (1984)
10. B.D.Bhattacharyya, Phys.Stat.Sol.(B) 74, 695 (1976)
11. D.L.Stein, Proc.Natl.Acad.Sc. USA, 82,3670 (1985)
12. H.Frauenfelder and R.D.Young, Molec.& Cellular Biophys. (1985,submitted)
13. B.D.Bhattacharyya, Proc.II Int.Symp.on Electron Micross; & Biophys. 49 B (1986)
14. P.W.Anderson,B.I.Halperin and C.M.Varma,Phil.Mag.25,1 ('72)
15. A.Warshel and D.Schlosser, Proc;Natl.Acad.Sc. USA, 78 5564 (1981)
16. A.Warshel and A.Lappicirella, J.Am.Chem.Soc.,103, 4664 (1981)

DETERMINATION OF RETINAL CHROMOPHORE STRUCTURE IN RHODOPSINS WITH RESONANCE RAMAN SPECTROSCOPY.

Richard A. Mathies,[1] S.O. Smith,[1] I. Palings[1] and J. Lugtenburg[2]

[1]Chemistry Department, University of California, Berkeley, CA 94720, USA
[2]Chemistry Department, University of Leiden, 2300 RA Leiden, The Netherlands

There are a variety of light-sensing pigments, called rhodopsins, which employ retinal or vitamin A as their principal chromophore. These include the visual pigment, rhodopsin, a light-driven proton pump called bacteriorhodopsin, and a light-driven chloride ion pump called halorhodopsin. In each system, light absorption drives the protein through a series of photochemical intermediates which involve structural changes of the retinal chromophore. These structural alterations have been studied in detail with FTIR, resonance Raman spectroscopy, and time-resolved resonance Raman. The vibrational frequencies and intensities in these spectra can now be quantitatively interpreted in terms of chromophore structure.

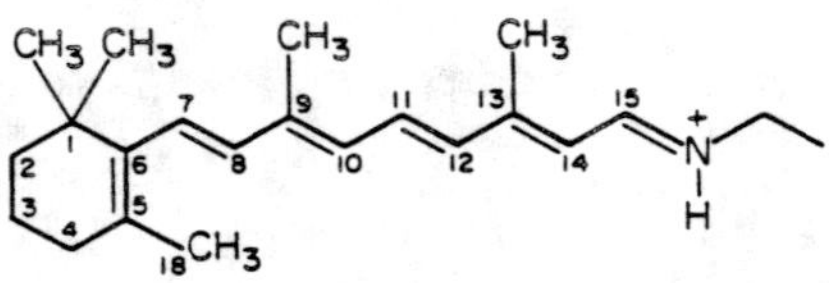

The first step in this analysis is the interpretation of the vibrational spectra of the retinal model compounds. The IR and Raman spectra of the retinal isomers and their isotopic derivatives have been assigned and the observed frequencies fit with normal coordinate calculations.[1-3] In general, isomerization from <u>trans</u> to <u>cis</u> causes a characteristic lowering of the modes involving C–C stretches that are adjacent to the isomerized C=C bond. The vibrational spectra of the Schiff base and protonated Schiff base of retinal have also been assigned.[4] Schiff base formation causes a significant increase of the C_{14}–C_{15} stretching mode when the electronegative oxygen is replaced by nitrogen. Protonation of the Schiff base causes an ~10 cm^{-1} frequency increase of all the fingerprint modes as a result of increased π-electron delocalization. This effect is felt most strongly at C_{14}–C_{15}.

These analyses have provided the bases for the development of methods for determining chromophore structure in pigments. For example, characteristic marker bands have been identified for determining the structure of the C_{13}=C_{14}, C_{14}–C_{15}, and C_{15}=N bonds.[5-7] The C_{13}=C_{14} configuration is identified by the characteristic frequency of the $C_{12}D$ + $C_{14}D$ in-plane rocking mode in the 12,14-D_2 derivative (~910 cm^{-1} for <u>trans</u> and ~940 cm^{-1} for <u>cis</u>).[7] The C_{15}=N configuration in protonated Schiff bases is identified by the coupling of the N-H rock with the C_{14}-

C_{15} stretch (strong coupling for C=N *cis* and weak coupling for C=N *trans*).[5] Also, the frequency of the C_{14}–C_{15} stretching mode is diagnostic for C_{14}–C_{15} conformation since s-*cis* isomerization causes an ~100 cm^{-1} lowering of this frequency.[6]

These methods have been used to study chromophore structure in bacteriorhodopsin. The primary photochemical step in bacteriorhodopsin is shown to be a C_{13}=C_{14} *trans* → *cis* isomerization with no change of C_{14}–C_{15} or C_{15}=N structure. This result is inconsistent with the model of Schulten and Tavan which predicts a C_{14}–C_{15} s-*cis* structure for the primary photoproduct.[8] The vibrational properties of the Schiff base moiety are strongly altered by the primary event, indicating that isomerization drives the Schiff base into a new hydrogen-bonding environment. This change of Schiff base environment may be involved in primary energy storage.

Raman studies of visual pigments have shown that the primary photochemical transition from rhodopsin to bathorhodopsin involves an 11-*cis* → all-*trans* isomerization with no change of C_{10}–C_{11} or C_{15}=N structure. These observations are inconsistent with the model of Liu and Asato, which predicts a 10-s-*cis* structure for the primary photoproduct.[9] The vibrational properties of the Schiff base moiety *do not* change significantly in the primary photoproduct, suggesting that energy storage is not associated with the Schiff base. Finally, the surprising lack of protein-induced shifts of the C-C fingerprint modes in rhodopsin compared with those of the 11-*cis* protonated Schiff base will be discussed in terms of proposed mechanisms for the "opsin shift" of the λ_{max}.

(1) B. Curry, A. Broek, J. Lugtenburg and R.A. Mathies, J. Am. Chem. Soc. 104, 5274 (1982).
(2) B. Curry, I. Palings, A. Broek, J.A. Pardoen, P.P.J. Mulder, J. Lugtenburg and R.A. Mathies, J. Phys. Chem. 88, 688 (1984).
(3) B. Curry, I. Palings, A.D. Broek, J.A. Pardoen, J. Lugtenburg and R.A. Mathies, Adv. Infrared Raman Spectr. 12, 115 (1985).
(4) S. Smith, A.B. Myers, R.A. Mathies, J.A. Pardoen, C. Winkel, E.M.M. van den Berg and J. Lugtenburg, Biop. J. 47, 653 (1985).
(5) S. Smith, A.B. Myers, J.A. Pardoen, C. Winkel, P.P.J. Mulder, J. Lugtenburg and R. Mathies, Proc. Natl. Acad. Sci. USA 81, 2055 (1984).
(6) S. Smith, I. Hornung, R. van der Steen, J.A. Pardoen, M. Braiman, J. Lugtenburg and R. Mathies, Proc. Natl. Acad. Sci. USA 83, 967 (1986).
(7) S. Smith, J. Lugtenburg and R. Mathies, J. Memb. Biol. 85, 95 (1985).
(8) K. Schulten and P. Tavan, Nature (London) 272, 85 (1978).
(9) R. Liu and A. Asato, Proc. Natl. Acad. Sci. USA 82, 259 (1985).

KINETIC STUDIES ON BACTERIORHODOPSIN BY RR SPECTROSCOPY

R. Diller and M. Stockburger

Max-Planck-Institut für biophysikalische Chemie, Am Faßberg,
D-3400 Göttingen, Federal Republic of Germany

Bacteriorhodopsin is a retinal-binding protein in the purple membrane of Halobacteria which acts as a light-driven proton pump. The chromophore is formed by a retinal molecule which is bound to the protein via a Schiff base linkage. Light absorption initiates a cyclic reaction process ("photocycle") in which the parent chromophore BR-570 is converted to the primary photo-intermediate K-590 which subsequently relaxes via various intermediate states (L-550, M-412, the numbers refer to the absorption maxima) to the original chromophore [1].

$$\text{BR-570} \xrightarrow{h\nu} \text{K-590} \xrightarrow{4\ \mu s} \text{L-550} \xrightarrow[\searrow H^+]{80\ \mu s} \text{M-412} \xrightarrow[H^+ \nearrow]{5\ ms} \text{BR-570}$$

Time-resolved RR spectroscopy is an ideal tool to study this photocycle which controls BR's biological function [2, 3]. Of particular interest is the reaction step between L-550 and M-412 during which a proton is removed from the chromophore and thus might be involved in the

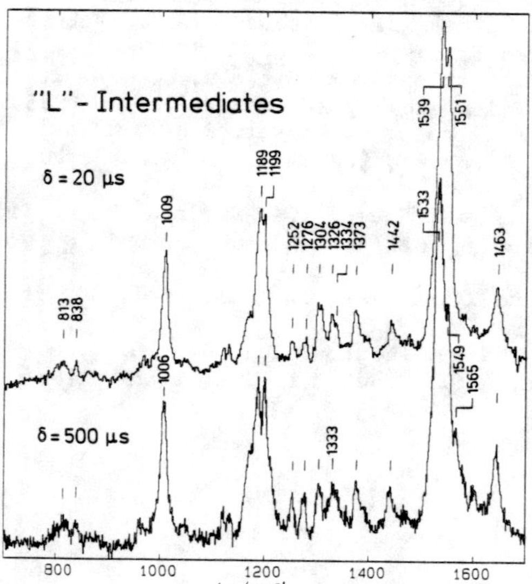

Fig. 1. RR spectra of "L" intermediates. Purple membrane in aqueous suspension is flown in a spinning cell across two parallel CW laser beams. Photolysis occurs in the first one and the intermediates are probed in the second one by RR scattering. The delay time, δ, between photolysis and probe event is given by the beam separation and the flow velocity. The spectra of L are derived by difference procedures. Photolysis at 647 nm; RR probe beam at 514 nm.

pump mechanism. We therefore have studied this reaction in detail. In a first attempt the temporal evolution of L and M was analyzed by band fitting on the basis of the frequency and shape of the strong C=C stretching bands which are characteristic of each species. Surprisingly it was found that an intermediate equilibrium between L and M is established which was ascribed to an oscillatory motion of a proton from the SB group to a proximate acceptor [4].

In this analysis it had been presupposed that only a single L intermediate is involved. In recent experiments, however, we found that this is not the case. This is demonstrated in Fig. 1 which shows spectra of L for delay times of 20 µs and 500 µs. The great similarity of the two spectra indicates that the L chromophore during this period does not undergo important structural changes (cis-trans isomerization, deprotonation). But on the other hand various spectral changes can be identified with certainty. The most significant one occurs in the C=C stretching region which therefore was studied for various delay times between 20 and 800 µs. In the two limits of this range two different species L(1) and L(2) can be clearly distinguished by their characteristic C=C bands (Fig. 2). From a preliminary analysis the following modification of the above reaction scheme is proposed.

$$BR \xrightarrow{h\nu} K \xrightarrow{4\ \mu s} L(1) \xrightarrow{80\ \mu s} M \underset{}{\overset{250\ \mu s}{\rightleftharpoons}} L(2) \xrightarrow{5\ ms} BR$$

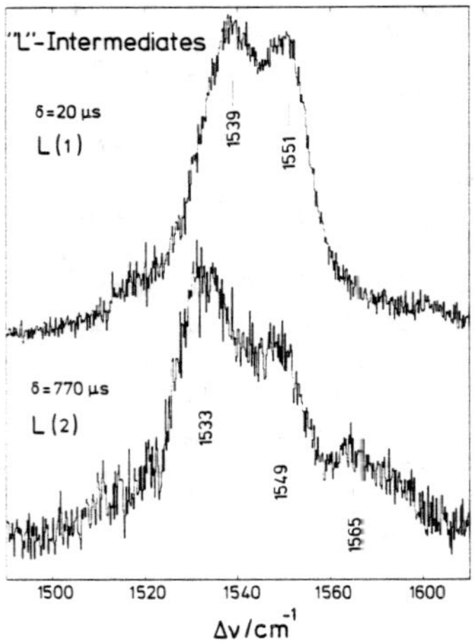

A possible explanation for the modification of L could be that a tyrosinate anion which has been found to appear nearly synchroneously with M [5] comes in close vicinity to the chromophore. This is supported by our previous observation that L(2) is preferentially formed at high pH [4]. The RR spectra thus provide an extremely sensitive probe for interactions of the chromophore with neighbouring groups which play an important role for BR's biological function.

Fig. 2. Spectra of two different L intermediates in the C=C stretching region. Conditions as in Fig. 1.

[1] W. Stoeckenius, R. H. Lozier, and R. Bogomolni, Annu. Rev. Biochem. 51, 587 (1982)

[2] S. O. Smith, J. Lugtenburg, and R. Mathies, J. Membrane Biol. 85, 95 (1985)

[3] M. Stockburger, T. Alshuth, D. Oesterhelt, and W. Gärtner, Advances in Spectroscopy 13 (1986) J. R. H. Clark and R. Hester, eds., John Wiley

[4] T. Alshuth and M. Stockburger, Photochem. Photobiol. 43, 55 (1986)

[5] J. H. Hanamoto, P. Dupuis, and M. A. El-Sayed, Natl. Acad. Sci. USA 81, 7083 (1984)

PICOSECOND CONFORMATIONAL INTERMEDIATES IN THE BACTERIORHODOPSIN PHOTOCYCLE.

T. L. Brack, D. Blanchard, L. Siemankowski, and G. H. Atkinson

Department of Chemistry and Optical Sciences Center, University of Arizona, Tucson, Arizona

The purple membrane of Halobacterium halobrium contains the protein bacteriorhodopsin (BR) which undergoes a cyclic reaction involving changes in the conformation and structure of the retinal chromophore. The value of BR as a model for the molecular dynamics associated with the retinal chromophore in visual pigments and for the associated proton and ion pumping is well established [1-3]. Although many aspects of both the BR dynamics and the structure of BR intermediates have been reported [2-3], significant parts of the molecular mechanism underlying its biochemical function remain either unknown or only partially characterized. The initial molecular changes occuring during the first few picoseconds after excitation are of particular interest since they involve the molecular mechanisms by which chemical energy is stored to drive the biochemical activity.

A variety of experimental methods have been successfully used to record the vibrational Raman spectra of BR intermediates [4-6]. The results presented here have been obtained with a two laser, pump-probe configuration derived from Nd:YAG pumping of two picosecond lasers [7]. This experimental approach provides the opportunity to independently control laser wavelengths, intensities, pulse duration, and focusing parameters. Such experimental versatility is fundamental to optimizing the initiation of the BR photocycle and the recording of picosecond time-resolved resonance Raman (PTR^3) spectra of the intermediates. Earlier results using single lasers to record PTR^3 spectra confirm this point [8,10].

The PTR^3 spectra of BR intermediates formed within 40 ps of excitation are of principle interest here. The time resolution of the measurements are 5 ps (autocorrelative pulsewidths of $\sim$7 ps). With 590 nm excitation (5ps) and resonance Raman probing between 590 nm and 610 nm (5ps), at least two different molecular transformations in retinal are observed to occur before 40 ps. Increased Raman intensity in the hydrogen-out-of-plane (800 - 1100 cm^{-1}) and fingerprint (1130 - 1240 cm^{-1}) regions are observed in the PTR^3 spectra recorded as a function of reaction time. These PTR^3 data, interpreted in terms of assignments based on isotopic studies of retinal [11-13], can be associated with two different molecular motions of the retinal chromophore. The changes in Raman band intensities appear as continuous functions of reaction time and can be assigned to twisting of the hydrocarbon chain, isomerization, and Schiff base deprotonation [14].

References

1. D. Oesterhelt and W. Stoeckenius, Proc. Natl. Acad. Sci. USA, 70, 289, (1973)
2. W. Stoeckenius and R. A. Bogomolni, Ann. Rev. Biochem., 51, 587, (1983)
3. R. H. Lozier, R. A. Bogomolni, and W. Stoeckenius, Biophys. J., 15, 955, (1975)
4. G. H. Atkinson, Advances in Infrared and Raman Spectroscopy, (R.E. Hester and R.J.H. Clark. eds.), North Holland Publ. London, 9, 1, (1981)
5. G. H. Atkinson, Time-Resolved Vibrational Spectroscopy, Academic Press, New York, N.Y. (1983)
6. G. H. Atkinson, in Chemical Dynamics: NATO Advanced Study Institute, (P. Rentzepis and C. Capellos eds.), Academic Press, New York, N.Y. (in press)
7. G. Rumbles, T. Brack, D. Blanchard, and G. H. Atkinson, Opt. Com., (submitted)
8. I. Grieger and G. H. Atkinson, Biochemistry 24 5660, (1985)
9. G. H. Atkinson, I. Grieger, and G. Rumbles in Time-Resolved Vibrational Spectroscopy, (A. Laubereau and M. Stockburger eds.), Springer-Verlag, Berlin, W. Germany, (1985)
10. G. H. Atkinson, T. L. Brack, I. Grieger, G. Rumbles, D. Blanchard, and L. Siemankowski, SPIE Proceedings (1986)
11. M. Braiman and R. Mathies, Proc. Natl. Acad. Sci. USA, 79, p. 403, (1982)
12. S. O. Smith, M. Braiman and R. Mathies, Time-Resolved Vibrational Spectroscopy, (G. H. Atkinson ed.), Academic Press, p. 219, (1983)
13. G. Eyring, B. Curry, R. Mathies, R. Fransen, I. Palings and J. Lugtenburg, Biochemistry, 19, p. 2410, (1980)
14. G. H. Atkinson, T. L. Brack, D. Blanchard, I. Grieger, G. Rumbles, and L. Siemankowski, in Time-Resolved Vibrational Spectroscopy, (G. H. Atkinson ed.), Gordon and Breach, New York, N.Y., (in press)

RAMAN STUDIES ON PIGMENTS RELATING TO PHOTOSYNTHESIS.

M. Tasumi, H. Hayashi, M. Fujiwara, H. Okamoto*, K. Iwata, and T. Noguchi.

Department of Chemistry, Faculty of Science, The University of Tokyo, Bunkyo-ku, Tokyo 113, Japan.
*Present address: Institute for Molecular Science, Okazaki 444, Japan.

The following studies have recently been performed in our laboratory on intact and metal-substituted chlorophylls, carotenoids in photosynthetic bacteria, partially deuterated all-*trans*-β-carotene, and tetradesmethyl-β-carotene.

1. Characterization of the Raman and infrared bands by substitution of the central metal.

Recently Fujiwara and Tasumi published a paper [1] on the downshifts of the Raman and infrared bands of chlorophylls *a* and *b* in going from five-coordinate state (one axial ligand to Mg) to six-coordinate state (two axial ligands). The downshifts were particularly notable in the C=C stretching region and were explained as a result of an expansion of the chlorin ring upon six coordination. It was then expected that the above bands showing the downshifts must be sensitive to substitution of the Mg atom with other metals. To examine this expectation experimentally, the Mg atom of chlorophyll *a* was substituted with Ni, Cu, Zn, and Ag, and the Raman and infrared spectra of these metal-substituted species were observed in the solid state and in solution. Nine series of metal-sensitive bands were found, as expected, in the region of 1700-900 cm^{-1} in both the Raman and infrared spectra, their frequencies changing in the order of Ni>Cu>Zn>Mg>Ag. The frequencies of these metal-sensitive bands were shown to be linearly correlated with the Ct-N distance (Ct is the center of the chlorin ring and N is the nitrogen atom of the pyrrole ring).

2. The intensity enhancement of the Raman band at 960 cm^{-1} and the torsion about the C=C bond in carotenoids *in vivo*.

Recently, Iwata, Hayashi, and Tasumi [2] studied the Raman spectra of carotenoids contained in chromatophores of *C. vinosum*, *R. palustris*, and *R. rubrum* and the same carotenoids extracted in acetone. A distinct difference between the Raman spectra of the *in vivo* and *in vitro* states was found in the intensity of a band at about 960 cm^{-1}; the intensity of this band was higher in the *in vivo* state. This band is assigned to the CH out-of-plane wag of the conjugated all-*trans* chain [3] (see below for a more detailed assignment), and its intensity is in fact very weak when carotenoid molecules are planar. The enhancement of the intensity of this Raman band is considered to reflect out-of-plane torsion about the C=C bond in the *in vivo* state, where carotenoid molecules exist as a complex with protein moieties. Thus, the intensity of the Raman band at about 960 cm^{-1} can be used as a marker of the nonplanarity of an all-*trans* carotenoid chain.

More recently, an inverse correlation was found to exist between the intensity of the Raman band at 960 cm^{-1} and the efficiency of the energy transfer from carotenoids to bacteriochlorophyll *a* in photosynthetic bacteria. The Raman spectra of chromatophores of the following photosynthetic bacteria were observed: (a) *R. sphaeroides* and its green mutant, (b) *R. capsulata*, (c) *R. gelatinosa*, (d) *R. rubrum*, (e) *R. palustris*, (f) *C. vinosum*. These bacteria can be classified into two groups according to the intensity of the 960 cm^{-1} band; group A consisting of (a), (b), and (c) shows the 960 cm^{-1} band with lower intensities, and group B consisting of (d), (e), and (f) with higher intensities. The efficiency of the energy transfer was estimated from the comparison of the excitation spectrum of fluorescence from bacteriochlorophyll and the absorption spectrum of carotenoids. The efficiency was high (80~90 %) for group A and low (30~40 %) for group B. These results suggest that the planarity of the carotenoid chain might be an important factor for ensuring the efficient energy transfer.

To establish the assignment of the 960 cm^{-1} band, partially deuterated β-carotene (11,12,11',12'-d_4) was prepared, and its Raman and infrared spectra were compared with the results of normal-coordinate calculations. For normal β-carotene (d_0 species) in solution, the 960 cm^{-1} band consists of two components at 967 and 955 cm^{-1}, the latter being a shoulder of the former. Comparison of the spectra of the d_0 species with those of the 11,12,11',12'-d_4 species has convincingly shown that the 967 and 955 cm^{-1} components of the d_0 species are due, respectively, to the in-phase CH out-of-plane wags of the $C_7=C_8$ (and $C_{7'}=C_{8'}$) and $C_{11}=C_{12}$ (and $C_{11'}=C_{12'}$) parts, where 'in-phase' means the simultaneous movements of the two hydrogens across the double bond to the same direction. Normal-coordinate calculations have been performed using the data for the d_0 and d_4 species as well as those for tetradesmethyl-β carotene.

1. M. Fujiwara and M. Tasumi, J. Phys. Chem. 90, 250 (1986).
2. K. Iwata, H. Hayashi, and M. Tasumi, Biochim. Biophys. Acta 810, 269 (1985).
3. S. Saito and M. Tasumi, J. Raman Spectrosc. 14, 310 (1983).

RESONANCE RAMAN STUDIES ON LIPOXYGENASE-INHIBITOR COMPLEXES.

William G. Gustafson and James T. McFarland

Department of Chemistry, University of Wisconsin-Milwaukee,
Milwaukee, WI 53201 U.S.A.

Lipoxygenases are the first enzymes in the metabolic pathway for the formation of leukotrienes, a class of compounds which is involved in a large range of physiological activity including inflammation response. Because of the potential physiological benefit of having potent enzyme inhibitors with dissociation constants in the micromolar range, compounds such as 4-nitrocatechol have been of great interest to investigators of this enzyme system. There are two 4-nitrocatechol-enzyme complexes; competitive inhibition is observed for the green inhibitor complex while an orange complex formed from the green complex exhibits noncompetitive inhibition.

We have investigated these two complexes along with model ferric complexes of 4-nitrocatechol by resonance Raman, RR, spectroscopy. The Figure below shows the RR spectrum of the green 1:1 complex, A, between ferric ion and 4-nitrocatechol, the green lipoxygenase-4-nitrocatechol complex, B, and 4-nitrocatechol (pH 4.0), C. Notice that the iron

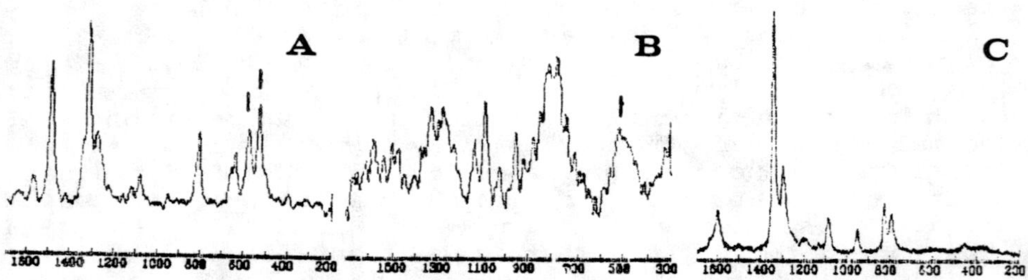

complex and enzyme complex of the inhibitor both exhibit vibrational bands in the 500 cm^{-1}-600 cm^{-1} region of the spectrum which are not present in the uncomplexed inhibitor. We have tentatively assigned two bands at 531 and 579 cm^{-1} to Fe-O stretching vibrations and take their presence to indicate direct inner sphere coordination between catechol oxygen and iron in both the enzyme and the model complex. There is also similarity between the RR spectra of the lipoxygenase-inhibitor complex and model system in the C-O stretch and O-H bend of the catechol and in the N-O stretch and C-N stretch of the nitro group indicating structural similarity. The orange form of the lipoxygenase-inhibitor complex shows little change in the Fe-O stretch, but substantial changes in the other vibrational modes indicating a difference in the structure of the two complexes. We are currently searching for an appropriate iron-inhibitor complex to serve as a model for the orange enzyme-inhibitor complex.

DETERMINATION OF FLAVIN OXIDATION STATE BY RESONANCE RAMAN SPECTROSCOPY.

Chris R. Lively, William G. Gustafson, and James T. McFarland

Department of Chemistry, University of Wisconsin-Milwaukee, Milwaukee, WI 53201 U.S.A.

We have been investigating the usefulness of resonance Raman (RR) spectroscopy for the investigation of flavoprotein intermediates. Figure I shows the RR spectrum of Glutathione Reductase, A, and a two electron reduced species, EH_2, B. The EH_2 spectrum is typical of the spectra

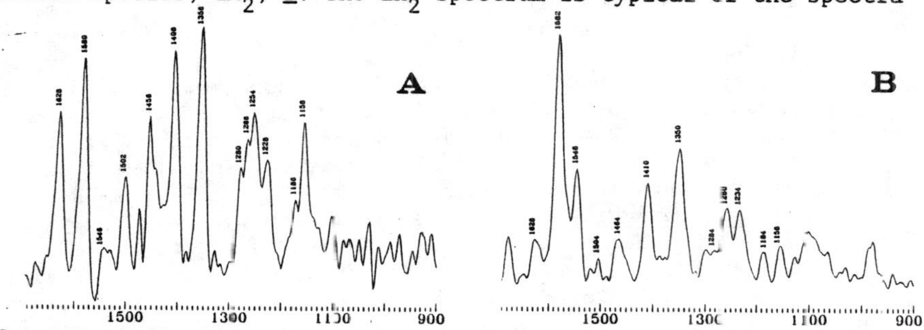

of oxidized flavin species; the excitation profile indicates that the charge transfer electronic band observed for this species may involve transfer of electrons from the thiolate anion of the reduced enzyme to the N-5 position of oxidized flavin. The RR spectrum of the four electron reduced EH_4 species appears to show band II at a frequency typical of a reduced flavin hydroquinone. Thus during reduction of Glutathione Reductase electrons flow to the cystine first and then to flavin.

The RR spectrum of the one electron reduced semiquinone of flavin has proven useful for the investigation of the hydrogen bonding of these intermediates. The RR spectrum of the blue neutral semiquinone of several proteins has been reported; Figure II shows the first RR spectrum of a red anionic semiquinone (Glucose Oxidase). We observe four vibrational bands in the red semiquinone; three of these are shifted approximately 14 cm^{-1} from the frequencies observed for the blue semiquinone of P-450 Reductase. This shift may result from the difference in overall charge of these species. The fourth band at 1340 cm^{-1} (VI) is shifted to lower frequency by 47 cm^{-1}; this band represents an N_1-C_2 stretching vibration in oxidized flavin. These data are consistent with increased hydrogen bonding at N_1 in the anionic semiquinone.

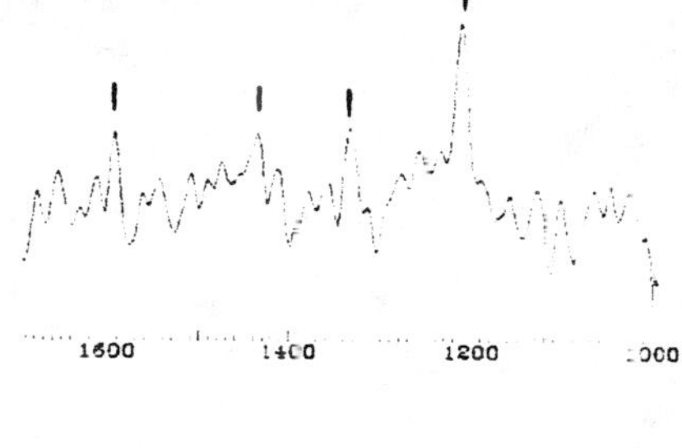

RAMAN AND RESONANCE RAMAN STUDIES OF SOME DEHYDROGENASES.

W. H. Hall, Jr., R. A. Price and P. W. Jagodzinski

Department of Chemistry, West Virginia University, Morgantown, WV 26506 USA

One common trait of a number of dehydrogenases is that they contain a subunit with a mass of approximately 40,000d. This subunit is present despite the broad range of reactions which the enzymes catalyze. The significance of this subunit as a link to the evolution of these proteins may be an important question. The similarity of these subunits is probably most important at the tertiary level since their primary and secondary structures show various degrees of homology between proteins.

We have investigated the secondary structures of a number of dehydrogenases using Raman spectroscopy and have found major differences among members of the group. We will discuss our investigations of alcohol dehydrogenases extracted from yeast (EC 1.1.1.1, YADH) and Thermoanaerobium brockii (TADH) as well as aldehyde dehydrogenase extracted from Saccharomyces cerevisiae (Baker's yeast, EC 1.2.1.5, ALDH). Our analyses of the secondary structures of these proteins will be presented and correlated with the degree of homology of primary structures where such are known. These dehydrogenases may be considered as models for the enzymes which are primarily responsible for the human metabolism of ethanol.

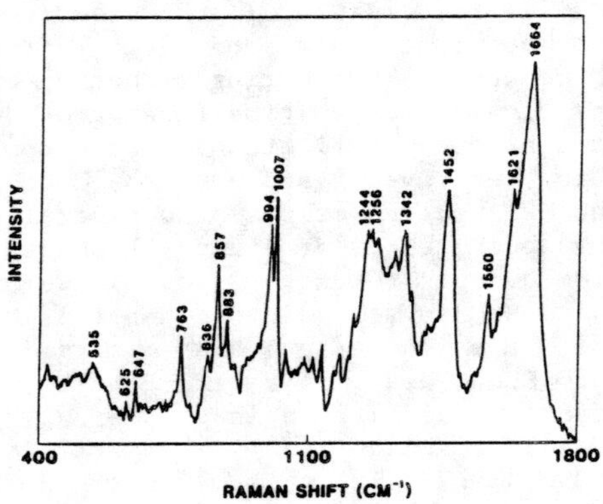

Fig. 1. Raman spectrum of YADH in 0.10 M phosphate buffer at pH = 7.50.

The data indicate that YADH, in contrast to liver alcohol dehydrogenase extracted from equine liver (EC 1.1.1.1, LADH), contains a larger percentage of alpha-helix than of beta-sheet [1]. The spectrum of YADH shown above is representative of the normal Raman spectra of many dehydrogenases. Using Raman techniques, LADH was found to contain 21% alpha-helix and 35% beta-sheet [2] while we have found YADH to contain 43% alpha-helix and 12% beta-sheet [1]. The relationship of this information with the possible evolutionary changes in dehydrogenases will be discussed in light of the possible connection between such changes and the homology of primary structure between YADH and LADH [3].

The study of the TADH was particularly revealing since relatively little spectroscopic study of this protein has been previously undertaken. This enzyme has a very broad substrate specificity and is extremely thermostable. The Raman data has been analyzed in terms of the residue environment and configuration and the secondary structure of this alcohol-aldehyde/ketone oxidoreductase [4].

Coupled with these structural studies of dehydrogenases we have used resonance Raman techniques to investigate the nature of their catalytic mechanisms at a molecular level. The observation of signals from extrinsic chromophoric probes within stabilized reaction intermediates has allowed us to propose mechanistic details essential for the formation of these intermediates. This has been particularly true in the case of the aldehyde dehydrogenase reaction. The details of these studies will be discussed in order to relate the importance of the subunit in dehydrogenases to their catalytic function.

REFERENCES

1. W. H. Hall and P. W. Jagodzinski, J. Raman Spectrosc., accepted for publication.

2. K. T. Yue, J.-P. Yang, C. L. Martin, D. L. Sloan and R. H. Callender, Biochem. Biophys. Res. Commun., 122, 225 (1984).

3. H. Jornvall, Eur. J. Biochem., 72, 425 (1977).

4. R. A. Price and P. W. Jagodzinski, manuscript in preparation.

RESONANCE RAMAN SPECTROSCOPY OF IRON-SULFUR PROTEINS

R.S.Czernuszewicz, S.Han, K.A.Macor, M.K.Johnson, M.W.Adams* and T.G.Spiro

Chemistry Department, Princeton University, Princeton, N.J. 08544, U.S.A.
*Exxon Research and Engineering Company, Annandale, N.J. 08801, U.S.A.

Significant progress in both instrumentation and sampling techniques has in recent years been achieved in this laboratory and improved resonance Raman(RR) data for 1Fe-, 2Fe-, 3Fe-, and 4Fe-center containing iron sulfur proteins, including assignments based on $^{34}S \rightarrow {}^{32}S$ and $^{54}Fe \rightarrow {}^{56}Fe$ isotopic substitution experiments have been obtained[1].

1). <u>4Fe MOLECULES.</u>

4Fe proteins contain [4Fe-4S] "cubes" consisting of interpenetrating Fe_4 and S_4 tetrahedra, the Fe corners of which are bound to cysteinate sulfur atoms. Their geometries are in good agreement with the $[4Fe\text{-}4S]^{2+}$ core structures of synthetic analogs, $[Fe_4S_4(SR)_4]^{2-}$[2]. A complete interpretation and assignments of $[4Fe\text{-}4S]^{2+}$ RR spectra, both as analogs and in proteins have been reached. Figure 1 compares RR spectra at 77K of oxidized <u>C. pasteurianum</u> (<u>Cp</u>) ferredoxin(Fd) and reduced <u>C. vinosum</u> HiPIP with those of the solution and solid $[Fe_4S_4(SCH_2Ph)_4]^{2-}$(A), all of which contain $[4Fe\text{-}4S]^{2+}$ cluster. They exhibit bands attributable to Fe-S stretching vibrations of both bridging (ca. 385,336,273,and 250cm^{-1}) and terminal (ca. 390 and 360cm^{-1}) sulfur atoms. <u>Cp</u> Fd_{ox} was reconstituted with either $^{34}S^{2-}$ or ^{54}Fe isotopes and their RR spectra compared with that of a native protein. RR spectra of A were complemented with IR data and the vibrational assignments assisted by ^{34}S substitutions in both cluster and terminal thiolate positions and by normal coordinate calculations. Appreciable splittings of 273(E^{br}), 250 (T_2^{br}) and 360cm^{-1}(T_2^{term}) bands were observed in the <u>Cp</u> Fd_{ox} and solid A spectra but not in those of <u>Cv</u> HiPIP and A in solution. Hence these two spectra were assigned in D_{2d} local symmetry in agreement with the compressed cube structure established by X-ray crystallography[3]. On the other hand, <u>Cv</u> HiPIP and A in solution display essentially T_d spectra; only five bands are observed instead of the eleven expected in D_{2d} symmetry.

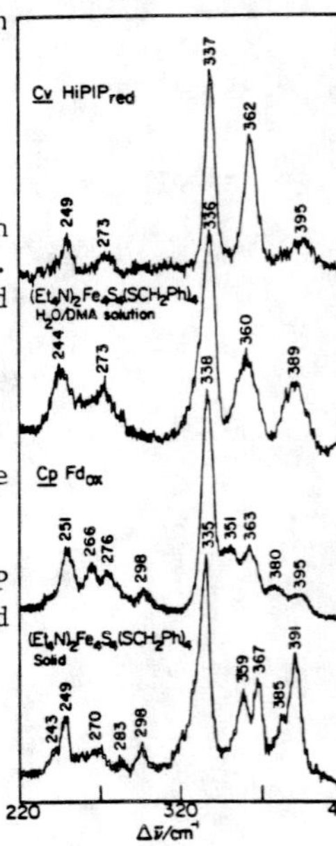

Fig.1.

2). <u>RUBREDOXIN.</u>

Detailed RR spectra were obtained from frozen solutions (77K) of oxidized <u>Desulfovibrio gigas</u>(D. gigas Rd_{ox}) and its ^{54}Fe substituted protein with excitation from 4880 to 5682Å. They display, as it is shown in Fig.2 five widely spaced bands in the Fe-S stretching region, 314, 324(sh), 348,

368 and 376cm^{-1}, instead of the two expected for tetrahedral FeS_4. The low-temperature difference Raman technique was designed and utilized to evaluate the ^{54}Fe isotope shifts [4]; all three components of the asymmetric $\nu_3(T_2)$ Fe–S sulfur stretch have been identified, at 376, 363 and 348cm^{-1}. The previous assignment of one of the ν_3 components to a 324cm^{-1} shoulder on the 314cm^{-1} ν_1 band is excluded by its lack of isotope shift. Numerous overtone and combination bands are observed, and a peak at 653cm^{-1} is assigned to a C-S stretch. The $\nu_2(E)$ and $\nu_4(T_2)$ S-Fe-S bending modes are located at 130 and 150cm^{-1}. The $\nu_2(E)$ band is enhanced with 4965Å excitation and its intensity is suggested to arise via its A_1 component in the D_{2d} effective optical symmetry of the chromophore. In contrast, the $\nu_4(T_2)$ mode is enhanced at 5682Å and it appears to be depolarized; a vibronic enhacement mechanism is suggested whereby the E component of this mode mixes the locally resonant $^6B_2$ charge transfer excited state with the nearby ^6E state.

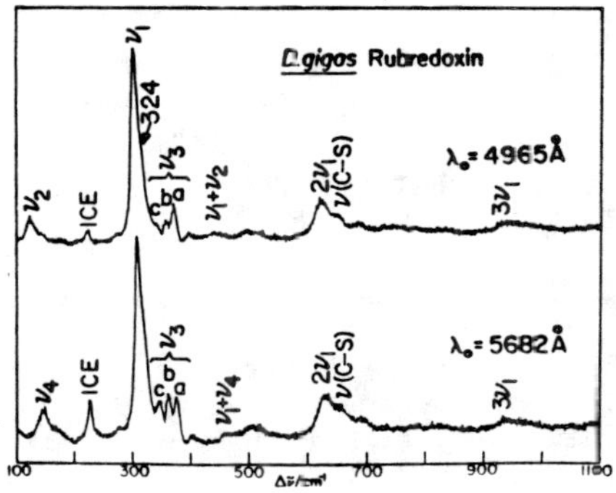

Fig.2

3). HYDROGENASES.

The hydrogenases(H_2-ases) are a group of iron-sulfur enzymes which catalyze the reversible activation of H_2[5]. Two distinct H_2-ases have been isolated from C. pasteurinum; H_2-ase I("bidirectional") and H_2-ase II("uptake"), both of which contain similar cubane-like [4Fe-4S] clusters, called F-centers. Bidirectional H_2-ase has two such centers while uptake has only one. In addition they each contain so called H-centers, whose structures are not well understood.

We report low frequency RR spectra of Cp H_2-ases obtained from frozen enzyme solutions with 4067Å excitation. With this increased resolution we were able to identify [4Fe-4S](F-centers) and [2Fe-2S](H-center) clusters in bidirectional H_2-ase and [4Fe-4S](F-center) and [3Fe-4S](H-center) clusters in uptake H_2-ase. Our results provide the first evidence for a major structural difference between the uptake and bidirectional enzyme H centers.

4). REFERENCES.

[1]. V.K.Yachandra,J.Hare,I.Moura,T.G.Spiro J.Am.Chem.Soc. 1983, 105, 6455.; V.K.Yachandra,J.Hare,A.Gewirth,R.S.Czernuszewicz,T.Kimura, R.H.Holm,T.G.Spiro J.Am.Chem.Soc. 1983, 105, 6462.;M.K.Johnson,R.S. Czernuszewicz,T.G.Spiro,R.R.Ramsey,T.P.Singer J.Biol.Chem. 1983, 258, 12771.; M.K.Johnson,R.S.Czernuszewicz,T.G.Spiro,J.A.Fee,W.V. Sweeney J.Am.Chem.Soc. 1983, 105, 6671.

[2]. J.H.Berg,R.H.Holm in "Iron-Sulfur Proteins"; T.G.Spiro,Ed., Wiley-Interscience, New York, 1982, Ch. 1.

[3]. C.D.Stout in "Iron-Sulfur Proteins"; T.G.Spiro,Ed., Wiley-Interscience, New York, 1982, Ch. 3.

[4]. J.F.Eng,R.S.Czernuszewicz,T.G.Spiro J.Raman Spectr. 1985, 16, 432.

[5]. M.W.W.Adams,L.E.Mortenson J.Biol.Chem. 1984, 259, 7045.

ULTRA-VIOLET (239NM) RESONANCE RAMAN SPECTROSCOPY OF AN ENZYME-SUBSTRATE INTERMEDIATE OF PAPAIN.

W.L. Peticolas, K. Bajdor, C.W. Wharton*, R.E. Hester**

Department of Chemistry and Chemical Physics Institute, University of Oregon, Eugene, Oregon 97403, USA; Department of Biochemistry, University of Birmingham, U.K.;** Department of Chemistry, Unlniversity of York, U.K.

A new technique is described for the study of intermediates formed in enzyme-substrate complexes using ultraviolet resonance Raman (UVRR) spectroscop. This technique uses an UV line from a YAG laser which is obtained by frequency doubling to 266nm and then frequency shifting by means of stimulated antiStokes Raman scattering to 239nm. The advantage of this line is that it falls in an absorption minimum for the aromatic residues, tyrosine and tryptophane, which permits the observation of resonance Raman spectra of enzyme-substrate intermediates whose absorption maxima fall within the range 225-245nm. In order not to decompose the sample and yet conserve it, a slow flow technique has been devised. Excellent UVRR spectra are obtained from the enzyme, papain, the substrate, N-acetyl-glycl-lysine methyl ester and the intermediate, an acyl enzyme which is a monothioester with an absorption band at 239nm. The slow flow technique which conserves sample uses a specially designed micro-mixing device which is fed from two hypodermic syringes which are driven by a constant displacement motor. In one syringe is the papain at a concentration of 2.5×10^{-4} M in a pH 7 phosphate buffer. In the other syringe is a 2.5×10^{-2} M solution of substrate also in the same buffer. When the two solutions are mixed, the substrate is present in a 100 fold excess over the enzyme. The mixed substrate--enzyme solution is introduced into a short glass capillary about 3 cm in length and 1 mm in internal diameter. A small pinhole is made in the side of the capillary and the laser beam is focused on this open pinhole. The Raman light is captured in back scattering geometry. The enzyme-substrate solution is moved in a slowly flowing stream through the capillary and out into a beaker. The solution does not flow out of the pinhole because of the high surface tension of water. In this way a continuous flow of sample is presented to the laser beam which pulses 10 times per second. We obtained very good spectra of the enzyme, the substrate and the enzyme-substrate mixture which showed clear evidence of additional Raman bands due to the formation of a chemical intermediate.

The top figure on the second page shows the UVRR of papain taken with 239nm laser light, the middle spectrum shows the UVRR spectrum of the substrate on the same scale (solid line), and expanded scale, (dotted line). The third or bottom spectrum shows the UVRR spectrum of the complex; the presence of a new band at 1680 cm^{-1} is seen which is ascribed to the carbonyl band of the acyl group bound at the active site.

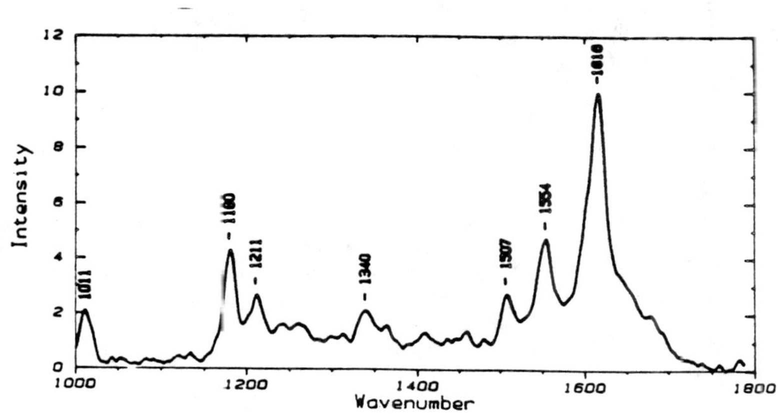
Intensity
Wavenumber
1011
1180
1211
1340
1507
1554
1616

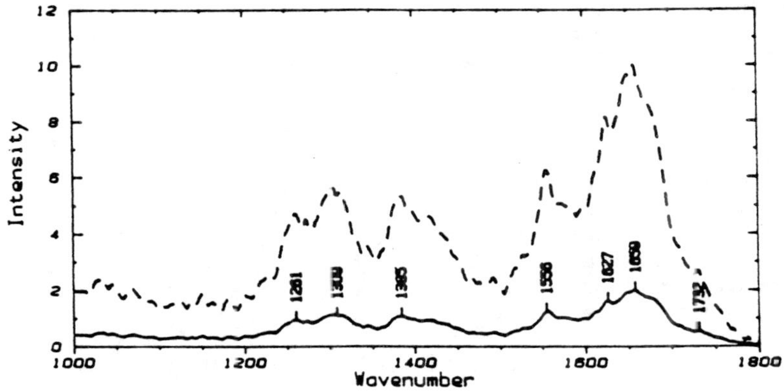
Intensity
Wavenumber
1261
1385
1556
1627
1659

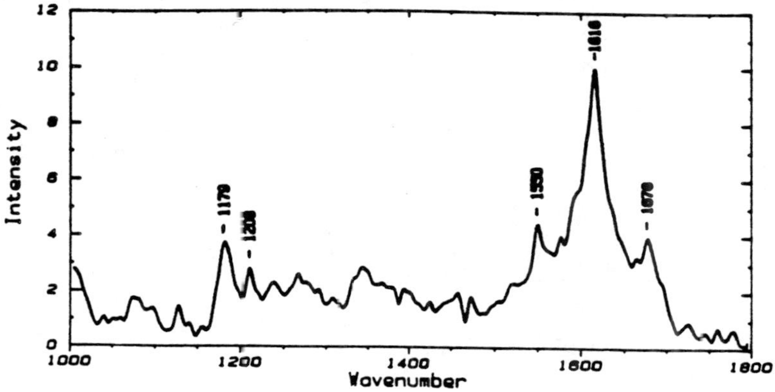
Intensity
Wavenumber
1178
1550
1616
1678

USING THE FURYLACRYLOLYL CHROMOPHORE AND UV RESONANCE RAMAN AND CLASSICAL RAMAN DIFFERENCE SPECTROSCOPY TO PROBE THE ACTIVE SITE ENVIRONMENTS IN PROTEINS

J. Wilson and W.L. Peticolas, Department of Chemistry, University of Oregon, Eugene, Oregon, 97403

Furylacrylolyl derivatives have been used in a number of earlier investigations to probe the active site of enzymes (1,2). Recently we have also begun to use chromophoric substates as a probe for the active site environment of proteins particularly enzymes (3,4). In our studies we have concentrated on the observation of the frequency carbonyl group of the substrate since an ester or amide is held in a double sp2 conformation with 5 or 6 atoms constrained to lie in a plane. Serine proteases such as chymotrypsin form an acyl enzyme intermediate (1). This intermediate is formed by means of an amide-ester exchange in which the amine portion of the amide substrate is split off and a simple ester is formed between the chromophoric acid from the substrate and the serine aliphatic hydroxyl group. In the past we have chosen chromophoric substrates in which the carbonyl vibration at 1660-1720 cm^{-1} is greatly enhanced in the RR spectrum (3,4). In our work, a linear plot is found to exist in a plot of the frequency, ν(C=O) of the carbonyl vibration versus the wavelength of maximum absorption, λ_{max} for the model ester in various solvents. When these parameters for the acyl enzyme intermediates are compared with the linear plot for the ester in various solvents it is found that the coordinates for the acyl enzyme do not lie anywhere near the line obtained for the solvent induced values. In this abstract we report the use of this approach with the FA-substrate. Although a very thorough study of this intermediate has been made previously (2), no comparison was made with the Raman spectra of the corresponding ethyl or methyl esters. One difficulty with the earlier studies is that the carbonyl bands are very week in the RR spectrum--indeed barely out of the noise level (2). Our RR spectrum of the FA-acyl enzyme using the 355nm line of the YAG laser showed many strong peaks but none in the carbonyl region. Consequently we have turned to classical Raman difference spectroscopy. Figure 1 shows the Raman difference spectrum of FA-chymotrypsin minus chymotrypsin. Here we see only one band at 1705 cm^{-1} which can be assigned to the carbonyl.

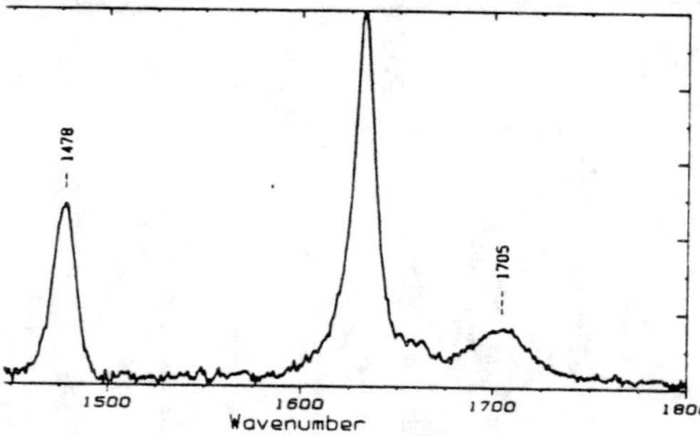

Fig. 1 R. Spectrum of FA chymotrypsin.

However when the classical Raman spectra were taken of the corresponding model compound, FA-ethyl ester in various solvents it was found that two carbonyl Raman bands were always seen (Figure 2). We assign these two bands to mixtures of the s-trans and s-cis conformations which seem to have equal populations in a varity of solvents. Figure 3 shows a plot of ν(C=O) vs. λ_{max} for the FA-ethyl ester (hexagons), the

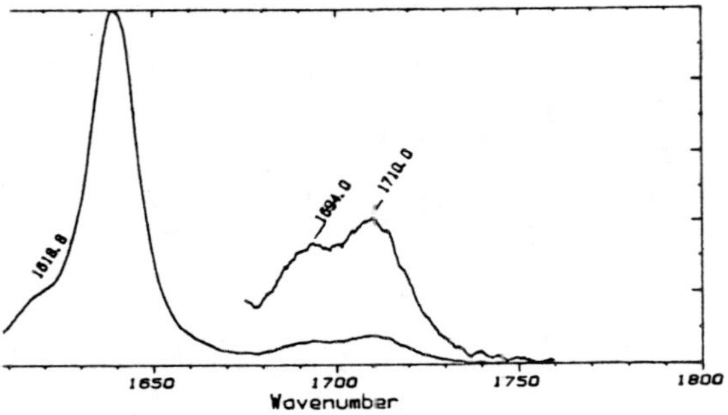

Fig. 2 R. Spectrum of FA-ethyl ester.

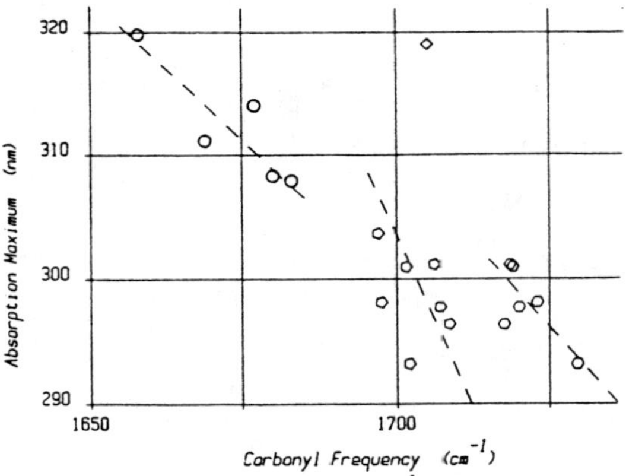

Fig. 3 Plot of λ_{max} vs $\nu_{C=O}$

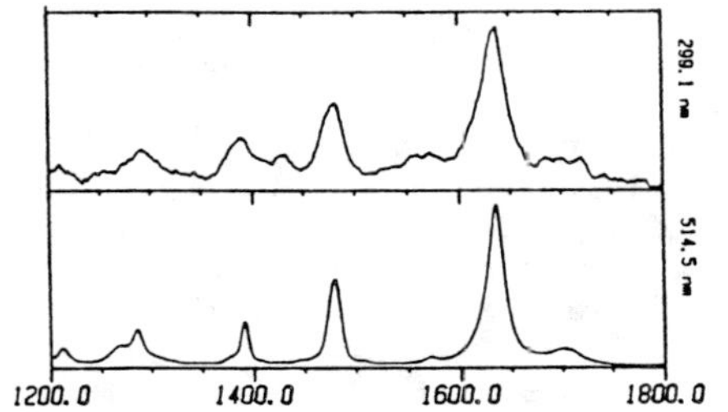

Fig. 4a Raman Spctrum of FA-choline

FA-aldehyde (circles) and the FA-chymotrypsin (diamond). Note that there are two lines for ethylester--one for each of the two carbonyl vibrational bands which are observed in each of the solvents but not in the FA-acylchymotrypsin. Again we see that the environment of the ester at the active site is different from that that can be induced in simple esters by solvent effects (4). Finally in figure 4 is shown the remarkable changes that take place in the UVRR of FA choline in aqueous solution as the exciting wavelength is changed from 514.5 to 217.9 nm. This indicates that this chromophore may find use in the far UVRR spectra as a probe of the environment of the active site of enzymes.

(1) S.A. Bernhard & S. Lau, Cold spring Harbor Symposia on Quantitative Biology, vol XXXI, 75 (1971)

(2) B.A.MacClement, R.G.Carriere, D.J.Phelps,& P.R. Carey Biochemistry,20,3438 (1981)

(3) P.V.Argade, G.K.Gerke, J.P.Weber, and W.L.Peticolas, Biochemistry,23,299 S(1984).

(4). J. A. Weber, P.Y.Turpin, S.A. Bernhard, & W.L. Peticolas, Biochemistry, 25, 1912 (1986)

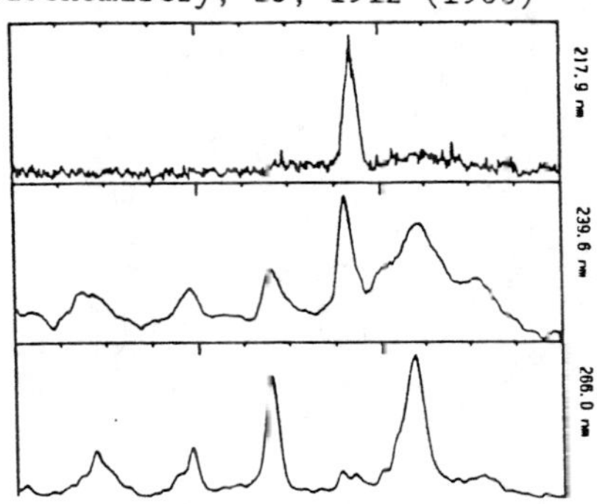

Fig. 4b UVRR spectrum of FA-choline

THE INFLUENCE OF pH AND Ca^{2+} CONCENTRATION ON THE TYROSINE DOUBLET INTENSITY RATIO IN THE BOVINE α-LACTALBUMIN RAMAN SPECTRUM

J.P. Lafaut, H. Van Dael

Interdisciplinair Research Centrum, K.U.Leuven Campus Kortrijk, B-8500 Kortrijk, Belgium.

α-Lactalbumin is a modifier protein in the milk lactose biosynthesis. Its amino acid sequence is homologuous with that of henn egg white lysozyme. In abscence of an X-ray determination of its three-dimensional structure, different techniques have been used to elucidate detailed molecular and conformational features (references in [1] and [2]). Fluorescence polarization studies of the tryptophan side groups have indicated that at low pH the protein undergoes a conformational transition to a more expanded structure. Similar transitions can be induced by thermal or chemical denaturation treatment. All these effects strongly depend on the Ca^{2+} concentration. Addition of Ca^{2+} to a decalcified expanded conformer leads to a more compact structure [3].

The tyrosine doublet intensity ratio $R = I_{850}/I_{830}$ in the Raman spectrum is known to be an useful tool to interprete the tyrosine environment and is suitable for conclusions on the protein conformation [4].

Our Raman measurements confirm the existence of an acid conformer, clearly distinguishable from the native state. The R value = 1.4 at neutral pH decreases to 0.6 at pH = 2 (fig. 1 and 2). In the Siamwiza theory [4], a low R value corresponds with a phenolic oxygen acting as a proton donor to a negatively charged acceptor. At this low pH however, practically no negatively charged groups are available for hydrogen bonding, so that the H-bonding at this pH, only can be understood if the protein undergoes a drastic conformational change whereby the COOH group is brought in the near vicinity of the Tyr residue.

A low R value also refers to a hydrophobic environment in the protein interior for the tyrosine moieties. In our case this theory leads to the surprising conclusion that the tyrosines move to a more hydrophobic position while the tryptophan groups are more exposed to the solvent. The ratio's obtained in experiments at higher temperatures (table 1) and in the presence of a chemical denaturant (table 2) are interesting and decisive checks of these results. In both these circumstances, where the protein evolves to a more unfolded structure, R decreases with temperature and with increasing amount of guanidine hydrochloride respectively.

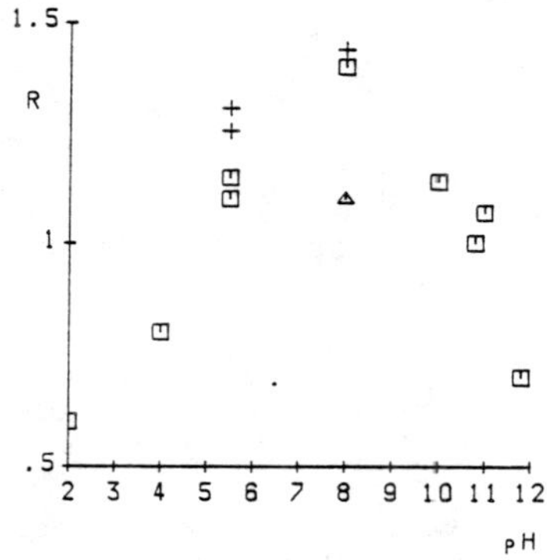

Fig. 1 : pH dependence of R for BLA with 0.3 Ca per α-lact ⊡, without Ca △ and with 1 Ca per α-lact (+)

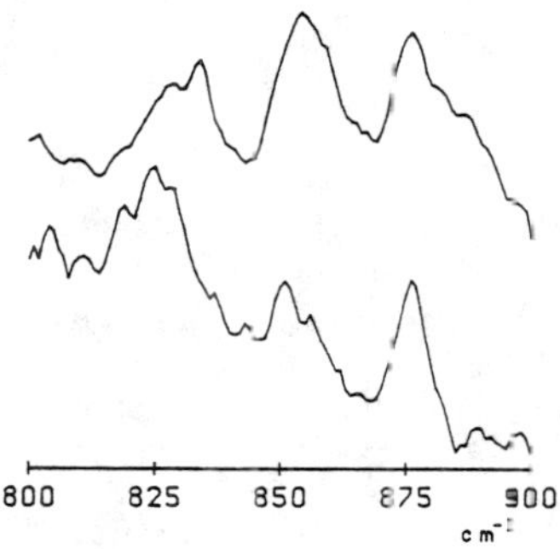

Fig. 2 : Spectra in the 800–900 cm^{-1} region at pH = 8 (top) and pH = 2 (bottom) at 23°C

T(°C)	R
5	1.11
25	1.10
35	1.08
45	1.15
55	0.97
65	0.92

Table 1 : Temperature dependence of R for apo-BLA at pH = 8

Concentration GuHCl (M)	R
0	1.30
1	1.27
2	1.08
3	1.04
4	0.88

Table 2 : Effect of GuHCl on R at pH = 8 and T = 23°C for BLA with 0.3 Ca

The removal of Ca^{2+} gives also rise to a lower R. This corresponds with a looser protein structure. Addition of Ca^{2+} on the other hand acts as a stabilizing agent for the protein.

References :

[1] L. Lindahl, H.J. Vogel, Anal. Biochem. 140, 394-402 (1984).

[2] E.A. Permyakov, L.A. Morozova, E.A. Bürstein, Biophys. Chem. 21, 21-31 (1985).

[3] I. Hanssens, H. Pottel, W. Herreman, F. Van Cauwelaert, Biochem. Biophys. Res. Commun. 119, 509-514 (1984).

[4] M.N. Siamwiza, R.C. Lord, M.C. Chen, T. Takamatsu, I. Harada, H. Matsuura, T. Shimanouchi, Biochemistry, 14, 4870-4876 (1975).

AMIDE I REGION OF RESOLUTION-ENHANCED RAMAN SPECTRA OF PROTEINS

D. Michael Byler and Heino Susi

U.S. Department of Agriculture, Eastern Regional Research Center
600 East Mermaid Lane, Philadelphia, PA 19118

Fourier deconvolution and second derivative techniques were applied to the 1550-1750 cm^{-1} region of the Raman spectra of the proteins lysozyme, immunoglobulin, insulin, β-lactoglobulin, and α-lactalbumin in the crystalline state. The amide I band was thereby resolved into three to six components. Some of these can be clearly assigned to conformations such as α-helices and β-strands.

The following generalizations appear to hold: helical sections exhibit a band close to 1657 cm^{-1}; in most proteins β-strands exhibit a band near 1670 cm^{-1} although in insulin the band appears at higher frequency (ca. 1682 cm^{-1}). A weak band close to 1638 cm^{-1} is also frequently observed. Turns show bands above 1680 cm^{-1}; a weak band sometimes observed near 1645 cm^{-1} is possibly associated with hydrated segments. Lysozyme, which contains a large proportion of unordered regions (ca. 30-40%), shows a strong band at 1683 cm^{-1} which may be associated with these segments. This is quite different from infrared spectra obtained in deuterium oxide solution, where the band for unordered segments is observed close to 1645 cm^{-1} [1]. These assignments must be regarded as tentative until more data are available.

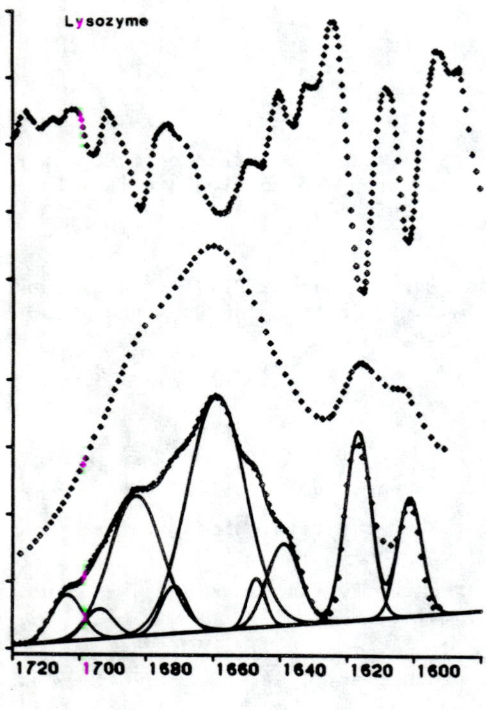

The Figure shows the second derivative spectrum, the original spectrum, and the deconvolved spectrum of lysozyme. (Relative intensity is plotted versus wavenumber in cm^{-1}.) The original spectrum was obtained at ca. 4 cm^{-1} spectral resolution and was treated with a thirteen point Savitsky-Golay smoothing function before resolution enhancement. The deconvolution procedure used is essentially the same as the one described by Kauppinen [2] and by Mantsch et al. [3]. The line shape function was assumed to be Lorentzian with a full width at half height (FWHH) of 18 cm^{-1}. The resolution enhancement factor K was set to 2.8. The figure demonstrates that resolution-enhanced Raman spectra of proteins in the amide I region can be obtained which are analogous to previously reported infrared results [1,4,5]. Assignments in the two cases differ, however, because (a) the Raman data are for solids while the IR are for D_2O solutions and because (b) selection rules and relative intensities are not the same for the two techniques. (Furthermore, the signal-to-noise ratio of conventional Raman spectra is markedly inferior to that of FTIR spectra.)

It is interesting to speculate about possible quantitative applications of this approach. The important questions concern (a) experimental intensity measurements and (b) the Raman scattering cross-section for segments of the peptide backbone with different conformations. Only a thorough study will provide clear answers. Nevertheless, if the area of the helix band component of lysozyme is measured as indicated in the Figure (the deconvolved spectrum was fitted with Gaussian components), we find that this component accounts for about 46% of the total area of the amide I band. Assuming that this number indicates the fraction of α-helix in the secondary structure of lysozyme, one can compare this number with previously reported values for the helical content of this protein: FTIR, 41% [1]; Raman (statistical correlations), 44-51% [6-7]; and X-ray, 45% [8]. Use of the Raman technique reported here to estimate the β-strand content of proteins is more difficult at present because the bands due to unordered chains, β-strands, and turns cannot yet be assigned with certainty.

1. D. M. Byler and H. Susi, Biopolymers, 1986, 25, 469-487.
2. J. K. Kauppinen, Spectrometric Tech., 1983, 3, 199-232.
3. H. H. Mantsch, H. L. Casal, and R. N. Jones, Adv. Spectrosc., 1986, 13, in press.
4. W.-J. Yang, P. R. Griffiths, D. M. Byler, and H. Susi, Appl. Spectrosc., 1985, 39, 282-286.
5. H. Susi and D. M. Byler, Biophys. Biochem. Res. Commun., 1983, 115, 391-387.
6. R. M. Williams, J. Mol. Biol., 1983, 166, 581-603.
7. M. Berjot, J. Marx, and A. J. P. Alix, J. Raman Spectrosc., 1986, 17, in press.
8. M. Levitt and J. Greer, J. Mol. Biol., 1977, 114, 181-239.

PROTEIN CONFORMATION BY LASER RAMAN SPECTROSCOPY: QUANTITATIVE DETERMINATION OF THE SECONDARY STRUCTURE FROM THE AMIDE 1 AND III BANDS BY THE REFERENCE INTENSITY PROFILES METHOD

A.J.P. Alix, M. Berjot, A. ElJoundi and J. Marx

Groupe de Spectroscopie Vibrationnelle des Systèmes Bio-Moléculaires, L.R.O., Faculté des Sciences, 51062 Reims Cedex, France

In this communication, we present a simple consistent method: the Reference Intensity Profiles Method (RIP-method) which is of use for the quantitative determination of the structural contents of proteins from the deconvolution of either their Raman Amide I band (1630 - 1700 cm^{-1}), or their Raman Amide III band (1220 - 1300 cm^{-1}).

Basically, the method uses as a model the Reference Intensity Profiles corresponding to the Raman spectral contributions to the Amide bands of 100% pure classes of secondary conformations. So, for a solid protein, the normalized Amide band intensities may be written as follows

$$A_\nu = \sum_{j=1}^{m} c_j R_\nu^j \quad ; \quad \sum_{j=1}^{m} c_j = 1 \qquad [1]$$

where c_j is the fraction of residues in the j-th conformation, R^j is the Reference Intensity Profile corresponding to the spectral activity of a pure j-th conformation and ν is the frequency which ranges from 1630 to 1700 cm^{-1} for the Amide I band, and from 1220 to 1300 cm^{-1} for the Amide III band.

Equation [1] is first used to generate the Reference Intensity Profiles: the c_j's are taken from crystallographic X-ray data and the A_ν are the Raman Amide bands data of a set of proteins and/or polypeptides. Afterwards, the same equation is used to compute the c_j's, i.e., the % contents of secondary structures for a given protein.

AMIDE I BAND ANALYSIS: THE 4-RIP METHOD (1)

4 Reference Intensity Profiles were set up from a chosen set of 17 proteins and/or polypeptides and correspond to 2 α-helices, 1 β-sheet and 1 undefined structure (containing all aperiodical conformations). The calculated correlation coefficient between X-ray data and the Raman estimates is higher than 0.96 for all conformations. For practical uses, the 4-RIP method was modified in order to take into account the contribution of the solvent (mainly due to the symmetric bending mode of water) in the case of the study of a protein in solution (1).

AMIDE III BAND ANALYSIS: THE m-RIP METHOD

The extension of the above method to the Amide III band is much more difficult as unfortunately, the amino acid side chains also contribute Raman scattering within the same frequency region. Two major problems are encountered: the right definition of the baseline of the spectrum and the subtraction of the residues contributions. Another problem is that, apparently, one has a less reliable structure-frequency relationship between the Amide III peak frequencies and the secondary sub-structures.

To generate the m-RIP (m is the number of Reference Intensity Profiles which is constrained to be minimum, according to statistical tests used in curve-fitting) we used the following steps: (i) recording of the spectra of all amino acids and of a large set of synthetic poypeptides and proteins (already analysed in their Amide I band region by use of the 4-RIP method, and of known X-ray data),

(ii) obtention of the treated Amide III bands (for instance, the baseline was adjusted by calculating at first the points of minimum intensity arising from the summation of the spectra of the different amino acid residues entering in the primary structure of the proteins),

(iii) obtention of the Reference Intensity Profiles (univoquely defined and common for all compounds) by curve-fitting,

(iv) assignment of the obtained reference profiles to the secondary sub-structures (in contrast to the Amide I band, the Amide III band is much more complex and can not be resolved into few components),

(v) applications to a set of proteins of known data in order to check the method used either alone or in conjunction with the 4-RIP method (Amide I band analysis).

In the search of the best results we used different kinds of shapes (hyperbolic, Lorentzian, Gaussian), different values for the bandwidths and the maximum frequencies as well as different structure criteria.

Some applications (especially the case of lysozymes) are presented in detail for the Amide I and the Amide III band analyses showing that both methods (4-RIP/m-RIP) correlate well together and give accurate estimates of % contents of secondary structures.

This work was partially supported by grant (CRE n° 855002 from INSERM: Institut National de la Santé et de la Recherche Médicale, France).

1. M. Berjot, J. Marx and A.J.P. Alix, J. Raman Spectrosc., in press;
A.J.P. Alix, M. Berjot and J. Marx, in "Spectroscopy of Biological Molecules, A.J.P. Alix, L. Bernard and M. Manfait editors, John Wiley & Sons, Chichester New York Toronto, Chapter IVb pp 149 - 154, august 1985.
2. A. ElJoundi, thèse doctorat, Reims, 1986.

POLARIZED RAMAN SPECTRA OF PROTEINS USING THE FOUR CHANNEL DEVIDED CELL TECHNIQUE.

Bernd M. Bussian[§] and Hans H. Eysel[$]

[§]Institut für Physikalische Chemie I, Technische Hochschule Darmstadt, Petersenstr. 20, D-6100 Darmstadt, FRG
[$]Institut für Anorganische Chemie, Universität Heidelberg, Im Neuenheimer Feld 270, D-6900 Heidelberg, FRG

There are only a few Raman spectra of proteins reported in literature which determine the two invariants of the classical Raman tensor. We demonstrate that additional and for protein chemistry quite useful information can be raised from Raman spectra. This is shown for some proteins.

Figs. 1a-c show the isotropic, the anisotropic and the total Raman spectra of henn egg white lysozyme. The interpretation of the intensity pattern of the isotropic and the anisotropic part of the spectra yields information of the symmetry of the vibration.

Characteristic vibrations of the ring containing side groups contribute only to the isotropic spectrum: Trp: 764, 880, 1013, 1582 cm^{-1}; Tyr: 840, 860, 120 cm^{-1}; Phe: 1008, 1200 cm^{-1}. Only a few modes show up in both isotropic and anisotropic spectra: Trp: 1345, 1368, 1556 cm^{-1}. The ring deformations around 1620 cm^{-1} of Trp, Tyr, Phe have strong anisotropic character because of local symmetry.

The CH_2-bending region around 1450 cm^{-1} which is sensitive to structural changes of the backbone and of the aliphatic side chains has intensity only in the anisotropic part.

The amide III region which has been widely used to trap structural changes is of higher intensity in the isotropic spectrum but, surprinsingly, resolution is better in the anisotropic part. This fact allows a better understanding of helix -> random coil transition due to thermal denaturation. Up to now sufficient and conclusive interpretation could not be made on the basis of unpolarized spectra.

The amide I region appears in the isotropic spectrum with much higher intensity. When using this region for a quantitative assignment of protein secondary structure types aromatic side groups around 1620 cm^{-1} must be subtracted. Up to now there is no way to do this accurately. Since the aromatic ring vibrations appear mainly in the anisotropic part an exact subtraction can now be done.

The experimental setup comprises an argon ion laser (514.5 nm), an electrooptical modulator, a devided spinning quartz cell, SPEX-1403 double monochromator and four channel photo counting collection electronics. Incident laser intensity is of ca. 600 mW, collection time is of some second per wavenumber, slit width corresponds to about 5 cm^{-1}.

This results in the following experimental advantages:

- small amounts of total sample solutions, less than 0.5ml
- small amounts of protein
- low protein contentration (less than 2% weight)
- spectra of solvent can be completely subtracted off
- no loss of enzyme activity has been observed

The proposed technique gives a spectrum of the 'pure' protein in solution. Therefore any buffer can be used and the sample can be reused for other experiments. Especially fields like protein design or site specific mutagenesis need a fast and reliable technique which probes even small amounts of proteins. We demonstrate that classical Raman can be emplcid as a routine technique to yield detailed information of specific structural changes.

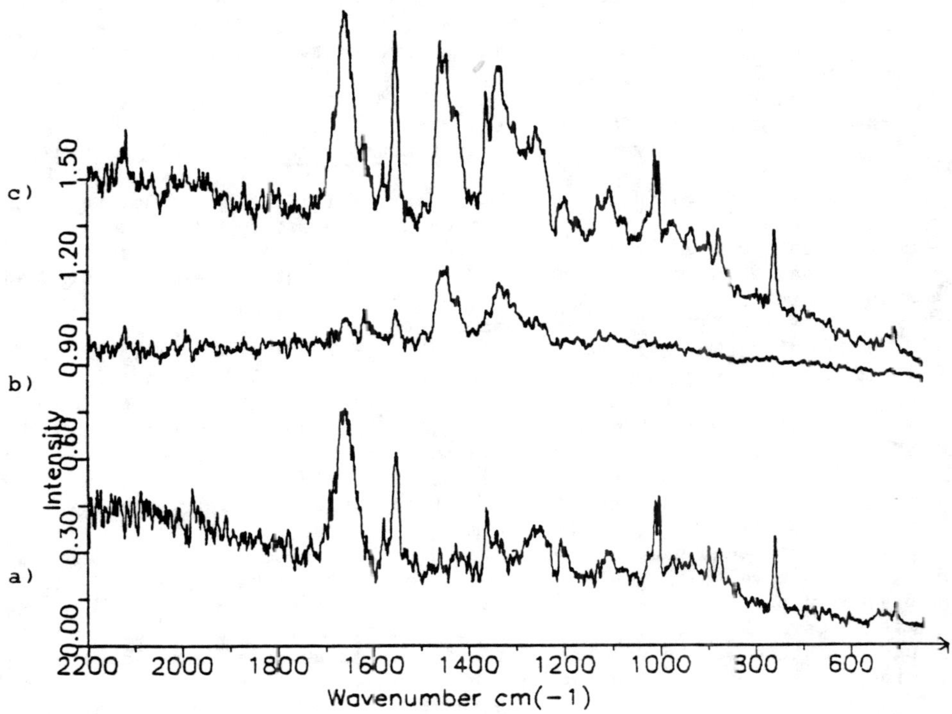

Fig. 1 a-c, polarized Raman Spectra of Lysozyme; a, isotropic part; b, anisotropic part; c, total spectrum

CHANGES IN THE LYSOZYME RAMAN SPECTRA AFTER IODINATION: RELATIONSHIP WITH ENZYMATIC ACTIVITY

J. Jacquot[b], J. Marx[a], B. Lambert[c], M. Berjot[a] and A.J.P. Alix[a]

a Groupe de Spectroscopie Vibrationnelle des Systèmes Bio-Moléculaires, L.R.O., Faculté des Sciences, 51062 Reims Cedex, France,
b Groupe d'Etude des Systèmes Mucociliaires, Laboratoire d'Histologie, Faculté de Médecine, 51095 Reims Cedex, France,
c Laboratoire de Biochimie, Faculté des Sciences, F-51062 Reims Cedex.

We recently analysed the laser Raman spectra of Human Airway Lysozyme (HAL, purified from pathological bronchial secretions) and of Hen-Egg White Lysozyme (HEWL) which showed marked intensity differences in the ranges characteristic of the aromatic side chain (Tyr, Trp), the peptide and bond vibrations. Moreover, HAL was characterized by a specific activity 3-fold higher than that of HEWL. The lytic activity was evaluated by measuring the initial rate of lysis of an <u>Micrococcus luteus</u> cell wall suspension used as substrate (1).

In order to go further in the investigation of the structure-activity relationship of both enzymes, we decided to modify the micro-environment of the active site of that molecules by iodinations of respectively 1 and 3 iodine atoms by molecule. Here are presented the first obtained results with iodinated-HEWL.

RAMAN SPECTROSCOPY: The Raman spectra of the enzymes, dissolved in 0.06 M sodium phosphate buffer 0.15 NaCl (pH 7.20) at a concentration of 10 mg/ml, were recorded in the wavelength region 500-1800 cm^{-1}. A Coherent Radiation Model 52Ar^+ laser, at power 1 W with the 488. nm line, was used with a small two-prism monochromator (ANASPEC 300S) to remove the background plasma lines. A cool photomultiplier EMI 9558 QB was used as a detector. The signal pulse is amplified, digitally counted and then stored in the computer system (ALCYANE, MBC France) which coordinates the scanning (steps of 2 cm^{-1}, counting time 2s/step) and the spectral data acquisition. Data accumulation (time averaging) was used to obtain the spectra and to improve the signal to noise ratio. Slit widths are 6 cm^{-1} and the frequency precision is ± 2 cm^{-1}. We like to point out that the baseline was adjusted by calculating the points of minimum intensity arising from the summation of the spectra of the different amino acid residues entering in the primary structure of the enzyme.

COMPARISON BETWEEN 1 I - HEWL and HEWL: Two very strong bands at 876 and 1078 cm^{-1} are dominent in the spectrum of 1 I-HEWL and are attributed to aromatic rings-substitute couplings (2). No major changes are observed in the 760, 1360 and 1552 cm^{-1} Trp-bands, and the 1008 and 1618 cm^{-1} Tyr, Phe-bands show only slight modifications. In contrast, the ranges characteristic of the backbone and peptide bond vibrations are significantly modified. Deconvolution of the Amide I band by use of the Reference Intensity Method (3) gives a significant increase of the α-helix structure, which is confirmed by the up-shift of the Amide III band and by the intensity increase of the $\nu(C_\alpha-C)$ backbone band at 940 cm^{-1}.

1I-HEWL: 46% α-helix + 18% β-sheet + 36% undefined structure,

HEWL: 39% α-helix + 20% β-sheet + 41% undefined structure.

That modification of secondary structure is not accompanied by a corresponding decrease in the bacteriolytic activity of the 1I-iodinated lysozyme compared to the normal lysozyme used as standard (ratio 95:100).

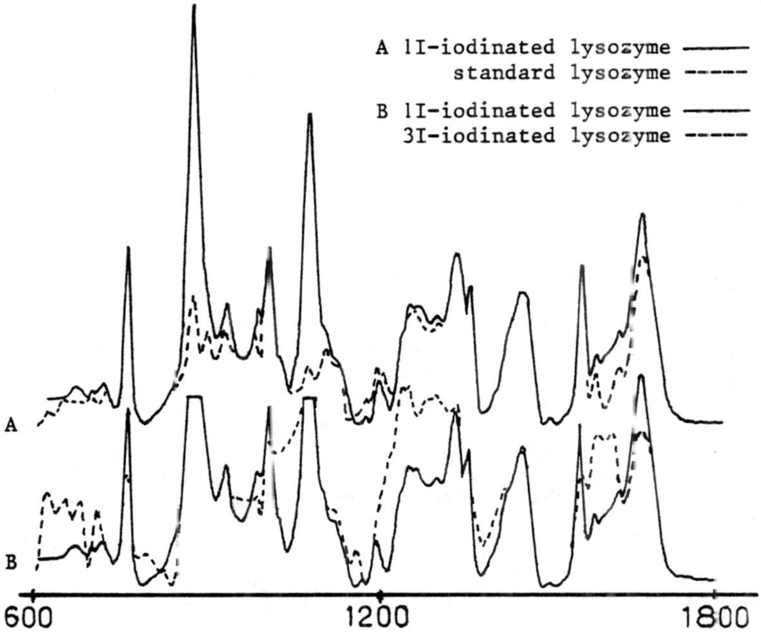

COMPARISON BETWEEN 3I and 1I IODINATED LYSOZYMES: Dramatic intensity changes are respectively observed in the Trp-bands at 1552 and 1578 cm^{-1} and in the Tyr-line at 1618 cm^{-1}. Also, the Amide I and III bands are affected (maximum frequencies, shapes and intensities). Deconvolution of the Amide I band gives a higher amount of β-sheet structure which agrees with the intensity increase of the 1240 cm^{-1} component of the Amide III band.

3I-HEWL: 45% α-helix + 28% β-sheet + 27% undefined structure.

Finally, we have shown that these marked changes (secondary structure, side chain vibrations) are correlated to a dramatic loss of specific activity (-70%) of 3I-HEWL compared to that of 1I-HEWL (-5%).

1. J. Marx, J. Jacquot, M. Berjot, E Puchelle and A.J.P. Alix, BBA, in press; J. Marx, M. Berjot, A.J.P. Alix, J. Jacquot and E. Puchelle, see Ref. 4, Chapt. IVb pp 169-172.
2. M.H. Baron, C. de Loze, T. Méjean, M.J. Coulange, P.Y. Turpin and L. Chinsky, J. Chim. Phys., 80, 729 (1983).
3. M. Berjot, J. Marx and A.J.P. Alix, J. Raman Spectrosc., in press; A.J.P. Alix, M. Berjot and J. Marx, see Ref. 4, Chapt IVb pp 149-154.
4. "Spectroscopy of Biological Molecules", A.J.P. Alix, L. Bernard and M. Manfait eds, J. Wiley & sons, Chichester New York Toronto (1985).

This work was supported by grants (CRE n° 855002 and 855019 from INSERM: Institut National de la Santé et de la Recherche Médicale).

THE EFFECT OF WATER ON THE SECONDARY STRUCTURE OF THE PROTEIN BOVINE SERUM ALBUMIN

B.A. Bolton and J.R.Scherer

Plant Development Quality Unit, Western Regional Research Center
800 Buchanan St., Albany, CA,94710

The structure and activity of proteins is known to be affected by water. An understanding of protein hydration requires information about the nature and strength of water-protein interactions, the extent of water/water interactions and the effect of hydration on the protein structure. The present investigation attempts to obtain this information from the stepwise hydration/dehydration of the globular protein bovine serum albumin (BSA).

The hydration characteristics of thin bovine serum albumin films(1.5 - 2.0 μ) were studied using the simultaneous application of integrated optics and Raman spectroscopy(1). By measuring the refractive index and thickness of the BSA film held at different humidities (0-100% RH), the water concentration within the film can be calculated using the Clausius-Mosotti equation. In the same experimental arrangement, Raman scattering from the laser beam propagating inside the films characterize the protein/water (D_2O or H_2O) interactions.

Information about the secondary structure of BSA and changes in conformation with water content is obtained from frequencies and/or integrated peak intensities of the amide I, amide III and the 900-1000 cm^{-1} skeletal vibration regions. At 0% RH we observe the amide I band at 1663 cm^{-1} (FWHH=57 cm^{-1}). Dry BSA film spectra of the amide III region (1320-1220 cm^{-1}) consists of highly mixed vibrational modes (Fig.1). A better resolved amide III band was obtained by subtracting a BSA solution (28% by wt. in D_2O) spectrum from the BSA film spectra. Using the integrated intensity of the phenylalanine band (1004 cm^{-1}) as an internal standard, these difference spectra exhibit a doublet at 1280 and 1237 cm^{-1} at 0% RH (Fig.1). The frequency of the amide bands occur in regions that are associated with α helix and random coil conformations. ORD measurements (2) have shown that BSA in solution contains 50-60% α helix and 50-40% disordered structure. A quantitative estimate of the two conformations in the dry film was determined from the amide III integrated intensities due to α helix (1260 - 1320 cm^{-1}) and random coil (1200 - 1260 cm^{-1}). The relative amounts were found to be 60% and 40% respectively.

As the water content within the film increases (higher % RH), the amide I band shifts to lower cm^{-1} and the band width decreases (100 % RH 1658 cm^{-1} & FWHH=49.5 cm^{-1}). The intensity of the skeletal stretching mode near 940 cm^{-1} increases by 43% from 0% RH to 100% RH. This observation has been associated with an increase helical content (3) which is consistent with the changes observed with the amide I band. Furthermore the amide III integrated intensities (1320-1260 & 1260-1200 cm^{-1}) indicate an increase in helical conformation (Fig.1). Our estimate of the two conformations at 100% RH from amide III intensities is 69.5% α helix and 30.5 % random coil. These spectral observations and a 41% increase in film thickness (100% RH) suggest that water molecules promote the mobility of the skeletal chains and reduce the amount of disorder in the wet sample.

Figure 2 shows spectra of a BSA film at various D_2O vapor pressures and D_2O neat. At 2% RH we observe a doublet at 2498 and 2416 cm^{-1} which continues to increase with increasing % RH. The relative peak intensities change with the higher cm^{-1} band becoming more intense. At 64% RH the band shape appears similiar to the liquid state but with a FWHH $\approx$ 50 cm^{-1} less. These results suggest that water absorbed below $\approx$ 40% RH binds more tightly to the protein than water to water in the liquid state.

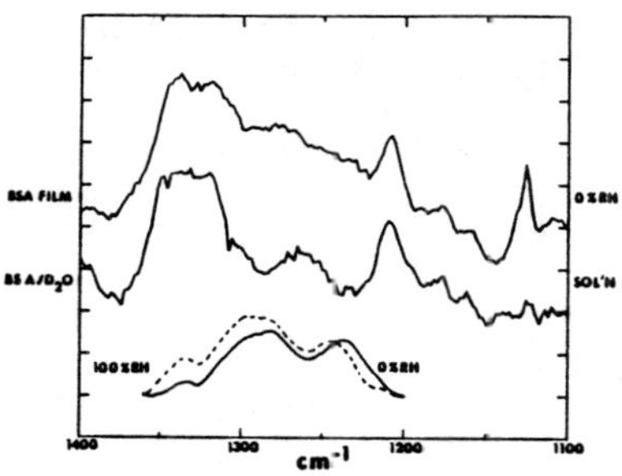

Fig.1 Amide III region for BSA sol'n, BSA film (0% RH) and the difference spectra (FILM-SOL'N) at 0% & 100% RH.

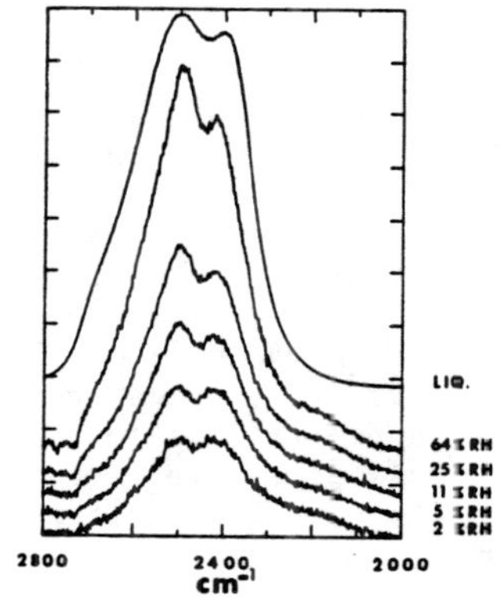

Fig.2 Spectra of OD stretching region at several relative humidities and in the liquid state

REFERENCES

1. J.R.Scherer and G.F.Bailey,J.Membrane Sci.,13,43(1983)
2. T.Peters,in "Advances in Protein Chemistry",37,161(1985)
3. V.J.C. Lin and J.L.Koenig, Biopolymers,15,203(1976)

RAMAN SPECTROSCOPIC STUDIES OF STRUCTURE IN POLYPEPTIDES

Samuel Krimm

Department of Physics and Biophysics Research Division
The University of Michigan, Ann Arbor, MI 48109

INTRODUCTION

The polypeptide chain is capable of adopting many types of spatial structures, or conformations. This diversity accounts for the wide range of specific biological functions that can be performed by protein molecules. Knowledge of such structure is therefore important to an understanding of function, and vibrational spectroscopy is inherently able to provide this kind of information.

Until recently, however, Raman (as well as infrared) spectroscopy has been used primarily as a tool for routine identification of known conformations, rather than as a means for deriving detailed insights into structure and interactions. The latter can only be obtained by combining experimental studies with the structure-specific predictive capabilities of normal mode analysis, which permits definitive correlations based on the complete spectral observations.

In recent years, we have been developing transferable vibrational force fields that permit the calculation of reliable normal modes of the polypeptide chain [1]. This has resulted in complete analyses of Raman spectra of peptides and polypeptides of known structure as well as specific predictions for molecules whose structure was uncertain [2]. In this paper, we review the development of these force fields and describe new applications of this approach to several structural problems.

POLYPEPTIDE FORCE FIELD

As discussed in detail elsewhere [2], our polypeptide force field was refined from N-methylacetamide and known structures of poly(glycine)I, β-poly(L-alanine), and α-poly(L-alanine). The chain has standard bond lengths and angles, conformational variability being determined by the dihedral angles $\phi(NC^{\alpha})$ and $\psi(C^{\alpha}C)$. A simplified general valence force field was used, with hydrogen bonds included explicitly and transition dipole coupling incorporated for amide I and amide II modes [3,4]. For the above molecules, observed Raman and infrared bands are reproduced with an average error of ~5 cm^{-1}.

In addition to good frequency agreement, this force field, together with *ab initio* dipole derivatives, provides quantitative reproduction of observed infrared intensities [5], and, from density of states derived from dispersion curves, can account for observed inelastic neutron scattering spectra [6]. It should, therefore, provide a reliable base from which to analyze vibrational spectra in terms of conformation.

SPECTRA AND CONFORMATION

If the force field is reasonably transferable, it becomes possible to predict the spectra of structures not yet studied by model compounds. This has been done for the parallel-chain pleated sheet structure [7]. Thus, we determine that, in distinction to the antiparallel-chain pleated sheet, only two amide I modes are expected: $\nu(0)$ at 1666 cm^{-1}, which should be strong in the Raman and weak in the infrared, and $\nu(\pi)$ at 1654 cm^{-1}, with the reverse intensity ratio. Subsequent studies of tripeptides with such a structure confirm these predictions [8].

A number of different chain conformations have been proposed for the transmembrane ion channel gramicidin A, an alternating L,D pentadecapeptide. Normal mode calculations on six plausible model structures permit the assignment, on the basis of the amide I Raman band, of a single-stranded helix to the active membrane species and double-stranded helical structures to the crystalline forms [9].

On the basis of circular dichroism spectra, it was proposed that polypeptides with charged side chains adopt locally ordered, so-called "extended-helix", rather than truly random, conformations in solution [10]. This is now further supported by a normal mode analysis of the strong skeletal stretch mode in the Raman spectrum near 900 cm^{-1} [11].

These, as well as previous [2], studies show that it is becoming possible to analyze the spectra of macromolecules such as polypeptides with a rigor approaching that heretofore applicable to small molecules, and that by doing so significant information on three-dimensional structure can be obtained.

This research was supported by NSF grants DMB-8517812 and DMR-8303610.

REFERENCES

1. S. Krimm, Biopolymers 22, 217 (1983).
2. S. Krimm and J. Bandekar, Adv. Prot. Chem., in press.
3. S. Krimm and Y. Abe, Proc. Natl. Acad. Sci. U.S.A. 69, 2788 (1972).
4. T.C. Cheam and S. Krimm, Chem. Phys. Lett. 107, 613 (1984).
5. T.C. Cheam and S. Krimm, J. Chem. Phys. 82, 1631 (1985).
6. V.K. Datye and S. Krimm, J. Chem. Phys., in press.
7. J. Bandekar and S. Krimm, Biophys. J. 49, 295a (1986); to be published.
8. J. Bandekar and S. Krimm, to be published.
9. V.M. Naik and S. Krimm, this Conference; Biophys. J., in press.
10. M.L. Tiffany and S. Krimm, Biopolymers 6, 1379 (1968).
11. P.K. Sengupta and S. Krimm, to be published.

HYDROGEN BONDING IN AMIDES AND PROTEINS. A LOW FREQUENCY STUDY (10-400 cm^{-1}).

O. Faurskov Nielsen*, I. Bigio**, J.-M. Berquier* and I. Olsen*

*Chem. Lab. V, University of Copenhagen, 5 - Universitetsparken, DK-2100 Copenhagen, Denmark. **Los Alamos National Laboratory, Los Alamos, New Mexico 87545, U.S.A.

Low-frequency Raman spectra are studied conveniently in the low frequency representation: $R(\bar{\nu}) = \bar{\nu}[1-\exp(-h\bar{\nu}c/kT)]I(\bar{\nu})$ where $\bar{\nu}$ is the wave number (in cm^{-1}) and h,c,k and T have their usual meanings [1-7]. A similar expression has been used for low frequency spectra of gases [8,9]. This contribution will present results for liquid N-methylacetamide and aqueous solutions of lysozyme.

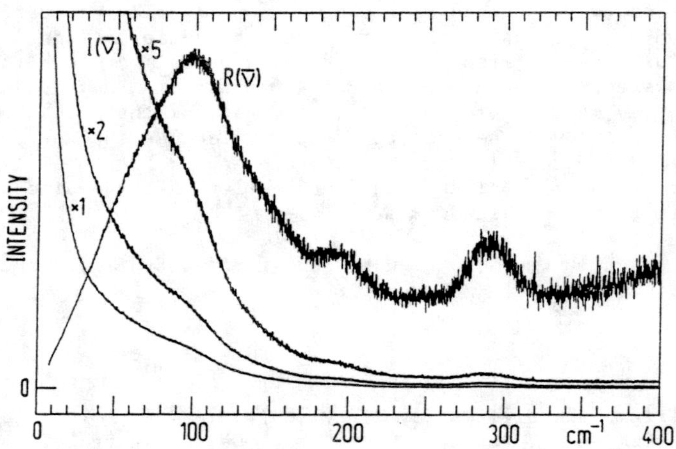

Figure 1. Raman spectra $I(\bar{\nu})$ of liquid N-methylacetamide at 30° C together with the $R(\bar{\nu})$-curve obtained directly from the experimental data (see text). I_{VH}-configuration in a horizontal scattering plane.

In figure 1 is given the conventional Raman spectrum of liquid N-methylacetamide at different gains together with the $R(\bar{\nu})$-curve calculated directly from the experimental data. A strong band with a maximum at ca. 100 cm^{-1} is easily observed in the $R(\bar{\nu})$-spectrum. This band corresponds to the band observed in the FIR-spectrum at the same frequency [10] .

R($\bar{\nu}$)-spectra of N-methylacetamide at 30, 60 and 90° C show that the maximum of the band at ca. 100 cm^{-1} shifts towards lower frequency with increasing temperature. Polarization measurements indicate that the band is depolarized. In accordance with similar observations for liquid formamide [11,12] the band at ca. 100 cm^{-1} is assigned to a mode involving atoms in the intermolecular hydrogen bonds in N-methylacetamide. This assumption is supported by results obtained for N-deuterated N-methylacetamide.

Aqueous solutions of lysozyme show a somewhat broader band around 100 cm^{-1}, which can be assigned to the hydrogen bonding in lysozyme. Spectra of 2, 5 and 10 % lysozyme solutions are investigated. Quantitative results are difficult to obtain due to an unavoidable fluorescense, but some conclusions can be drawn concerning the amount of water directly interacting with the protein.

The importance of nonlinear dynamics has been proposed for a coupling in proteins between the amide-I vibrations and low frequency vibrations [13]. Low frequency Raman spectra of N-methylacetamide irradiated with a CO-laser in resonance with the carbonyl-stretching vibration at 1660 cm^{-1} have been obtained. Observed changes are difficult to distinguish from an ordinary heating of the sample.

Acknowledgement. This work has been supported by the Danish Natural Science Research Council.

REFERENCES

1. P.-A. Lund, O. Faurskov Nielsen and E. Praestgaard, Chem. Phys. 28, 167 (1978).
2. M. Perrot, M.H. Brooker, and J. Lascombe, J. Chem. Phys. 74, 2787 (1981).
3. M.H. Brooker and M. Perrot, J. Chem. Phys. 74, 2795 (1981).
4. O. Faurskov Nielsen, P.-A. Lund, and E. Praestgaard, J. Chem. Phys. 75, 1586 (1981).
5. O. Faurskov Nielsen, P.-A. Lund, Lone S. Nielsen and E. Praestgaard, Biochem. Biophys. Res. Com. 111, 120 (1983) and references therein.
6. O. Faurskov Nielsen, D.H. Christensen, J. Aubard, and G. Dodin, J. Raman Spectrosc. 15, 134 (1984).
7. O. Faurskov Nielsen, D.H. Christensen, and E. Praestgaard J. Chem. Phys. 82, 1183 (1985).
8. J.E. Bertie and K.H. Michaelian, J. Chem. Phys. 77, 5267 (1982).
9. J.E. Bertie, H.H. Eysel, D.N.S. Permann, and D.H. Kalantar, J. Raman spectrosc. 16, 137 (1985).
10. K.Itoh and T. Shimanouchi, Biopolymers 5, 921 (1967).
11. O. Faurskov Nielsen and P.-A. Lund, Chem. Phys. Lett. 78 626 (1981).
12. O. Faurskov Nielsen, P.-A. Lund, and E. Praestgaard, J. Chem. Phys. 77, 3878 (1982).
13. G. Careri, U. Buontempo, F. Galluzzi, A.C. Scott, E. Gratton and E. Shyamsunder, Phys. Rev. B. 30, 4689 (1984).

VIBRATIONS OF TRYPTOPHAN AND TYROSINE SIDE CHAINS IN PROTEINS.

Issei Harada, Takashi Miura, Noriko Watanabe, and Hideo Takeuchi

Pharmaceutical Institute, Tohoku University, Aobayama, Sendai 980, Japan

In interpreting UV resonant as well as non-resonant Raman spectra of proteins, it is helpful to know the vibrational mode of each band. It is also important to find out correlations between spectral patterns of side chain vibrations and the environment. We examined the vibrations of tryptophan (trp) and tyrosine (tyr) side chains in detail including
(1) the normal coordinate treatments on indole, skatole, and *p*-cresol, and the isotopically substituted compounds,
(2) model studies on the 1360-1340 cm^{-1} doublet and the 880 cm^{-1} band of trp, with an application to the H → D exchange in lysozyme, and
(3) model studies on the tyr trio [1] at 1260, 1210, and 1180 cm^{-1}.

Normal coordinate calculations were performed for indole and eleven deuterated indoles. The obtained set of valence force constants reproduces infrared and Raman wavenumbers [2] with an average error of about 6 cm^{-1}. The force field was extended to skatole and the calculated wavenumber shifts on isotopic substitutions (^{15}N , 1-d_1, and 4-d_1) agree generally well with the observed ones of trp [3,4]. Normal vibrations of *p*-cresol, as a model of tyr, and seven deuterated analogs were treated.

Raman spectra of indole, skatole, tryptophan and its derivatives were studied with special attention to the bands around 1350 and 880 cm^{-1}. Although these bands in proteins are known to be sensitive to the environment, their origins and the spectrum-environment relationships have not been fully clarified yet. The 1350-cm^{-1} band is observed as a doublet or triplet depending on the compound and solvent, which was ascribed to Fermi resonance between one skeletal stretching fundamental and one or two combination bands of the out-of-plane bending vibrations. Removal of the resonance is seen for trp-4-d_1 and some solutions of skatole. The assignments of the component vibrations were made based on the shifts in wavenumber upon C_4-deuteration. Relations between the relative intensity, R = I_{1360}/I_{1340}, and environment of indole are, 0.65-0.92 in hydrophilic solvents including MeOH:H_2O (2:8), 1.02-1.11 in benzene and its derivatives, and 1.23-1.32 in saturated hydrocarbons. This classification is considered to apply to the trp side chain in proteins the environment of which is similar to that of indole in solution [5]. However, it must be pointed out that a trp side chain which gives a large R value is not necessarily inaccessible by water (vide infra). The bands shift down by 10 cm^{-1} on N-deuteration.

The 880-cm^{-1} band arises from a mixed mode of the benzene 12 like vibration and the NH motion. The wavenumber in model compounds is different depending on solutions and shifts down by about 20 cm^{-1} on N-deuteration. Accordingly, the band can be useful for studies of the environments of trp side chains in proteins. As a test of such application, we studied the H → D exchange of six trp side-chains in lysozyme [6]. The amount of deuterated trp was estimated by the intensity of the 1382-cm^{-1} band of trp-ND [6,7]. First of all, native lysozyme [pH 5.5, 22°C, to be termed (6H)] gives a peak at 877 with a shoulder at 872 cm^{-1} and the FWHM is 12.6 cm^{-1}. In a denatured structure [8M LiBr, pH 5.5, d-(6H)], the band at 877 cm^{-1} is symmetric and sharp (FWHM, 7.5 cm^{-1}). Similar to the report by Peticolas [6], 2.5 trp side chains were deuterated by 30 min after dissolution of the enzyme in D_2O (pD 5.5) at 22°C (3.5H), three by 4 hr (3H), four by 10 hr (2H), and five, by 100 hr after dissolution at 35°C (1H). Completely deuterated enzyme (6D) was obtained either by keeping the solution at 40°C for 125 hr (6D) or by dialysis of a 8M LiBr-D_2O solution. The following assignments were possible on the basis of the analysis of difference spectra: trp-1 and trp-2, 877; trp-3, 874; trp-4, 879; trp-5, 879; and trp-6, 871 cm^{-1}. There is no simple correlation between the wavenumber and accessibility of water to trp. However, the anomalously low value of trp-6 may reflect the peculiar environment of completely buried trp-6 (trp 28). The intensity of 1360-cm^{-1} peak decreases substantially during the first 30 min. It is strong in trp-3, -5, and -6, and weak in trp-4.

Three tyr bands at 1260, 1210, and 1180 cm^{-1} arise mainly from the CO stretch, the C_β-C_ϕ stretch, and the ϕ-CH bend, respectively. We studied Raman spectra of *p*-cresol, its deuterated analogs, and tyrosine derivatives in various solutions (10-500 mM). The CO stretching wavenumber was found to be sensitive to the state of phenolic hydroxyl group: e.g., for *p*-cresol, 1269 in DMSO, 1261 in H_2O, 1256 in CCl_4, and 1257 cm^{-1} in benzene. The peak intensity is weakest in H_2O, and becomes stronger in DMSO, benzene, and CCl_4, in that order. The $C_\beta C_\phi$ stretch gains intensity in aromatic solvents. It is interesting to note that in state III of cytochrome c, tyr 48 and tyr 67 are strongly H-bonded and also in aromatic environment, and actually, the CO stretch wavenumber is high and the $C_\beta C_\phi$ stretch is strong in the UV Raman spectra [1]. The former shifts down and the latter weakens on going from state III to state II where the tyrosines are exposed to water.

REFERENCES

1. R. A. Copeland and T. G. Spiro, *Biochemistry*, 24, 4960 (1985).
2. A. Lautié, M. F. Lautié, A. Gruger, and S. A. Fakhri, *Spectrochim. Acta*, 36, 85 (1980): M. F. Lautié, Dr. Thesis, Univ. Pierre et Marie Curie, Paris, 1978.
3. I. Harada and H. Takeuchi, in *Advances in Spectroscopy Vol. 13 - Spectroscopy of Biological Systems*, (Clark and Hester, eds.), 1986.
4. H. Takeuchi and I. Harada, *Spectrochim. Acta*, submitted.
5. I. Harada, T. Miura, and H. Takeuchi, *Spectrochim. Acta*, in press.
6. W. L. Peticolas, in *Raman Spectroscopy Linear and Nonlinear*, (Lascombe and Huong, eds.), Wiley, 1982, pp. 694-702.
7. H. Takesada, M. Nakanishi, A. Y. Hirakawa, and M. Tsuboi, *Biopolymers*, 15, 1929 (1976).

EXPERIMENTAL INVESTIGATION OF THE LATTICE MODES OF α-GLYCINE.

R. Ramnarine[1] and W. Sherman[2]

[1]Faculty of Engineering
University of the West Indies
St. Augustine, Trinidad, West Indies.

[2]Department of Physics, King's College,
Strand WC2, U.K.

The study of the vibrational modes of α-glycine and a search of any phase in the far infrared region is presented. α-glycine (NH_3CH_2COO) has a monoclinic crystal structure with space group $C^5 2h = P2_{1/n}$ (1), and $C^5_{2h} = P2_{1/b}$ a C_1 site group with four molecules per unit Bravais cell (2,3). Since α-glycine is a basic building block for protein, a study of its vibrational modes will lend greater understanding to the nature of the intermolecular forces in the protein chain.

The group theoretical analysis has shown that α-glycine has the following modes active :

$$\Gamma^{I.R.}_{Gly} = 5Au + 4Bu$$

These nine bands are expected to be active in the far infrared. Powdered α-glycine was mixed with polyethylene and made into discs of varying sample concentrations. The appropriate spectra were obtained on a Fourier Transform photospectrometer. The lattice modes in the far infrared were investigated with respect to temperature variation. The spectra of these modes were recorded (Fig. 1) and their values tabulated and compared with the values of previous workers (4,5) in Table 1.

Table 1: Comparison of the observed modes of α-glycine of this work With the calculated and observed values of other workers.

Species	Previous work		This work				
	γ-cal (7.12)	obs(7.21) 295K	γ-obs 295K	240K	190K	140K	80K
Au	53	-	54.91	58	59	60	81
	100R	-	74.45	75	76	77	78
	137T	132	90.32	91.54	92.8	94	95.21
	199R	202	176	177	179	183	184
	355R	352	361	363	365	366	368
Bu	140R	138	106	107	108	109	110
	148T	170	139	142	144	145	147
	211R	224	242		246		250
	354R	-	242		252		262

Temperature dependent curves were obtained in the 80K to 295 K range. These showed no evidence of temperature induced phase changes. Though all the bands showed an apparently linear frequency increment on cooling, one of the bands (242 cm) was found to split into two components.

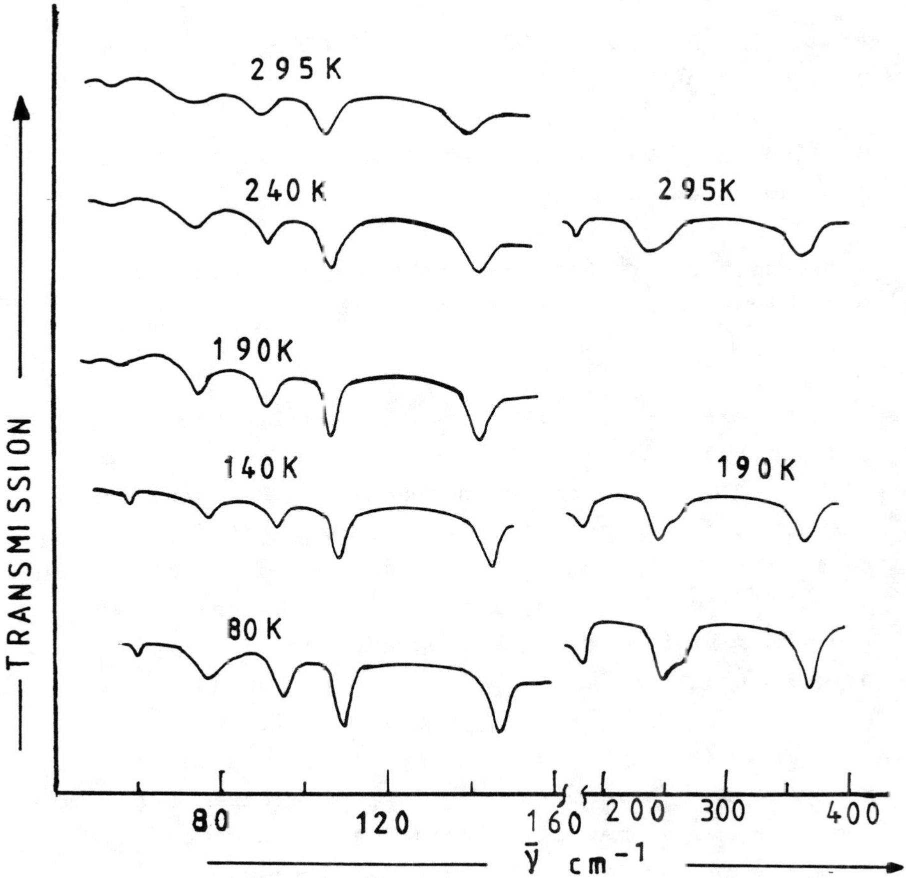

Fig. 1 Spectra of α-glycine at various temperatures

References

1. W.G. Fateley et al., "Infrared and Raman Selection rules for molecular and lattice vibrations: The correlation method." J. Wiley, 1982.
2. P.Q. Jonnson, A. Kuick, "Acta Cryst. B 28 p. 1827 (1939).
3. E. Albrecht, T.B. Corey, J. Am. Chem. Soc. 61, p. 1087.
4. E. Castelluci et al., "Proceedings of the 5th Int. Conf. on Raman Spectro 7B 1600 p. 550 (1976).
5. F. Sani Thesis (Ph.D.) University of Florence (1975).

RAMAN SPECTRA AND CONFORMATIONAL ANALYSIS OF N-ACYL-HOMOOLIGOGLYCINES.

Keijiro Taga, Takahiko Okuoka, Radhaballabh Debnath and Hirofumi Okabayashi

Department of Applied Chemistry, Nagoya Institute of Technology, Gokiso-cho, Showa-ku, Nagoya 466, Japan

Polyglycine exists in two crystalline modifications, e.g., polyglycine I (extended β form) and polyglycine II (helical form). The Raman spectra of both the form of polyglycine have been investigated and their normal coordinate calculations have been made by several authors (1-6).

In case of simple oligomers, it has been reported that the oligoglycine up to the pentamer assumes a planar form of polyglycine I, while hexamer and dodecamer seem to attain a polyglycine II type structure (2,6).

In the present work, we synthesized a series of N-acylglycine oligomers and its N-deuterated derivatives and made a systematic study on their structure in the crystals in the light of their Raman spectra.

Figure 1 shows the Raman spectra of a series of sodium salt of N-hexanoylglycine oligomers in the crystalline state in the region 1800–800 cm^{-1}. It is evident from the spectra that trimer and higher homologues have almost the same spectral feature which markedly differs from those of monomer and dimer. A detailed comparison of these spectra with the Raman spectra of polyglycine I and II shows that trimer and higher homologues resemble to the structure of polyglycine II (1,2). The Raman bands with asterisks in Figure 1 correspond to the bands which appear in the Raman spectrum of polyglycine II.

The observed frequency shift in the Raman spectra of the corresponding N-deuterated species and the IR spectra of both the normal and N-deuterated species also support the above view. Since the spectra of monomer and dimer differ from those of higher homologues, it seems that helicity in the chain occurs actually right from the trimer.

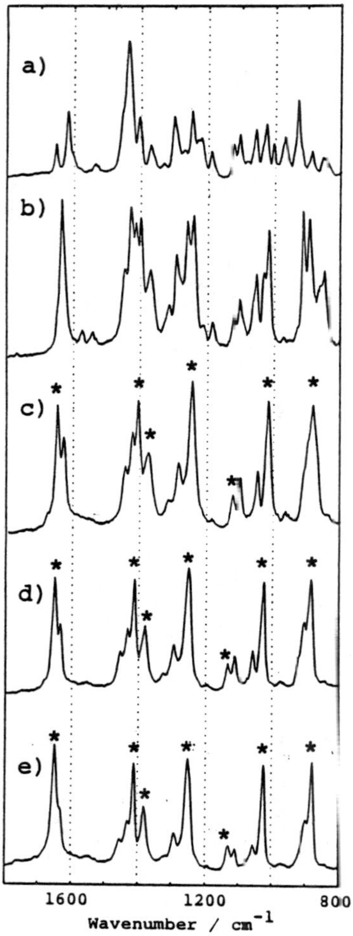

Figure 1. Raman spectra of sodium salt of N-hexanoyl oligoglycines, Hex(Gly)$_n$ONa, in the crystalline state. a) monomer, b) dimer, c) trimer, d) tetramer and e) pentamer.

REFERENCES

1. E. W. Small, B. Fanconi and W. L. Peticolas, J. Chem. Phys., 52, 4369 (1970).
2. M. Smith, A. G. Walton and J. L. Koenig, Biopolymers, 8, 29 (1969).
3. S. Suzuki, Y. Iwashita, T. Shimanouchi and M. Tsuboi, Biopolymers, 4, 337 (1966).
4. A. M. Dwivedi and S. Krimm, Macromolecules, 15, 177 (1982).
5. A. M. Dwivedi and S. Krimm, Biopolymers, 21, 2377 (1982).
6. V. D. Gupta, M. K. Gupta and K. Nath, Biopolymers, 14, 1987 (1975).

THE USE OF RAMAN SPECTROSCOPY TO SOLVE THE CONFORMATIONAL FEATURES OF SMALL PEPTIDES.

P. LAGANT, G. VERGOTEN, G.A. THOMAS*, W.L. PETICOLAS*

Faculty of Pharmacy - U. 279 INSERM, Rue du Professeur Laguesse - 59045 LILLE CEDEX - FRANCE
* Department of Chemistry - University of Oregon - EUGENE OR - 97403 - U.S.A.

Summary

For last years, Raman spectroscopic applications to folded peptides have been performed by us using experimental and theoretical investigations e.g. β turns (1,2) and C7C5 (3) conformations. We try at the present time to distinguish between γ and β turn conformations using normal coordinate treatments.

Experimental

The CBZ-Gly-Pro-L-Leu-gly spectra (H and D species were recorded by singlechannel Raman spectrometry using an Ar^+ ion laser as the exciting source at the University of Oregon.

The spectrum of N-acetyl D prolineamide which forms a γ turn (4) in $CHCl_3$ was recorded on a multichannel OMARS spectrometer (DILOR, LILLE FRANCE) using a 351 nm line on an UV Ar^+ laser.

Results

Normal coordinate treatments were performed onto two possible conformations of tBoc-glycyl-L-prolyl-glycyl-glycyl-OMe (tBoc-GpGGOMe) and tBoc-glycyl-L-prolyl-L-leucyl-glycyl-OMe (tBoc GpLGOMe) peptides.

* Typical features of the Raman spectrum of a β turn conformation are : (amide I at 1686, 1653 and 1610 cm^{-1}) and amide III at 1250, 1286 cm^{-1}. These last two bands shift upon N-deuteration to 951, 963 cm^{-1} while a 1271 cm^{-1} band appears.

However, according our previous work (1,2), the amide III vibrations are often strongly coupled with vibrations of CH_2 groups and N-deuteration would shift the 1286 cm^{-1} (Amide III + CH_2 twisting) to 1271 cm^{-1} (CH_2 groups of prolyl ring).

It is difficult however to predict the particular set of dihedral angles for this β turn since the amide I and III frequencies are near each other for the various kinds of β turns (I, II, III and mirror images).

* A normal coordinate treatment using Urey Bradley force field refined in our previous work (1,2,3) was done using the amide I and III frequencies of N-acetyl D-prolineamide, which is a typical turn (4). Interactions between nonbonded atoms and dipolar effects between C = 0 groups have been included.

For the N acetyl-D-prolineamide, amide I and III frequencies are summarized into Table I.

Table I

	Group ν: Streching, δ: Bending	Calculated frequency (cm^{-1})	Observed in N-acetyl-D-prolineamide
Amide I	$\nu(C=0)_4$ (gly-OMe)	1865	
	$\nu(C=0)_5$ (Tboc-gly)	1838	
	$\nu(C=0)_2$ (gly,C_7ring)	1698	1715 ? 1682
	$\nu(C=0)_1$ (Pro,C_7ring)	1631	1639
	$\nu(C=0)_3$ (gly,C_{11}ring)	1611	
Amide III	42 % $\delta(N-H)_2$ 30 % $\delta(C=0)_2$ 28 % $\nu(C-N)_2$ 25 % $\nu(C\alpha-CO)_2$	1275 (933 in N-deuterated compound)	1267
	19 % $\nu(C_\delta-N)_1$ 25 % $\delta(C=0)_1$ 8 % $\delta(NC_\alpha C)_1$ 33 % $C_\beta H_2$ Twist	1246	1248

Conclusion

Amide III vibrations are on lower frequencies for a γ turn than for a β turn conformation. However for complex molecules, it would be difficult to distinguish between these two kinds of structures according to results involving Electric circular dichroïsm (5) or other Raman investigations given by Bandekar and Krimm (6) for the γturn pentapeptide cyclo (D Phe L pro gly D ala Lpro).

References

1- "Vibrational normal modes of folded prolyl containing peptides. Application to turns".
P. LAGANT, G. VERGOTEN, G. FLEURY, M.H. LOUCHEUX-LEFEBVRE.
Eur. J. Biochem., 139, 149-154 (1984).

2- P. LAGANT, G. VERGOTEN, G. FLEURY, M.H. LOUCHEUX-LEFEBVRE.
Eur. J. Biochem., 139, 137-148, (1984).

3- P. LAGANT, G. VERGOTEN, G. FLEURY, M.H. LOUCHEUX-LEFEBVRE.
Int. J. Peptide Protein Res., 24, (1984), 543-552.

4- R.E. LONDON
Int. J. Peptide Protein Res. 14, (1979), 377-387.

5- KAWAI A., FASMAN G.
J. Am. Chem, Soc., 100, 3630-3632, (1978).

6- J. BANDEKAR, S. KRIMM.
Xth ICORS, Tokyo Japan, 122-123

RAMAN SPECTROSCOPIC STUDIES OF SOLID BOVINE SOMATOTROPIN AND RELATED ANALOGS: ESTIMATION OF SECONDARY STRUCTURE, TYROSINE ENVIRONMENT, AND DISULFIDE CONFORMATION.

Thomas J. Thamann

The Upjohn Company
Kalamazoo, MI

Raman spectra have been obtained for solid pituitary, recombinant, partially reduced, and completely reduced bovine Somatotropin (bSt), plus CNBr and tryptic digest fragments of bSt. The amide data indicate pituitary and recombinant bSt have similar structures of approximately 50% helix, 25% sheet, and 25% disordered conformation. Reduced bSt analogs in which one or both disulfide bridges were cleaved have progressively less order. The data indicate that extensive order in secondary structure is lost when the disulfide bridge connecting cysteines 53 & 164 is cleaved. The second disulfide connects a loop of 9 residues, and its cleavage results in only a minor loss of ordered conformation. This analog contains 6% less helix than native bSt, accompanied by an increase in disordered structure. The residues inclosed in this disulfide loop form about 5% of the total bSt helical segments, thus the unfolding of this partially reduced bSt analog is thought to occur predominantly in the area stabilized by the disulfied bridge.

Most of the CNBr-cleaved bSt fragments have secondary structures with large amounts of disorder (>50%). An exception is a fragment containing 149 amino acids and a disulfide bridge. This fragment has about 39% helix, 22% sheet, and 39% disorder, values not very different than native bSt. This fragment shows biological activity, although at a considerably lower level than native bSt. The activity of this fragment, and the high level of ordered secondary structure, is thought to be a result of the large size and the presence of one disulfide. Native bSt has 191 amino acids and only two disulfide bridges. As previously mentioned, the second disulfide bridge contributes little to the native secondary structure of bSt.

The tryptic digest fragment contains 38 amino acids and has also been found to exihibit biological activity. Secondary structural prediction algorithms indicate this segment of the protein to contain large amounts of helical content. The Raman amide data support this prediction by estimating the helical content to be about 57%, with about equal amounts of sheet (23%) and disorder(20%). CD data from aqueous solutions of this fragment indicate about 50% helix. Although this fragment is not large (38 residues) and possesses no disulfide bridges, the presence of biological activity may be the result of significant amounts of ordered stucture, which exists as a result of the large amount stable helix present.

Features of tyrosine environment were investigated by analysis of the relative intensities of the Fermi doublets, I(850)/I(830), for bSt and analogs. The Raman Fermi doublet data indicates that the tyrosines of native pituitary and recombinant bSt are in very similar, although not completely identical, environments. The doublet ratio for these molecules is about 1.0, which indicates that the majority of the six bSt tyrosines form hydrogen bonds by serving as weak acceptors of protons from residues that are mildly acidic. Varying amounts of residual water did not alter the results.

The Fermi doublet data for partially reduced bSt, in which one disulfide bridge has been broken, indicates that a change in tyrosine environment occurs, when the small loop (9 residues) bridge is cleaved. This is also manifested in UV spectral changes observed on reoxidation of the disulfide bridge. The portion of the protein encompassed by this disulfide bridge contains no tyrosine residues. There is a tyrosine at position 175, which is adjacent to the small loop. Tertiary molecular rearrangements encountered when the disulfide bridge is broken could affect any of the tyrosines. However, as previously mentioned, the secondary structure of this bSt analog varies little from native bSt. Larger secondary structural changes might be expected if tyrosines farther removed from the small loop area were being affected. Indeed, when both disulfide bridges are broken, large secondary and tertiary structural changes are encountered, and a large change in the Fermi doublet ratio (0.6) is noted.

The tyrosine doublet ratio for the tryptic digest fragment is interesting, because this portion of bSt contains only one tyrosine. As previously discussed, this fragment retains much of its native structure, thus its tyrosine environment is probably similar to that found for the analogous tyrosine in the native protein. The Raman data suggests that the tyrosine in the tryptic fragment is hydrogen bonded to a negative proton acceptor.

Raman data of the disulfide stretching regions for native bSt, plus selected analogs were examined. Native bSt exhibits prominent ν(S-S) bands at 488, 494, and 508 cm^{-1}. The bSt analog, that has the small loop disulfide bridge broken has peaks at 488 and 508 cm^{-1}, but lacks the 494 cm^{-1} band, indicating it may be due to the small loop disulfide stretch. Conversely, a CNBr fragment, that is composed almostly entirely of small loop residues (fragment D), has a vibration at 494 cm^{-1}, but lacks the 508 cm^{-1} peak. Completely reduced bSt, which possesses no disulfide bridges, lacks Raman bands at both 494 and 508 cm^{-1}. Thus, the data suggests that ν(S-S) for the large disulfide loop in native bSt is at 508 cm^{-1}, and the small loop ν(S-S) is at 494 cm^{-1}. These data suggest that the large loop disulfide bridge is of similar geometry to those found in many proteins, but that the small loop bridge is atypical and is in a somewhat strained configuration. The unusual geometry of the small loop bridge may be due to the small size of this protein loop.

UV RESONANCE RAMAN STUDIES OF AROMATIC AMINO ACIDS AND PROTEINS.

M. Ludwig and S. A. Asher

Department of Chemistry, University of Pittsburgh, Pittsburgh, PA 15260

The ultraviolet resonance Raman (UVRR) excitation profiles have been measured for the aromatic amino acids, phenylalanine, tryptophan, and tyrosine. Resonance excitation enhances Raman scattering from those vibrational modes that distort the ground state configuration towards the configuration of the excited state. The excitation profile maxima are red-shifted with respect to the absorption spectral maxima for each aromatic amino acid. These excitation profiles directly indicate the excitation wavelengths required to maximally enhance particular aromatic amino acid residues in proteins. Individual aromatic amino acids in environmentally distinguishable positions in a protein have slightly different transition energies, and therefore may be identifiable and distinguishable by proper selection of the excitation frequency. For example, surface tyrosyl residues in hydrophilic environments will have lower lying excited states due to extensive hydrogen bonding. These residues should be maximally enhanced with longer wavelength excitation than residues buried in the hydrophobic pockets within proteins. The frequency of certain vibrational bands, particularly those known to be influenced by substituents on the aromatic ring may also be indicative of the local environment. The relative intensities of other enhanced bands may also contain information concerning specific local environment. For example, the relative intensity of the peaks of the 830/850 cm^{-1} Fermi resonance doublet of tyrosine are known from normal Raman studies to be sensitive to hydrogen bonding.

Vibronic substructure is obscured in the absorption spectral measurements due to the breadth of the absorption spectral features. This vibronic substructure is amplified and is detectable in the resonance Raman excitation profiles.

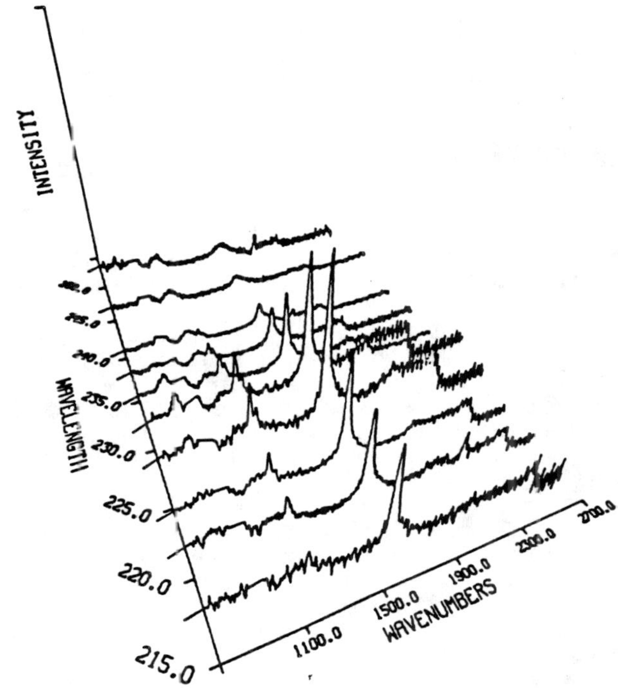

Figure Caption: UVRR excitation profile of tyrosine from 217 to 250 nm. Maximal enhancement of tyrosine occurs with 227 nm excitation. All solutions are .001 M tyrosine, 1.0 M sodium perchlorate, pH = 6.0.

SELECTIVE ENHANCEMENT OF PROLINE RAMAN SIGNALS WITH ULTRAVIOLET EXCITATION

Leland Mayne and Bruce Hudson
Department of Chemistry and Institute of Molecular Biology
University of Oregon, Eugene, Oregon 97403

Selective excitation of the resonance Raman spectrum of the peptide linkage to proline using 230 nm radiation is demonstrated. The far UV resonance Raman spectroscopy of protein components has been of considerable recent interest[1]. N-Methylacetamide has been studied as a model for the peptide bond, leading to further understanding of the nature of its vibrations and electronic states. The resonance Raman spectroscopy of the x-proline peptide bond is of particular interest because the structure around the nitrogen in this group makes it different from the normal secondary amide peptide bond in terms of its electronic absorption and vibrational spectra. Proline bonds are also of interest because of their implication in studies of protein renaturation kinetics[2].

We have studied several small model compounds containing proline residues. Figure 1 shows the Raman spectra of pro-gly (dashed) and gly-pro (solid) taken with 218nm radiation. We see that in the gly-pro case the absence of the amide proton causes the C-N stretch motion to be decoupled from the C-N-H bending. This decoupling leads to an almost pure C-N stretch amide II'-like mode at 1485 cm^{-1}. In a normal peptide these motions are coupled and give rise to the amide II and III modes. This is analogous to the case of N-methylacetamide where N-deuteration decouples these motions and concentrates intensity in the amide II' band. Preliminary studies of cyclic proline dipeptides indicate that the frequency of this band is sensitive to the state of cis-trans isomerization of the X-Pro linkage[3].

A significant problem that will have to be overcome for this technique to be of use in applications to proteins is that of detecting the resonances of one or a few X-Pro bonds amid the signals of the other more numerous peptide bonds. To this end we have measured the intensity of the amide II mode of Pro-Gly and the amide II'-like mode of gly-pro against the intensity of a sodium sulfate internal standard at a variety of excitation wavelengths. These results are shown in figure 2. As can be seen, the amide II'-like mode is enhanced relative to the amide II mode by a factor of about 30 at excitations near 230 nm due to the relative red shift of the X-pro absorption.

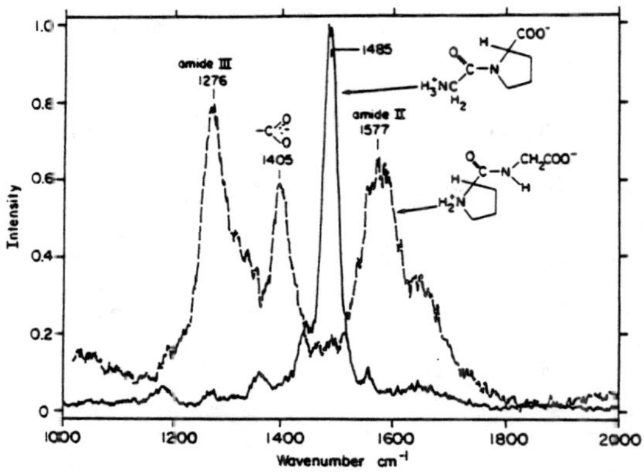

Figure 1. The resonance Raman spectra of prolylglycine (solid line) and glycyl proline (dashed line). Both spectra obtained with 218 nm excitation from 10mM solutions of the dipeptides.

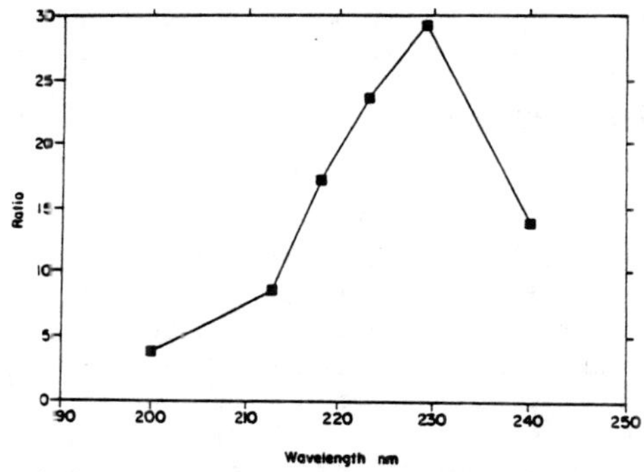

Figure 2. Plot of the intensity of the amide II'-like mode of gly-pro relative to the amide II mode of pro-gly. Intensities are based on peak heights.

Acknowledgement

This work was supported by NIH grant GM 32323 and NSF grant PCM 8308529.

(1) (a) Mayne, L. C.; Ziegler, L. D.; Hudson, B. J. Phys. Chem., **89**, 3395 (1985)
(b) Hudson, B.; Mayne, L. Methods in Enzymology, **130**, 331 (1986)
(c) Dudik, J. M.; Johnson, C. R.; Asher, S. A. J. Phys. Chem., **89**, 3805 (1985)
(d) Copeland, R. A.; Spiro, T. G. Biochem., **24**, 4960 (1985)

(2) Mui, P. W.; Konishi, Y.; Scheraga, H. A. Biochem., **24**, 4881 (1985)

(3) Mayne, L.; Ramahi, T.; Oas, T.; Hudson, B. Biophys. J., **45**, 322a (1984)

ULTRAVIOLET RESONANCE RAMAN STUDY OF ARGININE

Peter Hildebrandt and Thomas G. Spiro

Department of Chemistry, Princeton University, Princeton, NJ 08544

The development of high power pulsed lasers which can provide large photon fluxes in the ultraviolet (UV) has made it possible to probe the resonance Raman (RR) spectra of UV chromophores with high sensitivity. With respect to the application to biologically important molecules most studies have been focused on nucleic acid bases, aromatic amino acid side chains and the amide bonds of the polypeptides backbone which have strong absorption bands between 260 and 190nm.

Here we report a UV RR study on an aliphatic amino acid, arginine (Arg). Arg has a relatively weak absorption band peaking below 190nm which is ascribed to the $\pi \rightarrow \pi^*$ transition of the amidine group.

The RR spectrum excited at 192nm of Arg in an aqueous solution at pH=7 is dominated by three bands at 1640, 1175, and 1090 cm^{-1}. In D_2O they shift to 1615, 910 and 830 cm^{-1} accompanied by an intensity increase of the strongest band at 1615 cm^{-1}. The latter band is assigned to the C=N stretching vibration while the other two bands probably originate from vibrational modes including the NH_2 rocking and the C-N stretching motion of the amide chromophore. The relative intensities of the amidine vibrations with respect to an internal standard (sulfate) decrease with increasing excitation wavelength as expected from the decreasing absorbance which can be explained by the loss of the resonance enhancement. Besides the amidine bands only a small contribution of other Arg modes is observed in the spectra. These results show that the intensity of the RR bands is mainly provided by the $\pi - \pi^*$ transition of the amidine chromophore. However, even at 192nm excitation the resonance enhancement is relatively low compared to the aromatic amino acids. Therefore, a selective enhancement of the RR bands of Arg in proteins may only be possible at excitation wavelengths below 190nm.

Laser Raman spectra of β- alanine silver nitrate and β-alanine copper nitrate

T.J. Bhoopathy, S. Mohan and Feridoun Payami
Department of Physics,
Pachaiyappa's College,
Madras-600 030, India.

Amino acids readily form complexes with metallic salts. They exhibit very interesting physical properties such as optical activity, ferro electric phase transition at low temperature, Faraday rotation, dichrosim and good conductors of electricity. The aim of the present study is to investigate the laser Raman spectra of β-alanine silver nitrate and β-alanine copper nitrate to look for their molecular properties. β-alanine silver nitrate and β-alanine copper nitrate single crystals have been prepared in the laboratory by evaporation technique. The laser Raman spectra of the samples have been recorded using 4880 Å line of Ar^+ for excitation in the region 100-4000 cm^{-1} on Cary Model 82 grating spectrophotometer and a 4W argon laser. The vibrations of these crystals have been analysed in terms of the normal modes of β-alanine group and nitrate groups. As expected, a lowering of symmetry of molecular group (NO_3^-) has been observed. This has led to the splitting of the degenerate vibrations. The bonding properties of the above complexes are briefly discussed.

RAMAN SPECTROSCOPY STUDY OF pH AND PRESSURE INDUCED CONFORMATIONAL CHANGES IN CHYMOTRYPSIN

Karel HEREMANS and Luc HEREMANS

Department of Chemistry, Katholieke Universiteit Leuven, B-3030 Leuven, Belgium

Introduction: A fundamental property of enzymes is that the active conformation can be changed to an inactive state under the influence of parameters such as pH, salt, temperature and pressure. The study of the effect of pressure on proteins and other biomolecules has gained considerable interest in recent years (1,2,3). There is a fundamental reason to study the effect of pressure on molecules in solution. When one studies the temperature dependence of dynamical processes, one varies in fact two parameters: the energy of the molecules and the space available to the molecules. With pressure studies one only changes the space available for the molecules.

The activity versus pH profile of chymotrypsin catalyzed reactions is bell shaped with pKapp around 7 and 8.5 (4). On the acid side, the pK is related to the ionization of the imidazole nitrogen of histidine 57. On the alkaline side the ionization of the a-aminogroup of isoleucine 16 is involved. The changes that occur on the alkaline side are related to the activation process from chymotrypsinogen to chymotrypsin. In both cases the activity depends on the integrity of the ion pair between isoleucine 16 and aspartate 194. At neutral pH the enzyme exists in an equilibrium between an active and an inactive conformation. This equilibrium is subject to changes by pH (4), temperature (4) and pressure (5). The active conformation is controlled by the intactness of the saltbridge close to the active site. The changes in conformation can be followed by activity measurements, dye binding and circular dichroism. We have characterized the change in conformation with Raman spectroscopy.

Materials and Methods: The methodology is essentially as described (6). The pressure cell was constructed as described (7).

The amide I band was resolved into six protein secondary structure components by published procedures. Two methods were used. In the first (I), a linear combination of calculated reference spectra is fitted to the experimental (solvent subtracted) amide I spectra to estimate the contributions of the six components (8). In the second method (II), the amide I band of a protein is analyzed directly as a linear combination of amide I bands of proteins whose secondary structures are known (9).

Results and Discussion: In a previous report (6) we noted no significant changes in the linewidth and the frequency of tryptophan residues of chymotrypsin, measured at the 760 cm-1 ringvibration, between pH 7 (active conformation) and pH 10.5 (inactive conformation). Similarly, the frequency and linewidth for the tryptophan vibration in chymotrypsinogen is the same as in chymotrypsin. Small changes in the amide I vibration were detected. The shift in amide I frequency going from pH 7 to 10.6. shows a midpoint of transition corresponding satisfactorily with the pK of 3.5 detected by other methods (4). Interestingly, the frequency position for chymotrypsinogen is the same as for the inactive conformation of chymotrypsin. This corresponds to the observation that the activation process from zymoger to enzyme is related to the alkaline transition.

The amide I bands were further analyzed in terms of six components: a-helix, mono and bi-hydrogen bonded, parallel and antiparallel b-sheet, b-turn and undefined. With method I we find that no changes occur in a-helix content of the protein when the pH changes from 8 to 10. The contribution of antiparallel b-sheet diminishes, with a concomitant increase in parallel b-sheet and b-turn. Similar, but smaller changes are calculated with method II.

At 2 kbar, we note a decrease in the bi-hydrogen bonded helix and an increase in the antiparallel b-sheet. Under the pressure conditions no ionization occurs, in contrast to the pH induced changes. Further experiments are in progress to obtain the complete pH and pressure profile.

Recent high pressure Raman experiments on lysozyme indicated that above 5 kbar not only intramolecular but also intermolecular changes are important (10). The increased intermolecular interactions become evident from the observed precipitation of the denatured protein. This necessitates a new interpretation on the reported domain melting in lysozyme (2). The denatured protein shows a decrease in random coil and a-helix in contrast to the temperature and salt denatured protein. The side chains are more exposed to the solvent. It would be of interest to see whether chymotrypsin shows a similar behaviour.

<u>References</u>

1. K.Heremans, Ann Rev Biophys Bioeng (1982) 11, 1
2. G.Weber, H.G.Drickamer, Q Rev Biophys (1983) 16 89
3. P.T.T.Wong, Ann Rev Biophys Bioeng (1984) 13, 1
4. A.R.Fersht, J. Mol. Biol. 64, 497 (1972).
5. K.Heremans, F.Ceuterick, J.Snauwaert and J.Wauters in Techniques and Applications of Fast Reactions in Solution, ed. by W.J.Gettins and E.Wyn-Jones, Reidel, Dordrecht (1979) p.429
6. K.Heremans, L.Heremans, R.Audenaert and Y.Engelborghs, in Spectroscopy of Biolcgical Molecules, A.J.P.Alix, L.Bernard and M.Manfait. Eds. John Wiley & Sons, (1985) 187
7. P.T.T.Wong, Rev. Sci. Instrum. 56 (1985) 1417.
8. R.W.Williams and A.K.Dunker, J. Mol. Biol. 152 (1981) 783
9. R.W.Williams, J. Mol. Biol. 166 (1983) 581
10. K.Heremans, P.T.T.Wong, Chem Phys Letters, 118 (1985) 101.

chapter 2

NUCLEIC ACIDS and VIRUSES

NUCLEIC ACID STRUCTURES IN VIRUSES FROM CORRELATIONS ESTABLISHED BY RAMAN SPECTROSCOPY OF DNA AND RNA CRYSTALS.

George J. Thomas, Jr.

Department of Chemistry, Southeastern Massachusetts University, North Dartmouth, Massachusetts 02747 U.S.A.

Single crystal x-ray diffraction analysis has provided new insight into the molecular architecture of oligomeric DNA.[1] The x-ray structures obtained from crystalline DNA fragments also suggest probable structural models for DNA *in vivo*. Nevertheless, other methods are required to investigate directly the structures of nucleic acids for experimental conditions which are different than those of the crystal lattice. This is especially important for high molecular weight DNA in aqueous solution, and for assemblies of DNA with protein in viruses, chromatin and gene regulatory complexes. Raman spectroscopy is one method potentially well suited to the study of DNA structures in native macromolecular assemblies.[2] However, in order to realize the potential of the Raman method the spectral lines which are diagnostic of specific DNA conformations and specific base sequences must be reliably established beforehand.

In an empirical approach to this problem we have combined the results obtained from atomic-resolution DNA crystallography and laser Raman spectroscopy to assign vibrational frequencies and intensities to specific DNA backbone and nucleoside geometries in structures of the A, B and Z-DNA families.[3] The identification of distinct conformation markers has been aided by the use of a recently developed Fourier deconvolution method for enhancement of resolution in Raman spectra of biological molecules.[4] The transferability of single-crystal Raman data to samples of non-crystalline morphologies, including fibers, gels and dilute solutions, has been exploited to determine the preferred conformers of DNA for experimental conditions which sometimes vary significantly from those in the crystal and which may in certain cases more closely approximate those *in vivo*. This analysis has also been extended to RNA models, as well as to RNA/DNA hybrid structures. Single crystals and solutions containing the following nucleic acid sequences will be discussed: $[d(CGCGCG)]_2$, $[d(CGTACG)]_2$, $[d(CGCATGCG)]_2$, $[d(CCCCGGGG)]_2$, $[r(GCG)d(CGC)]_2$, $[r(GCG)d(TATACGC)]_2$ and $[d(CGCGTG)]_2$. The last of these is a miniature double helix incorporating the "wobble" base pair GT in place of the normal Watson-Crick GC pair at the penultimate base positions.

In applications to DNA fibers several new or unorthodox structures have been revealed, including structures for poly(dG)·poly(dC), poly(dA-dT)·poly(dA-dT) and poly(rA)·poly(dT) which show characteristics of the A-helix geometry. For the case of RNA/DNA hybrids, both oligonucleotides and polynucleotides appear to be capable of forming "mixed" structures in which A and B type conformers are simultaneously present.

In applications to viruses and other biologically active nucleic acid sequences, we find evidence of considerable heterogeneity in DNA secondary structure. For example, in the 17-base pair operator O_L1, which is the tightest binding site for the phage lambda cI repressor, more than one type of backbone conformation is indicated by the Raman spectrum. Microheterogeneity of DNA secondary structure is also indicated by the Raman scattering from chromosomes of isometric double-stranded (P22) and single-stranded (ØX174) DNA bacteriophages, as well as by the Raman markers from several filamentous DNA bacteriophages. In isometric RNA bacteriophages (R17, MS2) and plant viruses (CCMV, BDMV, TYMV), homogeneity of secondary structure is usually observed for the RNA genome. However, in the rod-shaped RNA virus TMV, at least two-thirds of the nucleosides exhibit an unorthodox conformation which is not found in any model system and which is not maintained in the absence of the viral protein coat.[5]

References

1. F. A. Jurnak and A. McPherson (Eds.), Biological Macromolecules and Assemblies, Volume 2: Nucleic Acids and Interactive Proteins, Wiley, New York, 1985, Chapters 1-3, pp.1-170.

2. G. J. Thomas, Jr. (1976) Appl. Spectrosc. 30, 483-494.

3. G. J. Thomas, Jr., J. M. Benevides and B. Prescott, in Biomolecular Stereodynamics III, R. H. Sarma and M. H. Sarma (Eds.), Adenine Press, Guilderland, New York, 1986 (in press).

4. G. J. Thomas, Jr. and D. A. Agard (1984) Biophys. J. 46, 763-768.

5. Supported by N.I.H. Grants AI11855 and AI18758.

A RAMAN SPECTROSCOPIC STUDY ON SEQUENCE DEPENDENT MODULATION OF DNA CONFORMATIONS

Yoshifumi Nishimura

Faculty of Pharmaceutical Sciences, University of Tokyo, Hongo, Bunkyo-ku, Tokyo 113, Japan

Recent X-ray crystallographic studies of DNA oligomers have shown that a DNA molecule in its crystalline state takes one of the polymorphic forms, A, B, or Z depending on its base sequence. DNA oligomers with CG sequences are crystallized into Z forms, those with GG/CC seqences into A forms. Although these DNA oligomers are considered to assume B forms in their aqueous solutions at low salt conditions, until now B-form crystal was found only for a dodecamer with a sequence of CGCGAATTCGCG, which showed the local variation within a B form depending on its sequence.

A DNA molecule with a defined sequence in solutions is now considered to take a characteristic conformation within a B family. Such sequence dependent characteristics might be recognized by sequence specific enzymes, such as restriction enzymes (1,2). In the present study our recent Raman spectroscopic investigations on various DNA molecules with different sequences are shown to reveal the sequence dependent DNA conformations in aqueous solutions at different conditions.

Our tactics to obtain DNA coformations in solutions is as follows.

i) In order to know the vibrational modes of nucleic acid base residues, we have calculated normal vibrations based on an ab initio MO method for uracil (3), adenine (4), guanine (5), and cytosine (6).

ii) To obtain the correlation between local conformations and Raman lines, we have examined Raman spectra of mononucleic acid crystals whose conformations were well established by X-ray crystallographic analyses (7,8).

iii) Based on these characteristic Raman lines of local coformations, we could reproduce the structures of A-, or Z-form nucleic acids whose conformations were well characterized (9-13).

iv) To obtain solution conformations of DNA polymers, we have examined Raman spectra of poly[d(G-C)]-poly[d(G-C)], poly(dG)-poly(dC), poly[d(G-meC)]-poly[d(G-meC)], poly(dG)-poly(dmeC), poly[d(A-T)]-

poly[d(A-T)], poly(dA)-poly(dT), and poly(rA)-poly(dT) (8,10,11,13-15).

v) Then we have studied solution conformations of various DNA oligomers with different base sequences: CGCG, GCGC, GGCC, CCGG, CGCGCG, ATAT, TATA, AATT, TTAA, GGAATTCC, CCTCGAGG, TTATATAA, GGGGCCCC, CGCGAATTCGCG, CGCGATATCGCG, CGCGTATACGCG, CGCGTTAACGCG, AATTAATTAATT (15-17).

vi) By examining these conformations, some correlations between DNA conformations and base sequences are obtained.

vii) To know the interaction between DNA and protein, Raman spectra of nucleosome core particles have been also investigated (18).

ACCKNOWLEDGEMENTS

The author would like to express sincere thanks to Prof. M. Tsuboi for his valuable discussions and sugesstions and to Prof. T. Hata for his kind cooperation on DNA oligomers.

REFERENCES

1. Y.Nishimura and M.Tsuboi,J.Biochem.,96,1807(1984)
2. Y.Nishimura,Advances in Biophysics,vol.20,59(1985)
3. Y.Nishimura,M.Tsuboi,S.Kato,and K.Morokuma,J.Am.Chem.Soc., 103,1354(1981)
4. Y.Nishimura,M.Tsuboi,S.Kato,and K.Morokuma,in "Raman Spectroscopy,Linear and nonlinear" eds. J.Lascombe and P.V. Huong,p703,John Wiley and Sons, Chichester(1982)
5. Y.Nishimura,M.Tsuboi,S.Kato,and K.Morokuma,Bull.Chem. Soc. Jpn.,58,638(1985)
6. Y.Nishimura and M.Tsuboi,Chem.Phys.,98,71(1985)
7. Y.Nishimura,M.Tsuboi,and T.Sato,Nucl.Acids Res.12,6901(1984)
8. Y.Nishimura,M.Tsuboi,T.Sato,and K.Aoki,Submitted for publication.
9. Y.Nishimura,M.Tsuboi,T.Nakano,S.Higuchi,T.Sato,T.Shida,S. Uesugi,E.Ohtsuka,and M.Ikehara,Nucl.Acids Res.,11,1579(1983)
10. Y.Nishimura,C.Torigoe,and M.Tsuboi,Biopol.,24,1841(1985)
11. Y.Nishimura,C.Torigoe,and M.Tsuboi,Nucl.Acids Res.,In press
12. Y.Nishimura,M.Tsuboi,S.Uesugi,M.Ohkubo,and M.Ikehara Nucle.Acids Res.,Symp.Ser.,No.16,25(1985)
13. M.Katahira,Y.Nishimura,M.Tsuboi,Y.Mitsui,and Y.Iitaka, Submitted for publication.
14. Y.Nishimura,C.Torigoe,M.Katahira,and M.Tsuboi,Nucl. Acids Res.,Symp.Ser.,No.15,147(1984)
15. Y.Nishimura and M. Tsboi,in "Advances in Spectroscopy: Biological Systems" eds. R.J.H.Klark and R.E.Hester,John Wiley and Sons,In press.
16. C.Torigoe,Y.Nishimura,M.Tsuboi,J.Matsuzaki,H.Hotoda, M.Sekine,and T.Hata,Submitted for publication.
17. M.Katahira,Y.Nishimura,M.Tsboi,J.Matuzaki,H.Hotoda,and T.Hata, To be published.
18. H.Hayashi,Y.Nishimura,and M.Tsuboi,Nucl.Acids Res.,In press

DNA STRUCTURES AND DNA-PROTEIN INTERACTIONS IN U.V. RESONANCE RAMAN SPECTROSCOPY.

P.Y. Turpin, L. Chinsky, A. Laigle and K. Otto*

Laboratoire de Physique et Chimie Biomoleculaire, Institut Curie, 11 rue P. et M. Curie, 75231 Paris Cedex 05, France, and *Twente University of Technology, Department of Applied Physics, P.O.B. 217, Enschede, Netherlands.

The UV resonance Raman excitation profiles of the various mononucleotides (1,2) have demonstrated the interest of exciting the DNA with a set of different wavelengths in order to optimize the observation of molecular changes induced by an interaction or correlated to some structure transitions. It was recently shown that base vibrations that are resonance-enhanced are highly sensitive to the conformation of DNA in solution(3), which adds complementary information to what is obtained in classical Raman spectroscopy from the phosphodiester backbone.

Poly(dG-dC) at high and low salt concentrations, excited with different wavelengths within the lowest energy absorption band, show resonance-enhanced lines (one for the purine and three for the pyrimidine) which may serve as marker bands for the B and Z structures. In the same way, large changes have been also observed in the B and Z forms of the polymer between H_2O and D_2O solutions. The 1484-1576cm^{-1} lines and the 1330-1350cm^{-1} doublet of guanine are good markers for the B form, while the 1486-1578cm^{-1} lines are good markers for the Z form, when Poly(dG-dC) is excited at 257nm. For an excitation at 280nm, the cytosine lines are in their turn observable, and their variations may be correlated to the H-D exchange. Hence it was possible to follow by multiple fast scanning the dynamics of hydrogen-deuterium exchange for each family, and relate them to the flexibility of the B and Z helices. It was found that the guanine proton exchange rate was faster than 5 seconds (which is the minimum time appreciable with our method) in the B form of Poly(dG-dC), while the half-rate of exchange of the guanine protons was around 12 minutes in the Z form. For the rate of exchange of the cytosine protons, it seems to be much slower, i.e. in the time range of a few hours. This seems to indicate that the flexibility of the Z helix is much lower than that of the B helix.

The thermal stability of the high salt d(m^5C-g)$_3$ was studied in resonance Raman with the 257nm excitation wavelength and in CD. At low temperature the Raman spectrum resembles that of Poly(dG-dC) in the Z conformation. As the temperature is increased the frequency and the intensity of several bands (1312-, 1482-, 1584- and 1632cm^{-1}) are modified, the variation of intensity versus temperature being biphasic. The same observation is made for $\Delta\varepsilon_{295}$ in the CD melting profile of the same sample. Analysis of these results suggests that the increase of temperature first induces a transition from the Z form to an intermediate stable form, which then melts at higher temperature. The Raman features of this intermediate form, obtained from a deconvolution method of the bands, suggest that it belongs to the left- handed family, but with changes in the stacking of the bases and the geometry of the phosphate groups as compared to the canonical Z form. The thermal transition follows the pathway:

Z-genus <--> left-handed intermediate form <--> denatured species.

The PolyrA- and PolydA interaction with the Gp32 protein of the T4 phage and with the Gp5 protein of the M13 bacteriophage was investigated with the 245nm excitation wavelength, for which essentially the adenine line intensities are resonance-enhanced, the protein contribution remaining weak. The unstacking effects of these proteins were compared to the temperature effects on the polymers, in following essentially the line intensity variations due to the base to base interaction changes. These variations are much larger in the case of PolydA than in the case of PolyrA and suggest to investigate further the excitation profile of the Raman hypochromism for both polynucleotides, as it was done before for PolyrU(4). Both proteins show important unstacking effects at 3°C, larger than a mere temperature effect at 40°C. The Gp5 interaction is more efficient than the Gp32 interaction at all temperatures investigated, but for both proteins no specific interaction seems to take place, which could give rise to variations in the frequency or intensity of a specific mode of vibration.

1) W.L. Kubasek, B. Hudson and W.L. Peticolas, P.N.A.S. 82, 1985, p. 2369
2) S. Fodor, R.P. Rava, T.H. Hays and T.G.Spiro, J.A.C.S. 107, 1985, p. 1520
3) B. Jollès, L. Chinsky and A. Laigle, J. Biomol. Struc. Dynam. 1, 1984, p. 1335
4) L. Chinsky and P.Y.Turpin, Biopolymers 19, 1980, p. 1507

RAMAN SPECTRA OF THE MODEL B-DNA OLIGOMER, $d(CGCGAATTCGCG)_2$ AND OF THE DNA IN LIVING SALMON SPERM SHOW THAT BOTH HAVE VERY SIMILAR B-TYPE CONFORMATIONS.

W.L. Kubasek, Y.Wang, G.A. Thomas, T.W.Patapoff, and W.L. Peticolas

Department of Chemistry, University of Oregon Eugene, Oregon 97403 USA

Raman spectra were obtained from aqueous solutions of the deoxy-oligonucleotide, $d(CGCGAATTCGCG)_2$ (I) which has been suggested as a model for B-type DNA conformation on the basis of X-ray diffraction studies. The Raman spectra were compared with the Raman spectra of aqueous solutions of several DNA's of natural origin. The Raman spectrum of (I) (66% GC) was compared with the Raman spectrum of the DNA from Micrococcus lysodeikticus (72% GC) and the two spectra were found to be identical in 50mM salt but with differences at 6M salt which indicates that (I) has a slight tendency to go into the Z form. The Raman spectrum of the dodecamer (I) in low salt was then used to analyze the Raman spectum of the DNA inside the head of living intact salmon sperm. Sperm spectra were taken using a newly developed intracavity laser Raman microscope system. Although the DNA in the sperm head is required by packing considerations to be in a highly compacted or condensed state, the Raman spectrum of the intact sperm is almost identical to that of the purified salmon sperm DNA in solution. Both are very similar to the Raman spectrum of the (I) in aqueous solulution if the difference in base composition is taken into account. Bandshape analysis of the backbone vibrations shows the existence of several bands which indicate a heterogeneity of furanose ring pucker in both the model dodecamer and the native DNA's. It is concluded that the furanose ring conformations in the purified salmon sperm DNA in solution are so varied that packing the DNA into the sperm head is carried out without significantly changing the DNA average backbone conformation.

The top Figure on the next page shows the spectra of intact salmon sperm (bottom spectrum) and purified DNA from the same sperm preparation (top spectrum). It is apparent that both the purified DNA and that in the intact Salmon sperm are in the B-family and the similarity of the Raman spectra of the purified and intact DNA is remarkably high since there is always the possibility that bands from other parts of the sperm head could be present in the Raman spectrum of the intact salmon sperm. The bands at 896 and 1450 cm^{-1} in the spectrum of the intact sperm head are assigned to membrane components of the sperm head. In addition to differences which are plainly due to other components, there is a slight change in the relative intensities of the Raman bands in the region 1257 to 1340 cm^{-1}. The middle and bottom figures show the region of the backbone vibrations of the intact Salmon sperm and (I). Both are in the B form and the slight differences in the base vibrations are probably due to differences in base composition.

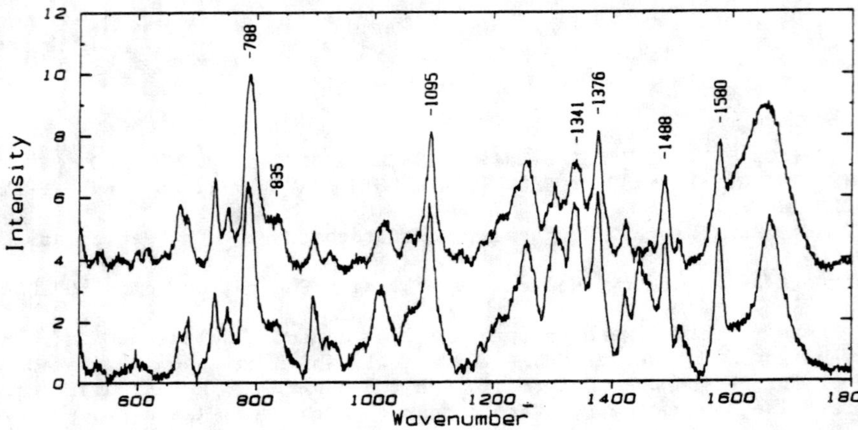
Intensity
Wavenumber
-788
-835
-1095
-1341
-1376
-1488
-1580

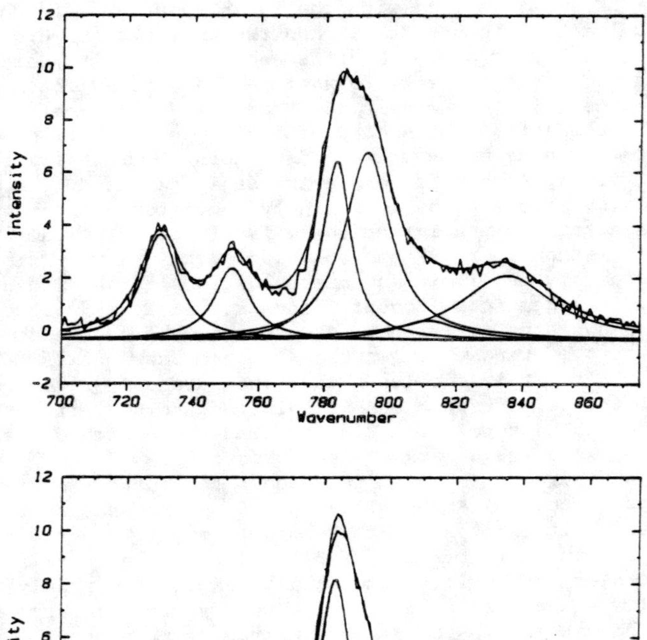
Intensity
Wavenumber

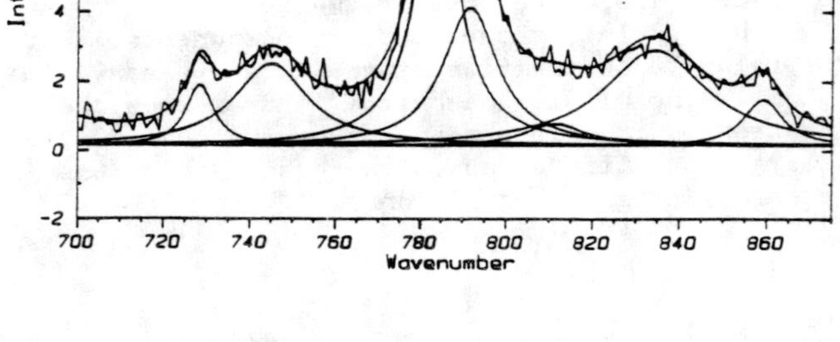
Intensity
Wavenumber

ETHANOL-INDUCED CONFORMATIONAL CHANGES IN POLYdA.POLYdT, POLYd(A-T), POLYdG.POLYdC AND POLYd(G-C) MONITORED BY RAMAN SPECTROSCOPY AND INTERPRETED BY BASE STACKING ENERGIES.

Y. Wang, and W.L. Peticolas

Department of Chemistry, University of Oregon, Eugene, Oregon 97403.

Recently it has been reported that the synthetic DNA, polydA.polydT in dilute (0.05M) solution at pH7 and 0.1M salt contains both the C3'-endo Raman marker band (at 816 cm^{-1}) as well as the C2'-endo marker band (at 837 cm^{-1}) (1). This was a novel finding since it is usually supposed that deoxynucleic acids have predominately C2'-endo ring pucker. Some evidence for polydispersity of ring pucker conformation was also found in the double helical poly(dA-dT)[1]. These results have been confirmed and extended by band shape analysis of the 850-800cm^{-1} region in these polymers [2]. In a parallel set of investigations, it has been shown that the alternating deoxy(purine-pyrimidine) homopolymers can go into the A form in which the C3'-endo marker band is extremely pronounced [3,4]. These latter measurments were made on fibers at 75% relative humidity or in fairly concentrated solutions (0.2M) where is is known that aggregation helps with the formation of the A-form of DNA [5]. The A-form DNA is a form in which the spine of hydration which has been found to exist in the minor groove of B-DNA is no longer present. It should be noted that in the absence of an excess of salts such as NaCl and MgCl2, the DNA transition is almost always between the A and B forms. In the presence of excess salt, one can obtain the Z form with polyd(G-C). The question naturally arises is: What is the conformation of the synthetic DNA's which have all purine in one chain and all pyrimidines in the other chain if at high concentrations and low hydrations. Recently it has been proposed on the basis of X-ray diffraction work that polydA.polydT can go into a heteronymous structure in which all of the furanose rings in one strand are in the C3'-endo and all of the furanose rings in the other strand are in the C2'-endo [5]. The existence of Raman marker bands for both ring forms [1,2] would indicate that there may be regions of this polymer in which this is the case but the intensity of the C3' marker band is probably not strong enough to warrant the interpretation of the entire chain being in this form. This is not too supprising since the X-ray fiber work must assume all of the furanose rings are in the same conformation which is probably too restrictive.

In an effort to shed light on this problem we have made a detailed Raman spectrographic study of all five polymers, polydG.polydC (I), polydG.polyd(m5C)(II), polyd(G-C)(III), polydA.polydT (IV), polyd(A-T) (V) in fiber form in alcohol-water mixtures as a function of the alcohol concentration. It has previously been shown that DNA goes into the A form at ethanol concentrations of 80% or higher[5]. In each case the window at 600-850 cm-1 was examined in detail and sujected to band shape analysis. What we find is rather interesting. Starting with (I) we find (frequencies are in wave numbers)

DNA	AQUEOUS SOLUTION (Dominating Frequencies)	80%ETHANOL(FIBER) (Dominating Frequencies)
	CG Containing Polymers	
I	686s 783vs 792s 812wb 835m	617m 666s 742s 784vs 794vs 811vs
II	680m 778vs 784m 810m none	614,641,665,687,742 779vs 813vsb
III	682s 784s 794vs 808vw 833m	615,642,667,684,741 786vs 815mb
	AT Containing Polymers	
IV	673s 731vs 754s 792vs 816m 843m;	615,645,667,674 731s 754s794s 821mb
V	673s 732vs 753s 794vs 814w 843w;	620,645,668,708,730s,748,778s,808vs

From the above table it is apparent that both polydG.polydC (I) and polyd(G-C) (III) are in the B form in dilute (0.05M) aqueous solution; both have a moderate band at 835cm^{-1} and a weak band at 810cm^{-1} which can only be obtained from band shape analysis [2]. Methylating the C in the 5-position causes the polydG.poly(5mC)(II) to go into a partial or predominately A form in dilute solution as evidenced by the moderate band at 810cm^{-1} and lack of any observable at 835-840cm^{-1}. In 80% ethanol I and II go into a pure A form in which the 813cm^{-1} band is the strongest in the spectrum, but the alternating copolymer d(GCGC....) (III) goes into a strange form with a moderately strong broad band at 815cm^{-1} and no disernable band at 835-40cm^{-1}. But with the AT containing polymers the situation is just reversed. The alternating copolymer d(ATATAT....) goes into an A-form with the 810cm^{-1} band the strongest in the spectrum, while in polydA.polydT the 816cm^{-1} band observable in the dilute aqueous solution gets much stronger and broader but the 843cm^{-1} band goes completely away. This observation correlates well with calculated stacking energies of stabilization for these polymers [6]. For AT polymers the side-by-side form is more stablized in the B form by stacking energies while for GC polymers the alternating form is the more stable B form.

1. G.A.Thomas & W.L.Peticolas, J.Am.Chem.Soc. 105, 986-999(1983).
2. R.M.Wartell & J.T.Harrell, Biochemistry, in press.
3. G.J.Thomas Jr. & J.M.Benevides, Biopolymers 24, 1101(1985).
4. Y.Nishimura, C.Torigoe, & M.Tsuboi, Biopolymers 24, 1841(1985).
5. R.Herbeck, T.J.Yu, and W.L. Peticolas, Biochemistry 15,2656(1976).
6. R.L.Ornstein,R.Rein,D.L.Breen, & R.D.MacElroy,Biopolymers 17, 2341(1978)

ULTRA-VIOLET RESONANCE RAMAN SPECTROSCOPY OF COMPLEXES OF CIS- AND TRANS-DICHLORODIAMMINEPLATINUM WITH GUANOSINE AS A FUNCTION OF PH.

K. Bajdor, Y. Wang and W.L. Peticolas

Department of Chemistry, University of Oregon, Eugene, Oregon 97403.

The use of pulsed ultraviolet lasers to obtain the Ultraviolet Resonance Raman (UVRR) spectra of nucleic acids components was first reported at ICORS-V, ten years ago (1). At that time it was only possible to reach the first excited electronic state at about 260nm. More recently in ICORS-VIII it was shown that using frequency doubled YAG lasers it is possible to obtain much shorter wavelengths (2) so that one can reach the second excited electronic state about 200-210nm. This is of interest in studying interactions with guanine because the carbonyl vibration is absent from the UVRR spectrum of guanine derivatives taken at 260nm (1) while it is present in the UVRR spectrum from the second excited state taken at 213nm or 218nm (2).

The interaction of the cis- and trans-dichlorodiaminoplatinum (II) (cisplatin and transplatin) with nucleic acid components is of interest because cisplatin is one of the most effective and extensively studied inorganic antitumor drugs while for reasones that are not fully understood, the transplantin isomer exhibits little or no antineoplastic activity. In each case it is known that the first Pt-Cl bond is replaced by Pt-N7 bond as the result of the reaction at the N-7 site in guanine. In a recent X-ray study, Dickerson et al.(3) showed that one Pt-Cl group of cisplatin reacted with the N7 position of guanine in a double stranded deoxyoligonucleotide containing alternating cytosine and guanine while the second interacted weakly, probably through a water bridge with the O6 group of guanine. These investigators found no evidence for crosslink across the double helix which has been reported to occur in solution in double stranded helices upon interaction with cisplatin (4). In our studies we wanted to see if there was evidence for interaction of platinum-O6 interaction after reaction of cis and transplatin with the N7 position of guanine in guanosine in solution. Since the Pt-Cl bonds are opposite each other in the transplatin and adjacent to each other in the cisplatin it seems impossible that the transplatin can interact with the O6 oxygen of the guanine but the cisplatin might possibly interact although this would require either interaction through water or hydroxide ion since the Pt-O6 distance would be quite large (This is what the Dickerson group found.) Thus by looking at the intensity and the frequency of the C=O6 carbonyl vibration of native guanosine and the 1-1 complexes of cisplatin and tansplatin with guanosine one could tell if there is evidence of an interaction in cis- and not in the trans- case.

Another feature of our study is the variation of pH. Guanosine and guanosine derivatives which have been substituted by Pt or methylated at the N7 position are known to exist in two completely different forms.

Below pH=7 the alkylated N7 position has a positive charge as in the ammonium ion which adds a positive charge to the molecule. Above pH=7 the N7-substituted guanine exists as a zwitterion. The hydrogen on the N1 position is lost, the electronic structure of the six-membered ring becomes isoelectronic with benzene and the C=O6 carbonyl becomes a single bonded C-O$^-$ with a negative charge on the oxygen atom. In view of this incredible change in structure with pH for N7-substituted guanosines it seems puzzling that no one has looked at the effect of pH of the interaction of these Pt compounds with guanine in DNA.

UVRR spectra of aqueous solutions of guanosine and the 1:1 complexes of cisplatin and transplatin with guanosine have been taken with 218nm line as a function of pH. At low pH all three exhibit a strong UVRR active band at 1662 cm^{-1} which is assigned to the C=O6 carbonyl vibration. At pH above 7 this band goes away and it is completely gone at pH=9. The band is strongest in the native guanosine, weaker in the 1:1 transplatin complex and very weak in the 1:1 cisplatin complex. From this we conclude that the carbonyl vibration is a mixed mode involving the vibration of the ring in addition to the vibration of the C=O double bond, substitution at the N7 position changes the ring electronic structure, and hence appears to be responsible for the decrease in the intensity of the transplatin substituted complex. The decrease in the intensity of the cisplatin-guanosine complex is assigned to an interaction of the Pt with the O6 of the guanosine through a hydroxide ion. This interpretation differs slightly from that of Dickerson. We base this conclusion on the fact that at pH = 6.5 and above the transplatin-guanosine complex becomes insoluble due evidently to the Pt-OH bond formation at higher pH leading to the formation of intermolecular Pt-OH hydrogen bonds which results in the formation of an insoluble precipitate. It seems likely that this Pt-OH group is also formed at higher pH's in the cisplatin complex. However, in this case the Pt-OH group does not lead to insolubility because of intramolecular hydrogen bonding to the O6 group of the guanosine. In conclusion, cisplatin reacts with guanosine first at the N7 position followed by interaction through a hydroxid ion to the carbonyl oxygen, Pt-OH$\cdots$O6.

1. W.L.Peticolas, in Proc. of the Fifth International Conference on Raman Spectroscopy ed by E.D.Schmid, J.Brandmuller, W.Kiefer, B.Schrader, and H.W.Schrotter. Hans Ferdinand Schulz Verlag D-7800 Freiburg im Breisgau, FRG, (1976). pp163-175.

2. L.D. Ziegler, S.P. Strommen, B.S. Hudson, and W.L. Peticolas, Raman Spectroscopy Linear and Nonlinear, Proc. of the Eighth Inter. Conf. on Raman Spectroscopy ed. by J.Lascombe and P.Huong (1982) pp707-708 John Wiley, New York.

3. R.M. Wing, P.Pjura, H.R. Drew, and R.E. Dickerson, (1984) EMBO Jurnal 3, 1201-1206.

4. S.J. Lippard, Science (Wash.) 218, 1075-1083(1982).

LASER RAMAN SPECTROSCOPY OF COMPLEXES OF NUCLEOTIDES WITH ACETAMIDE

M. Molina and P. Carmona
Instituto de Optica (C.S.I.C.). Serrano 121. 28006 Madrid

Interactions between nucleic acid and proteins play a central role in the biology of the living cell by controlling the processes of cellular metabolism, replication and development. In many cases the interacting proteins are bound only to highly specific nucleotide sequences or sites of the nucleic acid substrate (1). As a step toward studies of complexes between nucleotides and amino acid side chains we have been conducting investigations on complexes of 5'-UMP, 5'-AMP and their polynucleotides poly(rA) and poly(rU) with acetamide. Direct interactions through hydrogen bonds might be of great importance and therefore acetamide may provide a basis for the understanding of interactions between nucleotides and glutamine or asparagine.

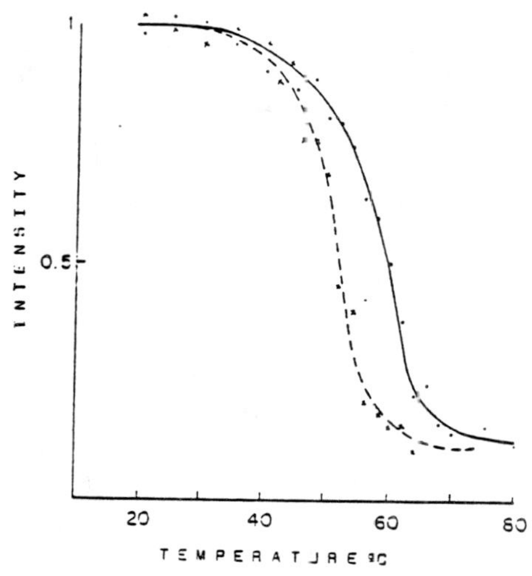

Fig. 1. Transition temperatures of (——) poly(rA).poly(rU) and (- -) poly(rA).poly(rU) + acetamide. Measured with Raman band at 814 cm−1.

Laser-excited Raman spectra of complexes of poly(rA).poly(rU) with acetamide reveal conservation of the backbone structure normally occurring in aqueous poly(rA).poly(rU). This is evident by the appearance of the phosphodiester group band near 820 cm−1 with no change from the Raman line of free poly(rA).poly(rU) at this frequency. However, the PO_2^- symmetric stretching band shifts to 1110 cm−1 (table 1). The position and half-width of this band appear to be affected by gross changes in the degree of hydration of nucleic acid fibers (2). Moreover, theoretical calculations demonstrated the lowering effect of hydration on the anionic P-O stretching frequencies (3). On the other hand, an intensity increase is observed for the band at 1345 cm^{-1} due to adenine. Concerning ribose the band located at 716 cm^{-1} has been assigned to the C_3-endo ring puckering of the ribofuranosyl moiety (4) and its intensity increases upon acetamide binding.

The complex formation is also detected by changes in melting properties (Fig. 1). The characteristic transition temperature T_m decreases when acetamide is bound to poly(rA).poly(rU), probably due to the fact that acetamide binds more tightly to coil than to helix. The phosphodiester Raman intensities versus temperature indicate that this binding is reversible.

Table 1. Principal Raman spectral changes of poly(rA).poly(rU) accompanying complex formation.

Frequency	Assignment	Effect of complex formation
716 sh	Ribose	Intensity increase
1102 m	PO_2^- symmetric stretch	Shift to 1110 cm^{-1}
1345 m	A ring	Intensity increase
1560 w	A ring	Intensity decrease

No significant spectral changes have been found for the aqueous solutions of the aforementioned mononucleotides and acetamide. This is in contrast to that occurring in the solid state, in which case infrared and Raman spectra show that most sensitive to complex formation are vibrational frequencies of poly (rA). Carbonyl bands in 5'-UMP and poly(rU) shift to lower frequencies upon acetamide binding, and their location in the infrared and Raman spectra suggests the existence of cyclic dimers.

REFERENCES

1. S. Casjens and J.King. Annu. Rev. Biochem. 44, 555 (1975).
2. S. C. Erfurth, P.J.Bond and W.L.Peticolas. Biopolymers 14, 1245 (1975).
3. E. B. Brown and W.L. Peticolas. Biopolymers 14, 1259 (1975).
4. B. Prescott, R.Gamache, J.Livramento and G.J.Thomas, Jr. Biopolymers 13, 1821 (1974).

RAMAN SPECTROSCOPIC STUDY OF THE BINDING OF COPPER (II) TO NUCLEOTIDES

P. Carmona*, P.V. Huong** and E. Gredilla*

*Instituto de Optica (CSIC), Serrano 121. 28006 Madrid.

**Laboratoire de Spectroscopie Infrarouge et Raman, Université de Bordeaux I, 33405 Talence.

Metal ions have profound effects on the structure and function of nucleic acids and mono– and polynucleotides. They alter the coding specificity of polynucleotides acting as templates for protein synthesis and are required for stabilizing the structure of tRNA. It is therefore of interest to determine the sites of attachment of the metal to these substances. On the basis of the recent assignments for the Raman bands of nucleobases and their D, ^{15}N and ^{13}C substitution products (1–5) we have analyzed the Raman spectra of the nucleotides 5'–AMP, 5'–GMP, 5'–CMP and 5'–UMP and their complexes with copper (II).

In purine mononucleotides, the Raman spectra show binding of the metal to a nitrogen atom in the imidazole ring. Binding to a nitrogen atom of the cytosine base in 5'–CMP has also been demonstrated by the spectral shifts of certain bands. However, copper was found to bind 5'–UMP near the carbonyl groups. Strong changes are also recorded for Raman bands characteristic of the $-PO_3^{2-}$, indicating a participation of the phosphate group in the ligands surrounding the copper ion (Table 1).

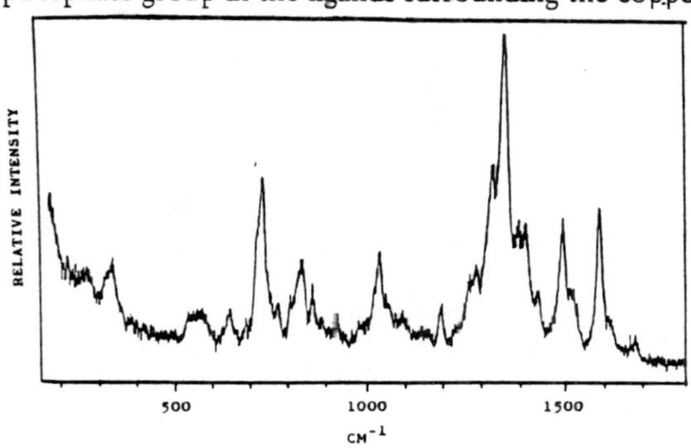

Fig. 1. Raman spectrum of Cu(AMP)

Table 1. Raman spectra of mononucleotides and their complexes with copper (II)

5'-$AMPNa_2$	Cu(AMP)	
1642 vw	1680 vw	NH_2 scis
1584 m	1608 sh 1587 m	ν_2 ring
1521 w 1510 w	1515 sh	ν_3 ring
1484 m	1493 m	ν_4 ring
1434 m	1434 w	δCH, ring
	1408 m	
1382 m	1382 w	δC(2)–H, ring
1337 vs	1347 vs	ν_6 ring
1312 m	1312 m	ν_7 ring (Pyr)
	1276 w	
1259 w	1259 sh	νC–NH_2
1237 w		ν_8 ring (Im)
1183 w	1183	δC(8)–H
1155 vw	1155 vw	ν_9 ring
1077 w	1090 vw	ν_{as}–PO_3^{2-}
989 m	1026 m	ν_s–PO_3^{2-}
912 vw	912 vw	δ–ring
877 vw	877 vw	ribose
841 w	855 w	γC(2)–H
800 m	828 800 sh	ν(–O–P–)
772 vw	772 vw	ν_{16} ring(Pyr)
731 s 717 sh	731 s 711 sh	ring br.
648 vw	648 w	ν_{12} δ–ring(Pyr) ribose
	582 w	
539 w	539 w	ribose δ–ring(Pyr)
387 w		τ NH_2
	343 m	
322 w	326 sh	δ C–NH_2
	275 w	Cu–ligand

5'-$GMPNa_2$	Cu(GMP)	
1705 w	1697	ν C=O
1647 w	1655 w	NH_2 scis
1588 vs	1587 vs	ν ring(Pyr)
1497 vs	1500 vs	ν ring
	1432 m	
1375 m	1365 sh	ν ring (Im)
1332 m	1337 s	ν ring
1180 w	1192 w	δ CH + ν ring
1075 w	1084 w	ν_{as}–PO_3^{2-}
990 m	987 w	ν_s–PO_3^{2-}
885 m	869 w	ribose
787 w	812 w	
679 s	675 s	δ–ring
647 w	647 w	ribose
590 w	590 w	δ–ring
507 w	507 w	δ–ring
	377 w	Cu–ligand
307 w		

5'–$CMPNa_2$	Cu(CMP)	
1658 sh 1621 m	1670 b, m	νC=C, νC=O
1536 m	1535 m	NH_2 scis, ν–ring
1462 w	1460 w	δ CH, ν C–N
1417 w		
1384 m	1394 w	δ CH, ν C–N
1304 vs	1292 vs	δ CH O.P.
1257 s 1227 s	1253 vs	ν–ring
1140 vw	1140 vw	
1067 vw	1050 vw	ν_{as}–PO_3^{2-}
1012 m 982 m	1007 m 987 m	ν_s–PO_3^{2-}
889 w	887 w	ribose
787 vs	792 vs	ν, δ–ring
653 w 635 w	655 vw	ribose
607 w	607 m	δ–ring
	567 w	δ C=O, δ C–N
546	540 sh	ribose
	478 w	Cu–ligand
420 w	420 w	ribose

5'–$UMPNa_2$	Cu(UMP)	
	1682 s	ν C_2=O, νC_4=O
1671 vs		
	1635 m	νC=C, νC=O...Cu
1467 w	1467 w	
1438 vw		δ NH
1401 m	1396 m	δ CH, ν–ring
1270 sh	1292 w	
1232 vs	1232 vs	δ CH
1125 w	1115 vw	ν–ring
1069 w	1052 vw	ν_{as}–PO_3^{2-}
1010 w 977 m	997 m	ν_s–PO_3^{2-}
905 w	887 w	ribose
825 w	820 sh	γ CH
776 s	787 s	δ–ring
661 w	637 w	ribose
643 w	607 vw	ribose
543 w	560 m	δ C=O, ribose
	462 w	γ C=O
424 w	424 w	ribose

REFERENCES

1. A.Y. Hirakawa, H. Okada, S. Sasagawa and M. Tsuboi. Spectrochim. Acta. 41A, 209 (1985).
2. M. Majoube. Biopolymers 24, 1075 (1985).
3. J.M. Delabar and W. Guschlbauer. Biopolymers 18, 2073 (1979).
4. Bandekar and G. Zundel. Spectrochim. Acta 39A, 343 (1983).
5. H. Susi, J.S. Ard and J.M. Purcell. Spectrochim. Acta 20A, 725 (1973).

SOFTENING OF THE LOWEST OPTIC MODES OF DNA.[a]

S.A. Lee, S.M. Lindsay, J.W. Powell[b], N.J. Tao, C. DeMarco and A. Rupprecht[c]
Physics Department, Arizona State University, Tempe, Az. 85287 USA.

Raman and Brillouin scattering, together with fiber diffraction and other physical studies, have been used to study fibers and films of Na-DNA over the A to B transition. Both low frequency (0 to 100 cm^{-1}) and high frequency (500 to 1000 cm^{-1}) spectra have been studied. Li-DNA (which does not undergo an A to B transition) was used as a benchmark sample. While there is some evidence of a soft mode for the transition (as predicted by Eyster and Prohofsky[1]) the dominant factor is interhelical contacts between adjacent double helices in the crystal (as first suggested by Herbeck et al.[2]). Formation of the contacts probably drives DNA into the A form, and we can account for the observed softening of low frequency modes with a model of these interhelical bonds.

The lowest lying optic modes[3,4,5] of Na-DNA and Li-DNA have been located using both conventional monochromators and tandem interferometry. VH polarized modes are seen at about 34 (Na-DNA) and 25 cm^{-1} (Na and Li-DNA), while a VV polarized mode is seen at 12cm^{-1} in Li-DNA (there is some evidence of a VH polarized mode at 15 cm^{-1} in Na-DNA). Both the 12 and 25 cm^{-1} modes are only seen with q parallel to the helix axis in highly crystalline samples. This selection rule is probably due to highly anisotropic polarizibility derivatives. Lattice dynamical calculations[4] indicate that the inner product of the eigenvector of the 12 cm^{-1} mode of B-form DNA with the order parameter for the A to B transition is nearly 0.9. Thus, this mode should soften if the A to B transition is mediated by it. This mode is difficult to study, and we have studied the related 25 cm^{-1} mode. The 25 cm^{-1} mode does soften with hydration in Na-DNA as the A to B transition is approached. However, the 25 cm^{-1} mode of Li-DNA also softens with hydration as the point of

a) This work was supported in part by NSF grant PCM8215433, ONR contract N00014-84-C-048 and EPA contract 68-02-4105.
b) present address: Max Planck Institute, 1 Heisenbergstrasse, 7000 Stuttgart 80, West Germany.
c) Arrhenius Laboratory, University of Stockholm, S-106, 91 Stockholm, Sweden.

decrystallization is approached (although not by as much as the related mode in Na-DNA). These results indicate that the soft mode theory of the A to B transition is incomplete.

Brillouin scattering data show that the dry material is held together by very strong interhelical bonds. The observed negative elastic anisotropy is in line with lattice dynamics estimates of the compressibilities parallel and perpendicular to the helix axis. Observation of this anisotropy requires <u>inter</u> double helical bonds which are comparable in strength to the intra helical bonds. The common factor underlying anomalies in the Brillouin spectra of both Li- and Na-DNA is decrystallization and not conformational transitions[6].

Fiber diffraction and high frequency Raman studies show that the rate of equilibration near a phase transition is a function of the degree of crystallinity of the samples. Furthermore, the intensity of the A form marker band at 807 cm^{-1} may be correlated with the extent of crystallinity: dry samples that are completely disordered contain <u>no</u> detectible A-DNA. Thus the A form of DNA is stabilized by interhelical contacts. X-ray fiber diffraction data indicate one near contact per turn of double helix involving an O4 of the phosphate groups of neighboring double helices in the A-DNA unit cell, and we propose a model of the interhelical bonding in Na-DNA based on a Coulomb interaction of these phosphates mediated by an excess Na^+ ion. This model accounts for the observed softening of the 25 cm^{-1} mode in terms of the measured separation of these charges on swelling.

1. Eyster, J.M. and Prohofsky, E.W., Biopolymers 13, 2505 (1974).
2. Herbeck, R., Yu, T.J. and Peticolas, W.L., Biochemistry <u>15</u>, 2656 (1976).
3. Urabe, H. and Tominaga, Y. , Biopolymers, <u>21</u>, 2477 (1982).
4. Lindsay, S.M., Powell, J.W., Prohofsky, E.W. and Devi-Prasad, K.V. in <u>Structure and Motion, Nucleic Acids and Proteins</u>, (1984), Eds. Clementi, E., Corongiu, G., Sarma, M.H. and Sarma, R.H. (Adenine Press, New York) pp. 531.
5. Lindsay, S.M., Powell, J.W. and Rupprecht, A. Phys. Rev. Lett., <u>53</u>, 1853 (1984).
6 DeMarco, C., Lindsay, S.M., Pokorny, M., Powell, J.W. and Rupprecht, A., Biopolymers, <u>24</u>, 2035 (1985).

TAUTOMERIC EQUILIBRIA AND VIBRATIONAL SPECTROSCOPY OF MATRIX-ISOLATED GUANINES.

M. Szczesniak*†, K. Szczepaniak*†, L. Hernandez† and W. B. Person†

*Institute of Physics, Polish Academy of Science, 02668 Warsaw, Poland, and †Department of Chemistry, University of Florida, Gainesville, FL 32611, USA

Guanine and the guanosine analogue, 9-methylguanine, have both been found[1,2] to exist in an inert environment in tautomeric equilibrium between keto and enol forms:

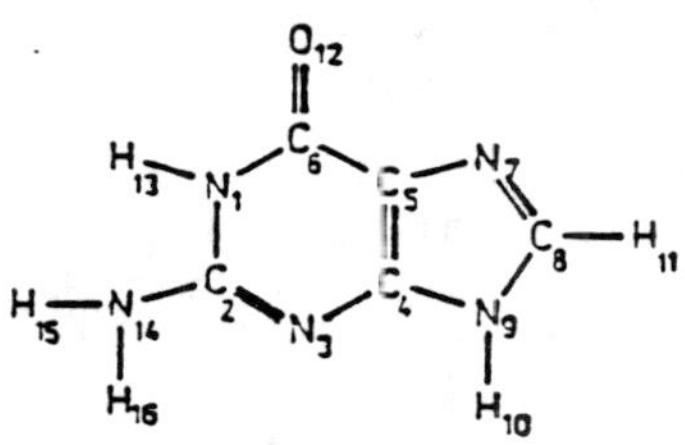

keto-amino

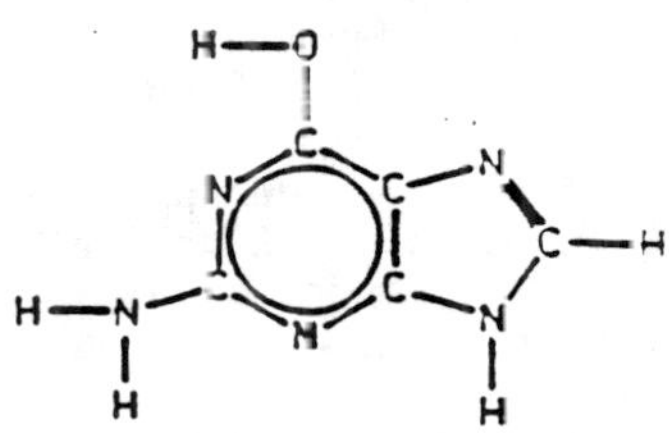

enol-amino

These findings have, of course, considerable implications for the understanding of mutagenesis, for example.

Our studies have confirmed and extended these initial observations[1,2] to include infrared and Raman studies of a number of other methylated guanine derivatives and to thioguanine compounds. Detailed analysis of the spectrum compared to *ab initio* calculations[3] of the infrared spectrum have led to assignments for these compounds and to some understanding of the effects of methyl substitution in the vibrational spectrum.

References

1. K. Szczepaniak and M. Szczesniak, J. Mol. Structure, to be published.
2. G. G. Sheina, E. D. Radchenko and Yu. P. Blagoi, Dokl. Acad. Nauk, 282 (1985) 1497 (Russ. ed.).
3. Z. Latajka, W. B. Person and K. Morokuma, J. Mol. Structure (THEOCHEM), 135 (1986) 253.

RAMAN SPECTROSCOPIC STUDIES OF ZEOLITE FORMATION.

Prabir K. Dutta and D. C. Shieh

Department of Chemistry, The Ohio State University, Columbus, Ohio 43210, U.S.A.

In the last thirty years, the development of aluminosilicate zeolitic materials has seen a phenomenal growth arising from the increasing range of industrial applications. However, as is true in the synthesis of all new materials, an understanding at the molecular level of the chemical transformations that result in zeolitic structures is practically non-existent. This accounts for the fact that only a few percent of the possible framework topologies have been synthesized.

We have been using Raman spectroscopy[1-3] along with NMR spectroscopy, IR spectroscopy and X-ray diffraction to study the molecular species involved in zeolite synthesis. The synthesis mixtures are considerably complex, consisting of soluble precursor species, amorphous gel and zeolite crystals. Figure 1 shows the evolution of the zeolite A structure as determined by Raman spectroscopy. The Raman bands suggests that the amorphous gel is predominantly composed of four-membered rings connected randomly which fold in specific ways depending on the Si/Al ratio, the cations and the solution species. These studies are providing new insight into the formation of these important materials.

1) P. K. Dutta and D. C. Shieh, Appl. Spect., 39, 343, (1985).

2) P. K. Dutta and B. DelBarco, J. Phys. Chem., 89, 1861, (1985).

3) P. K. Dutta and B. DelBarco, J.C.S. Chem. Comm., 1297, (1985).

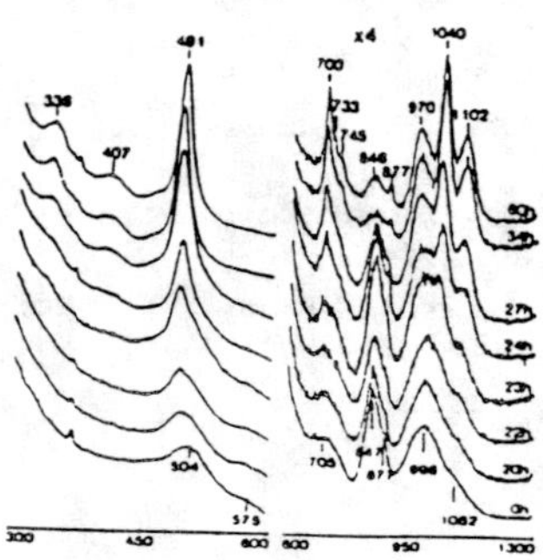

Figure 1.

NORMAL COORDINATE ANALYSIS OF THE NUCLEIC ACID VIBRATIONAL MODES

R. LETELLIER, M. GHOMI and E. TAILLANDIER

Laboratoire de Spectroscopie Biomoléculaire, U.F.R. Biomédicale de Bobigny, Université Paris XIII,
74, Rue Marcel Cachin 93 012 Bobigny cedex, France.

The main Raman and infrared peaks arising from the vibration modes of the four nucleosides appearing in DNA double-helical chains, have been assigned using a normal coordinate analysis * based on the Wilson GF-method (1).

The redundant internal coordinates have been removed by a $B.\tilde{B}$ method (2) respecting the local symmetry of the molecular sites in nucleosides. A valence force field has been used in order to construct the interatomic interaction matrix. The validity of this force field has been tested in the case of 5′-dGMP (3), 2′-deoxycytidine (4), 2′-deoxyadenosine and 2′-deoxythymidine (5).

The calculated results take account of the DNA vibrational characteristics observed experimentally in the 1550-1250 cm^{-1} and 800-600 cm^{-1} spectra regions . Moreover the spectral evolution upon the right left-handed transition observed in oligonucleotides and polynucleotides has been satisfactorily reproduced (6,7,8,9,10).

In a recent paper we have described the variation of the characteristic vibrational frequencies of dG and dC residues versus the glycosidic χ angle and the sugar-pucker conformation (4).

Similar calculations have already been done in the case of dA and dT residues. A complete set of oligonucleotide and polynucleotides with 5 different forms (A,B,C,D and Z) has been studied. So, on the basis of these calculations a reliable force field can be proposed for the four nucleosides found in DNA double-helical chains.

REFERENCES

1. WILSON E.B., DECIUS J.C. and CROSS P.C. in Molecular Vibrations (Mc Graw Hill - New York) (1955).

2. GUSONI M. and ZERBI G. J. Mol. Spectrosc. 26, 485-488 (1968).

3. GHOMI M. and TAILLANDIER E. Eur. Biophys. J. 12, 153-162 (1985).

4. LETELLIER R., GHOMI M. and TAILLANDIER E. J. Biomol. Struct. Dyn 3, 671-687 (1986).

5. LETELLIER R., GHOMI M. and TAILLANDIER E., (to be published).

6. THOMAS G.A., PETICOLAS W.L., Biochemistry 23, 3201-3207 (1984).

7. TSUBOI M., NISHIMURA Y., HIRAKAWA A.Y. and PETICOLAS W.L. (to be published).

8. BENEVIDES J.M., WANG A.H.J., VAN DER MAREL G.A., VAN BOOM J.H., RICH A., and THOMAS Jr. G.J. Nucl. Acids Res. 12, 5913-5925 (1984).

9. GHOMI M., TABOURY J.A. and TAILLANDIER E., Biochimie (Paris), 66, 87-94 (1984).

10. ADAM S., LIQUIER J., TABOURY J.A. and TAILLANDIER E., (in press).

* We would like to thank the scietific Council of the CCVR (Centre de Calcul Vectoriel pour la Recherche) for the computational facilities on Cray-1s computer.

chapter 3

LIPID BILAYERS and RELATED STRUCTURES

RAMAN SPECTROSCOPIC STUDY OF BOUNDARY LIPID IN APOLIPOPROTEIN A-1/DIMYRISTOYLPHOSPHATIDYLCHOLINE RECOMBINANTS

S. F. Bush[1] and I. W. Levin[2]

Chemistry Department, University of North Carolina at Charlotte, Charlotte, NC 28223[1] and the National Institutes of Health, National Institute for Arthritis, Bethesda, MD 20205[2]

Aqueous multilamellar dispersions of dimyristoylphosphatidylcholine (DMPC) normally undergo a sharp gel-to-liquid-crystal main phase transition at 23.8°C. Treatment of multilamellar dispersions with ultrasound will convert them nearly quantitatively into small unilamellar vesicles. These vesicles yield a broadened main phase transition centered about 22°C. Addition of a major protein component of serum high density lipoprotein, apolipoprotein A-1, results in the autoconversion of small unilamellar vesicles into discoidal complexes as shown by electron microscopy (Figure 1). These disk shaped particles are believed to be comprised of small pools of lipid material in a bilayer arrangement with the protein molecules forming an annulus for the lipid pool.[1]

Figure 1. Electron micrograph of discoidal complex of DMPC and apolipoprotein A-1. Magnification, 105.000 x.

We have used Raman spectroscopy to generate thermal phase transition profiles of both unilamellar vesicles and the discoidal complex resulting from apolipoprotein A-1 addition. The discoidal complex has a main transition at $\sim$27°C and a secondary transition at $\sim$47°C (Figure 2). These transitions are interpreted as belonging respectively to melting of bulk lipid acyl chains and protein boundary lipid acyl chains. Simple geometric considerations and the chemical analysis of the discoidal complex allow us to conclude that a single annular ring of DMPC lipid molecules constitutes the boundary lipid layer.

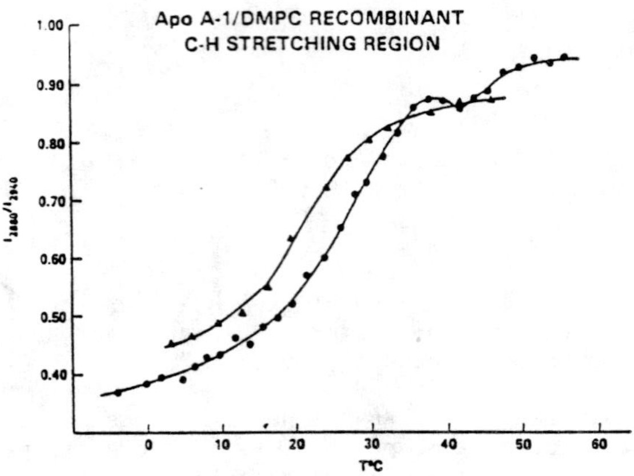

Figure 2. Temperature profiles of DMPC unilamellar vesicles (▲) and of DMPC/apo A-1 recombinant material (•) shown in Figure 1.

1. A. R. Tall and D. M. Small in Advances in Lipid Research Vol. 17, eds. R. Paolette and D. Kritchevsky (Academic Press, New York, NY 1980) pp. 1-51.

INTERACTIONS OF FERRICYTOCHROME c WITH MODEL PHOSPHOLIPID BILAYERS.

James S. Vincent[1] and Ira W. Levin[2]

[1]Chemistry Department, University of Maryland Baltimore County, 5401 Wilkens Avenue, Catonsville Maryland 21228
[2]Laboratory of Chemical Physics, NIADDK, National Institutes of Health, Bethesda Maryland 20892

The intermolecular perturbations of 1-α-dipalmitoylphosphatidylcholine (DPPC) multilayers induced by the extrinsic interactions of ferricytochrome c were studied using vibrational spectroscopic techniques. Raman spectra in the C-H stretching mode region (2800-3100 cm^{-1}) were observed over a temperature range of 5 to 55°C for approximately 100:1 mole ratio mixtures of DPPC and ferricytochrome c. The temperature profiles were measured for samples at pH 4 and 7 with and without 0.1 M phosphate buffer and also in the presence and absence of NaCl concentrations which ranged from 0.5 M to 1.0 M. The results of these studies were summarized by graphing the ratios of the intensities of the peaks near 2935 cm^{-1} and 2880 cm^{-1} and by plotting the total integrated intensity of the 2800-3100 cm^{-1} spectral interval both as function of temperature. Temperature profiles of DPPC in the absence of ferricytochrome c show evidence of the subtransition near 17°C, the pretransition from 33 to 35°C and the sharp gel state to liquid crystalline main transition (T_M) at 41.5°C. These transitions display no dependence upon buffer or pH between 4 and 7 and no salt concentration dependence below 1.0 M. However, in the presence of ferricytochrome c both the peak intensity ratios and total integrated intensities show changes indicative of ferricytochrome c interaction with the lipid bilayer. With increasing ferricytochome c concentration the subtransition observed in the ratio of peak intensities disappears; the pretransition becomes more prominent in the integrated intensity plot, apparently at the expense of the main transition which decreases in temperature and becomes much smaller in amplitude. The extent of these changes appear to be more dependent upon the microscopic concentration of ferricytochrome c (aggregation properties) than upon the buffer or salt concentration or the pH.

The ferricytochrome c was not distributed uniformly in the multilayers. The color in the capillary sample tube varied from white at the tip to red near the lipid-solution interface. It was virtually impossible to place the tube in the laser beam such that the ferricytochrome c concentration was constant from sample to sample. We assume that the decrease in T_M from 41.5°C to about 37°C is proportional to the increased microscopic ferricytochrome c concentration at the point of laser excitation. The subtransition around 15 - 17°C and the main gel to liquid crystalline phase transition at T_M = 37-42°C were observed in the peak intensity ratio (I_{2935}/I_{2880}) temperature profiles. The pretransition at approximately 30 - 33°C, as well as the main transition, was monitored by the total integrated intensity plots. The disappearance of the subtransition, using the peak ratio index, and the increased prominence of the pretransition along with the near disappearance of the main transition in the integrated intensity plot, with increased ferricytochrome c concentration are interpreted in terms of gel state lattice reorganization. An additional feature of the samples with relatively large ferricytochrome c concentrations was the increase in intensity of the lipid 2935 cm^{-1} feature. This arises from the chain termini experiencing a more polar environment. These ferricytochrome c interactions will be contrasted with studies involving negatively charged bilayers, as for example, phosphatidic acid assemblies.

RAMAN SPECTRA OF LIPID MEMBRANES AND HEAD GROUP HYDRATION.

J. R. Scherer, S. Kint, B. A. Bolton

Western Regional Research Center, Agricultural Research Service, U.S. Dept. of Agriculture, Albany, CA 94710

Liquid-crystal/gel phase transitions at temperature T_m of di-acylchain lipids in the presence of excess water have been the focus of many studies[1]. T_m has been found to depend on the length of acyl chains[2] and their degree of unsaturation[3]. Some information has been obtained on the relative degree of hydration of lipids in their gel and liquid crystal phases[4] and also on isothermal (usually ambient temperatures) absorption of water as a function of relative humidity (water activity)[5]. However, no measurements have been made to determine the effect of lowered water activity on T_m or other phase transitions such as the liquid crystal/Hex II transition. In the following we describe our first attempts to monitor the T_m phase transition of dimyrstoyl phosphatidylcholine (DMPC) by observing changes in spectroscopic parameters as functions of temperature and relative humidity.

An apparatus for observing Raman scattering from ~ 10-30 µg of aqueous lipid dispersions exposed to conditions of controlled humidity and temperature will be described. The sample environment is established by passing humidified N_2 over the surface of a sample contained in a horizontal slit shaped trough. The humidified N_2 is generated by mixing dry N_2 with water saturated N_2 at the same temperature as (and in close proximity to) the sample. The dry and saturated gas flows are set with computer controlled needle valves. The humidity levels can be set to within 1% with a stability of ~ 0.03% and widely different humidity levels can be established within a few seconds.

Figure 1.

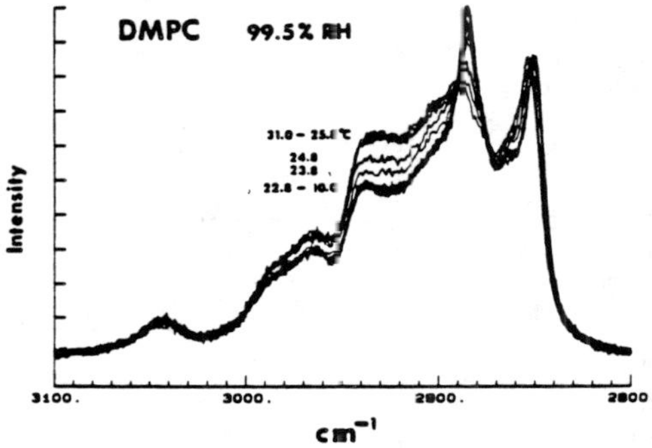

In Figure 1 are Raman spectra (CH streching region) of a DMPC dispersion held at 99.5% RH over the temperature range of 10°C to 31°C. These spectra, (obtained by increasing the temperature) show marked changes in the intensities of the 2935 cm^{-1} and 2883 cm^{-1} bands between 22.8°C and 25.8°C (T_m = 23.8°C) and only small changes outside this range. A 3 to 4 cm^{-1} shift of the 2850 cm^{-1} band also occurs over the same temperature interval. These spectra, which are similar to what has been observed for DMPC dispersions in exess water, show the eqivalence between measurements made just under 100% RH and measurements in excess water. The peak ht. ratio $R=I_{2935}/I_{2883}$ shows a large increase from 0.5 to 0.75 on going from gel phase to liquid crystal around 23.8°C and only small changes on either side of the main transition[2].

Figure 2

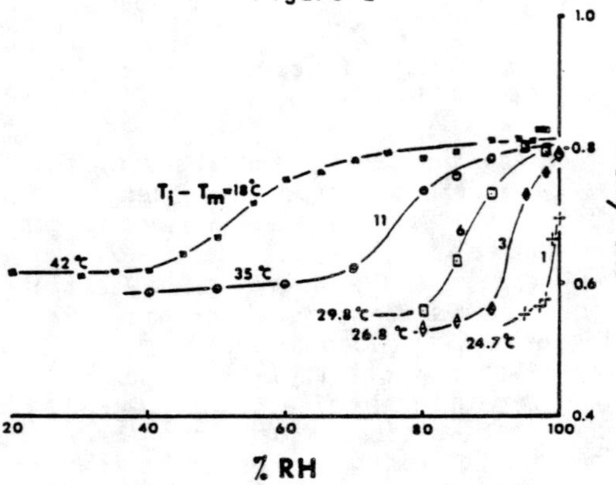

In Figure 2 we show the change in R as a function of relative humidity along isotherms that are 1,3,6,11 and 18°C above T_m. These isotherms, taken with increasing or decreasing humidity (reversible), show that the gel phase can be induced at temperatures above T_m by lowering the relative humidity. Furthermore, the phase transition becomes less cooperative at higher temperatures (lower humidity) and the resulting gel phase has higher disorder than the gel phase induced at lower temperatures.

Figure 3

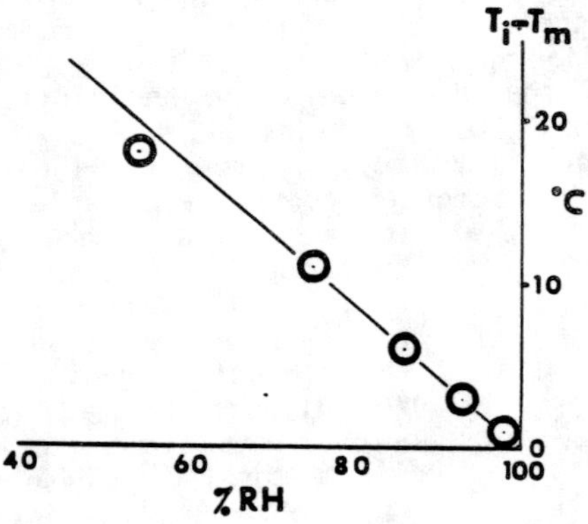

In Figure 3 we show the elevated temperature of the "phase transition" as a function of the humidity at which the "phase transition" occurs $(R=(R_{min} + R_{max})/2)$. There appears to be a linear correlation between the "transition temperature" and the relative humidity from 100% to 50% RH.

1. I. Levin, Advances in Infrared and Raman Spectroscopy (R. J. H. Clark and R. E. Hester, eds.) Wiley Heyden, Vol. 11, p 1, 1984.
2. C. H. Huang, J. Lapides and I. W. Levin, J.A.C.S. 22, 5926 (1982).
3. C. B. Berde, H. C. Andersen and B. Hudson, Biochemistry (1980) 19, 4279 (1980).
4. R. P Taylor, C. H. Huang, A. V. Broccoli and J. K. Chun. Arch. of Biochem. and Biophys. 187 197 (1978).
5. G. L. Jendrasiak and J. H. Harty, B.B.A. 337, 79 (1974).

RAMAN AND CALORIMETRIC STUDIES OF PHASE TRANSITIONS IN SEVERAL LIPID BILAYERS.

Nathan H. Rich[1)], Christopher Stevenson[1)], and Kevin M. W. Keough[2)]

(1) Department of Physics and (2) Department of Biochemistry, Memorial University of Newfoundland, St. John's, Newfoundland, Canada

We have undertaken studies of phase transitions in phospholid bilayers in which we will use both techniques of vibrational Raman spectroscopy and of differential scanning calorimetry. Our intent is that use of both methods on materials prepared under conditions as nearly equivalent as possible can ensure strict comparability of transition temperatures (and perhaps enthalpies), minimizing uncertainties caused by preparation differences among different laboratories. We choose first to test our measurements by experiments on well-studied materials: 1,2-dipalmitoylphosphatidylcholine (DPPC) and 1,2-dimyristoylphosphatidylcholine (DMPC). We then can study with confidence other lipids such as 1-palmitoyl - 2 - oleoylphosphatidylcholine (POPC), 1-palmitoyl-2-oleoylphosphatidylethanolamine (POPE), 1-oleoyl-2-palmitoylphosphatidylethanolamine (OPPE), 1,2-dioleoylphosphatidylcholine (DOPC), and 1,2-dipalmitoylphosphatidylethanolamine (DPPE) to systematically investigate effects on transition characteristics owing to presence of unsaturated carbon-carbon bonds, and the location of those bonds in the hydrocarbon chains, and to the effect of changing the headgroup of the lipids.

DPPC and DMPC were 99^{+}% grade synthetic, obtained from Calbiochem and Sigma Chemical Corporation, used without further purification. DPPC dispersions were prepared using weighed quantities of sample and doubly-distilled water. With DMPC saline solutions were used to adjust the ionic strength to improve dispersion. Samples were rehydrated at temperatures higher than that of the main transitions and maintained there prior to measurements. Calorimetric experiments were performed in Microcal MC2 and Perkin-Elmer DSC-2 instruments using standard procedures.

Raman measurements were made using Argon-ion lasers operating at 514.5 nm with power levels in the range 100 to 200 mw focused in the samples. Both single- and multi-mode operation were used. Scattered light was collected by a two lens system and dispersed by a Spex double monochromator using 1800/mm holographic gratings; spectral slit widths were typically 3 to 4 cm^{-1}. An RCA 31034 phototube (cooled) detected the signal and photon counting electronics was used with strip chart recording. Scan rates were typically 0.2 cm^{-1}/second - faster rates were used for surveys and substantially slower for detailed study of small spectral intervals. A polarization scrambler was used at the entrance slit for some preliminary polarization studies. We have used a number of kinds of sample cells and illumination geometries. Most successful is an arrangement using backscattering from samples held in vertical quartz tubes of small diameter, rotated at 100 RPM to minimize sample heating to ensure accurate temperatures and to prevent sample deterioration. The quartz tubes fit into a cylindrical hole bored through a massive brass jacket, in turn wrapped with copper tubing soldered to its surface, through which ethylene glycol/water mixture was pumped from a temperature regulated circulating bath. A small hole bored horizontally through the side of the brass cylinder admitted the focused laser beam for excitation and allowed scattered light to exit. Temperature inside the sample holder was monitored using calibrated thermistors. Temperature accuracy and stability are conservatively estimated to be better than 0.1°C. Spectra were typically taken at 0.5°C intervals with equilibration times of fifteen minutes at each new temperature. Spectra were readily reproducible in their temperature dependence, without difficulty in drift of backscatter or signal, indicating stability of the apparatus and of the sample. Reproducibility and consistency were maintained while decreasing sample temperature through the transition region, then subsequently raising the temperature.

We illustrate our Raman results with DMPC. It is known to have a primary transition between 22 and 25°C and a characteristic pre-transition 8 to 10 degrees lower. These both can be seen spectroscopically as changes in spectral contour in the C-C stretching region near 1100 cm^{-1}. (We observed corresponding effects in the C-H stretching region as well). We use the ratio of peak heights of the features at 1130 and 718 cm^{-1} as a function of temperature to indicate the transition behaviour, corrected by subtraction of background before ratioing. The 718 cm^{-1} peak used as reference is from C-N stretching in the head-group and its intensity is unchanged as the temperature varies through the bilayer transition. From our measurements one can readily see both main- and pre-transitions located at 22.2 ± 0.2°C and 15.2 ± 0.2°C, respectively. We have similar results for DPPC to test our methods. At the time of preparation of this manuscript our results are preliminary; we anticipate comparing calorimetric with Raman results on these and other lipids at the Conference.

We acknowledge support from NRC and NSERC of Canada.

VIBRATIONAL SPECTROSCOPIC STUDIES OF THE STRUCTURE OF CRYSTALLINE AND MEMBRANE BOUND GRAMICIDIN A

Vaman M. Naik[†] and S. Krimm[‡]

†Department of Physics, Northeastern University, Boston, MA 02115
‡Biophysics Research Division, Univ. of Michigan, Ann Arbor, MI 48109

Gramicidin A (GA), an atibiotic polypeptide isolated from Bacillus brevis, is a linear alternating L, D pentadecapeptide that facilitates the passive transport of alkali metal ions and water molecules across biological membranes and synthetic lipid bilayers. The transmembrane channel-forming species is known to consist of two molecules. For this active dimer of GA single and double helical structures have been proposed. Diffraction studies on GA crystallized in its native and various cation-bound forms (1,2) and low-resolution Fourier maps (3) are consistent with a helical molecule but no detailed crystal structure has been published. However, x-ray studies do show that the length of the dimer decreases from 32 to 26Å, and the diameter of the cylinder formed by the peptide main chain atoms increases from 5 to 6.8Å on binding Cs^+ (2). From ^{13}C NMR studies of single-stranded and synthetic selectively ^{13}C enriched, GAs, a head-to-head dimer of single-stranded molecule has been inferred for the active transmembrane channel (4).

We present here the results of infrared (IR) and Raman studies of crystalline native, crystalline Cs^+-bound, and vesicle-bound GA. We also present the results of normal vibration frequency calculations on the $\beta^{4.4}$, $\beta^{6.3}$, $\uparrow\downarrow\beta^{5.6}$, $\uparrow\downarrow\beta^{7.2}$, $\uparrow\uparrow\beta^{5.6}$, and $\uparrow\uparrow\beta^{7.2}$ structures. The details of the experimental work and the normal mode calculations are published elsewhere (5,6).

The calculated strong parallel-polarized amide I IR modes and their splittings, $\Delta\nu$ (i.e. the difference between the strong parallel-polarized IR band and other high intensity IR components, see ref. 5), together with the observed frequencies and splittings (7,8) are given in Table I.

In the IR spectrum of ion-free GA crystals there is a strong amide I band at 1638cm^{-1} accompanied by a shoulder at 1680cm^{-1}, corresponding to a $\Delta\nu$ of 42cm^{-1} (7). On binding Cs^+ or K^+ the strong band shifts down by 8cm^{-1}, to 1632cm^{-1}, and the high frequency band, which appeared at about 1680cm^{-1} as a shoulder in the native GA, now appears as a distinct band at 1685cm^{-1}, giving $\Delta\nu$=55cm^{-1}. In the Raman spectrum, the strong amide I band is observed at 1666cm^{-1} and this shifts up by 2cm^{-1} to 1668cm^{-1}, upon binding Cs^+ or K^+ (note that the Raman modes are observed near the mid-point of the range of frequencies remaining after the exclusion of the strong parallel-polarized IR mode, see refs. 5 and 6). The calculated frequencies of the strong (parallel-polarized) bands, their shifts on ion binding, and the magnitudes and changes in $\Delta\nu$ are in excellent agreement with an assignment of the $\uparrow\downarrow\beta^{5.6}$ structure to crystalline native GA and the $\uparrow\downarrow\beta^{7.2}$ structure to crystalline ion-bound GA. The assignment to parallel double-stranded helices can clearly be excluded because of their poor agreement between calculated and observed frequencies. The assignment to the single-stranded structures is also precluded on the basis of much smaller calculated $\Delta\nu$ for these structures and the poorer agreement with the observed strong IR and Raman modes.

Table I. Observed and calculated amide I frequencies (in cm^{-1}) for single- and double-stranded Gramicidin A structures

Structure	calculated			observed			State
	ν(IR)	$\Delta\nu$	ν(R)	ν(IR)	$\Delta\nu$	ν(R)	
$\beta^{4.4}$	1631	17	1663				
$\beta^{6.3}$	1643	9	1650	1640*	--	1654	GA-lipid
$\uparrow\downarrow\beta^{5.6}$	1636	30-33,39	1666	1638	42	1666	GA(cryst)
$\uparrow\downarrow\beta^{7.2}$	1632	42-54	1669	1630	55	1668	GA-Cs^{+}(cryst)
$\uparrow\uparrow\beta^{5.6}$	1656	13,27,49	1685				
$\uparrow\uparrow\beta^{7.2}$	1676	-6,6	1697				

*Estimated from observations on vesicles (8,9).

In the earlier IR studies on GA in lipid vesicles (8,9), there had been attempts to assign the structure based on the observed strong amide I IR mode. We have shown (7) that it is difficult to make such an assignment based on the IR mode alone, because similar calculated strong IR amide I frequencies are expected for $\beta^{6.3}$ and $\uparrow\downarrow\beta^{7.2}$ structures. However, the Raman mode is expected to be significantly different (see Table I). We have studied GA in vesicles by Raman spectroscopy (7). In the conducting state there is only one Raman amide I mode observed at $1654 cm^{-1}$, which can be assigned to the $\beta^{6.3}$ structure. This assignment is preferable to $\beta^{4.4}$, not only because of the better agreement with the Raman mode but also because of the much better agreement with the strong IR mode. Such a structure is also supported by the NMR studies (4).

Research supported by NSF grants PCM-8214064 and DMR-8303610.

REFERENCES

1. Koeppe II, R.E., Hodgson, K.O. & Stryer, L. (1978) J. Mol. Biol. 121, 41.
2. Koeppe II, R.E., Berg, J.M., Hodgson. K.O. & Stryer, L. (1979), Nature (London), 279, 723.
3. Koeppe II, R.E. & Shoenborn, B.P. (1984) Biophys. J. 45, 503.
4. Urry, D.W., Trapane, T.L. & Prasad, K.U. (1983) Science, 221, 1064.
5. Naik, V.M. & Krimm, S. (1986) Biophys. J. (in press).
6. Naik, V.M. & Krimm, S. (1986) Biophys. J. (in press).
7. Naik, V.M. & Krimm, S. (1984) Biochem. Biophys. Res. Comm. 125, 919.
8. Sychev, S.V. & Ivanov, V.T. (1982) in Membrane Transport, ed. A.N. Martonosi, Vol. 2, pp. 301-307, Plenum, New York.
9. Urry, D.W., Shaw, R.G., Trapane, T.L. & Prasad, K.U. (1983) Biochem. Biophys. Res. Comm. 114, 373.

MELITTIN-PHOSPHATIDYLCHOLINE COMPLEXES: A MODEL TO STUDY LIPID-PROTEIN INTERACTIONS.

M. Lafleur, J.-L. Dasseux and M. Pézolet

Département de chimie, Université Laval, Québec, Qc, G1K 7P4, Canada.

It is now well established that melittin has a marked affinity for phospholipids. This toxin causes the lysis of natural as well as synthetic membranes, and it has recently been shown that it can change drastically the lipidic arrangement by inducing under some conditions the formation of small lipid-melittin micellar complexes (1-2). Melittin displays an amphiphilic character and, therefore, is an interesting model peptide to study lipid-protein interactions. Due to the strong electrostatic interactions between the basic residues of the toxin and the polar head group charges of the phospholipids, melittin has a high affinity for anionic lipids. However, since the hydrophobic part of peptide tends to penetrate the apolar zone of lipid bilayers, the effect of the toxin is also modulated by the cohesion of the hydrophobic core of the membrane. We report here some Raman results on the effect of lipids chain length on the thermotropic behavior of phosphatidylcholine-melittin complexes.

In order to monitor the structure of melittin-phosphatidylcholine complexes, we have used the C-C and C-H stretching mode regions of their Raman spectra. The first region provides direct information about the intramolecular order of the lipid chains while the latter is sensitive to both the overall disorder and the lateral packing of the acyl chain matrix.

For melittin-dimyristoylphosphatidylcholine (DMPC) complexes, the toxin increases significantly the intramolecular disorder of the bilayer in the gel phase. The C-H stretching modes indicate that a large decrease of the intermolecular coupling as well as an augmentation of the overall disorder are induced by the presence of the toxin. In the liquid crystalline phase, melittin affects the molecular packing of the lipid without appreciable change of the number of gauche bonds. These effects are accompanied by a shift of the gel to liquid-crystalline phase transition toward lower temperature and by a significant loss of the cooperativity of the transition. For example, at a lipid to peptide incubation molar ratio of 10 (R_i=10), the transition occurs over a 15°C range (10 to 25°C), while at R_i=4 the phase transition is essentially abolished. All these results show unambiguously that melittin penetrates rather easily in the acyl chain core of DMPC bilayers, and can destroy the bilayer structure of this lipid.

The thermotropic behavior of dipalmitoylphosphatidylcholine(DPPC)-melittin complexes was also studied. In the gel state of the lipid melittin induces a drastic decrease of the h_{2880}/h_{2850} intensity ratio which indicates that the interchain vibrational coupling is significantly lowered. However, the overall and intramolecular orders of the lipid acyl chains, measured from the C-H and C-C stretching mode regions respectively, are much less affected by the presence of the toxin. As for DMPC, the gel to liquid-crystalline phase transition is shifted to lower temperature by melittin but always remains more cooperative; at R_i=2, a transition is still detectable.

From the results obtained from various physical techniques such as freeze fracture electron microscopy, quasi elastic light scattering and NMR spectroscopy, it was concluded that melittin destroys the multilamellar structure of phosphatidylcholine dispersions and induces the formation of small particles. Our Raman results provide valuable information on the structure of these particles. First, they are characterized by a rather weak intermolecular vibrational coupling which may be associated to their small size or to the presence of melittin in the lipidic structure. However, since for melittin-DPPC complexes the intramolecular order in the gel phase is slightly affected by the toxin and the phase transition is always detected, we believe that the lipids still exist as bilayers in the complexes and that melittin does not penetrate deeply in these bilayers. Therefore, we propose that, as for apolipoproteins(3), melittin induces the formation of small DPPC bilayer discs whose hydrophobic edges are shielded from the aqueous phase by a layer of the toxin. In such a "bicycle tire" like structure, the chains can remain in almost all-trans conformation but due to the small size of the discs, there is less intermolecular vibrational coupling. In the case of DMPC, the fact that the addition of melittin greatly increases the number of gauche bonds of the lipid acyl chains and that the phase transition is abolished at $R_i \leq 4$ suggests that in that case the small particles consist of mixed micelles of melittin tetramer swollen by a few DMPC molecules that are strongly perturbed. This difference of behavior between DMPC and DPPC can be associated with the greater cohesion of the hydrophobic core of DPPC bilayers due to the longer chains of the lipid.

References:

(1) J.-L. Dasseux. J.-F. Faucon, M. Lafleur, M. Pézolet and J. Dufourcq. Biochim. Biophys. Acta 775, 37-50 (1984).

(2) F. G. Prendergast, J. Lu, G. J. Wei and V. A. Bloomfield. Biochemistry 21, 6963-6971 (1982).

(3) J. P. Segrest. Chem. Phys. Lipids 18, 7-22 (1977).

POLARIZED UV RESONANCE RAMAN AND ABSORPTION SPECTRA OF MONOLAYERS ADSORBED AT THE AQUEOUS SOLUTION-CARBON TETRACHLORIDE INTERFACE.

T. Takenaka, J. Umemura, and N. Isono

Institute for Chemical Research, Kyoto University, Uji, Kyoto-Fu 611, JAPAN

Bilayer-forming amphiphiles have drawn much interest in anticiptation of their functionalization in aqueous solution. Shimomura et al. have reported the relationship between the spectral characteristics and the molecular alignment of azobenzene-containing, single-chain ammonium amphiphile represented by

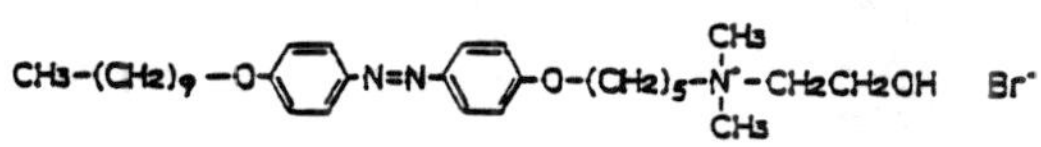

(abbreviated as $C_{10}AzoC_5N^+$). In the present work, we studied polarized UV resonance Raman and absorption spectra and molecular orientation of the $C_{10}AzoC_5N^+$ monolayers adsorbed at the aqueous solution-CCl_4 interface.

Both resonance Raman and absorption spectra of the adsorbed monolayers at the interface were measured by the total reflection method at the interface by using truncated pyramidal cells (placed upside down). For excitation of the monolayers in Raman measurements, the 350.7 nm line of a Coherent model Innova 90-K Kr^+ laser was used. This line gives rise to a resonance Raman scattering because the bulk aqueous solution of $C_{10}AzoC_5N^+$ has an absorption band at 345 nm in the liquid-crystalline state above the phase transition temperature (Tc=24°C) and a doublet band at 375 and 393 nm in the gel state below Tc as shown in Fig.1(A).

Fig.1(B) is the polarized absorption spectra of the $C_{10}AzoC_5N^+$ monolayer adsorbed from the aqueous solution in the gel state to the interface with CCl_4. Obviously, the spectra of the adsorbed monolayer are different in shape and wavelength from that of the aqueous solution in the gel state and resemble that in the liquid-crystalline state. Further, the spectra of the monolayer adsorbed from the liquid-crystalline solution are identical to those shown in Fig.1(B). Therefore, it is revealed that the adsorbed monolayer is in the liquid-crystalline state irrespective of the state of the bulk aqueous solution from which the molecules are adsorbed to the interface.

Fig.1(B) also shows a large parallel dichroism with respect to the plane of incidence. Since these bands are assigned to the π-π* transition of the trans-$C_{10}AzoC_5N^+$ molecule with the transition moment parallel to the long axis of the chromophor, the parallel dichroism suggests that the chromophors tend to orient normal to the interface.

In Fig.2, intensity of the parallel band ($A_{\parallel}$), dichroic ratio ($A_{\parallel}/A_{\perp}$), and wavelength of the parallel band maximum ($\lambda_{\parallel}$) are plotted against logarithm of concentration. Both $A_{\parallel}$ and $A_{\parallel}/A_{\perp}$ increase gradually with increase in concentration attaining constant values at 2 x 10^{-5}M. Further, $\lambda_{\parallel}$ decreases with increasing concentration and attain a constant value at the same concentration. These facts indicate that in the low concentration region, the number of adsorbed molecules increases with increasing concentration and comes to an equilibrium value at 2×10^{-5}M (Langmuir adsorption). At the same time, the orientation of the chromophors is gradually improved until the adsorption equilibrium is reached. Interactions between adsorbed moleucles also change with increase in concentration and become a stationary state at the same concentration.

Fig.3 represents polarized UV resonance Raman spectra of the adsorbed monolayer obtained after the equilibrium is established. The subscript // means that the exciting light is polarized parallel to the plane of incidence, i.e., perpendicular to the interface, and X and Y mean that the scattered light is polarized parallel and perpendicular to the interface, respectively. (A) is the spectrum of the interface and (B) is that of background. Consequently (C)=(A)-(B) is the spectrum of monolayer. In the $I_{/\!/Y}$ (C) spectrum, three bands are apparently observed at 1458, 1413, and 1145 cm^{-1} which are assigned to the phenyl, N=N stretching, and phenyl-N stretching vibrations, respectively, of the trans $C_{10}AzoC_5N^+$ molecules. In the $I_{/\!/X}$(C) spectrum, however, only a very weak peak is seen around 1460 cm^{-1}. This fact reveals that the chromophors are oriented almost perpendicular to the interface, being in good agreement with results of the absorption measurements.

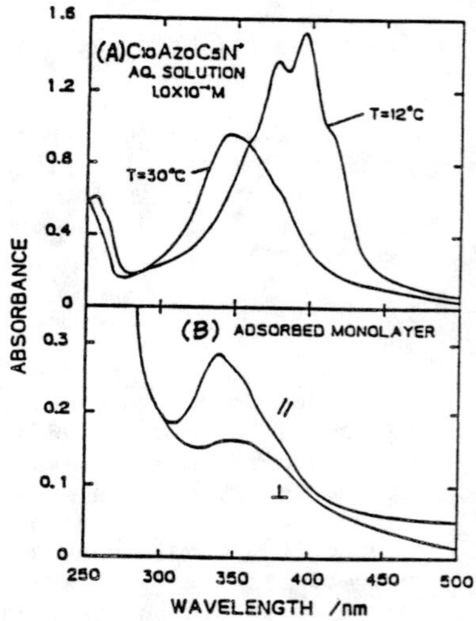

Fig.1. (A) Absorption spectra of bulk solution and (B) polarized spectra of adsorbed monolayers of C_{10}Azo C_5N^+.

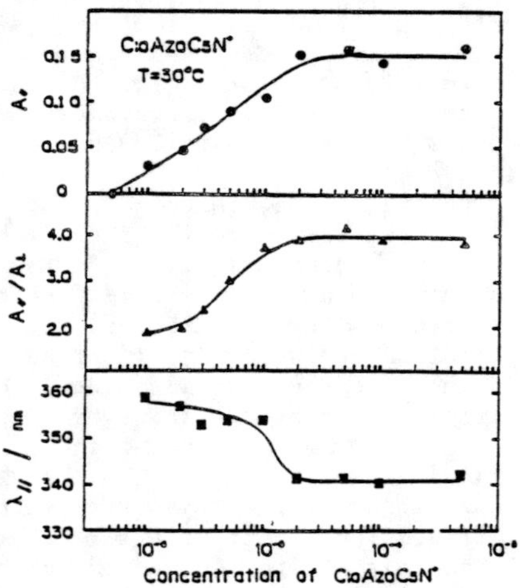

Fig.2. $A_{/\!/}$, $A_{/\!/}/A_{\perp}$, and $\lambda_{/\!/}$ as a function of logarithm of concentration.

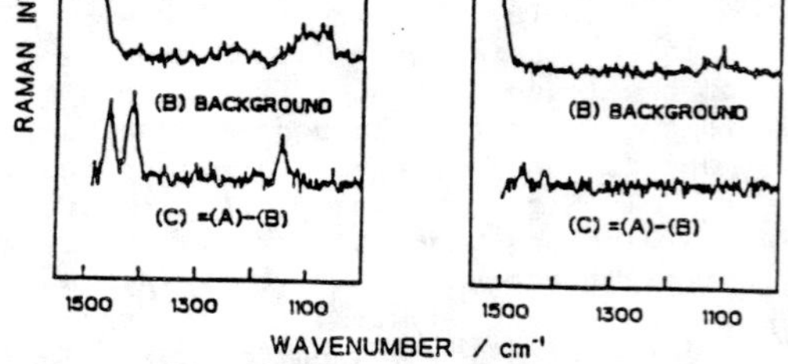

Fig.3. Polarized resonance Raman spectra of adsorbed monolayers of $C_{10}AzoC_5N^+$.

RAMAN SPECTROSCOPIC ESTIMATION OF SOLUBILIZATION EQUILIBRIA IN AQUEOUS POLYMER/SURFACTANT SOLUTIONS.

L. B. Shih, M. H. Luccas and R. J. Warner

Polymer Research Department, S. C. Johnson & Son, Inc., Racine, WI 53403

It has been shown that certain Raman stretching vibrational frequencies are sensitive to the solvent microenvironments and can be used as a quantitative probe to determine the partition of a solubilizate in surfactant solutions (1). In the present study, we have extended this technique to investigate the solubilization of benzyl-d_7 alcohol in aqueous solutions containing a comb-shaped copolymer of 1-octadecene-co-maleic anhydride in the presence and absence of sodium dodecyl sulfate, SDS, and as a function of the degree of neutralization of the copolymer. We have also applied FTIR technique to determine the conformational states of the copolymer and its interactions with the surfactant. The results suggest that the solubilization behavior in the polymer/surfactant solution can be explained in terms of the solution conformation of the copolymer and its association with the surfactant.

The procedure of analyzing the solubilizate spectrum in the polymer/surfactant solution is to fit the unknown spectrum as the linear combination of the reference spectra modeling the hydrocarbon and aqueous phases and to relate the fractional intensities to solubilizate concentrations in each phase. A typical example of decomposing the aromatic C-D stretching band of benzyl-d_7 alcohol in an aqueous solution containing 1 wt.% copolymer fully neutralized with LiOH and 0.05 M SDS is shown in Figure 1. The reference spectra are benzyl-d_7 alcohol in water and in heptane. The fitting procedure is described in detail in Reference 1.

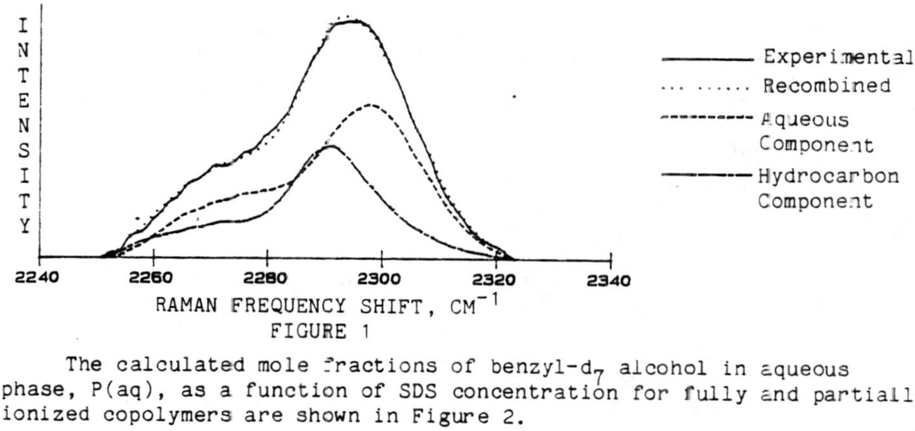

FIGURE 1

The calculated mole fractions of benzyl-d_7 alcohol in aqueous phase, P(aq), as a function of SDS concentration for fully and partially ionized copolymers are shown in Figure 2.

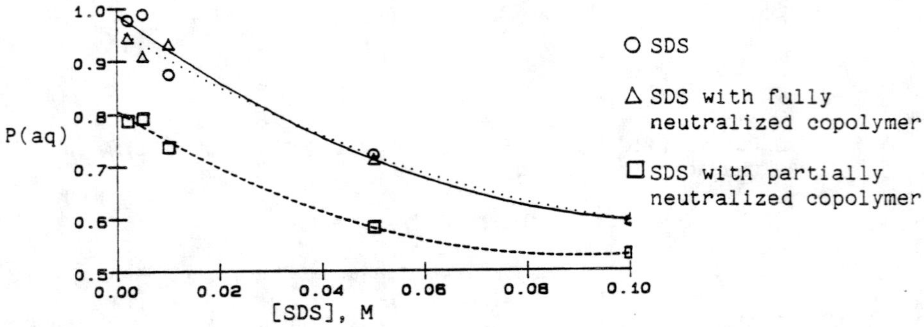

FIGURE 2

Figure 2 shows that the solubilization behavior is drastically different for fully and partially neutralized copolymers. The data suggest that there exists an additional hydrocarbon domain available for solubilization in the partially ionized copolymer solutions and that there are two different types of polymer/surfactant associations at different extents of copolymer head group ionization. Potentiometric titration data on amphiphilic polymers of this type have suggested polymer conformational change as a parameter of neutralization (2). Such conformational change is also supported by spectroscopic evidence. The FTIR asymmetrical COO^- stretching spectrum of freeze-dried samples shows a frequency shift from 1574 cm^{-1} at partial neutralization to 1569 cm^{-1} at full neutralization. The higher frequency at partial ionization is due to the formation of intramolecular hydrogen-bonding between the carboxylate ion and the -OH from the neighboring carboxylic acid. The hydrogen-bonding is absent when the head groups are completely ionized. Dynamic Light Scattering measurements also show more compact aggregates at full ionization. The hydrodynamic radii are 50Å and 110Å at full and partial ionizations, respectively. In the presence of SDS, the COO^- frequency is shifted to 1581 cm^{-1} for the fully ionized copolymer indicating the formation of mixed micelles. The COO^- frequency remains unchanged for partially neutralized copolymers in the presence of SDS. The data suggest that there is probably no association between SDS and the partially ionized copolymer and that the copolymer forms a hydrocarbon phase by itself and thus provides an additional hydrocarbon domain for solubilization.

In summary, we have applied Raman and FTIR techniques to determine solubilization behavior and interactions of a comb-shaped copolymer and a similarly charged surfactant in aqueous solutions. The data indicate that the copolymer assumes different conformations as a parameter of the extent of ionization through the formation of hydrogen-bonding at the ionic heads. Different solubilization behaviors can be explained in terms of different copolymer conformations and different types of polymer/surfactant associations.

REFERENCES:

1. L. B. Shih and R. W. Williams, J. Phys. Chem. in press.
2. P. Dubin and U. P. Strauss, J. Phys. Chem. 77, 1427, (1973).

DYNAMICS OF ALKYLAMMONIUM CHAINS IN $C_nH_{2n+1}NH_3Cl$ COMPOUNDS (n = 8,10). A SPECTROSCOPIC STUDY.

K.J.SCHENK (1) and M.REY-LAFON, Laboratoire de Spectroscopie Infrarouge, UA 124 CNRS, Université de Bordeaux I,33405 TALENCE, FRANCE.

Layer-structure compounds containing long alkyl chains, in particular n-alkanes and perovskite type compounds $(C_nH_{2n+1}NH_3)_2MX_4$ (M=Cd,Mn; X=Cl,Br), present structural phase transitions related to the dynamics of the chains and thus have been considered as models of bilayer lipid membranes. n-alkylammonium chlorides (C_nCl) also crystallize in a layer structure but the packing of the chains in the ordered phase is different. An X-ray study on $C_{10}Cl$(2) shows that the polar layers are constituted of N and Cl atoms arranged in a boat configuration. The adjacent chains which are linked through hydrogen bonds to the lattice are intercalated ; they are in the all-trans configuration and tilted with respect to the normal to the mean lattice plane. This solid undergoes four first order phase transitions. The lowest temperature one (phase I → phase II) involves an irreversible desintercalation of the chains probably due to the onset of a reorientational motion of the polar heads and to the occurrence of conformational disorder(2,3). When raising the temperature another disordered phase (III) is observed; a metastable phase(I') is obtained by cooling back to room temperature. Phases II and III have been shown to be isomorphous to the room temperature phases of C_8Cl(phase II) and C_6Cl(phase(III) respectively and a model of structure has been recently proposed for these two phases (4).

The aim of this work is to obtain a more precise description of the dynamics of the chains in the different phases. We applied infrared and Raman spectroscopy to $C_{10}Cl$ and three selectively bi-deuterated derivatives, on carbons 1,2 and 3 respectively, and to C_8Cl.

The spectra of the room temperature phase of $C_{10}Cl$ are typical of an ordered solid, as proposed from X-ray measurements(2). Two different NH...Cl bonds are observed and strong intermolecular interactions exist between chains in the all-trans geometry. A weakening of the molecular interactions as well as conformational defects appear within phase I when raising the temperature ; in phase II of $C_{10}Cl$ and C_8Cl, kink defects and gauche forms in the vicinity of the methyl group are observed. It can be deduced from the $r(CD_2)$modes intensities in the three deuterated derivatives of $C_{10}Cl$ that the C_1C_2 and C_3C_4 bonds are trans and about 40% of C_2C_3 bonds are gauche. To be compatible with the monoclinic structure containing double layers of non intercalated alkylammonium ions, one must assume an interconversion of two types of chains thus creating a b-glide plane.

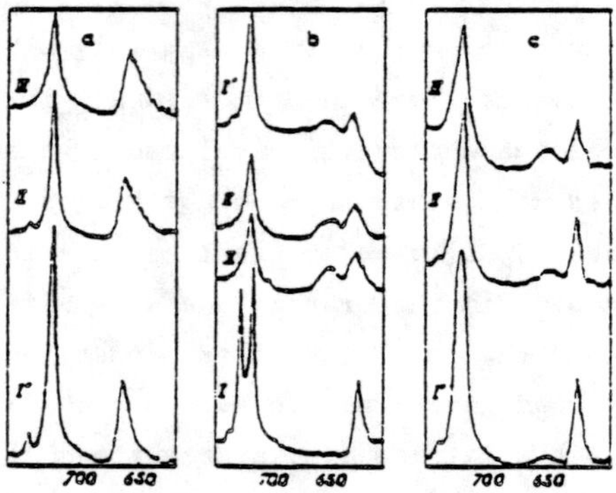

Fig. - IR spectra :(a)$C_{10}Cl$-1,1-d_2 ; (b) $C_{10}Cl$-2,2-d_2 ; (c) $C_{10}Cl$-3,3-d_2

In the high temperature phases (III),the chains do not interact and the hydrogen bonds are all equivalent.The number of gauche forms increases but the nature of the defects remains the same. The study of the bideuterated derivatives of $C_{10}Cl$ shows that C_1C_2 and C_3C_4 bonds again are trans and about 42% of C_2C_3 bonds are gauche; the chains beyond carbon 4 are flexible. This result agrees with the model of a "tree-like" structure proposed for phase III of C_6Cl (4) in which C_2C_3 bond and others beyond carbon 4 would be perpendicular to the polar lattice.

In phase I' at room temperature, the chains are still disordered and present kink defects localized in their middle part. The hydrogen bonds are equivalent. The disordering process in n-alkylammonium chlorides is thus different from that of perovskites; this can be related to the structure of the chlorine matrix.

(1) Permanent address : Institut de Cristallographie,Université de Lausanne,CH-1015-LAUSANNE - Switzerland.

(2) R.KIND et al., Phys.Rev., A26,1816 (1982)
(3) J.SELIGER et al.,J.Chem.Phys., 78, 2661 (1983).
(4) K.J.SCHENK and G.CHAPUIS, to be published.

RAMAN SPECTROSCOPIC STUDIES OF CHAIN INTERACTIONS AND MOBILITY IN LIQUID CRYSTALLINE PHASES OF CATIONIC SURFACTANTS.

W.F. Pacynko§, G.J.T. Tiddy* and J. Yarwood§

§ Department of Chemistry, University of Durham, South Road, Durham, DH1 3LE.

* Unilever Research, Port Sunlight Laboratory, Port Sunlight, Wirral, Merseyside, L62 4XN.

Three frequency regions have been employed to monitor intra- and intermolecular interactions of surfactant molecules using vibrational spectroscopy. In this work we have studied two of these in order to cast light on the nature of the forces which stabilize lyotropic mesophases of $C_{12}TACl$ (dodecyltrimethylammonium chloride). The C-C skeletal stretching region (1000-1200 cm^{-1}) where the vibrations, strongly intramolecularly coupled, are sensitive to chain geometry [1]. As the temperature is raised giving rise to increased numbers of gauche conformers the $I_{1130\ cm^{-1}}/I_{1080\ cm^{-1}}$ band ratio is expected to rise [1,2]. However, careful examination of this ratio over a wide concentration range shows that this ratio remains, within the experimental error, unchanged. It therefore seems unlikely that changes in gauche/trans conformation ratios change across the phase diagram for $C_{12}TACl$.

The C-H stretching region 2800-3000 cm^{-1} has also been used [1,3,4] to monitor gauche/trans conformational changes but more recently it has been shown [5,6] that relative intensities in this region are seriously affected by Fermi resonance (with CH bending modes) which may arise from either intramolecular coupling of bending modes along the chains or through intermolecular interactions involving lateral interactions from one chain to another. Thus the I_{2880}/I_{2850} intensity ratio may be affected, not only by differences in gauche/trans ratio between the crystal (all trans) and/or liquid (largely gauche) conformations, but also by the degree of dynamic interactions between adjacent molecules. Examination of this ratio for the crystalline surfactant - where an all trans configuration is expected [2] shows that the value is $\sim$1.7. On going to different liquid crystalline phases the ratio shows a step-like behaviour - being low (0.7-0.8) for isotropic phases (L_1, I_1, V_1) and much higher (1.1-1.4) in the anisotropic phases (H_1, L_α). This behaviour is unexpected in view of the indications that the gauche/trans ratio changes very little. The strong implication is that the lateral interactions change significantly on crossing the phase boundary between isotropic and anisotropic phases. Certainly any decrease in gauche conformations in order to accommodate micelle surface curvature changes (even though they may not be seen in the skeletal region) is expected to be a

THE BEHAVIOUR OF THE I_{2880}/I_{2850} RATIO FOR C_{12}TACl AT 30°C

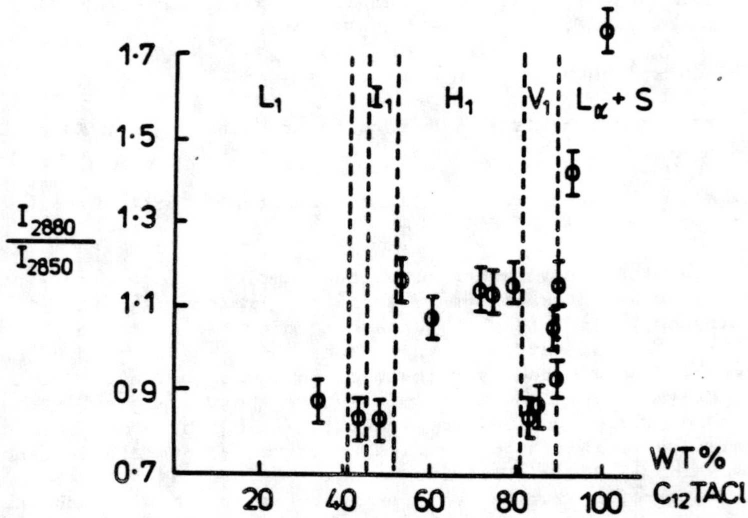

monotonic change. Associated studies on the gel phase of C_{16}TACl have shown that the ν(CH) intensity ratios are affected by chain mobility, so this may also be a factor for mesophases formed by C_{12}TACl.

[1] R. Faiman and D.A. Long, J. Raman Spect., 3, 371 (1975).

[2] R.G. Snyder, J. Chem. Phys., 47, 1316 (1967).

[3] A.M. Amorim da Costa, C.F.G.S. Geraldes and J.J.C. Texeira-Dias, J. Coll. Int. Sci., 86, 254 (1981).

[4] R. Mendelsohn, R. Dluhy, T. Taraschi, D.G. Cameron and M.H. Mantsch, Biochem., 20, 6699 (1981); Biochem., 17, 3944 (1978).

[5] R.G. Snyder, J.R. Scherer and B.P. Gaber, Biochem. Biophys. Acta, 601, 47 (1980).

[6] R.G. Snyder, Spectrochim. Acta, 34A, 395 (1978).

RAMAN BAND SHAPE ANALYSIS AND MOLECULAR INERTIAL MOTION IN ANISOTROPIC ALIGNED FLUIDS

N. Kirov,[+] I. Dozov,[+] M. Fontana[++] and B. Petroff[+]

[+] Institute of Solid State Physics, 1784 Sofia, Bulgaria

[++] Physics Department, University of Parma, Parma, Italy

Molecular reorientation in ordered fluids is usually considered as a stochastic process and the experimental data are treated in the framework of the strong collision or small step rotational diffusion models. Both models give in short time approximation an exponential decay of the rotational correlation function $\Phi^{l}_{mn}(t)$. Recently has been demonstrated /1/ that at $t \to 0$ $\Phi^{l}_{mn}(t)$ have Gaussian form and the stochastic models are not valid in this time domain.

Here we present a simple model of the molecular reorientational dynamics in ordered fluids. We assume that: 1) the molecules reorient through small but finite angular steps due to the collisions with other molecules, the avarage time between the collisions being τ; 2) between two collisions the molecules reorient inerially in the mean field of the other molecules; 3) during the time of the collisions τ_c ($\tau_c \ll \tau$) the molecules do not change their orientation but only their angular momentum $\vec{J}$; 4) the molecules are considered cyllindrically symmetric. Under these assumptions we obtain the following equations for the rotational correlation functions:

$$\Phi^{l}_{mn}(t) = \bar{\Phi}^{l}_{mn}(0) + \ddot{\Phi}^{l}_{mn}(0)(\tau t - \tau^2 + \tau^2 \exp(-t/\tau)) \quad (1)$$

where both $\bar{\Phi}^{l}_{mn}(0)$ and $\ddot{\Phi}^{l}_{mn}(0)$ are model independent functions of the order parameters $\langle P_2 \rangle$ and $\langle P_4 \rangle$ /1/. At $t \ll \tau$ eq. (1) reproduces the results reported in ref. 1 and at $\tau \ll t \ll$ $\Phi^{l}_{mn}(0)/\tau\ddot{\Phi}^{l}_{mn}(0)$ is equivalent to the small step rotational diffusion theory with $D_i^r=(kT/I_i)$, $i=\parallel,\perp$ where I_i is the inertial molecular moment.

In order to verify the validity of eq.(1) we selected the Raman band at 1603 cm^{-1}(benzene ring stretch vibration) of the aligned nematic liquid crystal 2-cyanobiphenyl 4' n-heptyl-oxybenzoate(the temperature interval of the mesophase is 60°-$109^{\circ}C$). We applied the approach developed by us/2/ which is based on the comparison of the Fourier transforms $a_{ik}(t)$ of the polarized and depolarized components of the band at three scattering geometries. For the picosecond time domain we obtain:

$$\ln\frac{a_{zx}(t)}{a_{zz}(t)} = (kT/I\tau)\left[t-\tau+\tau\exp(-t/\tau)\right]\times\left[6\frac{7+2.5\ P_2+8\ P_4}{7+5\ P_2-12\ P_4} - (8/3)\frac{7+5\ P_2-12\ P_4}{7+20\ P_2+8\ P_4}\right] \quad (2)$$

$\langle P_2\rangle$ and $\langle P_4\rangle$ have been calculated from the depolarization ratios assuming the uniaxiality of the polarizability tensor of the vibration 1603 cm^{-1}. At $60^{\circ}C$ they are 0.64 and 0.31 respectively. The experimental data are in good agreement with the eq.(2) and the best fit procedure gives τ = 0.2 and $I_\perp \cong 1.5\times10^{-43} kg.m^2$($60^{\circ}C$). The values of $I_\perp$ seems reasonable compared with $0.7\times10^{-43} kg.m^2$ reported in ref.3 for the smaller mesogen 4,4'-dimethoxyazoxybenzene(PAA). For the rotational diffusion coefficient we obtain $D_\perp^r$ = $7\times10^9 s^{-1}$ in good agreement with the data for other nematics/2/. We estimate also that the mean angle of the molecular rotation around short axes (rotation of the long axes itself) during one reorientational step is $(\varepsilon^2)^{1/2}\cong 0.04$rad. This small value is the first direct evidence that the small step rotational diffusion model gives an adequate description of the reorientation in the stochastic regime.

REFERENCES: 1.P.Pasini and C.Zannoni, Mol.Phys., 52, 749(1984)
2.N.Kirov, I.Dozov and M.P.Fontana, J.Chem.Phys., 83, 5267(1985
3.A.J.Martin, G.Meier and A.Saupe, Farad.Symp., 1A, 327(1971).

chapter 4

OTHER BIOLOGICAL and MEDICAL APPLICATIONS

RAMAN INVESTIGATIONS OF INTERACTIONS BETWEEN GLUTATHIONE AND ANTIMICRO-TUBULAR AGENTS

S. Turrell[1,2], T. Bouchaour[1-3], and I. Lesieur[4]

[1]L.A.S.I.R. (L.P. 2641) Universite Lille I, 59655 Villeneuve d'Ascq

[2]I.N.S.E.R.M. U279, 59800 Lille

[3]Laboratoire Physique, Universite Lille II, 59046 Lille

[4]Institut Chimie Pharmaceutique, 59045 Lille, France

Over the past few years the syntheses and structure-activity studies of several families of antimicrotubular compounds have been reported. These compounds, which inhibit <u>in-vitro</u> microtubular assembly, and hence eukaryotic cellular division, <u>apparently</u> react <u>via</u> the formation of a covalently-bound complex with a cysteine site <u>in tubulin</u>.[1,2]

In a previous spectroscopic study of a model systen, we found evidence that a covalert complex is indeed formed between cysteine and certain aminoacrylopherones.[3,4] Moreover, Raman frequency changes in the νC=C and νC=O bands of the drugs and νSH bands of systeine support pharmacological conclusions that the mechanism of such interactions involves the ketone αβ ethylenic structure of the drug and the SH group of cysteine. However, because of the amphoteric properties of cysteine in aqueous solution, we were forced to work at acidic pH, thus further altering an already overly simplified model.

In the present work, we have attempted a better simulation of biological conditions. By working with a tripeptide, glutathione (GSH) (Fig. 1a), we obtain cysteine in a more complex site and a system which can be studied at physiological pH. The νC=O and νC=C Raman bands for the drug employed, ILI (Fig. 1b), which occur at 1655 and 1623 cm^{-1}

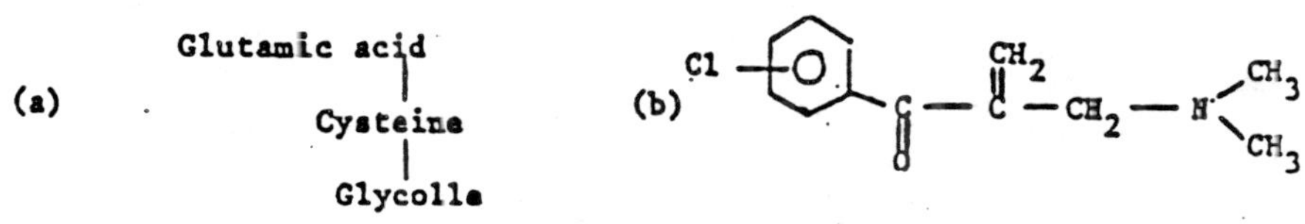

Fig. 1. Molecular Structures of (a) Glutathione (GSH) (b) ILI

respectively for the solid, combine to yield a single, broader component in the spectrum of the aqueous solution at 1645 cm^{-1}. Unfortunately, the spectrum of GSH also contains a weak band in this region (amide I at 1650 cm^{-1}. On the other hand, the bands of ILI do not present any interference with the NH and SH spectral regions of GSH which occur at 3100

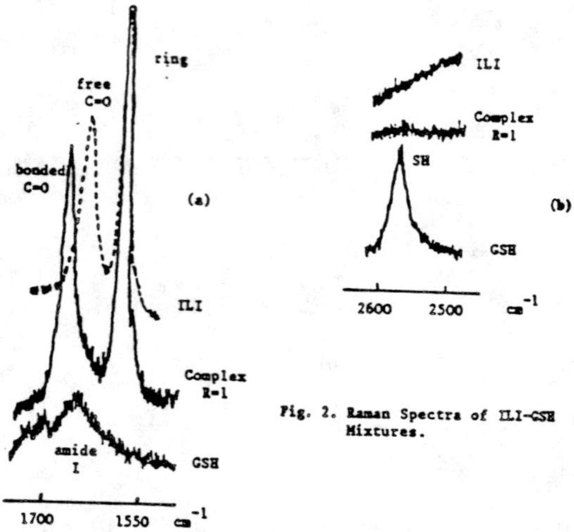

Fig. 2. Raman Spectra of ILI-GSH Mixtures.

and 2573 cm^{-1}, respectively.

Raman spectra of GSH-ILI mixtures of varying proportions of concentrations, R=[ILI] / [GSH], have been obtained. It is seen that while the υNH region remains virtually unchanged, the intensity of the SH band gradually decreases until it disappears at R=1 (Fig. 2b). No other appreciable changes appear in the GSH part of the spectrum. In the ILI spectral reigons, the ring vibration band remains unaffected, but the frequency of the υC=O band shifts to 1675 cm^{-1} when R=1. This latter is the same frequency as that observed for υC=O in ILY (a compound in this series of drugs which lacks the $=CH_2$ group). These frequency variations imply a marked decrease in conjugation caused by the formation of a stable complex.

Hence, even in an environment of larger biomolecules, these antimicrotubular compounds appear to react preferentially at the SH site of cysteine.

References:

1. K. Mann, M. Giesel, H. FAsold, FEBS Letters 92, 45 (1978).

2. J. Deinum, M. Wallin, C. Lagercrantz, Biochim. Biophys. Acta 671, 1 (1981).

3. I. Lesieur, D. Lesieur, M. Cazin, M.-L. Mallevais, A. Delacourte, M. Brunet, M. Luyckx, Arzneimitt. Forsch. 36(I), 20 (1986).

4. T. Bouchaour, S. Turrell, G. Fleury, M.-L. Mallevais, I. Lesieur, J. Mol. Structure, In Press.

METALLO-PROTEINS AND METALLO-NUCLEOTIDES

Pham V.HUONG *, P.CARMONA and J.LOROSCH.
Laboratoire de Spectroscopie Infrarouge et Raman (UA 124 CNRS)
UNIVERSITE de BORDEAUX I, 351, Cours de la Libération
33405 TALENCE FRANCE.

INTRODUCTION

Organometallic complexes play a basic role in biological media, not only for living functions but also for dis-functions responsible of many deseases.

The recent discovery of Copper-organic complexes is cancerous tissues from any origin and in blood of leukemic patients (1-3), intensifies the interest in the study of metallo-complexes with nucleotides and with proteins.

This paper concerns with Copper complexes with mono-nucleotides, poly-nucleotides and with some synthetic protein model molecules.

RESULTS AND DISCUSSION

1-Copper nucleotides complexes

The Raman spectra of complexes Cu^{II}AMP, Cu^{II}CMP, Cu^{II}GMP, Cu^{II}UMP, Cu^{II}Poly A, Cu ^{II}Poly C, Cu^{II} Poly G, Cu^{II} Poly U were recorded and analyzed (Fig.1).

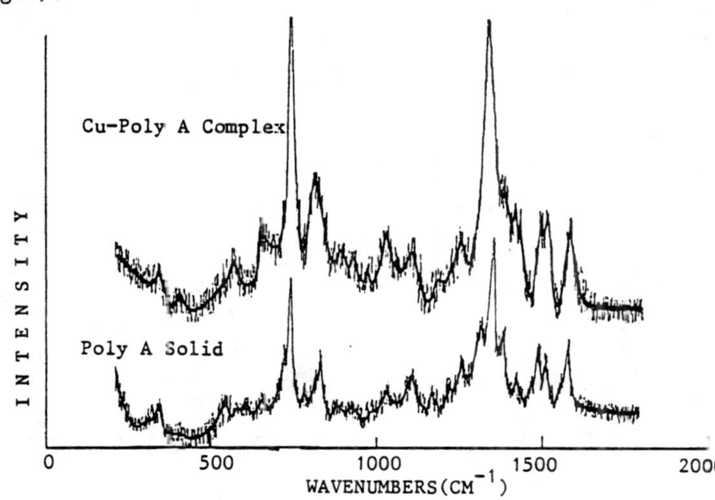

Fig.1 - Raman spectra of poly A and Cu^{II} poly A complex.

Two types of interactions were evidenced. For adenosine and guanosine phosphate series, the ligand is localized on the nitrogen atom N_7 of the imidazol cycle. With Cu^{II} GMP, the Cu-N is located at 374 cm^{-1}.

With UMP and CMP complexes, the metal is bound to the carbonyl group.

For poly-nucleotides series, the order is much lowered in the complexes in comparison to their parent macromolecules.

2-Copper-protein model complexes

Phenolate bridged binuclear Cu^{II} complexes were studied by resonance Raman spectroscopy to help the understanding of the properties of proteins such as hemocyanin.

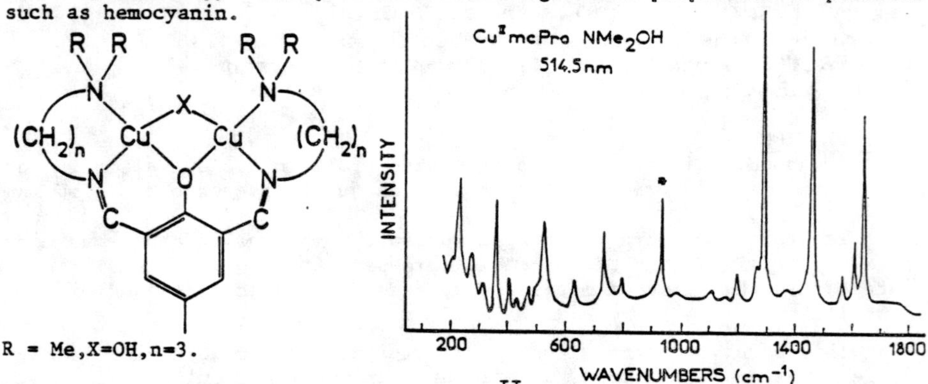

R = Me,X=OH,n=3.

Fig.2 - Resonance Raman spectrum of a Cu^{II} protein model complex.

The nature of ligands to Copper can be deduced. An increased double bond character of the C-O bond is observed. The Cu-O-C angle is wider in comparison to mononuclear complexes. Bands observed in the spectral range 200-450 cm^{-1} for all studied complexes were assigned to metal-ligand vibrations. The Raman excitation profile study leads to the evidence of a noticeable charge transfer from the phenolate towards the Copper atom (5).

3-Copper-organic complexes in cancerous tissues

Evidence of blue and green particles were shown in a large number of cancerous tissues. Their Raman spectra are very characteristic and are in-

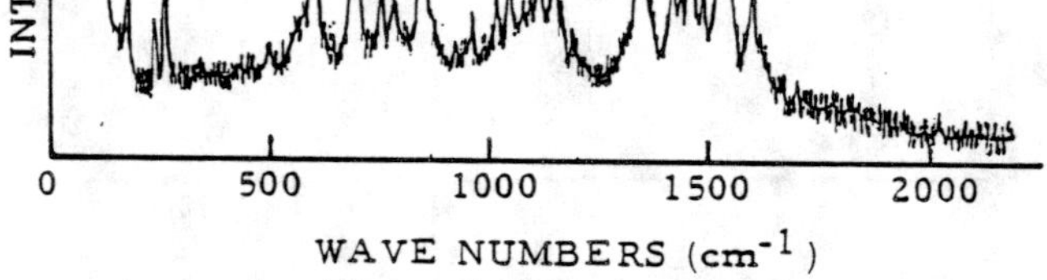

Fig.3 - Raman spectrum of a blue particle in an ovary-cancer.

dependant of the origin of the samples (Fig.3).These particles were identified as copper complexes and their copper-ligands Cu-O, Cu-N and Cu-S were assigned (6,7).

RAPID HEADSPACE ANALYSIS IN SEALED DRUG VIALS BY MULTICHANNEL RAMAN SPECTROSCOPY.

Laura Porter Powell[1] and Alan Campion[2]

American Cyanamid Company, Medical Research Division, Lederle Laboratories, Pearl River, New York[1]; Department of Chemistry, University of Texas, Austin, Texas[2]

Within seconds, intact drug vials are analyzed to verify the nitrogen purge. The analysis speed is achieved by multichannel Raman spectroscopy, together with the novel sample imaging technique depicted in the figure (side view).

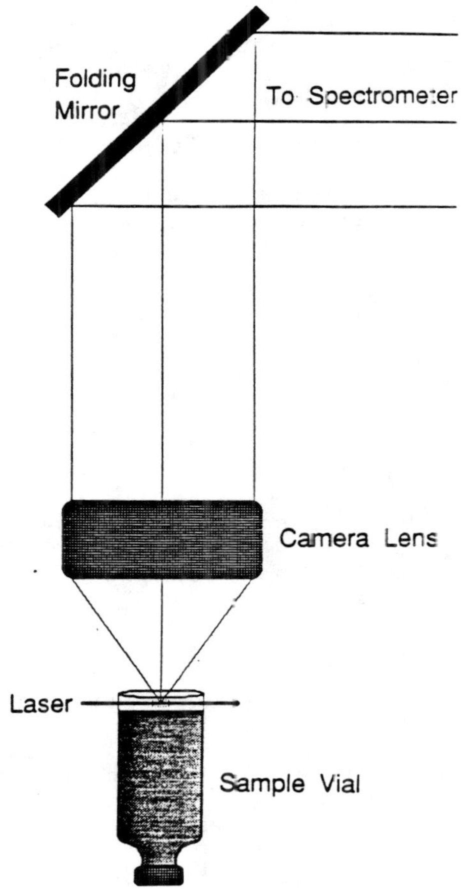

The ratio of the oxygen to nitrogen vibrational peak intensities is used to quantitate the residual oxygen content. Using multichannel detection any changes in experimental conditions occur for both peaks *simultaneously* and cancel out in the ratio.

In this imaging technique the horizontal image in the sample headspace is optically matched with the vertical spectrometer entrance slit. Wide spectrometer slits can be used without any peak overlap or interferences from drug or vial Raman peaks. Sample alignment is *simple and rapid*. Thermal degradation of the sample is eliminated because the laser beam only illuminates the headspace. The technique is useful for analyses in vials with highly scattering samples (powders), and in vials with a small headspace volume.

LASER RAMAN SPECTROSCOPIC STUDIES OF GALACTOSE CATARACT IN RATS

Cai Ming-Zhi*, Chen Yu-Ling**
Department of Chemistry, Tsing Hua University, Beijing, China*
Institute of Biophysics, Academia Sinica, Beijing, China**

In the past decade, the pathogenesis of sugar cataracts has been studied, mainly by biochemical and histopathologic methods.[1-5] To the best of our knowledge, no Raman spectroscopic studies on changes of lens protein structure and conformation in experimental galactosemic cataracts have been reported. This paper will describe some biochemical changes detected by Raman techniques during sugar cataract formation.

Materials and Methods

WISTAR rats, either sex, weighing about 100 g., were injected with 50% galactose solution, 30 g. per Kg into the abdominal cavity twice a day. After 21 days of injection, the lenses became opaque in varying degrees. The eye balls of galactosemic and normal rats were enucleated and the lenses were dissected. The lenses were stored in normal saline for Raman spectroscopic studies.

The Raman spectra of the intact lenses can be obtained by using a laser Raman spectrometer, which consists of a SPEX 1403 double monochromator, a RCA-type C-31034 photomultiplier tube, photon-counting electronics and a SPEX Datamate (microprocessor). All spectra were recorded at a scan speed of 4 cm^{-1}/sec with 900 μm slit width and 3 cm height, scan range: 450-3200 cm^{-1}. The spectra we recorded have not been smoothed. The excitation wavelength at 514.5 nm (150 mw) was provided by a SPECTRA PHYSICS Argon-Ion Laser model 2020.

Results and Discussion

The results showed the differences of Raman spectra between the normal and the galactose cataract lens of rats. The results also showed the different Raman spectra of galactose cataract lens at different stage, early cataract, moderately opaque lenses and mature cataract.

No significant changes in Raman pattern were noted in the range of 900 cm^{-1}-1700 cm^{-1}. The peaks at 1242 cm^{-1} (amide III) and 1668 cm^{-1} (amide I) appeared in both normal and galactose cataract lenses, indicating that the lens soluble proteins have trans parallel β-fold conformation, and no changes of secondary structure occurred in these proteins as well.

Dramatic changes of intensity ratio I_{858}/I_{832} were noted in all galactose cataractous lenses as compared to that of normal. This suggested that the hydrogen bond of Tyr residue might be modified during the process of protein aggregation.

In addition, the changes of I_{882}/I_{758} in galactosemic rats might indicate the same modifications of lens crystallins also occurred at Trp residues. Changes of I_{644}/I_{622} in galactosemic rats might demonstrate the changes of side chains of the lens protein at Tyr and Phe residues. Furthermore, changes of I_{2578}/I_{2730} showed the transformation of -SH into -S-S- bonds during crystallin aggregation in these cataractous lenses was very dramatic.

The results of Raman spectrosocpic studies in this paper have demonstrated the biochemical changes at the molecular level in sugar cataract formation.

References

1. Kinoshita, J.H.: Mechanisms initiating cataract formation. Invest. Ophthalmol. **13**, 713 (1974).
2. Reddy, V.N., Schwass, D., Chakrapani, B., and Lim, C.P.: Biochemical changes associated with the development and reversal of galactose cataracts. Exp. Eye Res. **23**, 483 (1976).
3. Unakar, N.J., Genyea, C., Reddan, J.R., and Reddy, V.N.: Ultrastructural changes during the development and reversal of galactose cataract. Exp. Eye Res. **26**, 123 (1978).
4. Datiles, M., Fukui, H., and Kinoshita, J.H.: Galactose cataract prevention with sorbinil, an aldose reductase inhibitor: a light microscopic study. Invest.Ophthalmol. Vis. Sci. **22**: 174 (1982).
5. Hu, T.-S., Datiles, M. and Kinoshita, J.H.: Reversal of galactose cataract with sorbinil in rats. Invest. Ophthalmol. Vis. Sci. **24**, 640 (1983).

chapter 5

SURFACES and SURFACE ENHANCED RAMAN SCATTERING

DOES SERS WORK WITHOUT ELECTROMAGNETIC FIELD ENHANCEMENT?

C. Pettenkofer and A. Otto

Physikalisches Institut III, Universität Düsseldorf, D-4000 Düsseldorf , Fed. Rep. of Germany

For the case of coldly deposited films, this question was already raised in ref. /1/. In the meantime, we have observed the quenching of the Raman signals of ethylene, ethane and pyridine by a submonolayer of oxygen and have excluded an explanation based on an electromagnetic field effect /2/.

The installation of an evaporation cryostat into our UHV system has allowed for the condensation of solid films of CO and ethane on silver films deposited at about 10 K, which serve as internal standards for the evaluation of the Raman enhancement.

Thermodesorption spectroscopy /3/ and "time resolved sticking spectroscopy" yields information which exposure of the porous films /3/ is needed to complete a monolayer. In the case of CO and silver films deposited at about 9 K (thickness about 200 nm), this exposure is about 40 L.

The SERS intensity of CO saturates at about 2 - 3 L exposure indicating a detectable enhancement for only a small fraction of a monolayer. The CO stretching line shifts eventually from 2143 cm^{-1} to about 2112 cm^{-1} /4/. This line is not observable for CO adsorbed on a film deposited at 305 K and cooled to 9 K, but rather a narrow line at 2139 cm^{-1} after very high exposures. Therefore, we attribute the SERS signal at 2112 cm^{-1} to CO adsorbed at sites of increased heat of adsorption /3/, as also corroborated by TDS.

During growth of the thick CO film, interferences of the reflected light of an extra small HeNe laser at 632.8 nm, of the Rayleigh scattered light at λ_L = 514.5 nm and the Stokes scattered light corresponding to a shift of 2112 cm^{-1} from λ_L = 514.5 nm (see fig. 1) allow a measurement of the thickness of the solid CO film. The quasi linear increase of the line intensity at 2139 cm^{-1} with CO film thickness for CO on silver films deposited at room temperature is shown in fig. 1. This intensity is compared to the intensity at 2112 cm^{-1} from a cold deposited film which saturated at 3 L CO exposure and which is modulated by interferences at λ_L = 514.5 nm and of the Stokes scattered light at λ = 577 nm. At high coverages of CO on coldly deposited silver the Raman signal of the thick condensed layer at 2139 cm^{-1} eventually becomes observable, its peak intensities for various CO film thicknesses on two coldly deposited Ag films are also displayed in fig. 1. Comparison of the integrated intensities of the "minority line" near 2112 cm^{-1} and the "thick condensed layer line" at 2139 cm^{-1} yields an enhancement for the first 3 L of CO on coldly deposited silver of about $6 \cdot 10^6$ x, with $0.25 \leq x \leq 4$, with respect to the Raman signal of condensed CO.

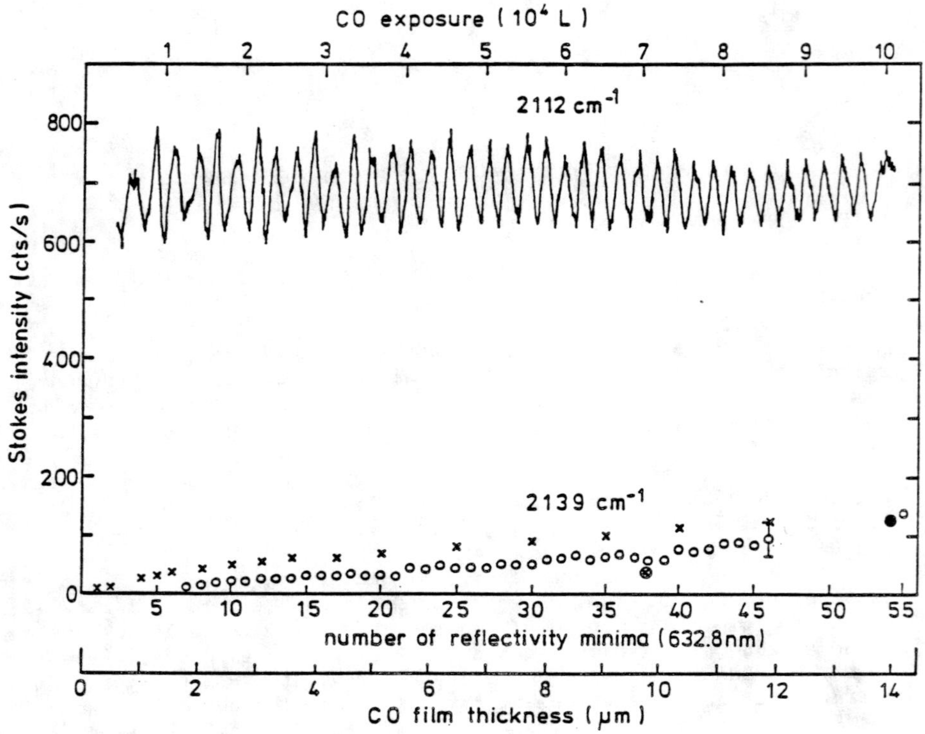

Fig. 1. Upper trace: peak intensity of CO stretch Raman band at 2112 cm^{-1} for CO on a silver film deposited at 9 K versus exposure and numbers of reflectivity minima at 632.8 nm converted into CO film thickness. Lower trace: peak intensity of CO stretch Raman band at 2139 cm^{-1} for CO on two silver films deposited at room temperature (x, o) and two silver films deposited at 9 K (⊗, ●).

We will present further results for analogous experiments with ethane C_2H_6, and preliminary results for CO on silver island films prepared in UHV.

We will confront our results with the prevalent philosophy of exclusive electromagnetic enhancement /5/.

/1/ A. Otto, I. Pockrand, J. Billmann and C. Pettenkofer in Surface Enhanced Raman Scattering, ed. by R.K. Chang and T.E. Furtak, Pergamon Press, 1982

/2/ Ü. Ertürk, C. Pettenkofer and A. Otto, J. Electron Spectrosc. Relat. Phenomena, in press

/3/ J. Eickmans, A. Otto and A. Goldmann, Surf. Science, in press

/4/ C. Pettenkofer and A. Otto, Surf. Science 151 (1985) 37

/5/ M. Moskovits, Rev. Mod. Physics 57 (1985) 783

THEORY OF SURFACE ENHANCED RAMAN SCATTERING.

R. Zeyher

Max-Planck-Institut für Festkörperforschung,
7000 Stuttgart 80, Federal Republic of Germany

Large enhancements of the cross section for inelastic light scattering from molecules have experimentally been found if the molecules were adsorbed on metal surfaces. This surface-enhanced Raman scattering has been observed for many molecules, i.e., CO, CN, Cl, pyridine, and a few substrates such as Ag, Cu, Au, Pb, Ni. It is now generally believed that surface roughness together with polarization and chemisorption effects play an important role for this effect.

I will review a general theory for surface enhanced Raman scattering based on a polariton description for the interaction of light with the molecule and the bounded metal. Taking only one or two electronic states on the molecule into account it will be shown that the general expression for the enhancement ρ of the cross section due to the presence of the metal assumes the simple form

$$\rho = \rho_I \cdot \rho_{II} \cdot \rho_{III} .$$

The three partial enhancement factors have a simple physical meaning:

a) ρ_I represents a local field effect due to polarization contributions of the bounded metal in the absence of the molecule. The bounded metal has electromagnetic modes which are localized near the surface and have large field amplitudes there. If an incident or scattered photon can excite these modes substantially, the molecule will be exposed to large effective fields. This occurs for a plane metal surface if the surface is rough. We will present results for ρ_I for the case where the roughness is due to a weak sinusoidal grating and for a randomly rough surface. The dielectric function of the metal (we consider Ag, Cu, and Au) is assumed to be local and is taken from the experiment. We discuss scattering from a randomly rough surface in some detail because this case represents a true and challenging many-body problem where multiple scattering of surface polaritons plays a decisive role even for a moderate strength of the roughness.

b) ρ_{II} represents also a local field effect due to polarization contributions because of the presence of the molecule. Real or virtual dipolar transitions on the molecule (between the ground state of the molecule and unoccupied states of the molecule or of the metal) induce image dipoles in the metal and thus change the electric fields. We represent simple model calculations for this effect assuming again a local dielectric function for the metal. This assumption however is rather delicate for ρ_{II} because, in contrast to the case a), ρ_{II} involves large momentum transfers of the order of $1/z$ where z is the distance between the molecule and the metal.

c) ρ_{III} is essentially due to chemisorption effects. The electronic levels of the molecule may shift and broaden because of the presence of the metal. The initial, final, and intermediate electronic states are thus changed compared with the free molecule and the continuum of unoccupied metal states can act as intermediate states. We present results for ρ_{III} within the Newns-Anderson model of chemisorption.

Raman Spectroscopy of Molecules Adsorbed on Solid Surfaces

Alan Campion
University of Texas at Austin

ABSTRACT

The ability to obtain Raman spectra of molecules adsorbed at low coverages on well-characterized single crystal surfaces without enhancement has opened many new exciting areas for investigation. In particular, it should be possible to connect the results of studies on model single crystal substrates with those on technical materials such as supported catalysts or electronic devices in order to gain insight into surface structures and the dynamics of interfacial processes. Raman spectroscopy has a number of important advantages as a surface spectroscopic probe; these include high spectral and spatial resolution, a wide spectral range, good sensitivity and the important capability of working with non-vacuum ambient phases. In this talk I will review the physical considerations that must be understood to design and interpret surface Raman experiments. Selection rules for Raman scattering by molecules adsorbed on single crystal metal surfaces will be discussed. I will describe how different geometries or angle-resolved experiments can be used to make symmetry assignments. New developments which increase the sensitivity of the technique will be discussed, as will the extension of the method to include semiconductor and dielectric substrates. Finally, experiments designed to elucidate molecular contributions to the total enhancement in surface - enhanced Raman scattering will be discussed.

DEVELOPMENTS IN THE USE OF RAMAN SCATTERING TO INVESTIGATE THE EXCITATION OF ADSORBATE-SUBSTRATE COMPLEXES

A. D. Brotman

University of Pennsylvania
Department of Physics
Philadelphia, PA 19104*

Over the last several years we have used measurements of Raman scattering excitation profile (RS-EP) spectra and the polarization dependence of the resonant Raman scattering (RS) by molecules adsorbed on smooth metal surfaces to investigate their valence electron excitations.[1-3] We have demonstrated that these measurements provide information on the energies and widths of electronic and vibrational excitations of the adsorbates-substrate (A-S) complex, on the nature (i.e., intra-molecular, inter-molecular or charge-transfer) of the A-S excitation and on the RS mechanism. Measurement of the RS-EP spectra provide information on the energy and width of the electronic excitations of the A-S complex. Examination of the symmetry and spatial loction of the vibration modes which exhibit resonance enhancement in the vicinity of the electronic excitation provides information on the vibration mode electronic excitation interaction (the RS mechanism). It was demonstrated by Campion that polarization studies can be carried out for the RS by molecules adsorbed on smooth metal surfaces (the RS by molecules adsorbed on rough surfaces is always depolarized).[4] The polarization dependence of the RS in the vicinity of the electronic excitation under investigation along with the information on the electron excitation vibration mode interaction can lead to an understanding of the nature of the excitation being probed.

The phthalocyanines are square, planar molecules which have their first singlet π to π^* electronic excitations in the visible. In metal phthalocyanines (MPc's) a metal atom is bonded to the central nitrogen atoms resulting in a molecule with D_{4h} symmetry. The x, y equivalence of the MPc's is responsible for the degeneracy of the x and components of the first singlet (Q) excitation. In the free base molecule the central hydrogen atoms break this symmetry and the Q_x and Q_y excitations are separated by ~750cm^{-1}. H_2Pc lies flat and forms strong bonds with metal surfaces.[5] It is expected that the lone pair electrons on the central nitrogen atoms play the predominate role in this bonding. The RS-EP spectra of H_2Pc adsorbed on a silver surface was measured in the vicinity of the Q excitation. We observed two peaks in the RS-EP spectra at 692 nm and 674 nm (a separation of ~400 cm^{-1}). The ring vibration modes exhibited resonance enhancement in the vicinity of these excitations. Polarization studies indicated that the x and y components of the transition dipole moment were predominantly responsible for the Raman intensity. The lack of a z component and strong interaction with ring vibration modes indicate that these excitations retain their intra-molecular π to π^* character and are shifted in energy through dielectric interaction with the substrate. The change in the separation of the Q_x and Q_y excitations is attributed to the bonding of the molecule to the surface through the lone pair electrons on the central nitrogen atoms resulting in weakening of the symmetry break N-H bonds.

---

* Present Address: AT&T Bell Laboratories, 2525 North 12th Street, Reading, PA, 19604

CV also has its first singlet π to π^* excitation in the visible. The excitation in the free molecule involves the transfer of charge between the central carbon atom and the peripheral di-methyl amine groups. The RS-EP spectra of CV adsorbed on gold exhibits a resonance peak for the central carbon breathing mode at 606 nm indicating that the excitation is intra-molecular in nature. Polarization studies, however, indicate a sizable zz component to the Raman tensor. In order to determine the origin of the zz component polarization resolved RS-EP spectra were measured in an orthogonal plane configuration so that only the zz contribution to the RS was obtained.[3] The RS-EP spectra, in this configuration exhibits no π to π^* resonance peak, only a steadily increasing Raman intensity as a function of wavelength is observed. This contribution is attributed to a charge transfer from the gold substrate to the adsorbed CV molecule. The absence of a π to π^* resonance peak in this configuration indicates that the CV molecule lies flat on the metal surface.

These studies demonstrate the usefulness of RS in the characterization of A-S complexes. RS is a virtual process involving contributions by all of the states of the system weighted by an energy denominator. By measuring the RS and its polarization dependence in the vicinity of a particular electronic excitation we pick out the contribution to the RS by that particular excitation (assuming that there are no other nearby excitations). Measurement of polarization resolved RS-EP spectra can, therefore, be used to provide information not only on the energies and widths of the electronic excitations of the A-S complex but also information on the mechanism responsible for the RS which helps determine the nature of the A-S electronic excitation. We believe that measurement of polarization resolved RS-EP spectra is a valuable addition to the array of tools available for the characterization of surfaces and that it will help provide a better understanding adsorbate-substrate interactions.

I would like to acknowledge useful discussions with E. Burstein, J. D. Jiang, H. Kobayashi and G. Pajer.

REFERENCES

[1] A. Brotman, E. Burstein and J. D. Jiang, Surf. Sci. 158, (1985).

[2] A. Brotman and E. Burstein, Physica Scripta 32, 385 (1985).

[3] J. D. Jiang, E. Brustein and H. Kobayashi, Proc. of the 10th Intern. Conf. on Raman Spectroscopy (1986).

[4] A. Campion, V. M. Grizzle, D. R. Mullins and J. K. Brown, J. Physique Colloq. 44 (1983) C10-341.

[5] J. C. Bucholz and G. A. Somorjai, J. Chem. Phys. 66, 537 (1977).

SURFACE ENHANCED SPECTROSCOPY - CALCULATION OF LOCAL FIELDS

J.I. Gersten[a], Abraham Nitzan[b]

[a] Department of Chemistry, City College of the City University of New York, New York 10031

[b] School of Chemistry, Tel Aviv University, Tel Aviv 69978, Israel

The discovery of surface enhanced Raman scattering has focussed attention on the role played local electromagnetic fields in this phenomenon as well as in other processes involving the electromagnetic response of adsorbed molecule or molecules imbedded in small clusters. After reviewing some such processes we describe in this talk a new numerical approach to calculating local fields. This approach is based on the Feynman path integral solution to the Poisson equation: The propagator associated with this equation, the solution $G(\vec{r},\vec{r}')$ to

$$\nabla\cdot\varepsilon(\vec{r})\nabla G(\vec{r},\vec{r}') = 4\pi\delta(\vec{r}-\vec{r}') \tag{1}$$

may be shown to be represented by

$$G(\vec{r},\vec{r}') = \frac{4}{\varepsilon(r')}\int_0^\infty dt \int D[\vec{r}(t)]\, e^{-S} \tag{2}$$

where $\int D[\vec{r}]$ is a path integral subjected to the conditions $\vec{r}(0) = \vec{r}'$ and $\vec{r}(t) = \vec{r}$, and where the "action" S is given by

$$S = \int_0^t L\left(\vec{r}, \frac{d\vec{r}}{dt}\right) dt$$

$$L = \frac{1}{4}\left(\nabla \ln\varepsilon\,(\vec{r}) + \frac{d\vec{r}}{dt}\right)^2$$

Given the spatial dependence of the dielectric function $\varepsilon(r)$, the path integral can be evaluated either on the "semiclassical" level (expanding to quadratic terms about the maximal path) or numerically by existing methods. In particular we shall examine evaluation of (2) by (for 1-dimensional problems) the technique of matrix multiplication and (for the general case) by Monte Carlo methods. Applications to local field calculations and optical properties of molecules near diffuse surfaces will be described.

"POLARIZATION RESOLVED" RESONANCE RAMAN SCATTERING BY CRYSTAL VIOLET ADSORBED ON SMOOTH METAL SURFACES: EVIDENCE FOR A CHARGE-TRANSFER EXCITATION*.

J.D. Jiang, E. Burstein and H. Kobayashi

Physics Department, University of Pennsylvania, Philadelphia, PA 19104

We report here results of a study of the polarization-resolved resonance Raman scattering (RS) by Crystal Violet Chloride (CV-Cl) adsorbed on smooth noble metal surfaces in ultrahigh vacuum. CV exists in aqueous solution as the carbonium (CV^+) ion which has its first $\pi-\pi^*$ singlet excitation in the visible at 590 nm. The CV^+ ion has a propeller-like (D_3 symmetry) structure. Theoretical calculations of the electronic structure, based on a simple planar model, indicate that the positive charge of CV^+ in the ground state is located predominantly at the central-carbon [1]. The $\pi-\pi^*$ excitation leads to an excited state in which the positive charge is located predominantly on the nitrogen atoms of the dimethyl amine groups. The Raman scattering excitation-profile spectrum of the totally symmetric central-carbon breathing mode of CV^+ adsorbed on a smooth evaporated Au film, for s-polarized incident radiation and unpolarized scattered radiation (s_i,u_s), exhibits a resonance peak at 606 nm (FWHM ~ 1100 cm^{-1}) [2]. The fact that the frequency of the resonance peak for the adsorbed molecules is not appreciably different from that of the "free" CV^+ ions in aqueous solution, and that the frequency of the central-carbon breathing mode (208 cm^{-1}) is not appreciably different from that of the "free" CV^+ ions (202 cm^{-1}), indicates that the resonance peak corresponds to the first singlet $\pi-\pi^*$ electronic excitation of CV^+ ions, and that, therefore, that the CV was adsorbed on the Au substrate as carbonium ions. Measurements of the polarization dependence of the resonance Raman scattering by the central-carbon breathing mode, in the vicinity of the first $\pi-\pi^*$ singlet excitation, showed that the Raman scattering by the central-carbon breathing mode in a (p_i,p_s) scattering configuration, (which can involve xx, xz and zz Raman tensor components, where z is direction of the surface normal) was appreciably stronger than in a (s_i,s_s) scattering configuration (which can involve only the xx, or yy component). This suggested that the contribution from the $\pi-\pi^*$ electronic excitations to the Raman scattering may have a sizeable xz or zz component.

In order to get further information about the relative magnitudes of the different components of the Raman tensor that contribute to the Raman scattering in the vicinity of the first $\pi-\pi^*$ excitation, we have carried out Raman measurements using a scattering-configuration in which the plane of incidence and the plane of scattering are orthogonal (Fig. 1). For this geometry, the (s_i,s_s) polarization configuration involves the xy component of the Raman tensor; the (p_i,p_s) configuration involves a linear combination of xy, xz and zz components; and the (p_i,s_s)

configuration involves the xx and xz components. A sizeable Raman scattering by the central-carbon breathing mode was observed only in the (p_i,p_s) polarization configuration. On this basis we may conclude that there is a sizeable zz component, and that the xy and xz components were either very weak compared to the zz component, or zero.

To ascertain the origin of the zz component, we carried out a measurement of the Raman scattering excitation-profile for the central-carbon breathing mode in the (p_i,p_s) orthogonal-plane configuration. The resultant spectrum does not have any $\pi-\pi^*$ resonance feature. Rather, it exhibits a steadily increasing scattering intensity with wavelength and, at 606 nm, the Raman scattering intensity is appreciably stronger than the corresponding intensity for the (s_i,u_s) parallel-plane configuration (Fig. 2). We tentatively attribute the sizeable zz component of Raman tensor to charge-transfer excitations from the Au to the adsorbed CV^+. We also conclude, from the absence of the $\pi-\pi^*$ resonance, that the adsorbed CV^+ ions lie flat on the surface.

We have also carried out measurements of the Raman scattering by CV adsorbed on smooth Cu and Ag surfaces. The Raman scattering intensities, and their polarization dependence, for CV adsorbed on Cu were similar to those for CV adsorbed on Au. On the other hand, we were not able to observe any Raman scattering by CV deposited on a smooth Ag surface. We tentatively believe this is due to the preferential adsorption of Cl^- ions by the Ag surface, and to the consequent low CV coverage on the Ag surface.

Our studies of the resonance Raman scattering of molecules adsorbed on smooth noble metal surfaces demonstrate the usefullness of Raman scattering excitation-profile spectra as a probe of the valence electron excitations of the adsorbate-substrate complexes that are formed. The polarization resolved Raman scattering data provide information, not only about the energies, widths and vibronic character of the electronic excitations, but also about the nature of the contribution to the Raman scattering from the different electronic excitations involved.

We wish to acknowledge valuable discussions with A. Brotman and G. Pajer.

*. Research supported in part by ONR and by NSF MRL Program at Penn.

1. F.C. Adams and W.T. Simpson, J. Mol. Spectrosc. 3, 363 (1959).
2. A. Brotman, E. Burstein and J.D. Jiang, Surf. Sci. 158, 1 (1985).

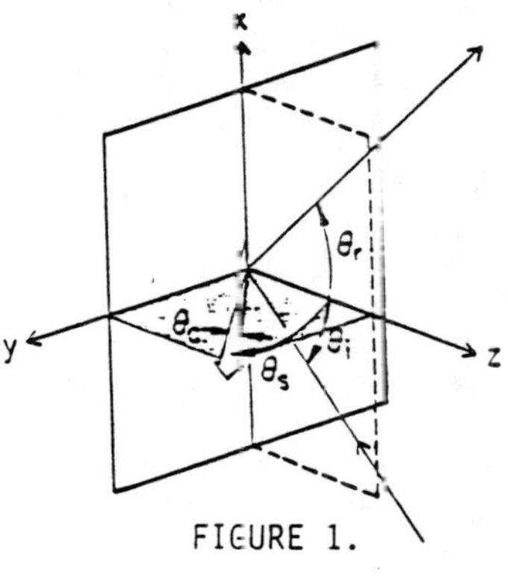

FIGURE 1.

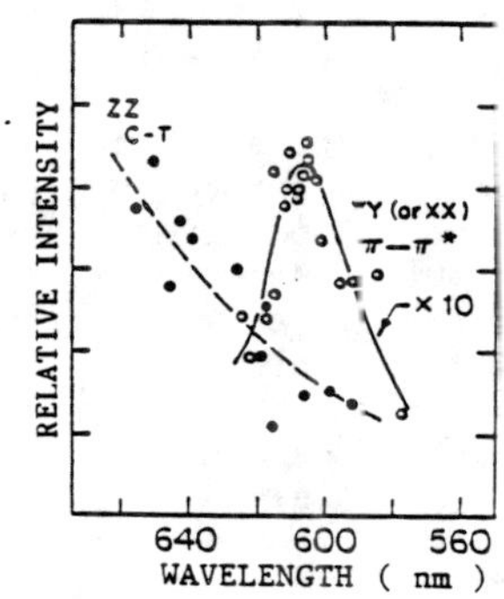

FIGURE 2.

SURFACE ENHANCED RESONANCE RAMAN SCATTERING (SERRS) FROM TRI-P-DIMETHYLAMINO PHENYL CARBONIUM ION (CRYSTAL VIOLET).

N. T. Liang, W. S. Tse and Y. C. Chou*

Institute of Physics, Academia Sinica, Taipei, Taiwan (11529), R.O.C.
* Department of Physics, National Tsing Hua University, Hsinchu, Taiwan, (300), R.O.C.

Surface enhanced Raman scattering (SERS) of molecules adsorbed on noble-metal surface has been extensively studied both experimentally and theoretically [1]. In the semiclassical treatment [2], the classical EM model successfully explained most of the observed SERS features. For molecules exhibiting resonance Raman scattering, where the intermediate excited state of Raman scattering is a real electronic state, one might expect completely different behaviors from the ordinary SERS when molecules were adsorbed on a rough metal surface, and the quantum mechanical effects should be considered [1,3,4]. Many authors study the SERS from crystal violet (CV) adsorbed on Ag surface [5,6,7].

Figure 1 shows the excitation profiles of 225 cm^{-1}, 920 cm^{-1} and 1590 cm^{-1} bands of CV adsorbed on the 50 Å silver film, where the ω^4 dependence and the spectra response was eliminated by normalizing the Raman intensities of adsorbed CV molecules against the 150 cm^{-1} band of $LiNbO_3$ crystal. A very sharp increase of Raman cross section was observed in the range from 4579 Å to 5682 Å of λ_{in}. Because CV molecules have absorption bands at 5500 Å and 5900 Å, this sharp increase might be due to multiplicative contribution from both electromagnetic effects of rough metal film and the electronic resonance of CV molecules. The decrease of the enhancement factor in the long wavelength side might be due to off resonance of both effects. The change in maximum position of the excitation profiles is believed to be originated from the resonance of molecular excitation.

In this work, we also measured the transmitted absorptive spectra of CV/50 Å Ag island film/glass system by using JASGO uvidec-505 apparatus. Figure 1(a) shows the absorptive spectrum of the difference between 50 Å Ag island film/glass/CV and glass/CV or glass only. In this system, CV molecules are adsorbed on glass surface and the absorption is small so that the absorptive spectrum is nearly flat. However, figure 1(b) represents the absorption spectrum of the interface between CV and 50 Å Ag island film which exhibit relatively strong and broad as compared to fig. 1(a) and fig. 1(c). One possible reason is that the silver island film with CV molecules adsorbed on it induce the localized surface plasmon resonance. Besides, the absorption band of CV molecules might become strong and broaden when adsorbed on Ag island film. The excitation profile of the Raman mode of 225 cm^{-1} band of CV adsorbed on Ag island film, as shown in figure 1, is consistent with the absorption band shown in figure 1(b).

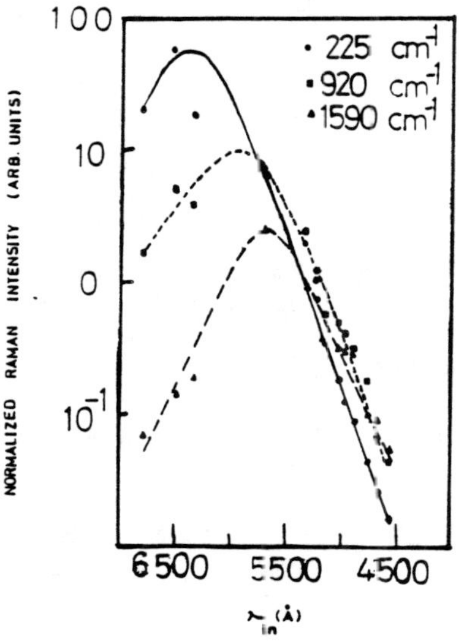

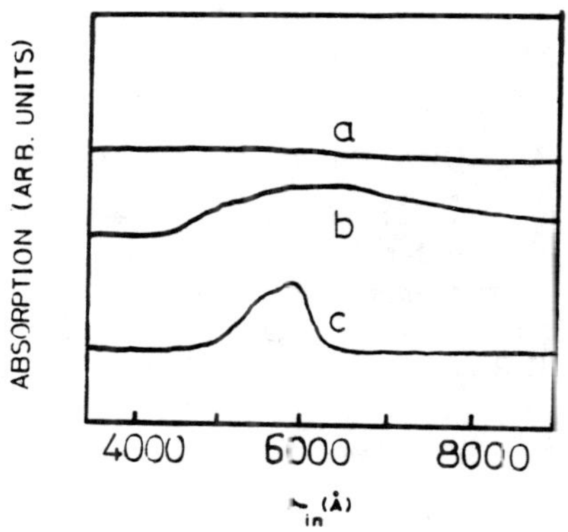

Figure 1. Excitation profiles of Raman modes at 225 cm^{-1} (closed circles), 920 cm^{-1} (closed squares), and 1590 cm^{-1} (closed triangles) of CV adsorbed on the 50 Å silver island film. The external reference standard was the 150 cm^{-1} Raman mode of the $LiNbO_3$ crystal. The ordinate is in logrithmic scale.

Figure 2. The absorptive spectra of (a) difference between 50 Å island film/glass/C V and glass/C V or glass only (b) difference between C V/50 Å Ag island films/glass and 50 Å Ag island film/glass/C V. (c) 0.01 mM C V. in CH_3OH.

References

[1] R. K. Chang and T. E. Furtak, Surface Enhanced Raman Scattering (Plenum Press, New York) 1982.
[2] J. Gersten and A. Nitzan, J. Chem. Phys. 73 (1980) 3023.
[3] W. L. Peticolas, L. A. Nafie, P. Stein, and B. Fanconi, J. Chem. Phy Phys. 52 (1970) 1576.
[4] D. A. Weitz, S. Garoff, J. I. Gersten and A. Nitzan, J. Chem. Phys. 78, 5324, (1983).
[5] R. P. van Duyne, in : Chemical and Biochemical Applications of Lasers, Vol. 4, ed. C.B. Moore (Academic Press, New York, 1979).
[6] T. Watanabe and B. Pettinger, Chem. Phys. Lett., 89, (1982) 501.
[7] E. Burstein, G. Burns and F. H. Dacol. Solid State Commun. 46 (1983) 595.

THE ROLE OF THE CAVITY MODEL IN SERS.

H. Seki
Almaden Research Center, 650 Harry Road, San Jose, California 95120-6099

The first clear Raman spectra of pyridine adsorbed on a silver electrode was reported by Fleischmann et al., in 1974[1] and it is almost ten years since the realization that we had a novel phenomenon[2,3]. Nevertheless a clear concensus regarding the essential mechanism of the SERS process has yet to be achieved. This lack of complete understanding has prevented SERS from living up to the expectation as a general purpose surface analysis tool. Recently we have seen impressive developments in the IR reflection absorption spectroscopy (IRRAS), especially at the electrode-electrolyte interphase due to the pioneering work of Bewick and his collaborators [4]. There is also the work started by Campion investigating the possibility of measuring normal Raman scattering of submonolayer order films[5]. These two developments may suggest that the need to complete our understanding of SERS is greatly reduced. It can be seen that if the possibilty of unknown enhancement exists this introduces uncertainty to the normal Raman scattering signal. Thus the need to understand and control enhancement processes at the surface remains the same, if not greater, as the the detecting techniques become more sensitive. In fact these developments provide an improved opportunity to gain further insight into the SERS process.

Here we discuss the Cavity model of SERS[6,7] which has been reasonably successful in explaining the SERS behaviour of pyridine on Cold Ag films, i.e. the low temperature anneal[8], the high temperature anneal[9,10], the extreme short ranged SERS effect[10], the loss of SERS upon annealing to room temperature[9,10] and the lack of SERS from deliberately formed adatoms and adclusters[11-13]. Recently a number of papers which discuss and add plausibility to the Cavity model for SERS have appeared[14-25]. Wirgin and Lopez-Rios have presented a waveguide model in which a set of narrow deep grooves simulates the crevices in the cold silver films. It is found that the amplitude of standing waves at resonant frequencies peak to give enhancements of several hundred inside the groove[14,15], larger than enhancements obtained for surface plasma polaritons. Liver, Nitzan and Gersten approached this from an interesting perspective, i.e. that of aggregates of small metal spheres and finds that the field within voids surrounded by several close spheres can lead to very high enhancements[16]. This result is interesting since it ties in with the SERS of Ag and Au sols where the greatest enhancements are observed when there is aggregation[17]. Chew and Kerker have calculated the field within a small dielectric spheroid embedded in silver near a flat surface[18] and find that enormous enhancements can be obtained.

More details on the characterization of the cold silver films and their relation to SERS have been discussed by Albano et al[19]. Experiments using deposition of Xe underlayers prior to the adsorption of pyridine on the cold Ag film have clearly shown that one can fill up the cavity with Xe and hinder the possibility of SERS[20]. It is also reported that when the cold Ag film with the adsorbed pyridine is overcoated with optimum monolayer amounts of additional Ag the SERS effect is increased supporting the idea that a geometry of an embedded spheroid may be the most optimum for enhancement[21]. The difference in the enhancement for Cold films deposited on a smooth substrate and a rough one[10] suggests that there may be two levels of enhancement, one with surface plasmon polaritons and then its interaction with the localized surface plasmons of the cavity[21]. The normal surface Raman measurements of pyridine adsorbed on carefully prepared Ag single crystal surfaces have demonstrated that chemisorption does not necessarily lead to enhancement[22] which certainly suggests, that in the case of pyridine on silver, the SERS mechanism is entirely due to collective electron resonance effects.

The models used for the theoretical estimates of the enhancement are generally greatly simplified for the sake of analysis. The attempt to examine the surface of an annealed Cold silver by tunneling microscopy demonstrates that the actual surfaces in experiments are quite complex[23]. Nevertheless, the fact that these seemingly different models all predict huge enhancements points to the uniqueness of the cavity model from the protrusive type geometries often considered in the collective electron resonance mechanism. There is need for more theoretical analysis. A good model has to explain the marked quenching of SERS when more than monolayer order of the adsorbates are deposited[24] or when submonolayer amounts of other metals are deposited on SERS active situations[25]. Note how the latter contrasts with the enhancement seen with the additional deposition of Ag[21].

REFERENCES

1 M. Fleischmann, P.J. Hendra and A.J. McQuillan, Chem. Phys. Lett., 26 (1974) 163-166.
2 M.G. Albrecht and J.A. Creighton, J. Am. Chem. Soc., 99 (1977) 5215-5217.
3 D.J. Jeanmaire and R.P. van Duyne, J. Electroanal. Chem. 84 (1977) 1-20.
4 A. Bewick, K. Kunimatsu, B.S. Pons, and J.W. Russell, J. Electroanal. Chem., 160 (1984) 47.
5 Alan Campion, J. Vac. Sci. Technol., B3 (1985) 1404.
6 E.V. Albano, S. Daiser, G. Ertl, R. Miranda, K. Wandelt and N. Garcia, Phys. Rev. Letters, 51 (1983) 2314.
7 H. Seki and T. J. Chuang, Chem. Phys. Letters, 100 (1983) 393.
8 H. Seki, Solid State Commun., 42 (1982) 695.
9 I. Pockrand and A. Otto, Solid State Commun., 37 (1981) 1159.
10 H. Seki, J. Chem. Phys., 76 (1982) 4412.
11 J. Billmann, G. Kovacs and A. Otto, Surf. Sci., 92 (1980) 153.
12 T.H. Wood, Phys. Rev., B24 (1981) 2289.
13 H. Seki, J. Electron Spectry., 29 (1983) 413.
14 A. Wirgin and T. Lopez-Rios, Opt. Commun., 48 (1984) 416.
15 T. Lopez-Rios and A. Wirgin, Solid State Commun., 52 (1984) 197.
16 N. Liver, A. Nitzan and J. I. Gersten, Chem. Phys. Letters, 111 (1984) 449.
17 C.G. Blatchford, J.R. Campbell and J.A. Creighton, Surface Sci., 120 (1982) 435.
18 Herman Chew and Milton Kerker, J. Opt. Soc. Am., B2 (1985) 1025.
19 E.V. Albano, S. Daiser, R. Miranda and K. Wandelt, Surface Sci., 150 (1985) 367 and 386.
20 H. Seki, T.J. Chuang, M.R. Philpott,E.V. Albano and K. Wandelt, Phys. Rev., B31 (1985) 5533.
21 H. Seki, T.J. Chuang, J. F. Escobar and E.V. Albano, Surface Sci., 158 (1985) 254.
22 A. Campion and D. R. Mullins, Surface Sci., 158 (1985) 263.
23 A. Humbert, J. K. Gimzewski and B. Reihl, Phys. Rev., B32 (1985) 4252.
24 H. Seki, J. Electroanal. Chem., 150 (1983) 425.
25 T. Lopez-Rios, Y. Gao and G Vuye, Chem. Phys. Letters, 111 (1984) 249.

COVERAGE DEPENDENCE OF SURFACE ENHANCED RESONANCE RAMAN SCATTERING AND SURFACE FLUORESCENCE

A. Gerolymatou and B. Pettinger

Fritz-Haber-Institut der Max-Planck-Gesellschaft
Faradayweg 4-6, D-1000 Berlin 33, West-Germany

An extremely low dye concentrations (10^{-10}– 10^{-8} M/l) is apparently optimal for a very great surface enhancement. This holds in particular for rhodamine 6G and crystal violet. In aqueous solution, the former is highly fluorescent with a fluorescence quantum yield of $\eta = 0.2$, whereas the latter is basically non-fluorescent showing pre-resonance Raman effects with

$$d\sigma/d\Omega_{pre-RR} = 6 \cdot 10^{-26}\ [cm^2 sr^{-1}].$$

Inspite of their destinct optical properties, these dyes exhibit, if adsorbed at Ag-colloids, a very intense surface enhanced resonance Raman scattering (SERRS) with a cross sections of about

$$d\sigma/d\Omega_{SERRS} = 10^{-19}\ [cm^2 sr^{-1}].$$

Thus a pure surface enhancement of ca. $2 \cdot 10^6$ is evident for the latter case.

Further, we give evidence for a remarkable coverage dependence of the surface enhancement: With decreasing dye concentration it rises roughly with $1/\vartheta$, reaches a maximum and decays somewhat (see fig.1). In turn, with increasing dye coverage the droping enhancement for Raman scattering is accompanied with increasing amount of surface fluorescence (SFL) by adsorbed rhodamine 6G. This shows the <u>competitive character</u> of the various types of surface (enhanced) light scattering.

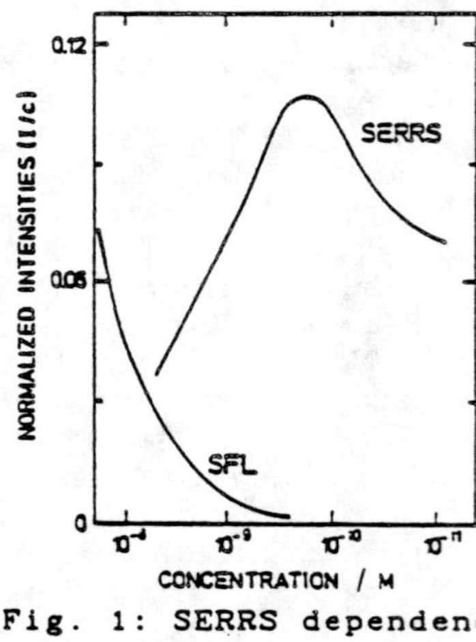

Fig. 1: SERRS dependence of concentration.

These results can be explained with an additional energy transfer induced enhancement.

QUANTUM-MECHANICAL MODEL FOR SURFACE ENHANCED RAMAN SCATTERING: AN ADDITIONAL ENERGY TRANSFER INDUCED GAIN

B. Pettinger

Fritz-Haber-Institut der Max-Planck-Gesellschaft
Faradayweg 4-6, D-1000 Berlin 33, West-Germany

The surface enhanced Raman intensity for molecules at spheres is expressed in terms of measurable quantities such as the spherical Rayleigh polarizability and the (resonance) Raman tensor for the adsorbed molecules. The total surface enhancement (without the measurable resonance Raman contribution) is the product of two or three terms, depending on the molecular resonance conditions (off or on):

$$\overline{F}_{off} = 0.0127(R^{-12}\varepsilon_{eff}^{-4}\overline{\alpha_M^4})\eta_{dd} \quad ; \quad \overline{F}_{on} = 0.0127(R^{-12}\varepsilon_{eff}^{-4}\overline{\alpha_M^4})\eta_{dd}\, g^{ET}.$$

with $g^{ET} = 0.11\,(R^{-6}\varepsilon_{eff}^{-2}\overline{\alpha_M^2})\,\eta_{dd}^{-1}$

The first term is equivalent to the classically derived electromagnetic enhancement (gain ≈ 10000); the second means that the efficiency of the dipole - dipole coupling is significantly reduced due to a competitive dipole - multipole and dipole - electron-hole pair coupling (loss ≈ 0.01); and the third term reveals an additional energy transfer induced gain for excitable molecules (gain ≈ 10000). Using the concept of competitive second order processes such as Rayleigh, Raman, dephasing and relaxation processes, it is shown that the ET-induced gain stems from an enhanced participation on the molecular excitation rate, e.g. from the suppression of usually occuring dephasing and relaxation processes. This explains why a highly fluorescent molecule (quantum yield ≈ 0.2) may change, if properly adsorbed, into a Raman scatterer yet shows the same peak intensity. This means a Raman cross section as high as:

$$d\sigma/d\Omega \approx 2\cdot 10^{-19}\ cm^2 sr^{-1}.$$

Since this ET-induced enhancement depends only on the strength of the molecule→metal coupling but not on its type, it is a general phenomenon. The important consequence is that the pure surface enhancement can be much greater for "coloured" species than for transparent ones. In so far an old controversy about the surface enhancement meachanisms is solved.

ELECTROMAGNETIC SHORT RANGE ENHANCEMENT OF SURFACE-ENHANCED RAMAN SCATTERING.

T. Lopez-Rios and Y. Gao.

Laboratoire d'Optique des Solides, Université Pierre et Marie Curie, 4, place Jussieu, 75252 Paris Cédex O5, France.

The short range enhancement of surface enhanced Raman scattering (SERS) is still a matter of discussion. We have already shown that very thin deposits of Pd (1,2), Al (3), Au, Cu or Si (4) quench the SERS of pyridine adsorbed on cold or island Ag films. For cold Ag films covered by very thin Pd films, we have thoroughly investigated the SERS of pyridine as a function of the pyridine coverages. We found that superficial Pd films strongly reduce the SERS of the first monolayer of pyridine and to lesser extent the SERS of the following monolayers as shown in fig. 1. Curve a corresponds to the intensity of the ν_1 mode (1006 cm^{-1}) as a function of pyridine exposures (uncorrected) for an uncoated Ag cold film. Curve b is for a cold Ag film coated with 2.5 A of Pd before pyridine exposures. Curve c gives the ratio a/b. Samples and experiments were made

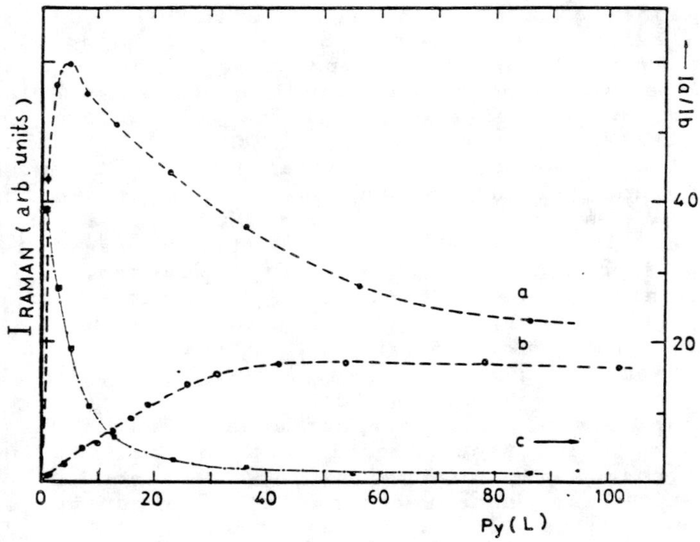

Fig. 1 Raman intensity of breathing mode ν_1 of the pyridine as a function of coverage at T = 140 K for a bare Ag cold film (a) and a Ag cold film coated by 2.5 A of Pd (b). The curve (c) gives the ratios of (a) and (b).

at T = 140 K. For thicker Pd films we obtain the same kind of results, but the intensity of the SERS is even lower and the difference with respect to the uncoated surface is more pronounced. Induced changes, due to pyridine adsorption, in the optical reflectivity of cold Ag films and cold Ag films covered by one monolayer of Pd present similar curves as shown in fig. 1.

Combining Auger and Raman experiments, we confirm the hypothesis already brought forward that large Raman enhancements take place within the pores of Ag cold films (5,6). We found, on the one hand, that Pd does not migrate into the pores and, on the other, that Pd supresses the SERS of molecules located in the pores thus excluding the possibility of a chemical origin of the observed quenching of SERS by the Pd deposits.

In other experiments, we have mesured, for $\hbar\omega_L$ = 2.41 eV, the intensity of the Raman scattering of pyridine adsorbed on cold Cu films (T = 140 K) coated with continuous Ag layers of different thickness (10 to 100 A). We found that the SERS intensity increases with pyridine exposures up to a saturation value L_{sat} beyond which the Raman intensity remains constant. L_{sat} increases linearly with the thickness of the Ag layer.

Our results can be understood in terms of localized surface plasmons of large wave-vectors. Within the pores and in between the interacting particles, the electromagnetic fields have a multipolar character and are highly concentrated at the interacting metal surfaces (7, 8, 9). In these regions the electromagnetic fields are strongly modified by very thin metal deposits (10). On the outer side of the surface, the electromagnetic fields are less short ranged and less modified by the foreign deposits. In conclusion, our experiments can be understood in terms of the excitation of surface plasmons, those with large wave-vectors giving rise to short range enhancements, those with shorter wave-vectors presenting a longer spatial range.

References

1. Y. Gao, T. Lopez-Rios, Phys. Rev. Lett. 53, 2584 (1984).
2. T. Lopez-Rios, Y. Gao, J. Vac. Sci. Technol. B3, 1539 (1985).
3. T. Lopez-Rios, Y. Gao, G. Vuye, Chem. Phys. Lett. 111, 249 (1984).
4. Y. Gao, Thesis (1986) University of Paris and results to be published.
5. E.V. Albano, S. Daiser, G. Ertl, R. Miranda, K. Wandelt, N. Garcia. Phys. Rev. Lett. 51, 2314 (1983).
6. H. Seki, T.J. Chuang, M.R. Philpott, E.V. Albano, K. Wandelt, Phys. Rev. B31, 5533 (1985).
7. P.K. Aravind, A. Nitzan and H. Metiu, Surf. Sci. 110, 189 (1981), P.K. Aravind and H. Metiu, Surf. Sci. 124, 506 (1983).
8. M. Inoue, K. Ohtaka, J. Phys. Soc. Japan 52, 3853 (1983).
9. F. Claro, Phys. Rev. B30, 4989 (1984).
10. A. Wirgin private communication.

SURFACE-ENHANCED RAMAN SPECTROSCOPY OF GUANINE DERIVATIVES

E. Koglin and J.-M. Séquaris

Institute of Applied Physical Chemistry, Nuclear Research Center (KFA), D-5170 Juelich, P.O. Box 1913, Federal Republic of Germany

The knowledge of strucutural characteristics of the nucleotides, nucleosides, and bases in aqueous solution provides fundamental information for the understanding of the conformation of nucleic acids. Modified bases in DNA cause mutagenic, carcinogenic or cytotoxic effects on the cell. A large group of mutagens and carcinogens are alkylating agents of which the electrophilic groups react in particular with guanine sites of DNA.

For the interpretation of vibrational spectra of methylated nucleic acids it is therefore necessary to measure precise frequencies of the guanine base and the methylated guanine derivatives. Recording normal Raman spectra of guanine and its derivatives under physiological salt conditions is impossible because of their very low solubility in water ($< 5 \times 10^{-4}$ g in 100 g H_2O). However, vibration spectra of these molecules have been obtained by means of colloid-surface-enhanced Raman spectroscopy (Colloid-SERS)[1,2]. The purpose of this contribution is to show the recent advances in the application of SERS from guanine derivatives adsorbed on a silver electrode surface as a function of the adsorption potential (Electrode-SERS). These Electrode-SERS investigations of guanine and its derivatives N_1, N_3, N_7, N_9-methylguanine have shown that the scattering intensity depends strongly on the surface charge, the concentration and the laser excitation wavelength. High resolution vibration spectra are obtained from very diluted aqueous solutions down to 10^{-6} M near the potential of zero charge (E_s = -0.8 V). To get an idea of the sensitivity of SERS spectroscopy, Fig. 1 compares the Colloid-

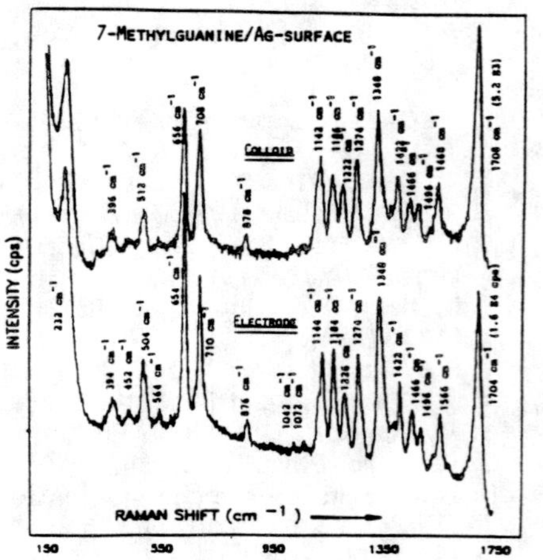

Fig. 1: SERS spectrum of 7-Me-Gua adsorbed on Ag colloids (above): λ_{ex} = 514.5 nm, power 20 mW, concentration 3×10^{-6} M. SERS spectrum of 7-Me-Gua adsorbed on an Ag electrode (below): λ_{ex} = 514.5 nm, power 10 mW, E_s = -0.80 V, pH 8, concentration 9×10^{-6} M

and Electrode SERS spectrum of 7-methyl-guanine adsorbed on a highly posi- tively charged surface.
This communication also compares the results of SERS on colloids to SERS investigations on silver electrode surfaces, attempting to clarify the mo- lecular orientational specifity and the short-range sensitivity of SERS. The observed frequencies and relative identities of the Electrode-SERS bands of guanine and it methylated derivatives are given in Table 1.

TAB. 1 RAMAN BAND FREQUENCIES (cm^{-1}) AND TENTATIVE VIBRATIONAL MODE ASSIGNMENT OF GUANINE AND ITS METHYLATED DERIVATIVES ADSORBED ON A SILVER ELECTRODE SURFACE (E_S = -0.6 V).

NOTE: W, M,S, VS INDICATE WEAK, MEDIUM, STRONG AND VERY STRONG INTENSITIES. SH = SHOULDER; STR = BOND STRETCHING, D = ANGLE DEFORMATION, B = BENDING, R = ROCKING

GUA	1 ME-GUA	3 ME-GUA	7 ME-GUA	9 ME-GUA	MODE ASSIGNMENT
226 SH	224 SH		240 SH	216 SH	AG/O: N
332 W	342 M				
378 W		372 M		366 W	$C_0(B) + C_2N_{10}(B)$
	394 M		404 M	416 W	$C_2N_{10}(B) + N_9R(B)$
456 W	474 W	468 W			
518 W	516 W	508 W	502 M	506 M	$CC_6N(D) - CH_3C(D) + CC_5C(D)$
				536 M	$N-CH_3(B)$
		558 W		556 S	$N-CH_3(B)$
<u>656 VS</u>	<u>648 VS</u>	<u>632 S</u>	<u>652 VS</u>	<u>654 VS</u>	<u>RING BREATHING</u>
		<u>654 VS</u>	<u>692 M</u>		
	710 W	690 W			$N_9R(STR) + C_2N_{10}(B) + C_5N_7C(D)$
		730 W	726 W		
	806 M				
852 M	862 M	825 M		848 M	$C_5N_7(STR) - C_2N_{10}(STR) + NC_2N(D)$
	928 M				
958 M		950 M		958 M	$N_1C_2(STR) + C_2N_3(STR) - N_9C_4(STR)$
	970 S				
	1022 M				
1046 W		1050 W	1042 W	1058 W	
	1090 W	1094 W	1080 W		
1148 M	1142 W	1152 W		1146 W	$C_6N_1(STR) + C_5N_7 - NH_2(R)$
			1080 S	1182 W	
1222 S	1212 S	1226 M	1230 W	1224 M	$N_9R(STR) + N_3C_4(STR) + C_4C_5(STR)$
	1238 S				
<u>1272 M</u>	<u>1272 M</u>	<u>1274 M</u>	<u>1282 S</u>	<u>1280 S</u>	$C_8H(B) + C_8N_9(STR)$
1328 S		1330 M		1332 S	
1352 W	1354 M	1350 W	<u>1352 VS</u>	1354 S	$C_8N_9(STR) + N_7C_8(STR)$
<u>1382 VS</u>	1378 M	<u>1382 VS</u>		1386 S	$N_1H(B) + C_2N_{10}(STR)$
1454 SH	1458 M	<u>1404 VS</u>	1422 M		$C_5N_7(STR) + C_5C_6(STR) + C_8N_9(STR) - N_9C_4(STR)$
<u>1468 VS</u>			1468 S	<u>1460 VS</u>	$N_7C_8(STR) + C_4C_5(STR) + C_8H(B)$
1532 M	<u>1528 VS</u>	1518 M	1492 SH		$N_9C_4(STR) + N_7C_8(STR) + N_1C_2(STR) + C_8H(B)$
	1548 SH		1530 M	1536 M	
1598 M		1584 M		1596 M	$N_3C_4(STR) + N_9C_4(STR) + N_9C_4C(D)$
1692 M	1702 S		1704 SH		$C_6 = O(STR) - C_5C_6(STR) + C_4C_5(STR)$

The detailed analysis of these investigations can be summarized as follow:

1. Ring-breathing vibrations are very sensitive to substituents: 656 cm^{-1} in guanine; splitting in 3-Me-Gua, 7-Me-Gua (c.f. Tab. 1) and deuterated 9-Me-Gua (616 cm^{-1}, 648 cm^{-1}). 582 cm^{-1} in C(6)-Cl-Gua, 620 cm^{-1} in O(6)-Me-Gua, 636 cm^{-1} in C(8)-OH-Gua, 690 cm^{-1} in C(8)-Br-Gua.
2. The scattering intensity of these ring-breathing modes is so high that after an application of silver colloid solution on HPTLC silica gel plates, used for thin-layer chromatography, in situ identification of chromatogram spots is possible at the nanogram level[3].
3. The sterical interpretations of these investigations indicated that the strong enhancement is caused by the short range mechanism for distances smaller than 5 Å from the surface.

References

1. J.-M. Séquaris, J. Fritz, H.H. Lewinsky, E. Koglin, J. Coll. Interf. Sci., <u>105</u>, 417 (1985)
2. E. Koglin, J.-M. Séquaris, Topics in Current Chemistry, in press, 1986
3. J.-M. Séquaris, E. Koglin, in preparation

SURFACE-ENHANCED RAMAN SPECTROSCOPY OF MACROMOLECULES.

R. L. Garrell, K. D. Beer, M.-A. Tadayyoni and A. M. Ahern

Department of Chemistry, University of Pittsburgh, Pittsburgh, PA 15260

Polymeric coatings of metal surfaces are used to protect the surface from degradation or to modify the surface for specific applications. The extent of surface coverage, and the physical and chemical interactions between the coating and surface, define the utility and durability of the system. Surface-enhanced Raman spectroscopy (SERS) has been shown to be an effective technique for monitoring adsorption on metal surfaces, for assessing adsorbate orientation, and for probing the effects of various environmental factors on the adsorbate-substrate interactions. Few macromolecules have been studied by SERS, however. We report here results of SER spectroscopic studies of polystyrene and poly(4-vinylpyridine) on roughened silver electrodes. Our results show that SERS can be used to monitor polymer adsorption and desorption at the electrode-electrolyte interface, and that SERS can discriminate between chemically similar adsorbed species. These results suggest the feasibility of using SERS as a probe of more complex (e.g. block copolymer or polymer blend) macromolecular adsorbates in order to detect preferential adsorption or sequestering of certain sites to the metal-polymer interface.

Polystyrene (PS) was chosen for its general utility as a metal coating material, and for its lack of lone pairs that would be expected to interact strongly with the Ag surface. The polymer used was free of charged end groups, surfactants and ions (1). The SER spectra for PS on Ag were obtained as a function of applied potential. In general, the band positions and relative intensities are consistent with those for bulk polystyrene, and there does not seem to be any clear orientational preference for the aromatic rings. As the potential is made more negative (approaching the PZC) there is increased adsorption on the surface, and small changes in the relative intensities are observed. The increased adsorption near the PZC can be attributed to increased hydrophobic interactions between the less charged surface and the uncharged polymer. These relative intensity changes may result from potential-dependent changes in polymer conformation. It should be noted that all of the observed potential-dependent changes are completely reversible. These spectra demonstrate the feasibility of using SERS to probe polymer-metal interfaces, even in the absence of specific (e.g. lone pair) adsorbate-substrate interactions.

Poly(4-vinylpyridine) (P4VP) was chosen for study because of its utility as an electrode modifer (2). SERS was used to study the effect of pH and applied potential on the polymer-surface interactions. 4-ethylpyridine (4EP) was used as a model compound to aid in the SERS band assignments. Potential-dependent SER spectra of partially protonated P4VP are shown in Figure 1. As with PS and most organic materials, adsorption increases as the potential is made more negative. At potentials down to -400 meV, only neutral (non-protonated) P4VP segments are adsorbed. At more extreme negative potentials, a new band at 1672 cm^{-1} is observed which is attributed to protonated P4VP. In the case of the monomeric analog, 4EP, SERS of $4EPH^+$ can only be observed if it is trapped during the roughening procedure. In the case of the polymer, it is likely that the protonated moieties are forced onto the surface as the neutral polymer becomes increasingly adsorbed, and as coulombic attractions begin to be important. All potential-dependent changes shown here are reversible, but occur over a very long time scale compared with the monomer. This is not surprising, since desorption of the polymer must be concertive. The SERS results for P4VP indicate that the potential window (useful potential range) for the polymer-coated gold electrode is much narrower than for the uncoated surface. It is also clear that the pH and applied potential are strong determinants of the nature and extent of polymer-surface interactions. SERS is thus an extremely useful analytical tool for assessing the useful working range for modified electrodes. We are continuing work on these polymers and related redox copolymer and block copolymer coatings in order to gain insight into the relative adsorptivity of polymers, and their utility as electrode modifying materials.

1. Schaaf, V. R. and Garrell, R. L., to be published.
2. Anson, F. C. and Oyama, J. Electrochem. Soc., 27 (1980) 248.

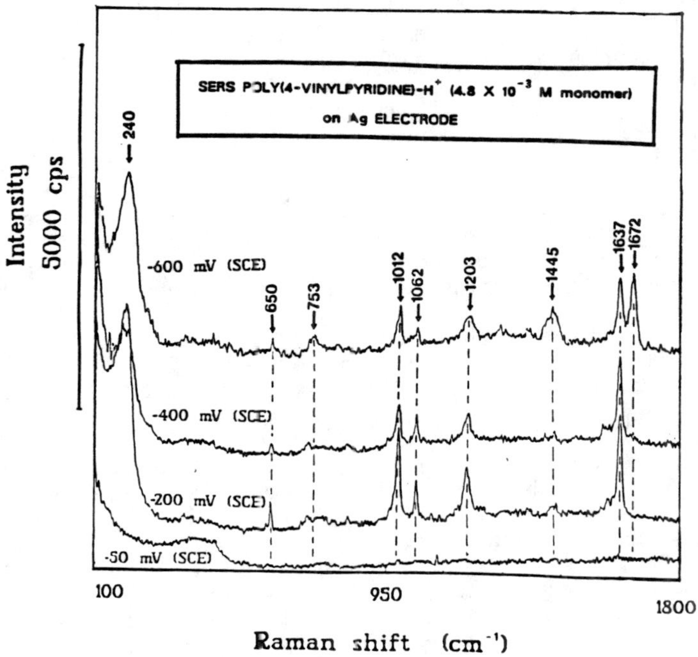

Fig. 1

ADSORPTION OF ANILINE IN ACIDIC SOLUTIONS STUDIED BY SERS

H. Shindo and C. Nishihara

National Chemical Laboratory for Industry
Tsukuba Research Center, Yatabe, Ibaraki 305, Japan

The authors have been applying Surface Enhanced Raman Scattering(SERS) to the study of adsorbates during electrochemical reduction of nitrobenzene in aqueous solutions[1]. In neutral and alkaline solutions aniline was detected on a Ag electrode in the most cathodic potential range. From the change in intensity of Raman bands of respective symmetry species upon adsorption, it was concluded that the molecule lies flatly on the electrode[2].

In this paper adsorption of aniline in acidic conditions are reported, where the molecule dominantly takes the protonated form in the solution phase. The sample solutions were prepared by dissolving aniline(5mM) in a dilute hydrohalogenic acid(HX; X=Cl, Br, I). Oxidation of Ag for SERS activation was performed at +0.2, 0.0 and -0.2V vs. SCE for Cl^-, Br^- and I^- solutions, respectively. Raman spectra were observed with a double dispersion spectrometer(JASCO R-800) with 514.5nm excitation(50mW).

The Raman spectrum of anilinium bromide is shown in Fig.1 (a) as a reference. The assignment was obtained by studying ir and Raman spectra of anilinium halides, including deuterated compounds, in solid and solution states. As the NH_3 group is considered to rotate freely around the C-N axis in the solution state, the molecule nearly has the C_{2v} symmetry. For the notation of symmetry species σ_v was chosen to be perpendicular to the phenyl ring to facilitate comparison with aniline.

When aniline was adsorbed at -0.6V from a 1M HBr solution, the spectrum shown in Fig.1(b) was observed. The sharp bands (ν_1, ν_{12}, ν_{18a}) clearly indicate that the adsorbate is anilinium ion. Marked enhancement of a_{2u} bands(ν_{16a}, ν_{10a}) again excludes the end-on geometry. When the potential was stepped to -0.85V, ν_{12} band of neutral aniline appeared at 995 cm^{-1}. The system is not in equilibrium since generation of hydrogen occurs continuously.

In a 0.1M HCl solution co-adsorption of neutral and protonated aniline was observed at around -0.5V where chloride ion starts to desorb quickly[3]. At -0.6V the molecule is present only in the neutral form on the surface. The neutral form was not observed for HI solutions. These facts strongly suggest that the anilinium ion adsorbs on the electrode surface by Coulomb's interaction with halogen ions.

For HI solutions marked enhancement was observed for ν_{19b} and ν_{8b} bands together with structured N-H stretching bands in 3000-2800cm^{-1} range. This fact suggests a different geometry of adsorption.

[1] H. Shindo et. al., Proc. ICORS-IX, Tokyo 1984, p.710.
[2] H. Shindo, J. Chem. Soc. Faraday I, 1986, in press.
[3] H. Wetzel et. al., Chem. Phys. Lett., 1981 78 392.

Fig.1 Raman spectra of $C_6H_5NH_3^+$ in solid and adsorbed states. (a); anilinium bromide(powder),(b)(c); -0.6 and -0.85V, respectively. 1M HBr solution, (d); -0.7V HI solution.

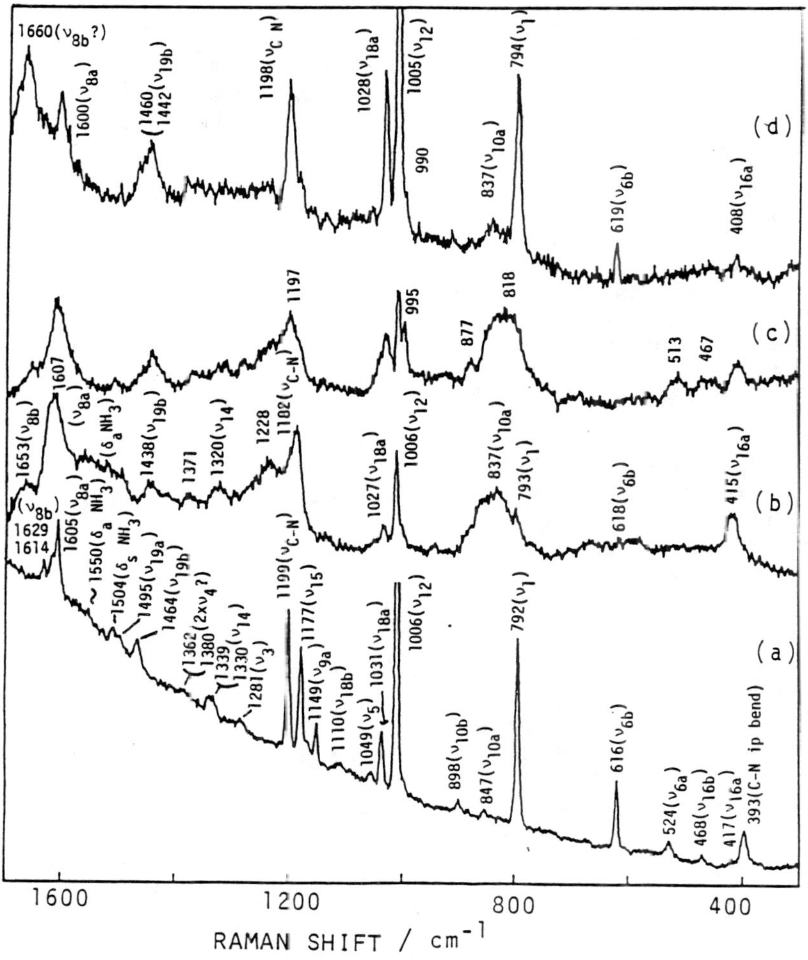

SURFACE-ENHANCED RESONANCE RAMAN SPECTRA OF TRANSITION METAL α-DIIMINE COMPLEXES.

Trevor J. Dines[a] and Robert D. Peacock[b].

a. Dept. of Chemistry, The University, Dundee, DD1 4HN, Scotland;
b. Dept. of Chemistry, The University, Glasgow, G12 8QQ, Scotland.

Studies of surface enhanced Raman scattering (SERS) have been mainly concerned with organic adsorbates, very little attention has been directed towards SERS investigations of inorganic molecules. As part of a research programme concerning SERS spectroscopy of transition metal complexes we have measured the Raman spectra of some α-diimine complexes, especially those involving the ligands 2,2'-bipyridyl (bpy) and 1,10-phenanthroline (o-phen), in the presence of colloidal silver. These species were chosen for initial study because they yield good quality Raman spectra[1] and are photochemically stable.

Ag sols were prepared according to the method of Creighton *et al.*[2] These were yellow, showing a single visible extinction band at 390 nm, and remained stable for several weeks. Addition of aqueous solutions of the adsorbates in the concentration range 10^{-7} - 10^{-4} M caused coagulation of the sols, as evidenced by changes in colour (the sequence of colour change is yellow, orange, red, violet, blue, green and grey), and the development of a shoulder at ca. 500 nm in the extinction spectrum.

The figure shows the surface-enhanced resonance Raman spectrum (SERRS) of 5×10^{-7} M $[Fe(bpy)_3]^{2+}$ in the presence of Ag sol, together with the resonance Raman (RR) spectrum obtained from a 10^{-3} M solution, both spectra were recorded using 514.5 nm excitation. The only difference between these two spectra is the appearance of a band at 1011 cm^{-1} in the SERRS spectrum, all other band wavenumbers are identical to those in the RR spectrum. The relative intensity distribution remains unchanged but the signal-to-noise ratio has increased 10-fold. Thus, the surface enhancement factor (SEF) is estimated to be 5×10^5. Since the SERRS spectrum does not show any evidence of chemisorption it is postulated that the enhancement arises from an electromagnetic mechanism resulting from an interaction of the adsorbate with the electrical double layer surrounding the colloidal Ag particles. No evidence was found for photodissociation or surface carbon formation in any of the spectra recorded. The SERRS spectra of $[Fe(bpy)_2(CN)_2]$ and $[Fe(bpy)(CN)_4]^{2-}$ are also very similar to their RR spectra and it is therefore concluded that the enhancement mechanism is the same as that for the $[Fe(bpy)_3]^{2+}$ ion.

RR spectroscopy of $[Ru(bpy)_3]^{2+}$ is seriously hampered by the luminsecence which arises from excited triplet states. We have found

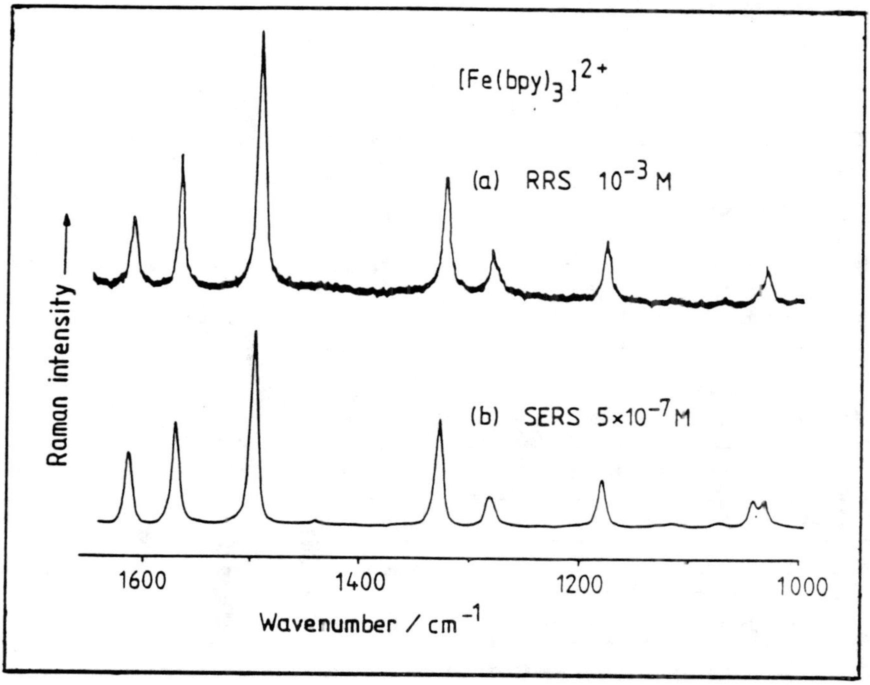

that in the presence of Ag sol the luminescence is almost completely quenched and good quality SERRS spectra can be obtained using 514.5 nm excitation. RR spectra can only be obtained with shorter wavelength excitation (e.g. 457.9 nm), where the resonance enhancement is weaker. As was found for the Fe complexes, the SERRS and RR spectra are very similar and a substantial improvement in signal-to-noise ratio is observed.

1. R. J. H. Clark, P. C. Turtle, D. P. Strommen, B. Streusand, J. Kincaid and K. Nakamoto, Inorg. Chem., 16, 84 (1977).

2. J. A. Creighton, C. G. Blatchford and M. G. Albrecht, J. Chem. Soc. Faraday II, 75, 790 (1979).

SPIN- AND REDOX EQUILIBRIA OF CYTOCHROME C ADSORBED AT THE Ag/ELECTROLYTE INTERFACE PROBED BY SURFACE ENHANCED RESONANCE RAMAN SPECTROSCOPY (SERS)

Peter Hildebrandt* and Manfred Stockburger

Max-Planck-Institut f. biophysikalische Chemie, Abt. Spectroskopie, D-3400 Goettingen, F.R.G.
*present address: Department of Chemistry, Princeton University, Princeton, New Jersey 08544

The application of Surface Enhanced Raman Spectroscopy (SERS) to complex biological molecules raises the question if their native structural and functional properties are modified after adsorption. In this work we have investigated this problem for cytochrome c(cyt) which acts as an electron carrier in the respiratory chain of organism. We have used different types of Ag/electrolyte interfaces (colloidal Ag at room and low temperature, Ag electrode) where the adsorbed cyt gives rise to an intense vibrational spectrum of the heme chromophore due to the combination of the SER and the molecular resonance Raman (RR) effect (SERS) [1-4].

Spin state equilibria: While in both the oxidized and the reduced form of the native enzyme (cyt^{3+}, cyt^{2+}) the central iron is in the low spin (LS) configuration, adsorption at the Ag/electrolyte interface leads to a partial transition to the high spin (HS) state in cyt^{3+} and cyt^{2+}. This is concluded from the appearance of the corresponding HS state marker bands in the SER spectra. The intensity ratio of the HS and LS state marker bands (e.g. ν_{10} at 1624 (cyt^{3+}HS) and 1637/1642cm^{-1} (cyt^{3+} -LS);Fig.1) changes reversibly with the temperature indicating thermal spin equilibria for the adsorbed cyt^{3+} and cyt^{2+}. The partial LS→HS transition is not accompanied by irreversible distortions of the heme-protein interactions since, for example, the characteristic "acid" transition of the native cyt also occurs in the adsorbed state. On the other hand denaturation of the adsorbed cyt obviously takes place at negative potentials below -0.7V (vs. SCE) on the Ag electrode.

The LS→HS transition of the adsorbed cyt is a general effect. It is also observed in many other heme proteins and iron porphyrin compounds which have different axial ligands or protein environments. A possible explanation for this phenomenon will be discussed.

Redox equilibria: The redox reactions of the adsorbed cyt were studied at the Ag electrode by varying the potential. From the coexistence of both spin states in cyt^{3+} and cyt^{2+} it can be concluded that all four states are involved in the electron transfer process. Therefore it is necessary to determine their relative concentrations as a function of the potential which, in principle, can be obtained from the relative intensities of the characteristic oxidation- and spin-state marker bands ν_4 and ν_3 in the SER spectra. However, since these bands are partially overlapping (Fig.2) they have to be disentangled by band fitting procedures. For this purpose SER spectra were recorded at higher spectral resolution. In these experiments a rotating Ag electrode was used in order to reduce thermal heating by the exciting laser beam.

Evaluation of the data by band fitting revealed that the redox potential for the HS state is more negative than for the LS state. However, in both cases the data could not be fitted to the Nernstian equation for a one-electron transfer. An interpretation for this behavior will be presented. Furthermore, the implications of our SER study on cyt for the biological redox process will be discussed.

References:

[1] Cotton, T.M., Schultz, S.G., van Duyne, R.P.; J. Am. Chem. Soc. 1980, 102, 7960.

[2] Hildebrandt, P., Stockburger, M. In: "Spectroscopy of Biological Molecules", Alix, A.J.P., Bernard, L., Manfait, M., eds.; Wiley, New York, 1985, p.25.

[3] Smulevich, G., Spiro, T.G.; J. Phys. Chem. 1985, 89, 5168.

[4] Hildebrandt, P., Stockburger, M.; submitted to J. Phys. Chem.

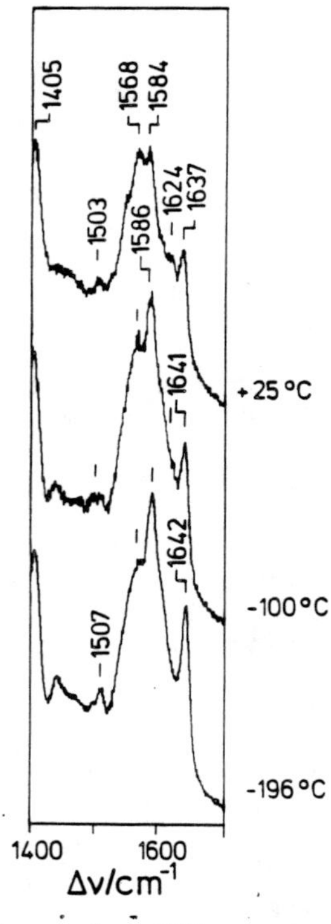

Figure 1: SER spectra of cyt^{3+} adsorbed on Ag glycerol sol at different temperatures (λ_o = 514nm)

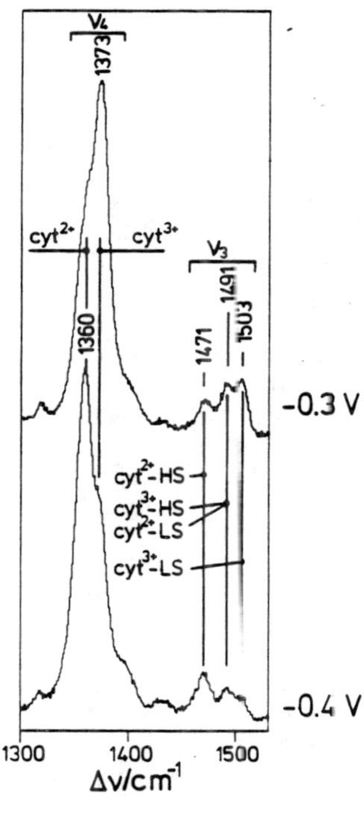

Figure 2: SER spectra of cyt adsorbed on the rotating Ag electrode at different potentials (λ_o = 413nm)

SERS AS A PROBE OF SURFACE GEOMETRY EFFECTS ON THE POLYMERIZATION OF ACRYLIC ACID ON SILVER

J.S. Suh and K.H. Michaelian

Energy, Mines & Resources Canada, CANMET,
Edmonton Coal Research Laboratory,
P.O. Bag 1280, Devon, Alberta, CANADA TOC 1E0

Surface-enhanced Raman scattering (SERS) spectroscopy provides a powerful means for the study of surface chemical reactions, since these reactions result directly in changes in SERS spectra of the adsorbed molecules. In this paper we report surface geometry effects on the polymerization of acrylic acid (AA) adsorbed on silver colloid surfaces using SERS spectroscopy.

Some evidence of polymerization of AA is found in the CH vibrations in the SERS spectrum (Fig. 1): The band at 2928 cm^{-1} corresponds to CH stretching of the polymer, and the band at 1348 cm^{-1} is due to polymer CH_2 wagging. Moreover, the band at 1164 cm^{-1} can be assigned to the polymer skeletal vibration. These results lead to the conclusion that some AA is polymerized on the silver surfaces.

The νCH bands of AA adsorbed on the surface at three different overall concentrations are shown in Fig. 2. The SERS spectra were observed immediately after adding one drop of adsorbate to the colloid and shaking. At high concentration (Fig. 2, upper), only the νCH bands corresponding to the monomer are observed initially (upper left); as time passes, the νCH band of the polymer appears (upper right). At medium concentration (Fig. 2, middle), the νCH bands corresponding to monomer and polymer were both observed. The relative intensity of the polymer νCH band increased with time. At low concentration (Fig. 2, bottom), only the polymer νCH band was observed.

The surface geometry of AA can be determined by observing the intensity of νCH bands in the SERS spectra. In the SERS spectrum of AA at high concentration, the strong νCH bands of the monomer mean that the ethylene plane of AA is at least tilted with respect to the surface, since in the flat down geometry the CH stretching vibrations have pure tangential Raman polarizability components and they are almost absent in SERS spectra [1]. In the tilted geometry of AA, no polymerization takes place. In the SERS spectrum at low concentration, only the νCH band corresponding to the polymer is observed, although the strong ν(C=C) band at 1632 cm^{-1} may signify the existence of unpolymerized AA on the surface. This means that the unpolymerized AA is lying down flat on the surface at low concentration. From these observations, we can conclude that polymerization takes place only when the ethylene plane of AA lies down flat on the surface. In the sample of high or medium concentration, polymerization of AA with passage of time is either due to the surface geometry change of AA from standing up to lying down flat on the surface, or simply the surface geometry change of AA accompanied by the aggregation of colloid particles. Anyway, in the flat down geometry of AA on the surface, the π-electrons of the ethylene group can easily interact with the electrons of the metal surface. This interaction probably brings about some intermediate state of polymerization.

REFERENCE

1. M. Moskovits and J.S. Suh, J. Phys. Chem., 88, 5526 (1984).

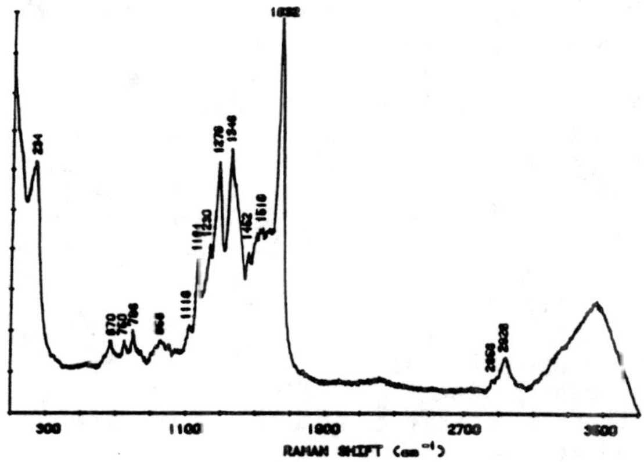

Fig. 1. SERS spectrum of acrylic acid adsorbed on silver colloid surfaces. The overall concentration is about $6x10^{-4}$ M.

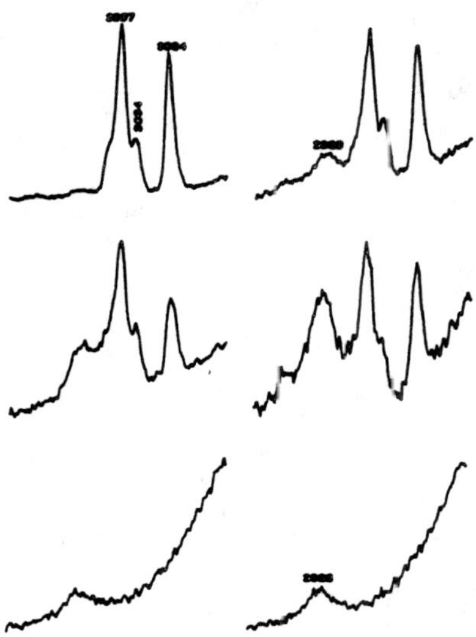

Fig. 2. The CH stretching region of the SERS spectra of acrylic acid at overall concentration of $3.6x10^{-3}$ M (upper), $2.1x10^{-3}$ M (middle) and $2.1x10^{-4}$ M (bottom). The spectra in each row were observed sequentially for the same sample.

CHARGE-TRANSFER BAND AND SERS MECHANISM FOR PYRIDINE-Ag SYSTEM

Haruka Yamada, Hisao Nagata and Kazuhiro Toba

Department of Chemistry, Kwansei Gakuin University, Nishinomiya 662, Japan

As the main mechanisms of SERS, electromagnetic(EM) effect and charge-transfer (CT, or chemical) effect have generally been accepted. The CT mechanism, however, has been supported mostly by indirect evidence and the CT levels have been detected only for a few cases by EELS or optical reflectance measurements. Several attempts to observe the CT band have so far been made without success. The present work is a challenge to observe a CT band for pyridine adsorbed on Ag film under suitable conditions.

The Ag island film of ca. 4 nm in mass thickness, which was prepared on glass or quartz plates by vacuum evaporation (10^{-5} torr, 300 K), showed an extinction maximum at ca. 500 nm. After pyridine was adsorbed on the Ag film at 300 K from the vapor and pumped out to remove excess or pysisorbed species, only monomolecularly chemisorbed species were left on the Ag surface. The transmission spectrum of this sample showed two extinction maxima, overlapping each other, one of them at ca. 450 nm was ascribed to the shifted Ag band by coating pyridine and the other at ca. 600 nm to a new band produced by the pyridine-Ag interaction.

After the Ag film having a maximum at ca. 500 nm as shown in Fig. 1(a) was heated in H_2 (450 K, for 1 h), a rather sharp maximum at ca. 410 nm (b) was observed, which was probably due to the Ag ellipsoidal particles. When pyridine was adsorbed on the Ag (heated in H_2) film, a new peak at ca. 600 nm was clearly observed with the band at 410 nm [Fig. 1(c)]. The new peak coincides well with that found on the Ag film without heating in H_2. By heating the sample to desorb pyridine the band at ca. 600 nm disappeared (d), but it did appear again as a result of readsorption of pyridine. The new band at ca. 600 nm is, therefore, confirmed to be the pyridine-Ag interation band.

Clear Raman spectrum of pyridine adsorbed on the Ag film was observed as shown in Fig. 2, which is identified as SERS of pure chemisorbed (N-bonded) pyridine, with the enhancement factor up to 10^6. The stronger the spectrum, the larger the 600 nm absorption band. The SERS spectrum was stronger on the non-heated Ag film than on the heated one. The excitation profile measured for the 1010 cm^{-1} band appeared overlapping almost with the 600 nm band. Since the N-bonded pyridine does not absorb in visible region, the 600 nm band should be ascribed to the CT band of Ag-pyridine.

We repeated the measurements on various Ag films of 2 ~ 6 nm thickness. In order to observe the CT band clearly, the Ag film having a rather sharp extinction peak and less absorption at the longer wavelength side had to be used. When water or ethanol vapor was adsorbed on the same Ag films, no absorption peak was observed in 450 ~800 nm region. If the 600 nm band for pyridine-Ag is caused from impurity of the material, SERS based on that impurity should be detected. Since we observed only the typical SERS of N-bonded pyridine, the 600 nm band should not be assigned to any impurity. It does not seem to be caused by aggregation of the Ag particles too, because of its reversible appearance by adsorption and desorption of pyridine regardless of the adsorption time.

The number of molecules adsorbed on the Ag film can be approximately estimated by measuring the π- π^* band of pyridine at 250 nm to be $(2 \sim 15) \times 10^{15}$ molecules/cm^2, which is considerably larger than the value on flat surface. The molar extinction coefficient of the CT band is obtained to be $10^3 \sim 10^4$. The observed CT band at ca. 600 nm (2.1 ev) is compared with the CT bands by EELS; 1.9 ev for pyridine-Ag(111) and 2.5 ev for pyridine-Ag evaporated.

Based on these experiments, the contribution of each effect to the total SERS enhancement factor 10^6 can be roughly estimated; $\sim 10^3$ for CT effect, 5 ~ 50 for increased number of adsorbed molecules, and $10 \sim 10^2$ for EM effect.

The reason why the CT band had not been detected as optical absorption so far might be considered as follows: (i) For the colloidal systems, adsorption of pyridine on Ag always gives rise to aggregation of Ag particles which shows strong extinction in visible region and obscures the CT band. (ii) For vacuum systems, most of the adsorbed pyridine molecules are often physisorbed at low temperatures. Since the energy required for chemisorption can be provided by laser irradiation, SERS is easily observed, but optical absorption or reflection is hard to be detected.

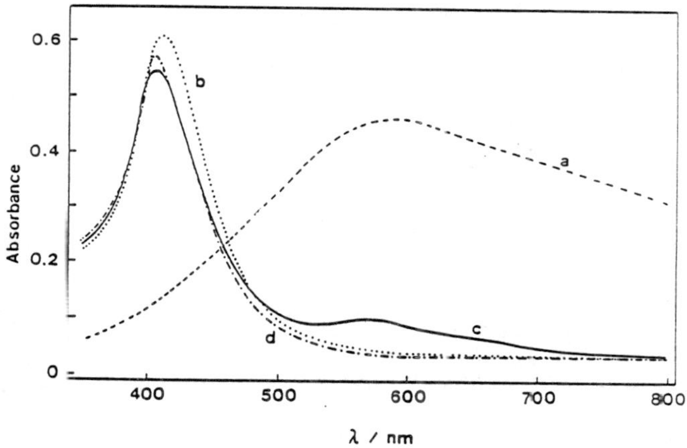

Fig.1. Absorption spectra of; (a) Ag film evaporated, (b) Ag film heated in H_2, (c) pyridine adsorbed on (b), (d) heated (c) to desorb pyridine.

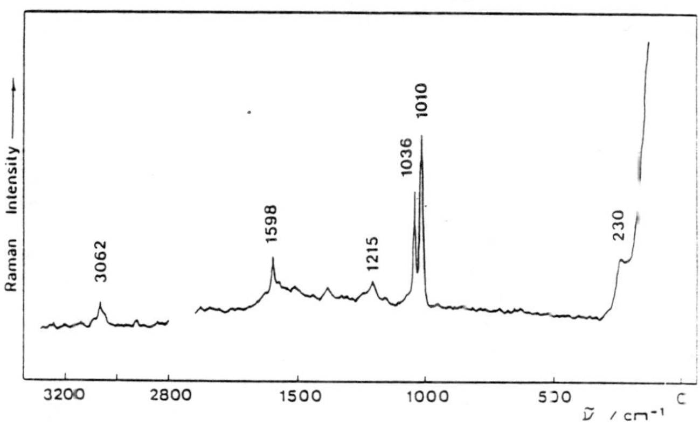

Fig.2. SERS of pyridine adsorbed on Ag film (514.5nm ex.)

SURFACE CHARACTERIZATION OF ION-IMPLANTED POLYETHYLENE FILMS BY THE USE OF SERS EFFECT

H. Ishida, G. Katagiri, H. Fukuda, S. Tomita, K. Yoshida and A. Ishitani

Toray Research Center, Inc.
1-Chome, Sonoyama Otsu, Shiga 520, Japan

Surface enhanced Raman scattering(SERS) phenomenon has attracted the interest of analytical chemists because of its potential as a high-sensitive probe for surface analyis[1]. We have recently succeeded in observing the enhanced Raman scattering from the outermost surface layer of carbon materials such as highly oriented pyrolitic graphite, glassy carbon, diamond and carbon fibers by evaporating Ag island films (Ag overlayer method)[2]. In this method, we can expect the enhanced Raman scattering from the surface of substrate materials due to the excitation of surface plasmon localized on Ag island film. In the present paper, we have applied the same method to the surface characterization of ion-implanted polyethylene films. The surface structure of ion-implanted polyethylene films was parallelly characterized by XPS, EELS and SIMS in order to confirm the Raman results. Surface morphology of Ag island films was examined by FE-SEM.

EXPERIMENTAL

The surface of polyethylene films(190μm thickness) were irradiated with 60keV O_2^+ and 150keV C^+ ions at the ion doses ranging from $2x10^{15}$ to $5x10^{16}$ ions/cm^2. The Ag island films were prepared by the vacuum evaporation method. The Raman spectra of the polyethylene films were recorded with the use of the p-polarized 5145Å line of Ar ion laser with the incident angle of 70°.

RESULTS AND DISCUSSION

Fig. 1 shows the Raman spectra of O_2^+ ion implanted polyethylene films. The Raman intensity of a broad band newly observed around $1550cm^{-1}$ increases with increasing the ion dose. This band is in good agreement with that of evaporated amorphous carbon(α-carbon). The SIMS depth profile analysis of the implanted polyethylene film indicates that the amorphous carbon layer exists within a few thousands Å depth from the surface.

Fig. 2 shows the Raman spectra of O_2^+ ion-implanted polyethylene films covered with Ag island films. The Raman scattering intensity of polyethylene substrate is negligible compared to the enhanced Raman bands. These enhanced Raman spectra agree closely with that of diamond-like carbon(i-carbon) which is prepared recently by chemical vapor deposition and ion beam sputtering methods[3]. The similar Raman spectra were observed from the surface of C^+ ion-implanted polyethylene films.

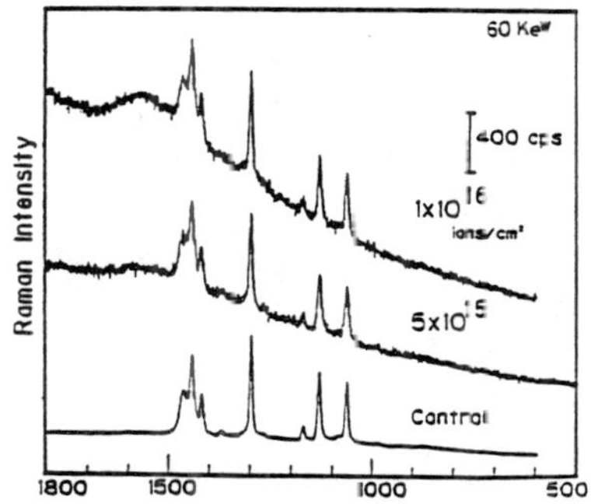

Fig.1
Raman Spectra of O_2^+ Ion Implanted Polyethylene Films

Fig. 3 shows the XPS spectra and C1s electron energy loss spectra (X-EELS) of O_2^+ ion implanted polyethylene films. The surface oxygen contents are less than 7 atomic %, much lower than expected from the O_2^+ doses. The X-EELS spectrum of control polyethylene shows an energy loss peak around 20eV due to the plasmon of sp3 valence electrons. The shift of the loss peak toward the higher energy by the ion implantation can be well explained by the formation of i-carbon.

The presence of diamon-like carbon layer has been revealed in the outrmost surface layer of the ion-implanted polyethylene films by the Ag overlayer method. The result here seems to be of great interest also from the viewpoint of surface modification of organic materials.

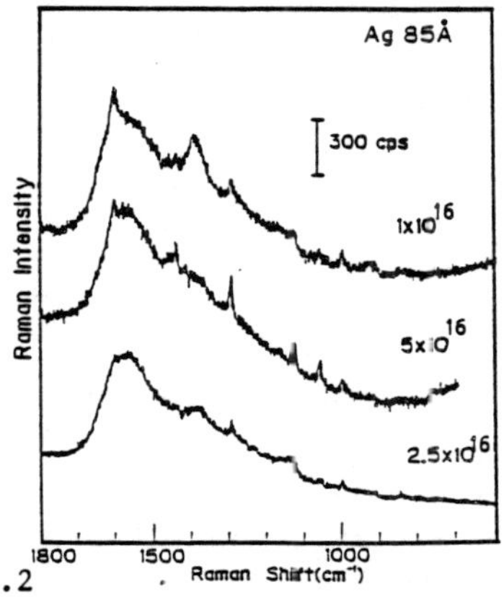

Fig.2
Raman Spectra of O_2 Ion Implanted Polyethylene Films Covered with Ag Island Films

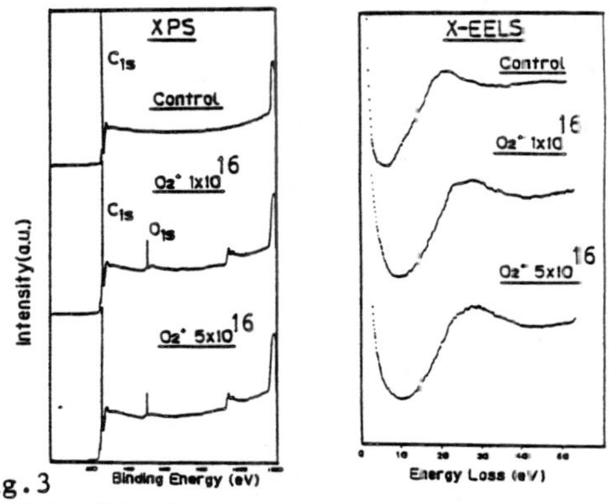

Fig.3
XPS and C_{1s} Energy Loss(X-EELS) Spectra of O_2^+ Ion Implanted Polyethylene Films

REFERENCES

1) H. Ishida and A. Ishitani, Appl. Spectrosc., 37, 450 (1983).
2) H. Ishida, H. Fukuda, G. Katagiri and A. Ishitani, Appl. Spectrosc., March-April issue (1986).
3) S. Matsumoto, Y. Sato, M. Tsutsumi and N. Setaka, J. Mater. Sci., 17, 3106(1982).

STRUCTURES OF ADSORBED 2,2'-BIPYRIDINE AND PYRIDINE ON Ag COLLOIDS STUDIED BY SERS SPECTROSCOPY.

Munsok KIM and Koichi ITOH

Department of Chemistry, School of Science and Engineering, Waseda University, Shinjuku-ku, Tokyo 160, Japan

Surface enhanced Raman scattering (SERS) spectroscopy can be used to moniter structural changes of adsorbed molecules on metal-liquid interfaces (e.g., electrodes in electrolytes and colloids suspended in a liquid). Recently, we reported the potential dependence of the SERS spectra of 2,2'-bipyridine adsorbed on the Ag electrode from its 0.1 M KCl, KBr and KI solutions [1,2] and proved that a series of transformations are taking place between four kinds of molecular states (Type I - IV) of the adsorbed 2,2'-bipyridine on the electrode surface. In this paper, we extend the SERS study to 2,2'-bipyridine (and also pyridine) adsorbed on Ag colloids in slightly alkaline and acidic solutions in the presence of a variety of anions (Cl^-, I^-, SO_4^{2-} and NO_3^-).

Aqueous Ag colloids was prepared by reducing silver nitrate with sodium borohydride as previously reported [3].

As the pH of the colloidal solution was altered from 8.5 (Fig. 1(A)) to 2.8 (Fig. 2(B)) with HCl, the SERS spectra of 2,2'-bipyridine showed drastic changes in their features, indicating that 2,2'-bipyridine molecules on Ag colloids take on two different types of adsorbateds. One type of adsorbate, which is characterized by SERS bands in Fig. 1(A), can be found out throughout the entire pH range from 8.5 to 2.8. This adsorbate was identified as the type I species found for the Ag electrode sysytem, taking on a coplaner cis conformation and has a molecular state almost identical with that of an argentous complex, Ag(2,2'-bipyridine)$_2$NO$_3$. In acidic solutions, the other type of adsobate, which is characterized by the SERS bands in Fig. 1(B) except those at 1011 and 765 cm^{-1}, appears and increases with lowering the pH. This adsorbate has an identical spectroscopic feature with that of the type III species found for the Ag electrode system. Further studies on the effect of both the pH and co-existing anions (i.e., Cl^-, I^-, SO_4^{2-}, NO_3^-) on the formatin of the second speices revealed that both halide and hydrogen ions play a crucial role in the formation of the adsorbates. In the bulk acidic solution at pH 2.8, 2,2'-bipyridinium ion is a predominant species. The SERS bands due to the adsorbed species from the solution cannot be ascribed to the 2,2'-bipyridinium ion. Thus we can conclude that in this case 2,2'-bipyridine molecules, halide ions and hydrogen ions are adsorbed toghether at active sites (or adatoms) on the colloidal silver surface, forming a strongly chemisorbed surface complex.

Figure 2 show the pH dependence of the SERS spectra of pyridine adsorbed on Ag colloids. On lowering the pH from 5.9 to 4.5 by HCl, the 1037-cm^{-1} band at pH 5.9 (Fig. 2(A)) decreases remarkably its intensity and a new band appears at 1025 cm^{-1} (Fig. 2(B)). On the other hand, in acidic solutions without Cl^-, where the pH was adjusted with H_2SO_4, the positions of the SERS bands at pH 5.1 (Fig. 2(C)) remain unchanged even at pH 3.6 (Fig. 2(D)). Therefore, it was found that halide ions play an essential role in the appearance of the 1025-cm^{-1} band. In the Ag electrode system, both strongly chemisorbed pyridine [4] and adsorbed pyridinium chloride [5] give rise to a distinct SERS band around 1025 cm^{-1}. From the similality in spectral feature of the pH dependence observed in Figs. 2(A) and (B) to the results of ref. [4], the 1025-cm^{-1} band in Fig. 2(B) can be ascribed to pyridinium chloride adsorbed on the surface of the Ag colloids. Although the Cl^- ion plays an impotant role in the formation of the type III of 2,2'-bipyridine and the 1025-cm^{-1} species of pyridine, the proposed structures of both species are quite different, which can be explained in terms of their coordination properties.

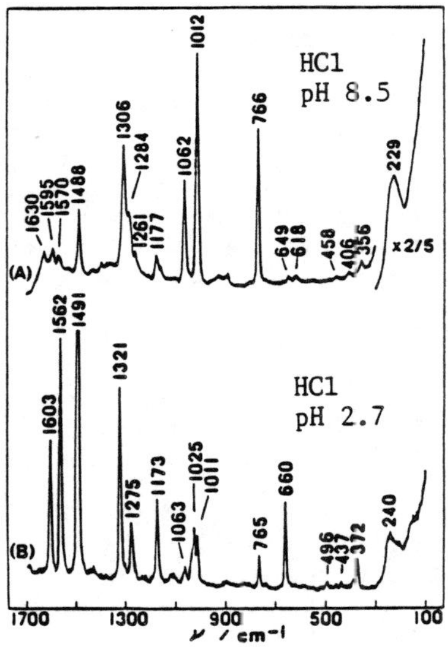

Fig. 1 SERS spectra of 2,2'-bipyridine adsorbed on Ag colloids. Excitation with 514.5 nm (20 mW).

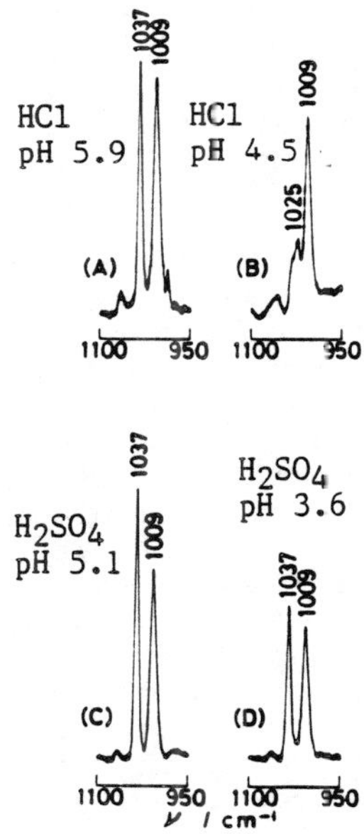

Fig. 2 SERS spectra of pyridine adsorbed on Ag colloids. Excitation with 514.5 nm (40 mW).

References

1. M. Kim and K. Itoh, Chem. Lett. 1984, 357.
2. M. Kim and K. Itoh, J. Electroanal. Chem. 188, 137 (1985).
3. J.A. Creighton, C.G. Blatchford and M.G. Albrecht, J. Chem. Soc. Faraday II 75, 790 (1979).
4. For example, M. Fleischmann and I.R. Hill, J. Electroanal. Chem. 146, 353 (1983).
5. For example, H. Chang and K.C. Hwang, J. Am. Chem. Soc. 106, 6586 (1984).

SURFACE ENHANCED RESONANCE RAMAN OF HAEM PROTEINS ON COLLOID SILVER

J. de Groot and R.E. Hester

Department of Chemistry, University of York, York YO1 5DD

Surface enhanced resonance Raman (SERR) spectra are reported for oxyhaemoglobin and cyanomethaemoglobin adsorbed on aqueous silver colloids, with excitation wavelengths in resonance with Soret and β bands of oxyhaemoglobin. The SERR spectra observed give rise to Raman bands at wavenumbers which correspond with the resonance Raman bands of oxyhaemoglobin. Raman excitation profiles of the SERR bands of oxyhaemoglobin in the region 457.9 nm to 568.2 nm indicate that some SERR bands are in resonance at ~490 nm, which may be due to the formation of a μ-oxo-dimer as reported by Spiro and Smuleuich.(1) However, the excitation profiles indicate that most of the SERR bands are in resonance at ~540 nm, which is highly indicative of the β band absorption of oxyhaemoglobin, suggesting the possible coexistance of denatured and native oxyhaemoglobin. The excitation profiles also indicate that the polarized modes in the resonance Raman spectra of oxyhaemoglobin are anomalously enhanced at longer wavelengths. This behaviour is similar to that seen in SERR spectra of cytochromes as reported by Stockburger and Hildebrandt.(2) We also have obtained SERR spectra of cyanomethaemoglobin and found these to be consistent with resonance Raman spectra of the native form. However, as with oxyhaemoglobin, cyanomet-haemoblogin gives rise to anomalously high intensities of the bands associated with the symmetric modes. This suggests that major structural changes may be occurring in the haem protein upon addition to the silver colloid, although the strong Fe-CN bond is unlikely to lead to formation of a μ-oxo-dimer in this case.

(1) T.G. Spiro and G. Smuleuich, J. Phys. Chem., in Press.
(2) M. Stockburger and P. Hildebrandt, J. Phys. Chem., in Press.

COORDINATION CHEMISTRY OF THIOSEMICARBIZIDE ON COPPER AND SILVER SURFACES - A SERS STUDY.

J. Leahey, Y. G. Lee and B. H. Loo

Department of Chemistry, The University of Alabama in Huntsville, Huntsville, AL. 35899

Thiosemicarbizide (tsc) exhibits interesting coordination chemistry with metal ions. It can bond through S, or through both S and N atoms. Both of these possibilities have been observed in metal complexes of tsc. The molecule exists in two conformational forms, cis (I) and trans (II). The trans configuration is reported in monodentate complexes of AgCl(tsc) and $HgCl_2(tsc)_2$, where x-ray structural determination shows that coordination is through the S atom [1,2]. Other metal complexes of tsc, $ZnCl_2(tsc)$, $ZnCl_2(tsc)_3$, $HgCl_2(tsc)$, $Ni(tsc)_2$, $Ni(tsc)_2SO_4 3H_2O$, $Cu(tsc)_2Cl_2$, $Cu(tsc)Cl_2$, are reported to have bidentate coordination through S and N atoms [3-10].

We have previously used SERS to study several structurally related compounds of tsc on Cu and Ag surfaces (thiourea, thioacetamide, N,N-dimethyl-thioformamide). We found that the ν(C=S) bands of these molecules showed significant shifts when adsorbed on Cu and Ag [11]. The observed frequency shifts were similar to those reported for their organometallic analogs in which S-metal bonding occurred. We concluded that these molecules were bonded through their S atoms to the metal surfaces.

Table 1 - Vibrational frequencies (in cm^{-1})

Solid		SERS		Complexes			Assignments[f]
IR[a]	Raman[b]	Ag	Cu	$HgCl_2X^c_2$	$HgCl_2X^d$	$NiCl_2X^e_2$	(PED.%)
1618	1614	1584		1600	1628	1650	δ NH_2(70)νCN(23)
1530	1530	1534	1531	1545	1596	1618	ν CN(50)νNH(18)
1480	1500	1497	1474	1502	1548 1456	1585 1445	δ NH(43)νCN(29) δ NH_2(12)
		1404 1343	1395				
1283	1274	1303	1295	1302	1380	1350	δ NH(27)νCN(21) ρ NH_2(18)νCS(15)
1160	1170	1164		1188	1196	1160	ν NN(61)$\rho$$NH_2$(26)
995	1008	1013	1096	1016	1038	1005	ρ NH_2(35)νNN(22)
			1014		999		ν CN(22)
800	807	797	799	796	680	707	ν CS(35)$\omega$$NH_2$(28) δ NCN(13)
	638	615	599				
503	507	503	501	496	600	615	δ NCN(44)νCS(20) δ CS(19)
408	406	407	403	408	444 368	438 380	δ CS(68)
206	288	310	289	315	302	290	δ CNN(72)δNCN(21)

PED = Potential energy distribution. X=TSC

ν = strech; δ = deformation; ω = wagg; ρ = rocking

[a,f] Ref. 9; [b,c,d] Ref. 10, Raman; [e] Ref. 8, IR.

For tsc complexes, the predominantly ν(C=S) at 800 cm^{-1} band shows a small decrease (10 cm^{-1}) in the monodentate coordination and a large decrease (100 cm^{-1}) in the bidentate coordination [8-10].
Based on our SERS results (Table 1), we conclude that the tsc molecule is S-bonded to Cu and Ag surfaces and the adsorbed molecule assumes the trans configuration.

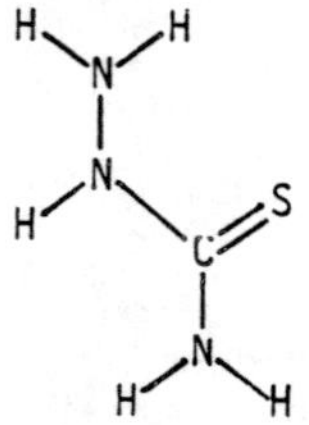

(I)

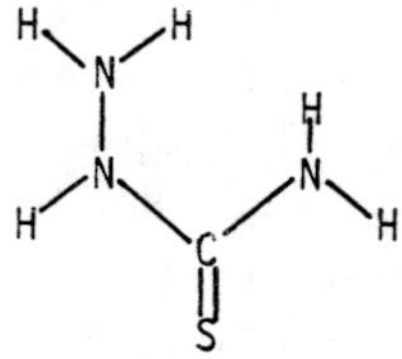

(II)

Acknowledgement: The authors thank International Copper Research Association for the support of this work.

References

1. G. C. Gasparire, A. Mangia, A. Musathi and M. Nardellei, Acta Cryst., 24, 367 (1968).
2. C. Chieh, Can. J. Chem., 55, 1583 (1977).
3. J. M. Campbell, Trans. Chem. Rev., 15, 279 (1975).
4. C. Chieh and D. H. Cowell, Can. J. Chem. 55, 3898 (1977).
5. M. Cavalca, M. Nardelli and G. Brianchi, Acta Cryst., 13, 688 (1968).
6. R. Gronbaek and S. E. Rasmussen, Acta Chem. Scand., 16, 2325 (1962).
7. L. Cavelca, M. Nardelli and G. Fava, Acta Cryst., 15, 1139 (1962).
8. K. Geetharani and D. N. Sathyanarayana, Aust. J. Chem., 30, 1617 (1977).
9. D. N. Sathyanarayana, K. Volka and K. Geetharani, Theochem. 71, 517 (1981).
10. C. Chieh, L. P. C. Lee and C. Chiu, Can. J. Chem., 56, 2526 (1978).
11. B. H. Loo, Chem. Phys. Lett., 89, 346 (1982); B. H. Loo and Y. G. Lee, Chem. Phys. Lett., 112, 580 (1984); B. H. Loo, Y. G. Lee and Z. Yazid, Chem. Phys. Lett., 114, 405 (1985).

SURFACE ENHANCED RAMAN SCATTERING (SERS) OF BENZOIC ACID ADSORBED ON COLLOIDAL SILVER PARTICLES.

M.Pagannone, B.Fornari and G. Mattei

C.N.R., I.M.A.I.-Area della Ricerca di Roma, P.O.Box 10, Monterotondo Scalo (Roma), Italy.

In this work we report our results on SERS (1) from benzoic acid adsorbed on aqueous silver sol particles. By carefully avoiding the disturbing effect due to graphitic carbon contamination frequently observed at about 1350 and 1550 cm^{-1}, we have obtained reliable results throughout the investigated frequency region. We give a complete assignment of the SER spectrum and discuss the structure and the orientation of the chemisorbed species.

Our SER spectrum of benzoic acid adsorbed on silver sol, see Fig. 1,

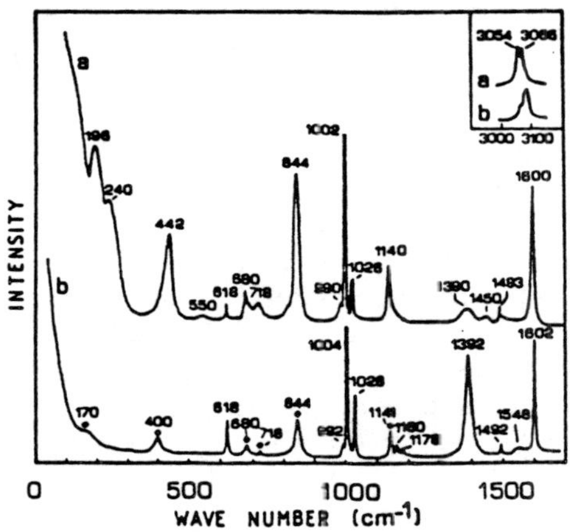

Fig.1. SER spectrum of benzoic acid adsorbed on silver sol (a) in comparison with the Raman spectrum of a saturated aqueous solution of sodium benzoate (b). In the insert the C-H stretching frequency region is shown.

has a stronger similarity to the Raman spectrum from a saturated aqueous solution of sodium benzoate, to which comparison is then made hereafter, than to those from either solid and dissolved (in alcohol) benzoic acid or solid sodium benzoate. There are no bands above 1600 cm^{-1} attributable to the acid C=O stretching mode. At frequencies higher than about 600 cm^{-1} there is almost a one to one correspondence between the bands in the two spectra. All but those at 1450 cm^{-1} in the SER spectrum and at 1548 cm^{-1} in the benzoate solution have the same frequency position to within 2 cm^{-1}. In Fig.1(b) asterisks mark the benzene ring modes known to be substituent sensitive.

The salient features of our SER spectrum can be summarized as follows. All of the bands can be reasonably assigned on the basis of the extensively studied vibrational spectra of benzoates (2,3). The modes arising mainly from vibrations in the benzene ring are only little affected by adsorption and preserve their intensity, exception made for the B_2 mode at 618 cm^{-1}. The largest enhancement is experienced by those modes that draw their intensity from mixed ring and substituent motions. A prominent band appears at 240 cm^{-1}. The symmetric OCO stretching band ν_S is rather small, even smaller and shifted to lower frequencies by as much as 100 cm^{-1} as compared to the free benzoate ion in solution is the asymmetric mode ν_A. Finally, the C-H stretching vibrations are present with appreciable intensity.

On the basis of the present results we can draw the following conclusions about the nature, the adsorption geometry and the orientation of the surface species. Undoubtedly benzoic acid adsorbs on Ag sol particles in the form of benzoate ion, as evidenced by the absence of the C=O stretching mode and by the strong similarity between the SER and the solution spectra. From the metal-ligand band at 240 cm^{-1} and from the remarkable increase in relative intensity of the modes involving motions in the substituent group we obtain strong evidence that bonding to the surface occurs largely through the carboxylate group.

However, the actual orientation of the adsorbate must be different from an "end-on" configuration with the benzene ring standing up. As a matter of fact, most of the results summarized above can be qualitatively explained by assuming a bidentate chelating bonding structure (4), in which the two carboxylate oxygens bond to the same Ag surface site and the orientation is such that the three molecular axes x, y and z are all canted with respect to the silver surface. Some additional interaction is likely to occur between the surface and such an inclined benzene ring (see the small broadening and shift to lower frequencies of the ring modes). This could be the reason why the prominence expected for the SER modes arising from the groups strongly interacting with the surface is not acquired (in addition to the mixed modes already mentioned) also by the symmetric stretching ν_S (localized mainly on the COO^-group) and why the stiffening (shift to higher frequencies) characterizes the C-C' bending modes only. The correlation between the adsorbate orientation and the intensity of the SER bands involving motions in the ring (e.g., decrease of the B_2 mode at 618 cm^{-1}, unchanged relative intensities of both the C-H in plane and the C-H out of plane motions) is in keeping with the surface selection rules (5). The same consideration can be made as far as the intensity of ν_S and ν_A is concerned.

References:

(1) Surface Enhanced Raman Scattering, R.K.Chang and T.E.Furtak, eds. (Plenum, N.Y.,1982).
(2) J.H.S.Green, Spectrochim.Acta 33A, 575 (1977).
(3) M.Machida, A.Kuwae, Y.Saito and T.Uno, Spectrochim.Acta 34A, 793 (1978).
(4) K.Nakamoto, Infrared and Raman Spectra of inorganic and Coordination Compounds (Wiley, N.Y., 1978).
(5) See, e.g., M.Moskovitz and J.S.Suh, J.Phys.Chem. 88, 5526 (1984).

EVIDENCE FOR A PHOTON-DRIVEN CHARGE-TRANSFER MECHANISM IN THE SURFACE ENHANCED RAMAN SCATTERING FROM DABCO AT A SILVER ELECTRODE

D. Guzonas, D.E. Irish, and G.F. Atkinson

Guelph-Waterloo Centre for Graduate Work in Chemistry, Waterloo Campus, Department of Chemistry, University of Waterloo, Waterloo, ON, Canada, N2L 3G1.

DABCO (1,4-diazabicyclo[2.2.2.]octane) is a saturated diamine whose cage-like structure and high symmetry give it unusual spectroscopic properties and make it an interesting SERS probe molecule. Recently we have proposed [1] that the preferential enhancement of four A_1' modes of DABCO and their relative intensities in the SERS spectra can be explained by a photon-driven charge-transfer (PDCT) mechanism [2] which contributes one to two orders of magnitude to the total SERS intensity. In this paper we present the results of further experiments, carried out on polycrystalline silver electrodes in the electrochemical SERS apparatus described in [1], which we believe add credence to this interpretation.

One prediction of the PDCT mechanism is that the electrode potential at which the SERS intensity maximizes, E_{max}, should exhibit a dependence on the excitation photon energy which is indicative of the direction of charge transfer (metal-to-molecule or molecule-to-metal); this dependence has been observed for a number of adsorbates [3]. A plot of E_{max} vs. photon energy for DABCO has a negative slope, indicating that the charge transfer is from the molecule to the metal. We have extended these experiments in order to examine the effect of the supporting electrolyte. The SERS spectra of DABCO adsorbed on a silver electrode in different supporting electrolytes (and therefore adsorbed on electrodes which have been roughened using very different ORC's) do not show significant differences in band positions or relative intensities; however, the behavior of the SERS intensity as a function of the electrode potential is altered, and this is reflected in the behavior of E_{max}. For example, as the supporting electrolyte is changed from 0.1 M $NaC\ell$ to 0.1 M NaF and then to 0.1 M $NaC\ell O_4$ i.e., as we move to a less specifically adsorbed anion, a shift in E_{max} to more positive potentials is observed, which is large for red excitation [Fig. 1], and thus the slope of the plot approaches the theoretical predicted value of unity [2]. This is possibly due to a reduction of the role of local potential variations at the electrode surface when the anions present undergo little or no specific adsorption.

Both the PDCT and the electromagnetic models of the enhancement predict a peak in the excitation profile of an adsorbate. Thus excitation profiles were measured at constant electrode potential using the available lines of Ar^+ and Kr^+ lasers. The integrated SERS intensities were corrected for ν^4 dependence and instrumental variations by using the internal standards $C\ell O_4^-$ and NO_3^-, which were found not to exhibit SERS under the conditions of these experiments: the measured excitation profiles for DABCO were independent of the internal standard. Excitation profiles were measured for two A_1' bands: ν_4, the C-C stretch at 986 cm^{-1} and ν_5, the C-N symmetric stretch at 794 cm^{-1} [Fig. 2]. One can see that the excitation profiles for the two bands are significantly different, ν_4 intensity passing through a maximum near 600 nm and ν_5 intensity peaking near 510 nm. This is difficult to reconcile with an electromagnetic enhancement mechanism, since, in the absence of "chemical" effects, the excitation of two lines 200 cm^{-1} apart should be nearly identical [4]. We therefore suggest that the observed SERS excitation profiles for DABCO on Ag are further indications of the PDCT mechanism, the maxima resulting from tuning the excitation wavelength into the CT resonance at constant electrode potential.

Finally, we note that the SERS intensities of different bands of DABCO do not maximize at the same electrode potential, the 794 cm^{-1} band always peaking at more positive potentials compared to the 986 cm^{-1} band. In terms of the PDCT model this behaviour is consistent with the maximizing of the excitation profiles of the 794 cm^{-1} band at shorter wavelengths compared to the 986 cm^{-1} band; at this time no explanation for either behaviour can be offered.

1. D.E. Irish, D. Guzonas, and G.F. Atkinson, Surface Sci. 158 (1985) 314.
2. A. Otto, in: Light Scattering in Solids, Vol. 4, Eds. M. Cardona and G. Güntherodt (Springer, Berlin, 1984).
3. J.R. Lombardi, R.L. Birke, L.A. Sanchez, I. Bernard, and S.C. Sun, Chem. Phys. Letters 104 (1984) 240.
4. H. Metiu, Progr. in Surface Sci. 17 (1984) 153.

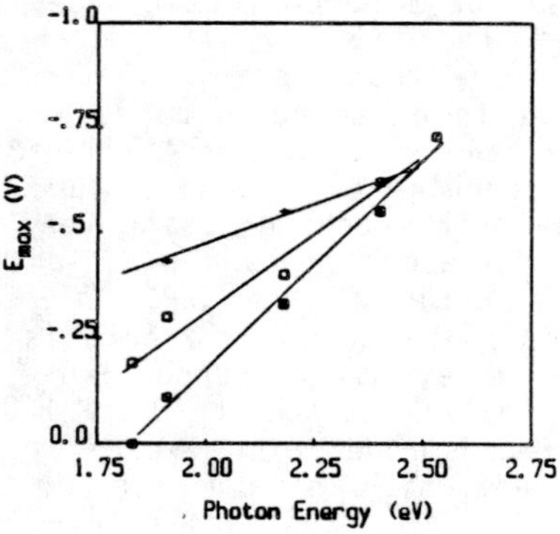

Figure 1

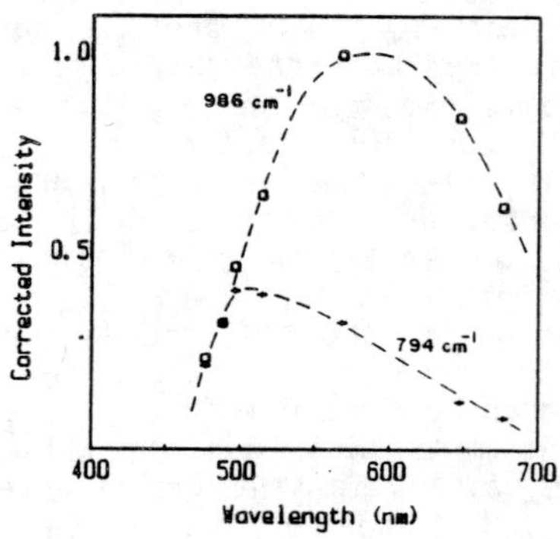

Figure 2

Figure 1: E_{max} for the 986 cm^{-1} band versus photon energy for three supporting electrolytes: + $C\ell^-$; ▫ F^-; ▪ $C\ell O_4^-$.

Figure 2: SERS excitation profiles for 0.1 M $KC\ell$, 0.9 M $NaC\ell O_4$, 0.5 M DABCO at a Ag electrode.

IN-SITU IDENTIFICATION DURING THE ELECTROCHEMICAL POLARISATION OF Ni AND NiMo ALLOY IN SULFURIC SOLUTIONS.

P. Delichère, A. Hugot-Le Goff, N. Yu

Physique des Liquides et Electrochimie, 4 place Jussieu 75005 PARIS

The role played by the molybdenum in the improvement of the electrochemical corrosion strength of iron and nickel can be approached by the study of the composition of films present in the different ranges : activity, passivity, transpassivity, secondary passivity.

In very concentrated sulfuric solutions, one knows that a thick porous NiO layer appears in the passive range. Its Raman spectrum is plotted on figure 1a, together with the spectrum of a NiO single crystal (1b. Peak at 725 cm^{-1} has never been reported in the literature), and of $Ni(OH)_2$ (1d) and Ni_2O_3 (1c) powder samples. On figure 2, Raman spectra obtained during the polarisation of Ni in 1M sulfuric solution at the corrosion potential (oxide-free surface, 2b), in the passive range (2a) and in the transpassive range (2c) are plotted together with the spectrum of $NiSO_4$, $7H_2O$ (2d). In the passive range, the main peak of NiO appears between the two sulfuric solution peaks (450, 595 cm^{-1}) and a salt is present in the transpassive range (as in activity, where the same type of spectrum has been found again). In the secondary passivity range, an other spectrum allows to assume that Ni_2O_3 is present.

Raman spectroscopy has shown that, during the electrochemical polarisation of Ni-10% Mo alloy, the NiO existence range extended at the detriment of the range of existence of salt, which is in agreement with the reducing of the d.c. current. These results illustrate the contribution that the Raman spectroscopy can bring in the field of electrochemistry, where insitu direct chemical informations are almost impossible to obtain by the usual techniques, taking into account that the passive films have a thickness which does not pass beyond 3-4 nm.

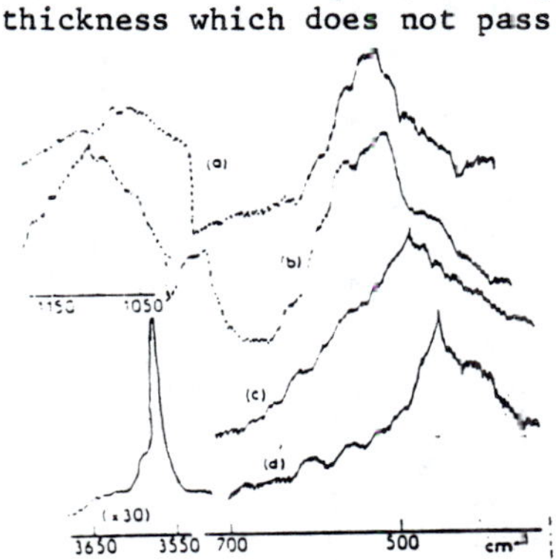

Figure 1

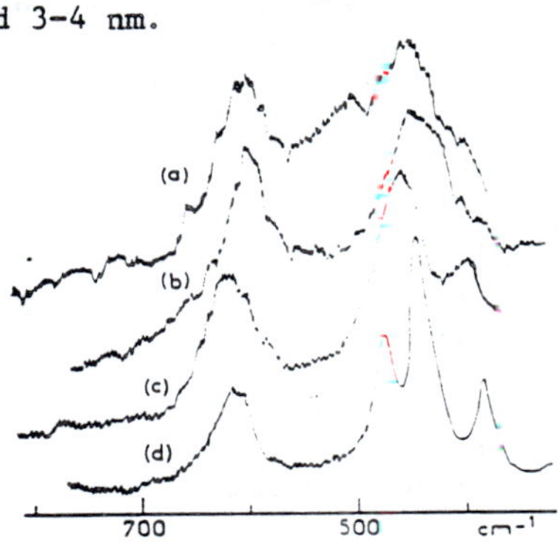

Figure 2

RAMAN SPECTROSCOPIC STUDIES OF $Cu(HCO_2)_2$ SURFACE COMPLEXES

H. Jeziorowski and B. Moser

WIM, D-8058 Erding, FRG

Formate complexes are formed on metal and metal oxide surfaces by reaction with formic acid, by CO2/H2 coadsorption and by CO adsorption on wet surfaces. The elucidation of the mechanism of film formation on materials used in electrical devices is of particular interest.

This study deals with the characterization of formate complexes formed on the surface of metallic copper under different conditions.

RESULTS AND DISCUSSION

Spectrum a in Fig. 1 is the Raman spectrum of copper formate crystallites obtained by the surface reaction of metallic copper with an aqueous solution of formic acid.

With regard to the frequencies as well as to the relative intensities, the marked Raman bands are in excellent agreement with the detailed vibrational assignment in the Raman spectrum of the formate ion in aqueous solution /1/.

In the vibrational spectra of water molecules in aquo metal complexes the region between 1000-300 cm-1 is associated with the librations and the translational modes. In particular, covalently bound water is indicated by the appearance of libration modes in the range of 1000-700 cm-1. The metal-water vibrations in aquo metal complexes are observed in the region between 530-230 cm-1, e.g. a polarized aquo copper Raman line characterizing covalently bound hydration spheres was found at 440 cm-1 /2,3/.

In a way similar to these results, the other observed features in spectrum a of Fig. 1 are attributed to frustrated rotational and translational modes of covalently bound water in the copper formate crystallites.

The result revealed by this spectrum corresponds to the ligand field theory: Because monocarboxylic acids are poorer coordinating ligands than water, Cu2+ forms a planar aquo complex with two weaker bound formate ligands perpendicular to this plane which is a distorted octahedral complex as expected from the Jahn-Teller effect for an ion with d9 electronic configuration.

Spectrum b in Fig. 1 was obtained from copper formate crystallites formed by the reaction of concentrated formic acid and metallic copper. No Raman bands attributable to vibrations of water molecules appear. The Raman bands of the formate ion undergo changes in frequencies and intensities compared to the bands in spectrum a due to stronger interaction of the formate ion with the cation.

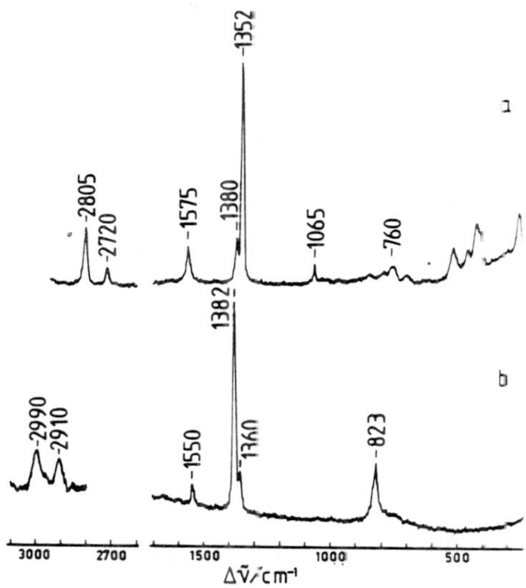

Fig. 1

The assignment of the symmetric COO stretching and the neighboring CH deformation modes to A1 and B1 species has been discussed. Ito attributed the band at 1350 cm-1 to the A1 mode and the band at 1380 cm-1 to the B1 mode, whereas Newman reversed this assignment /4/. In IR, EEL and IET spectroscopic studies of the adsorption of formic acid on different metals and metal oxides both assignments have been applied to the formed formate species /5/.

We have attributed the band at 1352 cm-1 and at 1382 cm-1 in the Raman spectra in Fig. 1 to the symmetric COO stretching vibration implying that its position depends on the coordination of the formate ion to the copper ion. This assignment is supported by deuteration studies. The Raman band of the CD deformation appears at 1030 cm-1 and the symmetric COO stretching vibration at 1348 cm-1 in different frequency ranges.

The observed Raman spectra show differencies due to the various configurations of copper formate. The dehydrated copper formate Cu/Cu(HCO2)4/ and the copper formate tetrahydrate /Cu(H2O)4/(HCO2)2 are well characterized by the vibrational frequencies of the formate ion and the modes below 500 cm-1 attributed to librations and translations of the formate group and the water molecule.

REFERENCES

/1/ K.Ito and H.J. Bernstein, Can.J.Chem. 34(1956)170.
/2/ A.da Silveira and M.A.Marques, Mol.Phys. 9 (1965)271.
/3/ R.E.Hester and R.A.Plane, Inorg.Chem.3(1964)768.
/4/ R.Newman, J.Chem.Phys. 20(1952)1663.
/5/ B.A. Sexton, Surf.Sci. 88(1979)319.

SURFACE-ENHANCED RAMAN SPECTRA FROM THIN FILMS OF CTAB ON NONMETALLIC SURFACES.

D. B. Parry and A. L. Dendramis

The Procter & Gamble Company, Miami Valley Laboratories, Cincinnati, Ohio

An area of surface analysis of great practical importance that can be successfully addressed by Surface-Enhanced Raman Scattering (SERS) is that involving nonmetallic surfaces. Although, to date, SERS has been used mostly for studies of liquid/metal interfaces (primarily in cases where the metal is one of Ag, Cu or Au), more SERS reports are beginning to appear in the literature where the substrate of interest is nonmetallic in nature.[1-3] Our past work in this area[4] has shown that SERS, via deposition of Ag-overlayers (of the appropriate roughness) on the interface of interest, can serve as a useful identification tool for a variety of molecules deposited as thin films on Si-wafers. In the work reported here we have used SERS to detect thin films of a typical cationic surfactant, CetylTrimethylAmmonium Bromide (CTAB), deposited on a variety of nonmetallic surfaces. Our data provide further evidence of the technique's versatility in this important area of surface science.

Fig. 1a shows the SERS spectrum obtained from a 390 Å-thick film of CTAB (as measured by ellipsometry) deposited on a Si-wafer. This spectrum bears a strong resemblance to that of CTAB at the Cu/H_2O interface[5], and it implies a CTAB molecular orientation similar to the one described in ref. 5. If the experiment is conducted under high-vacuum conditions (experimental details are published elsewhere[4]), the resultant SERS spectrum has a drastically different appearance, with improved S/N and more fine structure visible (Fig. 1b). If the vacuum is released (spectrum assumes the pattern in Fig. 1a) and then reapplied, the spectrum recovers most of its original (1b) features (Fig. 1c), implying that simple orientation changes are responsible for the observed spectral differences. The similarity of the spectrum taken under ambient pressure with that of CTAB at the Cu/H_2O interface prompts us to suggest that the orientation change is induced by the presence of H_2O molecules next to the polar head groups of the surfactant species.

A useful reference point for thin film spectral identification is provided by the SERS spectrum of CTAB in its "bulk" powder form (Fig. 2b). This is obtained by packing the CTAB solid into an indented cylindrical sample holder and depositing a Ag-overlayer of the appropriate roughness. Although similar to the normal Raman spectrum of solid CTAB (Fig. 2a), the resultant SERS spectrum has certain distinct features (e.g. the reduced relative band intensities in the C-C stretching and CH_2 twisting regions) that may need to be taken into account in attempts aimed at correlating spectra of thin CTAB films with a specific molecular orientation on the surface.

SERS spectra of CTAB are highly dependent, in appearance, on the film's thickness[4], but highly reproducible for films of comparable thickness. The spectra also appear to vary with the nature of the substrate supporting the CTAB film, as is illustrated in Fig. 3, where spectra of CTAB films on polystyrene (3a) and hydroxyapatite (3b) substrates are shown. In each of these cases the spectrum reported was taken by using the high-vacuum approach, as it consistently provided better quality spectra. Bands marked with an asterisk belong to the substrate part of the interface. The S/N ratio is more than acceptable for identification purposes and the relatively broad range of substrates employed points out the general applicability of the technique for nonmetallic interfacial systems. Efforts are currently under way for the purpose of explaining the observed spectral patterns in terms of CTAB's binding mode and orientation on the surface.

References

1. K. Shoji, Y. Kobayashi and K. Itoh, Chem. Phys. Letters 102, 179(1983).
2. A. Takenaka, S. Takeuchi, Y. Kobayashi and K. Itoh, Surface Sci. 158, 359(1985).
3. H. Ishida, T. Kawai and A. Ishitani, Pittsburgh Conference, Atlantic City, NJ (1986), Paper No. 873.
4. D. B. Parry and A. L. Dendramis, Appl. Spectroscopy, July-August issue (1986).
5. A. L. Dendramis, E. W. Schwinn and R. P. Sperline, Surface Sci. 134, 675 (1983).

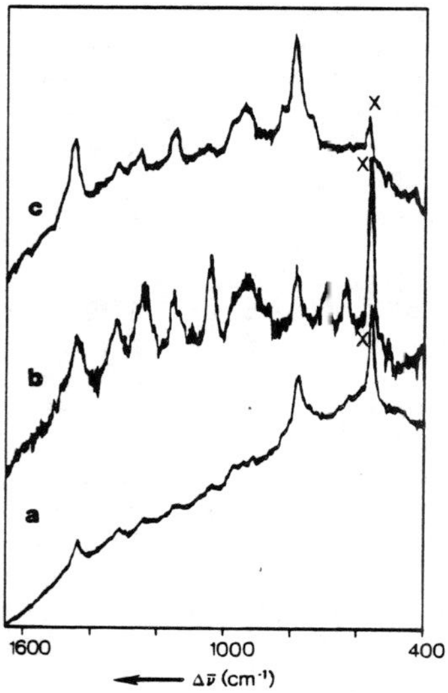

Fig. 1 SERS spectra of a 390 Å CTAB film on Si (see text for details).

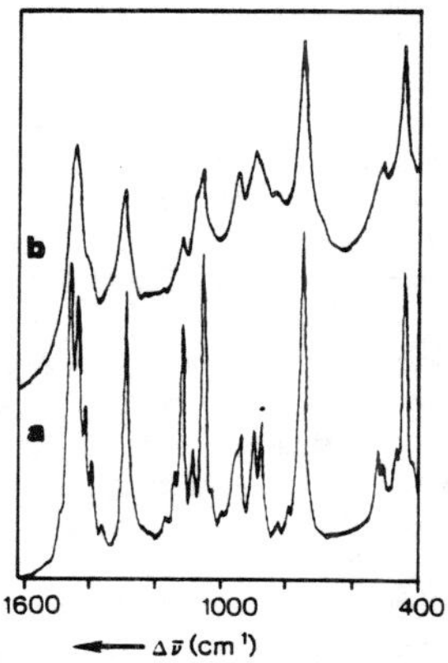

Fig. 2 Raman spectra of CTAB powder: a) Normal Raman b) SERS, via Ag-overlayer.

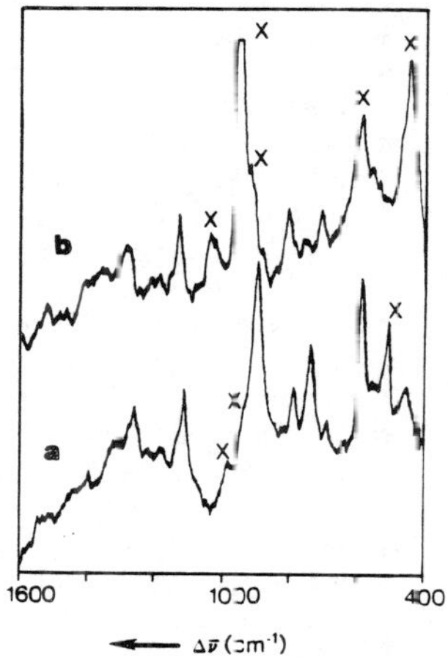

Fig. 3 SERS spectra of CTAB films (<900 Å) on: a) polystyrene b) hydroxyapatite

CHARACTERIZATION OF THE SILVER AGGREGATES RESPONSIBLE FOR THE ENHANCEMENT OF RAMAN SPECTRA ON SILVER HALIDES CRYSTALS AND COATED ELECTRODES

Guilment J., Régis A., Corset J. and Riveccie M.

LASIR, CNRS, Thiais, France.
Kodak-Pathé, Chalon-sur-Saône, France.

In an effort to clarify the nature and size of silver roughness responsible for the enhancement of the Raman spectra of species adsorbed on a silver electrode, we have already observed such enhanced spectra on silver halide coated silver electrode either for pyridine or for pyridinium ions formed when pyridine is adsorbed (1 to 4) or for the halide depending low frequency modes (2, 3, 4) with or without pyridine adsorbed.

Through excitation of Raman spectra at 77K of preliminary irradiated pure silver halide crystals or halide coated silver electrodes, well resolved fluorescence bands are observed even at incident energy lower than that of the AgX band gap. The relative intensities and frequencies of these fluorescences are depending on the "history" of the AgX matrices (X = Cl or Br) : preparation methods, further chemical treatments, irradiations at room temperature.

These weak and resolved fluorescence bands (fig. 1) are related to the size distribution of small silver aggregates generated at low concentration in the silver halide matrices. The formation of silver specks in crystals upon irradiation at room temperature can be monitored by absorption spectroscopy. These aggregates are preexisting in the electrochemically prepared silver halide coatings or are formed through illumination at room temperature.

The Raman spectra show low frequency bands probably due to resonance Raman diffusion of species involving aggregates. These signals are superposed over the second order spectrum of the AgX matrix and the fluorescence background spectra. The second order Raman spectra of the silver halide is observed alone before irradiation of the crystals. After a room temperature irradiation, two type of new signals appears : 1) some bands are depending on the nature of the halide and will be discussed in relation to the nature of the holes either generated through crystal irradiation or resulting from halide anion adsorption on silver halide coated electrodes ; 2) the other bands are thought to be due to silver aggregates as already discussed by us (1 to 3) and others (5). Simultaneous evolution of fluorescence and enhanced Raman spectra will be discussed in reference to the already proposed models : electron-hole pairs, excitons, latent image formation.

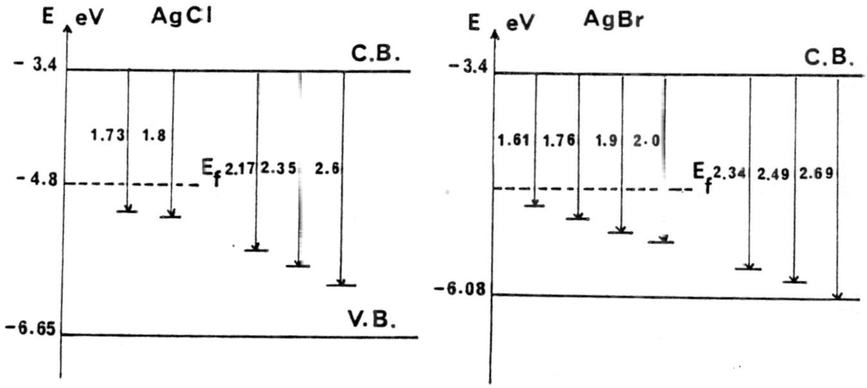

Fig. 1. Fluorescences observed at 77K for irradiated crystals at 273 K.

References

(1) A. Régis, J. Corset, Chem. Phys. Letters, **70** 2, 305, 1980.
(2) A. Régis, P. Dumas, J. Corset, J. of Electron Spectroscopy and Related Phenomena, **30**, 203, 1983.
(3) A. Régis, P. Dumas, J. Corset, Chem. Phys. Letters, **107** 6, 502, 1984.
(4) G.F. Atkinson, D.A. Gujones, D.E. Irish, Chem. Phys. Letters, **75** 3, 557, 1980.
(5) D. Roy, T.E. Furtak, Chem. Phys. Letters, **124** 4, 299, 1986.

VOLTAGE-INDUCED ENERGY LEVEL SHIFTING OF ADSORBED MOLECULES IN SERS

Gu Ren-ao Leng Yong-zhang Zhu Han-ru Ling De-hong
Suzhou University,Suzhou,PRC

A focus in the continuing debate concerning the origin of Surface-enhanced Raman Scattering (SERS) is: the "classical electromagnetic enhancement" and "chemical enhancement"which is more important. The former is nearly independent of the adsorbed molecules, the latter considers the interaction between adsorbates and adsorbents and it is sensitive to the presence of "active site " on the metal and details of the molecule-metal bond. Although many experimental results led to a wide acceptance of the fact that EM enhancement is responsible for "part of " the strong enhancement of the order of 10^4-10^6, it is sure that there are sufficient evidence which is not fully understood within the frame of the EM enhancement. Experimental findings stimulate us to explore the chemical effect,the model of charge-transfer(CT) excitation found most favorable among the various items for "chemical enhancement". When the incident photon energy $\hbar\omega$ is equal to the energy separation between the participating state $\hbar\omega = \Delta E = |E_{mol} - E_{met}|$. Furtak[1] considers the large gradient in the electrostatic potential perpendicular to the metal surface and explains the role of voltage-induced shifting of charge-transfer excitations in surface-enhanced Raman scattering partly. For further investigation of the origin of SERS,the experiment of voltage-induced shifting of charge-transfer excitation is made and the idea of voltage-induced energy level shifting of adsorbed pyridine molecules at Ag electrode is introduced.

The SERS spectra were recorded with a JASCO R500 Raman spectrometer using 488.0, 514.5, and 632.8 nm lines from Ar^+ and He-Ne lasers, operated at about 50 mw, 50 mw, 20 mw respectively. The sample is 0.05 M pyridine + 0.1M KCl aqueous solution. The Ag electrode is pretreaed by a double-step ORC from -0.60V to +0.60V where the potential is held for 5Sec. and then Turn to -0.60V. All potentials are versus the saturated calomel electrode (SCE). For a given Raman band, voltages are scanning from 0.0V to -1.0V, o.oV to-1.0V and o.oV to -1.5V respectively.Repeat the procedure in the exactly similar way for the other Raman bands.

The experimental results are shown in Table 1. From Table 1, we get following results:

Table 1. The values of potential at which spectral intensity reachs maximum

$h\omega_i$ \ U_{Imax}(V)	$U_{1215cm^{-1}}$	$U_{1035cm^{-1}}$	$U_{1008cm^{-1}}$
2.54eV	-0.73	-0.54	-0.60
2.41eV	-0.79	-0.61	-0.65
1.96eV	-0.83	-0.72	-0.77

1. U_{Imax} of different Raman bands varies with incident energy $\hbar\omega$;
2. The ratio $\hbar\Delta\omega/e\Delta U$ doesn't equal to unity;
3. The ratio $\hbar\Delta\omega/e\Delta U$ varies with different Raman bands;
4. The slope of curve $\hbar\omega$ versus U_{app} isn't a constant.

It is believed that the maxima of intensity are the result of voltage -induced changes in the efficiency of charge-transfer excitations and modification of electronic resonance of the molecule-metal system.

From Ueba's CT excitation theory[2], the Raman intensity of the adsorbed molecules is enhanced when the incident photon energy satisfies the condition:

$$\hbar\omega + E_f(U) = E_a$$

Where E_a denotes final state of the CT excitation of the coupled molecule-metal system, and $E_f(U) = E_f o - eU$. Changing the incident photon energy, the following formula is derived

$$e\,(U_1 - U_2)/\,\hbar(\omega_1-\omega_2) = 1$$

It doesn't fit our experiment results. Later, Furtak noticed that the local electrostatic potential at the surface of the Ag is incompletely screened by the electrolyte, but it cann't explain the difference of ratio $\hbar\Delta\omega/e\Delta U$ of different Raman bands. Therefore, we consider the interaction between adsorbed pyridine molecules and adatoms of silver when pyridine molecules chemisorbed at Ag electrode and the energy level shifts of adsorbed molecules vary as voltage applied, then introduce a corrective term $f_i(eU)$ which describes the excited vibrational energy level shifts caused by applied voltage. The Raman intensity is enhanced when the incident photon energy satisfies the condition:

$$\hbar\omega + E_f(U) = E_a - f_i(eU)$$

where $E_f(U) = E_f o + (d\Delta E/dV_{app})U$, $(d\Delta E/dV_{app})$ described that the local electrostatic potential at the surface of the Ag electrode is incompletely screened by electrolyte.

With different incident photon energy, we get following formula:

$$\hbar(\omega_1-\omega_2)/e(U_1-U_2) = d\Delta E/deV_{app} + (f_i(eU_1)-f_i(eU_2))/e(U_1-U_2)$$

The value of $\hbar(\omega_1-\omega_2)/e(U_1-U_2)$ is not unity and varies with different excited vibrational energy level of adsorbed molecules, i.e. different Raman band. Therefrom we explain the experimental results qualitatively.

REFERENCES:

1. T. E. Furtak, et al. Chem. Phys. Letters 95(1983)328.
2. H. Ueba, Surface Sci. 131(1983) 347.

TEMPERATURE AND ELECTRODE POTENTIAL DEPENDENCY OF THE CO-ADSORPTION OF PYRIDINE AND METHYLPYRIDINE ON A SILVER ELECTRODE.

Hiroshi Saito, Yasunari Ikezawa, and Genjiro Toda

Department of Chemistry, College of Science, Rikkyo (St. Paul's) University, Nishi-Ikebukuro, Toshima-Ku, Tokyo, 171 Japan

In a study of the dependence of the SERS intensity on the bulk concentration of pyridine, it was found that a decrease in coverage was accompanied by deactivation of the SERS active sites [1]. Hence the coverage could not be determined from the SERS intensity. However, the coverage of co-adsorption can be determined from the SERS intensities if the total coverage is almost unity[2].

In the present work, the dependence of the SERS intensities on the bulk concentration of pyridine (py) and methylpyridine (mpy) was investigated for various temperatures and electrode potentials. The dependence of the adsorption constants thus obtained on the temperature and on the potential was discussed.

0.1 M KCl solution was used as the supporting electrolyte. The concentration ratio was changed in situ by replacing a part of the py solution with mpy solution, the total concentration being maintained at 50 mM. By ORC procedure, a silver electrode was oxidized at 0.2 V /SCE for 3 sec., then reduced at each observed potential. For excitation of the Raman spectra a Jeol He-Ne laser (6328 Å, 50 mW) was used.

Fig. 1 shows the SERS spectra at various concentration ratios of py and 3-mpy. It was assumed that the band intensities of the breathing mode of the adsorbed species of d_5-py (975 cm^{-1}) and 3-mpy (1033 cm^{-1}) were proportional to the number of the adsorbed molecules, and that their co-adsorption could be approximated by Langmuir isotherm:

$$Cpy \cdot Kpy = \theta py/(1-\theta py-\theta mpy) \text{ and } Cmpy \cdot Kmpy = \theta mpy/(1-\theta py-\theta mpy).$$

The equilibrium constant ratio of adsorption (Kmpy/Kpy) was determined by means of the method of least squares. The isotherms calculated with the adsorption constant ratio gave a good fit to the observed isotherms. It was confirmed for the binary system that the total coverage had been maintained at unity for the temperature and potential ranges used.

The dependency of the adsorption constant ratio on the temperature was examined for py and 3-mpy over the temperature range from 10 to 40°C. Neither of the breathing band positions shifted with change in temperature. In general, the equilibrium constant was represented by the terms of entropy and enthalpy, and the following equation was derived in this case of co-adsorption:

$$\ln(K_{mpy}/K_{py}) = -(dG_{mpy}-dG_{py})/RT$$
$$= (dS_{mpy}-dS_{py})/R - (dH_{mpy}-dH_{py})/RT,$$

where dG, dS, and dH are the standard adsorption free energy, entropy, and enthalpy, respectively. The curve of ln(Kmpy/Kpy) vs. 1/T is shown in Fig. 2. The values of (dSmpy-dSpy) and (dHmpy-dHpy) were 37 J $mol^{-1}K^{-1}$ and 8.3 kJ mol^{-1}, respectively; and in this case the heat of adsorption of py was larger than that of 3-mpy. The dependency of these thermodynamic parameters on the potential was examined.

Fig. 3 shows the dependency of the Kx/Kpy on the electrode potential, where x designates 2-, 3-, or 4-mpy. The adsorption constants of the three kinds of mpy were relatively larger than that of py as the potential was changed from -0.3 V to -0.7 V. Thus it became clear that the strength of adsorption was closely related with the interaction between the electron density of the nitrogen atom of the pyridine-ring and the electrode potential.

Moreover, as shown in this paper, by means of SERS, the constant of adsorption on a solid electrode can be obtained as a function of the temperature and of the electrode potential.

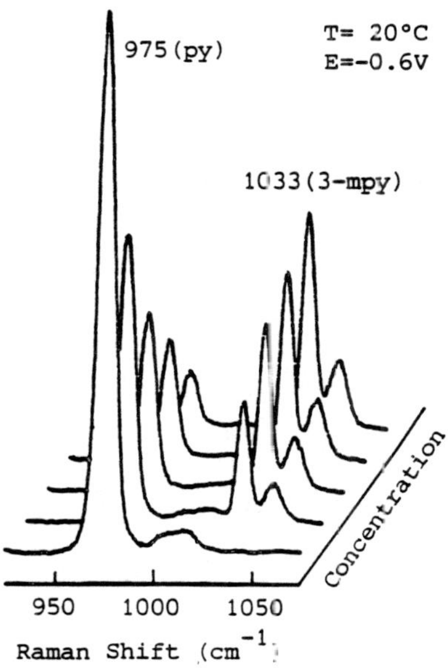

Fig. 1 Concentration dependency of SERS spectra

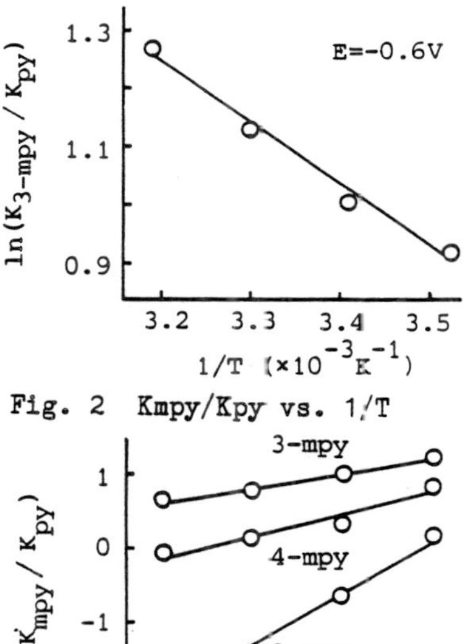

Fig. 2 Kmpy/Kpy vs. 1/T

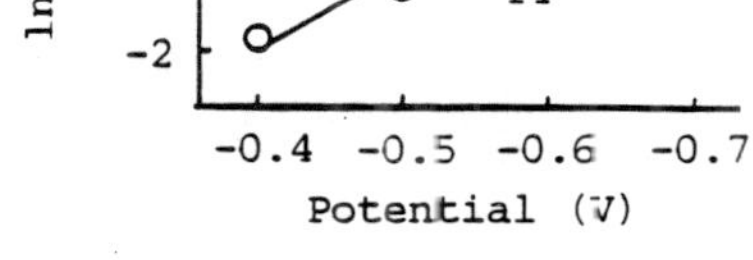

Fig. 3 Kmpy/Kpy vs. potential
T = 20°C

REFERENCE

1. H.Saito, Y.Ikezawa, and G.Toda, Chem. Phys. Letters 110 (1985) 291
2. Y.Ikezawa, H.Saito, K.Matsui, and G.Toda, to be published.

SIMPLE, REPRODUCIBLE CHEMICAL PROCEDURE FOR PREPARING SERS-ACTIVE SILVER FILMS ON GLASS SUBSTRATES.

Therese M. Cotton, Fan Ni, and Qiao Feng

Department of Chemistry, University of Nebraska-Lincoln, Lincoln, NE 68588-0304

There are three general types of substrates used to observe surface enhanced Raman scattering (SERS) from molecules adsorbed onto silver. These include electrodes, colloids or sols, and vacuum deposited films. Through judicious choice of experimental parameters, each of these substrates can produce enhancement of the normal Raman intensity of adsorbates by six orders of magnitude. However, each substrate suffers from at least one drawback. For example, sols are easily prepared, but are very unstable and tend to aggregate upon addition of the adsorbate. In addition, sols are chemically complex and undesirable reactions can occur with the adsorbate. For example, sodium borohydride, the reducing agent commonly used, is a strong reductant and can reduce the substrate if it is present in excess. On the other hand, if insufficient reductant is added, silver ion will be present at the sol surface and it may interact with the adsorbate to form a silver complex. Thus, the surface potential of the sol is difficult to control. The variability in SER spectra reported in the literature for adsorbates on silver sols is at least partially due to differences in the preparation procedure and surface potential. The use of silver electrodes or island films prepared by vacuum deposition both require more complex procedures and the availability of the necessary equipment (potentiostat or high vacuum deposition apparatus). A simple, rapid and highly reproducible method for preparing silver surfaces for SERS is reported here. The application of this procedure to the determination of trace organics in water is demonstrated.

Silver films were deposited from Tollens reagent onto glass strips (10mm x 3 mm) cut from frosted microscope slides. Each of the experimental variables (concentration of $AgNO_3$, temperature of the reduction bath, and reduction time) were optimized to produce maximal enhancement in the adsorbate signal. 4,4'-bipyridine was chosen as a model compound for optimizing the experimental parameters, since it had previously been shown to produce strong SER spectra on silver [1].

The strongest SER signals were observed for silver films prepared from 2% $AgNO_3$ solutions. Briefly, the procedure involves three steps. In the first step, NaOH is added to the $AgNO_3$ solution to produce a precipitate of silver oxide. In the second step, NH_4OH is added to redissolve the precipitate and form the ammonia complex of Ag(I). In the final step, the silver complex is reduced with glucose in a water bath at 55°C for 2 minutes. Following this step, the film is washed with distilled water, after which it may be stored in distilled water for several days without loss of activity. SER spectra were recorded by immersing the Ag film in a 5 mm tube containing several hundred microliters of the adsorbate solution.

The above experimental procedure resulted in uniform deposits of Ag on the glass surface, as indicated by scanning electron microscopy (SEM). The particle size ranged from 500-1000 Å and the enhancement factor for 4,4'-bipyridine was determined to be 10^5. Higher concentrations of $AgNO_3$ solutions or longer reduction times produced a thicker deposit of silver with highly aggregated particles. These films gave much lower enhancement factors, as would be expected from electromagnetic enhancement theory.

The detection limit for 4,4'-bipyridine was determined to be 1 x 10^{-7}M using 40 mW at 514.5 nm. The spectra were recorded on an optical multichannel analyzer (OMA-2) interfaced to a Spex Triplemate monochromator/spectrograph. The total spectral acquisition time was 20 seconds.

The above films gave very reproducible spectra (standard deviation ± 3% for a 100 µM solution). A plot of SER intensity versus the logarithm of the concentration of bipyridine solution resulted in two linear regions. For the concentration range from 0.1 - 1.0 µM the slope was much smaller than for the concentration range from 1.0 µM to 1.0 mM. In the first region of the plot, or for low solution concentrations, the bipyridine adsorbs with the aromatic rings parallel to the Ag surface. As the solution concentration is increased, the preferred orientation of the rings is perpendicular, with one of the ring nitrogens bonded to the Ag surface. According to EM theory, modes perpendicular to the silver surface should be most strongly enhanced. The change in intensity of ring modes with the concentration of adsorbate solution support an orientational change in the bipyridine at the silver surface.

The chemically deposited silver films were also found to be conductive and, hence, can be used as electrodes. We have used cyclic voltammetry for determining the redox properties of adsorbates. The ability to control the surface potential at these films is another advantage over silver sols.

The results to be presented illustrate that the chemically deposited films are ideal for analytical detection of trace organic species. A large number of slides can easily be prepared. Calibration curves using standard solutions of the analyte should be prepared from the same batch of slides, since the precision between two separate preparations is not as good as within a single preparation.

In summary, the procedure described represents a significant improvement over the use of silver sols as SERS-active substrates. The silver films are much cleaner chemically than sols, since all of the chemicals used in their preparation are washed away prior to the addition of the adsorbate solution. The films are also stable with time - only a small decrease in SER signal intensity was observed after several days. Finally, potential control is achievable with these films.

[1]. Cotton, T.M. and Vavra, M. Chem. Phys. Lett. 106, 491 (1984).

ENHANCED RAMAN SCATTERING FROM THIN CHROMIA FILM SUBSTRATES

P. R. Graves,

Materials Development Division, AERE Harwell, Didcot, Oxon, England

Much work has been recently reported in the literature on the fabrication of silver diffraction grating type devices. The aim of this is to generate a surface with periodic structures such as posts(1) or a sinusoidal profile(2), onto which metallic silver may be deposited. These surfaces will greatly enhance the Raman scattering cross-section of molecules absorbed onto them. However, no paper has yet reported enhancement of Raman scattering from the substrate itself. This paper reports the first observation of SERS from the substrate itself, a rough chromia (Cr_2O_3) surface. Rough chromia surfaces were produced by oxidising polished chromium metal coupons in a furnace. Results indicate that oxygen adsorbs at chromium (V) atom sites rather than chromium (III) sites on the surface.

RESULTS AND DISCUSSION

Figure 1 shows the excitation dependence of SERS from an oxidised chromium metal coupon onto which a 90nm layer of silver was evapourated. The evaporation rate and layer thickness of the silver were both found to be important to observing strong SERS. Deviation of the 878 cm^{-1} intensities from the $1/\lambda^4$ scattering rule and the broad linewidth are taken as indications that the large Raman signal is due to SERS. The chromia lattice has the space group D_{3d}^6 and group theory predicts 7 Raman active vibrational modes(3), six of which have been observed(4). However, none of these modes have frequencies coincident with the Raman bands shown in Figure 1. Therefore it is concluded that the enhanced Raman signal is only from the surface of the oxide and not from the bulk.

BAND ASSIGNMENTS

Chromia (VI) oxyanions have Cr–O vibrations in the 750–1000 cm^{-1} region and CrO_3, which is substantially covalent, has vibrations in the 800–1000 cm^{-1} region(5), as does $K_2Cr_3O_{10}$(5). Therefore the 787 cm^{-1} vibrational mode is assigned as a Cr–O stretch. Formation of $Cr(VI)-O_x^{3-}$ species on the surface is unlikely but $Cr(V)-O^{2-}$ and $Cr(V)-O_2^{2-}$ species have previously been suggested on the basis of infra-red studies of Cr_2O_3 surfaces(6). Adsorption of O_2^{2-} would lead to a ν_{O-O} symmetric stretch and O_2 libration if π-bonded, and a ν_1, stretch and δ_2 deformation if σ-bonded via an oxygen. For these reasons the mononuclear $Cr(V)-O^{2-}$ site is a better interpretation of the Raman spectra, since only a single vibrational mode is expected.

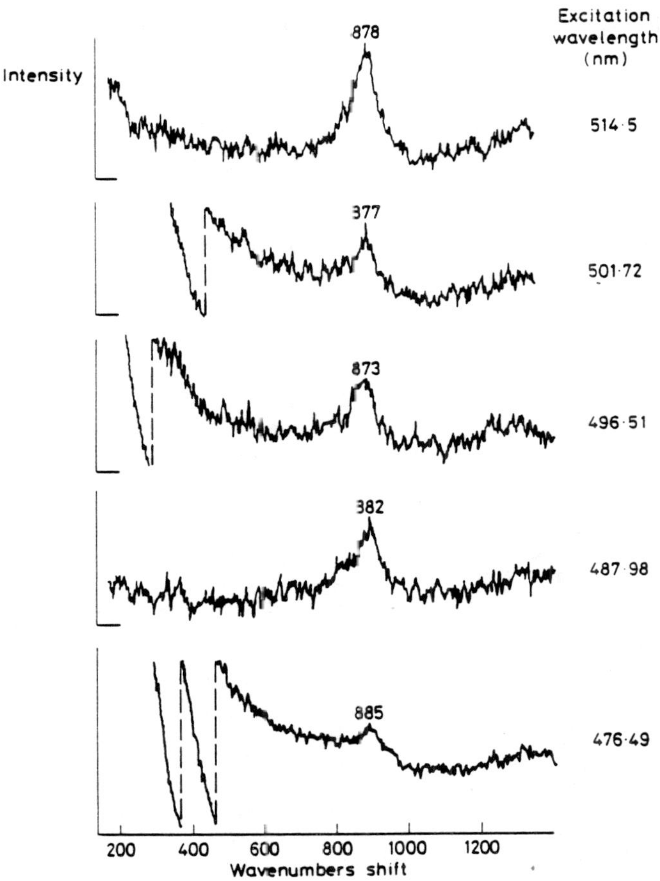

Figure 1 Excitation dependence of SERS from an evaporated silver film deposited onto an oxidised chromium metal surface. Spectra were obtained with 0.1mW laser excitation and an 160s acquisition time.

REFERENCES

1. M. Meyer, A. Wohaun, T. Vo-Dinh, J.Phys.Chem., (1985), 89, 1843-1846.
2. P.F. Liao, in "Surface Enhanced Raman Scattering", 379 - 390, ed. R.W. Chang, T.E. Furtak, (Plenum 1982)
3. S.P.S. Porto, R. S. Krishnan, J.Chem.Phys., (1967), 47, 1009.
4. I. R. Beattie, T. R. Gilson, J.Chem.Soc. (A), (1970), 980-87.
5. R. Mattes, Zeit. fur anorg. allge. Chemie (1971), 382, 163-176.
6. M. Che, A. J. Tench, Adv. in Cat., 32, 51, (1983).

VIBRATIONS OF KINKS AND FOLDS: UNDERSTANDING THE PHASES OF THE CYCLOALKANES

Victoria L. Shannon*, Herbert L. Strauss, Robert G. Snyder

Department of Chemistry, University of California,
Berkeley, California, 94720 USA

The Raman and infrared spectra of several members of the series of cycloalkanes, cyclo-$C_{12}H_{24}$ through cyclo-$C_{22}H_{44}$, have been investigated as a function of temperature. The even members of this series (cyclo-$C_{14}H_{28}$ and above) crystallize in a compact conformation with two extended chains connected by a "fold" sequence of trans and gauche bonds. These folds are very much like the folds found in the crystalline lamella of polyethylene. With the aid of normal coordinate calculations, and by comparing the Raman and infrared spectra of these molecules, we have been able to find many characteristic vibrations of the specific sequences of trans and gauche bonds involved in the folds.

Prior to melting, the medium-sized cycloalkanes (except cyclo-$C_{12}H_{24}$) undergo a solid-solid phase transition in which the enthalpy change is considerably larger than that at the melting transition. For cyclo-$C_{22}H_{44}$, for example, the ΔH at the solid-solid phase transition is four times the ΔH at the melting transition. In order to interpret the changes that take place for the various cycloalkanes on going into the high temperature solid phase and into the melt, we have studied the relative intensities of several of the vibrational bands which are characteristic of specific sequences of trans and gauche bonds.

For cyclo-$C_{22}H_{44}$, our studies support the idea that a gtg' kink is introduced into one of the long extended chain segments at the solid-solid phase transition. It is uncertain if any "new" gauche bonds are introduced into the ring upon melting, but the spectra indicate that, at the very least, the gauche bonds already present are redistributed. Both intra- and intermolecular interactions are important in determining the conformations of cyclo-$C_{22}H_{44}$ in the high temperature solid phase and in the melt.

For cyclo-$C_{16}H_{32}$, changes in the vibrational spectra seem to indicate that no "new" gauche bonds are introduced into the ring at the solid-solid phase transition. The spectra are consistant with the idea that the gauche bonds already present are redistributed in the ring at each phase transition. It is likely that conformational rearrangements take place at the solid-solid phase transition and at the melting transition for all of the other cycloalkanes that we have studied.

*Present address: Dept. of Chem. University of Oregon, Eugene, OR 97403

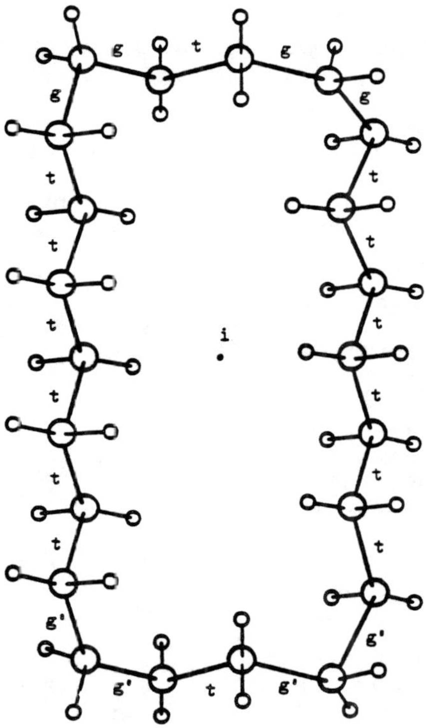

An example of a cyclo-$C_{22}H_{44}$ molecule in the low temperature solid phase. The molecule crystallizes with a pair of extended chains connected by the fold sequence ...t(ggtgg)t... at each end.

INTERFACIAL STUDIES OF SILVER-AQUEOUS ELECTROLYTES BY OPTICAL SECOND HARMONIC AND DIFFERENTIAL CAPACITANCE.

Hossein M. Rojhantalab, and Geraldine L. Richmond

Department of Chemistry
University of Oregon, Eugene, 97403

Second harmonic generation (SHG) is a nonlinear optical effect which has unique sensitivity to surface structure and interfacial electrostates. Since SHG is electric dipole forbidden in centrosymmetric media, it is a powerful probe when measurements require a clear distinction between bulk phase and interfacial properties[1,2].

The second harmonic signal for an electrode can be expressed by the nonlinear polarization $P^{nl}(2\omega)$ as the sum of the second order and third order terms:

$$P^{nl}(2\omega) = \alpha\, E(\omega) \times H(\omega) + \beta\, E(\omega)(\nabla . E(\omega)) + \gamma\, E_{dc}(E(\omega).E(\omega)) + \gamma'\, E(\omega)(E_{dc}\cdot E(\omega))$$

Recent SHG work on silver electrode-alkali halide electrolytes have shown some promising trends in elucidating the nature of the second harmonic signal at the interfaces[3,4]. These studies showed a strong dependence of the harmonic signal on the nature of the silver electrode surface, the anion adsorbed, and correlated the intensity of the second harmonic signal with excess charge density, σ_m, on the metal surface,

$$I(2\omega) \propto (\sigma_m)^2$$

In order to understand the various contributions to SHG at electrode surfaces, we have compared the optical second harmonic signal with the differential double layer capacitance for adsorption of anions on polycrystalline silver electrode. These studies focused on potentials within the ideally polarizable region in binary mixtures of various electrolytes. The effects of the adsorbed species on the nature and intensity of the SHG in binary mixtures of electrolytes of constant ionic strengths were obtained by applying linear potential sweep voltammetry. The surface concentrations of specifically adsorbed anions in each case as measured from capacitance-potential data were correlated with the intensity of the second harmonic signal. These results will be presented for smooth and electrochemically roughened silver electrodes.

Acknowledgement:

The financial support from the National Science Foundation for this work (che#853008) is greatly appreciated.

References:

1. C.H. Lee, R.K. Chang, and N. Bloembergen; Phys. Rev. Lett. 18, 167(1967).
2. C.K. Chen, T.F. Heinz, D. Ricard, and Y.R. Shen; Phys. Rev. B 27, 1965(1983).
3. G.L. Richmond; Chem. Phys. Lett. 106, 26(1985), 110, 571(1985).
4. R.M. Corn, M. Romagnoli, M.D. Levenson and M.R. Philpott, Chem. Phys. Lett. 106, 30(1984); J. Chem. Phys. 81, 4127(1984).

OPTICAL STUDIES OF ELECTROCHEMICAL INTERFACES: A COMPARATIVE ANALYSIS BY SECOND HARMONIC GENERATION AND PHOTOACOUSTIC METHODS

Geraldine L. Richmond[a] and Pamela Chu[b]

[a]Department of Chemistry
University of Oregon, Eugene, OR 97403

[b]Department of Chemistry
University of California, Berkeley, CA 94720

Recent experiments from our laboratory have demonstrated the use of SHG in measuring interfacial interactions at the electrochemical junction.[1] In addition to monitoring oxidative film growth on and removal from the surface, reversible adsorption and desorption of ions and molecules occurring at potentials within the ideally polarizable region have also been investigated. In the course of these experiments on both smooth and electrochemically modified silver electrodes, questions have arisen with respect to the potential dependence of the linear and nonlinear optical properties of the interface. To address these questions and to understand further the mechanisms leading to SHG enhancement, we have coupled our SHG measurements with photoacoustic detection methods. These are the first studies of this kind which directly correlate the optical absorptive properties of the cycled interface with the SHG response during electrode polarization.

This presentation will focus on silver electrode surfaces polarized in various simple aqueous electrolyte solutions. The experimentation for these studies is shown below in figure 1 in which photoacoustic detection[2] has been integrated into the optical and electrochemical measurements. The periodic heating of the electrode surface due to optical absorption results in periodic pressure changes in the electrochemical cell as detected by the coupled transducer. As with SHG, this signal shows a strong potential dependence which can be compared with various perturbations at the electrode surface. These induced effects include surface roughening, surface film formation, double-layer polarization and molecular adsorption. The photoacoustic results show a strong correlation with SHG efficiency including enhancement effects. Two examples of our results are shown below. In figure 2, a comparison between the SHG and photoacoustic behavior is shown for an electrode polarized within the ideally polarizable region. The different SH potential dependence observed as a result of surface roughening which has been discussed in earlier work[3] follows the potential dependence of the photoacoustic response. In figure 3, the optical coupling to the surface is enhanced with the addition of pyridine. Similar enhancement in this region is seen in the SHG response. The wavelength dependence of these studies will be presented and correlated with other optical measurements.

This work has been supported by the National Science Foundation (CHE#8513008) and the Petroleum Research Fund of the American Chemical Society(16905-AC5).

1. G. L. Richmond, Langmuir, 2 (1986) 132; Surf. Sci. 147 (1984) 115; Chem. Phys. Lett. 106 (1986) 26.
2. C. K. N. Patel and A. C. Tam, Rev. Mod. Phys. 53 (1981) 517.
3. G. L. Richmond, Chem. Phys. Lett., 110 (1984) 571.

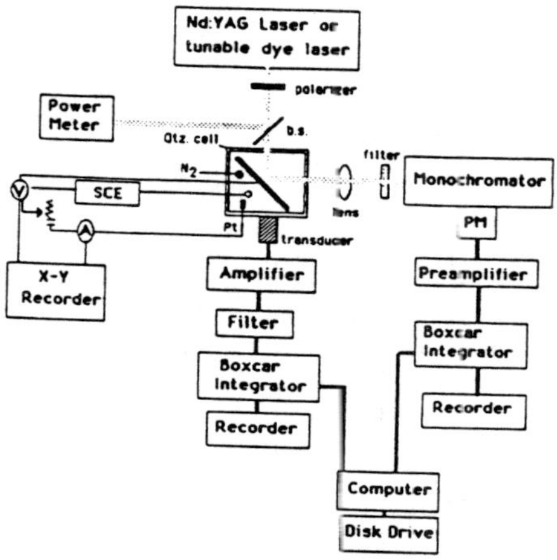

Fig. 1. Experimental apparatus.

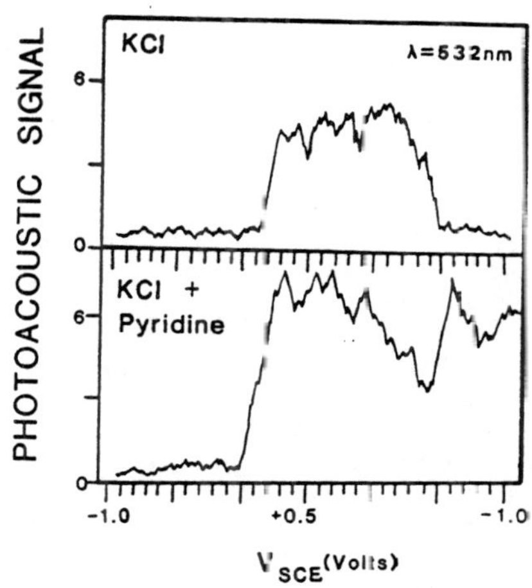

Fig. 2. Polycrystalline silver in 0.1M K_2SO_4 before and after one oxidation-reduction cycle(ORC). Top: SHG response with 2ω = 532nm. Bottom: Photoacoustic response.

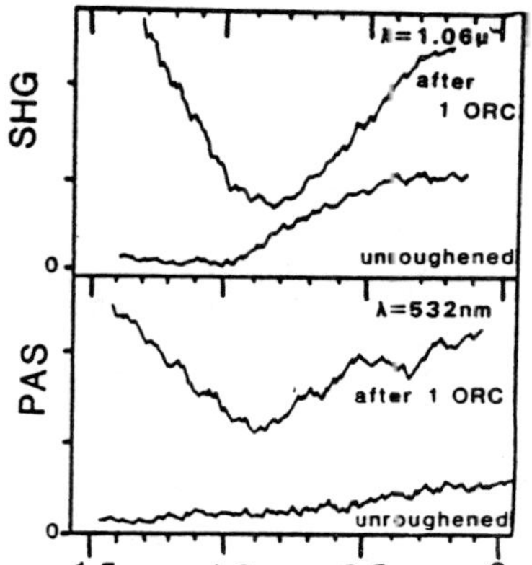

Fig. 3. Photoacoustic response for silver cycled in 0.1M KCl. In the lower figure, 10^{-4}M pyridine has been added.

STATIC AND TIME-RESOLVED NONLINEAR OPTICAL STUDIES OF SEMICONDUCTOR INTERFACIAL PROPERTIES

J. M. Robinson and G. L. Richmond

Department of Chemistry
University of Oregon
Eugene, OR 97403

Optical second harmonic generation (SHG) is a highly sensitive probe of the surface structure and reactivity of bulk isotropic materials. Static measurements of SH intensity determine, for example, surface symmetry and orientation of adsorbed species. Use of this technique as a real time probe of interfacial processes such as electrochemical reactions or trapping in surface defects requires an ultrafast pulsed laser system. These new time-resolved measurements potentially provide information about interfacial kinetics and dynamics on nanosecond and subnanosecond timescales. The semiconductors investigated are amorphous selenium thin films, isomorphic selenium-tellurium mixtures, and transition metal hexagonal layer compounds.

Selenium thin films are isotropic and we have readily observed SH signals from them. The effects of layer thickness and angle of incidence of pump radiation on SH intensity are interesting parameters for the static measurements. The amorphous form converts to the grey metallic form upon heating to 100 C. Transient SHG studies of the crystallization process may prove informative.

The bandgap of amorphous selenium is about 800 nm whereas that of crystalline tellurium is about 4000 nm. Selenium-tellurium mixtures with increasing proportions of tellurium effectively tune the bandgap to longer infrared wavelengths. Investigations of SH intensity for a single pump frequency as a function of bandgap will be discussed in terms of the effects of resonance and absorption.

The hexagonal layer compounds studied are single crystal transition metal dichalcogenides including WSe_2, $MoSe_2$, and $ZrSe_2$. These materials easily cleave along the van der Waals surface between two layers of unit cells and produce strong SH signals under ambient conditions. Useful as semiconductor electrodes, slow transient photocurrent measurements suggest that surface states of these materials trap electrons. Time-resolved SHG studies should help unravel the detailed kinetics at the semiconductor interface.

This work has been supported by the National Science Foundation (CHE#8451346) and the Petroleum Research Fund of the American Chemical Society (16905-ACS).

TENTATIVE INVESTIGATION ON MECHANISM OF SERS -EXPERIMENTAL STUDY OF ELECTRODE SURFACE

Zhu Hanru,Leng Yongzhang,Gu Renao and Ling Dehong

Suzhou University, Suzhou, Jiangsu, China

Much achievements have been acquired in the elucidation of the mechanism of SERS since it was discovered, but the relationship between the SERS intensity and the thickness of surface layer of electrode has not yet been determined clearly[1,2]. Although it was known that the electrochemecal pretreatment of electrode is necessary to produce intense SERS signal[3], the affection of various factors of pretreatment on the SERS intensity depends on the condition of different experiments, therefore, the results obtained are not able to be compared each other in the same standard. We report here an experiment and its results indicating the dependence of the SERS intensity on the thickness of electrode surface layer.

In our experiment, the 2-electrode—Ag(99.99%) and SCE—electrochemecal cell was used. After the active surface layer of Ag electrode was formed by exerting a positive rectangular potential pulse on it for oxidization and then changing its potential to -0.6v(vs. SCE) for reduction, the intensity of the 1008cm^{-1} band of pyr adsorbed on Ag electrode was measured while the potential of Ag electrode was maintained at -0.6v. The observed results are the following:

1. The polished electrode surface(smooth surface) can hardly produce obvious SERS signal. We take this scattering background as the measuring zero point.
2. If the electrode surface was properly pretreated, the maximum enhancement factor of Raman intensity amounted to 10. However, if the pretreatment was inadequate eg. too short oxidizing time and too low positive potential, the SERS intensity was only of the order of the scattering background mentioned above.
3. The SERS intensity depends strongly on the oxidizing time and positive oxidizing potential. Within a certain extent, when the oxidizing time was fixed, the SERS intensity enhanced with the increase of the positive oxidizing potential, and this enhancement can reach a "saturated state". The results are shown in Fig.1. When the oxidizing potential was fixed, the SERS intensity depends on the oxidizing time in a similar way.

According to item 3, in order to elucidate the relationship between the SERS intensity and pretreatment process, it is reasonable to consider associated affection of the oxidizing potential and time on the SERS intensity. Changing the potential and time while fixing their product, we have not obtained the regular correspondence between the product and the SERS intensity, thus our attention was attracted by the pretreating current instead of pretreating potential. In our experiment, the surface area of Ag electrode is 2.4mm^2. The oxidizing potentials and corresponding average of oxidizing current are listed in Table 1. It has been found that the SERS intensity is regularly related to the product of the oxidizing current and oxidizing time: corresponding to various combinations of a product the nearly same SERS intensity was obtained. (For simplisity, this product will be called "transferred charge" below). Some results are listed in Table 2. Because the Ag electrode was sufficiently reduced under the potential of -0.6v, the relationship between the SERS intensity and "transferred charge" reflects the dependence of the enhancement factor of Raman intensity on the thickness of the active surface layer of electrode.

With the concept of "transferred charge" Q, we obtained the following results: When Q<0.02mc, the SERS intensity was not larger than the scattering background produced by unpretreated electrode surface; in the range of Q=0.02mc—0.66mc, the more the Q is, the stronger the SERS intensity gets; in the range of Q=0.66mc—1.7mc, the SERS intensity was approximately constant. Using Q and surface area of electrode, it is easy to estimate the thickness D of the active surface layer of electrode. According to our experiment results, we conclude that in Ag/pyr+Kcl system, the enhancement factor of Raman intensity depends on the thickness D of the active surface layer. In the range of D=0—10Å, the enhancement factor is not larger than 1.(vs. scattering background of smooth surface). In the range of D=10Å—300Å, the enhancement factor enhance sharply and nonlinerly with the increase of D. In the range of D=300Å—800Å, the enhancement factor is approximately constant. We have not done the experiment for D>800Å.

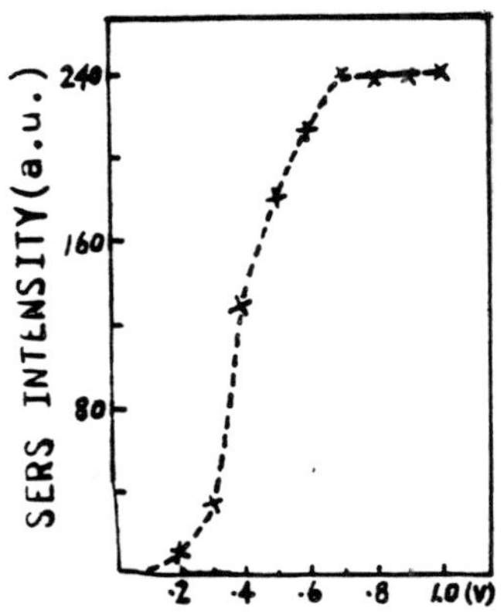

Fig.1. Intensity of signal vs. pretreatment potential(each pretreatment time is 5s.).

TABLE 1

POTENTIAL(10v)	1	3	4	5	7	10
CURRENT(A)	5	26	67	90	135	190

TABLE 2

POTENTIAL(10v)	2	3	4	5	6	7
TIME(s)	50	25	10	7	6	5
CHARGE(mc)	.65	.65	.67	.63	.66	.68
INTENSITY	230	228	225	208	232	240

REFERENCES

1. J.R.Kirtley,S.S.Jha and J.C.Tsang, Solid State Commun.,35(1980)509.

2. S.H.Macomber,T.E.Furtak and T.M.Devine, Sur.Sci.,122(1982)556.

3. M.Fleischmann et al.,Chem.Phys. Lett.,26(1974)163.

chapter 6

THIN FILMS

RAMAN SPECTROSCOPY OF SURFACES AND LAYERED STRUCTURES WITH OPTICAL WAVEGUIDES AND SURFACE PLASMONS.

J. D. Swalen

IBM Almaden Research Center
San Jose, California 95120

Advances in Raman spectroscopy of thin films have led to new and better ways to probe organic, inorganic and metallic surfaces and layered structures. The small number of absorbing species present, however, causes a problem of sensitivity and hence a low signal-to-noise ratio. Various optical techniques have recently been devised to enhance the optical fields at specific interfaces or within a specific thin film layer. Two of these techniques are guided optical waves using integrated optics and surface electromagnetic waves, such as surface plasmons. Some of the details that can be obtained from analyses of the observed spectra are the identification of specific groups, their packing and orientation, and interfacial interactions.

Initial investigations of thin films with integrated optics techniques showed that this method was convenient and accurate for determining film thickness and refractive index, at a specific wavelength. A corollary to this is that by making measurements at a series of wavelengths, one, in effect, measures the dispersion of the real part, $n(\lambda)$, and the absorption spectrum, $k(\lambda)$, from the imaginary part of the index of refraction. Experimentally, light is coupled into a optical waveguide conveniently by a prism or a grating. A change in angle changes the propagation vector so as to match those for the propagating modes at a given polarization. These modes for non-absorbing films are defined by an eigenvalue equation. For a given film thickness only a specific number of modes can propagate and the minimum thickness below which no modes are supported is roughly about a third of the wavelength of light used for propagation, i.e., a fraction of a micron for visible light. A number of methods used to find the solution to the eigenvalue equation will be discussed, as well as some of the experimental details.

From the optical properties of a film, the optimal conditions for recording the Raman spectrum of the whole film, of a layer on the surface or from a layered region within the film can be determined. Since Raman cross sections are much smaller than infrared or visible absorption spectral cross sections, our first goal was to determine how thin a polymer film we could observe with the waveguide geometry. Enhancements were realized because the optical fields are higher in a waveguide, a large number of sites are irradiated for Raman scattering and the scattering streak imaged onto the slit of a monochromator can completely fill it. In addition, the quality of detectors is continually improving and computer control of the spectrometer allows the averaging of many scans, improving the signal-to-noise ratio. As a consequence we were able to measure the ring modes of a 80nm film of poly (styrene) on a glass (Corning 7059) waveguide with an argon ion laser operating at 488nm with a signal-to-noise ratio of about 4:1. Adding the resonance Raman effect from a dye molecule to the enhancement from the waveguide, the Raman spectrum of a single L-B monolayer of a cyanine dye molecule on a glass waveguide was clearly observable. Not only could we see the ring modes of the cyanine molecule but we could measure wavelength shifts of several wavenumber units from interactions with the surface. Observations made with each polarization further showed that the electronic transition dipole moment of the interacting group is oriented parallel to the plane of the film. To study an interfacial region between two layers, a film of poly (styrene) was deposited onto a film of poly (vinyl alcohol). The Raman spectrum depended on the specific waveguide mode and the intensities agreed with the calculated optical field contours. Mathematical deconvolution of our signals and subtraction techniques further showed that some scattering came from interfacial species. Recent experiments of Langmuir-Blodgett films of cadmium arachidate on waveguides have shown that we can observe the polarized Raman spectrum to give us information about the conformation and orientations of the groups. In comparison with our infrared spectral results, where a two-step melting process occurs, the Raman bands further indicated that up to 95°C there was no evidence for any trans to gauche conformational changes in the aliphatic chains. Apparently the chains disorder by tilting before the head groups melt.

Although optical waveguides have many advantages for spectroscopy, surface plasmons have certain distinct and unique advantages. As such we have used both approaches depending on the specific situation. Matching E and H fields for radiation at an interface we see that surface electromagnetic waves require both a dielectric function which is strongly negative in some spectral region and a p (TM) polarized wave. Highly reflective metals, such as silver, gold, aluminum, or copper satisfy this relation over most of the visible spectral region. Other metals, such as the alkali or alkaline earths, also are appropriate, except they require more care as to environment. A dispersion relation (ω vs. k_z) for surface plasmons exists and for large k ω is asymptotic to a frequency somewhat below the frequency for bulk plasmons. Still further below this surface plasmon frequency, the propagation vector, $k(\theta)$, becomes very sensitive to the optical properties adjacent to the surface, e.g., overcoatings of a few nanometers can result in an angular shift in θ of the order of a degree. Absorption in a film leads to a further angular shift, a broadening of the reflectivity versus angle curve, and a reduction in the depth of the minimum, assuming close to optimal conditions exist.

The high optical fields at a metallic surface propagating surface electromagnetic waves can lead to significant Raman scattering. This has been proposed as one of the mechanisms for surface enhanced Raman scattering (SERS). For example, a 10nm film of poly (styrene) is clearly observable at the surface plasmon angle and shows a strong angular dependency, following that of the surface plasmon. This effect is about 30 times greater than comparable waveguide results. L-B monolayers have also been investigated by surface plasmons and they show a strong amplification from coupling energy from the surface mode to the Raman mode. In fact, as one changes the angle, the surface mode moves through the various Raman bands, adding increased intensity as it goes. Coupling to molecular fluorescence can also occur. Here, as the surface wave energy passes through fluorescence bands, each is increased in intensity. Energy is exchanged between the surface mode, the molecular levels, and those of the metal. Clearly these transfers, as with the quenching of fluorescence and Förster type dipole-dipole energy transfer, can be used both for informative and practical uses.

I wish to thank the many colleagues who have collaborated on much of the work: J. F. Rabolt, J. P. Rabe, N. E. Schlotter, M. R. Philpott, W. Knoll, and M. Jurich.

RAMAN SPECTROMETRY OF INTERFACES AND ULTRA-THIN FILMS.

R. Dupeyrat, M. Harrand, and M. Masson

Departement de Recherches Physiques, Universite Pierre et Marie Curie,
4, Place Jussieu, 75252 Paris Cedex 05, France

Numerous practical problems are related to the properties of interfaces and ultra-thin films: we can say lubrication, adhesion, which are industrial problems, but also hemocompatibility, cell-membrane permeability, which are biological ones. For their comprehension and for their solution these problems need a large knowledge of molecular interactions. Two types of study will be described.

Non-spectroscopic studies: Generally workers are interested in forces and we shall give two examples. A) For a long time Israelachvili and co-workers (1) have been studying solid liquid interfaces, measuring the forces between the two walls of a liquid film as a function of their distance (i.e. the thickness of the liquid film). Their results related to intermolecular forces are, at first very surprising because in the range of the thickness of ten molecular dimensions the measured forces (attractive and repulsive) vary approximately periodically with the thickness (period of the order of one molecular dimension). These forces decrease asymptotically towards the continuum van der Waals attractive curve. This result, whatever the explanation can be, gives an idea of the thicknesses which are involved in the interactions. Therefore when studying interactions between two media by other methods it is necessary that the thickness can vary in the same range. B) Toshev and Ivanov (2), by the way of hydrodynamics and thermodynamics, using the shape of meniscus, have given indications on the structure of black films.

Spectroscopic studies: A) Infrared studies: Cholet (3) has studied by infrared absorption the modification of the orientation of successive Langmuir-Blodgett layers of docosanoic acid deposited on metals. He has observed a variation of the tilting of the molecules from zero degrees for the first layer to 23 degrees after the fifth layer. Rabolt and Swalen (4) have studied films of polymer (PMMA) 180 A thick by infrared reflection and have described the modification of the spectrum with the nature of the support. B) Raman studies: Takenaka and Nakanaga (5) were the first to obtain the spectrum of a monolayer between two liquids. Their film was resonant in order to obtain an "enormous enhancement of the Raman intensity". The monolayer of methyl orange solution was produced at the interface of carbon tetrachloride and aqueous solution of a cationic surfactant. The illumination was obtained by evanescent wave. The authors have then determined the orientation of the dye molecules in one monolayer. Other experiments were performed by Rabolt and Swalen (6) with films obtained by a spinning machine or by Langmuir-Blodgett deposition.

Several experimental devices used wave-guide technique or plasmon excitation. In the case of solid polymer films produced by a spinning machine the thickness can be obtained by measuring the coupling angle. In the case of solid films deposited by the Langmuir-Blodgett method the thickness is easily determined by the number of monolayers.

We have obtained (7) the spectra of films formed by multilayers from one to nine layers and we have observed a modification of the Raman spectra in the same range of thickness as studied by Israelachvili. It seems that nobody has published Raman spectra of liquid films of different thicknesses. The reason is if it is easy to have, in a box, a sample with ultra-thin solid film, it is not the case with a liquid sample because it is difficult to obtain the stability of a film during the experiment. Nevertheless this way would be interesting in order to verify and perform the hypothesis made with force measurements because Raman can give information on molecular conformations.

Future projects: In the future the Raman spectra of ultra-thin liquid films will be realized for the case of a film between two solid media, using classical Raman spectroscopy, but in the case of a free film (between two gases) we need a pico second pulse and a convenient electronic device to record the spectrum before the explosion of the film, and this will be more difficult.

1. CF. bibliography in Israelachvili's book: Intermolecular and Surface Forces, Academie Press (1985).

2. B.V. Toshev and I.B. Ivanov, Colloid and Polym. Science 253, 558.

3. P.A. Cholet, J. Chem. Phys. 64, 3 (1976).

4. J.F. Rabolt, M. Jurich, and J.D. Swalen, Applied Spectroscopy, 39, 269 (1985).

5. T. Takenaka and T. Nakanaga, J. Phys. Chem. 80, 475 (1976).

6. J.F. Rabolt and J.D. Swalen, Applied Spectroscopy (1985).

7. M. Masson, M. Caillaud, Z-N. Peng, and R. Dupeyrat, Optics Communications 53, 33 (1985).

UNENHANCED RAMAN SPECTROSCOPY OF THIN ORGANIC FILMS ON SURFACES.

N. E. Schlotter

Bell Communications Research, Murray Hill, New Jersey

Advances in detector technology and sampling methods permit acquisition of unenhanced Raman spectra of monolayers on surfaces. Combining the improvments discussed below makes it possible to examine monolayers in detail as well as extensions to studies of surface interactions and chemistry.

For the detection of photons, array-based systems have a multiplex advantage that is overwhelming in comparison to single channel systems. Under optimum conditions intensified optical multichannel array (OMA) detectors have been able to measure spectra from a monolayer or less of molecules with large Raman cross sections. These experiments have been done on metallic substrates but have been shown to be unenhanced Raman spectra with signals as small as one count per second detected [1,2]. Another approach has been to use a charge coupled device (CCD) camera. The CCD camera has a several advantages over the OMA detector. The quantum efficiency of the CCD camera can approach 0.75 versus 0.068 for the OMA or, in terms of sensitivity the CCD camera "sees" nearly every photon while the OMA needs approximately 15 photons per count. Additionally, the dark current of the CCD camera is extremely low; most of the noise occurs at readout (approximately 15 electrons versus 4000 counts per second for the intensified OMA). There are several disadvantages such as a slow readout and a useful chip dynamic range of about 50,000 - 100,000 electrons. Experiments that require time resolution or have high background signal levels can take advantage of the OMA's gating speed and higher readout rate. Figure 1 shows thin film spectra recorded using a CCD camera. The samples are built up from monolayers of cadmium stearate (CdSt) prepared on a Langmuir - Blodgett (LB) trough before transfer to the solid fused silica substrate. Each CdSt monolayer is approximately 25 angstroms thick and the structure is initially a two dimensional crystalline array. The spectra have been normalized for the number of layers and offset for clarity. Further details can be found in reference [3].

In addition to signal collection, signal generation can be improved by using integrated optical techniques. The waveguide Raman sampling method described in references [4] and [5] increases both the power at the sample and the sampling volume for a slit geometry. This technique has been used recently to study multilayer stacks of LB films [6] with a scanning double monochromator and photon counting system. Also, temperature studies can be done on the waveguide by controlling the temperature of the flow of dry nitrogen gas in the cell.

FIGURE 1.

Unenhanced Raman Spectra of Cadmium Stearate LB films on fused silica substrates.

Conditions:

500 mW, 5145 Å, 360 sec. exposure, Binned 40 by 2 pixels.

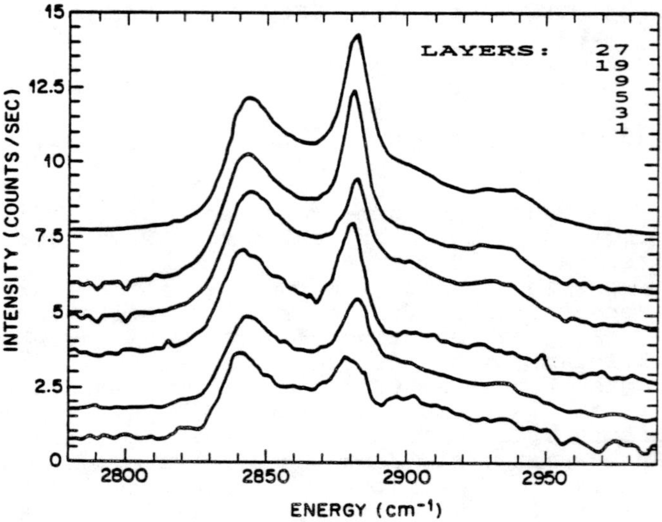

By combining waveguide Raman sampling with the above array detection systems single monolayers of LB films can be studied by Raman spectroscopy on a more routine basis. This allows the polarized Raman spectra to be collected and partial orientation analysis applied to obtain detailed structural information concerning molecules on surfaces. Preliminary results will be presented.

REFERENCES

1. Vicki M. Hallmark and Alan Campion, Chem. Phys. Lett., 110, 561-564 (1984).
2. David R. Mullins and Alan Campion, Chem. Phys. Lett., 110, 565-570 (1984).
3. S. B. Dierker, C. A. Murray, J. D. LeGrange, and N. E. Schlotter, in press in Chem. Phys. Lett.
4. J. F. Rabolt, R. Santo, N. E. Schlotter, and J. D. Swalen, IBM J. Res. Develop., 26, 209-216 (1982).
5. N. E. Schlotter and J. F. Rabolt, J. Phys. Chem., 88, 2062-2067 (1984).
6. J. P. Rabe, J. F. Rabolt, and J. D. Swalen, Bul. Am. Phys. Soc., 31, No. 3, EX 5, March Meeting, Las Vegas, NV, 1986.

MOLECULAR ORIENTATION IN THIN FILMS BY RAMAN SPECTROSCOPY USING INTEGRATED OPTICAL TECHNIQUES.

John F. Rabolt
IBM Almaden Research Center
San Jose, CA 95120

The difficulty of studying ultra thin (1-5 microns) films by Raman scattering techniques is well documented. It arises from the low concentration of material present and the inability to obtain large optical field intensities within this minuscule sampling volume. The use of integrated optical techniques in conjunction with Raman scattering to investigate micron[1] and submicron[2] films has greatly alleviated these problems. Recently[3] polarized Raman measurements of Langmuir-Blodgett multilayered (200 angstroms in thickness) molecules have been obtained. A comparison of the polarization data obtained from both $CH_3(CH_2)_{18}$ COOH and $CF_3(CF_2)_8(CH_2)_{10}$ COOH multilayers indicate that while the former is oriented almost normal to the substrate surface, the latter is considerably inclined. Results of high temperature studies using the waveguide Raman technique suggest considerable differences in their disordering mechanisms.

[1]Y. Levy, C. Imbert, J. Cipriani, S. Racine and R. Dupeyrat, Opt. Commun. 11,66 (1974)

[2]J. F. Rabolt, R. Santo, N. E. Schlotter and J. D. Swalen, IBM J. Res. Dev. 26,217

[3]J. Rabe, J. F. Rabolt and J. D. Swalen (unpublished results)

CHARACTERIZATION OF ORDER IN LANGMUIR-BLODGETT MONO- AND MULTILAYERS BY MULTICHANNEL RAMAN SPECTROSCOPY USING A CCD DETECTOR.

S. B. Dierker

AT&T Bell Laboratories, Murray Hill, New Jersey 07974

Langmuir-Blodgett (LB) films have become particularly attractive model systems for elucidating the relationship between structure and function of ultrathin organic films. Raman spectroscopy is an especially promising nondestructive technique for studying LB films, since, as a vibrational spectroscopy, it is sensitive to the internal and local molecular structure, as well as the chemical nature of the molecule and its environment. It has not been widely applied to thin organic films, however, due to the weakness of their Raman scattering cross sections. In order to circumvent these limitations, we have developed a high sensitivity Raman scattering instrument based on a charge-coupled-device (CCD) multichannel detector. The sensitivity of the instrument is such that, in a few minutes, we can obtain Raman scattering spectra of monolayer LB films with a signal to noise ratio of $\gtrsim$ 15 without relying on field or resonance enhancement of the signal. We report here initial results of studies of thin films of cadmium stearate (CdSt) deposited on fused silica and hydroxylated silicon substrates. A fuller discussion of this work is presented in Ref. 1.

The novel high sensitivity detector we use is a liquid nitrogen cooled RCA SID501 CCD array consisting of 520 x 312 30 μm square pixels. The quantum efficiency of the chip is over 70 % between 3500 and 7000 Å , and it has a dark current of $<$ 0.01 electrons/pixel/sec and a read out noise of 50 electrons. An extended discussion of our system is given in Ref. 2. The CCD chip is placed at the focus of a SPEX Triplemate spectrometer. Raman spectra were taken with 500 mW of 5145 Å radiation and the scattered radiation was collected with F/2.0 collection optics in 360 second integrations. The Raman data images were averaged vertically along the chip, perpendicular to the dispersion, to obtain the final Raman spectra.

Langmuir-Blodgett multilayers of fatty acid salts were deposited on fused silica plates and highly polished hydroxylated silicon wafers using a Joyce-Loebl trough. All depositions were performed at 22° C at a dipping speed of 15-30 mm/minute and at a surface pressure of 20-30 dyne/cm. The structure of fatty acid salt LB films is one in which the alkyl tails are arranged in layers and the head groups and alkyl tails have a certain degree of lateral packing order. The tails themselves are most ordered when they exist in the all *trans* planar zig-zag conformation. We have focused on the C-H stretching region which is particularly sensitive to changes in lateral and conformational order.

In Figure 1 are displayed representative spectra of various thickness (1 - 27 layer) CdSt LB film samples on fused silica substrates. The Raman intensities were found to be directly proportional to film thickness and are divided by the number of layers in Fig. 1.

Vibrational spectra of the C-H stretching region are complex due to Fermi-resonance (FR) interactions[3] between the symmetric methylene C-H stretching mode, d^+, at 2843 cm^{-1}, and the manifold of overtones of the CH_2 bending modes, $\delta\,(CH_2)$. The CH_2 bending modes undergo significant intermolecular coupling[3] and their FR interaction with δ^+ is thus dependent on the lateral packing order. The antisymmetric methylene C-H stretching mode[3], d^-, at 2882 cm^{-1}, is forbidden by symmetry from entering into FR interactions and is thus insensitive to lateral order. However it is sensitive to chain conformation. Consequently, the ratio of the peak intensity at 2882 cm^{-1} to that at 2843 cm^{-1} is a convenient parameter which is sensitive to both the lateral packing of extended alkyl chains and to conformational order within the chains. The value of this ratio ranges from 0.7 for completely melted hydrocarbon chains to 1.5 for vibrationally decoupled all-trans chains to 2.2 for a highly ordered crystalline lattice.[4]

The spectra shown in Fig. 1 exhibit significant differences as a function of layer thickness. I_{2882}/I_{2843} increases rapidly with increasing thickness, rising from 0.95 for a monolayer up to 1.25 for 5 layers and then less rapidly to 1.45 for 19 and 27 layers. Ratios found for spectra acquired from different positions on a given film varied by $\sim$ 10%, indicating the presence of a significant degree of structural inhomogeneity across the films. More detailed analysis of the spectra shows that the increase in I_{2882}/I_{2843} is accompanied by a small broadening of the d^+ fundamental, an increase in intensity of the d^+ sidebands and a narrowing and consequent increase in the peak height of the d^- peak. Broadening of the d^- peak in the first few layers was not accompanied by a shift to higher energy, indicating the presence of metastable static gauche defects rather than the dynamic conformational disorder characteristic of the liquid[4]. These results show that the lateral packing is somewhat disrupted and there is a higher population of static gauche defects in the first few layers deposited.

An additional effect of the lateral packing order is to induce mode splittings[5] due to the crystal field symmetry of the unit cell. Raman measurements of our samples in the 1450 cm^{-1} CH_2 bending region are consistent with either orthorhombic, or hexagonal subcell packing. An orthorhombic subcell packing in the thicker films is supported by the shape[5] of the d^+ fundamental. The decrease in linewidth of the d^+ fundamental with decreasing layer thickness indicates a tendency for hexagonal packing in the first few layers.

The lattice disruption in the first few layers, characterized by a greater number of static gauche defects and the tendency for hexagonal packing, could arise from substrate surface roughness and/or mismatch of the lattice parameter of the LB film with the amorphous substrate. As successive layers are deposited the effects of the substrate surface is diminished so that subsequent layers should be more regular. However, it is intriguing to speculate that the structural differences between 1 to 3 layer thin films and thick films has its origin in the reduced dimensionality of the thin films, independent of coupling to the substrate. Further studies of the trends reported here, in order to ascertain whether LB monolayers are an example of a system exhibiting two-dimensional phase transitions and melting, are certainly warranted.

In conclusion, we have demonstrated the capability of monitoring structural changes in organic LB films as thin as one layer ($\sim$ 25 Å) by using a CCD detector and multichannel Raman spectroscopy. It is expected that these results will stimulate new studies of order-disorder transitions in LB films and also lead to studies of liquid crystal phases of Langmuir monolayers floating on water.

I would like to acknowledge collaboration with Cherry Murray of AT&T Bell Laboratories in the experiments and design of the system, collaboration with Jane LeGrange of AT&T Technologies and Nick Schlotter of Bell Communications Research (BCR) in the Langmuir Blodgett film experiment, helpful discussions with D. L. Allara of Bell Communications Research and J. A. Tyson of AT&T Bell Laboratories, and software development by R. Lee of Lassen Research.

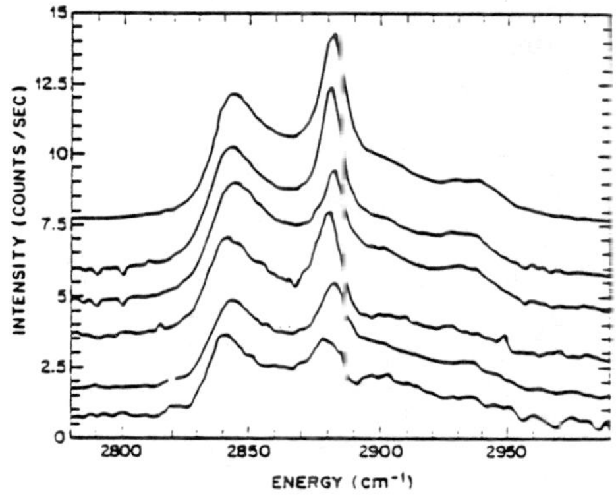

Figure 1 Raman spectra of Cd Stearate films on fused silica, top to bottom: 27, 19, 9, 5, 3, and 1 layer. The curves have been shifted vertically for clarity of display.

REFERENCES

[1] S. B. Dierker, C. A. Murray, J. D. LeGrange, and N. E. Schlotter, Chem. Phys. Lett., to be published.

[2] C. A. Murray and S. B. Dierker, J. Opt. Soc. Amer., to be published.

[3] R. G. Snyder, H. L. Strauss, and C. A. Elliger, J. Phys. Chem. *86*, 5145 (1982).

[4] B. P. Gaber and W. L. Peticolas, Biochimica et Biophysica Acta. *465*, 260 (1977).

[5] R. G. Snyder, S. L. Hsu, and S. Krimm, Spectrochimica Acta A *34*, 395 (1978).

DAVYDOV SPLITTING IN THE RAMAN SPECTRA OF THIN SOLID FILMS.

Ricardo Aroca, Carol Jennings, Ah-Mee Hor*, Rafik O. Loutfy*

Department of Chemistry and Biochemistry
University of Windsor
Windsor, Ontario, Canada
N9B 3P4

*Xerox Research Centre of Canac
2660 Speakman Drive
Mississauga, Ontario, Canada
L5K 2L1

INTRODUCTION: The crystal structure and molecular orientation in thin organic films are known to correlate with physical properties such as: carrier mobility, electrophotographic performance, photovoltaic effect and optical properties. In a recent report[1] we have studied depolarization ratios of the Raman bands in phthalocyanine(Pc) films, and have show that the results provide information about the molecular orientation with respect to the substrate plane. It was also found that vibrational bands presented a multiplet structure due to correlation field (Davydov) splitting[2] in the molecular crystal. In this report, we present experimental evidence of Davydov splitting in evaporated films of chloroaluminum Pc (AlPcCl). Films were grown simultaneously onto two different substrates: quartz plate (A) and a clean face of NaCl (B) crystal. Raman and fluorescence spectra of A and B samples show significant differences that seem to indicate an induced orientation in the film grown onto NaCl crystal. Similar experiments are now in progress for a series of Pc derivatives.

EXPERIMENTAL: The AlPcCl was purchased from Eastman Chemicals and purified by a train sublimation procedure. Thin films were evaporated from a tantalum boat onto a quartz plate and NaCl crystal in a Balzers vacuum system at 10^{-6} Torr. A film thickness of 200nm was monitored with an Inficon XTM quartz crystal oscillator. Polarization measurements were carried out using a polarization rotator mounted on the 2000 argon and krypton ion lasers. Scattered light was analyzed using a linear polarizer and scrambled before entering the Spex 1403 spectrometer. The experimental geometry has the xz plane as the scattering and incidence planes. There are four polarized components of the scattered light that can be measured in this geometry: SS, SP, PP and PS, where S indicates that the electric field is perpendicular to the scattering plane, while P polarization is parallel to the plane of incidence. The figure shows the resonance Raman spectra of samples A and B obtained when the laser beam and the normal to the substrate form a 30° angle.

RESULTS AND DISCUSSION: The absorption spectrum of AlPcCl shows maxima at 350nm (Soret band) and from 654 to 760nm (Red band). Excitation with the 647.1nm laser line produces RR scattering and a strong fluorescence. The fluorescence of film A (on quartz substrate) is structureless with a 799nm maximum. P-polarized excitation light gives rise to a twofold higher fluorescence intensity than the S-polarized excitation. This could be related to a specific structure of the crystal absorption spectrum for polarized light[1]. Davydov splitting of Raman bands are clearly observed for vibrational fundamentals. Splitting with a small magnitude were separated by selective polarization since multiplet components belong to different symmetry types. For instance, the pyrrole stretching is observed with components at 1339cm^{-1} and 1345cm^{-1}; isoindole stretching vibrations are observed as a series of doublets: 1494cm^{-1} and 1500cm^{-1}; 1450cm^{-1} and 1455cm^{-1}; 1430cm^{-1} and 1441cm^{-1}. The strongest Raman band observed with excitation at 488nm (outside the red band and close to the Soret band) is the pyrrole stretch at 1540cm^{-1}. Polarization ratios[1] for this band indicate that it belongs to a totally symmetric vibration of A_1 type. It is shown in the figure that in the RRS this band is accompanied by an intense component at 1526cm^{-1}. In particular, the 1526cm^{-1} line dominates the spectrum of sample B(NaCl substrate), and the intensity ratio of these two components is sensitive to changes of the experimental angle formed by the laser beam and the normal to the substrate plane. From the number of components in the Davydov splitting it can be concluded that there are two AlPcCl molecules in the unit cell of the molecular crystal. The differences between the spectra of films grown on quartz and NaCl crystal could be explained assuming identical crystal structures but in different orientations, or that the films do have different space groups. Complementary information from electron diffraction[3] studies would be necessary to answer this question.

REFERENCES:

1. A.S. Davydov, Theory of Molecular Excitons. Plenum Press, N.Y. 1971
2. R. Aroca, C. Jennings, A.M. Hor, R.O. Loutfy, J. Phys. Chem. (in press).
3. R.S. Kirk, Mol. Crystals 5, 211 (1968).

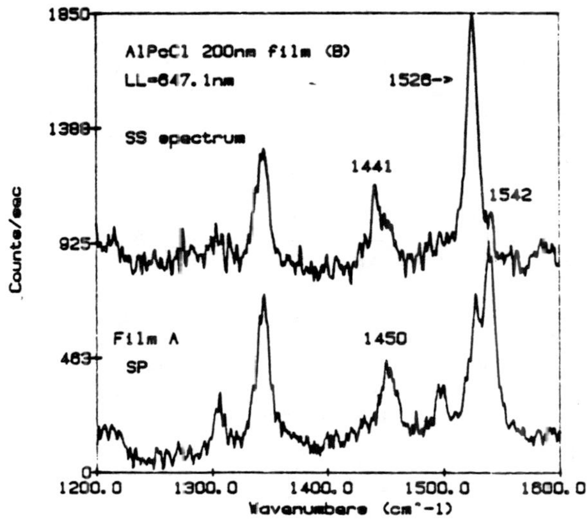

HIGH-TEMPERATURE IN-SITU RAMAN SPECTROSCOPY OF SURFACE FILMS IN AQUEOUS ENVIRONMENTS.*

C. A. Melendres, J. J. McMahon,+ and W. Ruther

Argonne National Laboratory
Argonne, Illinois 60439, U.S.A.

The application of laser Raman spectroscopy to the study of corrosion films on metal surfaces has been amply demonstrated. We have now extended use of the technique for "in-situ" measurements in high-temperature/high-pressure aqueous environments. Construction of the experimental facility and initial results on the identification of anodic corrosion films on lead formed in very dilute sulfate solution from 25 to 280°C and ∿100 atm pressure are presented.

The cell for spectroscopic and electrochemical studies is shown in Figure 1. The cell body (A) is made of Ti and measures about 2 1/4" dia x 3" long. Provision is made to admit the laser beam through a 3/8" dia x 7/16" thick sapphire window (S_1) and to collect the scattered light through a second window (S_2), which is 7/8" dia x 3/8" thick. The corrosion specimen under study is in the form of a 3/16" diameter disc and used as the working electrode (W). A Pt counter electrode and a Ag/Ag_2SO_4 reference electrode are fitted into ports B and C, respectively. The electrolyte solution is pumped from an external reservoir through port D and exits through port C. Port E is used either as an outlet for scattered light, for inserting a thermocouple, or the counter electrode. The cell is heated by two heater cartridges (H) which are imbedded in the cell body. Thermocouples are similarly mounted. All seals are made of teflon or metal reinforced teflon ring seals available commercially. Electrical isolation of the electrodes is achieved through the use of heat-shrinkable teflon sheaths (e.g.,T). A viewing port (F) fitted with a sapphire window is also provided to facilitate positioning of the working electrode. In addition, a specially designed shaft seal (G) allows movement of the working electrode for optical alignment. Details of construction of the cell, the electrodes, and the recirculation system will be presented. Excitation was by Ar^+ laser using the 5145 Å line at power levels between 50 and 200 mW. Standard electrochemical equipment consisting of a potentiostat and a universal programmer was used.

Electrochemical and laser Raman spectroscopic measurements were undertaken on Pb in 7 x 10^{-4} M Na_2SO_4 (pH = 6.2). The inset in Fig. 2 shows a cyclic voltammogram of Pb at 280°C taken at a scan rate of 1mV/sec. Typical active dissolution and passivation behavior are observed even at this temperature. Figure 2 shows a spectrum obtained "in-situ" at 280°C for the passive film produced on anodization at -0.10 V vs. Ag/Ag_2SO_4. The observed spectrum corresponds to $PbSO_4$. Assuming that the cathodic wave III corresponds wholly to the reduction of the surface film, the integrated charge under the wave indicates a film thickness of about 1000-3000 Å. Formation of a surface film of composition $Pb_4O_3SO_4$ was also found to occur on standing at open-circuit at about -0.90 V. The observation of a phase identified as Pb_2OSO_4 appears to correlate with the occurrence of wave I in the cyclic voltammogram at 280°C and lower temperatures.

*Work supported by the U.S. Department of Energy, Office of Basic Energy Sciences, Division of Materials Science, under Contract W-31-109-Eng-38.
+Permanent address: Williams College, Williamstown, MA 01267.

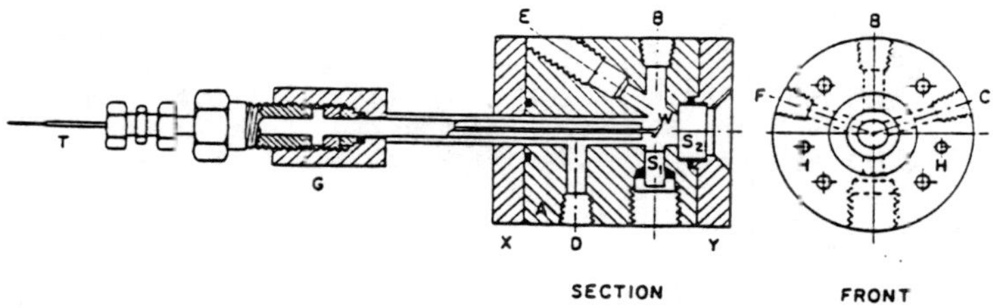

Fig. 1. High-Temperature/High-Pressure Laser Raman Spectro-electrochemical Cell

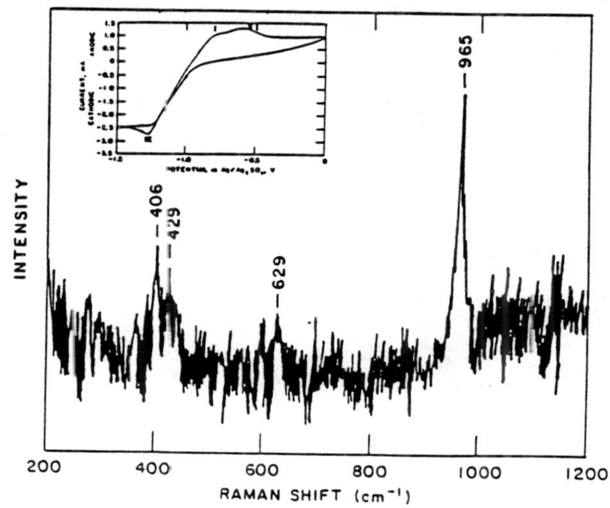

Fig. 2. "In-situ" Laser Raman Spectrum of Surface Film formed on Pb in 6×10^{-4} M Na_2SO_4 at 280°C
Ar^+ laser, 5145 A° line, P ∿ 100 mW
Inset - Cyclic Voltammogram of Pb at 280°C.

STUDY OF THIN TITANIUM OXIDE FILMS BY INTERFERENCE-ENHANCED RAMAN SPECTROSCOPY.

A. Hugot-Le Goff

Physique des Liquides et Electrochimie, 4 place Jussieu 75005 PARIS

The films grown by anodic oxidation of titanium present two different crystallographic organisations, in function of their oxidation voltage. If it is higher than 50V, the films are composed of anatase. At lower voltages, they are microcrystallised, each aggregate having the local order of anatase, and containing about 20-25 atoms.

These thin films give a Raman spectrum quite different from the spectrum of anatase given by films grown at higher voltage (figure 1a). As seen on figure 1b, the spectrum of thin films present a bump at 350 cm^{-1}, but not the two peaks at 520 and 398 cm^{-1}.

We have found again this particular structure on a native oxide (thickness : 7nm) obtained by leaving at the atmosphere exposure, during a long period, an electropolished titanium sample, in spite of the fact that the native oxide is generally assumed to be rutile. This spectrum must be characteristical of very thin films, or, more generally, of films having only a local order. It can be obtained using a long accumulation time (2h30).

The enhancement observed when the film was in optical interference conditions with the laser wavelength, made easier the obtainement of spectra on thin films. We have plotted on figure 2 the reflectance of oxidised titanium measured at 514.5 nm (laser wavelength). Of course, it passes by interferences maxima and minima, in function of the thickness of film, d. We have plotted on the same figure the intensity of the main Raman peak (630 cm^{-1}), which passes by a maximum exactly when the film is in interference condition. The enhancement ratio is about 20 for the first interference condition, and 10 for the second. It is an interesting possibility to study a material present in low amounts at the surface of a metallic electrode.

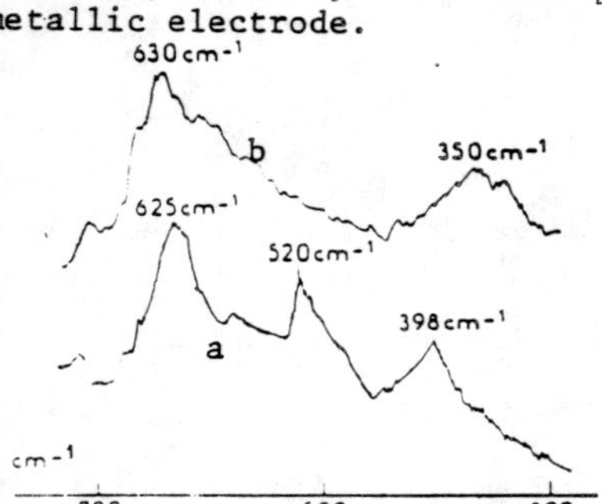

Figure 1

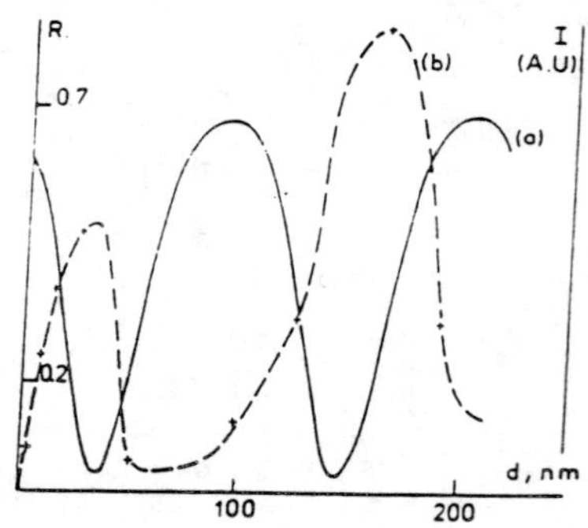

Figure 2

ELECTROCHROME BEHAVIOR OF WO_3 FILMS STUDIED BY RAMAN SPECTROSCOPY.

P. Delichère, P. Falaras, A. Hugot-Le Goff

Physique des Liquides et Electrochimie, 4 place Jussieu 75005 PARIS

WO_3 is an "electrochrome" material, i.e., its optical absorption index in the visible range increases when the material is polarised in electrolytic solutions at low cathodic potential. The coloration is obtained in about 500 ms. It is due to the insertion in the oxide, of cations present in the solution : H^+ in sulfuric medium, or Li^+, Na^+ in PC medium where lithium or sodium perchlorate has been added. The formation of a bronze during the coloration (destroyed during the bleaching) was generally postulated.

The films are porous and, if grown at voltages lower than 58V, they are amorphous. At higher voltages, they are formed by the hexagonal modification of WO_3.

On figure 1, the development of the Raman spectrum of an hexagonal film is followed when the coloration goes on in a PC + $LiClO_4$ solution. The two peaks characteristical of this WO_3 modification decrease and then desappear, and a pic characteristical of terminal groups W=0 (950 cm^{-1}) appears and grows. In this case, the inserred element is not put in definite lattice sites, but in desordered places, just leading to the weakening of the WO_3 crystal polarisability tensor. The appearance of the W=0 peak is related to the breakening of lattice continuity.

On figure 2, the coloration has been obtained in a sulfuric medium and one has to deal with a quite different mechanisme since it leads to the formation of hexagonal tungstic acid, WO_3 1/3 H_2O.

But in no case, any bronze spectrum has appeared.

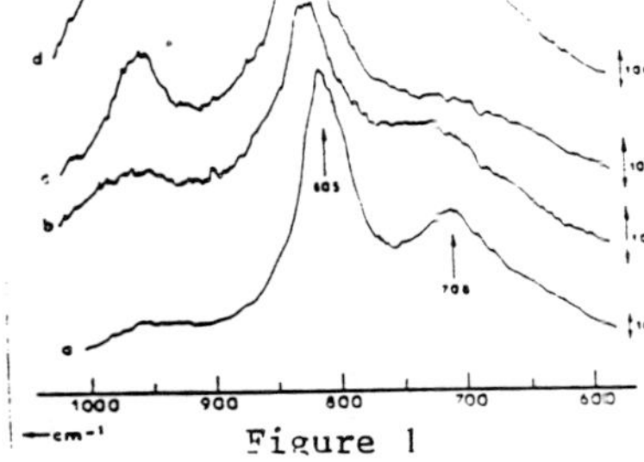

Figure 1

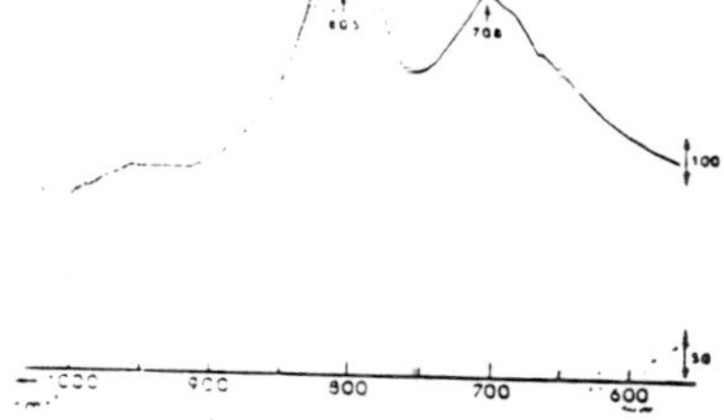

Figure 2

IN-SITU RAMAN SPECTROSCOPY FOR THE STUDY OF IRON PASSIVITY.

A. Hugot-Le Goff, C. Pallotta

Physique des Liquides et Electrochimie, 4 place Jussieu 75005 PARIS

The nature of the films present during the electrochemical passivity of iron and steels in acidic solutions is a really controversial topic in the field of electrochemistry. It is essential to do this identification in-situ during the sample polarisation. In-situ indirect method, like ellipsometry, did not give infaillible results. The presence of the γ-modification of oxide or oxihydroxide was often assumed in the literature.

The iron sample was left under polarisation in the passive range in 1N sulfuric acid solution, during several hours. The reflectance of the iron surface decreased immediately after the polarisation, attesting the presence of an absorbing layer in front of the electrode, but no Raman spectrum is observed. After 8 hours, a slight increase is seen in the reflectance (corresponding to a change of the absorption index of the layer ?) and the spectrum given on figure was obtained. It corresponds to a mixture of α-Fe_2O_3 and Fe_3O_4. This mixture is perfectly stable and can be studied out of the solution. Its thickness, measured by direct observation of nuclear reaction, is 10 nm (the thickness of a "true" passive film, ellipsometrically measured after about 1 hour of polarisation is 3nm).

We have analysed these data by comparison with electrochemical kinetics results. The lack of spectrum during the first hours of polarisation cannot be related to an insufficient film thickness since the spectrum cannot be restored by extending the duration of data accumulation. It is due to the "gel" nature of the passive film, the ageing of which involves its crystallisation. Anyhow, neither γ-, nor α-FeOOH can be present during the passivity. These two forms give intense Raman spectra and would be identified while amorphous, β- and δ- modifications of FeOOH have not.

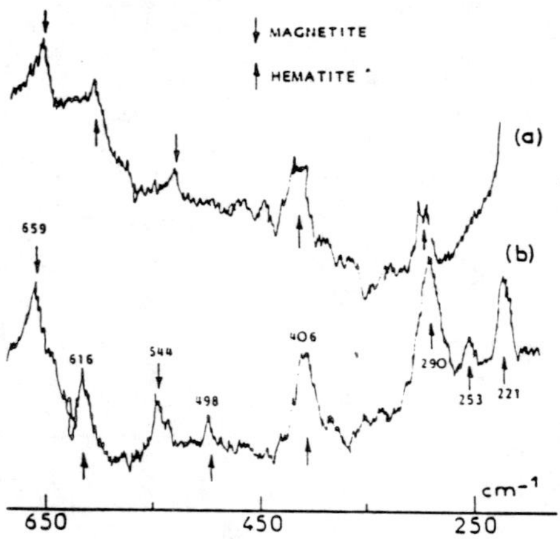

chapter 7

SEMICONDUCTORS

RAMAN SCATTERING BY MAGNETIC EXCITATIONS IN DILUTED MAGNETIC SEMICONDUCTORS.

S. Rodriguez and A. K. Ramdas

Department of Physics, Purdue University, W. Lafayette, IN 47907, U.S.A.

A semiconductor AB where A and B are, respectively, group II and VI elements, upon being alloyed with a compound MB where M is an element of the iron group, forms part of a class of materials usually called diluted magnetic semiconductors (DMS). If x is the atomic concentration of M, the formula of a DMS is written $A_{1-x}M_xB$. A typical example is $Cd_{1-x}Mn_xTe$ which can form homogeneous solid solutions for x ranging from 0 to 0.75. These materials have a unique combination of semiconducting and magnetic properties. $Cd_{1-x}Mn_xTe$ crystallizes in the zinc-blende structure for $0 \leq x \leq 0.75$ in which at random cation sites Cd is replaced by Mn atoms. For $x < 0.17$, $Cd_{1-x}Mn_xTe$ is paramagnetic for all temperatures. If $x > 0.17$ there is a magnetically ordered phase due to antiferromagnetic coupling between neighboring Mn^{2+} ions below a critical temperature $T_c(x)$. For example, $Cd_{1-x}Mn_xTe$ with $x = 0.75$ is in a type 2 antiferromagnetic phase below 40K.

Raman scattering studies provide significant information on the nature of the magnetic excitations of these materials. Transitions in which a quantum of angular momentum is exchanged between the system and the radiation field give rise to Raman lines whose shifts equal the energy of the magnetic excitations in which the total spin of the material changes by $\hbar$. These transitions are in one-to-one correspondence with those observed in magnetic resonance experiments but are carried out in the optical rather than in the microwave region of the electromagnetic spectrum.

The electric polarization $\vec{P}$ is a functional of the magnetization as well as of the other variables describing internal modes of motion. Expanding $\vec{P}$ in a power series in $\vec{M}$, invoking the Onsager reciprocity relations, the lack of absorption in the frequency region of interest, and the cubic symmetry, the component of $\vec{P}$ linear in $\vec{M}$ is $\vec{P}^{(1)} = i\,G\,\vec{M} \times \vec{E}$, where G is a constant.

The scattering cross section for a first order Raman process is proportional to $|(\hat{\varepsilon}' \times \hat{\varepsilon}) \cdot \vec{M}|^2$, where $\hat{\varepsilon}$ and $\hat{\varepsilon}'$ are unit vectors along the directions of polarization of the incident and scattered radiation, respectively. Thus Raman scattering does not occur when the polarizations of the incident and scattered light are parallel. With incident radiation circularly polarized propagating parallel to the z-axis selected along $\vec{H}$ and $\vec{E} = (\hat{x} \pm i\hat{y})E_o\, e^{-i\omega t}$, $\vec{P}^{(1)}$ is proportional to $\mp \hat{z}\, E_o \exp(-i\omega t \pm i\omega_R t)$, where $\omega_R = \omega_{PM} = -\gamma H$, in the paramagnetic phase and $\omega_R = \omega_M^+(-\omega_M^-)$ in the ordered phase. Thus, in this geometry, there is a Stokes (anti-Stokes) line with polarization $(\hat{\sigma}_+, \hat{z})$ and an anti-Stokes (Stokes) line with $(\hat{\sigma}_-, \hat{z})$

where $\hat{\sigma}_+$ and $\hat{\sigma}_-$ denote positive and negative helicites, respectively. In a similar way, it can be shown that the ω_{PM} and $\omega_M^+(\omega_M^-)$ Stokes (anti-Stokes) and anti-Stokes (Stokes) lines also occur in the geometries $(\hat{z},\hat{\sigma}_-)$ and $(\hat{z},\hat{\sigma}_+)$, respectively. The two cases described above are, of course, related to each other by time reversal symmetry.

In DMS the following phenomena of interest have been observed:

(i) Raman scattering in the paramagnetic phase.[1] There is a pronounced feature observed in the presence of a magnetic field only. The Raman shifts $\hbar\omega_{PM}$ correspond to transitions between adjacent sub-levels of the Zeeman split ground state $^6S_{5/2}$ of the Mn^{2+} ion. The g-factor is $g = 2.01 \pm 0.02$ in agreement with the expectation for Mn^{2+}.

(ii) Raman scattering in the magnetic phase.[1] The Raman line at ω_{PM} observed in the paramagnetic phase of $Cd_{1-x}Mn_xTe$ broadens and shifts to higher frequencies as the temperature approaches the critical temperature $T_c(x)$. Below $T_c(x)$ the excitations responsible for the Raman lines are described as spin waves. The spin wave excitation is split by the magnetic field into two components M_+ and M_- observed as Stokes and anti-Stokes lines in the $(\hat{\sigma}_+,\hat{z})$ geometry, respectively. In the $(\hat{\sigma}_-,\hat{z})$ geometry they are reversed. In the absence of a magnetic field M_+ and M_- coalesce into a line M giving the energy of the long wavelength spin waves in zero magnetic field. The continuous evolution of ω_{PM} into $M_\pm$ as the temperature is lowered shows that these excitations involve a change of angular momentum equal to $\hbar$. It thus corresponds to a single spin wave.

(iii) Electrons bound to donors.[2] The large exchange coupling between the 3d electrons of the Mn^{2+} ions and the conduction electrons gives rise to a large Zeeman splitting of the conduction electrons. Spin-flip Raman lines have been observed which are attributed to transitions between the Zeeman levels of the ground state of donors in DMS. The Raman shift, $\hbar\omega_{SF}$, in the paramagnetic phase exhibits a temperature dependence and saturation characteristic of the magnetization of Mn^{2+} in the external magnetic field. The exchange field on the donor electrons predominates over the Zeeman splitting caused by the applied magnetic field. At sufficient concentration x, and $T > T_c(x)$ a zero field line is observed due to magnetization caused by Mn^{2+}-donor-electron exchange (bound magnetic polaron).

(iv) Magnetic field and temperature tuning of resonant Raman scattering in diluted magnetic semiconductors.[3] A resonant enhancement of the Raman line in $Cd_{1-x}Mn_xTe$ associated with LO phonons in combination with the spin flip of Mn^{2+} is observed. The resonance condition is fulfilled when the large magnetic-field-induced exciton splitting allows one of its components to be brought into coincidence with the Raman line. The exciton splittings are due to the large effective g factors of the conduction and valence bands arising from the strong Mn^{2+}-band-electron exchange interaction. The associated temperature tuning is also observed in a constant magnetic field. The feasibility of magnetic field or temperature tuning of resonance conditions for Raman scattering in DMS is a result of the sensitive manner in which excitonic splittings depend on magnetic field, temperature, and concentration of magnetic ions.

---

[1]A. Petrou, *et al.*, Phys. Rev. B27, 3471 (1983).

[2]D. L. Peterson, *et al.*, Phys. Rev. B32, 323 (1985).

[3]D. L. Peterson, *et al.*, Phys. Rev. B31, 7932 (1985).

RAMAN SCATTERING STUDIES OF OPTICALLY EXCITED GaAs

Heinz Nather

Physikalisches Institut, Teil 4, Universität Stuttgart
Pfaffenwaldring 57, D-7000 Stuttgart 80, Fed. Rep. Germany

present address: Telefunken electronic GmbH, Abt. T2/E12
Theresienstrasse 2, D-7100 Heilbronn, Fed. Rep. Germany

Absorbed laser light can generate an electron-hole plasma (EHP) in intrinsic semiconductors. These plasmas are characterized by the presence of more than one mobile species (multicomponent plasma). The interaction between the ensemble of carriers (electrons, light-, heavy-holes) and the elementary excitation of longitudinal phonons (LO) as well as interaction among the carriers is responsible for the collective plasmon-LO-phonon modes (PLP) and single-particle-excitations (SPE), as seen in the inelastic light scattering experiments. The correlation of the EHP parameters, density and temperature with energy and lineshape of the Raman-spectra allow the EHP to be investigated in an extended range of these parameters. The present studies are concerned with an application of Raman-spectroscopy to low and high temperature EHP at different density levels.

Because of the resonant character of the scattering cross-section, only scattering around the $E_0 + \Delta_0$ energy gap is considered. The excitation energy of the photo-created EHP, however, changes between the fundamental gap and $E_0 + \Delta_0$ in order to vary the electronic temperature of the EHP. In this case the dominating scattering contributions fall into two categories: 1. collective excitations (PLP) due to charge carrier density fluctuations with parallel polarization and 2. single particle scattering (SPE) which couples to the laser radiation via spin density fluctuations that show perpendicular polarization.

Figure 1a shows a typical polarized spectrum $\bar{x}(yy)x$ for a highly optically excited GaAs sample. This spectrum is assigned to charge density fluctuations and shows two well defined bands at the energies of the coupled PLP-modes. The spectral lineshape is particularly pronounced in Fig. 1b (dots) after subtraction of the broad split-off band luminescence background at 1.91eV. A quantitative analysis of the spectrum, i.e. the calculation of the lineshape, requires the total longitudinal dielectric function $\varepsilon(q,\omega)$ including the intraband contributions of electrons, light- and heavy-holes. Furthermore, one has to include the intervalence band transitions between the light- and heavy-hole bands /1/. The solid line in Fig. 1b corresponds to a best fit by variation of EHP density and temperature. The linewidth as well as the relative intensities of the two coupled modes are simultaneously fitted by the theory. An analysis of the different contributions to the cross-section indicate that the inter-valence band transitions of photo-excited holes have a substantial influence on the damping of the coupled modes, although the peak position is well approximated by the simple Drude expression, which neglects any damping. The simple relationship between the EHP density and the energy of the coupled modes distinguish this scattering process and is therefore best suited to measure the plasma density.

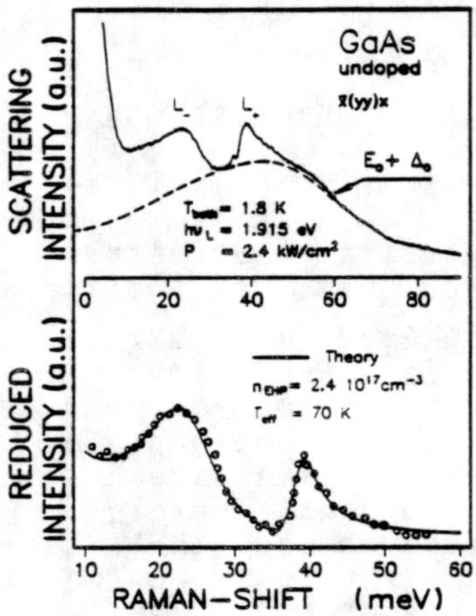

Fig. 1 Coupled plasmon-LO-phonon mode spectrum of photo-excited GaAs.

Typical examples for depolarized spectra $\bar{x}(yz)x$ are shown in Fig. 2 after subtraction of the luminescence background. Similar to the case of n-type GaAs, such spectra are caused by spin density fluctuations of the free electrons and reflect directly the single particle excitations of the EHP. In a first order approximation for the scattering cross-section only the electronic dielectric function has to be taken into account. The best fits are shown by the solid line in Fig. 2. It should be pointed out that the plasma density and temperature determined from the SPE spectra agree very well with the values obtained from the coupled mode analysis and photoluminescence fittings.

In conclusion, we can state that the inelastic light scattering has proven to be an accurate and versatile tool in measuring the equilibrium properties of photo-excited EHP. From the spectral features of high temperature EHP we deduce novel correlation of EHP-density and temperature with recombination channels.

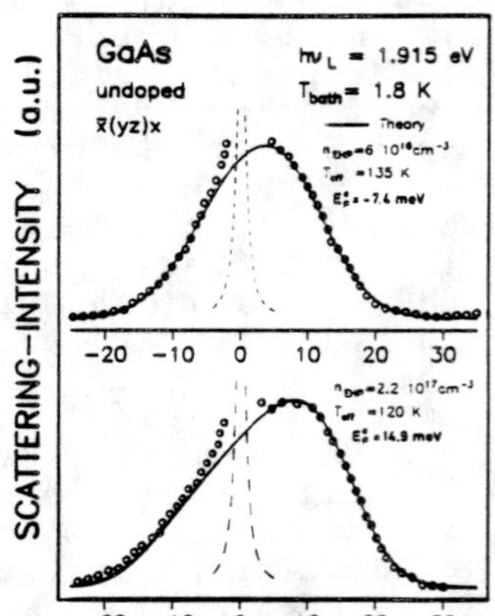

Fig. 2 Single-particle spectra of GaAs.

1/ P.M. Platzman, P.A. Wolff, in "Waves and Interactions in Solid State Plasmas", Academic Press, New York 1973.

SUB-PICOSECOND RAMAN STUDIES OF CARRIER RELAXATION IN GaAs

J. C. Tsang and J. A. Kash
IBM Thomas J. Watson Research Center, P. O. Box 218, Yorktown Heights, New York 10598

The advent of sub-picosecond optical sources raised the possibility of sub-picosecond time resolved studies of dynamic processes in semiconductors.[1,2] Carrier relaxation times in III-V semiconductors are determined by the carrier-phonon and carrier-carrier scattering times which are generally less than 1 ps.[3] Therefore, direct quantitative studies of hot carrier relaxation became feasible. We have shown [4] that the combination of these sub-picosecond sources and new, more sensitive optical detectors now make it possible to study these phenomena using time resolved Raman spectroscopy and high energy hot luminesence as probes. The dynamics of the initial relaxation of optically injected carriers with average excess energies far greater than any phonon energy can be directly studied through the time dependence of Raman scattering from the phonons generated by the relaxing carriers and through measurements of the hot luminesence for energies far above the band gap. We demonstrate that these probes can effectively complement the more traditional optical probes[5] in detailing the dynamics of hot carriers in semiconductors such as GaAs.

It is difficult to achieve sub-picosecond temporal resolution in optical studies by electrical means. However, the pump-probe technique where one beam is used to excite a sample and a second to study the response of the sample as a function of time after the excitation can readily measure changes on the sub-picosecond time scale. The temporal resolution of the pump-probe technique is solely limited by the pulse width of the excitation source and the ability to measure the delay between two optical pulses. In our Raman experiments, the pump beam is used to excite carriers which relax to the band edges through the excitation of small wavevector LO phonons.[5] The probe beam excites Raman scattering from both the thermal and the non-equilibrium, carrier generated phonons, as a function of the delay between the pump and the probe. The variation of the LO phonon scattering with the delay between the pump and the probe provides a unique time resolved view of the process by which energy is transferred from the carriers to the lattice through the excitation of LO phonons. In the case of hot luminesence from band to band transitions, the bi-molecular nature of the recombination process involving free holes and electrons whose concentrations are not recombination limited [6] means that the intensity of the emission varies as I^2 where I is the intensity of the excitation source. If the injected carrier densities are high enough so that carrier-carrier scattering can quickly randomize the carrier distribution in momentum space, then the the density of electrons and holes separated by a given energy at a given time determines the intensity of the emission at that energy and time. von der Linde et al. showed that correlation type measurements can be performed on this emission providing information on the cooling of the carriers.[7] While the analysis of this data is very difficult for times greater than 10 ps, high energy emission where the correlation times is below 2 to 3 picoseconds can be analyzed.

Both the Raman and the non-equilibrium luminesence measurements are severely limited by the low efficiencies of the scattering and recombination processes. When these are combined with the maximum average powers available from the 76 MHz repetition rate, 0. ps width, pulsed dye laser source used in this experiments (<40 mw) the powers availabl for the probe beams (<8 mw) and the spectral widths of the probe (30 cm^{-1}) for sub picosecond temporal widths, the signal levels of interest can be an order of magnitude smalle than those detectable by a conventional Raman scattering system with a single channe photomultiplier detector. Microchannel plate photodetectors with resistive anodes have quantum efficiency comparable to that of a conventional single channel photomultiplier, dark count per channel at least two orders of magnitude smaller and the capability to simul taneously obtain data over 400 channels.[8] The enhanced capabilities of these detectors mak it possible to perform sub-picosecond time resolved Raman scattering experiments fron opaque semiconductors.

The time required for the generation of the non-equilibrium LO phonon distribution b the optically injected hot carriers is short compared to the phonon lifetime. This results in non-equilibrium population of LO phonons. The dispersion of the conduction bands of thes materials is such that the hot phonons generated by the relaxing carriers are first orde Raman active. The increase in the intensity of the LO phonon scattering for injected carrie densities below $10^{17}cm^{-3}$ has been measured as a function of the delay between the pum and the probe. From this, we have been able to directly deduce that the generation time c an LO phonon by a hot electron is about 160 fs.[4] This result is based on a simple pictur where the optically injected carriers are non-interacting and do not modify the phono dispersion. The introduction of injected carrier densities in excess of $2x10^{17}cm^{-3}$ results i a breakdown of this picture. Modifications of the dynamic behavior of the excess phono population can arise from finite wave vector effects due to the the back scattering geometr used in these experiments and the screening of the Raman active LO phonon. When the in jected carrier density is well above 10^{18} cm^{-3}, the intensity of the pump induced LO phono scattering no longer varies as the product of the pump and probe intensities. This is due t screening of the LO phonons by the optically injected carriers.

When the injected carrier concentration exceeds $10^{17}cm^{-3}$, broad band anti Stoke emission has been observed more than .9 eV above the band gap. This emission is due t recombination of non-equilibrium thermalized carriers. The correlation time for th excitation of this emission by a pump and a probe beam depends strongly on the energy o the emission. For energies far above the band gap, the correlation time can approach 1 ps This correlation time can be related to the rate at which energy is transferred from th electron distribution to the lattice and we find the correlation time is consistent wit electron-phonon interaction times of the order of 0.2 ps.

ACKNOWLEDGEMENT: We thank Professor S. S. Jha of the Tata Institute for Fundamental Research for many stimulating conversations.

REFERENCES

1. C. V. Shank, R. L. Fork, R. F. Leheny and J. Shah, Phys. Rev. Lett.42, 112 (1979).
2. J. L. Oudar, A. Migus, D. Hulin, G. Grillon, J. Etchepare and A. Antonetti, Phys. Rev. Lett. 53, 384 (1984).
3. E. M. Conwell and M. O. Vassell, IEEE Trans. Electron. Devices 13, 22 (1966).
4. J. A. Kash, J. C. Tsang and J. M. Hvam, Phys. Rev. Lett. 54, 2151 (1985).
5. J. Shah, Solid State Electronics 21, 43 (1978).
6. D. von der Linde, J. Kuhl and H. Klingenburg, Phys. Rev. Lett. 44, 1505 (1980).
7. D. von der Linde, J. Kuhl and E. Rosengart, J. Lumin. 24/25, 675 (1981).
8. J. Tsang in Dynamics on Surfaces, ed. by B. Pullman et al. (Reidel, Dordrecht, 1984). p. 379.

TIME RESOLVED CARS MEASUREMENTS OF COHERENT EXCITATIONS OF SEMICONDUCTORS.

W. E. Bron

Indiana University, Bloomington, IN 47405

Generation of coherent optical phonons is achieved through coherent Raman excitation (CRE) and detection is (for the most part) obtained through time resolved coherent anti-Stokes Raman scattering (TRCARS).[1,2] The coherent phonon state most closely resembles a damped classical harmonic oscillator, and is found to be driven by forces involving a Raman interaction and by forces associated with the longitudinal component of the nonlinear polarization.

It is observed that there are two contributions to the TRCARS signal $I_c(\Delta t)$; one due to the nonlinear response of the lattice, $I_c^l(\Delta t)$ and one due to the nonlinear response of the electronic system, $I_c^e(\Delta t)$, of the solid. As has been demonstrated in ref. (1), the contribution due to I_c^e is observed only as long as the temporal overlap exists among the three laser beams (~ 50 ps), whereas I_c^l can be observed for times of the order of hundreds of picoseconds. Hence, the lattice component represents free fall dephasing of the coherent phonon state. Moreover, the temperature dependence of the dephasing time, $T_2/2$, of LO phonons is observed[1,2] to correspond to three-phonon anharmonic processes within a temperature range from 0 to 150K. In addition, it is observed[1,2] that the temperature dependence of $T_2/2$ as measured via $I_c^l(\Delta t)$ is the same as that observed in traditional measurements of the linewidth of the spontaneous (incoherent) Raman scattering intensity.

It is also possible to extract from $I_c(\Delta t)$ all the components of the complex third order electronic susceptibility[2,3] $\chi^{(3)}$. For example, the components for GaP under our experimental conditions are found (in units of 10^{-10} esu) to be: $\chi'_{1221} = 2.3 \pm 0.3$, $\chi''_{1122} = -0.8 \pm 0.8$, $\chi'_{1221} = 2.9 \pm 0/2$, $\chi''_{1221} = -0.7 \pm 0.3$, $\chi'_{1111} = 1.7 \pm 0.5$ and $\chi''_{111.} = -0.3 \pm 1.1$. The corresponding values for ZnSe (in units of 10^{-11} esu) are: $\chi'_{1122} = 9.1 \pm 1.6$, $\chi''_{1122} = -4.0 \pm 1.4$, $\chi'_{1221} = 6.9 \pm 0.5$, $\chi''_{1221} = -1.7 \pm$

2.0, $\chi'_{1111} = 9.2 \pm 1.9$ and $\chi''_{1111} = -4.6 \pm 3.0$. Nonzero values of χ'' (the imaginary components) imply that real multiphoton absorption of laser light must have occurred.

The LO phonons under observation possess an energy $\hbar\omega \approx 50$ meV and wavevector $|\vec{q}| \approx 3000\ \mathrm{cm}^{-1}$. Thus electrons which scatter from these phonons can gain or lose energy and momentum of these amounts. But this turns out not to be possible for the following reasons. The important relevant parameter is the curvature of the conduction band surface. For example, if one assumes a very simple parabolic conduction band for, say, GaP and an effective mass of $m^* = 0.29\ m_e$ one finds that LO phonons which can interact with an electron must possess $|\vec{q}| \approx 10^5\ \mathrm{cm}^{-1}$ or higher. This holds for electronic states which are up to of the order of 1eV above the band minimum. Thus, $T_2/2$ as measured for LO phonons of $|\vec{q}| \approx 3000\ \mathrm{cm}^{-1}$ is not influenced by the presence of hot electrons. Nor is it influenced by the presence of an electron-hole plasma, at some conduction band minimum, which is formed as the result of the decay of the hot electrons through optical phonon emission. This has indeed been confirmed experimentally as long as the plasma concentration is less than $\sim 5 \times 10^{17}\ \mathrm{cm}^{-3}$, which is the case here.[4] However, it is observed[5] that the plasma does interfere with the generation of the coherent state through its influence on the nonlinear polarization. In fact, it is observed that optical phonon generation through CRE goes through a maximum, and actually is totally destroyed as the concentration of electron-hole plasma increases.

The examples, presented above, of coherent nonlinear excitation in semiconductors due to interaction with laser light, are but a few which illustrate the potential of these techniques. More investigations on the possible existence of stimulated phonon decay, phonon breakdown, phonon renormalization effects, and free-electron-phonon interactions are either underway or planned for the future.

The author acknowledges the collaboration of Drs. B. K. Rhee and J. Kuhl, and financial support through ARO DAAG29-83-K-0091.

REFERENCES

1. J. Kuhl and W.E. Bron, Physica 117B/118B, 532 (1983), Solid State Commun. 49, 935 (1984).
2. W.E. Bron, J. Kuhl, B.K. Rhee, submitted for publication.
3. B.K. Rhee, W.E. Bron, and J. Kuhl, Phys. Rev. B30, 7358 (1984).
4. J.E. Kardontchik and E. Cohen, Phys. Rev. Lett. 42, 669 (1979)
5. B.K. Rhee and W.E. Bron, submitted for publication elsewhere.

STUDY OF DYNAMICS OF PHOTOEXCITED ELECTRON-HOLE PLASMAS IN GaAs BY PICOSECOND TIME-RESOLVED RAMAN SCATTERING.

Yihe Huang and Peter Y. Yu

Department of Physics, University of California and Materials and Molecular Research Division, Lawrence Berkeley Laboratory, Berkeley, CA 94720.

The dynamics of hot carriers excited by picosecond and femtosecond laser pulses have been of great interest recently. So far techniques for studying these hot carriers have been limited to absorption, reflection and emission. Raman scattering has the potential of obtaining more complete information on the density and distribution function of the hot carriers than the above techniques. We have succeeded in obtaining time-resolved Raman spectra from photoexcited hot carriers in GaAs with picoseconds resolution using a pump-probe technique.

The sample, a high purity layer of GaAs grown by liquid phase epitaxy, is cooled to 77 K in an optical dewar. The output of a synchronously pumped modelocked dye laser, consisting of a train of 4 picosecond (psec) long pulses separated by 12 nsec, is split into two beams. One beam (A) containing 90% of the laser energy is used to excite hot carriers in the sample while the weaker probe beam (B) is suitably delayed and probes the hot carriers by Raman scattering. A system of shutters controlled by a microcomputer is used to eliminate the Raman scattering from the individual beams and to extract only the Raman signal arising from the scattering of beam B by the carriers excited by beam A. By varying the delay between zero and <30 psec the evolution of the hot carriers can be monitored.

From the Raman scattering by the coupled longitudinal optical (LO) phonon-plasmon mode the maximum density of the hot carrier was determined to be about 5×10^{18} cm^{-3}. The hot carrier density decreases by roughly a factor of six in 20 psec due to expansion into the bulk. The expansion velocity is estimated to be $< 10^{7}$ cm/sec.

The lineshape of the Raman spectra due to single particle scatterings is compared with existing theories. It is found that theories which neglect carrier relaxation fail to explain the experimental lineshapes. A recent theory proposed by Hertel and Appel which includes carrier relaxation gives the most satisfactory fit to the time-dependent Raman spectra. From these fits of the experimental spectra with the theory of Hertel and Appel the temperature and relaxation time of the hot carriers have been determined as a function of time. At zero time delay the hot carrier temperature was found to be about 800 K. After 20 psec the carriers cooled to about 400 K. During this time interval the carrier relaxation time increased from 3.5 femtoseconds (fsec) to 12 fsec. These relaxation time are consistent with carrier-carrier scattering times in GaAs determined recently by femtosecond laser spectroscopy at considerably lower hot carrier densities.

The carrier cooling rate deduced from our Raman scattering experiment is about one hundred times smaller than the theoretical value based on emission of LO phonons. To resolve this difference between theory and experiment we have determined the LO phonon temperature from the ratio of the Anti-Stokes to Stokes Raman intensities. The LO phonon temperature was found to be about 800 K also at zero time delay. This suggests that within the laser pulse width of 4 psec the hot carrier and the LO phonons have already reached a quasi-thermal equilibrium so the cooling rate of the hot carriers is now determined by the lifetime of the hot LO phonons. The observed hot carrier cooling rate can be explained satisfactorily by this hot phonon effect.

ACKNOWLEDGEMENT

This work was supported by the Director, Office of Energy Research, Office of Basic Energy Sciences, Materials Sciences Division of the U.S. Department of Energy under Contract Number DE-AC03-76SF00098.

THEORY OF TIME-RESOLVED RAMAN SPECTROSCOPY OF NON-EQUILIBRIUM CARRIERS AND PHONONS IN SEMICONDUCTORS

Sudhanshu S. Jha

Tata Institute of Fundamental Research, Bombay 400005, India

Recent experimental developments [1] in time-resolved Raman scattering from GaAs, in the presence of photo-excited non-equilibrium carriers and LO-phonons, generated as well as probed by very short sub-picosecond laser pulses, have lead to renewed interest in the theory [2] of inelastic light scattering from non-equilibrium electronic and ionic excitations in semiconductors. Such non-equilibrium distributions are produced when light from a sub-picosecond laser pulse with photon energy above the band gap is absorbed by a semiconductor. In GaAs, the fastest dynamics involving electron-electron collisions and electron-LO phonon scattering among the various types of carriers is that of the highly excited electrons in the conduction band. For high enough excited conduction electron density $n_e(t) \simeq 5 \times 10^{17}\ cm^{-3}$, these carriers equilibriate among themselves quite fast, with the initial effective electron temperature $T_e(t)$ being quite different than the lattice temperature T_L. These carriers also relax rapidly towards the conduction band minimum, the electron -LO-phonon scattering time being of the order of 200 femtoseconds for electrons of energy 0.5 ev above the bottom of the conduction band, but rising to a few nanoseconds for electrons close to the bottom of the band. Initially, the population of LO-phonons $n_{ph}(t)$ in the sample, therefore, starts rising quite fast above the thermal equilibrium value, only to saturate and decay to its equilibrium value (with a lifetime of about 6-7 picoseconds) when the conduction electrons no longer produce these phonons on such a fast time scale. Spontaneous Raman scattering with sub-picosecond pulses and powerful multi-channel detection is capable of resolving this dynamics of changing distributions of electrons and phonons as a function of time.

The theoretical calculation of Raman scattering efficiency from the electronic as well as ionic excitations in semiconductors has been extended [2] in a very general way to the nonequilibrium case, which differs considerably from the case of usual equilibrium doped-semiconductors. For a given time-dependent carrier density $n_e(t)$, LO-phonon population $n_{ph}(t)$ and the pulse width of the incident Raman probe beam t_p, there may or may not be any steady state during the scattering process. For very short t_p, there is no energy conservation, and the coupled LO-phonon-plasmon collective mode structure and the single-particle electronic excitation density are functions of time, since the longitudinal dielectric function $\varepsilon(q,\omega,t)$ is now also a function of time.

It can be shown in general that the number of Raman scattered photons per unit solid angle, per unit frequency, for given incident and scattered unit polarizations $\vec{e}_i$ and $\vec{e}_s$, is determined by

$$\frac{dN_s(t)}{d\Omega d\omega_s} = \frac{\omega_s}{\omega_i}\left(\frac{n_i c}{2\pi}\right) \sum_f \sum_i |\langle f|\chi_{si}|i\rangle|^2 \, g_{fi}(t) \qquad (1)$$

$$g_{fi}(t) = \left|\int_0^t d\tau \, e^{-\Gamma_{fi}\tau} \, e^{-i(\omega_{fi}-\omega)\tau} \, b^{(o)}_{i\to f}(t-\tau)\right|^2 \qquad (2)$$

where $\langle f|\chi_{si}|i\rangle$ is the effective Raman transition susceptibility for excitation of the initial state i to the final state f of the system, $\hbar\omega_{fi} = \hbar(\omega_f-\omega_i)$ is the energy difference between these states, ω is the photon frequency shift $\omega_s-\omega_i$, Γ_{fi} is the phenomenological damping coefficient for the transition, n_i is the incident photon density and $b^{(o)}_{if}$ denotes the initial state amplitude of the system for transition to the final state. For simplicity, the trivial time-dependence of the occupation number of phonons, plasmons, etc. is kept in the transition matrix element $\langle f|\chi_{si}|i\rangle$, so that the nontrivial temporal evolution of the scattering efficiency is determined by the line-width function $g_{fi}(t)$.

For the observation time $t_p \ll 1/\Gamma_{fi}$ and slowly varying b_{if}, the expression (2) leads to the Raman scattering linewidth of the order of π/t_p. This implies that for proper observation of Raman modes of frequencies in the range of 100 cm^{-1}, one cannot use probe pulse of duration shorter than 200 femtoseconds or so. When the damping is appreciable, i.e. for times $t_p \gg 1/\Gamma_{fi}$, the expression (2) leads to the Lorentzian shape with the usual linewidth Γ_{fi} characteristic to the mode.

For scattering from LO-phonons, all the electrons in the system in the valence as well as conduction bands take part in the process, the number of nonequilibrium carriers being only a small part of the total. Since in this case the lifetime $1/\Gamma_{fi} \simeq 7$ picoseconds, for picosecond and sub-picosecond pulses, the linewidth is determined by the pulse width alone. The scattering signal is approximately proportional to

$$\frac{dN_s(t)}{d\Omega d\omega_s}(\text{LO-phonon}) \sim |R_{\mu si}|^2 \begin{bmatrix} n_{ph}(t)+1 & \text{Stokes} \\ n_{ph}(t) & \text{Anti Stokes} \end{bmatrix} \qquad (3)$$

for the Stokes and anti-Stokes processes, respectively, from one mode.

By calculating approximate expressions for $\varepsilon(\vec{q},\omega,t)$ for the nonequilibrium carriers, it can be shown that the collective plasma mode is not well defined here when $qV_e(t)$ is large. For qV_e greater than the plasma frequency, because of heavy Landau damping, all electronic excitations are mostly single particle type for typical values of $q \sim 8\times10^5 cm^{-1}$ being used in experiments. Because of this, the coupled LO-phonon-plasma modes can be seen in Raman experiments only in the region where the mode is phonon like. All these results seem to be borne out by recent experiments, for large hot electron average velocities $V_e(t)$.

[1] J.A. Kash, J.C. Tsang and J.M. Hvam, Phys. Rev. Lett. 54, 2151 (1985) ; H. Nather and G. Quagliano, J. Lumin. 30, 50 (1985); C. Collins and P.Y. Yu, Phys. Rev. B 30, 450 (1984).

[2] S.S. Jha, J.A. Kash and J.C. Tsang (to be published).

GREENS FUNCTION THEORY FOR THE DISORDER INDUCED RAMAN SPECTRA IN CdHgTe SYSTEM.

D. N. Talwar[+] and Michel Vandevyver[*]

[+]Department of Marine Sciences, Texas A&M University at Galveston, Texas.

[*]Department de Physico-Chimie, CEA-CEN de Saclay, Gif-sur-Yvette, France.

INTRODUCTION:

In recent years, potential device applications have spurred the theoretical and experimental research efforts to elucidate the intrinsic properties of the mixed II-VI compounds. CdHgTe is an important group of alloy systems that has established itself as the most significant material for infrared (i.r.) detection.[1)] Raman spectroscopy is a useful tool for characterization of the thin crystalline films; crystal orientation, imperfections, alloy composition and superlattice structure can be studied non destructively on the vibrational spectra.[2)]

Quite recently, Amirtharaj et. al.,[3)] have investigated the polarization dependent Raman scattering from the <100>, <110> and <111> surfaces of p-$Cd_xHg_{1-x}Te$ ($x \sim 0.2$, $E_g = 100$meV) and HgTe (zero band gap) single crystals. Besides the normally allowed $\vec{q}=0$ longitudinal (LO) and transverse (TO) optical phonon peaks (*two* mode behavior), the authors of Ref. [3] have also observed several other features arising from the deviation of the perfect crystallinity (e.g., disorder-induced Raman active "impurity-mode" or "clustering mode", etc.).

THEORETICAL CONSIDERATIONS:

From the existing i.r. data of impurity modes and using the well known Greens function (G.F.) theory,[4)] we are able to estimate the nature of nearest neighbor (n.n.) force perturbation caused by Hg (and Cd) impurity in *Cd*Te (and *Hg*Te) system. To treat the impurity induced Raman scattering problem in zinc-blende type crystals, we follow the off resonance theory of Xinh[5)] and define the intensity of linearly polarized Raman scattering for unit solid angle Ω and for unit frequency as:

$$I(\omega_i:\omega_s:\Omega) = \frac{\hbar.\omega_i^{L}}{2\pi^2 c^{3}} \sum_{\substack{\alpha\alpha' \\ \beta\beta' \\ \gamma\gamma'}} \sum_{\substack{\ell\kappa \\ \ell'\kappa'}} n_\alpha n_{\alpha'} E_\beta E^*_{\beta'} n(\omega_i-\omega_s) P_{\alpha\beta',\gamma}(\ell\kappa) \times P_{\alpha'\beta',\gamma'}(\ell'\kappa') \lim_{\xi\to 0} \mathrm{Im} U_{\gamma\gamma'}(\ell\kappa;\ell'\kappa':z), \quad (1)$$

where ω_i and ω_s are the frequencies of the incident and scattered light; n_α is the α component of a unit polarization vector parallel to the scattered light; E_β is the β component of the incident field; $U(\equiv G^0(1-PG^0))$ is the Greens function matrix of the imperfect crystal (related with the G.F. matrix for the perfect system G^0 and the perturbation matrix P caused by the defect); $n(\omega_i-\omega_s)$ is the thermal occupation number; z stands for

$\omega^2 - i\xi$, the limit $\xi \to 0$ being understood and ω is near the absolute value $(\omega_i - \omega_s)$.

NUMERICAL COMPUTATIONS, RESULTS AND CONCLUSIONS:

In actual numerical calculations for the scattering tensor (using Eq. 1), the effect of an impurity is assumed significant only up to the n.n.'s. From the Group theoretical arguments, it is found that in zinc-blende type crystals, only A_1, E and F_2 type vibrations of the point group symmetry T_d are Raman active, and in the G.F. theory one can separate out the contribution of each mode to the scatterint intensity.[6] For *Hg*Te:Cd system, we report the results of our calculations for the Raman intensities and compare it with the existing data (Ref. [3]) (see Fig. 1).

To explore the possibility for the origin of Ω_2 structure observed in i.r. (near $\sim$105cm^{-1}) and in Raman (near $\sim$108cm^{-1}) measurements, our Green's function calculations[4] in the F_2 irreducible representation suggest the possibility of a gap mode of an antisite (e.g., Hg substituted for Te) defect in HgTe. Another feature near $\sim$135cm^{-1}, although absent in the i.r. spectra, however shows a distinct signature of A_1 character in Raman scattering and in all the three faces of the $Hg_{0.8}Cd_{0.2}Te$ samples. The authors of Ref. [3] have assigned it as a "cluster-mode", our calculations on the other hand suggest the possibility of an impurity mode of A_1 character. Since the mode lies in the region where the density of phonon states of HgTe are very low whether it is a gap or resonance mode, require further experimental investigations. Similar calculations for $Cd_{1-x}Mn_xTe$ and $Cd_{1-x}Zn_xTe$ systems have also been performed and the results will be reported elsewhere.

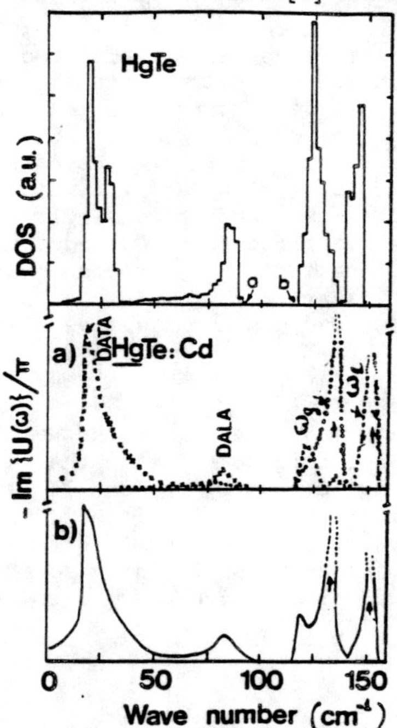

Fig. 1. Calculated DOS of the perfect HgTe (Ref. 4) (upper panel). a) Imaginary part of the imperfect G.F. for *Hg*Te:Cd system in the ....A_1 and -x-x-x- F_2 irreducible representation. b) same as a) but for A_1+ F_2+ E irreducible representation. ω_g - gap mode; ω_ℓ: localized mode.

REFERENCES:

1. B.L. Sharma and R.K. Purohit, Semiconductors Hetrojunctions (Pergamon, Oxford, 1974), pp. 24.
2. See for example, "Physics and chemistry of Mercury Cadmium telluride (MCT)", J. Vac. Sci. Techn. 21, 1 (1982); A1 (1983); A3 (1985).
3. P.M. Amirtharaj, K.K. Tiong, P. Parayanthal, F.H. Pollack and J.K. Furdyna, Proc. Int. Conf. on Semiconductors, San Francisco, 1984.
4. D.N. Talwar and M. Vandevyver, J. Appl. Phys. 56, 1601 (1984).
5. N.X. Xinh, Ph.d. thesis, University of Paris, (1966).
6. D.N. Talwar, M. Vandevyver and M. Zigone, J. Phys. C13, 3775 (1980).

HIGH PRESSURE RAMAN SCATTERING STUDY OF PHASE TRANSITION IN $Cd_{1-x}Mn_xTe$.

Akhilesh K. Arora*, D. U. Bartholomew, D. L. Peterson# and A. K. Ramdas

Physics Department, Purdue University, West Lafayette, IN 47907
#Kodak Research Laboratory, Eastman Kodak Company, Rochester, NY 14650

The semiconductors CdS, CdSe and CdTe undergo a phase transition from the low pressure cubic zinc-blende or wurtzite, to the high pressure NaCl phase at hydrostatic pressures below 35 kbars. The alloy of CdTe with MnTe, which forms the Diluted Magnetic Semiconductor $Cd_{1-x}Mn_xTe$, also exhibits the same transition. However, the transition pressure for the zinc-blende phase ($0 < x < 0.7$) is not found to depend strongly on the composition; besides, the transition is associated with a large volume change.[1] Recent optical absorption measurements[2] show that the band gap is smaller in the NaCl phase. In the present work we report the first high pressure Raman scattering investigation of this phase transition in $Cd_{1-x}Mn_xTe$ ($0 < x < 0.7$).

Raman spectra at high pressure were recorded using a gasketed diamond anvil cell in the back scattering configuration. Ruby fluorescence was used for measuring pressure.[3] Measurements were made while increasing, as well as decreasing, the pressure.

$Cd_{1-x}Mn_xTe$ is known to show a two mode behavior.[4] Apart from zone center CdTe-like and MnTe-like transverse optic(TO) and longitudinal optic(LO) phonons, the one-phonon density of states, consisting of peaks corresponding to zone boundary phonons, are also observed in the first order Raman spectrum because of the disorder in the alloy. The intensities of all the modes decrease as one approaches the transition pressure and finally disappear above it, consistent with the absence of the first order Raman spectrum in the NaCl phase. Frequencies of all zone center phonons increase linearly with pressure whereas the peak at 42 cm^{-1} due to TA(X) shows a noticable softening. This mode has been identified as the soft mode for the cubic zinc-blende to beta-Sn metallic phase transition in a number of II-VI and III-V semiconductors.[5]

Crystals of all compositions undergo the phase transition at pressures between 25 and 35 kbars, showing a weak dependence of the transition pressure on composition. In each case the reverse transition occurs at a lower pressure, indicating hysteresis. Transition pressure decreases as x increases. The reverse transition also shows a similar trend.

In pure CdTe the phonon frequencies, while decreasing pressure, match with the ones with increasing pressure, whereas in the alloys there is a marked difference between the frequencies while increasing and decreasing pressure. CdTe-like LO phonon frequencies, when the crystal is brought back to room pressure after the phase transition, are higher than those before the transition. The MnTe-like LO phonon shows an opposite trend. Thus the LO-TO splitting of CdTe-like phonon increases whereas that of MnTe-like phonon decreases as a result of cycling through the phase transition.

The LO-TO splitting is determined by the oscillator strength of the mode which in turn depends on the Born transverse dynamic charge.[5] Hence the present results indicate that there is a redistribution of the Born transverse dynamic charge between the CdTe-like and MnTe-like modes when the crystal is cycled through the phase transition. The quantitative estimates of the total oscillator strength from the observed phonon frequencies, before and after transition, show that it remains unchanged, indicating that the redistribution of the Born transverse dynamic charge is such that the total charge remains unaltered. The redistribution of the charge could, perhaps, be due to the introduction of defects in the process of cycling through the transition.

---

[*]On Leave from Materials Science Laboratory, Reactor Research Centre, Kalpakkam 603102 INDIA.

[1]K. Strössner, S. Ves, W. Dieterich, W. Gebhardt and M. Cardona, Solid State Commun. 56, 563 (1985).

[2]W. Shan, S. C. Shen and H. R. Zhu, Solid State Commun. 55, 475 (1985).

[3]A. Jayaraman, Rev. Mod. Phys. 55, 65 (1983).

[4]S. Venugopalan, A. Petrou, R. R. Galazka, A. K. Ramdas and S. Rodriguez, Phys. Rev. B25, 2681 (1982).

[5]R. Trommer, H. Müller, M. Cardona and P. Vogl, Phys. Rev. B21, 4869 (1980).

INTERFERENCE BETWEEN ALLOWED AND FORBIDDEN SCATTERING IN ZINCBLENDE-TYPE SEMICONDUCTORS.

M. Cardona, W. Kauschke, and N. Mestres

Max-Planck-Institut für Festkörperforschung, Heisenbergstrasse 1, D-7000 Stuttgart 80, Federal Republic of Germany

In the last ICORS, one of us reported on an interesting phenomenon involving constructive and destructive interference between allowed and forbidden resonant Raman scattering by LO phonons in zincblende-type semiconductors.[1] The effects had been observed for the $E_o+\Delta_o$ resonance of GaAs and also for the E_1 resonance of InSb. These phenomena appear on [001] faces, either polished or epitaxially grown by MBE. In the $z(xx)\bar{z}$ configuration only forbidden scattering is observed, while for the $z(xy)\bar{z}$ configuration pure allowed scattering appears. The scattering in the former configuration is strongly resonant at all interband gaps (E_o, $E_o+\Delta_o$, E_1, $E_1+\Delta_1$) while the latter resonates strongly at E_o, E_1, and $E_1+\Delta_1$ but not at $E_o+\Delta_o$. In the configurations $z(x+y,x+y)\bar{z}$ and $z(x-y,x-y)\bar{z}$ both, allowed and forbidden scattering, appear. They interfere constructively for one of these configurations and destructively for the other (Fig. 2 of Ref. 1). Through a fit of these structures with theoretical expressions, rather detailed information on the real and imaginary parts of the allowed and forbidden Raman tensors is found. One also finds accurate information on the relevant electron-phonon coupling constants.

Here we investigate the generality of these phenomena. We have extended the GaAs observations of the LO-phonon $E_o+\Delta_o$ resonance to $Ga_{1-x}Al_xAs$ alloys. This resonance and the allowed-forbidden interferences are seen for x up to 0.1. They yield rather accurate information on the energy of the $E_o+\Delta_o$ gap and its broadening vs. x. The forbidden scattering was decomposed in Ref. 1 into two components, one which interferes coherently with the allowed scattering and an incoherent one, attributed to scattering by ionized impurities. The same procedure is followed here for the alloys. It is shown here that alloy scattering in $Ga_{1-x}Al_xAs$ does not enhance the incoherent component to the forbidden scattering.

We have also observed the $E_o+\Delta_o$ LO resonances and related interferences in two other zincblende-type semiconductors: InP and GaSb. The former has a spin-orbit splitting $\Delta_o = 0.11$ eV, much smaller than that of GaAs (0.34 eV), while for GaSb $\Delta_o = 0.75$ eV is much larger. Nevertheless, analogous interference phenomena are observed in all three cases. We also discuss the E_1 and $E_1+\Delta_1$ resonances investigated in GaSb. Particularly striking are the resonant phenomena for scattering by nearly pure plasmons ($\omega_p >> \omega_{LO}$). In this case we see an allowed (cross polarizations) scattering which we attribute to coupling via the first order electrooptic coefficient (interband Fröhlich interaction). We also see a forbidden scattering which strongly resonates at E_1 and $E_1+\Delta_1$. Both of these scattering mechanisms interfere completely coherently. A quantitative interpretation of these phenomena will be given.

[1]J. Menéndez and M. Cardona, Pure and Appl. Chem. 57, 181 (1985).
J. Menéndez and M. Cardona, Phys. Rev. Letters 51, 1297 (1983).
J. Menéndez, L. Viña, M. Cardona, and E. Anastassakis, Phys. Rev. B 32, 3966 (1985).

LASER-INDUCED CRYSTALLIZATION AND MEDIUM RANGE ORDER IN AMORPHOUS $GeSe_2$

Kazuo Murase, Koichi Inoue and Koji Kawamoto
Department of Physics, Faculty of Science, Osaka University
Toyonaka 560, Japan

A layered crystalline form is stable in c-$GeSe_2$ at STP [1], whereas there are two forms in c-GeS_2 i.e., a three dimensional form (low temperature phase)[2] and a layered form (high temperature phase)[3]. In the Raman spectra of the glassy phase, it is acknowledged that the g-A_1 band (199 cm^{-1}) is primarily determined by the symmetric breathing motions of chalcogen atoms at $GeSe_{4/2}$ tetrahedra. The higher energy side band at 216 cm^{-1} is called the companion A_1 line (A_1^c). Models for the A_1^c line have been subjected to controversy for many years. Proposed origins are more or less from modified fragments of the layered c-$GeSe_2$[4]. To investigate the glass forming mechanism and medium range order in the amorphous or glassy state, laser-induced-crystallization experiment is made at room temperature in evaporated a-$GeSe_2$ films with thickness 6000A. A time resolved Raman spectroscopy is utilized to observe the laser-induced-structural change in $GeSe_2$. The laser light (5145 A) is used either for inducing, or probing the structural change. The light is focused onto a film surface area with a typical diameter of 30-50 μm with excitation power of 10-50 mW. The scattered light is detected by a silicon photodiode array with image intensifier through a monochromator with prefiltering stage by a pair of gratings.

At a high power density region, the sharp 210 cm^{-1} A-line is observed to grow appreciably even within a minute, which is due to the layered crystal formation. The crystalline part becomes dominant within 10 min.. In an intermediate power level, the sharp 210 cm^{-1} and 200 cm^{-1} lines grow, before or after, in a few minutes. These spectral profiles are more or less similar to the previous authors'[5,6] measured by a convensional Raman spectrometer in bulk glasses. At a lowest power region, we found that only the 210 cm^{-1} crystalline appears after a "latent period" of ~15 min. and is still growing even after 180 min. as shown in Figure 1 (a).

We deduce the temperature of the crystallized part from the line shift with reference to the temperature-shift-rate of the bulk crystal line position, ~0.017 $cm^{-1}K^{-1}$. Even at the highest power level, the temperature under illumination is below the recrystallization temperature T_c. The intermediate temperature is about 230°C which is below the glass transition temperature T_g~390°C. For the lowest power level, the temperature should be ~100°C. The fact that only the layered crystal grows very much below T_g under a gentle photoexcitation naturally supports the existence of layered crystalline fragments in the glassy or amorphous phase.

We reinterpret the origin of the 200 cm^{-1} line which is always associated with a new subsidiary band at 90 cm^{-1} as a new crystalline phase which may possibly be of the 3D form. We have clearly found the 200 cm^{-1} line with several new subsidiary bands including the 90 cm^{-1} line in a crystalline phase which was grown in $GeSe_2$ film after annealing at 420°C (between T_g and T_c) in Ar gas, where we have also observed 2D phase in a different part of the specimen. The new crystalline part in our crystallized film by annealing seems to be more transparent than the 2D part. In the 2D part we have observed the 210 cm^{-1} crystal A-line with a broad band centered at 216 cm^{-1}, whereas no distinct band near 200 cm^{-1}. A low energy shift of the broad line position with increasing laser intensity (or temperature) is about two times faster in comparison with the 210 cm^{-1} peak. This broad 216 cm^{-1} peak may be due to a paracrystal which is topologically similar to a fine layered crystal, and might be related to the A_1^c line in the glass phase, as is suggested in Figure 1(b), though we need further study on these problems.

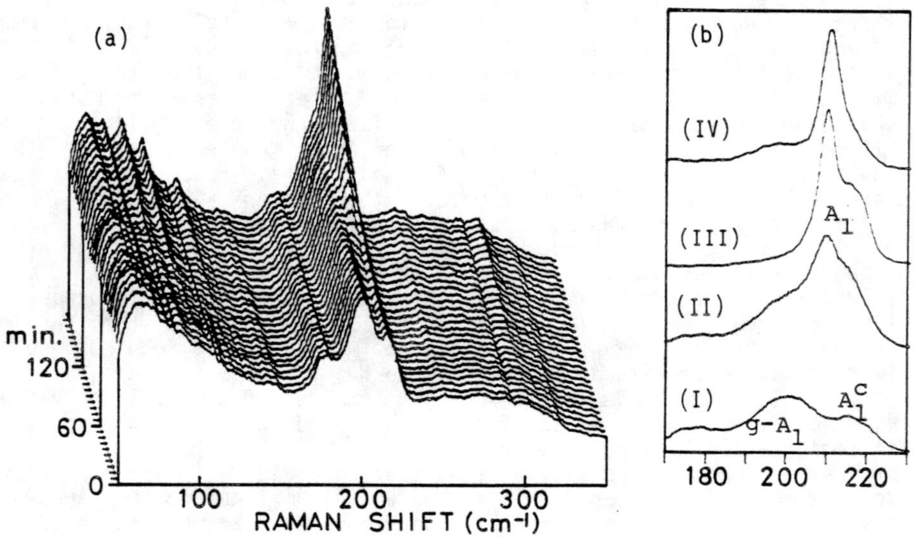

Figure 1 (a) time dependence of Raman spectra in the lowest power region; the first amorphous (I) and the last partially crystallized spectra (II) are shown in (b) where, (III), the 2D spectra of a crystallized film by annealing at 420°C with a broad band at 216 cm^{-1}, and (IV), a spectrum of ground c-$GeSe_2$ powder in fresh water which has g-A_1 band and c-A-line with a high energy tail (A_1^c ?).

This work was supported by the Grant-in-aid for Scientific Research from the Ministry of Education Science and Culture.

1. Von G. Dittmar and Herbert Schafer, Acta Cryst. B32, 2726 (1976).
2. Von G. Dittmar and Herbert Schafer, Acta Cryst. B32, 1188 (1976).
3. Von G. Dittmar and Herbert Schafer, Acta Cryst. B21, 2060 (1975).
4. K. Murase and T. Fukunaga, Mat. Res. Soc. Proc. 61, 102 (1986).
5. J. E. Griffiths, G. P. Espinosa, J. P. Remeika and J. C. Phillips, Phys. Rev. 25, 1272 (1982).
6. E. Haro, A. S. Zu, J. F. Maorhange and M. Balkanski, Phys. Rev. B32, 969 (1985).

THIN FILM REACTIONS AT THE ALUMINUM-GALLIUM ANTIMONIDE INTERFACE.

G. P. Schwartz, R. G. Susnow, G. J. Gualtieri, and W. A. Sunder

AT&T Bell Laboratories, Murray Hill, New Jersey 07974

In the present work surface reflection Raman scattering has been utilized to probe interfacial chemical reactions occurring at the interface between thin films of aluminum on GaSb substrates. In the Al-Ga-Sb system thermodynamics favors the formation of a solid solution of (Al,Ga)Sb with the subsequent release of metallic Ga. Nevertheless, previous studies of the Al/GaAs system have shown that for relatively low reaction temperatures kinetic factors often favor the formation of a metastable phase such as AlAs(1). An additional consideration of the work concerns the question of whether crystalline or amorphous products are forming at the metal/semiconductor interface.

The GaSb substrates were chemically etched according to Kodama et. al.(2) and inserted into a UHV chamber with Auger capability and an MBE Al cell. Following the chemical etch, the residual oxides on the substrate were flash evaporated at 530C; Auger spectra showed that all oxides were totally removed and only vestiges of carbon remained. The substrates were cooled to room temperature and Al films ranging from 100 to 300A were deposited at a typical rate of 1A/sec. The films were then thermally aged in vacuo. After the latter step, the substrates were removed and placed in a glass cell with flowing argon and examined via Raman scattering using 5145A excitation. The electric vector of the incident light was polarized in the scattering plane and the polarization of the scattered light was not analyzed.

For the primary reaction products of interest, i.e. AlSb or alloys of (Al,Ga)Sb, their compositions can be obtained directly from a measurement of their optical mode frequencies. Lucovsky et. al.(3) have shown that (Al,Ga)Sb alloys display two-mode behavior and obtained the four optical mode frequencies (LO1,LO2,TO1,TO2) as a function of composition in analytical form. Neither Al nor Ga can be detected in Raman scattering because FCC metals are symmetry forbidden to be Raman active. In addition to the interfacial product modes, we also typically detect the LO and TO frequencies from the substrate because our optical penetration depth exceeds the thickness of the reacted interfacial layer. Some caution is required to react a significant portion of the initial Al film during thermal aging because the penetration depth in Al is of the order of 70A for an excitation wavelength of 5145A.

Our measurements have identified the following patterns. Under thermal aging conditions where the Al overlayer is not fully reacted, the primary interfacial reaction product is nearly pure AlSb. Based on the analytical mode frequencies tabulated in ref.(3), the composition is 95±5% AlSb. A typical set of reaction conditions which yields this metastable interfacial phase would be a 300A Al film aged at 510C for 1.5 hours. If one continues to age the film beyond the point where the Al is fully consumed via reaction, then we detect an interdiffused surface layer of (Al,Ga)Sb whose lineshape indicates that a concentration gradient exists normal to the interface.

Reaction sequences for temperatures as low as 300C have also been examined. For a 100A Al film reacted in vacuo for 65 hours we detect a broad band centered in the region between the TO and LO modes of AlSb. If the sample is reinserted after the initial measurement back in the vacuum system and heated to 450C for 30 minutes, two sharp peaks are then detected at 318 and 338 cm^{-1}, thus indicating that the interfacial layer initially formed was amorphous AlSb and that the brief second aging cycle served to crystallize some of this material. For aging temperatures intermediate between 300 and 500C we typically observe superimposed amorphous and crystalline bands assigned to nearly pure AlSb.

References

(1) G. P. Schwartz and A. Y. Cho, J. Vac. Sci. Technol. 19(3), 607 (1981).
(2) M. Kodama, J. Hasegawa, and M. Kimata, J. Electrochem. Soc. 132, 659 (1985).
(3) G. Lucovsky, K. Y. Cheng, and G. L. Pearson, Phys. Rev. B 12, 4135 (1975).

RAMAN STUDY OF THE PHONON SPECTRA IN $Hg_{0.83}Cd_{0.17}Te$ FROM 10K TO 250K.

M.L. Bansal and A.P. Roy

Nuclear Physics Division, Bhabha Atomic Research Center
Bombay 400 085, India

$Hg_{(1-x)}Cd_xTe$ is a narrow gap semiconductor or semimetal for x<0.2 and is therefore an interesting system for investigating electron-phonon interaction. Mooradian and Harman (1) have measured phonon frequencies at 1.6 K over a range of concentrations. Amritharaj et al. (2) have reported polarized Raman spectra from $Hg_{0.8}Cd_{0.2}Te$ at 77 K. These authors find a 'defect' mode at 108cm^{-1} and additional modes around 135cm^{-1}. We present here a study of the line width of the HgTe like TO phonon, over a temperature range of 10K-250K in $Hg_{(1-x)}Cd_xTe$ with a nominal composition of $x \approx 0.17$. In addition we show that the 'defect' mode of Ref. 2 is most likely a combination mode.

Disc like samples (ϕ=5mm, t=0.35mm) were mounted with Cryocan grease on the cold finger of a Displex system. The geometry was such that only TO modes could be observed. The beam from a HeCd laser (λ4416$\AA$, 60mW) was focused using a cylindrical lens to form a spot of 60μx2.5mm, on the sample, in the back scattering configuration. Typical spectra recorded for $x \approx 0.17$ at different temperatures are shown in Fig. 1.

We find that the instrument resolution corrected width (FWHM) of the TO phonon increases from 4cm^{-1} at 10K to 13cm^{-1} at 250K but the phonon frequency does not show any softening. This behavior is not expected from usual anharmonic interactions. In $Hg_{(1-x)}Cd_xTe$, interband transitions between the light hole valence band and heavy hole valence band span a continuum of energy overlapping phonon frequencies and these processes become progressively important with increasing temperature. Electron-phonon interaction is therefore expected to play an important role leading to broadening of the phonon as observed. It may be noted that in $Hg_{(1-x)}Cd_xTe$ the energy gap between the conduction band and valence band can also be made to vary around the optical phonon energy by choosing suitable compositions. But our measurements for $x \approx 0.11$ and $x \approx 0.22$ suggest that direct transitions across the gap do not affect phonon line shape in any important way (1).

We see from Fig. 1 that the band at 102cm^{-1} decreases in intensity as temperature is lowered and is practically absent at 10K. The intensity variation of the 102cm^{-1} band indicates that it is a combination mode (3) and not a 'defect' mode as suggested in Ref. 2. Similar intensity variation has also been observed in the IR spectrum of pure HgTe by Grynberg et al. (4).

The authors thank Shri D. Sahoo and Shri R.V. Srikantiah for providing the crystals grown by them at BARC, Trombay.

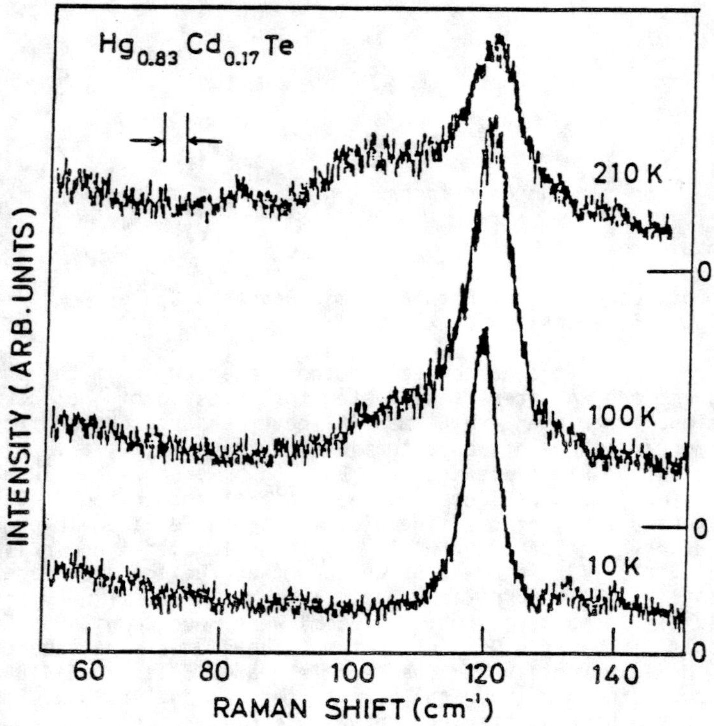

Fig. 1. Raman spectra of $Hg_{(1-x)}Cd_xTe$ for x=0.17, at various temperatures. For x≈0.11 and 0.22 the spectra show similar behavior.

References:

1. A. Mooradian and T.C. Harman in Proceedings of the Conference on the Physics of Semimetals and Narrow Gap Semiconductors, Dallas, ed. D.L. Carter and R.T. Bate (Pergamon, N.Y., 1970), p. 297.

2. P.M. Amritharaj, Kwong-Kau Tiong, and Fred H. Pollak, J. Vac. Sci. Technol. A1(3), 1744 (1983).

3. J. Baars and F. Sorger, Solid State Commun., 10, 875 (1972).

4. M. Grynberg, R. Le Toullec, and M. Balkanski, Phys. Rev. B, 9, 517 (1972).

chapter 8

CONFINED RESONANCES

RAMAN SCATTERING IN OPTICAL FIBERS.

R.H.Stolen

AT&T Bell Laboratories, Holmdel, NJ 07760

An optical fiber confines and guides light over distances limited only by absorption and scattering. There is a tremendous intensity enhancement of spontaneous Raman scattering as scattered light is collected over the entire length of the guide which could be many kilometers. This increased interaction length also drastically reduces the threshold power for stimulated Raman scattering as well as reducing power requirements for Raman amplifiers and lasers.(1)

A simple fiber consists of a guiding core surrounded by a lower index cladding. more complicated fibers can contain multiple core and cladding layers.(2) Typical core diameters for single-mode fibers are 3 um, 5 um, and 8 um for operation at wavelengths of 500 nm, 1000 nm, and 1.3-1.5 um respectively. Typical core diameters for multimode operation fall betwen 50 and 100 um. Most fibers do not maintain the state of polarization of the light although polarization-preserving fibers can be made which preserve linear polarization along the fiber.(3)

The light gathering efficiency of fibers has been used to investigate weak Raman features in both glass and liquid-core fibers.(4) The enhancement in Raman scattered intensity is usually sufficient to make forward scattering measurements possible. With care, forward scattering measurements have been made to 2 cm^{-1} of the exciting line in fused silica fibers and backwards scattering measurements to 0.03 cm^{-1}.(5)

The Raman interaction can also be used to amplify a Raman-shifted signal if it is injected along with the exciting light which is called the pump. A Raman Stokes signal is amplified exponentially with a gain equal to the product of a Raman gain coefficient times pump intensity and length. The gain coefficient is proportional to the spontaneous cross-section. The pump intensity is approximately the power divided by the fiber core area. The length falls somewhere between the actual fiber length and the absorption length.(1)

Raman gain is under investigation as an amplifier for lightwave communications(6) and as an undesirable source of crosstalk in wavelength-multiplexed communications systems(7). Raman amplification is also used in reentrent fiber loops to keep optical pulses propagating dozens of passes around the loop.(8) With sufficient pump power, no injected signal is necessary. Spontaneous scattering can be amplified to pump depletion. The process can be repeated sequentially along the fiber producing colorful bands of Raman shifted light at the fiber output.

The addition of feedback to a fiber with Raman gain produces a Raman laser which can be tuned over the broad Raman gain curve. Both cw and synchronously-pumped pulsed lasers have been made in linear and ring configurations. Tuning can be by internal prisms or etalons or by time-dispersion tuning in repetitively pulsed lasers.(1) Practical application of fiber Raman lasers has so far been limited by hardware problems and interference from competing nonlinear effects such as stimulated Brillouin scattering and intensity-dependent spectral broadening.(1) However, improvements in diode lasers and inline fiber optical components may provide the versatility necessary to produce practical fiber Raman lasers.

Recent interest centers around Raman amplification of short optical pulses in a fiber. This is particularly important in connection with studies of optical solitons where Raman amplification can compensate for fiber loss and permit the propagation of a non-spreading optical pulse over enormous distances.(9) Stimulated scattering from short pulses becomes a complicated problem because of the walk-off imposed by group-velocity dispersion. In the short pulse work we appear to be entering a regime in which it is more useful to view the Raman interaction strictly in the time domain rather than as a Raman spectrum.

1. R.H.Stolen, "Fiber Raman Lasers", Fiber and Integ. Opt., 3, 21 (1980).
2. S.E.Miller and A.G.Chynoweth,Eds. "Optical Fiber Telecommunications", (Academic Press,1979).
3. S.C.Rashleigh and R.H.Stolen, "Preservation of Polarization in Single-Mode Fibers", Laser Focus (May, 1983).
4. G.E.Walrafen and J.Stone, Appl. Spectrosc. 26, 585 (1972); 29, 337 (1975).
5. K.B.Lyons, P.A.Fleury, R.H.Stolen, and M.A.Bosch, Phys. Rev. B, 26, 7123 (1982).
6. J.Hegarty and N.A.Olsson, Electron. Lett., 21, 290 (1985); Y.Aoki, S.Kishida, k.Washio, and K.Minemura, Electron. Lett., 21, 191 (1985).
7. A.R.Chraplyvy, Electron. Lett., 20, 58 (1984).
8. E.Desurviere, M.Digonnet, and H.J.Shaw, Opt. Lett., 10, 83 (1985).
9. L.F.Mollenauer, J.P.Gordon, and M.N.Islam, IEEE J. Quantum Electron, QE-22, 157 (1986).

RAMAN - MIE - SCATTERING FROM LEVITATED DIELECTRIC MICROSPHERES

W. Kiefer* and R. Thurn#

*Institut für Experimentalphysik, Karl-Franzens-Universität Graz, Universitätsplatz 5, A-8010 Graz, Austria

#Physikalisches Institut der Universität Bayreuth, Postfach 3008, D-8580 Bayreuth, Federal Republic Germany

Recently there has been a great deal of interest in investigations of the light-scattering properties of dielectric spheres. Studies of the wavelength-dependent features of this type of scattering are of particular interest because of the sharp structural resonances that occur as the Mie size parameter $x = 2\pi a/\lambda$ (a = radius of sphere, λ = wavelength in the surrounding medium) is varied.

These structural resonances have been predicted for incident and/or inelastically scattered wavelengths which correspond to the natural modes of dielectric spheres. The first observation of morphology-dependent resonances in Raman scattering from dielectric microparticles has been published by Owen *et al.*[1]

Here we report on the observation of such structural resonances in the Raman spectra of single dielectric microspheres. The experiments were performed applying a straightforward sample arrangement[3] where the particle is excited by laser radiation while free in space without any supporting sample holder. This was achieved by employing the technique of optical levitation by radiation pressure, which has been pioneered by Ashkin[3] some years ago.

Pronounced structural (Lorenz-Mie) resonances were observed in the Raman spectra of optically levitated glass spheres of diameters ranging between about 10 and 40 μm and of liquid droplets (water/glycerol mixture) of about the same size. As example, we show in Fig. 1 the spectrum of the inelastically scattered light from a dielectric liquid sphere of ca. 15μm radius. In this figure, we display in the lower field *A* part of the Raman spectrum of a 1:6 water/glycerol mixture which is obtained from the bulk material. Curve *B* of Fig. 1 represents an observed Raman spectrum of an optically levitated single droplet of the same mixture as for *A*. In the upper field of Fig. 1 we show a theoretical calculation of a synthetical Raman spectrum which one expects for a dielectric sphere of refractive index $n = 1.45$ and a radius of $a = 14.422$ μm, if the Lorenz-Mie formalism[4] is applied. For further details regarding the theoretical calculations we refer to Reference 2. Fairly good agreement between the experimental and the theoretical spectrum is obtained. The occurence of the sharp structural resonances in these spectra are due to the fact, that the Raman light scattered inside the sphere is more or less continuous over the region of interest, which means, that there are always frequencies present, which coincide with the frequencies of the structural resonances of the particular sphere. Since at these frequencies the wavelength of the Raman emission corresponds to a resonance condition of the microparticle, the Raman emission will be enhanced at that wavelength.

In this talk it is attempted to give a general overview on problems that arise when Raman spectra of microparticles in the Mie size range are investigated.

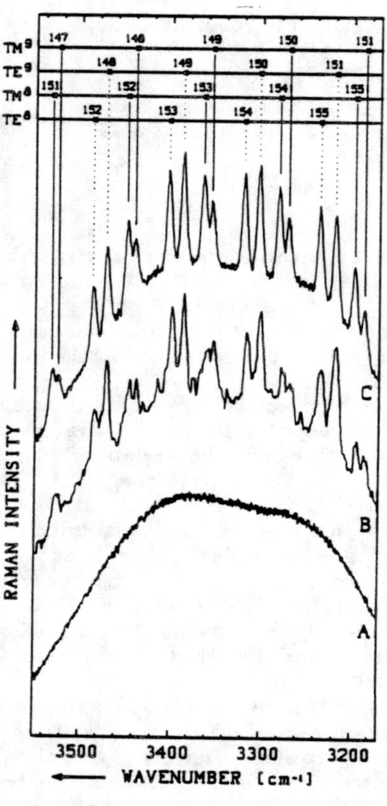

Fig.1: *A*, O-H stretching region of the Raman spectrum of a liquid 1:6 water-glycerol mixture; *B*, observed Raman spectrum of an optically levitated liquid drop of the same mixture as for *A*; *C*, theoretical curve applying the Lorenz-Mie formalism

REFERENCES

1. J.F. Owen, P.W. Barber, and R.K. Chang, in *Microbeam Analysis 1982*, K.F.J. Heinrich (San Francisco Press) 1982, p. 255.
2. R. Thurn and W. Kiefer, Appl.Spectrosc. **38**, 78 (1984); J.Raman Spectrosc. 15, 411 (1984); Appl.Opt. **24**, 1515 (1985).
3. A. Ashkin, Phys.Rev.Lett. **24**, 156 (1970).
4. M. Kerker, *The Scattering of Light and Other Electromagnetic Radiation* (Academic, New York, 1969).

ENHANCED RAMAN SCATTERING BY DIELECTRIC SPHERES

M. Inoue

Institute of Applied Physics, University of Tsukuba, Sakura, Ibaraki 305

Observation of surface-enhanced Raman scattering has stimulated activities in classical electromagnetic (EM) theory and the effects of the surface plasmon excitation have been investigated for various systems [1]. In this paper, we show another kind of resonance phenomena ; excitation of resonance states (virtual bound states of EM waves) of dielectric materials. [2]

Figure 1 shows various quantities as functions of ω in the case of a dielectric sphere of radius a subject to an incident EM plane wave of unit strength. In the lower half of the figure, solid line shows the electric field intensity averaged over the volume of the sphere, while the broken line shows the normal component strength averaged over the surface. The upper half shows the scattering wave phase shift for M (solid curve) and N (dashed curve) fields [3] for ι=1, 2, and 3. Sharp increase of phase shift at certain frequencies is regarded as the appearance of resonance states. When excited, these resonance states will enhance the local electric field. In fact the field intensities are seen to peak at those resonance frequencies. Note that both M and N fields contribute to the field intensity, while normal component is due only to N field.

Figure 2 is a similar plot for a two dimensional array of dielectric spheres in x-y plane. Square lattice is assumed with lattice constant d such that a/d=0.4. The incident wave vector lies in the x-z plane with incident angle 20 deg. This time, the specular reflectivities for s- and p-polarized light are shown in the upper half. In this case, the character of the resonance modes is found to be completely different from those of a single sphere. As the spheres are rather closely packed, a dielectric slab with periodic modulation is the appropriate reference system. Then, localized modes of the slab couples with the incident field by the umklapp processes due to its periodic structure. When the incident field excites the resonance modes, the reflectivity attains its peaks, and the local field shows sharp maximum. An interesting thing is that the average intensity is enhanced one hundred times as large as that of the incident field. This phenomenon is promising as a means of observation of the enhanced Raman scattering not only from the adsorbed molecules on the surface, but also from inside of the sphere; bulk Raman scattering. Furthermore, as the intensity is enhanced inside the sphere, this phenomenon is also applicable to the nonlinear optics of semiconductors.

The nature of these resonance modes of dielectric spheres will be discussed also for clustered spheres, a sphere on a substrate and two dimensional array of spheres.

The author is grateful to Dr. K. Ohtaka and Dr. T. Takemori for their valuable contributions.

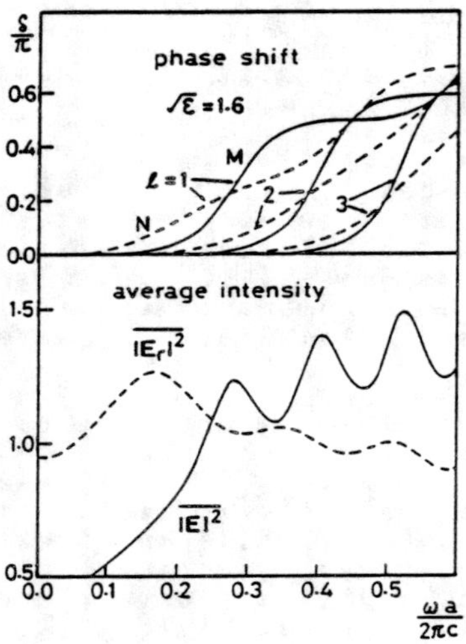

Fig.1

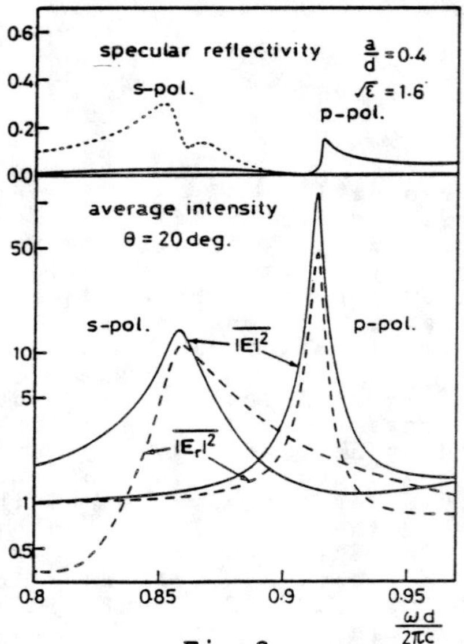

Fig.2

Fig.1 Phase shift and averaged electric field intensity as a function of frequency in the case of a single sphere.

$$\overline{|E|^2} = \int |E(r)|^2 dr/(4\pi a^3/3),$$

$$\overline{|E_r|^2} = \int |E_r(a+0)|^2 d\Omega/(4\pi).$$

Fig.2 Specular reflectivity and averaged electric field intensity for a two dimensional square array of dielectric spheres.

REFERENCES

[1] M. Moskovits: Rev. of Mod. Phys. 57, 783 (1985)

[2] K. Ohtaka and M. Inoue: Phys. Rev. B25, 677 (1982), M. Inoue, K. Ohtaka and S. Yanagawa: Phys. Rev. B25, 689 (1982), K. Ohtaka and M. Inoue: J. Phys. C15, 6463 (1982), M. Inoue and K. Ohtaka: Phys. Rev. B26, 3487 (1983); J. Phys. Soc. Jpn. 52, 1457 (1983).

[3] J. A. Stratton: Electromagnetic Theory (McGraw Hill, New York, 1941).

STIMULATED RAMAN SCATTERING SPECTRA FROM SINGLE WATER DROPLETS

Johannes H. Eickmans and Richard K. Chang

Section of Applied Physics and Center for Laser Diagnostics
Yale University, New Haven, Connecticut 06520, USA

Due to hydrogen bonding, the spontaneous Raman spectra of bulk water (H_2O and D_2O) are the broadest (~400 cm^{-1}) of all liquids. Interpretations of such spectra have led to two different models [1] for the structure of water, a mixture model and a continuum model. The stimulated Raman scattering (SRS) of bulk water has also been used as supporting evidence for the two models [2,3]. However, the low single-pass Raman gain, associated with the broad spontaneous Raman profile, can lead to complications which are inherent to the SRS process. The large fluctuations in the observed SRS spectra of water in a 10 cm cell could result from: (1) self-phase modulation, which can cause spectral broadening, (2) self-focusing, which can disrupt the thermal equilibrium within the filaments, and (3) Raman noise, which dictates the SRS emission within the Raman gain profile.

The input intensity necessary to achieve SRS threshold for an individual water droplet (with radius of 20-50 μm) is ~10 times lower than that for water in a 10 cm optical cell. The liquid-air interface of the droplet acts as an optical cavity for the internally generated Raman emission at discrete wavelengths which correspond to the morphology-dependent resonances (MDR's) of a sphere [4]. For a fixed mode order, the separations of the MDR's with different mode numbers are in the 10 cm^{-1} range. However, when different mode orders are considered, the separations of the MDR's are less than 1 cm^{-1}, well below the resolution of our optical multichannel analyzer (OMA). Because of the combination of 10 times lower input intensity to achieve SRS threshold and the 10^3-10^4 times shorter length, self-phase modulation broadening and self-focusing can be eliminated for a droplet. Furthermore, the fluctuations in the SRS emission from a droplet should be less than those from water in a cell because in a high optical feedback system SRS is less dominated by Raman noise and more representative of the Raman gain. SRS from individual water droplets may provide useful spectroscopic information related to the structure of liquid water.

A linear stream of highly monodispersed water droplets was generated with a modified Berglund-Liu piezoelectric vibrating orifice. The incident laser pulse from a Q-switched Nd:YAG laser (532 nm emission, 15-30 ns duration, and 0.4 mJ/pulse energy) was focused onto a single droplet. The SRS was collected at 90° relative to the incident beam direction and the vertically flowing droplets and detected with the OMA.

Typical spontaneous Raman spectra from D_2O, H_2O, and a mixture of 25% D_2O in H_2O in a 1 cm optical cell are shown in Figs. 1(a) and 1(b). Spontaneous Raman spectra from water droplets exhibit sharp structures commensurate with the MDR's [5]. The single-shot SRS spectra from individual D_2O and H_2O droplets (~93.5 μm in diam) are shown in Fig. 1(c). The 2395 and 3220 cm^{-1} droplet peaks have been classified as the respective symmetric vibrational modes of D_2O and H_2O molecules which are in-phase with the other hydrogen bonded (HB) molecules. The corresponding peaks with larger vibrational frequencies (2530 and 3435 cm^{-1}) have been classified as symmetric vibrational modes which are out-of-phase with the other HB molecules.

For droplets containing 25% D_2O in water, the in-phase peaks are characteristically absent in all the SRS spectra [see Fig. 1(d)]. Similarly,

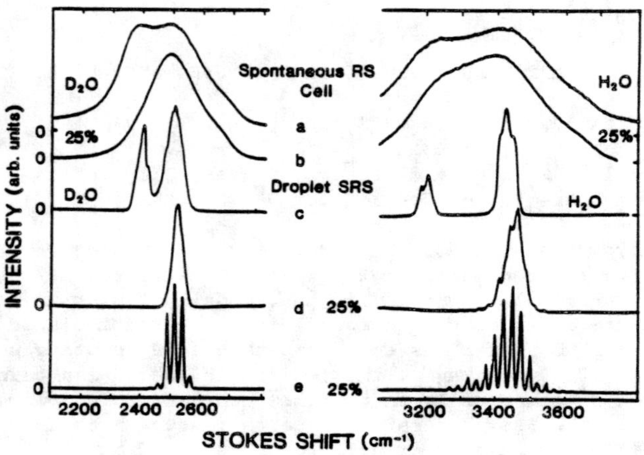

Figure 1

the in-phase peaks are missing for droplet mixtures containing 50% or 75% D_2O. The addition of bond breaking anions (e.g., 2-6 M $NaClO_4$) to pure D_2O or H_2O droplets also causes the in-phase SRS peaks to disappear without the appearance of the non-hydrogen bonded (NHB) peaks. A broad SRS peak for H_2O droplets containing high concentrations of ClO_4^- was detected at a higher frequency (e.g., 3500 cm^{-1} for 4 M $NaClO_4$). The 550 cm^{-1} vibrational mode reported in the SRS from bulk water [6] was not observed from any of the droplets (pure or mixed). Unlike the SRS from aqueous solutions in an optical cell, the SRS from droplets was considerably more reproducible.

By detuning the droplet radius (as little as 0.1%) so that the input laser wavelength no longer corresponds to a MDR (i.e., at an input resonance), the field intensity within the droplet is reduced [7,8]. Consequently, even though the incident intensity remains unchanged, the SRS gain is reduced and therefore only those Raman wavelengths corresponding to that mode order which provides the largest optical feedback can achieve SRS threshold. Figure 1(e) shows SRS spectra consisting of MDR's corresponding to one mode order but different mode numbers when the input wavelength or radius is not on an input MDR. The smooth spectra of Figs. 1(c) and 1(d) correspond to the on-input MDR case, where many unresolved MDR's (with different mode orders and numbers) have achieved SRS threshold.

We gratefully acknowledge the partial support of this work by the Air Force Office of Scientific Research (Contract No. F49620-85-K-0002) and the Army Research Office (Contract No. DAAG29-85-K-0063).

REFERENCES

[1] Structure of Water and Aqueous Solutions, W.A.P. Luck, ed. (Verlag Chemie und Physik, Weinheim, 1974).
[2] G.E. Walrafen, in ref. 1, p. 301.
[3] M. Sceats, S.A. Rice, and J.E. Butler, J. Chem. Phys. 63, 5390 (1975); ibid., 64, 2701 (1976).
[4] J.B. Snow, S.-X. Qian, and R.K. Chang, Opt. Lett. 10, 37 (1985).
[5] R. Thurn and W. Kiefer, Appl. Opt. 24, 1515 (1985).
[6] O. Rahn, M. Maier, and W. Kaiser, Opt. Commun. 1, 109 (1969).
[7] J.F. Owen, R.K. Chang, and P.W. Barber, Opt. Lett. 6, 540 (1981).
[8] P. Chýlek, J.D. Pendleton, and R.G. Pinnick, Appl. Opt. 24, 3940 (1985).

SPECTRA OF STIMULATED RAMAN EFFECT FOR ANHARMONIC COUPLING OF THE POLARITON AND STOKES WAVES ON A SLAB.

V. Schulz and L. Merten
Dep. of Physics, Westf. Wilhelms-University of Münster
4400 Münster, West Germany.

An optical arrangement is treated in which two electromagnetic waves are normally incident onto a plane surface of a slab of the thickness d of a piezoelectric cubic crystal (two-beam method). The first wave, called pump wave, couples the polariton and the Stokes wave, and the second wave excites the coupled waves in consequence of the boundary conditions. Only the coupling with pure transverse polaritons is regarded /1/. We assume that the amplitude E_L of the field strength of the pump wave is large compared with the field strength of the exciting wave and can be treated as a constant (parametric approximation).

The frequency of the pump wave is generally chosen from the visible frequency region. The frequency of the exciting wave can be taken either from the polariton frequency region in the infrared or from the Stokes frequency region in the visible. Two "reflected" and two "transmitted" waves are generated by the exciting wave /2/, /3/.

The spectra of the reflectivity, the transmissivity and of the transformation formula are derived and evaluated for GaP, where the transformation formula is defined by the ratio of the intensity of the excited wave in the Stokes frequency region in the visible and the incident wave in the polariton frequency region in the infrared and vice versa.

The neglect of the coupling with the anti-Stokes wave leads to qualitatively similar results as in /2/, but has the advantage that quantatively similar results can be derived for the frequency region, in which a stimulated effect is possible /4/. An example for a reflectivity spectrum is illustrated in figure 1.

Figure 1. $E_L = 10^5 V/cm$

$d = .01$ cm

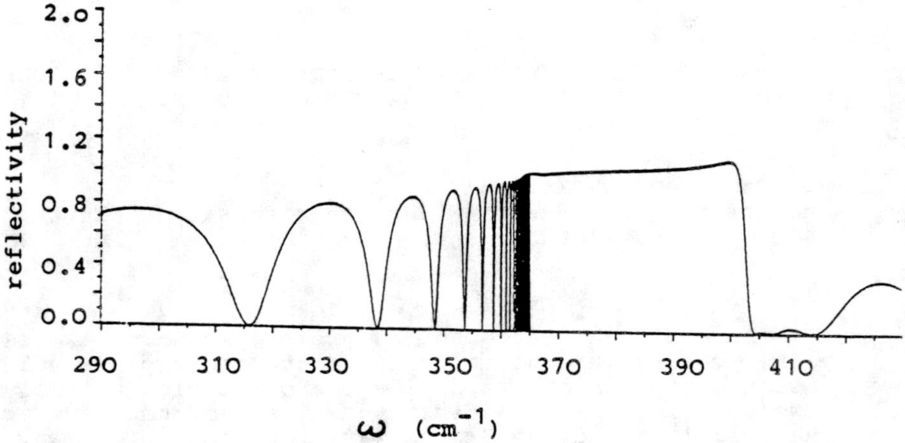

References

/1/ V. Schulz, L. Merten, phys. stat. sol.(b) 115, 225 (1983).

/2/ L. Merten, V. Schulz, J. Raman Spectrosc. 17, 93 (1986).

/3/ L. Merten, V. Schulz, to be published.

/4/ V. Schulz, L. Merten, phys. stat. sol.(b), in print.

chapter 9

SUPERLATTICE STRUCTURES

RAMAN SCATTERING FROM PHONONS IN SEMICONDUCTOR SUPERLATTICES

M. V. Klein
Department of Physics, University of Illinois at Urbana-Champaign 1110 W. Green St., Urbana IL 61801

Experimental results on the Raman spectra from artificial semiconductor superlattices have been available in the last few years. The status of the results from several research groups has recently been reviewed and compared with the predictions of simple models.[1] That work will be summarized here. The most popular samples have been GaAs-AlAs or GaAs-$Al_xGa_{1-x}As$ superlattices grown by molecular beam epitaxy or metal-organic vapor phase epitaxy, and the emphasis has been on understanding the role of superlattice layering on the vibrational properties of the samples. The samples are usually grown along the [001] direction. If the transferred wave-vector **q** is also in that direction, as it is for the back-scattering geometry, then the dynamics of the superlattices is particularly easy to model. Each atomic plane vibrates in phase, and linear chain models work well. In the low frequency region new Raman-active modes are observed from folding of virtual-crystal longitudinal acoustic modes due to mini-zone formation. The modes are weakly affected by the interfaces and possess a simple dispersion relation. Finite q effects are readily observed--unusual for Raman scattering. In the region of optical phonon frequencies simple behavior is also observed, namely strong confinement of the longitudinal optical modes. For binary superlattices a linear chain model fits well. For superlattices with alloy barriers effective-mass-like theory works reasonably well for the phonon modes.

When **q** has a component along the interfaces, theoretical treatments become much more complex, but some simple models still give good first approximations to the experimental results. In the optical regime slab modes are seen confined to one of the two layers in the superlattice repeat unit. These modes have frequencies given by quantum well models, and their transverse or longitudinal nature is exactly opposite to what is found in the bulk as long as the component of **q** along the interfaces is small compared with π divided by the layer thickness. The reason for this is decoupling of the wells that prevents a macroscopic electric field from existing in the direction of the layers.

Another class of optical modes in superlattices is interface modes. They result from coupling among surface phonons localized at each interface and can be understood using simple notions from electrostatics. They have been observed experimentally primarily under conditions of wave-vector violation, presumably due to defects or interface roughness. The first studies of resonance Raman behavior of these modes and slab modes have been undertaken. Better theoretical models for these phonons are needed.

This work was sponsored by the NSF under Grant DMR 8506674.

[1] For a review article see M. V. Klein, "Phonons in semiconductor superlattices," *IEEE J. Quantum Electron.*, Sept., 1986, to be published.

RESONANCE RAMAN STUDIES OF FOLDED ACOUSTIC PHONONS IN GaAs-AlAs SUPERLATTICES.

T. A. Gant[a], M. V. Klein[a,b], T. Henderson[c], and H. Morkoç[c]

[a]Department of Physics, University of Illinois at Urbana-Champaign (UIUC), 1110 W. Green St., Urbana, IL 61801

[b]Materials Research Laboratory and Coordinated Science Laboratory, UIUC

[c]Department of Electrical Engineering and Coordinated Science Laboratory, UIUC

Folded acoustic phonons are a well known phenomenon in GaAs-AlAs superlattices and several Raman scattering studies of them have been published[1,2]. We have made a resonance Raman study of folded acoustic phonons in a sample (50 Å GaAs - 20 Å AlAs), and have observed different orders of folded phonons resonating at different laser energies. The figure shows Raman spectra at three different laser energies. The top curve was taken far from resonance. Four orders of folded phonons, decreasing in intensity, are visible. Each order occurs as a doublet with a splitting of about 5 cm^{-1}. The relative intensities of the different orders can be roughly predicted by a theory outlined in Colvard et. al.[1] The relative intensity of the peaks within the doublet depend upon the phonon wavevector and the symmetries of the phonon states.

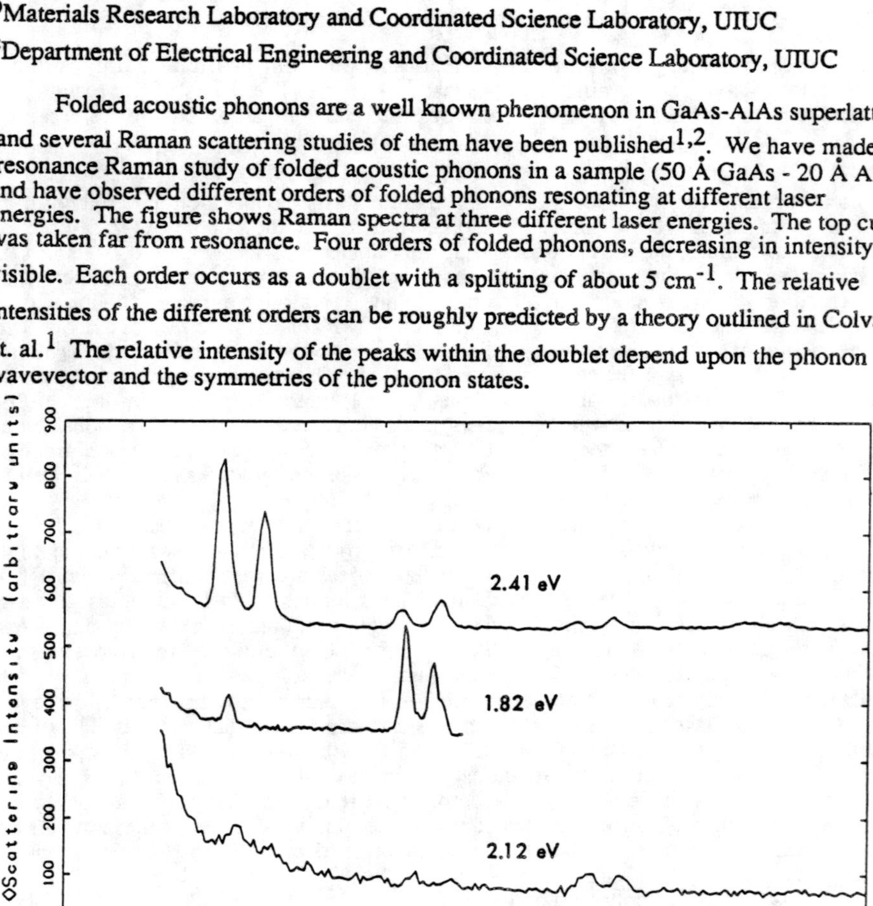

chapter 9

SUPERLATTICE STRUCTURES

In the middle curve the laser is near the $E_0+\Delta_0$ gap of the GaAs layers. The second order folded phonons are stronger than the first order, and the asymmetry of the doublets is enhanced. In the bottom curve the laser is near one of the higher subbands. The first three orders of folded phonons are all seen clearly. The second order is the weakest, and the third order is comparable in intensity to the first order.

The interaction between electrons and folded acoustic phonons depends upon the symmetries of the electron and phonon states. The electrons are confined in the GaAs well. The phonons are nearly plane waves with wavevector $2\pi m/d$ for an m'th order folded phonon, where d is the superlattice period. The strong asymmetry of the doublets in the spectra at 1.82 eV is not understood at this time.

[1]C. Colvard, T. A. Gant, M. V. Klein, R. Merlin, R. Fischer, H. Morkoç, and A. C. Gossard, Phys. Rev. B31, 2080 (1985)

[2]J. Sapriel, J. C. Michel, J. C. Tolédano, R. Vacher, J. Kervarec, and A. Regreny, Phys. Rev. B28, 2007 (1983)

RAMAN INVESTIGATION OF CONFINED OPTIC PHONONS AND THEIR ANNEALING EFFECTS IN GaAs-AlAs SUPERLATTICES.

Shu-Lin Zhang[a*], Dean H. Levi[a], T. A. Gant[a], Miles V. Klein[a], J. Klem[b], and H. Morkoç[b]

[a]Department of Physics and Materials Research Laboratory, University of Illinois at Urbana-Champaign (UIUC), 1110 W. Green St., Urbana, IL 61801

[b]Department of Electrical Engineering and Coordinated Science Laboratory, UIUC

We have observed longitudinal optical phonons confined either in the GaAs layers or the AlAs layers for the same GaAs-AlAs superlattice sample. The assignment of the confined orders has been made. For an un-annealed sample, the highest order we observe is the ninth one for the GaAs layer and the fifth one for the AlAs layer, the first such report as far as we know.[1-3] The dispersion relations obtained by our experiments fit the theoretical dispersion curves reasonably well for GaAs confined optic phonons as well as for AlAs modes.[4]

The shape changes of phonon well versus the annealing time are studied using the frequency shifts of the confined phonons after annealing at 850° C. We discovered that the phonon well is not a square well for the un-annealed sample, in agreement with the distribution of Ga atoms in the GaAs layer obtained by x-ray diffraction experiments. The variation of phonon well shapes and the position of phonon energy levels in the well are deduced. The results are compared with the other experimental results such as x-ray diffraction, Raman scattering experiments on $Ga_{1-x}Al_xAs$ alloys, etc.

The thickness of the GaAs layer and of an interfacial region, d_1 and d_0 respectively, have been calculated from Raman data on confined optic phonons. The d_1 and d_0 from Raman data have been compared with that from x-ray data and with theoretical calculations by Yip and Chang.Their results cannot fit our d_1 and d_0.

We recorded a spectral peak between the first and third order confined phonons of GaAs.It's frequency decreases with increasing annealing time. A discussion of the properties of this novel peak will be given.

*Permanent address: Department of Physics, Peking University, Beijing, P. R. China

1. C. Colvard, T. A. Gant, M. V. Klein, R. Merlin, R. Fischer, H. Morkoç, and A. C. Gossard, Phys. Rev. B<u>31</u>, 2080 (1985)
2. A. K. Sood, J. Menéndez, M. Cardona, and K. Ploog, Phys. Rev. Lett. <u>54</u>, 2111 (1985)
3. B. Jusserand, F. Alexandre, D. Paquet, and G. Le Roux, Appl. Phys. Lett. <u>47</u> , 301 (1985)
4. S. K. Yip and Y. C. Chang, Phys. Rev. B<u>30</u>, 7037 (1984)

OPTICAL PHONONS IN GaAs-AlGaAs SUPERLATTICES

A.K. Sood[+] and M. Cardona[*]

+ Materials Science Laboratory, Indira Gandhi Centre for Atomic Research, Kalpakkam - 603 102, India
* Max-Planck Institut für FKF, Heisenberg Strasse 1, 7000 Stuttgart 80, Federal Republic of Germany

This paper discusses Raman scattering by optical phonons confined to the GaAs layers and also by interface phonons in short period GaAs-AlAs superlattices.

When the optical phonon branches of the two constituent materials do not overlap in energy as is the case for GaAs-AlAs superlattices, the modes cannot propagate along the superlattice (SL) axis and become confined in either layer. The confined longitudinal optic (LO) phonons labelled by $m = 1,2,3,...$ belong either to the B_2 or A_1 irreducible representations of the D_{2d} point group symmetry of the SL grown along (001). The B_2 phonons (odd m) have off-diagonal Raman tensor and A_1 phonons (even m) have diagonal Raman tensor. The B_2 phonons were observed in Raman spectra[1,2] recorded with laser energy away from resonance. We showed[3,4] in Raman experiments on short period GaAs-AlAs superlattices that A_1 phonons which interact with electronic system via Fröhlich interaction dominate in resonance while B_2 phonons (interacting via deformation potential interaction) are stronger in off-resonance case. Figure 1 shows off-resonance and near-resonance Raman spectra of 20 Å GaAs-60 Å AlAs SL at 10 K for polarized and depolarised scattering configurations[3]. The results can be understood[3,4] by realising the symmetry properties of the electron-phonon interaction potentials and the wavefunctions of the quasi two-dimensional exciton with which the incident or the scattered light is in resonance.

Although forbidden by symmetry, a series of confined transverse-optic (TO) phonons in GaAs layers were observed[3] in GaAs-AlAs superlattices under extreme conditions of resonance. A quantitative explanation of the observed selection rules[3] for TO as well as LO phonons is still lacking.

Based on a linear chain model for the phonons in the SL, the frequency of modes confined in GaAs is found[1,2,5] to be the same as bulk GaAs phonons at wavevector $q = m\pi/(n_1+1)a$ instead of $q = m\pi/n_1a$ as expected from the correspondence with the slab-mode frequencies. The observed confined LO and TO phonon frequencies have been plotted[3] vs. $q = m\pi/n_1a$ (solid circles) and $q = m\pi/(n_1+1)a$ (solid squares) in fig.2.

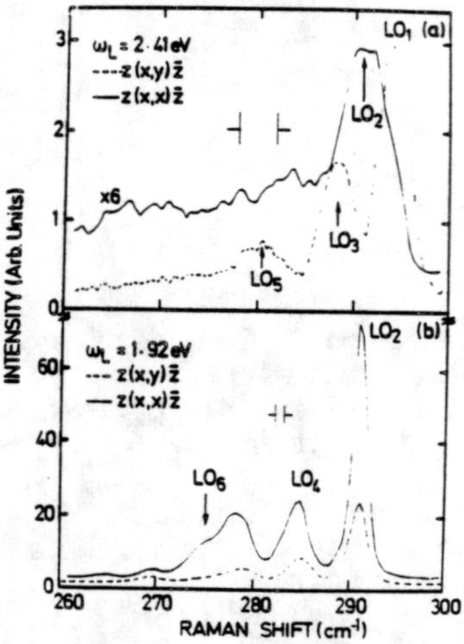

Fig.1 Raman spectra
(a) off-resonance,
(b) near-resonance

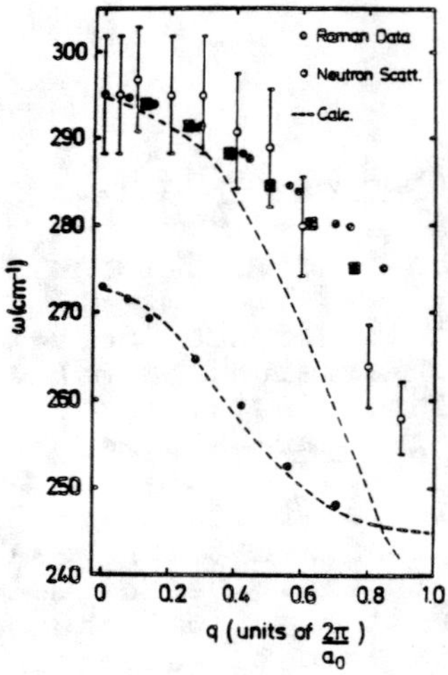

Fig.2 Observed LO_m and TO_m frequencies vs. q.

Also shown in the figure 2 are the neutron scattering data and calculated dispersion curves for the bulk GaAs. Here n_1 is the number of GaAs monolayers, a is monolayer thickness and m an integer.

We, however, note that there is a considerable discrepancy between the observed confined LO phonon frequencies when plotted against $q = m\pi/(n_1+1)a$ and the bulk GaAs phonon dispersion curve. A theoretical calculation to explain this discrepancy is desirable.

Interface vibrational modes propagating parallel to the interfaces and for which the vibrational amplitude is different from zero in both layers were also observed[6] in resonance Raman experiments on the short period GaAs-AlAs superlattices.

References

1. B. Jusserand, D. Paquet and A. Regremy, Phys. Rev. B30, 6245 (1984)
2. C. Colvard, T.A. Grant, M.V. Klein, R.Merlin,R.Fischer, H. Morkoc and A.C. Gossard, Phys. Rev. B31, 2080 (1985)
3. A.K. Sood, J. Menendez, M. Cardona and K. Ploog, Phys. Rev. Lett. 54, 2111 (1985)
4. A.K. Sood, J. Menendez, M. Cardona and K. Ploog, Phys. Rev. B32, 1412 (1985)
5. B. Jusserand and D. Paquet, unpublished
6. A.K. Sood, J. Menendez, M. Cardona and K. Ploog, Phys. Rev. Lett. 54, 2115 (1985)

RESONANT RAMAN SPECTROSCOPY OF EXCITONS AND PHONONS IN QUANTUM WELLS.

J. E. Zucker

AT&T Bell Laboratories
Holmdel, New Jersey 07733

Recent developments in semiconductor growth techniques have made possible a new class of materials in which alternating thin layers of two different semiconductors form an artificial superlattice. These heterostructures possess unique physical properties which are distinct from those of the constituent materials [1]. We have used resonant Raman scattering to study excitons, optical phonons and exciton-phonon interactions in GaAs-AlGaAs quantum well heterostructures. In our structures layer thicknesses are of the order of 100 Å so that the electronic spectrum is strongly modified by quantum size effects. The quantization results in a series of excitons that are quasi-two-dimensional in the plane of the GaAs layers. We are able to obtain the transition energies and spectral widths of these excitations from the resonant behavior of the Raman scattering intensity[2]. The polarization properties of the scattering reveal the reduced symmetry of the vibrational modes involved and show that the optical phonons which couple most strongly to quantum well excitons are themselves confined to the GaAs layers [3]. In addition these investigations have provided new examples of the types of information which can be obtained by Raman spectroscopy. We have used resonant Raman scattering in GaAs quantum wells to identify exciton-phonon scattering channels [2] and also have demonstrated the method as a probe of exciton localization for motion perpendicular to[4] and in the plane of[5] the quantum well layers.

REFERENCES

1. A. C. Gossard, IEEE Transactions on Electron Devices, ED-31, 1667 (1984).
2. J. E. Zucker, A. Pinczuk, D. S. Chemla, A. Gossard and W. Wiegmann, Phys. Rev. Lett. 51, 1293 (1983).
3. J. E. Zucker, A. Pinczuk, D.S. Chemla, A. Gossard and W. Wiegmann, Phys. Rev. Lett. 53, 1280 (1984).
4. J. E. Zucker, A. Pinczuk, D. S. Chemla, A. Gossard and W. Wiegmann, Phys. Rev. B 29, 7065 (1984).
5. J. E. Zucker, A. Pinczuk, D. S. Chemla, and A. Gossard, Bull. Am. Phys. Soc. 31, 450 (1986).

RAMAN SCATTERING FROM AMORPHOUS SUPERLATTICES

L. Ley and P. Santos

Max-Planck-Institut für Festkörperforschung, Heisenbergstrasse 1, D-7000 Stuttgart 80, Federal Republic of Germany

We have prepared compositional superlattices based on amorphous silicon (a-Si:H) and amorphous silicon nitride (a-SiN_x:H) by the glow discharge decomposition of appropriate gas mixtures and with periods ranging from 50 to 400 Å. The periodic modulation of the elastooptic constants leads to a folding of the dispersion curve for LA phonons travelling in a direction perpendicular to the layers. In a conventional Raman set-up we observe in backscattering geometry a doublett with an average Stokes shift that varies inversely proportional to the superlattice period. We identify the two peaks with transitions into the second and third branch of the folded LA phonon dispersion curve. The dispersion of these branches was determined by varying the superlattice period and the laser frequency. The result is well described by a continuum theory that takes the differences in the acoustical parameters of a-Si:H and a-SiN_x:H into account. Gaps open up at the new zone boundaries that are directly observed in Brillouin scattering experiments.

BRILLOUIN SCATTERING IN METALLIC SUPERLATTICES

M. Grimsditch

MST Division, Argonne National Laboratory,* Argonne, Illinois 60439

The elastic properties of superlattices has received considerable attention since the first reported enhancements, "supermodulus effects", found by Hilliard *et al.*[1] A recent review[2] summarizes these early experiments and the experiments which supported and contradicted these first findings. Most of these experiments are difficult and require the removal of the superlattice film from the substrate on which they were deposited, a factor which may explain the inconsistency of some of the results. Here we report the results of Brillouin scattering from surface waves in metallic superlattices. This technique is convenient since it requires no sample treatment after deposition and probes an excitation which depends only on the properties of the superlattice and not on the nature of the substrate.

We recall that the elastic properties of a medium composed of layers of two different materials should depend only on the fraction of each material and not on the thickness of the layers[3,4,5] (valid in the long wavelength limit). Contrary to this expectation in Fig. 1 we show the sound velocity of surface waves in V/Ni superlattices which shows a pronounced minimum at ~ 20 Å modulation wavelength.[6] Similar behavior has also been found for Nb/Cu[7] and Mo/Ni[8] superlattices. In the Mo/Ni system, which was also studied extensively by x-rays, it can be seen that the increase below 20 Å is concomitant with the disappearance of the superlattice structure and hence can be understood on the basis of a structural change. Above 20 Å however, well defined layering exists and the change in sound velocity is unexpected. The existence of the softening observed by Brillouin scattering in all these systems has recently been confirmed[6] by ultrasonic surface wave experiments on V/Ni in spite of the large role played by the substrate in determining the surface wave velocity.

A partial explanation for the phonon softening in Mo/Ni was presented by a molecular dynamics calculation in Ref. 9 as due to the ~2% lattice expansion observed with x-rays; it does not address, however, the origin of the lattice expansion which is yet to be explained.

In Fig. 2 the surface wave velocity in Au/Cr superlattices is shown.[10] Contrary to all other systems previously studied by Brillouin scattering, this system shows a pronounced increase in surface wave velocity. This behavior can also be understood on the basis of the lattice expansion and contraction perpendicular to the layers measured by x-rays. It has been possible to ascertain that the Au layers contract while the Cr layers expand as the layers are made thinner, and this explains, at least qualitatively, both the increase and decrease observed in the sound velocity. This argument can be made somewhat more quantitative by writing, simply by analogy with the Murnaghan equation of state,

$$C = C_o \left(\frac{a_o}{a}\right)^{3B'} \qquad (1)$$

where C is an elastic constant, a is the lattice constant and B' is a constant that ranges between 4 and 7 for almost all materials. Arbitrarily choosing B' as 5, using the expression[5] for the elastic constant of the superlattice in terms of those of Au and Cr, using the measured expansion of Cr and contraction of Au and Eq. 1, a structure similar to that shown in Fig. 2 is reproduced.

It is apparent therefore that the elastic anomalies observed in superlattices are due to strains induced in the layers; however the origin of these strains are not yet known.

1. W. M. C. Yang, T. Tsakalakos and J. E. Hilliard, J. Appl. Phys. 48, 876 (1977); T. Tsakalakos and J. E. Hilliard, J. Appl. Phys. 54, 734 (1983); G. E. Henein and J. E. Hilliard, J. Appl. Phys. 54, 728 (1983).
2. I. Schuller, to be published in the Proc. of the IEEE Ultrasonic Symposium, San Francisco, 1985.
3. S. M. Rytov, Akust. Zh 2, 71 (1956) [Sov. Phys. Acoust. 2, 68 (1956)].
4. A. Kueny and M. Grimsditch, Phys. Rev. B 26, 4699 (1982).
5. M. Grimsditch, Phys. Rev. B 31, 6818 (1985) and M. Grimsditch and F. Nizzoli, ibid (to be published).
6. R. Danner, R. P. Huebener, C. S. L. Chun, M. Grimsditch and I. K. Schuller, Phys. Rev. B (accepted for publication).
7. A. Kueny, M. Grimsditch, K. Miyano, I. Banerjee, C. M. Falco and I. K. Schuller, Phys. Rev. Lett. 48, 166 (1982).
8. M. R. Khan, C. S. L. Chun, G. P. Felcher, M. Grimsditch, A. Kueny, C. M. Falco and I. K. Schuller, Phys. Rev. B 27, 7186 (1983).
9. I. K. Schuller and A. Rahman, Phys. Rev. Lett. 50, 1377 (1983).
10. P. Bisanti, M. Brodsky, G. Felcher, M. Grimsditch and L. Sill (to be published).

* Work supported by the U.S. Department of Energy, BES-Materials Sciences, under Contract #W-31-109-ENG-38.

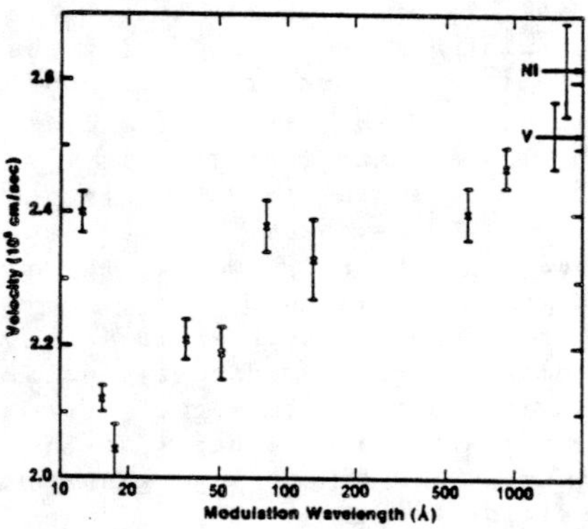

Fig. 1. Velocity in V/Ni

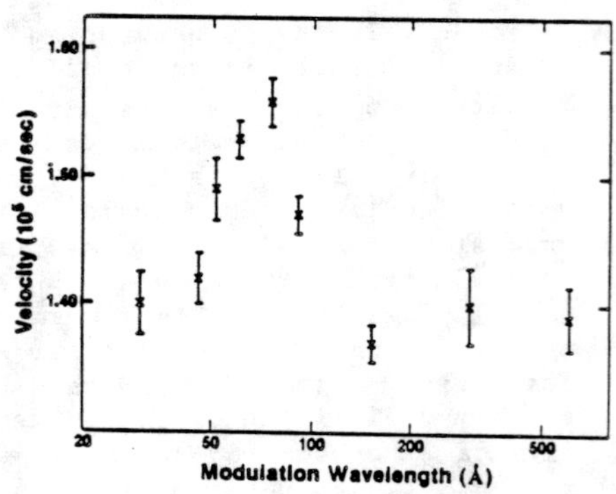

Fig. 2. Velocity in Au/Cr

INELASTIC LIGHT SCATTERING BY FREE CARRIERS IN SEMICONDUCTOR SUPERLATTICES.

A. Pinczuk

AT&T Bell Laboratories, Murray Hill, N.J. 07974

Inelastic light scattering is an extremely effective experimental method to probe the elementary excitations of free carriers in semiconductor superlattices. Sensitivity to low density (typically 10^{11} - 10^{12} cm^{-2}) two dimensional electron or hole gases occurs with the use of photon energies that are in resonance with optical energy gaps that involve the states of free carriers.[1,2] Light scattering studies in GaAs-(AlGa)As quantum-well heterostructures have revealed single particle and collective behavior associated with the reduction in dimensionality.[3,4,5] Major features of the spectroscopy of inter-subband excitations of two-dimensional electron gases are understood.[6] In the case of two-dimensional hole gases the spectra of intersubband excitations are much more complex because of the degenerate character of the valence states of zincblende-type semiconductors that leads to separate, but coupled, light and heavy valence subbands. The light scattering spectra of two-dimensional hole gases yield subband spacings and their lineshapes reveal the non-parabolicity of the valence subbands.[7] The first determination of the dispersion of plasma oscillations in GaAs-(AlGa)As multiple quantum wells has been carried out by inelastic light scattering.[8] These experiments have shown a nearly linear acoustic-like dispersion. The discrete character of the plasmon dispersion in finite multilayer systems, predicted by recent theories,[9] has been confirmed by the light scattering experiments.[10,11]

[1] E. Burstein, A. Pinczuk and S. Buchner, Physics of Semiconductors 1978, B.L.H. Wilson editor (The Institute of Physics, London, 1979) p.1231.

[2] E. Burstein, A. Pinczuk and D. L. Mills, Surface Science 98, 451 (1980).

[3] G. Abstreiter and K. Ploog, Phys. Rev. Letters 42, 1308 (1979).

[4] A. Pinczuk, H. L. Störmer, R. Dingle, J. M. Worlock, W. Wiegmann and A. C. Gossard, Solid State Commun. 32, 1001 (1979).

[5] A. Pinczuk, J. M. Worlock, H. L. Störmer, R. Dingle, W. Wiegmann and A. C. Gossard, Solid State Comm. 36, 43 (1980)

[6] A. Pinczuk and J. M. Worlock, Surface Science 113, 69 (1982).

[7] A. Pinczuk, D. Heimann, R. Sooryakumar, A. C. Gossard and W. Wiegmann, to be published in Surface Science.

[8] D. Olego, A. Pinczuk, A. C. Gossard and W. Wiegmann, Phys. Rev. B25, 7876 (1982).

[9] J. K. Jain and P. B. Allen, Phys. Rev. Letters 54, 2437 (1985).

[10] G. Fasol, N. Mestres, H. P. Hughes, A. Fischer and K. Ploog, submitted for publication.

[11] A. Pinczuk, M. G. Lamont and A. C. Gossard, to be published in Phys. Rev. Letters.

COHERENT ANTI-STOKES RAMAN SCATTERING IN NARROW GAP SEMICONDUCTOR SUPERLATTICES

H. Pascher

Experimentalphysik I, Univ. Bayreuth, D-8580 Bayreuth, F.R.G.

Due to the large nonparabolicity of the energy bands narrow gap semiconductors exhibit considerably high nonlinear susceptibilities. In a superlattice (SL) this nonparabolicity is further increased [1]. The energy gap of $Pb_{1-x}Sn_xTe$ with Sn-contents x between 0.12 and 0.17 varies between 113 meV and 94 meV. Thus using CO_2 laser radiation the susceptibilities are further increased by interband resonances. This is also true for the SL systems $CdTe/Hg_{1-x}Cd_xTe$ and $PbSe/Pb_{1-x}Eu_xSe$ with suitable x-values. Therefore coherent anti Stokes Raman scattering (CARS) using two CO_2 lasers is a very well suited method to observe Raman allowed transitions in such samples, even if epitaxial films have a thickness as small as 1 μm. In a magnetic field the strongest of these transitions are due to the spin resonances of the free carriers (elektrons or holes). Thus CARS is a very precise method to determine the effective g-values.

But there are several difficulties arising: (i) The large non resonant amount of the susceptibility results in complicated lineshapes which have to be analysed. (ii) One- and two-photon interband transitions between Landau levels also give resonances in the CARS-signal which have to be separated from the Raman transitions. (iii) Large laser intensities cause considerable heating of the carriers. Since the polulation difference of the spin levels enters the scattering cross section the lineshapes are intensity dependent. (iv) Thin superlattices can only be irradiated with the light propagating parallel to the sample normal. The polarization selection rules for spin-flip Raman scattering force the sample to be held in Voigt configuration into the magnetic field. Therefore the magnetic field is parallel to the SL layers. In that case the energy of a carrier in a Landau subband depends on the position of the center of the cyclotron orbit in real space. But envelope function calculations [2] show that this dependence is equal for the two spin subniveaus of one Landau level. Therefore the spin transitions should show sharp lines in contrast to cyclotron resonance- or magnetooptical interband transitions in Voigt configuration.

PbTe and PbSnTe have direct gaps a the L-point of the Brillouin zone. The bands have a many valley structure consisting of four ellipsoids oriented along the [111] axis. With the magnetic field along the [110] direction there are two ellipsoids with their axis perpendicular to the magnetic field (B) and two with their axis oriented with 35.26^0 with respect to B. Thus in this configuration magnetooptical resonances occur at two different magnetic fields.

Using two simultaneously Q-switched CO_2 lasers we have measured coherent Raman scattering in magnetic fields up to 7T with sample temperatures of 1.5K in a number of different PbTe/$Pb_{1-x}Sn_xTe$ superlattices which were grown by hot wall epitaxy [3]. Figure 1 shows registrations of the CARS-intensity as a function of the magnetic field of a SL and of the two reference samples of PbTe and $Pb_{1-x}Sn_xTe$, respectively. The special features of the different lineshapes are discussed in detail in [4]. In the figure the spin resonance positions as deduced by comparison of the measured with calculated lineshapes are marked by arrows. In figure 2 the resultant effective g-factors are plotted. In the inserts of the figures the structure of the superlattice is drawn scematically.

The (n-type) SL shows almost the same g-values as $Pb_{1-x}Sn_xTe$ which are much higher than those of PbTe. This fact demonstrates that the electrons in the SL are confined within the $Pb_{1-x}Sn_xTe$ layers and not in PbTe as assumed by other authors [5].

The linewidth of the SL is the same as of the $Pb_{1-x}Sn_xTe$ sample. This demonstrates the correctness of Kriechbaum's [2] calculations. The spin splitting in the SL is not dependent on the position of the cyclotron orbit of the electron with respect to the PbTe/PbSnTe interfaces.

The decrease of the g-value with increasing magnetic field is larger in the SL with respect to $Pb_{1-x}Sn_xTe$ (see fig. 2). This shows the larger nonparabolicity of the bands in the SL.

Acknowledgements: Discussions with Prof. Dr. G. Bauer are gratefully acknowledged. We thank H. Clemens for growing the samples. This work was supported by the Deutsche Forschungsgemeinschaft.

References:

[1] W. L. Bloss, L. Friedmann, Appl. Phys. Lett. 54, 1023 (1982)
[2] M. Kriechbaum, Proc. Int. Winter School, Mauterndorf (1986), ed. G. Bauer, F. Kuchar, H. Heinrich, to be published
[3] H. Clemens, E. J. Fantner, G. Bauer, Rev. Sci. Instr. 54, 685 (1983)
[4] H. Pascher, Appl. Phys. B 34, 107 (1984)
[5] K. Murase, S. Shimomura, S.Takaoka, A. Ishida, H. Fujiasu, Superlattices and Microstructures 1, 177 (1985)

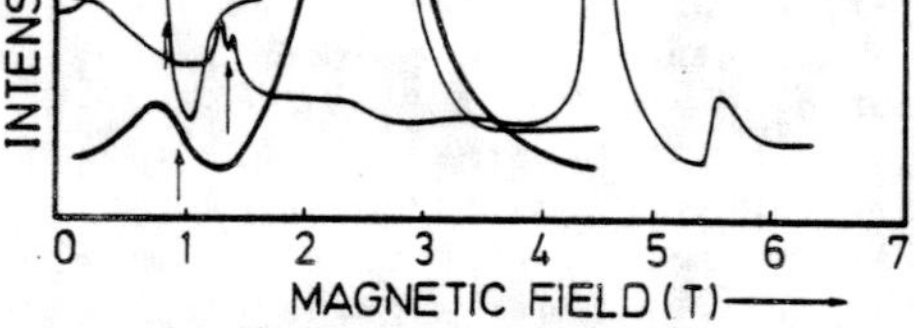

Fig.1: Four wave mixing intensity as a function of the magnetic field for a PbTe/PbSnTe SL, a n-PbTe and a PbSnTe sample for comparison. T=1.5K; B//[110]

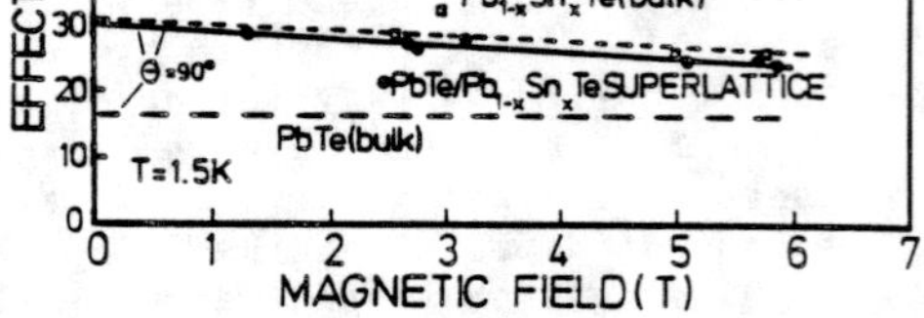

Fig.2: g-factors of a SL, PbTe and PbSnTe for several magnetic fields (different combinations of laser frequencies) o:SL; :bulk PbSnTe Θ:angle between valley axis and B

chapter 10

LAYERED STRUCTURES

RESONANT RAMAN SCATTERING AT THE INTERBAND ABSORPTION EDGE IN GRAPHITE-H_2SO_4 INTERCALATION COMPOUNDS

P.C. Eklund[a], G.D. Mahan[b], J.G. Spolar[b], J.M. Zhang[a], E.T. Arakawa[b], and D.M. Hoffman[a]

[a]Dept. of Physics and Astronomy, University of Kentucky, Lex., KY 40506
[b]Solid State Division, Oak Ridge National Laboratory, Oak Ridge, TN 37831

We report on resonant Raman scattering observed in the quasi two dimensional (2D) metals graphite-H_2SO_4. A previous account of this work has appeared recently [1]. A peak or resonance in the Raman cross section σ for phonon emission is observed when the laser frequency is centered on the threshold for interband electronic absorption. A series of four compounds were studied: $C_p^+(HSO4^-)(H_2SO_4)_x$, where p=60,48,28,21. The principal effect of decreasing p is to lower the Fermi level in the valence band(s), thereby shifting the threshold for interband absorption to higher energy. The E_{2g} graphitic intralayer phonon(~1600 cm^{-1}) used to monitor the resonance scattering is perturbed only slightly by changes in p. A theoretical model is presented which is found to be in good agreement with the experimental observations. We conclude that the effect should be observable in any Raman-active metal exhibiting a well-defined interband absorption threshold.

The samples were prepared by the electrochemical intercalation of plates of highly oriented pyrolytic graphite submerged in concentrated H_2SO_4 (~96%, Baker). Raman scattering and reflectivity measurements were carried out in situ in the electrochemical glass cells used to prepare the compounds. Optical dielectric constants ε_1, ε_2 were determined from a Kramers-Kronig analysis of the reflectance data. Details of the experimental procedure are available elsewhere [1]. The Raman scattering experiments were carried out in a backscattering geometry with TM photons incident on a cleaved sample c-face, the beam making a 45° angle to the c-axis. Since the phonons have E_{2g} symmetry, no analyzer was used. The data were corrected for the instrument transmission and the laser power.

In Fig. 1 we display both ε_2 and σ vs. ω_1, where ω_1 is the incident photon frequency, for the p = 60,48,28,21 compounds. The ε_2 and σ data appear, respectively, as solid lines and open circles. The dashed lines are guides to the eye. The ε_2 data are characteristic of a metal: high ε_2 for low ω due to intraband absorption, followed by a threshold for interband absorption. σ is seen to be correlated with $|d\varepsilon_2/d\omega_1|^2$. Slight misalignments are observed between σ and $d\varepsilon_2/d\omega$. We believe these are due to small differences in the samples used to study separately σ abd ε_2. The width of the interband threshold in ε_2 is observed to be comparable to the width of the resonance in σ which suggests a common origin. Calculations [2] indicate this width is consistent with the lifetime of the excited electron due to electron-electron interactions.

The differential Raman cross section is:

$$\frac{d^2\sigma}{d\omega_2 d\Omega} = r_0^2 \; (\omega_2/\omega_1)\,|M|^2\; \Psi(\omega_1,\omega_2) \tag{1}$$

where r_o is the classical electron radius, and $\Psi(\omega_1,\omega_2)$ is the transmission-absorption factor. Using $\varepsilon_1,\varepsilon_2$ determined experimentally, we find that $\Psi(\omega_1,\omega_2)$ is essentially constant in the region where $|M|^2$ is large. The matrix element M is calculated using the standard three-step scattering process and involves explicit calculation of the A•p and electron-phonon matrix elements. In Fig. 2 we display the frequency dependence of the calculated (solid line) and experimental values (Δ) for the p=28 compound. As can be seen, theory and experiment are in good agreement.

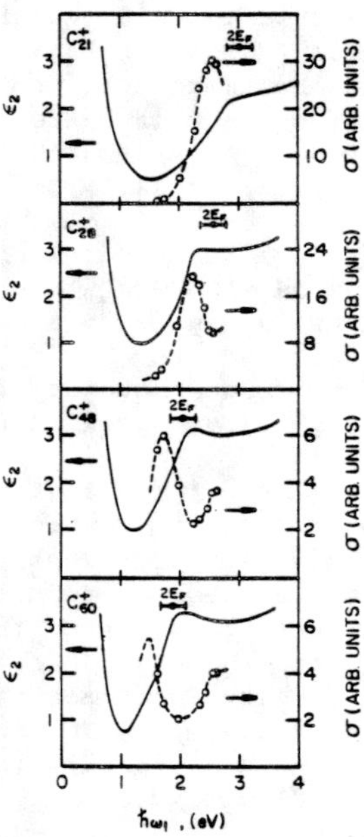

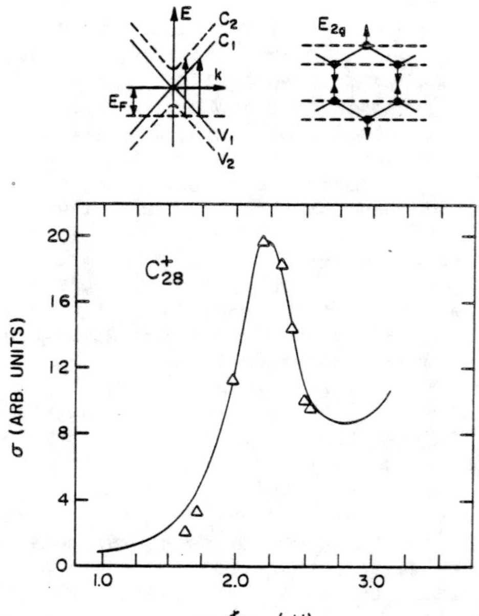

Fig. 1(left). ε_2, σ vs. $\hbar\omega_1$ for the p=60,48,28,21 compounds. The position of E_F is determined from Drude analyses of ε_2. Fig. 2 (right). Calculated and experimental (Δ) values for the Raman cross section σ vs. laser frequency ω_1. Above left: energy bands near E_F for the p=28,21 compounds (solid) and p=60,48 compounds (dashed and solid); Above right: E_{2g} graphtic intralayer phonon.

REFERENCES:

1. P.C. Eklund, G.D. Mahan, J.G. Spolar, J.M. Zhang, D.M. Hoffman and E.T. Arakawa, Solid St. Commun. 57 (8), (1986), pp. 567-570.
2. K.W. -K. Shung, Phys. Rev. B, in press.

ION SIZE AND SUBSTRATE RIGIDITY IN MIXED LAYERED SOLIDS

S. A. Solin[a], S. Lee[a], W. Leonard[b] and T. J. Pinnavaia[b]

Departments of Physics (a) and Chemistry (b)
Michigan State University
East Lansing, MI 48824
USA

The physical properties of layered solids are, to a large measure, determined by the rigidity of the constituent layers with respect to transverse distortions.[1] Thus, graphite which is composed of thin monatomic layers[2] can be characterized as "floppy" while layered aluminosilicate clays such as vermiculite which are composed of thick multiatomic layers[3] can be characterized as "rigid".

In recent x-ray and Raman studies[4] of the mixed layered silicate clays, $M_{1-x}M'_x$-V where $0 \leq x \leq 1$, M and M' are alkali ions and V repesents vermiculite, we showed that the dependence of the c-axis normalized basal spacing $[d_n = (d_{obs}-d_{min})/(d_{max}-d_{min})]$ on composition, x, was nonlinear and exhibited a sharp drop with increasing x at x ~ 0.4 as shown in the dashed curve of Fig. 1 for M = Cs and M' = Rb. This result was quite unexpected because the rigidity of the silicate layer relative to that of graphite, for instance, should allow the larger ion to prop the layers apart even for values of x ~ 1. In addition, corresponding measurements of $Cs_{1-x}Rb_xC_8$[5] which are shown as the dotted line in Fig. 1 revealed a more gradual decrease in basal spacing but at a higher value of x, opposite to expectations based on simple rigidity arguments.[4]

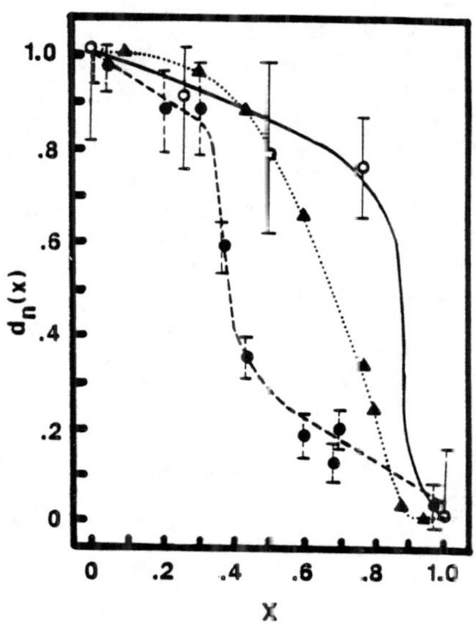

Fig. 1. Composition dependence of the normalized basal spacing, $d_n(x)$, of $Cs_{1-x}Rb_xC_8$ (▲), $Cs_{1-x}Rb_x$-V (●) and $[Me_4N^+]_{1-x}[Me_3NH^+]_x$-V (○).

It has been shown[4] that the silicate layer in vermiculite is indeed relatively rigid to transverse distortions but that it is very floppy vis-á-vis longitudinal in-plane perovskite-like rotations of the SiO_4

tetrahedra which comprise the bounding tetrahedral sheets of the silicate layers. These tetrahedral rotations cause the hexagonal cavities bounding the intercalated ions to locally expand or contract to accommodate those ions without appreciable transverse distortion of the layers. When the size difference between the guest ions is small as in the case of Cs and Rb, for which $\Delta r = r_{Cs} - r_{Rb} = 0.21 A$, the collapse of the silicate layers in the regions between the large ions is constrained by the small ions, not by the intrinsic stiffness of the layer itself.

To explore the intrinsic layer stiffness it is necessary to employ guest ions with a value of Δr.that is considerably larger than that for the Rb-Cs pair. Accordingly, we prepared a series of mixed -layer vermiculites with M = tetramethylammonium and M' = trimethylammonium, $(Me_4N^+)_{1-x}(Me_3NH^+)_x$-V, $0 \leq x \leq 1$, for which $\Delta r = 0.5A$. The details of the ion exchange methods used to prepare these clay intercalation compounds are given elsewhere.[6] The normalized basal-spacing of mixed layer tetramethylammonium-trimethylammonium vermiculite is shown in Fig1 as a solid line. The data of Fig. 1 were obtained from (001) x-ray diffraction measurements which confirmed that the specimens were homogenious.

It is clear from Fig. 1 that the normalized basal spacing of the tetramethylammonium-trimethylammonium system is dominated by the spacing induced by the larger ion which, with minor diminution persists up to x ~ 0.75 at which point it exhibits a quasidiscontinuous decrease similar to that exhibited by the Cs-Rb system at x ~ 0.4. Indeed the larger basal spacing may persist to x = 1 in agreement with the simple rigidity concepts alluded to above. [We expect to study this point in more detail using a series of samples with $0.7 \leq x \leq 1.0$.] The negative slope of $d_n(x)$ for the tetramethylammonium-trimethylammonium system in the range $0 < x < 0.75$ clearly indicates the some transverse layer distortion or "sagging" occurs ie. the vermiculite layers are not infinitely rigid. Also note that the tetramethyl and trimethylammonium ions are very large compared to the hexagonal pockets of the vermiculite layers between which they are nestled. Therefore the in-plane tetrahedral distortions which modify the hexagonal cavities to effectively compensate for the small ion size differences in the in the Cs-Rb system will have a much smaller effect on the normalized basal spacing exhibited by tetramethylammonium-trimethylammonium vermiculite. Instead it is the intrinsic rigidity of the vermiculite layer itself, not the size of the smaller ion, which limits the collapse of the layers in the region between the larger ions.

This work was supported by the US National Science Foundation under grants DMR 85-17223 (SAS) and CHE 83-06583 (TJP).

References

1. S.A. Solin, J. Molecular Catalysis 27, 291 (1984).
2. S.A. Solin, Adv. in Chemical Physics 49, 455 (1982).
3. R.E. Grimm, Clay Minerology (McGraw-Hill, New York, 1986).
4. B.R. York, S.A. Solin, N. Wada, R. Raythatha, I.D. Johnson, and T.J. Pinnavaia, Solid State Comm. 54, 475 (1985).
5. D. Billaud and A. Hérold, Bull. Soc. Chim. Fr. 11, 2402 (1974).
6. W. Leonard, T.J. Pinnavaia, S. Lee and S.A. Solin, to be published.

LONG-WAVELENGTH PHONONS IN METALLIC LAYERED TRANSITION METAL DICHALCOGENIDES.

J.C. Irwin and W.G. McMullan

Dept. of Physics, Simon Fraser University, Burnaby, B.C., Canada V5A 1S6

INTRODUCTION

This paper describes the application of a valence-force-field (VFF) model (1-3) to the metallic 2H layered dichalcogenides. This model includes force constants describing bond angle bending as well as bond stretching and has been found to provide better agreement with experiment than nearest neighbour models that employ central forces only (1,2). We have used a VFF model (model 2) developed by Feldman (3) because of the excellent agreement it provided with both Raman and neutron scattering data in 2H-$TaSe_2$ and 2H-$NbSe_2$ (3). We have used this model to assist in the interpretation of Raman scattering data in both the pristine and intercalated compounds of 2H-NbS_2, 2H-TaS_2 and 2H-$NbSe_2$. Analytic expressions have been obtained for the squares of the normal mode frequencies at Γ (4) in terms of four parameters that can be determined from the measured frequencies of the four Raman active modes of the 2H-compound.

RESULTS

Raman scattering experiments have been carried out on the metallic compounds 2H-NbS_2, 2H-TaS_2 and intercalated Ag_xTaS_2 (4). Recently, similar experiments have been carried out on 2H-$NbSe_2$ and $Ag_{0.33}NbSe_2$. Samples of $Ag_{0.33}NbSe_2$ were prepared by electrointercalation in a 0.1M solution of $AgNO_3$ in glycerin. The $Ag_{0.33}NbSe_2$ samples have been found to be stage-2 compounds with the Ag-atoms occupying sites with tetrahedral symmetry (5). For temperatures T such that 200K $< T <$ 380K, the Ag atoms form an ordered arrangement on a $2a_0 \times 2a_0$ lattice. As the temperature is lowered below 130K, a $2\sqrt{3}a_0 \times 2\sqrt{3}a_0$ ordering is observed to coexist with the $2a_0 \times 2a_0$ superlattice (6).

Zone boundary phonons in the intercalated compounds should thus be zone folded to the center of the reduced zone of the intercalated crystal (7). The $2a_0 \times 2a_0$ superlattice should result in phonons at the M point of the original zone becoming Raman active after intercalation. A $2\sqrt{3}a_0 \times 2\sqrt{3}a_0$ superlattice would also result in phonons from the K point (and a point midway between Γ and K) being folded into the center of the new zone.

The Raman spectrum of $Ag_{0.33}NbSe_2$ obtained at 15K contains many new features and these are attributed to the formation of an Ag-atom superlattice. The phonon frequencies at M and K have been calculated using the VFF model and a comparison of the measured and calculated frequencies is shown in Table 1. As can be seen, good agreement is obtained between the calculated and measured values.

TABLE 1. Phonon frequencies of $Ag_{0.33}NbSe_2$ obtained from Raman spectra and values calculated using the VFF model. All frequencies are given in cm^{-1}.

Experiment	Calculated	Zone Point Assignment
93	93,99	K
103	108	M
115	113	M
129	139	M
136	133	E_{1g}
150	147,151,155(4)	K
197	198	K
207	211	M
216	216	M
(234,239)	(231,238,242)	A_{1g},E^1_{2g},M
258	249	K
(265,272)	269	M

Unfortunately, the above comparison is not particularly selective given the large number of calculated values that are available. In an effort to make the comparison more rigorous, the symmetry properties of the zone boundary phonons have also been investigated. These properties can be compared to the results of Raman spectra obtained with alternately parallel and crossed polarizations. Good agreement is also obtained in this procedure, thus lending additional confidence in the assignments shown in Table 1.

SUMMARY

A relatively simple form of a VFF model has been used to analyze the long-wavelength phonons in both pristine and intercalated metallic 2H-layered dichalcogenides. The model has provided phonon frequencies that are in good agreement with experimental observations on several different compounds. The results are consistent with the assumption that additional phonons observed in the intercalated crystals arise from a zone-folding mechanism. It is conceivable, given additional model refinements and well-characterized compounds, that the Raman spectra of the pristine and intercalated compounds could provide sufficient data to enable a reliable determination of the phonon dispersion relations of the pristine compound.

REFERENCES

1. W. Wakabayashi, H.G. Smith and R.M. Wicklow, Phys. Rev. B12, 659 (1975).
2. G. Lucovsky, J.C. Mikkelson, Jr., W.Y. Ziang, R.M. White and R.M. Martin, Phys. Rev. B14, 1663 (1976).
3. J.L. Feldman, Phys. Rev. B25, 7132 (1982).
4. W.G. McMullan and J.C. Irwin, Can. J. Phys. 62, 789 (1984).
5. O.S. Rajora and A.E. Curzon, preprint (to appear in Phys. Stat. Sol.).
6. A.E. Curzon and O. Rajora, Phys. Stat Sol.(a) 87, 157 (1985).
7. M. Plischke, K.K. Bardhan, R. Leonelli and J.C. Irwin, Can. J. Phys. 61, 397 (1983).

INTRA- AND INTERLAYER RAMAN MODES IN LAYER SILICATES.

N. Wada

Schlumberger-Doll Research, Old Quarry Road, Ridgefield, CT 06877, U.S.A.

Layer silicates are composed of two basic two-dimensional sheets; one is a Si_2O_5-tetrahedral sheet and the other is an Al- or Mg- octahedral sheet. According to the stacking sequence of the two basic sheets and isomorphous substitutions, layer silicates can be classified into many different groups and species.[1] Of particular interest among the layer silicates is vermiculite, which can be cation-exchanged and hydrated to form one- or two-water layers between the adjacent silicate layers. Single-crystal vermiculite gives rise to resolution-limited sharp x-ray Bragg peaks (at least, the correlation length along the c-axis is more than 400 Å), which typically yield a 10, 12.5 or 15 Å c-axis repeat distance for zero, one or two water layer hydration states, respectively. (An exact c-axis repeat distance also depends on interlamellar cations.)

In this paper, we report Raman scattering results in low-frequency regions (less than 250 cm^{-1}) from single-crystal vermiculite .[2] Na- and Sr-vermiculites were prepared using Mg-vermiculite from Llano, Texas by immersing the crystals in a NaCl or $SrCl_2$ water solution for a couple of months at 60 ° C. The samples were confirmed to be well-defined by both (00*l*) x-ray and neutron scattering.

Figure 1 shows Raman spectra taken from Na-vermiculite in three different hydration states. As can be seen from this figure, an anhydrous Na-vermiculite exhibits peaks at 24, 102 and 196 cm^{-1}. The Raman peak at 24 cm^{-1} can be assigned to an interlayer shearing mode. Because of the weak interlayer interaction and the large c-axis repeat distance (c~20Å), the frequency in Na-vermiculite is very low. Note that no corresponding peaks were found in Na-vermiculite at other hydration states. Because of the interlamellar water molecules, the interlayer shearing frequency becomes even lower so that the peak can not be distinguished due to the laser stray light background.

We also examined an anhydrous Sr-vermiculite and found the frequency of the interlayer mode to be 21 cm^{-1}. We emphasize that the interlayer shearing mode should be quite sensitive to the interlamellar species, since the intralayer structure of the silicate layers is relatively rigid. The frequency difference between Sr- and Na-vermiculite can be understood in terms of the areal density of the interlayer cations. The number of Na^+ atoms per unit area is half that of Sr^{2+}. This means that the interlayer shearing force constant of Sr-vermiculite should be less than that of Na-vermiculite.

The Raman peak at 196 cm^{-1} shows drastic changes, depending on the hydration states. This mode may be an intralayer B_2 mode,[3] where only the surface oxygen atoms in the tetrahedral sheets are involved. The mode is then expected to be sensitive to the interlamellar in-plane structure. When Na-vermiculite is in the two-water layer hydration states, Na^+ cations occupy positions above the tetrahedral bases because of electrostatic interactions between interlayer cations and substituted atoms in the host. This hydration state exhibits a broad peak at 164 cm^{-1}. However, when the hydration state becomes one-layered, the cations are situated over either the tetrahedral bases or the ditrigonal cavities. With this the mode of the two-water layer state becomes perturbed and splits into two modes, namely at 137 and 199 cm^{-1}. Further hydration causes all the cations to sit in the ditrigonal cavities. This yields a relatively high-frequency Raman peak at 196 cm^{-1}, since the intralayer oxygen interactions are stiffened by the presence of Na-atoms in the ditrigonal cavities.

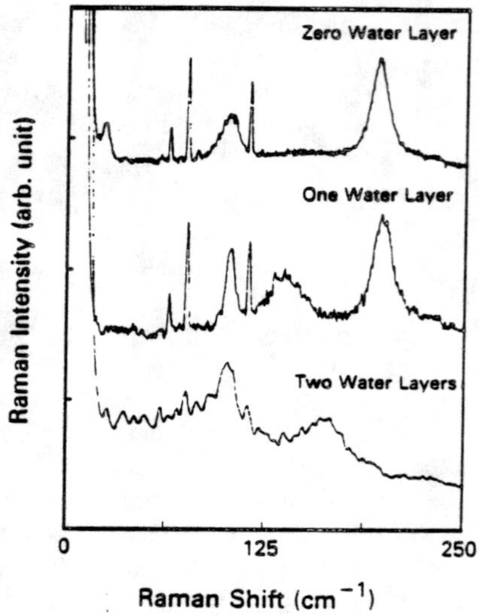

Fig. 1 Raman spectra taken from Na-vermiculite in different hydration states. A 5145 Å Ar laser cylindrically-focused beam was used. Sharp peaks at 66, 77 and 117 cm^{-1} are laser plasma lines.

Finally, we mention that these experimental results should contribute to a better understanding of the inter- and intralayer interactions in layer silicates, where one can study quasi-two dimensional phase transitions and other interesting physical phenomena.

ACKNOWLEDGEMENTS

This work was done in collaboration with Masatsugu Suzuki and W. A. Kamitakahara.

REFERENCE

1. *Crystal Structures of Clay Minerals and Their X-ray Identification,* edit. by G.W. Brindly and G. Brown (Mineralogical Society, 1980)

2. B. R. York, S. A. Solin, N. Wada, R. Raythatha, I. D. Johnson and T. J. Pinnavaia. Solid State Comm., **54** (1985) 475

3. M. Ishii, T. Shimanouchi and M. Nakahira, Inorg. Chim. Acta, 1 (1967) 387

RAMAN SCATTERING BY AMPLITUDE MODES AND SOFT MODES IN IODIZED TRANSITION-METAL TETRACHALCOGENIDES.

Tomoyuki Sekine

Institute of Physics, University of Tsukuba, Sakura-mura, Ibaraki 305, Japan

A new family of linear-chain compounds $(MSe_4)_nI$ (M=Nb or Ta; n=2, 3 or 10/3) has attracted much attention, because they show intriguing phase transitions. When n=2 and 10/3, these compounds exhibit metallic natures and undergo charge-density-wave (CDW) phase transitions. When n=3, on the other hand, these compounds show semiconducting property and undergo displacive-type structural phase transitions. In this paper we report the observation of amplitude modes in the CDW states of $(TaSe_4)_2I$ and $(NbSe_4)_{10/3}I$ and of soft modes in $(NbSe_4)_3I$ by Raman scattering.

1. $(TaSe_4)_2I$ and $(NbSe_4)_{10/3}I$

The CDW phase transitions were observed at T_c=263 K in $(TaSe_4)_2I$ and at T_c=289 K in $(NbSe_4)_{10/3}I$ by our resistivity measurements. Figure 1 shows the temperature dependence of the low-frequency parts of the a(cc)$\bar{a}$ spectra in $(TaSe_4)_2I$ [1,2] and $(NbSe_4)_{10/3}I$ [2], where a=[100] and c=[001]. A peak was found at about 90 cm^{-1} with very strong temperature dependence in $(TaSe_4)_2I$. A similar peak was observed at about 100 cm^{-1} in $(NbSe_4)_{10/3}I$. Their frequencies scarcely shift, but their Raman intensities decrease with increasing temperature and they disappear above T_c. A remarkable fact is that they broaden extremely with approaching T_c. This behaviour is very similar to that of the amplitude mode in KCP [3]. Kurihara [4] calcu-

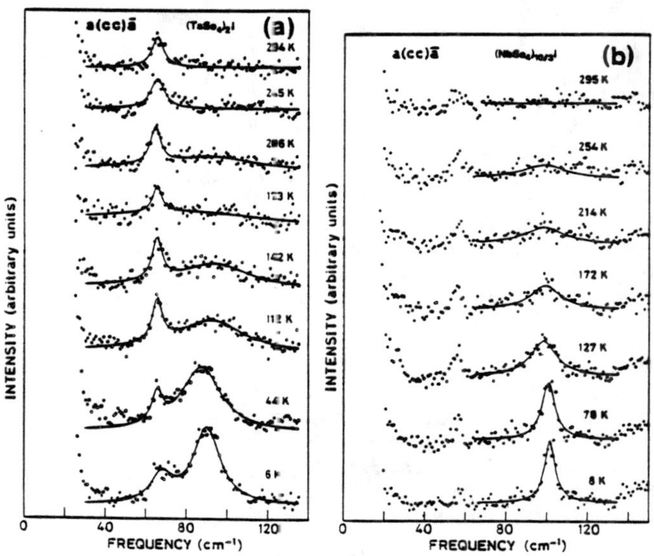

Fig. 1 The a(cc)$\bar{a}$ Raman spectra: (a) $(TaSe_4)_2I$ and (b) $(NbSe_4)_{10/3}I$.

lated the damping constant of the amplitude mode by using a nonlinear amplitude-phase interaction Hamiltonian. His result explains the temperature dependence of the damping constants of the 90 cm^{-1} peak in $(TaSe_4)_2I$ and the 100 cm^{-1} peak in $(NbSe_4)_{10/3}I$ [1,2]. We assign them to the amplitude modes in the CDW states.

2. $(NbSe_4)_3I$

In $(NbSe_4)_3I$ a structural phase transition has been found at 274 K by X-ray diffraction [5]. Since it is not due to the formation of CDW, it is unique in linear-chain compounds. We have observed a soft mode which is related to this phase transition [6]. This fact shows clearly that this phase transition is of displacive type. Figure 2 shows the low-frequency parts of the $a'(cc)\bar{a}'$ Raman spectra in $(NbSe_4)_3I$, where a'=[110]. We observed three peaks at 28, 58 and 81 cm^{-1} at 6 K. The strongest peak at 81 cm^{-1} softens with increasing temperature, and it, therefore, can be assigned to the soft mode which is related to the structural phase transition at 274 K [6]. A mode repulsion between the soft mode and the 58 cm^{-1} peak was observed at high temperatures [6]. On the other hand, the 28 cm^{-1} peak is sharp at low temperatures, but it broadens and softens with increasing temperature. It was not observed above 107 K. This result suggests the existence of another displacive-type structural phase transition around 100K in $(NbSe_4)_3I$.

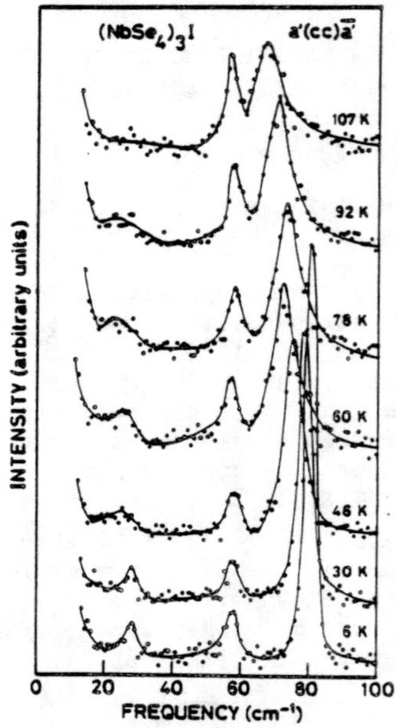

Fig. 2 The $a'(cc)\bar{a}'$ Raman spectra in $(NbSe_4)_3I$.

Gressier et al. [7] have done a structural analysis at 128 K by X-ray diffraction of $(NbSe_4)_3I$, and they pointed out that the crystal symmetry changes from P4/mnc to $P\bar{4}2_1c$ at 274 K. Our recent study of the X-ray diffraction [8] shows that new Bragg peaks appear at $(0,0,\ell)$, where $\ell=2n+1$, around 90 K. The structural analysis at 30 K indicates that the space group of the crystal is $P\bar{4}$. Therefore we assign the 28 cm^{-1} peak to a soft mode which is related to the low-temperature structural phase transition. In conclusion we found that the displacive-type structural phase transitions occur successively at 274 K and around 100 K in $(NbSe_4)_3I$.

References

[1] T. Sekine et al., Solid State Commun. 53, 767 (1985).
[2] T. Sekine et al., Proc. 2nd Int. Conf. on Phonon Physics, ed. by J. Kollár et al. (World Scientific, Singapore, 1985), p. 735.
[3] E.F. Steigmeier et al., Solid State Commun. 17, 1447 (1975).
[4] S. Kurihara, J. Phys. Soc. Jpn. 48, 1821 (1980).
[5] M. Izumi et al., Solid State Commun. 51, 191 (1984).
[6] T. Sekine et al., Solid State Commun. 52, 379 (1984).
[7] P. Gressier et al., Mat. Res. Bull. 20, 539 (1985).
[8] M. Izumi et al., to be published in Synthetic Metals.

STRUCTURAL PHASE TRANSITIONS IN PEROVSKITE-TYPE LAYER COMPOUNDS.

M.COUZI

Laboratoire de Spectroscopie Infrarouge (UA 124 CNRS), Université de Bordeaux I, 33405 TALENCE-CEDEX, FRANCE.

The perovskite-type layer compounds, with general formula AMX_4 or A_2MX_4 are built up from infinite sheets made of corner-sharing MX_6 octahedra; the spaces between layers are occupied by the A cations. These compounds have attracted a great deal of attention recently, because of their peculiar properties due to the two-dimensional character of their structure. In particular, they undergo numerous structural phase transitions involving different types of mechanism, and we have studied such phenomena by means of Raman scattering experiments.

In such compounds like $RbAlF_4$ and NH_4AlF_4, very classical examples of soft-mode induced (displacive) phase transitions and of purely order-disorder transitions are observed, respectively. In $RbAlF_4$, the two-dimensional perovskite arrangement results in a rather loose network, and the existence of octahedra disorder is evidenced in the high-temperature phases by the observation of a partial breakdown in the $\vec{k} = 0$ selection rules. With NH_4AlF_4, octahedra disorder is not observed, probably because of the very symmetrical hydrogen-bond network, but the NH_4^+ cations give rise to order-disorder processes (reorientable pseudo spins model) explaining the phase transition. A change-over in the transition regime from 2D to 3D is shown to occur around 170 K, whereas T_c = 150 K (3D transition).

In the series of alkylammonium derivatives, $(C_nH_{2n+1}NH_3)_2MX_4$ (M = Mn, Cd ... X = Cl, Br), the phase transitions are governed by the reorientational dynamics of the chains, coupled to different tilts of the octahedra within the layers, via the NH...X hydrogen bonds. For instance, the phase sequence observed with $(CH_3NH_3)_2MCl_4$ compounds (M = Mn, Cd) corresponds to successive freezings of reorientational motions of the $CH_3NH_3^+$, from a fully disordered tetragonal phase to an ordered monoclinic one.

Disorder induced scattering coming from the octahedra-layers is also observed on the Raman spectra ; the peculiar selection rules noticed for these additional features are interpreted in terms of fluctuating layer domains, whose instantaneous symmetry is lower than the averaged structure derived from diffraction methods. It is shown that the above mentioned fluctuations are essentially bidimensional. The P,T phase diagram of $(CH_3NH_3)_2MCl_4$ compounds, established by means of Raman spectroscopy, reveals the existence of a new high pressure ordered phase (HP). The results show that the HP structure results from the condensation of a shearing mode (antitranslations of the layers against each other) of very low frequency ; such a mechanism is coupled to an important compression of the interlayer distance.

The long chain alkylammonium compounds give rise to additional phase transitions resulting from a partial melting of the carbon chains. In the alkylammonium series, $NH_3(CH_2)_nNH_3MX_4$, such a phenomenon is generally not observed, because both ends of the same chain is hydrogen bonded to successive octahedra layers. However, for a number of crystals in this family, Raman scattering experiments show the existence of "twisted" chains, whose population relative to all-trans states increases with increasing temperature. Thus, at high temperature, a trans $\leftrightarrow$ twisted conformational disorder is superimposed on the orientationally disordered state with all-trans configuration ; complex mechanisms are then associated to the observed phase transition. Furthermore, in the case of $NH_3(CH_2)_5NH_3CdCl_4$, it is shown that beyond a threshold concentration, the twisted chains induce a structural phase transition to a disordered high temperature phase with monoclinic symmetry, whereas the ordered low temperature phase (with only all-trans chains) is orthorhombic. A model is proposed to account for the unusual behaviour of this system.

STUDY OF CHARGE-DENSITY-WAVE PHASE TRANSITIONS IN LAYERED TRANSITION-METAL DICHALCOGENIDES 4Hb-TaS_2 AND 4Hb-$TaSe_2$ BY RAMAN SCATTERING.

Tsuneo Nakashizu, Tomoyuki Sekine, Kunimitsu Uchinokura+ and Etsuyuki Matsuura

Institute of Physics, University of Tsukuba, Sakura-mura, Ibaraki 305, Japan
+Department of Applied Physics, The University of Tokyo, Hongo, Bunkyo-ku, Tokyo 113, Japan

The layered transition-metal dichalcogenides 4Hb-TaS_2 and 4Hb-$TaSe_2$ have charge-density-wave (CDW) phase transitions. The undistorted unit cell of 4Hb-compounds consists of four layers of alternating octahedral (OH) and trigonal-prismatic (TP) coordinations. Incommensurate (IC)-commensurate (C) CDW transition occurs in the OH layer at T_d=315 K in 4Hb-TaS_2 and at T_d=400 K in 4Hb-$TaSe_2$. This CDW transition resembles the IC-C CDW transition of 1T-compounds, although T_d of 4Hb-compounds is lower than that of 1T-compounds. Moreover normal (N)-IC CDW transition occurs at T_0 in the TP layers (22 K in 4Hb-TaS_2 and 75 K in 4Hb-$TaSe_2$). It resembles the N-IC CDW transition of 2H-compounds, although T_0 of 4Hb-compounds is lower than that of 2H-compounds. The Raman spectra of 4Hb-compounds are mainly divided into high- and low-frequency parts. The former comes from the optical phonon branches and the latter from the acoustic ones. In this report, we study Raman spectra of 4Hb-TaS_2 and 4Hb-$TaSe_2$.

Figure 1a shows the temperature dependence of the Raman spectra of 4Hb-TaS_2 obtained by a photon-counting system. The high-frequency spectrum above 200 cm^{-1} observed in the C phase is very similar to the corresponding spectrum of 1T-TaS_2 and has no corresponding component observed in 2H-TaS_2. The intensity of the high-frequency spectrum decreases as the temperature increases toward T_d=315 K. This result suggests that this spectrum apparently comes from the OH layer. Moreover the frequencies of the phonon peaks in this spectrum do not change. On the other hand, the frequencies of the phonon peaks in the low-frequency spectrum depend on temperature and the spectrum changes its shape especially near T_d. The strongest peak with A_g symmetry shows softening as the temperature increases. We assigned it to one of the soft modes which is related to the CDW transition of the OH layer. Using optical-multichannel detector we have found that not only this peak but also two A_g-symmetry peaks in its lower-frequency side strongly depend on temperature [1]. The temperature dependence of the frequencies of these peaks clearly showed anticrossing. We have attributed it to the LA-TA coupling [1]. Moreover we have found two other soft modes of E_{1g} and E_{2g} symmetries which anticross with other same-symmetry modes [1]. Since the center of inversion is in the OH layer, all these soft modes are amplitude modes [1]. The space group of the C phase was determined to be C_{6h}^2 from the number and symmetry of the soft modes. A rigid-layer mode was observed at 19 cm^{-1} (see 146 K spectrum in Fig. 1a) and the interlayer-shear-force constant was estimated to be 2.6x10^3 dyn/cm.

Figure 1b shows the temperature dependence of Raman spectra of 4Hb-$TaSe_2$. The spectra reveal more complicated feature than those in Fig. 1a. Soft modes which are related to the CDW transition of the OH layer are also observed (one of them is the strongest peak in Fig. 1b). They are the amplitude modes. The number and symmetry of the soft modes of the OH layers are the same as those of 4Hb-TaS_2, which indicates that the space group of the C phase is C_{6h}^2 [2]. We observed anticrossings with respect to these soft modes. We observed high-frequency spectrum which comes from the optical modes of the OH and TP layers. Intensity of the mode near 190 cm^{-1} decreases as the temperature increases toward T_d=400 K, which suggests that this mode comes from the OH layer. This mode corresponds to the high-frequency mode of 1T-$TaSe_2$. The 220 cm^{-1} (E_{2g}) and 234 cm^{-1} (A_g) modes at 45 K correspond to the high-frequency modes of 2H-$TaSe_2$. The relative intensity of the 234 cm^{-1} mode against the 220 cm^{-1} mode increases with decreasing temperature. The temperature dependence of the intensity of the A_g mode is larger than that of the E_{2g} mode. The temperature dependence of these modes is similar to the corresponding modes of 2H-$TaSe_2$ [3]. At room temperature a broad band near 150 cm^{-1} was observed. This band has temperature dependence. It is assigned to the two-phonon overtone Raman spectrum which comes from the Kohn-anomaly (KA) phonon branch of the TP layer with the distortion wave vector [2]. The frequency of the lower edge of this band corresponds to twice the minimum frequency of the KA mode (shown by arrows in Fig. 1b). Its temperature dependence is weaker than that of 2H-$TaSe_2$, which enabled us to explain the fact that T_0=75 K in 4Hb-$TaSe_2$ is lower than T_0=122 K in 2H-$TaSe_2$ [2]. We observed a broad peak below 50 cm^{-1} at 45 K in Fig. 1b. It softened with increasing temperature toward T_0=75 K [2]. We assigned it to a soft mode which is related to the CDW transition of the TP layer. A rigid-layer mode was observed at 16.5 cm^{-1} and the interlayer-shear-force constant was estimated to be 2.72x10^3 dyn/cm.

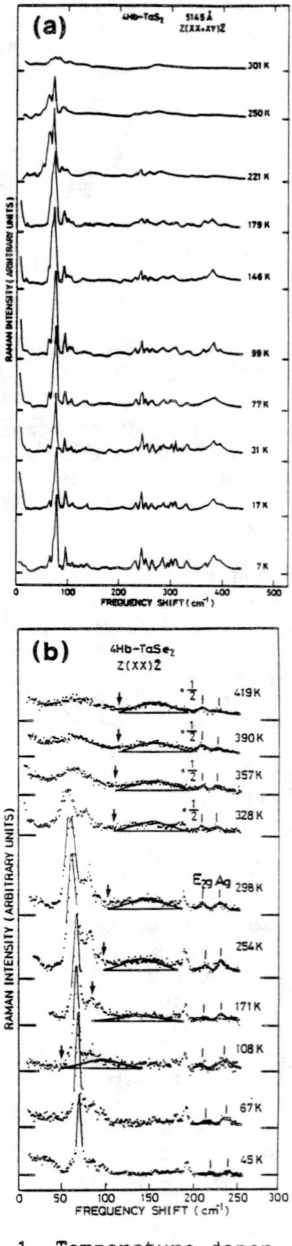

Fig. 1 Temperature dependence of Raman spectra of 4Hb-TaS_2 (a) and 4Hb-$TaSe_2$ (b) obtained by a photon-counting system.

References

[1] T. Nakashizu, T. Sekine, K. Uchinokura and E. Matsuura, Phys. Rev. B29, 3090 (1984).

[2] T. Nakashizu, T. Sekine, K. Uchinokura and E. Matsuura, J. Phys. Soc. Jpn. 55, 672 (1986).

[3] N. Nagaosa and E. Hanamura, Phys. Rev. B29, 2060 (1984).

chapter 11

SOLID STATE: INORGANIC SYSTEMS

RAMAN SCATTERING FROM LOCALIZED ELECTRONIC EXCITATIONS IN METALLIC RARE EARTH COMPOUNDS

E. Zirngiebl, S. Blumenröder, G. Pofahl, G. Güntherodt

II. Physikalisches Institut+, Universität Köln, 5000 Köln 41, FRG

Recent developments in applying Raman scattering to investigate localized electronic excitations in metallic rare earth (RE) Kondo-type and intermediate valence (IV) compounds are reviewed. Raman spectroscopy in these metallic systems unlike other normal metals, is very rewarding because of their detailed and rich localized electronic structure, including atomic spin-orbit- and crystalline-electric-field-split levels.
We focus on the following two selected highlights:

I. Crystalline-Electric-Field Excitations

The precise determination of the crystalline-electric-field (CEF) level scheme of RE-compounds is fundamental for the understanding of their thermodynamic, magnetic and elastic properties. Moreover, the determination of the CEF level scheme is indispensable in order to identify the Kondo-type or IV anomalies.
We have investigated the CEF excitations of some RE-B_6. An overview of the results is given in Fig.1. The Γ_8-Γ_7 CEF excitation of CeB_6 has been determined near 372 cm^{-1}. The Γ_8 ground state shows a splitting of 10 cm^{-1} which manifests itself in the spectra below 20 K by an abrupt shift of the mean CEF excitation energy towards higher frequency. The CEF level scheme of NdB_6 could be determined due to the high resolution of Raman spectroscopy, i.e. by revealing two CEF excitations near 95 cm^{-1}. Their temperature dependence clearly revealed the excited quartet ($\Gamma_8^{(1)}$) and doublet (Γ_6) states at 92 cm^{-1} and 190 cm^{-1}, respectively.

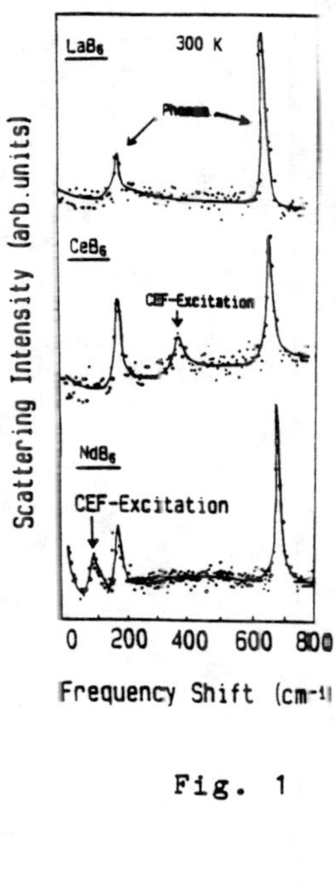

Fig. 1

II. Intra- and Inter-Configurational Excitations

Intermediate valence (IV) in RE-compounds is usually considered as a quantum-mechanical mixing of two atomic 4f configurations ($4f^n \longleftrightarrow 4f^{n-1} + e^-(E_F)$). We have been able to show spectroscopically the validity of this idea for several IV compounds ($EuPd_2Si_2$, $EuCu_2Si_2$, SmB_6 and $CePd_3$) by detecting the existence of the atomic spin-orbit split (J) multiplet level excitations (intraconfigurational excitations), and moreover by identifying transitions between the two 4f configurations participating in the valence fluctuation process (interconfigurational excitations). Fig.2 shows as an example electronic Raman scattering in the IV compound $EuPd_2Si_2$ at 145 K. The intraconfigurational excitations of the Eu^{3+} $4f^6$ (^7F_J) configuration are indicated by the dashed vertical lines; the interconfigurational excitations are shifted by E_x to lower energies compared to the intraconfigurational excitations. (As a reference we show in Fig. 2 (top) the electronic Raman scattering of Sm^{2+} $4f^6(^7F_J)$ in semiconducting SmSe). From the spectroscopic data we were able to extract the mixing width or lifetime broadening energy $k_B T_f$ of the J multiplet levels and the interconfigurational excitation energy E_x between the two 4f configurations. T_f and E_x are necessary input parameters for the description of the temperature dependence of the thermodynamic properties of IV compounds in the framework of the ionic model.

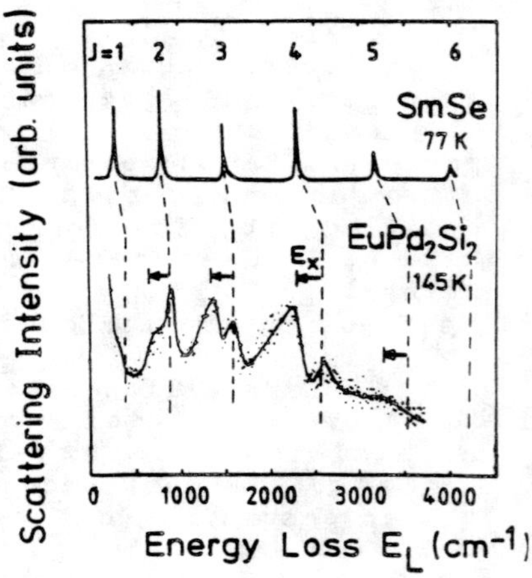

Fig. 2

In conclusion we would like to emphasize that light scattering has become a valuable complementary tool to neutron scattering in the field of Kondo- and IV- compounds /1/. The advantages of Raman spectroscopy are high resolution (0.5 meV), strict symmetry selection rules, high (< 1000 meV) energy losses observable, small usable sample size (1 mm^3) and independence of rare earth isotopes.

+ Work supported by Deutsche Forschungsgemeinschaft, SFB 125

/1/ G. Güntherodt, S. Blumenröder, B. Hillebrands, R. Mock, E. Zirngiebl, Z. Phys. B 60, 423 (1985)

MAGNETIC ORDERING INDUCED RAMAN SCATTERING IN MPS_3 (M = Fe, Ni) LAYERED COMPOUNDS

M. Balkanski, M. Jouanne, M. Scagliotti[+] and G. Benedek[++]

Laboratoire de Physique des Solides associé au CNRS, Université Pierre et Marie Curie, 4, Place Jussieu, 75252 PARIS CEDEX 05, France

+ Dipartimento Materiali, CISE S.p.a., Casella postale 12081, 20134 MILANO, Italia.

++ Università degli Studi si Milano, Dipartimento di Fisica, Sezione Stati Aggregati e Biomolecole, Via Celoria 16, 20133 MILANO, Italia.

Strong spin dependant inelastic light scattering in the phonon spectrum region is observed in $FePS_3$ (1). This compound is a good exemple of 2D-Ising type antiferromagnet with the honeycomb lattice. It exhibits around 118 K a magnetic phase transition from the paramagnetic phase to an ordered phase in which ferromagnetic linear chains are coupled antiferromagnetically in the ab plane. Similar magnetic properties are found in $NiPS_3$.

Enhancements of some phonon peaks are observed below the Néel temperature and this effect is found to be independent from the exciting wavelength.

In the specific case of MPS_3 antiferromagnetic layered compounds the investigations of the integrated intensity, band width and frequency of the Raman peaks suggests that the anomalies observed below the Néel temperature are due to direct phonon-magnon interaction. Both the internal vibrations of the P_2S_6 units and the phonons involving the vibrations of metal cations in the MS_6 units are sensitive to the magnetic order. In the antiferromagnetic phase, phonons are activated by a phonon-magnon interaction due to a d-electron transfer between neighbouring transition metal ions.

Anomalous intensity enhancements and abrupt appearance of new bands can be observed in the phonon Raman spectra of magnetic crystals at the onset of the magnetic ordering. The modulation of the spin-orbit coupling by phonons as well as the modulation of the exchange coupling and/or of the energy transfer in the transition metal compounds act as electron-phonon coupling and provide additional terms to the phonon Raman tensor. Zone boundary phonons can be activated by the magnetic ordering if the magnetic superstructure can assure the momentum conservation. Also magnetic ordering introducing new selection rules for which normal modes become Raman active.

For all these scattering processes the Raman intensity depends on the temperature like a spin correlation function. From the temperature dependence of the peak intensities the spin correlation function of

$FePS_3$ is evaluated and found to be in satisfactory agreement with the available neutron scattering data. However the spin-dependent and the spin-independent scattering have generally different selection rules and resonant behaviors so that the study of the low temperature phonon Raman spectrum in magnetic crystals can provide new details on the lattice dynamics.

(1) M. Scagliotti, M. Jouanne, M. Balkanski and G. Ouvrard, Solid State Communications 54, 291 (1985).

SOFT-MODE SPECTROSCOPY IN FERROELECTRICS BY HYPER-RAMAN SCATTERING.

H. Vogt

Max-Planck-Institut für Festkörperforschung, Heisenbergstr.1, 7000 Stuttgart 80, Federal Republic of Germany

Ferroelectric crystals are known to be particularly useful for optical second-harmonic generation (SHG). Their outstanding figures of merit are intimately related to the ferroelectric lattice distortion which in many cases can be interpreted as a frozen zone-center optical phonon.[1] When the frozen phonon is released in the paraelectric phase above the Curie temperature, the coherent SHG due to the static ferroelectric lattice distortion transforms into spontaneous hyper-Raman (HR) scattering by the thermally activated soft mode. Although any phonon of appropriate symmetry can participate in the HR process, the soft mode is expected to display the most easily detectable HR line because its HR polarizability directly evolves from the large second-order susceptibility of the ferroelectric phase.

We have studied HR scattering by the Raman-forbidden odd-parity soft modes in the cubic phases of oxygen perovskites in order to obtain the temperature dependence of the soft-mode parameters.[2-6] The source of the exciting radiation was an acousto-optically Q-switched Nd-YAG laser with a pulse repetition rate of 5 kHz, a pulse width of 150 ns, and an average power of 1 - 3 W. Spectra were taken by the conventional single-channel method using a double monochromator and a gated

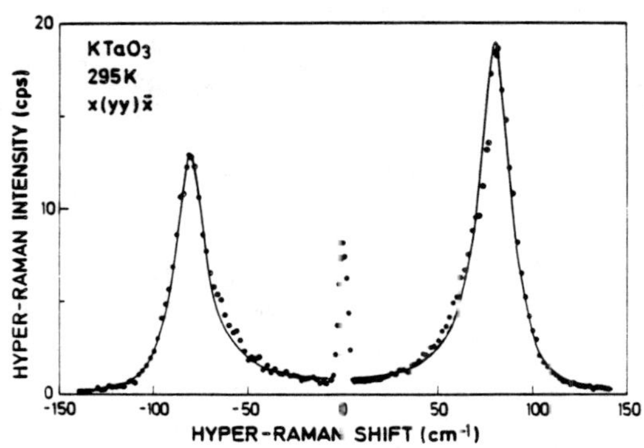

Fig.1: Stokes and anti-Stokes HR line of the soft mode in $KTaO_3$ at room temperature. The HR intensity in photon counts per second is plotted as a function of the wave-number shift from the second harmonic of the laser. Closed circles : experiment; solid line : classical-oscillator fit.

photon counting system. Typical results are shown in Figs. 1 and 2. Applying the fluctuation-dissipation theorem,we have deduced the imaginary part of the dielectric function which can directly be compared with far-infrared and millimeter-wave data. Since the hyper-Rayleigh intensity at exactly twice the laser frequency could only result from crystal imperfections disturbing the inversion symmetry and therefore was comparatively weak, we have been able to study the dielectric response down to wave numbers of about 2 cm^{-1}.

We intend to discuss our experimental findings in terms of recent theories of the paraelectric-ferroelectric phase transition in oxygen perovskites. A point of special interest will be the case of $BaTiO_3$ and $KNbO_3$ where soft-mode concepts compete with order-disorder models.

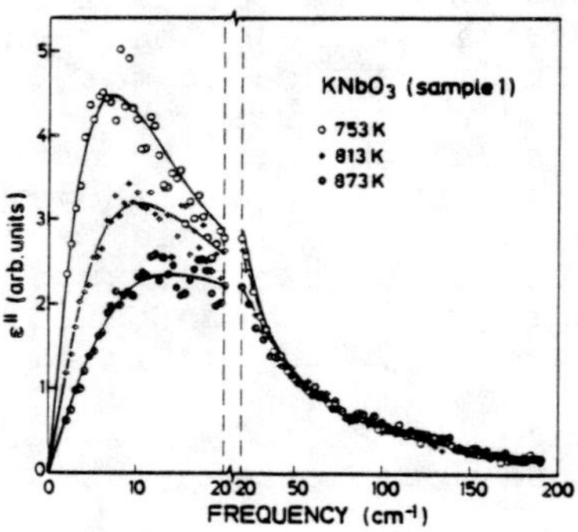

Fig.2: The relative imaginary part of the dielectric function of cubic $KNbO_3$ as deduced from HR spectra by means of the fluctuation-dissipation theorem. The solid lines represent single-oscillator fits involving three parameters. The soft mode is heavily overdamped.

REFERENCES

1. M.E. Lines and A.M. Glass, Principles and Applications of Ferroelectrics and Related Materials, Clarendon Press, Oxford, 1979, p. 467
2. H. Vogt and G. Rossbroich, Phys. Rev. B 24, 3086 (1981); ($SrTiO_3$)
3. H. Vogt, J.A. Sanjurjo, and G. Rossbroich, Phys. Rev. B 26, 5904 (1982); ($BaTiO_3$)
4. H. Vogt and H. Uwe, Phys. Rev. B 29, 1030 (1984);($KTaO_3$)
5. G. Kugel, H. Vogt, W. Kress, and D. Rytz, Phys. Rev. B 30, 985 (1984); ($KTa_{1-x}Nb_xO_3$)
6. H. Vogt, M.D. Fontana, G.E. Kugel, and P. Günter, to be published; ($KNbO_3$)

DIRECT MEASUREMENT OF WAVE VECTOR DEPENDENT POLARITON ENERGY VELOCITY AND DEPHASING TIME IN NH_4Cl

F. Vallée, G.M. Gale and C. Flytzanis

Laboratoire d'Optique Quantique, Ecole Polytechnique
91128 Palaiseau, cédex, France

The creation and evolution of short pulses of propagating collective excitations in a crystal is related to some fundamental problems in condensed matter physics. Up till now the relevant information has been derived indirectly for instance, in the case of vibrational excitations, by studying the frequency dependence of the dielectric constant in the far infrared. Recently Auston et al [1] demonstrated a technique to generate ultrashort far infrared pulses and use them to study the dynamics of low frequency infrared active vibrational motion in crystals ; the selectivity of this technique is however limited.

Here we show that it is possible to create and follow the evolution in space of a pulse of propagating Raman active polariton using an adapted time resolved coherent anti-Stokes Raman scattering (CARS) technique. Its first demonstration we report here concerns the ν_{4T}-polariton in Ammonium Chloride (NH_4Cl) whose dispersion curves and Raman spectra [2] have been extensively studied and furthermore indicate a strong coupling with many-phonon states.

The measurements were performed with a CARS system driven by a passively mode-locked Nd^{3+}/glass laser, providing three independent picosecond pulses ω_L , ω_S and ω_P one of which (ω_L) is continuously tunable ; the ω_L and ω_S-pulses create by coherent Raman scattering, at a certain instant and region in the sample, a pulse of the polariton mode whose propagation direction is fixed by the phase matching requirement : its subsequent propagating and loss of coherence are followed at a different point along the propagation direction by delayed coherent anti-Stokes scattering of a probe pulse at frequency ω_P ; with this technique we can follow the polariton propagation over more than one millimeter in the sample and gain direct access to the energy propagation characteristics ; a variation of the energy velocity by a factor 20 was measured over a spectral region of 200 cm^{-1} which is in agreement with the variation estimated from the polariton dispersion curves.

In addition the polariton dephasing time T_2 was determined as a function of the wave vector over a large portion of its dispersion curve and as a function of temperature ; globally T_2 decreases with decreasing wave vector and this decrease is rapid near the phase transition temperature T_C = 242°K but slows down considerably as the temperature is lowered well below T_C.

This above technique has many advantages over other indirect techniques and in particular high selectivity in space and frequency.

REFERENCES

[1] D.H. Auston and K.P. Cheung, J. Opt. Soc. Am. B 2, 606 (1985)

[2] V.S. Govelik, O.P. Maximov, G.G. Mitin, M.M. Sushchinsky, Sclid. State Com. 21, 615 (1977)

RAMAN SPECTRA OF CRYSTALLINE GERMANIUM AND SILICON TETRABROMIDES

W. S. Tse and G. Yur

Institute of Physics, Academia Sinica, Taipei, Taiwan (11529), R.O.C.

The present work reported a comprehensive study of the Raman spectra of $SiBr_4$ and $GeBr_4$ of the molecular fundamental modes as well as the lattice vibrational modes. This should lead to a better understanding of the reported crystal structure of $SiBr_4$ and also the most probable structure of the unknown $GeBr_4$ by comparing to other tetrabromide spectra.

Typical spectra of crystalline $SiBr_4$ and $GeBr_4$ samples at temperatures between 80K and 250K are shown in Figures 1 and 2. Frequencies of maximum intensity are also listed in Table 1. The temperature dependent Raman spectra show that both compounds exhibit only one phase between 80K and the melting.

Table 1. Raman frequencies (cm^{-1}) of crystalline $SiBr_4$ and $GeBr_4$ at 80K.

	$\nu_1(A_1)$	$\nu_2(E)$	$\nu_3(T_2)$	$\nu_4(T_2)$	Lattice
$SiBr_4$	246	91	467,471,486	133	26,32,34 39,42
$GeBr_4$	233	79,84	319,325,329 334,337	108,110	26,29,35 40,48

As shown in Fig. 1, there are some similarities in spectral band shape and relative intensities between the Raman spectra of $SiBr_4$ in this work and previous Raman work on CBr_4 (I) [1] where CBr_4 (I) at 320K forms the disordered "plastic" crystal in which the center of mass of the tetrahedral molecules occupy the sites of a face-centered cubic lattice. For space group I43m with 2 molecules per unit cell, the molecules (T_d molecular group) are at T_d sites and the group analysis [2] predict no site splitting and a maximum of two components in each fundamental is expected as a result of factor-group splitting. Therefore, there are two Raman active components of ν_1, ν_2, ν_3 and ν_4 respectively. The spectra of both compounds, $SiBr_4$ at 80K and CBr_4 (I) at 320K, show only one sharp peak for ν_1, ν_2 and ν_4 modes. For ν_3 mode (triply degenerate stretching) only one broad peak present in CBr_4 (I) spectra and three resolved peaks in $SiBr_4$, this is because in CBr_4 (I) the temperature is so high (320K) such that all resolved peaks may merged into one broad band due to temperature anharmonic effect. In this work, we also detect no isotopic splitting in totally symmetric stretching ν_1 mode and factor-group splitting of the triply degenerate bending ν_4 mode as observed by Creighton and Sinclar [2]. In the orientationaly disordered CBr_4 (I) phase the molecules are rapidly reorienting about a lattice point. The lattice Raman spectrum is a broad wind compare to the five sharp peaks in the lattice spectrum of $SiBr_4$ at 80K. This feature of sharp peaks in $SiBr_4$ is probably an indication of an ordered cubic phase. The above discussions show that the fine structure pattern for $SiBr_4$ and CBr_4 (I) are very similar indicated that they both may belong to the same cubic space group.

There appear to be no published crystal data for $GeBr_4$. A shown in Fig. 2, the two spectra of $GeBr_4$ and $SnBr_4$ [3] including both lattice and stretching modes, are very similar in band shapes and relative intensities. The factor-group analysis based on a monoclinic system with four molecules per unit cell seems to apply reasonably well to the Raman spectra of crystalline GeBr indicating its crystal structure may be similar to that of crystalline $SnBr_4$ of $P2_1/C$ system.

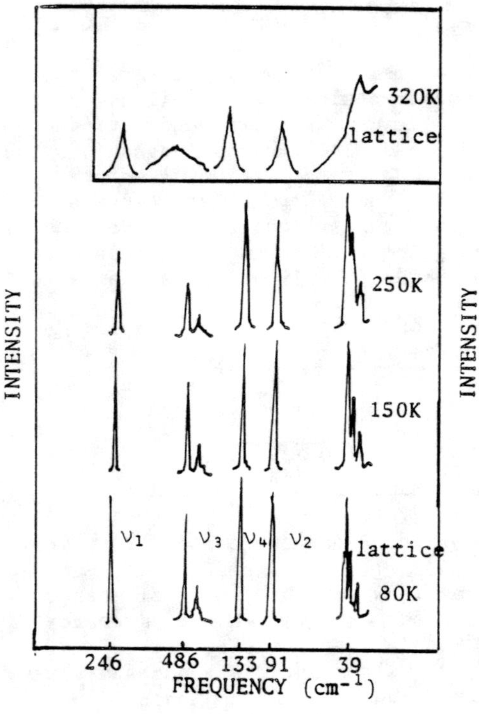

Figure 1. Temperature dependent Raman spectra of crystalline $SiBr_4$. The inserted spectra of crystalline CBr_4 (I) [1] is used for comparison.

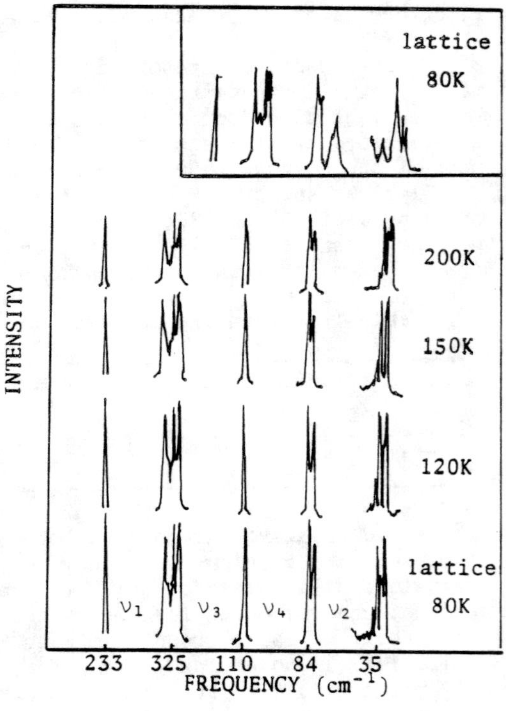

Figure 2. Temperature dependent Raman spectra of crystalline $GeBr_4$. The inserted spectra of crystalline $SnBr_4$ [3] is used for comparison.

References

(1) W. S. Tse, N. T. Liang, W. W. Lin and P. Y. Chiang, Chem. Phys. Lett., 119, 67 (1985).

(2) J. A. Creighton and T. J. Sinclar, Spectrochimic Acta, 35A, 137 (1979).

(3) W. S. Tse and N. T. Liang, Ann. Rept. Inst., Phys. Academia Sinica, 15, 81 (1985).

RAMAN SPECTRA OF HIGH-TEMPERATURE PHASE CHANGES IN SCANDIUM FLUORIDES

S. Hafsi[1], S. Turrell[1], M.-C. Dhamlincourt[1], P. Barbier[2], J. Champarnaud[3]

[1]L.A.S.I.R. (L.P. 2641 C.N.R.S.) and [2] Laboratoire de Cristallochimie (U.A. 452), Universite de Lille I, 59655, Villeneuve d'Ascq [3]Laboratoire de Chimie Structurale (L.A. 320 C.N.R.S.), Universite de Limoges, 87060 Limoges, France.

At high temperatures the scandium hexafluorides, M_3ScF_6 (M=K, Rb, Tl) have a high, $(NH_4)_3AlF_6$-type symmetry (O_h^5, Fm3m) with one molecule per unit cell.[1] This structure consists of a lattice of isolated ScF_6^{-3} octahedra whose interstices are filled with monovalent cations. At ambiant temperature this cubic symmetry is deformed due to packing strains and the octahedra are either deformed, or twisted or tilted with respect to the C_4 axis.

In the present communication we present results of some structural studies for the Rb and Cs compounds, both of which have tetragonal D_{4h}^5 cells at room temperature. DSC measurements indicate one phase transition for Rb_3ScF_6 (at 234°C) and three for Cs_3ScF_6 (at 101°, 152°, and 196°C). Corresponding crystallographic results for Cs_3ScF_6 yield transitions at 90°, 160°, and 185°C. Raman spectra, some of which are presented in Fig. 1, were obtained of powdered samples over the temperature range 25°-230°C. For Rb_3ScF_6, one of the components of the "$A_g^{?}$-mode" doublet, which is due to the presence of two octahedra in the O_{4h} cell, decreases in intensity and finally disappears around 190°C. For the cesium compound, the second component also disappears around 95°C, but in addition, band-shape changes suggest that other structural variations occur around 155°C and 195°C. No other spectral changes are observed above 200°C.

This work is being continued to complete the Raman intensity data and to correlate these results with crystallographic studies in order to explain the origin of these phase transitions.

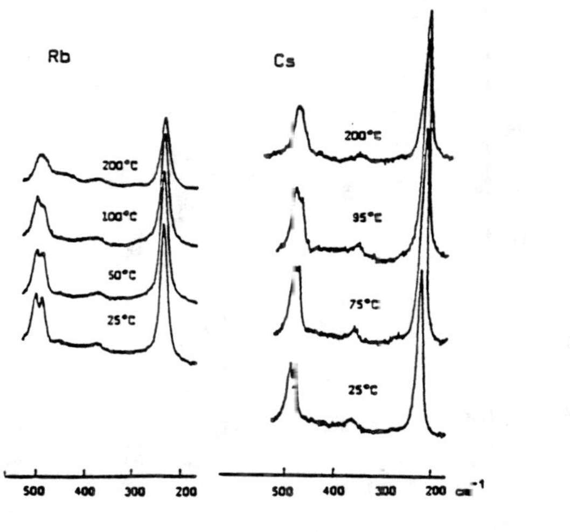

Fig. 1. Raman Spectra of M_3ScF_6

Reference:

1. J.-C. Champarnaud-Mesjard, Doctoral Thesis, Universite de Limoges, (1978).

MULTICHANNEL RAMAN STUDIES OF BINARY ALKALINE BORIDES

Z. Yahia, S. Turrell, M.-C. Dhamlincourt, and G. Turrell

L.A.S.I.R. (L.P. 2641) Universite de Lille Flandres Artois, Bat. C5, 59655 Villeneuve d'Ascq Cedex, France.

Recently, boron compounds have commanded much attention because of their chemical and physical properties (chemical reactivity, magnetic and electric characteristics). The present communication is based on our study of the crystal dynamics of a series of binary alkaline borides MB_6 (M=Ca, La, Sr, Sm)[1], which present much spectroscopic interest because of their high symmetry. All previous attempts to obtain Raman spectra of these compounds were unsuccessful due to their dark coloration. However with the development of multichannel detection systems coupled with data accumulation, a very low level of excitation intensity can be employed, thus avoiding decomposition of these light-sensitive materials.

In the present work, Raman spectra of powdered samples were obtained over the range 896 to 86 cm^{-1} using a Dilor MICRODIL 28 Raman microspectrometer.[2] Excitation was at 514.5 nm at a power of 80 mw.

Crystallographic studies have shown that these structures derive from a cubic O_h^1 (Pm3m) cell consisting of a three-dimensional skeleton of B_6 octahedra in which the interstices are filled by metal atoms.[3,4] Factor group analysis predicts $\Gamma=\underline{A_{1g}}+\underline{E_g}+F_{1g}+\underline{F_{2g}}+2F_{1u}+F_{2u}$ (where — indicates Raman activity and --- IR activity).

Portions of the preliminary Raman spectra for the first three members of the series are shown in Fig. 1. It is apparant that the band assigned to species A_{1g} is appreciably cation dependent, thus reflecting

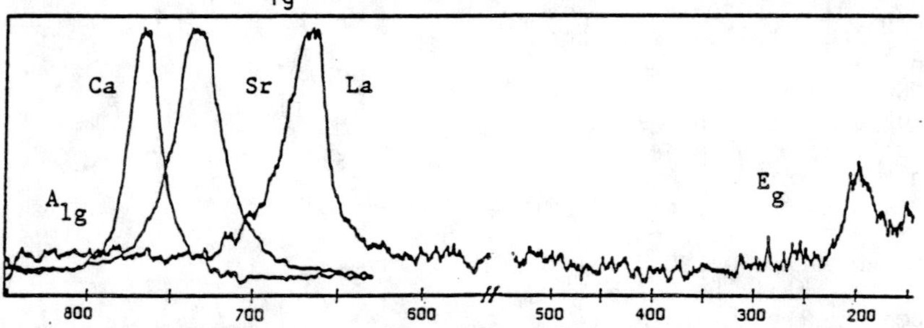

Fig. 1. Raman spectra of MB_6 (M=Ca, Sr, La)

the prevailing opinion that there is a marked electron transfer between the boron lattice and the metal ion in the case of boron-rich borides.[5] Work is now in progress to complete the Raman data and to obtain diffuse reflectance IR spectra of these compounds.

The authors would like to thank Professor J. Etourneau, Universite de Bordeaux, for providing the compounds used in this study.

References:

1. R. Naslain, Doctoral Thesis, University of Bordeaux, 1967.
2. J. Barbillat, P. Dhamlincourt, M. Delhaye, J. Phys.(Paris)45, 255 (1984).
3. J. Etourneau, P. Hagenmuller, Philos. Mag.B, 52(3), 589 (1985).
4. R. Kiessling, Acta Chem. Scand. 4, 209 (1950).
5. N. Greenwood, R. Parish, P. Thornton 20, 441 (1966).

THE INTENSITIES AND HIGH-PRESSURE BEHAVIOUR OF RAMAN BANDS IN RHOMBOHEDRAL AXF_6 CRYSTALS (A = K,Cs,X = Sb,As)

A.M. HEYNS

Department of Chemistry, University of Pretoria, PRETORIA 0002, S. AFRICA

$LiSbF_6$(1), $CsSbF_6$(2) and $TlSbF_6$(3) all have rhombohedral structures at ambient conditions, and it can be assumed that they have either the same or closely related structures. $KAsF_6$ II, is also rhombohedral, belonging to the space group $R\bar{3}-C_{3i}^2$ (4), and is therefore isostructural with $CsSbF_6$(2). The rhombohedral cell is slightly distorted from $R\bar{3}m-D_{3d}^5$-symmetry, which represents the densest packing that can be conveniently accommodated at moderate temperatures and pressures for these compounds.

The high-pressure Raman spectra of crystals of $CsSbF_6$ and rhombohedral $KSbF_6$ have been obtained in a diamond-anvil cell and are compared with the ones of $TlSbF_6$ reported elsewhere(6). These results are summarized in Table 1.

Table 1: The Raman Bands and their Pressure Dependences in Rhombohedral $KSbF_6$*, $CsSbF_6$ and $TlSbF_6$(6).

	$KSbF_6$ III			$CsSbF_6$ II			$TlSbF_6$		
Mode	cm^{-1}	$d\upsilon/dP$	γ	cm^{-1}	$d\upsilon/dP$	γ	cm^{-1}	$d\upsilon/dP$	γ
$\upsilon_1(Ag)$	660	0.41	0.51	652	0.33	0.10	648	0.25	0.08
$\upsilon_2(Eg)$	580	0.59	0.24	574	0.34	0.12	566	0.53	0.19
$\upsilon_{5a}(Eg)$	301	0.60	0.48	290	0.57	0.39	289	0.50	0.34
$\upsilon_{5b}(Ag)$	284	0.28	0.23	283	0.25	0.18	276	0.18	0.13
$\upsilon_{lib}(Ag)$	-	-	-	50	-	-	-	-	-
$\upsilon_{lib}(Eg)$	-	-	-	65	2.33	7.1	70	1.88	5.3

* The Raman Bands of rhombohedral $KSbF_6$ are reported at 4 kbar.

In these spectra, all the predicted Raman-active bands have been observed in $CsSbF_6$ II, in the remaining rhombohedral compounds only one librational mode has been observed in $TlSbF_6$ whereas no librational modes could be observed in $KSbF_6$ III. The splitting of υ_5 into Ag- and Eg-components under C_{3i}-symmetry is clearly evident in these spectra and the pressure dependences of the Ag-components are substantially less than those of the Eg-ones. A linear correlation is also found between the extent of the Ag-Eg splitting of υ_5 and the radius of the cation in these rhombohedral lattices showing that the interactions between the cations and anions are directly proportional to the distortion of the anion. Increasing pressures will also contribute to larger cation-anion interactions, thus explaining the fact that υ_5-increases for each of these compounds with increasing temperatures.

The polarized Raman spectra of single crystals of rhombohedral $CsSbF_6$ and $KAsF_6$ have also been obtained. In the case of $CsSbF_6$ these results clearly show that the librational mode is split into Ag- and Eg-components in agreement with $R\bar{3}$-C_{3i}^{2}-symmetry in direct contrast to the one Raman active component Eg predicted under $R\bar{3}m$-D_{3d}^{5}-symmetry. These results for $CsSbF_6$ II are summarized in Table 2.

Table 2: The frequencies (cm^{-1}), assignments, observed and calculated intensities of the Raman bands in $CsSbF_6$

			Observed						Calculated					
Mode	cm^{-1}	S_ℓ^P	I_{xx}	I_{yy}	I_{zz}	I_{yz}	I_{xy}	I_{xz}	I_{xx}	I_{yy}	I_{zz}	I_{yz}	I_{xy}	I_{xz}
υ_1	652	S_1(Ag)	100	100	100	-	-	-	*	90	95	-	-	-
υ_2	574	S_2(Eg)	6	6	-	7	7	7	*	5	-	4	3	7
υ_{5a}	290	S_3(Eg)	22	22	-	9	21	9	18	21	-	*	19	8
υ_{5b}	283	S_4(Ag)	14	16	52	-	-	-	13	16	*	-	-	-
υ_{lib}	50	Ag	<1	<1	<1	-	-	-	-	-	-	-	-	-
υ_{lib}	65	Eg	<1	<1	-	<1	<1	<1	-	-	-	-	-	-

*Shows that the observed relative intensities in these polarization directions were used to calculate the parameters.

Equations are obtained which represent the polarizability tensors ($\alpha(Qjk)$) in terms of the electro-optical parameters for the SbF_6^-- and AsF_6^--ions of D_{3d}-symmetry. The Raman tensors α (S_ℓ^P) can now be calcu= lated for the internal SbF_6^-- and AsF_6^--modes in $CsSbF_6$ and $KAsF_6$ by means of a well-known equation given in reference 6. The tensors for the SbF_6^-- and AsF_6^--ions are then used to predict the relative intensities of the Raman bands in the single crystal spectra and an excellent agree= ment is obtained between the calculated and observed values as is shown in Table 2. $\partial\alpha/\partial_r$-values are calculated for the totally symmetric stretching modes of these ions using the well-known delta function potential model equation of Long and Plane (9). The derivatives of the molecular polarizability with respect to symmetry coördinates $\partial\alpha/\partial_{S_1}$ are then obtained and the values for SbF_6^--, AsF_6^-- and PF_6^--ions compared and contrasted.

References:

1. J.H. Burns, Acta Cryst. 15, 1098 (1962).
2. M.M. de V. Steyn, A.M. Heyns and R.B. English, J. Cryst. and Spectrosc. Research 14, 505 (1984).
3. G.M. Begun and A.C. Rutenberg, Inorg. Chem. 6, 2212 (1967).
4. G. Gafner and G.J. Kruger, Acta Cryst. B30, 250 (1974).
5. U. Kristen, Proc. of the Fifth International Conference on Raman Spectroscopy, p.666, Shulz Verlg., Freiburg (West Germany) 1976.
6. S. Montero, Phys. Status Solidi B65, 133 (1974).
7. T.V. Long and R.A. Plane, J. Chem. Phys. 43, 457 (1965).

RAMAN SCATTERING IN $PbMo_6S_8$, $SnMo_6S_8$, AND $Cu_{1.8}Mo_6S_8$.

R.T. Demers, D.J. Holmgren, M.V. Klein, and D.M. Ginsberg

Department of Physics, University of Illinois at Urbana-Champaign, 1110 W. Green St., Urbana, Illinois 61801

We have grown single crystals of the high critical field, high transition temperature superconductors $SnMo_6S_8$,[1] $PbMo_6S_8$, and $Cu_{1.8}Mo_6S_8$, commonly known as Chevrel phases. The crystals were prepared by melting deoxygenated, sintered slugs in a pressurized RF Bridgman furnace at temperatures as high as 1750°C.

We prepared copper-Chevrel crystals of cubic shape, about 2 to 3 mm on a side. The lead- and tin-Chevrel crystals are smaller, about 0.5 mm on a side. The copper-Chevrel crystals were large enough to be separated from the solidified melt, and could be oriented by x-ray Laue techniques. The tin- and lead-Chevrel crystals could not be separated, so we polished a surface of the solidified melt to prepare a number of single crystal grains of arbitrary orientations for optical measurements. We prepared additional tin-Chevrel crystals about 0.7 mm on a side by a vapor growth technique.

We measured the Raman spectra of these Chevrel crystals.[2] The space group of the crystals is R3 (No. 148). Group theoretical techniques show that the spectra of the crystals should reveal 14 modes (7 modes belonging to the A_g representation and 7 pairs of two-fold degenerate modes belonging to the E_g representation). Room temperature spectra taken on three of our crystals are shown in the

Figure. The peaks at 66, 77 and 521 cm^{-1} correspond to plasma lines from the argon laser. We note the strong (anticipated) similarity in the spectra of the three compounds.[3] Because of their position near a center of inversion, we expect little Raman activity of the Sn and Pb atoms. The off-center delocalization of the Cu atoms, however, may be manifested in the "shifts" of peak frequencies near 110 and 285 cm^{-1} in the Sn and Pb systems to 117 and 310 cm^{-1} in the Cu system.

1. Review article: Ø. Fischer, Appl. Phys. 16, 1 (1978).
2. K. Yvon in Current Topics in Materials Science, Volume 3, edited by E. Kaldis (North-Holland, Amsterdam, 1979).
3. K. Yvon, Sol. State Comm., 25, 327 (1978).

This work was supported by the National Science Foundation under grants DMR85-01346 (D.J.H. and D.M.G.) and DMR84-06473 (R.T.D. and M.V.K.).

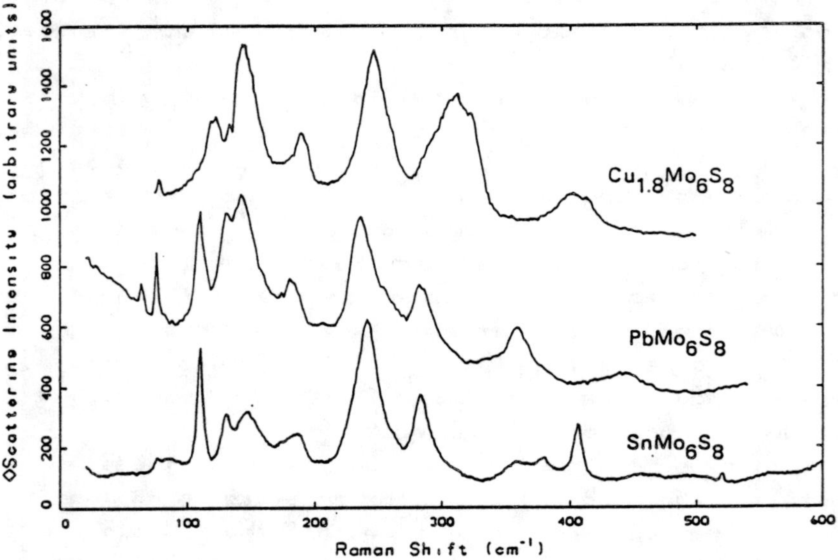

Figure: Raman Spectra of $Cu_{1.8}Mo_6S_8$, $PbMo_6S_8$, and $SnMo_6S_8$. The top and middle data are scaled up by factors of 2.5 and 4. From top to bottom, the spectra are shifted up by 800, 300, and -100 units.

RAMAN SCATTERING INVESTIGATION OF LOW-TEMPERATURE PHASE TRANSITIONS OF $LiTlC_4H_4O_6 \cdot H_2O$ AND $LiKSO_4$.

K.E.Haller

Institute of Physics, Estonian SSR Acad. Sci.,
202400 Tartu, Riia t. 142, Estonian SSR, USSR.

The polarized Raman spectra (RS) of lithium thallium tartrate (LTT) and $LiKSO_4$ in low-temperature phases have been investigated for the first time.

1. LTT, isomorphous to Rochelle salt, has an orthorhombic symmetry $D_2^4 = P2_12_12_1$ (z=4) at room temperature and at $T_c \approx 10K$ it undergoes a second-order phase transition (PhT). LTT appears to be a perspective object for studying the critical fluctuations arising near PhT, since the prolonging of the pyroeffect and the spontaneous polarization P_s are taking place in the paraelectric phase up to ∿100K /1/.

The RS of LTT have been measured in two phases (at 2K, 33K) and the temperature dependence of the low-frequency RS region (up to 250 cm^{-1}) has been studied in detail at 2-120K. An analysis of RS, with allowance made for literature data about the dielectric constant ε and P_s, shows an isomorphous dispersed PhT in LTT. The frequencies and depolarization ratios of lines practically do not change at T_c (<1%, <5%, respectively), the number of modes being the same at $T \gtrless T_c$. A partial redistribution of line intensities is observable at the temperatures 10-100K (for instance, with the increase of T the line at 136 cm^{-1} becomes weaker and broader, whereas the 30 cm^{-1} one becomes stronger). The low-frequency lines are intensive and sharp ($\Gamma \sim 2$-3 cm^{-1}), but most of the high-frequency ($\nu > 600$ cm^{-1}) ones are broad ($\Gamma \sim 30$-40 cm^{-1}) in the temperature region studied. What has been said is in good accordance with the data of /2/ to conclude that PhT and P_s are due to the ordering of O-H bonds. The reconstruction of O-H bonds gradually taking place in the low-symmetry unit of tartrate does not change the symmetry D_2 (z=4) of the elementary cell as a whole. The rise of ε by about two orders at PhT is due to a little change (∿1%) of the L-T splitting of a large number (222) of polar optical modes. Other literature data about the temperature-dependent physical characteristics of LTT are in accordance with this mechanism as well.

Note also that a damage (darkening) of the LTT crystal surface occurs when high-excitation densities are used. This effect has a threshold character ($\sim 10^4$ W/cm^2) and it takes place at all temperatures from 2 to 300K. Probably, it is due to the photoconversion of Tl^+ into Tl^o.

2. It is of great interest to study the influence of the dynamical motions of the crystal lattice on the quality of mobile ions. In this connection $LiKSO_4$ is a good object for studying, because the substitution of K with Na leads to a change of a number of physical properties of this crystal, including the rise of the ionic conductivity by about two orders /3/.

$LiKSO_4$ has a hexagonal symmetry C_6^6 = P6 (z=2) at room temperature and it undergoes as a minimum two PhT at low temperatures: at $T_1 \simeq 190K$ and at $T_2 = 38K$. The vibrational spectra of $LiKSO_4$ at $T < T_2$ have been studied here for the first time. An analysis of RS shows an increase of the number of lines in connection with the removal of the "accidental" degeneracy at T_2; the volume of an elementary cell is not changed at T_2. The symmetry of low-temperature phase is higher than C_i and does not include any rotational axes higher than C_2. The halfwidths of all lines are comparable ($\Gamma \sim 10 \div 14$ cm^{-1}) at $T > T_1$, but at $T < T_2$ the following changes take place: i) in the regions of external vibrations and librations of SO_4^{2-} (<210 cm^{-1}), the translational vibrations of Li^+ and the bending vibration of SO_4^{2-} (370-490 cm^{-1}), sharp ($\Gamma \sim 4$ cm^{-1}) lines can be observed; ii) in the regions of bending (620-650 cm^{-1}) and stretching (1000-1210 cm^{-1}) vibrations of SO_4^{2-} there are sharp ($\Gamma \sim 5$ cm^{-1}) and broad ($\Gamma \sim 20$ cm^{-1}) maxima. Based on literature data and on what has been said above one can conclude that the freezing of Li^+ occurs in different positions of the elementary cell below T_2. This leads to an "inhomogeneous" broadening of the internal modes of SO_4^{2-}, which is additional to the librational one (ν(libr.) $\ll$ ν(int.)).

/1/ V.K. Novik, S.N. Drozhdin, G.D. Gavrilova, Sov. Sol. State Phys., 24, N.5, 1555, 1982 (in Russian).

/2/ B. Brzhezina, F. Smutny, B. Janouchek, V. Marechek, Izv. AN SSSR, ser. phys., 33, 297, 1969 (in Russian).

/3/ D.Teeters, R. Frech, Phys. Rev. B26, 4132, 1982; B26, 5897, 1982.

RAMAN SPECTRA OF SOLID COMPLEXES FORMED OF TRIHALIDES OF GROUP V ELEMENTS WITH TRIHALIDES OF GROUP III ELEMENTS.

Jacob Shamir[1] and Rafael Rafaeloff[2]

[1]Department of Inorganic and Analytical Chemistry, The Hebrew University 91904 Jersualem, Israel. [2]Department of Chemistry, Soreg Nuclear Research Centre, Yavne, Israel.

The trihaldies of group V elements are of covalent nature both in solution as well as in the solid state. When these react with trihalides of group III of similar covalent structure, solid complexes are formed.

Two structural categories of these compounds have to be considered and each of them can be subdivided into two additional types.

One structural category is of a complex with a covalent bonding which can be either (a) a donor-acceptor type, in which the lone pair of electrons of the central atom of Group V forms a coordination bond with the central atom of Group III

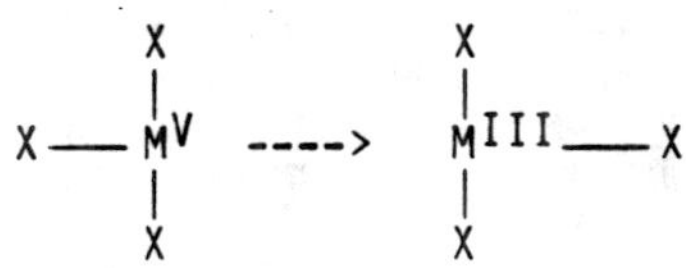

or (b) a dimeric structure with two bridging halogens:

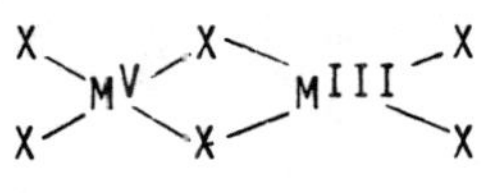

The second category is a complex with ionic bonding, formed by transferring a haldie anion from the more basic halide to the more acidic one:

c) $M^{V}X_3 + M^{III}X_3 \rightarrow [M^{V}X_2]^+[M^{III}X_4]^-$

or the reverse reaction:

d) $M^{III}X_3 + M^{V}X_3 \rightarrow [M^{III}X_2]^+[M^{V}X_4]^-$

The occurence of reactions (c) and (d) would depend on the relative acidities and basicities of the reactants involved.

The expected vibrational spectra for these different covalent and ionic structures have been discussed previously.[1]

The expected spectrum of a covalent complex is essentially a superposition of the spectra of the two reactants, with some additional lines resulting from the vibrations of the newly formed bonds in the complex.[2]

Thus, the vibrational spectrum of such a covalent complex would consist of more lines than the sum of those observed in the spectra of the separate free reactants. The observed spectrum of the complex would therefore be rather complicated.

On the other hand, the spectrum of an ionic complex may be simpler and consist of fewer lines. The ionic species present in the complex would generally be with one of the common coordination numbers 2,4 or 6, with a higher degree of symmetry; a tetrahedral or octahedral structure for coordination numbers 4 or 6. As a result, more degeneracies of the various fundamentals may be expected and the total number of observed vibrations would be fewer than in a related covalent complex.

Five compounds, $SbCl_3 \cdot GaCl_3$, $SbBr_3 \cdot GaBr_3$, $SbBr_3 \cdot AlBr_3$, $BiCl_3 \cdot GaCl_3$ and $BiBr_3 \cdot GaBr_3$ were prepared and their Raman spectra recorded. These spectra were interpreted on the basis of ionic structures, basically of the general formula $[N^{V}X_2]^+[M^{III}X_4]^-$, in agreement with the conclusions of the NQR measurements.[3,4]

The observed Raman spectra were simple with few lines and are best interpreted on the basis of ionic structures of the complexes. This interpretation is in agreement with the conclusions derived from the NQR measurements,[3,4] which is an entirely different physical method for structure determination.

In all these spectra the vibrations of the tetrahedral anion can be well assigned, their frequencies and relative intensities being in good agreement with literature data. Three additional lines are observed in these spectra, which are assigned to two stretching vibrations and one bending mode of the $M^{V}X_2^+$ cation.

However, a distinct difference seems to prevail between the antimony complexes and the related bismuth complexes. Whereas the antimony (III) halides donate a halide anion to form a monomeric cation SbX_2^+ with C_{2v} symmetry, the bismuth (III) halides seem to form, by a similar reaction, a polymeric chain-like cation $(BiX_2^+)_n$. In this chain-like cation,

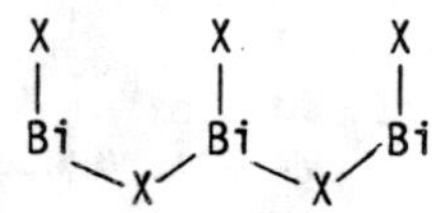

two different bismuth-halogen bonds are present: one is a terminal bond ($Bi—Cl_t$) and the other a bridging bond ($Bi—Cl_b$).

REFERENCES:

1) J. Shamir, Proc. 6th Int. Conf. Raman Spectrosc. 1, 61 (1978).
2) W.B. Person, in Spectrososcpy and Structure of Molecular Complexes, edited by J. Yarwood, p. 29, Plenum Press, New York (1973).
3) T. Okada, K. Yamada, H. Ishihara and H. Negita, Chem. Lett. 785, (1985).
4) T. Okada, K. Yamada, H. Ishihara and N. Negita, Bull. Chem. Soc. Jpn. 50, 3136 (1977).

RAMAN SCATTERING STUDY OF THE PHASE TRANSITIONS IN $TlAlF_4$

A. Bulou. M. Rousseau. J. Nouet

Laboratoire P.E.C.. Associé au C.N.R.S. (U.A. n° 807).
Faculté des Sciences du Mans 72017 LE MANS Cedex - France.

Many $A^{I}M^{III}F_4$ compounds have a structure closely related to the structure represented figure 1a where MF_6 octahedra are centered in tetragonal cells with A^+ ions located at the corners. These materials consist of MF_6 octahedra sheets (fig. 1b) and exhibit two dimensional properties.

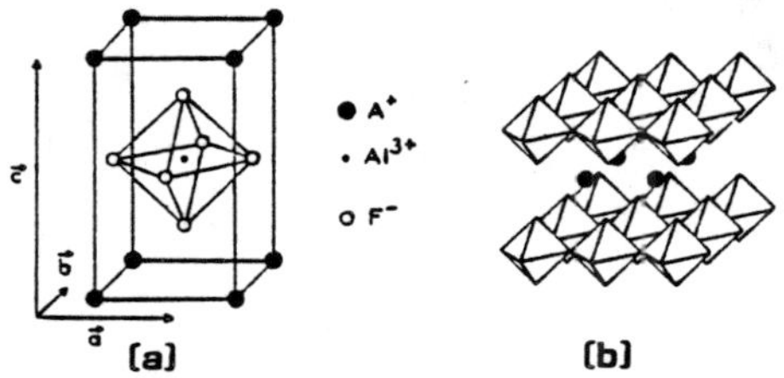

Figure 1 : Aristotype structure of $TlAlF_4$ (a) and evidence of AlF_6 octahedra sheets (b).

In a recent paper we have shown that the compound $TlAlF_4$ undergoes two structural phase transitions at high temperature |1|.

The structural phase transitions (SPT) occur at $T_{C1} = 241°$ C (weakly first order SPT) and at $T_{C2} = 162°$ C (second order SPT). The structure described figure 1 is observed from T_{C1} to the melting point (phase I). The space group is P 4/mmm with Z = 1 formula unit per cell. In the temperature range T_{C1} to T_{C2} (phase II) the AlF_6 octahedra undergo a tilting about |001| axis. alternated for the successive sheets. The space group is I 4/mcm with Z = 4 formulae units per cell. Below T_{C2} (phase III) new octahedra tilts appear about |100| and |010| axes together with Tl^+ displacements in (001) planes. The cell is the same as in phase II and the space group is I 2/a. In an extended Glazer notation |2. 3 | the sequence of transitions is :

$$a^o a^o c^o \rightarrow a^o a^o c^- \rightarrow a^-_a a^-_a c^-$$

Both transitions could be related to the freezing of a soft phonon at the A point (½ ½ ½) in the Brillouin zone of the aristotype tetragonal cell (phase I).

Raman scattering has been used to confirm these results and to check the displacive character of the transitions. According to the space groups of the three phases group theory analysis gives the following symmetry for the Raman active modes :

phase I	$A_{1g} + E_g$
phase II	$2A_{1g} + B_{1g} + 2B_{2g} + 4E_g$
phase III	$7A_g + 8B_g$

The Raman scattering investigations have been performed with a DILOR Z24 spectrometer on a microscope equiped with an heating system. The crystals are grown by the Bridgman technique.

The room temperature Raman spectrum of $TlAlF_4$ is given fig. 2. Eight lines are observed on the figure at 22, 38, 64, 92, 110, 226, 344 and 520 cm^{-1}. Moreover three very low intensity lines are observed at 258, 404 and 442 cm^{-1}. This spectrum clearly evidences the symmetry lowering with respect to the ideal structure where two lines should be observed. When heating the crystal the L_2 line frequency decreases and it coalesceswith the Rayleigh line in the vicinity of T_{C2}. Its temperature behaviour can be described by a law $\nu \sim \sqrt{T - T_C}$ (fig. 3) where $T_C = 163 \pm 2°$ C, in good agreement with the preceeding determinations |1|. Then the L_2 line corresponds to the soft mode of the III-II transition and its symmetry is A_g. At T_{C2} the L_1 line also disappears without significant frequency variation. Above T_{C1} only the L_6 and L_8 lines are still present. By comparison to $RbAlF_4$ |4| they can be attributed to E_g and A_{1g} modes respectively involving Fax vibrations only. In phase I it can be noted that the L_3, L_4, L_5 lines vanishing givesrise to an intense signal in the frequency range below 150 cm^{-1}. A similar effect already has been observed in $RbAlF_4$ |4|.

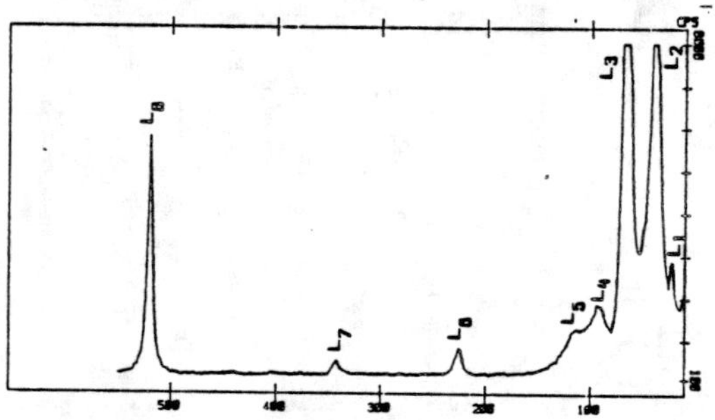

Figure 2

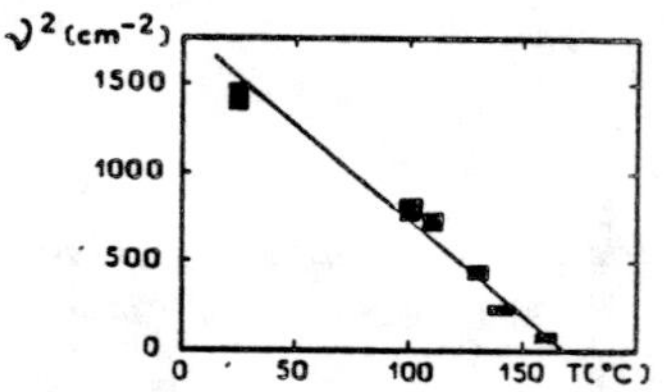

Figure 3

CONCLUSION

This work clearly evidences the structural change of $TlAlF_4$. The number of Raman active modes is consistent with the prediction according to the structural studies. It is shown unambiguously that the T_{C2} transition is displacive. The T_{C1} transition is evidenced by the vanishing of several modes but in this case no soft phonon is observed. This can be related to the fact that the tilt angle around the |001| axis is not temperature dependent |1|.

REFERENCES

[1] A. Bulou and J. Nouet (submitted to J. Phys. C)

[2] A. Bulou and J. Nouet J. Phys. C : Solid State Phys., 15 183 (1982)

[3] R. Deblieck, G. Van Tendeloo, J. Van Landuyt and S. Amelinckx Acta Cryst. B41, 319 (1985).

[4] A. Bulou, M. Rousseau, J. Nouet, P.L. Loyzance, R. Mokhlisse and M. Couzi J. Phys. C : Solid State Phys. 16, 4527 (1983).

COMPARISON OF LONGWAVELENGTH OPTICAL PHONONS IN $ThSe_3$ AND USe_3

G. Nouvel, H. Noël*, A. Zwick and M.A. Renucci

Laboratoire de Physique des Solides (U.A. 74) 118, Route de Narbonne
31062 Toulouse Cédex (France)

*Laboratoire de Chimie du Solide et Inorganique Moléculaire
Avenue du Général Leclerc, 35042 Rennes Cédex (France)

The uranium trichalcogenides UX_3 (X = S, Se, Te) display intriguing magnetic properties [1] in relation to the 5f electrons that have stimulated recent Raman investigations [2]. Whole electronic excitations through exchange coupling between neighbouring U^{4+} ions, expected to manifest themselves both in US_3 and in USe_3 below 50 K [3], were clearly observed in the former compound [2]. Unfortunately, the relative intractability of USe_3 due to strong light absorption and to photodecomposition made the Raman experiments rather difficult. As a first step towards a better insight into the properties of USe_3, a preliminary phonon assignment is achieved in this paper by comparison with the vibrational Raman spectra measured from $ThSe_3$, a non magnetic and isostructural compound.

The UX_3 compounds crystallize in the "$ZrSe_3$-type" monoclinic structure [4]. The unit cell consists of two molecular units belonging to adjacent chains, running parallel to the b axis and linked together to form layers in the (a, b) plane. The group theoretical analysis predicts 12 k = 0 Raman active modes, 8 belonging to Ag symmetry and 4 to Bg.

Typical polarized Raman spectra from $ThSe_3$ and USe_3 are shown in Fig. 1. Ten lines can be unambiguously detected in the spectra from $ThSe_3$. They lie respectively at 45, 62, 79, 107, 133, 148, 153, 166, 187 and 282.5 cm^{-1}. Fig. 1 exemplifies the similarity of corresponding spectra from the two selenides, with regard both to the relative intensities and to the wavenumbers. Indeed, no significant changes in wavenumbers are expected on the basis of only changes in atomic masses (U = 238, Th = 232). This gives the opportunity to confirm two phonons lines at 44 and 154 cm^{-1} in USe_3, although they can hardly be detected. Moreover, as Raman intensities depend on variations of the electronic susceptibility induced by the atomic displacements, Fig. 1 suggests that the modes responsible for first order light scattering in the two selenides have nearby eigenvectors. Then, assignment of Raman lines in $ThSe_3$ to phonons of specific symmetry will achieve quite straightforwardly symmetry assignment of phonon lines in USe_3, till prohibited by various shortcomings. Fig. 2 displays the three analyzed spectra of $ThSe_3$. In spite of the low scattering efficiency of the off-diagonal Raman tensor component, three among the four Bg modes are identified at 45, 62 and 107 cm^{-1} in $ThSe_3$, with two of their counterparts at 44 and 96 cm^{-1} in USe_3. All of the remaining structures are attributed to Ag modes. Lattice dynamical calculations, based on an interatomic force field model previously developped for the isostructural MX_3 compounds (M = Ti, Zr, Hf) [5], account well for the Raman data from $ThSe_3$ and USe_3 They give evidence of Se-Se pairing within the chains and support a rather strong coupling between the chains, even if the lattice dynamics are markedly influenced by chain symmetry.

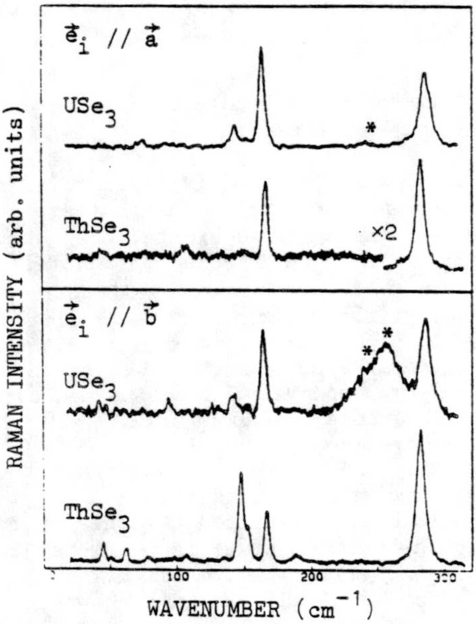

Fig. 1.- Room temperature polarized spectra from USe_3 and $ThSe_3$, excited respectively with λ = 488 nm and λ = 454.5 nm (Asterisks mark structures originating from amorphous and crystalline Se left behind by oxydation).

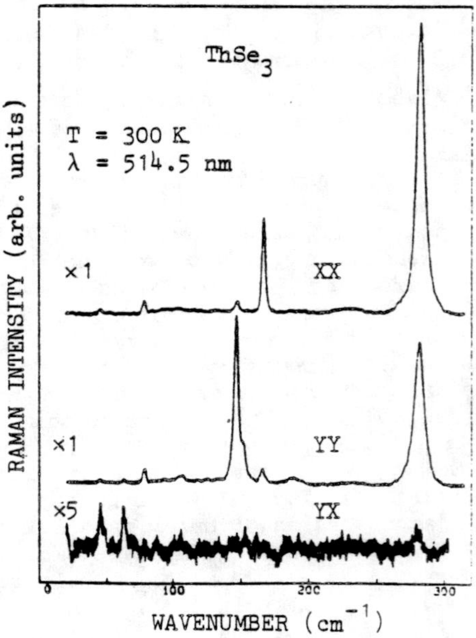

Fig. 2.- Room temperature analyzed spectra from $ThSe_3$. Ag modes appear in XX and YY polarizations, Bg modes in YX (≡ XY).

[1] B. Janus, W. Suski and A. Blaise, Proc. IV Int. Conf. on Crystalline Electric Field and Structural Effects in f-electron systems, Wroclaw 1981, Plenum Press, N.Y., p. 539 (1982).
[2] A. Zwick, G. Nouvel, M.A. Renucci, G. Mischler, W. Suski and D.J. Lockwood, Sol. State Commun. 53, 783 (1985).
[3] G. Nouvel, A. Zwick, H. Noel, M.A. Renucci and D.J. Lockwood, unpublished.
[4] S. Furuseth, L. Brattas and A. Kjekshus, Acta Chem. Scand. Ser. A29, 623 (1975).
[5] A. Zwick, Thèse, Toulouse 1983.

BRILLOUIN SCATTERING STUDIES OF SUPERIONIC $Ba_{1-x} La_x F_{2+x}$ (x =0.393)

P.E. Ngoepe* and J.D. Comins
Department of Physics, University of the Witwatersrand, Johannesburg 2001, South Africa.

Recently developed models of superionic conductors with the fluorite structure include the concept of transient vacancy - stabilized clusters of the 222 or analogous type. Their binding energy is supposed to reduce the effective anion Frenkel defect formation energy, thus leading to the diffuse superionic transition at temperature T_c.[1-3] The presence of clusters reduces the likelihood of additional formation of clusters nearby, thus suppressing further defect creation. In an analogous way impurity - stabilized clusters, formed in fluorites doped with trivalent cations, are considered to bind interstitial anions. This leads to a reduction in T_c with increasing dopant concentration.[4]

The present paper concerns Brillouin scattering measurements of $Ba_{1-x} La_x F_{2+x}$ (x = 0.393) and pure BaF_2 at high temperatures in the superionic region. The frequency shifts in the Brillouin scattered light $\Delta\omega_B$ corresponding to acoustic phonon modes in certain high symmetry directions are related to the elastic constants or their combinations.[5] The results on $Ba_{1-x} La_x F_{2+x}$ (x = 0.393) show a number of features. The transition temperature T_c is reduced as compared with pure BaF_2 as shown by the elastic constant C_{11}. The disorder is generated more rapidly, but then reduces considerably. This is similar to the behaviour observed in Brillouin measurements on PbF_2. The results appear consistent with the models of superionic behaviour in fluorites discussed above[1-4]. Elastic constant C_{44} also displays a reduction followed by a further quasi-linear stage.

A comparison of the results with those obtained on $Ba_{1-x} La_x F_{2+x}$ (x = 0.209 and x = 0.492) using neutron scattering[6,7] suggest that the reductions of elastic constants observed here correlate better with the observed increase in specific heat[7] rather than the disappearance of short-range correlations between impurity-stabilized 222-type clusters.

References

1. Clausen K, Hayes W, Hutchings M T, Kjems K J, Schnabel P and Smith C 1981 Solid State Ionics 5 589.
2. Catlow C R A and Hayes W 1982 J. Phys. C : Solid State Phys. 15 L9
3. Catlow C R A 1983 Solid State Ionics 8 89.
4. Catlow C R A, Comins J D, Germano F A, Harley R T, Hayes W and Owen I B 1981 J. Phys. C : Solid State Phys. 14 329.
5. Catlow C R A, Comins J D, Germano F A, Harley R T and Hayes W 1978 J. Phys. C : Solid State Phys. 11 3197.
6. Kjems J K, Anderson N H, Schoonman J and Clausen K 1983 Physics 120B 357.
7. Anderson N H, Clausen K, Kjems J K and Schoonman J 1986 J. Phys. C : Solid State Phys. to be published.

* On leave from: Department of Physics, University of the North, Sovenga 0727 South Africa.

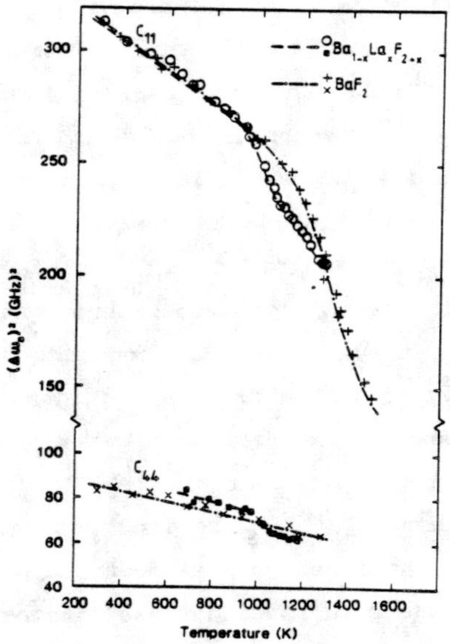

HIGH TEMPERATURE BRILLOUIN SCATTERING STUDIES IN SUPERIONIC CONDUCTORS: CUBIC ZIRCONIA AND LANTHANUM FLUORIDE.

P.E. Ngoepe* and J.D. Comins
Department of Physics, University of the Witwatersrand, Johannesburg 2001, South Africa.

Previous work on cubic zirconias provides evidence for superionic conduction above 1300K associated with disorder on the anion sublattice.[1,2] In LaF_3 there is evidence in favour of Schottky disorder: above 1100K there is a striking increase in the specific heat.[3]

Measurements of elastic constants have provided useful information concerning structural and defect properties of superionic conductors.[4] Elastic constants or their combinations are related to the Brillouin frequency shifts $\Delta\omega_3$ for the relevant acoustic modes.[4,5] Here the 514.5nm argon laser line was used to illuminate the specimens within an optical furnace. A triple-pass Fabry-Perot interferometer system was used to analyse the scattered light.

Studies of cubic zirconia stabilized with 9.4, 15, 21 and 24 mol% Y_2O_3 have been peformed. In all cases a behaviour of elastic constants was found which strongly resembles that found in several fluorites, e.g. CaF_2, SrF_2 etc. This involved a quasi-linear decrease in elastic constants with increasing temperature, attributed to lattice anharmonicity, followed by a softening of certain elastic constants above about 1300K, the temperature associated with the onset of superionic behaviour. In particular, C_{11} is affected considerably, but C_{44} undergoes at most a small change. This appears to be characteristic behaviour for the fluorite structure in which anion disorder is created.[4]

In LaF_3, studies of acoustic phonon modes in the [100] and [011] propagation directions show a softening near 1100K which corresponds to the substantial increase in specific heat attributed to the formation of lattice vacancies.(3)

References

1. Morinaga M and Cohen J B 1979 Acta Cryst A35 789.
2. Morinaga M and Cohen J B 1980 Acta Cryst A36 520.
3. Lyon W G, Osborne D W, Flotow H E, Grandjean F, Hubbard W N and Johnson G K 1978 J. Chem. Phys. 69 167.
4. Catlow C R A, Comins J D, Germano F A, Harley R T and Hayes W 1978 J. Phys. C : Solid State Phys. 11 3197.
5. Laiho R, Lakkisto M and Levola T 1983 Phil. Mag. A47 235.

* On leave from : Department of Physics, University of the North, Sovenga
0727, South Africa.

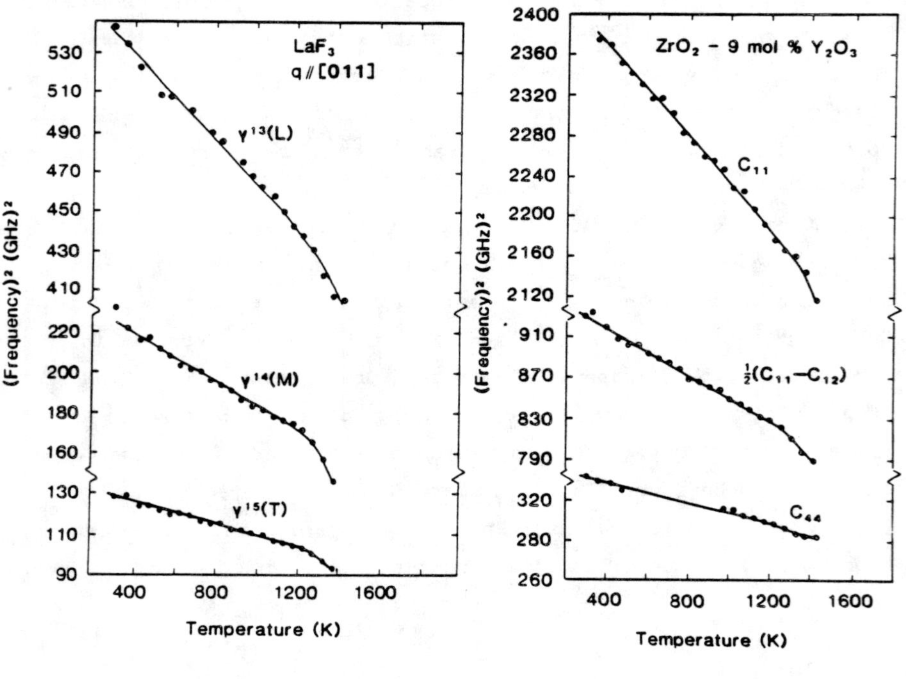

HYDROGEN IN $Fe_{90}Zr_{10}$ MAGNETIC GLASS - A RAMAN STUDY

M. S. Mathur*, H. A. Marzouk!, and J. S. C. McKee
Department of Physics, University of Manitoba, Winnipeg CANADA R3T2N2

The study of the hydrogenation of magnetic glasses has been of some interest recently [1]. Wronski *et al* and Boliang [2] *et al* have pointed out that charging of the iron-zirconium glasses with hydrogen converts them from paramagnetic to ferromagnetic. Whether the hydrogen induced ferromagnetic order is produced by bonded or by the non-bonded hydrogen, is not known. Present investigation of the H^+ bombarded $Fe_{90}Zr_{10}$ ribbon surface by Raman scattering has been undertaken to get some answers. $Fe_{90}Zr_{10}$ glass ribbon surface implanted with 30 KeV H^+ to a fluence of 10^{18} and 5×10^{19} H^+/cm^2 are characterized by the unenhanced surface Raman scattering. The Raman spectrum of the unimplanted and implanted $Fe_{90}Zr_{10}$ ribbon surface confirms the formation of the oxide layer on the surface. The presence of Raman bands due to OH and H-O-H stretching vibrations also confirms that the impinging H^+ react with this oxide layer making the above mentioned bondings. In addition, the bands due to metal-hydrogen stretching vibrations seem to suggest some sort of bonding between the constituents of the magnetic glass studied and the impinging H^+. At best in the H^+ implanted $Fe_{90}Zr_{10}$ surface, hydrogen is staying in the bonded state.

---

* on leave/! at the University of Kentucky, Department of Electrical Engineering, Lexington, Kentucky 40506 U.S.A.

---

[1] Z.S. Wronski, A.H. Morrich and A.W. Stewart. Phys. Lett. 101A, 5.6 (1984)

[2] Yu Boliang, D.H. Ryan, J.M.D. Coey, Z. Altourian, J.O. Strom-Olsen and F. Razavi, J.Phys. F. 13 L217 (1983)

TWO-MAGNON AND PHONON RAMAN SCATTERING IN $Cd_{1-x}Mn_xF_2$ MIXED CRYSTALS.

P.Ciepielewski[a/], I.Gregora[b/], V.Vorlicek[b/] and Z.Kaliński[a/]

[a/] Institute of Physics, Polish Academy of Sciences
02-668 Warsaw, Al.Lotników 32/46, Poland

[b/] Institute of Physics, Czechoslovak Academy of Sciences,
Prague, Na Slovance 2, CSSR

The $Cd_{1-x}Mn_xF_2$ mixed crystals provide a very interesting system for the study of phonon and magnon behaviour since the respective pure compounds, CdF_2 and MnF_2, exhibit markedly different properties, due to their different structure and lattice symmetry.

Two sets of technologically available samples were investigated within the temperature interval of 9-300K. The first set, in the composition range $0 \leq x \leq 0.26$ was cubic (CdF_2-like) whereas the second one, with $0.9 \leq x \leq 1$, had rutile structure (MnF_2-like).

Two Raman peaks have been observed in the spectra of the first set. The position of the stronger peak, near $320cm^{-1}$ shows a slight shift towards higher frequencies with increasing x. Such behaviour indicates a coupling between the T_{2g} Raman mode of pure CdF_2 ($320cm^{-1}$) and the A_{1g} Raman mode of pure MnF_2 ($341cm^{-1}$). The second Raman peak, less than ten times weaker, has emerged in the vicinity of $400cm^{-1}$ increasing its strength in a non-linear manner with increasing of x. This peak shows a shift towards higher frequencies with increasing x, too. The linear extrapolation of its position yields $395cm^{-1}$ for pure CdF_2 ($x \rightarrow 0$) which compares favourably with the frequency of the T_{1u}(LO) mode in CdF_2. Moreover, the variation of its position with x is in agreement with that of the LO phonon calculated from FIR reflectance spectra. These findings indicate that the peak probably originates in the activation of the LO component of the T_{1u} triplet, Raman inactive in pure CdF_2 crystals.

In the second set the Raman scattering arising from excitation of magnon pairs and phonons has been measured. The results show that the two-magnon peak drops in frequency and increases in width with increasing cadmium concentration. Similar behaviour is induced by increasing the temperature. The experimentally observed linewidths, lineshapes and frequency shifts, as functions of x, are in an agreement with the predictions of both the Isirg cluster model and the mean field model. In particular, the frequency shift agrees better with the theory than in the $Mn_{1-x}Zn_xF_2$ case. This fact is probably due to special properties of cadmium which introduces much larger distortion into the MnF_2 lattice than, e.g., zinc or iron. It is also in agreement with the difference in the percolation limit between $Cd_{1-x}Mn_xF_2$ ($x_o \approx 0.5$ from EPR measurements) and ordinary rutile structure mixed crystals like $Mn_{1-x}Zn_xF_2$ ($x_o = 0.76$).

The two-mode behaviour of the E_g mode, observed in $Mn_{1-x}Zn_xF_2$ and $Fe_{1-x}Zn_xF_2$ mixed crystals, occurs in $Cd_{1-x}Mn_xF_2$ too, as evidenced by a shoulder on the high-frequency flank of the Raman line, increasing with increasing cadmium concentration.

COHERENT ANTISTOKES RAMAN SCATTERING BY OBLIQUE POLARITONS IN THE LITHIUM IODATE CRYSTAL

K.B.Petrosyan and K.M.Pokhsraryan

Research Institute of Condensed Matter Physics, Yerevan State University, Kievyan 1a, 375028 Yerevan, USSR

Recently the methods of coherent antistokes Raman spectroscopy (CARS) are widely used in solid state spectroscopy.

The principle of polariton CARS is based on the excitation of polaritons by two rather powerful laser waves with frequencies ν_1 and ν_2, the difference between them being equal to the polariton frequency ν_p. If direct to the medium the probe laser wave with frequency ν the scattering by coherently excited polaritons at the stokes or antistokes frequency $\nu_{s,a} = \nu \mp (\nu_2 - \nu_1)$ takes place.

In the present work the first results of experimental study of oblique polaritons by CARS methods are reported.

As a source of tunable coherent radiation the optical parametric oscillator(OPO) on the KDP crystal pumped by second harmonic of picosecond YAG:Nd^{3+} laser was used. The coherent excitation of oblique polaritons realized by fundamental wave of YAG:Nd laser with frequency ν_1 =9396,7cm^{-1} and signal wave of OPO with frequency ν_2 tunable in the region 9396,7-11905cm^{-1}. As a probe wave the second harmonic of YAG:Nd^{3+} laser with frequency ν =18793cm^{-1} is used.

The 15mm long LiIO crystal(cut in the xz plane at the angle ϑ =30° to the optical Z axis) is used as a Raman medium. The antistokes wave of probe radiation was registered.

In the chosen scheme the waves ν_2 and ν were propagated perpendicular to the crystal input face, while the wave ν_1 -at the definite angle β. For the each value of $\Delta\nu = \nu_2 - \nu_1$ the value β between excited waves $\vec{k}_1$ and $\vec{k}_2$ into the crystal were determined from the four-photon phase-matching condition $\Delta\vec{k}_a = \vec{k} + \vec{k}_2 - \vec{k}_1 - \vec{k}_a$. When the frequency of biharmonic pump $\Delta\nu = \nu_2 - \nu_1$ occurs in the oblique polariton range simultaneously with the four-photon process $\nu_a = \nu + \nu_2 - \nu_1$ two-cascade three-photon processes may take place: at the first cascade the polariton is excited and at the second one - the antistokes photon $\nu_a = \nu + \nu_p$. The highest efficiency of these process is achived when
as well as the phase-matching condition
are carried out. Here $\vec{q}$ is defined from the condition $\Delta\vec{k}_a = 0$ while $\vec{k}_p$ - from oblique polaritons dispersion equation[1].

The calculations show that in the chosen CARS scheme a coherent excitation of oblique polariton with frequency ν_p = 783,5cm^{-1} (k_p=13899cm^{-1}) is possible.

The dependence of relative CARS-signal intensity $I_a/I_1I_2I \sim |\chi^{(3)}_{eff}|^2$ on $\Delta\nu = \nu_2 - \nu_1$ in the range 750–800 cm^{-1} was measured. In this frequency range two maximums took place. The first maximum corresponds to $\Delta\nu = 769\ cm^{-1}$, which is one of the frequencies of E_1 optical phonon [2]. Presence of this maximum is apparently connected with the resonance contribution of E_1 phonon to the "straight" four-photon process. The second maximum falls at $\Delta\nu = 783\ cm^{-1}$, which rather exactly coincides with the calculated value of coherent excited oblique polariton frequency.

The oblique polaritons characteristic pecularity - the dependence of frequency ν_p on direction $\vec{k}_p$ allows to investigate the CARS-signal intensity dependence on mismatch at fixed values $\nu_2 - \nu_1 = \nu_p$ and $\vec{k}_2 - \vec{k}_1 = \vec{k}_p$.

The CARS-signal intensity dependence on crystal rotation angle φ in the XZ plane at fixed values $\Delta\nu = 769$ and $783\ cm^{-1}$ was measured (keeping $\Delta\vec{k}_a = 0$). As was expected at $\Delta\nu = 769\ cm^{-1}$ the CARS-signal intensity does not change but at $\Delta\nu = 783\ cm^{-1}$ the obtained dependence had the form of Lorentzian function with maximum at $\varphi = 0$. According to De Martini [3] the full width of this function gives the absorption coefficient of the polariton wave. In our case the absorption coefficient at the frequency $\nu_p = 783\ cm^{-1}$ was $(42 \pm 4)\ cm^{-1}$.

Note that the violation of the condition $\Delta\vec{k}_p = 0$ brings to the decrease $|\chi^{(3)}_{eff}|^2$ in 1,3 times.

Thus the obtained results show that the study of anisotropic noncentrosymmetric crystalls by the CARS methods allows to define several characteristics of oblique polaritons.

R E F E R E N C E S

[1] W.S. Otaguro, E. Wiener-Avnear, S.P.S. Porto, J. Smith, Phys. Rev. B, 6, 3100 (1972).

[2] W. Otaguro, E. Wiener-Avnear, C.A. Arguello, S.P.S. Porto, Phys. Rev. B, 4, 4542 (1971).

[3] J.P. Coffinet, F. De Martini, Phys. Rev. Lett., 22, 60 (1969).

11-34

LOW TEMPERATURE BEHAVIOUR OF THE LITHIUM MODES IN MONOCLINIC Li_2SO_4

E.Cazzanelli, A.Fontana and F.Rocca

Dipartimento di Fisica - Universita' di Trento and
Centro CNR di Trento , 38050 Povo (TRENTO), ITALY

The compound lithium sulfate and some of other lithium containing mixed sulfates exhibit an interesting property in their high temperature phase: plastic behaviour (related to the orientational disorder of the molecular anions) together with a fast ion diffusion for the cations. The analysis of the Raman spectra between room temperature and the 583°C monoclinic-cubic phase transition [1,2,3] suggests the occurrence of pretransitional phenomena involving the rotational degree of freedom of the sulfate ions, as well as the interaction of the lithium cations with the internal vibrations of the sulfates. In fact, the high frequency of the lithium vibrations, respect to the modes of the others cations, makes more difficult in principle the usual separation between lattice and molecular modes, justified in most of the cases.

In a previous work [1] the lithium modes have been assigned via the isotopic substitution; their frequency range is between 300 and 450 cm^{-1}. The theoretical factor group analysis predicts 6 lithium modes in the A_g representation and 6 in the B_g one: the room temperature spectra reveal only 4 modes in each symmetry, with a very weak intensity and a large bandwidth, making difficult any further analysis. For this reason we have performed Raman measurements in a lower temperature range (90-300 K) for both the symmetries.

The spectra were collected on monocrystal anhydrous sample, grown from a solution in our laboratory with the same method of ref.[1]. The orientation of the sample corresponds to the scattering geometry Z(YY)X for the A_g and Z(YZ)X for the B_g representations. The spillover from the forbidden symmetries was less than 10% , due probably to a slightly depolarization of the incident light at the sample surface.

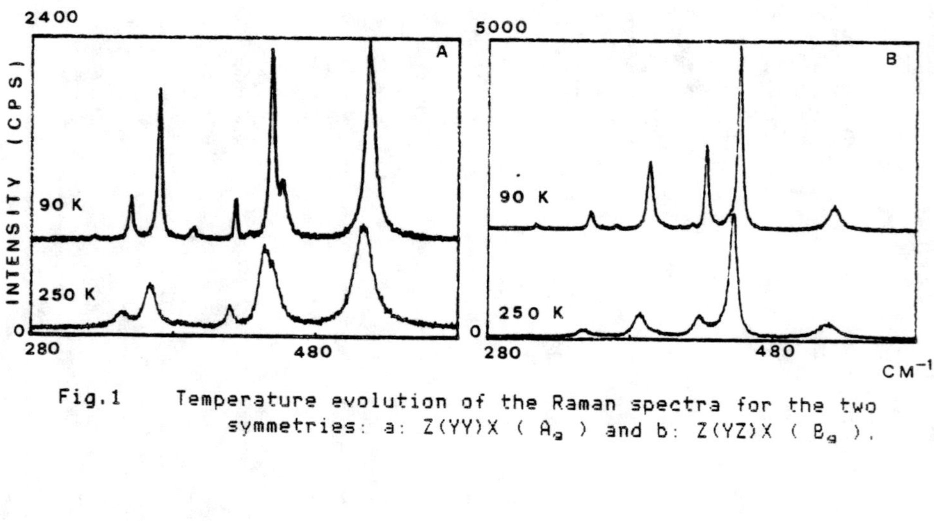

Fig.1 Temperature evolution of the Raman spectra for the two symmetries: a: Z(YY)X (A_g) and b: Z(YZ)X (B_g).

It is evident the broadening and the shift toward lower energy of all the peaks, consistent with the trends found in the higher temperature range [3]. A more interesting fact is, however, the anomalous decrease of the intensity for some peaks. A quantitative description of these trends has been obtained by fitting all the peaks by analitical functions (gaussians or lorentzians, depending from the bandwith).

An example of this behaviour is shown in the figure 2 , where the intensity of the B_g lithium mode, centered around 430 cm^{-1} at 90 K, has been plotted as a function of the temperature. The reported intensity values were normalized by the internal standard provided by the total-symmetric stretching mode of the sulfate group, which is temperature independent; furthermore the trivial population factor was divided out.

Fig.2:
Plot of the normalized intensity (see text) of the B_g lithium mode at 433 cm^{-1} vs. T.

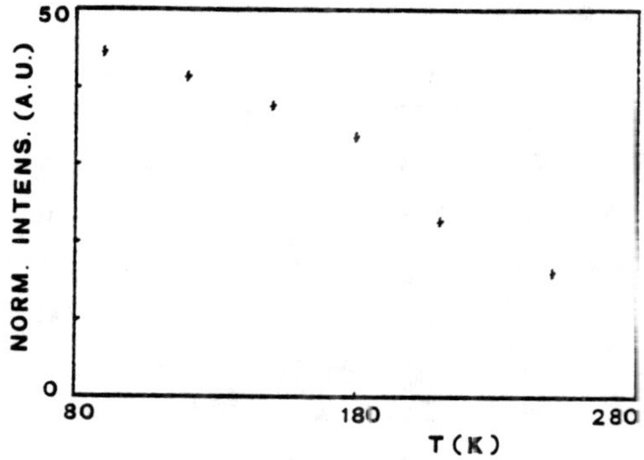

The expected behaviour for a B.E. normalized first order Raman peak would be a constant function of the temperature. Instead we have found for some peaks an intensity decrease of more than a factor two, which cannot be explained by systematic errors in the fitting procedure.

This decrease can be related to a continuus change of the crystallografic parameters (atomic distances and bonding angles) or even some kind of order-disorder evolution originating a more symmetric configuration of the lithium cations. Direct informations about these phenomena can be obtained by X-ray or neutron scattering analysis, as well as calorimetric investigation or dielettric loss measurements, to detect eventual phase transitions. Unfortunately no such data are available on anhydrous lithium sulfate, due to the difficulty to get monocrystalline samples. In similar salts ($LiKSO_4$) ferroelectric phase transitions are reported at temperatures around 200 K [4]. Our first Raman evidences suggest the need for a complete analysis of the phenomena occurring in this temperature range. A more extensive study about the behaviour of the Raman spectra is in progress and it will be presented elsewhere.

REFERENCES

[1] E.Cazzanelli and R.Frech, J.Chem.Phys.79, 2615 (1983)
[2] R.Frech and E.Cazzanelli, Solid State Ionics9&10, 95 (1983)
[3] E.Cazzanelli and R.Frech J.Chem. Phys.81, 4729 (1984)
[4] Sandhya Bhakay-Tamhane et alii , Solid St.Comm.53, 197 (1985)

RAMAN SPECTRA AND DIRECTIONAL DISPERSION IN $LiNbO_3$ AND $LiTaO_3$.

Guoxiang Lan, Huafu Wang, Xichen Yang and Bing Li

Department of Physics, Nankai University, Tianjin, P.R.China

$LiNbO_3$ and $LiTaO_3$ belong to the space group C_{3v}^6 in ferroelectric phase with the same structure. The vibrational symmetries of the optical modes at Γ point are $4A_1+9E+5A_2$. The A_1 and E modes are polar phonons, and there are 13 directional dispersion branches of extraordinary phonons. The optically active vibrational modes of these two crystals have been studied in the past, but the mode assignment is not conclusive. The directional dispersion of extraordinary phonons in $LiNbO_3$ has been determined by Claus et al.(1) with back scattering method, and that in $LiTaO_3$ by Penna et al.(2) with forward scattering techniques.

In the present work we have found a satisfactory right-angle scattering geometry. The YZ-plane constitutes the scattering plane. The Phonon wave vectors make an angle θ with the optic axis. (1) SP configuration: the polarization of incident light is perpendicular to the plane of scattering, and that of scattered light parallel to the plane of scattering, (2) SS configuration: the polarization of both incident and scattered light is perpendicular to the plane of scattering. Raman spectra of extraordinary phonons can be obtained with SS configuration and that of ordinary phonons with SP configuration. Thus Raman spectra of extraordinary phonons are completely separated from that of ordinary phonons and all of the extraordinary phonons always appear in the same series of Raman spectra, hence the directional dispersion can be determined with certainty. The Raman spectra of extraordinay phonons in $LiNbO_3$ and $LiTaO_3$ are shown in Fig.1 and Fig.3, respectively. From these Raman spectra, 13 directional dispersion branches of extraordinary phonons of $LiNbO_3$ and $LiTaO_3$ are determined without ambiguity, and sketched in Fig.

2 and Fig.4, respectively.

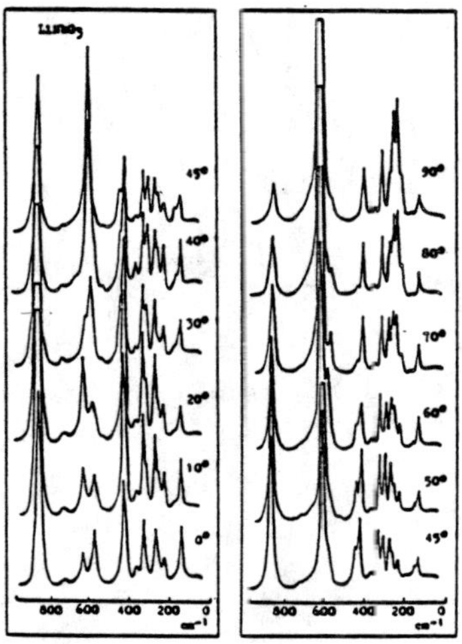

Fig.1 Raman spectra of extraordinary phonons in $LiNbO_3$

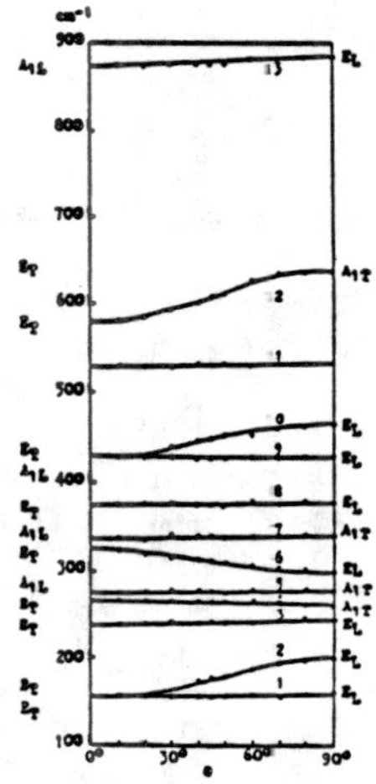

Fig.2 Directional dispersion branches of extraordinary phonons in $LiNbO_3$

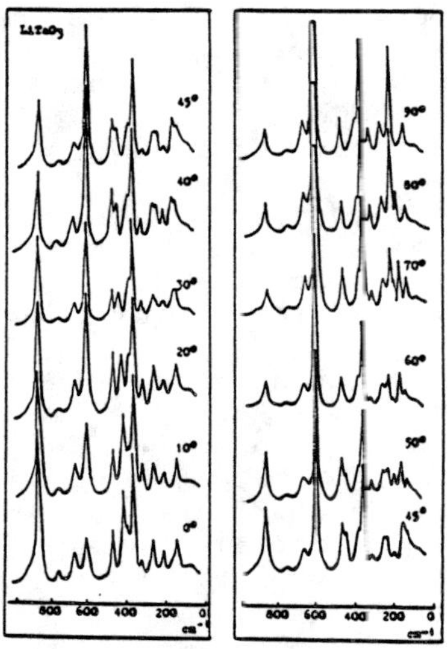

Fig.3 Raman spectra of extraordinary phonons in $LiTaO_3$

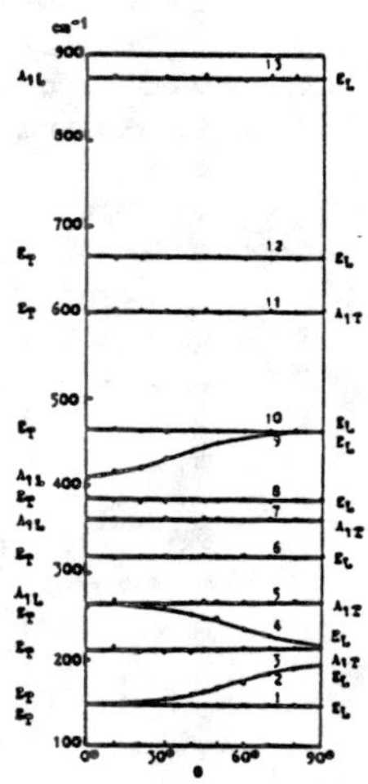

Fig.4 Directional dispersion branches of extraordinary phonons in $LiTaO_3$

References

(1) R. Claus et al., Z. Naturforsh, 27a,1186(1972).

(2) A. F. Penna et al., Phys. Rev. B13, 4907(1976).

RESONANCE RAMAN SCATTERING IN $CdGa_2(S_xSe_{1-x})_4$.

P.P. Lottici, A. Parisini and C. Razzetti

Dipartimento di Fisica dell'Universitá and GNSM-CNR, CISM-MPI
Via M. D'Azeglio 85 - 43100 - PARMA - Italy

Tetragonal compounds of the $A^{II}B_2^{III}X_4^{VI}$ type and their mixed phases have interesting applications in nonlinear optical devices (1).

$CdGa_2(S_xSe_{1-x})_4$ has a tetragonal defect chalcopyrite structure, derived from the chalcopyrite by the inclusion of an ordered array of vacancies. The primitive unit cell of $A\ B_2X_4$ compounds contains seven atoms: the vibrational spectrum at k=0 is given by 3 A-symmetry modes, 6 E-symmetry doubly degenerate modes and 6 B-symmetry modes. Common vibrational features of the tetragonal defect chalcopyrite compounds have been evidenced (2). In particular, one observes in Raman scattering strong breathing-like modes of A symmetry and zincblende like nearly degenerate E and B modes at the top of the spectrum. These last modes correspond to the relative motion of the cation and anion sublattices as a whole. Raman scattering measurements on $CdGa_2(S_xSe_{1-x})_4$ in the whole composition range (3) show a two-mode behaviour for all the observed phonons. A MREI (Modified Random Element Isodisplacement) model has been developed to describe the composition dependence of the zincblende-like modes in a nearly-cubic approximation (4). The model is completely defined by the macroscopic physical quantities of the end members and it is free from fitting parameters.

In a previous paper, resonant Raman scattering results have been reported for single crystals of the end member $CdGa_2Se_4$ (5). A resonance enhancement of the E(LO) zincblende-like phonon at 278 cm^{-1} in "forbidden" geometry was found, together with a weak resonance for the breathing A-symmetry mode at 141 cm^{-1}. In the present study we focus on the zincblende like optical modes arising from Se in $CdGa_2(S_xSe_{1-x})_4$.

The mixed crystals of the $CdGa_2(S_xSe_{1-x})_4$ family have been grown by a I_2-CVD-LTVTP procedure (1) at the MASPEC-CNR Laboratory, Parma, Italy. Raman scattering experiments were performed on unoriented samples using an Ar laser, a SPEX 1401 double monochromator and photon counting techniques. Resonance conditions may be attained up to x=0.5. To obtain absorption corrected values, a linear interpolation scheme has been used to obtain the unknown absorption curves for the mixed phase from those of the ternary crystals.

The ratio between the cross section for the zincblende like Se-derived E-B optical mode and the breathing (Se) A symmetry mode as a function of exciting laser energy is reported in Fig.1. We see that the resonance of the high frequency modes, through their longitudinal component, is stronger than the nonpolar A modes.

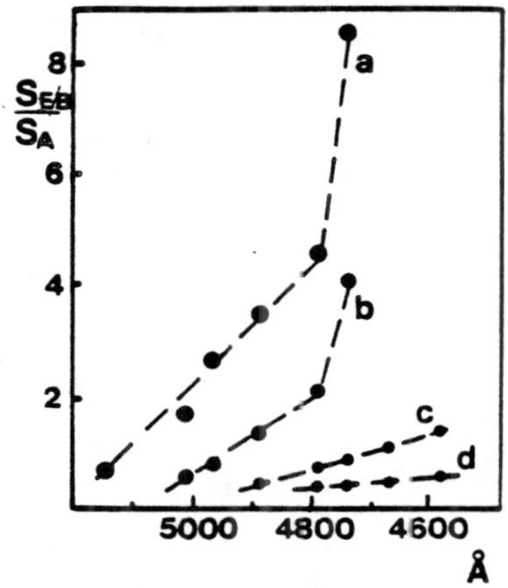

Fig.1 - Ratio between the absorption corrected Raman cross section of the highest frequency E-B zincblende-like optical modes and the cross section of the A symmetry breathing like mode vs. laser exciting frequency.
a) $CdGa_2Se_4$ - b) $CdGa_2S_{0.5}Se_{3.5}$
c) $CdGa_2S_{1.5}Se_{2.5}$
d) $CdGa_2S_2Se_2$.

In polar crystals, in addition to the deformation potential scattering mechanism, one has to account for the macroscopic electric fields associated withe the longitudinal phonons. In resonance conditions the LO phonons often are present in "forbidden" geometries, i.e. in scattering geometries where they should not be observable. This breakdown of the Raman selection rules, observed in $CdGa_2Se_4$ (5), has been ascribed to the finiteness of the phonon wave vector and then to the activation of an intraband scattering by the Frohlich Hamiltonian between electrons and longitudinal optical phonons which in resonance conditions may become very large (6). The same mechanism could be responsible for the selective resonance enhancement of the zincblende-like polar modes in $CdGa_2(S_xSe_{1-x})_4$ mixed crystals.

REFERENCES

1. C. Razzetti, P.P. Lottici and L. Zanotti
Mater. Chem. Phys. 11, 65 (1984)
2. P.P. Lottici and C. Razzetti
J.Phys.C: Solid State Phys. 16, 3449 (1983)
3. P.P. Lottici and C. Razzetti
J. Mol. Structure 115, 133 (1984)
4. A. Parisini and P.P. Lottici
phys. stat. sol.(b) 129, 539 (1985)
5. A. Parisini, P.P. Lottici, C. Razzetti and R.Bacewicz
phys. stat. sol.(b) 134, 279 (1986)
6. R.M.Martin
Phys. Rev. B 4, 3676 (1971)

DYNAMIC PROCESSES IN THE PYROELECTRIC SINGLE CRYSTALS OF URANYL NITRATE HEXAHYDRATE THROUGH RAMAN SPECTROSCOPY.

H. D. Bist,[*†§] T. S. Little,[*§] and J. R. Durig[*§]

[*]Department of Physics, Indian Institute of Technology, Kanpur, India.
†Institute for Molecular Science, Okazaki-444, Japan.
§Department of Chemistry, University of South Carolina, Columbia, SC 29208

Temperature dependence of the photoluminescence of uranyl nitrate hexahydrate was used recently to infer the occurrence of a phase transformation around 130K which has been associated with the twin phenomena of generation of pulsed pyrovoltage and pyroluminescence reversibly during its cooling and warming cycles.[1] Laser Raman spectroscopy is one of the most convenient techniques for investigating the temperature dependent dynamics of delicate crystals like uranyl nitrate hexahydrate, which are mildly deliquescent but become easily dehydrated under low pressures. The internal modes of the uranyl ion, the nitrate ions and the water molecules, along with their external lattice vibrations, will be discussed in order to explore the detailed nature and mechanism of the suggested phase transition[1].

The uranyl nitrate hexahydrate crystal near 300K belongs to a non-centrosymmetric space group $Cmc2_1$ with four molecules per unit cell (z = 4).[2] The primitive cell has z' = 2. Of the two available sites $C_s(2)$ and $C_1(4)$ in this space group, the former is permissible for occupancy of the various structural units of the crystal. If the uranyl O-U-O axis[2] is taken as the z-axis and the x-axis as bisecting the bidentate nitrate bonds in a rectangular coordinate convention, the optical translatory, $\Gamma(OT)$, and optical libratory, $\Gamma(OL)$, modes of the chromophoric group are distributed as: $\Gamma(OT) = A_1 + B_2 + A_2$; and $\Gamma(OL) = A_1 + B_2 + 2A_2 + 2B_1$. All the A_2, A_1, B_1 and B_2 modes are Raman active. On any structural phase transformation the crystalline field around the constituent groups changes.

The Raman spectra were recorded using a Cary model 82 spectrophotometer with 514.5 nm excitation from a Spectra Physics model 171 argon ion laser. A single crystal of uranyl nitrate hexahydrate, sealed hermetically in a pyrex tube, was cooled in a Cryogenics Technology Inc. Spectrim cryostat equipped with a Lake Shore Cryotronic model DTL-500 high-precision temperature controller.

At 300K the photoluminescence of the uranyl hexahydrate crystal predominates, and it diminishes with decrease in temperature in a non-exponential fashion. This behavior indicates that the occurrence of room temperature photoluminescence, at least partially, is due to factors other than the thermal population in the vibrationally excited states of this piezoelectric crystal.

Throughout the temperature range 300-26K, the ν_1 mode of the uranyl ion exhibits the highest intensity in all the scattering spectra. At 300K the ν_3 mode (~935 cm^{-1}) is observed with about 0.9% of the ν_1 intensity which supports at least a $C_{\infty v}$ structure with UO bond lengths of 1.770 and 1.745A.[2] A very weak hump at ~250 cm^{-1} also supports the previous assignment of the ν_2 mode in uranyl nitrate hexahydrate (250 ± 3 cm^{-1}).[3]

The temperature dependence of the intensity of the photoluminescence is conspicuous on 514.5 nm Ar^+ excitation and the case is similar for the ν_1 band observed in the Raman scattering. Their intensity ratio at different temperatures corroborates our earlier suggestion of a phase transformation in uranyl nitrate hexahydrate in the temperature range 130 ± 20K.

Since the bidentate character of the nitrate group is established,[2] the characteristic pattern of nitrate bands in the Raman spectra could be a diagnostic tool (for such bondings in actinyl compounds) with well separated doublets at 1488 and 1466, 1056 and 1044, and 756 and 721 cm^{-1} for its ν_3, ν_1 and ν_4 modes, respectively. The three types of H-bonding for the water molecules[2] are exhibited by a relatively sharp doublet (3600 and 3514 cm^{-1}), a broad band (3350 cm^{-1}) and a broader peak (3150 cm^{-1}) at 300K. The bending mode of water at 1620 cm^{-1} is weak and so are its librational modes in the 600 to 300 cm^{-1} region.

In spite of the expected heavy mixing in the low frequency region, the main bands could be assigned to the translatory modes of the uranyl ion (32, 45 and 54 cm^{-1}), the nitrate ions (69, 95 and 111 cm^{-1}) and the water (129, 173 and 208 cm^{-1}). These assignments indicate that the three constituent ions which have very different masses (~270, 62 and 18 amu, respectively) oscillate in a mutually created symmetrical field in the equatorial plane of the uranyl ion. The libratory modes are weaker in the Raman spectra at all temperatures.

Thanks are due to Professor S. Nagakura, Institute of Molecular Science, Okazaki, Japan, for providing a visiting assignment to one of us (HDB) and to the National Science Foundation of the United States of America for travel grant INT-84-11610.

[1]H. D. Bist, T. Nishiya, M. Baba and I. Hanazaki, to be published.
[2]J. C. Taylor and M. H. Mueller, Acta Cryst., 19, 536 (1965).
[3]H. D. Bist, J. Mol. Spectrosc., 27, 542 (1968).

A VARIABLE TEMPERATURE RAMAN STUDY OF THE VIBRATIONAL MODE IN NH_4I.

Y. Xu, W.F. Sherman and G.R. Wilkinson

Department of Physics, King's College London, London, England.

There are four (and possible five) solid phases[1] of NH_4I in the temperature and pressure range 0<T<300K, 0<p<100kbar, see Figure 1. As a preliminary to a variable pressure study being planned for NH_4I we started to check the zero pressure variable temperature data for this material. We are particularly interested in the part played by the torsional oscillations as the order-disorder transitions are approached and we therefore paid particular attention to this mode, which has generally been referred to as ν_6. Somewhat to our surprise, the behaviour of bands due to ν_6 and $2\nu_6$ in the Raman and $2\nu_6$ and $3\nu_6$ in the infrared showed evidence of a second order phase transition at about 110K and zero pressure. It is this data that we describe here.

From our earlier study[2] of NH_4Br we had suspected that the break-down of rotational order might proceed in at least two stages, with the dramatic (λ-type) transition taking place at relatively high temperature as the short-range order is destroyed. However at lower temperatures the long-range order might well be less than complete.

Full long-range order could be lost either by the onset of random mis-orientations or by the appearance of a systematic long-range disordering, i.e. an incommensurate phase. As the other modes studied so far show no obvious signs of a phase transition in the 110K region indciated by the bands due to the ν_6 mode, it seems fairly certain that the short-range order is not significantly effected by temperature changes in this region. However the ν_6 data presented in Figure 2 clearly suggest a transition of some sort. At this stage we are not prepared to make any firm statements about the nature of the transition beyond the contention that it represents the onset of a loss of long-range order. We will shortly be commencing the high pressure spectroscopic study of this material and we are considering the possibility of careful X-ray measurements to complement the spectroscopic study.

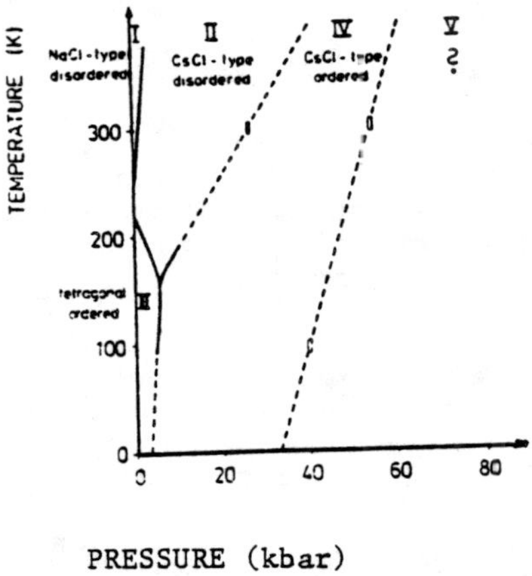

FIG. 1. Phase diagram of NH_4I

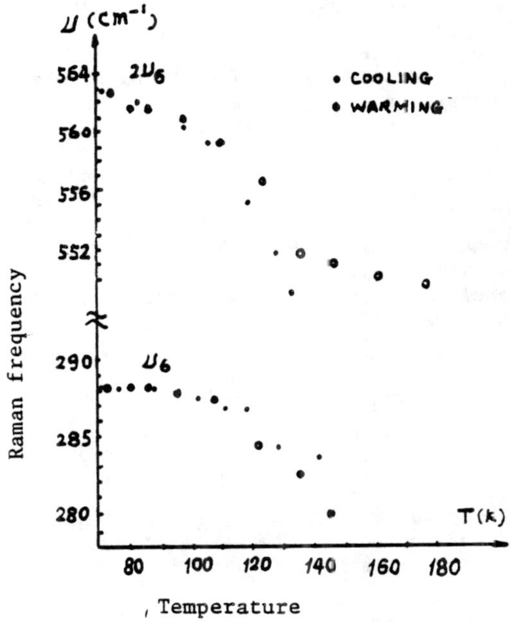

FIG. 2. Frequencies shits of $2\nu_6$ and ν_6 from Raman spectra.

REFERENCES

[1] A.M. Hens, K.R. Hirsch and W.B. Holzapfel, J.Chem.Phys. 73, 105, (1980).

[2] A.H. Abdullah and W.F. Sherman, J.Phys.E.Sci.Instrum. 13, 3735 (1980).

11-44

THEORY OF RAMAN SCATTERING IN ALKALI HALIDES.

G.D.Mahan* and K.R. Subbaswamy+

*Oak Ridge National Laboratory and the University of Tennesee, +University of Kentucky

A new mechanism is suggested for the Raman scattering in alkali halides based upon the electro-optic effect.[1-2] This effect is based upon the nonlinear, or hyperpolarizability, of the individual ions. This process gives a good account of the magnitude and polarization dependence of the Raman scattering. It appears to be as important, and perhaps more important, than the old mechanism of Born and Bradburn which used the nonlinear bond between ions.[3-4]

First order Raman scattering is forbidden in alkali halides because each ion sits at a site with inversion symmetry. Light scattering is caused by a much weaker second order process whereby the photon scatters by making two phonons. The experiments observe a continuous spectra which is adequately represented by the joint density of two phonon states.[5] Another experiment is to apply a strong dc electric field F_0 to the crystal which breaks the inversion symmetry. Then one finds a first order Raman cross section proportional to F_0^2. This experiment on alkali halides was first reported a year ago by Krantz and Luty.[6]

The electro-optic contribution is due to the nonlinear, or hyperpolarizability of the ions. We use Buckingham's notation[7] for the two hyperpolarizabilities γ and B. Both are fourth rank tensors, but the four indices are not written except when necessary. The induced dipole moment p and quadrupole moment q due to electric F and quadrupole Q fields are given by

$$p = \alpha_d F + (1/3!)\gamma F^3 + (1/3)BFQ$$

$$q = \alpha_q Q + (1/3!)BF^2$$

where the indices have been suppressed. The top formula for the induced dipole moment is the basis of our theory of the electro-optic contribution to Raman scattering. The first term $\alpha_d F$ gives the Rayleigh scattering in terms of the dipole polarizability α_d but does not contribute to the Raman scattering.

The easiest experiment to discuss theoretically is the field induced first order Raman scattering (FIFORS). It is well known from early theories that the cross section is proportional to $F_0{}^2$. However, we show that the electro-optic term gave a simple expression for the cross section which agrees with the experiment. For this case there is no quadrupole field Q, and the B coefficient is not involved. Instead, there is a field F_q from the phonon of wave vector q, a field F_ℓ from the laser, and the dc field F_0.

$$F = F_o + F_\ell + F_q$$

Taking the cube of this field produces six terms with the triple cross product

$$F^3 = 6\, F_o\, F_\ell\, F_q + \ldots$$

$$p_\alpha = \gamma_{\alpha\beta\mu\nu}\, F_{o\mu}\, F_{\ell\nu}\, F_{q\beta}$$

There are also local field corrections to this expression which are treated in detail but have little influence upon the answer. Here we neglect them to keep the discussion simple. Then the answer for the cross section for FIFORS is

$$d\sigma/d\Omega = (\omega/c)^4 \left| \gamma_{\alpha\beta\mu\nu}\, \eta_\alpha\, \eta_\beta\, F_{o\mu}\, F_{q\nu} \right|^2$$

where $\eta_{\alpha,\beta}$ are the polarizations of the initial and final photon. This rather simple expression relates the cross section to F_o^2, γ,[8-9] and the field due to the long wavelength optical phonon. Using experimental values for γ, we show that the magnitude of the cross section is in excellent agreement with the experimental result. Furthermore, from the properties of the γ-tensor, one can make simple and definite predictions regarding the polarization dependence of the FOFIRS. The word "polarization" is meant to apply to both the phonons and the scattered radiation.

We also develop the electro-optic theory of the second order Raman scattering. Here the B-term is included along with the γ-term. This introduces the problem that there are no experimental values for B. There is one calculation of B for several alkali halides.[10] Using these values, we find that the B-term was more important than the γ-term for the two-phonon scattering. Here again the magnitudes of the theoretical cross section agree with the data although few published data sets provide absolute values for comparison.

Research supported by NSF Grants DMR 85-01101 and DMR 82-16212 and by the Distinguished Scientist Program at The University of Tennessee and Oak Ridge National Laboratory. ORNL is operated by Martin Marietta Energy Systems, Inc. under contract DE-AC05-84OR21400 for the US Department of Energy.

References

1. K.R. Subbaswamy and G.D. Mahan, Phys. Rev. B32, 5433 (1985)
2. G.D. Mahan and K.R. Subbaswamy, Phys. Rev. (in press)
3. M. Born and M. Bradburn, Proc. Roy. Soc.(London) A188, 161 (1946)
4. J.R. Hardy and A.M. Karo, The Dynamics and Statics of Alkali Halide Crystals (Plenum, New York, 1979)
5. T.P. Sharma, R.D. Kirby, and S.S. Jaswal, Phys. Rev. B9, 1971 (1973)
6. M. Krantz and F. Luty, Phys. Rev. B31, 2599 (1985)
7. A.D. Buckingham, Adv. Chem. Phys. 12, 107 (1967)
8. P.D. Maker and R.W. Terhune, Phys. Rev. 137, A801 (1965)
9. C.C. Wang and E.L. Baardsen, Phys. Rev. 185, 1079 (1969); B1,2827 ('70)
10. P.W.Fowler and P.A. Madden, Phys. Rev. B31, 5443 (1985)

THE 2-DIMENSIONAL RAMAN MAP OF THE STRESS INDUCED MONOCLINIC PHASE IN PSZ.

Jiro Mizuno, Seishi Kudo and Hideo Hasegawa

Toyota Central Research and Development Laboratories, Inc.,
Nagakute, Aichi, Japan

It has been generally accepted that the stress induced phase transformation from the metastable phase (D_{4h}^{15}: tetragonal) to the stable phase (C_{2h}^{5}: monoclinic) plays a major role in the high toughness of the partially stabilized zirconia (PSZ).

Clarke and Adar[1)] have first applied Raman microprobe analysis to measurement of the transformed zone close to the fracture cracks of the ceramics containing PSZ.

We investigated the two dimensional distribution of monoclinic phase, induced by stress field around Vicker's indent of Y_2O_3-PSZ, in order to examine the effect of indentation temperature on the phase transformation.

The samples of sintered PSZ, of ZrO_2-2 mol% Y_2O_3 and 3 mol% Y_2O_3 (named 2Y- and 3Y-PSZ, respectively), were composed of the tetragonal phase with crystallite size of 0.3-0.4 µm. Vicker's indents were notched on the samples at the temperature between 20°C and 580°C, using the high temperature micro hardness tester by 500 g weight.

The new system of Raman microprobe, JASCO model RPM-1000, equipped with an X-Y step-scanning sample stage and with a multichannel detector was used to obtain the 2-dimensional maps of a monoclinic fraction. The Ar ion laser beam of 488 nm with the output power of 400 mW was focused on the sample by the microscope objective lens (×100, N.A. = 0.9) with a diameter of 4 µm. The backscattered light was leaded to the filtering monochromator (f = 20 cm), next to the main monochromator (f = 100 cm) and finally to the intensified multichannel detector of a diode array (TN-6122: 1024 ch/1 inch). Raman spectrum up to 480 cm^{-1}, with resolution of 10 cm^{-1}, was obtained at each sampling point by the exposure time of about 25 seconds (Fig. 1). The sample was scanned stepwise at intervals of 5 µm. The scan and aquisition were automatically controlled, and the set of spectra was stored on the 22 M bite hard disk.

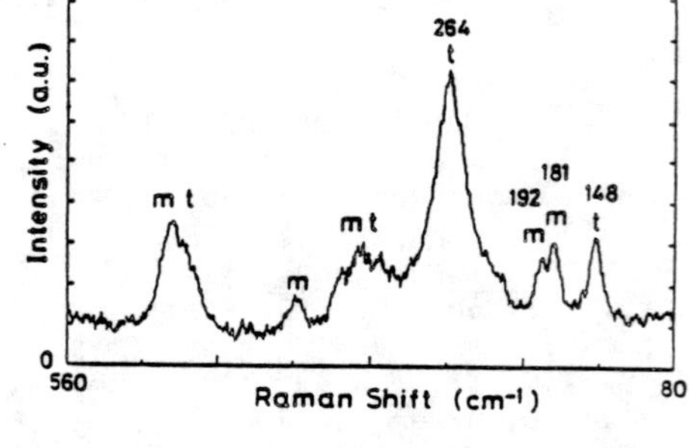

Fig. 1 Typical spectrum at a sampling point around the edge of the indent: 2Y-PSZ, indented at 20°C.

The monoclinic fraction, C_m, at each sampling point was calculated by the next formula from the each spectrum:

$$C_m = \frac{I_m^{181} + I_m^{192}}{F(I_t^{148} + I_t^{264}) + (I_m^{181} + I_m^{192})} ,$$

where I is the intensity of the Raman peak after background correction, and m, t and numerical number represent monoclinic phase, tetragonal phase and Raman shift, respectively. The correction factor, F, was put to be unity, in this case.

Two examples of the maps are shown in Fig. 2. It is noticed that C_m takes the maximum value, i.e. C_m (max), around the edge of each indent, and that it is rather small at the center of each indent, i.e. C_m (center). Both of C_m (max) and C_m (center) are shown in Fig. 3 to take maximum values at the indentation temperature of 100-300°C in each sample.

The physical meanings of these results will be discussed.

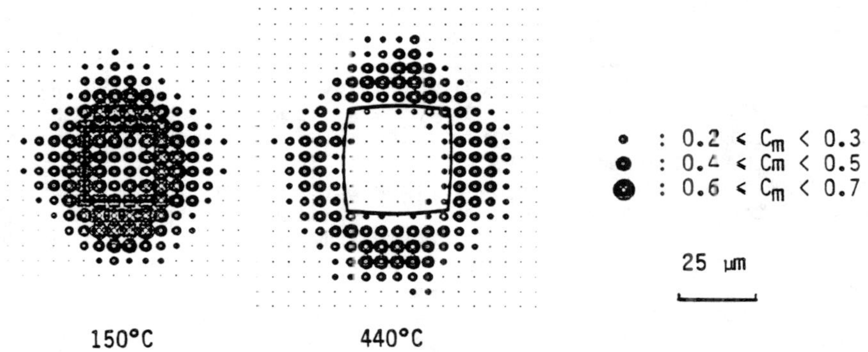

Fig. 2 Raman maps of monoclinic fraction, C_m, of 2Y-PSZ samples. The edge of an indent is shown by the solid line.

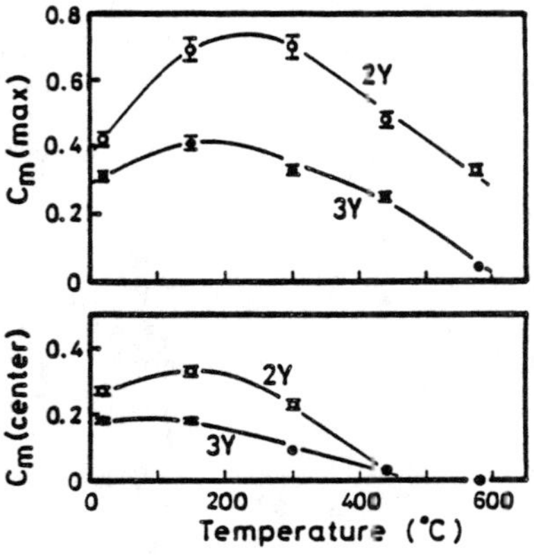

Fig. 3 The variation of C_m with the indentation temperature for 2Y- and 3Y-PSZ.

REFERENCE

1. D. R. Clarke and F. Adar, J. Am. Ceram. Soc. 65 [6], 284-288 (1982)

RAMAN MICROPROBE STUDY OF PHASE TRANSFORMATIONS IN CERIA STABILIZED ZIRCONIA.*

G. M. Begun, P. F. Becher, and E. F. Funkenbusch†

Oak Ridge National Laboratory, P. O. Box X, Oak Ridge, TN 37831

Transformation-toughened multiphase ceramics containing metastable tetragonal ZrO_2 are being widely investigated for use as technologically important materials. The ceramic samples are strengthened by the presence of the metastable tetragonal ZrO_2 phase which when subjected to a tensile stress, converts to the higher volume monoclinic phase which inhibits crack propagation in such materials. Koramidas and White[1] and Benner and Nagelberg[2] obtained the Raman spectra of amorphous, tetragonal, monoclinic, and cubic zirconia. The tetragonal and monoclinic forms are easily characterized and the ratio of tetragonal to monoclinic crystalline forms present in a sample can be determined by measurement of several of the non-overlapping Raman peaks assigned to the two species. The development of the Raman microprobe has made it possible to determine the microscopic state of ZrO_2 present in ceramic samples and to study its variation in the neighborhood of indentations and cracks. Clark and Adar[3] established that the ratio of monoclinic ZrO_2 to tetragonal ZrO_2 was equal to the sum of the intensities of the Raman peaks at 181 and 191 cm^{-1} divided by the sum of the intensities of the peaks at 148 and 264 cm^{-1}.

We have applied this method to samples of ceria stabilized ZrO_2 (10-15 mol % CeO_2) to determine the influence of CeO_2 content on the ability to trigger the transformation. Representative samples were impressed with a Diamond Pyramid Hardness (DPH) indentor with a maximum load of 30 Kg. The samples were then examined with a Ramanor Microprobe using 488.0 nm radiation from an argon-ion laser. Figure 1 shows some of the Raman traces made at various distances perpendicular to the side of a 30 Kg DPH indentation. The tetragonal (t) to monoclinic (m) phase transformation proceeds by cooperative shear, forming extended bands in which neighboring tetragonal grains transform. The results show that less extensive transformation occurs along the extension of the indent diagonal, then along the normal to the center of the side face of the indent. Figure 2 shows a plot of the results obtained (% monoclinic, assuming the samples to contain only the monoclinic and tetragonal forms, vs the distance (μm) from the edge of the indent. Data is given for samples with 10.0, 12.5, and 15.0 mol % CeO_2. Increasing the CeO_2 content suppresses the t→m transformation. The 15.0 mol % sample shows very little transformation, while the 10.0 and 12.5 mol % samples exhibit the effect mentioned above with more transformation taking place along the line normal to the indent. Figure 3 shows the results on the same samples which were carefully polished to remove 20-25 μm of the surface after indentation in order to expose the interior surface. Examination of the interior surface at 15 mol % shows that the t→m transformation has occurred; t→m→t transformation occurs on the free surface due to stress relaxation. Similar effects are indicated by the broader maxima exhibited by the interior surface results obtained for 10 and 12.5 mol % CeO_2. The zone shape, as well as the crack geometry, is altered by the change in constraining forces (free vs interior surfaces).

---

*Research sponsored by the Division of Materials Sciences, Office of Basic Energy Sciences, U.S. Department of Energy under contract DE-AC05-84OR21400 with Martin Marietta Energy Systems, Inc.

†3M Company, Saint Paul, MN 55101.

1. V. G. Keramidas and W. B. White, J. Am. Ceram. Soc. 57, 22 (1974).
2. R. E. Benner and A. S. Nagelberg, Thin Solid Films 84, 89 (1981).
3. D. R. Clarke and F. Adar, J. Am. Ceram. Soc. 65, 284 (1982).

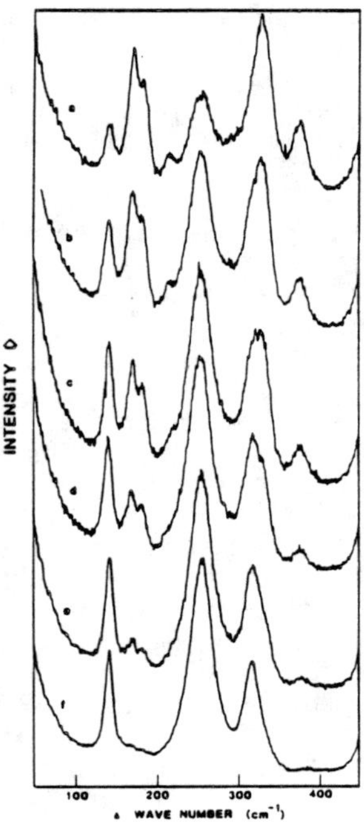

Fig. 1. a→f; 0, 60, 90, 140, 200, 300 μm from indent, respectively.

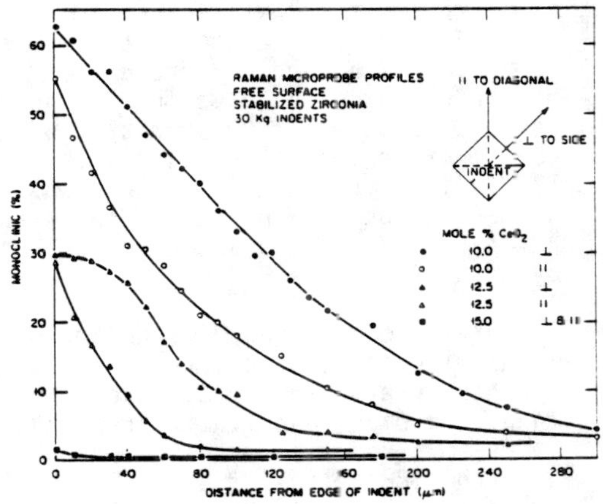

Fig.2

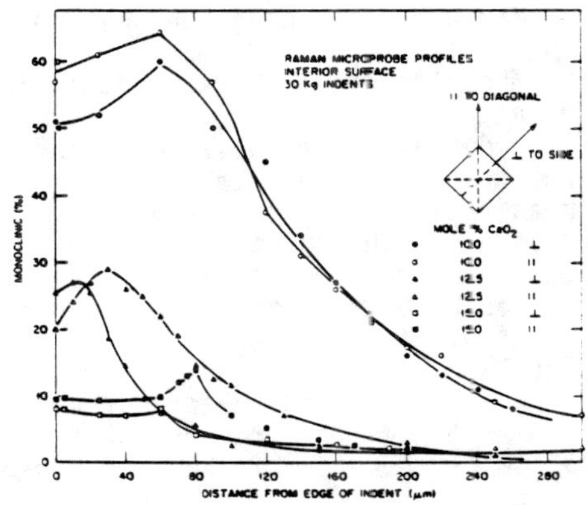

Fig.3

EVALUATION OF TRANSFORMATION ZONE SIZE OF Y-TZP BY RAMAN MICROPROBE

Gen Katagiri, Mikako Teramura, Hideyuki Ishida, Takaki Masaki* and Akira Ishitani

Toray Research Center, Inc. *Toray Industry Inc.
1-1, 1-Chome Sonoyama, Otsu, Shiga 520, Japan

Y-TZP(Yttria Stabilized Tetragonal Zirconia Polycrystal) with mostly tetragonal structure is gathering interest because of exceptionally high strength and toughness for ceramics. The stress induced transformation of Y-TZP from tetragonal to monoclinic structure is widely accepted as a cause for high bending strength and fracture toughness. Raman microprobe has been proved to be useful for investigation of the transformation[1], because of its ability of distinguishing different crystal structures with the μm-order spatial resolution. In this work, the transformation of Y-TZP by indentation and fracture was studied in relation to Y_2O_3 and Al_2O_3 concentrations. The size of the transformed zone determined directly by Raman microprobe was also compared with the theoretical values presented by Evans[2].

Y-TZP containing Al_2O_3 along with 2.0 to 4.0mol% Y_2O_3, was sintered by hot isostatic pressing at 1400°C for 1.5hr[3]. Vickers hardness indents were made with 20kg load on the surface of a polished test piece. Then, the piece was fractured by the 3 points bending test.

Raman spectra were measured using Raman microprobe, MOLE(Jobin Yvon), with the 4880Å line of an Ar ion laser.

The content of the monoclinic structure is presented by equation, $m = (I_M^{178} + I_M^{188})/(2I_T^{146} + I_M^{178} + I_M^{188})$, using the peak intensities of Raman bands of the given wavenumbers corresponding to two structures.

Monoclinic concentration , m, is mapped in Fig. 1. Transformed regions are observed around the indents and along the fracture edge. Both of the degree and the size of transformation is found to be largest in 2.0mol% Y_2O_3 sample. They decrease monotonously with the increase of Y_2O_3 concentration. It is noteworthy that the transformation is scarcely observed along the fracture edge in 4.0mol% Y_2O_3. The complicated contours around the cracks indicate that the stress in the corner of the indents is relieved by both of the generation of cracks and the stress induced transformation.

Successive measurements of Raman spectra on spots of different distance from edges of fracture or an indent give a monoclinic concentration profile as shown in Fig. 2. The transformation zone size (d), an important parameter in the theories of transformation-toughning, can be directly obtained from Fig. 2.

The fracture toughness, K_{IC} , is expected to be proportional to the product of the square root of the transformation zone size, $\sqrt{d}$, and the volume fraction of tetragonal phase, V_f, according to the theory by Evans[2].

Fig. 3 shows the experimentally obtained plot of K_{IC} against $V_f\sqrt{d}$. Although K_{IC} increases with increase of $V_f\sqrt{d}$, the expected linear relationship is not observed.

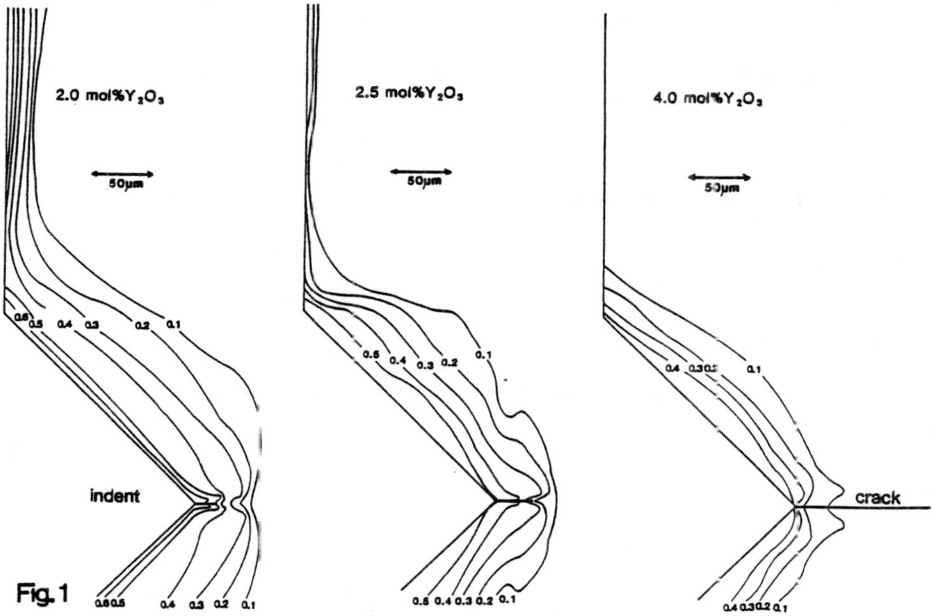

Fig.1

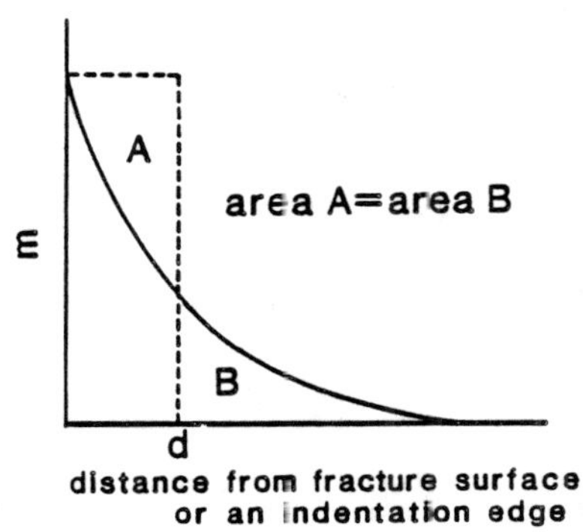

Fig.2

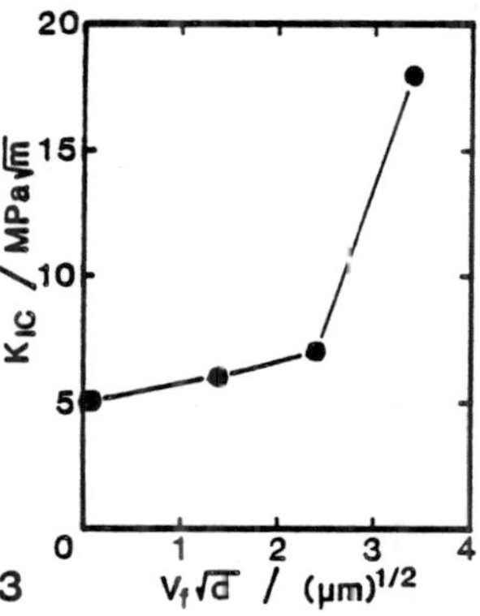

Fig.3

References

1) D. R. Clarke and F. Adar, J. Am. Ceram. Soc., 64, 284 (1982)

2) R. McMeeking and A. G. Evans, J. Am. Ceram. Soc., 65, 242 (1982).

3) T. Masaki, J. Am. Ceramic Soc., to be published in 1986.

DISTRIBUTION AND CHARACTERISATION OF HIGH TEMPERATURE AIR CORROSION PRODUCTS ON IRON-CHROMIUM ALLOYS BY RAMAN MICROSCOPY.

Derek J Gardiner, C John Littleton, K Mark Thomas* and Kenneth N Strafford.

School of Chemical and Life Sciences, Newcastle upon Tyne Polytechnic, Newcastle upon Tyne, UK, NE1 8ST.

* British Gas Corporation, London Research Station, Michael Road, London, UK, SW6 2AD.

Raman microscopy has been used to study the nature and distribution of corrosion products formed on iron and iron-chromium alloys in air at high temperatures. Fe and Fe-Cr alloys containing 2, 5, 14 and 18% Cr were oxidised at 400, 600 and 850°C for 2 hours, in addition samples of each alloy were oxidised for 24 hours at 400°C to obtain thicker scales at this temperature. The corroded samples showed varying distributions of the oxides Fe_2O_3, Fe_3O_4, Cr_2O_3 and $FeCr_2O_4$. Fe_2O_3 and Fe_3O_4 were formed exclusivly on the pure iron and the 2% and 5% chromium alloys at all temperatures and on the 14% chromium alloy at 400°C. The 14% and 18%Cr alloys formed scales containing Cr_2O_3 and $FeCr_2O_4$ at the higher temperatures (600 and 850°C). Examples of small regions of Fe_2O_3 being formed within $Cr_2O_3/FeCr_2O_4$ scales are suggested as possible indications of breakaway corrosion initiation sites.

CONFORMATIONAL ORDER IN LEAD(II) CARBOXYLATES FROM SKELETAL OPTICAL AND LONGITUDINAL ACOUSTIC RAMAN MODES

A.M.Amorim da Costa
Dept. Chemistry - The University
3000 Coimbra - Portugal

INTRODUCTION

Carboxylates of many divalent metal ions are known to exhibit one or two liquid crystalline phases between the solid and the isotropic liquid. For lead(II) carboxylates with chain lengths up to 12 carbon atoms, Adeosun and Sime (1) tentatively assigned the phase sequence as crystal ⟶ G (scmetic) ⟶ V_2 (cubic isotropic) ⟶ Liquid. For the longer chain carboxylates the V_2 phase was suggested to be absent.

In the present work we consider the conformational order information derived from the analysis of skeletal optical and longitudinal acoustic Ramàn modes.

Lead(II) carboxylates were prepared as described elsewhere(1), purified by repeated recrystallisations from benzene, and their purity checked by Pb, C and H elemental analysis. The measurements of the Raman spectra have been also already reported (2).

RESULTS AND INTERPRETATION

i) Skeletal optical modes

Relative to the frequencies, there are no differences between the Raman spectra of the n-carboxylic acids and the correspondent lead(II) carboxylates in the 1000 - 1150 cm^{-1} spectral region. However, from n-alkanes to carboxylic acids (or carboxylates) there is a small decrease (4-6 cm^{-1}) in the frequency of the in-phase C-C stretching, and an increase in frequency of the band usually described as a combination of skeletal stretching and methylene rocking. The frequency of this band increases with the number of carbon atoms in the hydrocarbon chain, either in n-alkanes or in carboxylic acids and carboxylates. In solid carboxylates, such a line appears as a weak shoulder in the case of decanoate, and is pratically undetectable in the case of hexanoate and octanoate.

As the hydrocarbon chain increases there will be a wider distribution of structures which include several gauche rotations, corresponding to an increase of conformational disorder. Thus, we conclude that the substitution of the methyl end group in the n-alkanes chain by the carboxylic group leads to some conformational disorder. As pointed out by Gaber and Peticolas (3) and Pink-Green-Chapman (4), the extent of this conformational disorder can be followed by the intensity ratio between the band described as a combination of skeletal stretching and methylene rocking and the band assigned to in-phase C-C stretching. This intensity ratio can be taken as a measure of the change in the concentration of gauche bonds with conformational changes. The inflexion point of the S-shaped curves which describe that intensity ratio as a function of temperature corresponds to the transition temperature from the scmetic phase to the isomorphous cubic phase for the carboxylates up to 12 carbon atoms, or to the liquid, for the longer carboxylates. The conformational changes involved in the transition from the all-trans crystalline state to the scmetic state is not enough to cause an abrupt change in that intensity ratio.

Quantitatively, taking that intensity ratio as being almost linearly related to the concentration of trans/gauche bonds (3), from the solid to the scmetic phase of the studied carboxylates there is a decrease in the trans concentration of about 20% and from the scmecic phase to the cubic (or the liquid) phase a decrease of about 60%.

In the case of decanoate, where seven gauche bonds in each chain are possible,simple Boltzmann statistics,using an energy difference of 2.1 kJ between gauche and trans states and placing no restriction on the possibility of gauche-gauche sequences,indicate a maximum gauche population of about 34%,at 373K. Combination of this with the Raman results suggests an average of ca. 1.9 gauche bonds per chain. This value is slightly higher than the figures obtained from molecular dynamics calculations or from single chain model.

ii) <u>Longitudinal Acoustic Modes (LAM)</u>

The intensity of the fundamental LAM decreases as the temperature increases. This decrease of intensity with temperature is gradual from the solid to the liquid without any discontinuous change upon the transition from the crystalline solid state to the scmetic phase or from this to the liquid.

The changes in the LAM frequency from n-alkanes to carboxylic acids and lead carboxylates indicate that there is some perturbation on the accordion-like motion of the chain on replacing the methyl group by the carboxylic group, both in the acid and in the salt forms. This perturbation does not seem to be linear to the all-trans chain length. Actually, for chain lengths longer than 12 carbon atoms, the perturbation effect of the end group in the acid is very similar to that in carboxylates, with the LAM frequency being greater than the LAM frequency of the corresponding n-alkanes. It seems that in the rotation of the chains as rigid rods,the carboxylic end group is responsible for a chain twisting which shortens the all-trans chain length.

For chain lengths up to 12 carbon atoms,it appears that the carboxylic end group does not affect so significantly the rotation of the whole chain.In this case,the LAM frequency of the n-alkanes is greater than the LAM frequency of the carboxylic compounds. These observations are consistent with the thought that shorter chains rotate as rigid rods while for sufficiently long chains the rotation of one end can leave the other end unaffected.

REFERENCES

1.S.O.Adeosun and S.J.Sime,Thermochim.Acta,<u>17</u>,351(1976);id.,<u>27</u>,319(1978).
2.A.M.Amorim da Costa,C.F.G.C.Geraldes and J.J.Teixeira-Dias,J.Raman Spectroscopy,<u>13</u>,56(1982).
3.B.P.Gaber and W.L.Peticolas,Biochim.Biophys.Acta,<u>465</u>,260(1977).
4.D.A.Pink,T.J.Green and B.Chapman,Bipchemistry,<u>19</u>,349(1980).

PHOTOELASTIC CONSTANTS OF ARAGONITE FROM RAMAN INTENSITY ANALYSIS OF CALCITE

by

N.Rajeswara Rao and L.D.Wahegoankar,
Dept. of Physics, Osmania University, Hyderabad-7, India.

Recently, Swarnakumari and Rajeswara Rao[1] have described a procedure to evaluate photoelastic constants of calcite from its Raman Intensity analysis. They have also published an erratum[2]. They have derived

$$\Delta\alpha = A'Urle = P'e \quad .....1$$

$$P' = \epsilon P \epsilon \quad .....2$$

Thus, the polarizability tensor α is related to the elasto-optic constants P and the dielectric constants ϵ, and the strains e. A' is transpose of A containing functions of the electrooptical parameters (eops) of the bonds C-O and Ca-O. A is related to the sqare of I, the Raman Intensity.

$$I = LA \quad .....3$$

L,U, r and l are characteristic of the normal coordinate treatment of molecules described in Ref 1. L and A are closely related the crystal structure and can be calculated for any crystal. If the eops of C-O and Ca-O of calcite are taken to be the same as those of aragonite, the values obtained from calcite can be used to calculate the characteristic functions of aragonite and using eq 1, the photoelastic constants of aragonite can be evaluated. Table 1 gies these values.

Table 1

e_{xx}	e_{yy}	e_{zz}	e_{yz}	w_x	e_{xz}	w_y	e_{xy}	w_z
0.0599	0.0767	-0.4190						
0.o502	0.0687	-0.4186						
0.0391	0.0590	0.3448						
			0.0420	0.0421				
					- 0.0426	0.0426		
							-0.005	-.002

It may be appreciated that the strain tensor is not symmetric as was explained by Nelson et al[3].

The symmetry of the P tensor of aragonite differs from that of calcile in many respects. These differences can be easily appreciated if we examine the close connection between the symmetry of the polarizability tensor and the derived P tensor from eq 1. The details are shown in table 2

The symmetry of the calcite shows that its E_g modes

contain both internal and rotational type vibrations. This results in the occurance of $P_{14} = -P_{24}$ in calcite. In aragonite, the rotational modes separate out into different species. Hence, these P elements do not occur. $(P_{11}-P_{12})/2 = P_{66}$ in calcite is a result of xx = -yy = xy in E_g. But in aragonite the degenerate modes separate out into A_{1g} and E_{1g} Hence, this equality is spoiled. P_{44} remains approximately equal to P_{55} in aragonite as its deviation from cylidrical symmetry is small.

Table 2.

	Raman Modes	Raman activity	Photoelastic consts
Calcite	1 A_{1g}	xx = yy ≠ zz	$P_{11}=P_{22}$ $P_{12}=P_{21}$; $P_{13}=P_{23}$
	4 E_g	xx = -yy = xy	$(P_{11}-P_{12})/2=P_{66}$
		xz = yz	$P_{44}=P_{55}$
			$-P_{24}=P_{14}$
Aragonite	4A_{1g}	xx ≠ yy ≠ zz	$P_{11} \neq P_{22}$ $P_{12} \neq P_{21}$
			$P_{13} \neq P_{23}$
	4 B_{1g}	xy	$(P_{11}-P_{12})/2 \neq P_{66}$
	1 B_{2g}	yz	P_{44}
	1 B_{3g}	xz	P_{55}

In the above table, xx,yy etc under Raman activity denote polarizability tensors.

$P_{11} + P_{12}$ is directly a function of the total symmetric line in calcite. Since the deviation from cylindrical symmetry is not large in argonite, a similar relationship holds good between the intensity of the total symmetric line and $P_{11} + P_{12}$. Also, $P_{11} + P_{12}$ is nearly equal to $P_{21}+P_{22}$ inspite of the fact that the crystal has lost its cylindrical symmetry.

References:

1. G.Swarna Kumari and N.Rajeswara Rao, Physical Rev.B Vol. 27, 15 April 1983.
2. G.Swarna Kumari and N.Rajeswara Rao, Erratum ibid, Vol. 29, Sept,1985.
3. D.F. Nelson,P.D. Lazay and M.Lax, Phys.Rev.,B Vol 6, 3109,(1972)

RAMAN SPECTRUM OF NaCN.

T.P. Sharma and R.M. Sharma

Department of Physics, Institute of Advanced Studies, Meerut University, Meerut 250005, India.

Sodium cyanide crystal exists in different phases. At room temperature $(CN)^-$ ions are disordered and the crystal has NaCl structure. At 288K the crystal transforms into a body centered orthorhombic structure. We report here theoretical interpretation of Raman spectrum of NaCN at room temperature based on lattice dynamics. We have calculated phonon dispersion curves of NaCN by using deformation dipole model/1/ after adding the contribution of rotational-vibrational coupling/2/. The model parameters are listed in table 1. The theoretical phonon dispersion curves are shown FIG.1. We find that these are in good agreement with the experimental data/3/.

Raman intensity in a disordered crystal is given by/4/

$$I(\omega) = \frac{n(\omega)+1}{\omega} \sum_j \rho_j(\omega)\{a_j+b_j[1-(q_0\xi)^{-1}\arctan(q_0\xi)]\}. \quad (1)$$

At room temperature, for a completely disordered crystal, the second term vanishes and we have

$$I(\omega) = \frac{n(\omega)+1}{\omega} \sum_j \rho_j(\omega)a_j. \quad (2)$$

The Raman intensity is determined by using this experession, where we have assumed a_j as constant for all the phonon branches. The results are plotted and compared with experimental Raman spectrum of NaCN at 296K/5/ in FIG.2. We find that experimental Raman spectrum resembles with the histogram. We,therefore, conclude that besides the libration motion of $(CN)^-$ ion (ω_{Lib} = 200 cm^{-1})/5/ one phonon process

Table 1. Model parameters

Paramter	Value
Z	0.96
A	10.5985
B	-1.07366
γ_+	0.06224 x 10^{-11} esu
γ_-	-0.61438 x 10^{-11} esu
δc_{11}	-1.4x10^{11} dyne cm^{-2}
δc_{44}	-0.9x10^{11} dyne cm^{-2}

contributes to the Raman spectrum of NaCN.

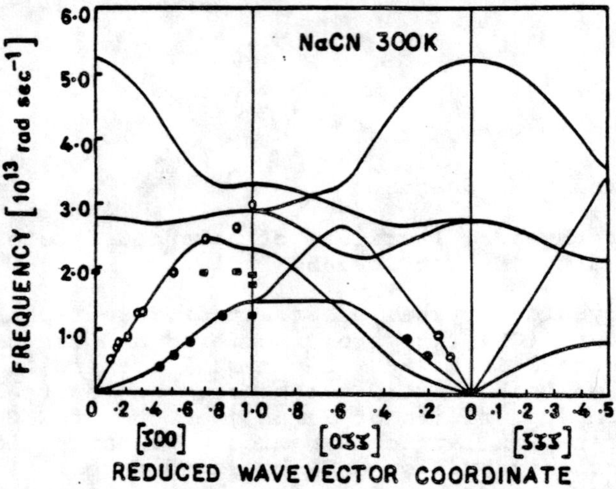

FIG.1. Phonon dispersion curves of NaCN; circles and rectangles represent experimental data/3/.

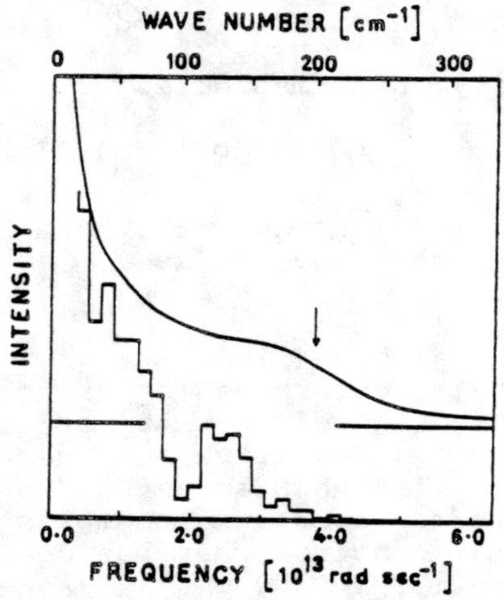

FIG.2. Raman spectrum of NaCN at 296K/5/. Histogram represents theoretical results based on one phonon process.

1. J.R. Hardy, Phil. Mag. 6, 27 (1961); 7, 315 (1962).
2. Strauch et al, Solid State Commun. 30, 559 (1979).
3. Rowe et al, J. Chem. Phys. 62, 4551 (1975).
4. W. Dultz, J. Chem. Phys. 65, 2812 (1976).
5. W. Dultz, H. Krause, and L.W. Winchester Jr., J. Chem. Phys. 67, 2560 (1977).

RAMAN SPECTRUM OF KCN.

R.M. Sharma and T.P. Sharma

Department of Physics, Institute of Advanced Studies, Meerut University, Meerut 250005, India.

At room temperature KCN crystal has a rocksalt structure with $(CN)^-$ ions disordered. We report here our theoretical results of Raman spectrum of KCN at room temperature based on its lattice dynamics. We have modified the deformation dipole model/1/ by incorporating the effect of rotational-vibrational coupling/2/ in it. The model parameters are listed in table 1. The lattice dynamical results are shown in figure 1 where the dotted lines are theoretical results without including the effect of rotational-vibrational coupling. We find that the continuous lines where we have included the effect of rotational-vibrational coupling are in better agreement with the experimental data/3,4,5/. Using these lattice dynamical results we have determined the Raman intensity by the following expression/6/ for a completely disordered crystal,

$$I(\omega) = a\frac{n(\omega)+1}{\omega} \sum_j \rho_j(\omega). \qquad (1)$$

The theoretical results are plotted in the form of histogram and compared with the experimental Raman spectrum of KCN/7/. We find that the histogram resembles with the experimental Raman spectrum of KCN confirming the contribution of disorder allowed one phonon process. The small peaks found in the histogram are not resolved experiment-

Table 1. Model parameters

Parameter	Value
Z	1.0
A	12.4784
B	-1.165
γ_-	$.17726 \times 10^{-11}$ esu
γ_-'	-1.89869×10^{-11} esu
δc_{11}	-1.5×10^{-1} dyne cm^{-2}
δc_{44}	$-.5 \times 10^{-1}$ dyne cm^{-2}

ally at room temperature. Two distinct peaks at 168 cm^{-1} and 190 cm^{-1} were observed besides the central band at 120 cm^{-1} in low temperature Raman spectrum of KCN/6/. We conclude that the Raman spectrum of KCN in high temperature phase is a combined effect of libration mode of $(CN)^-$ ion/6,7 and disorder allowed one phonon process.

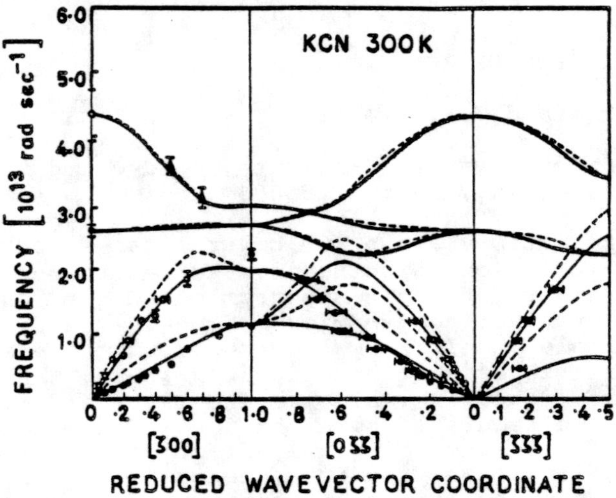

FIG.1. Phonon dispersion curves of KCN; circles and triangles represent experimental data/3,4,5/.

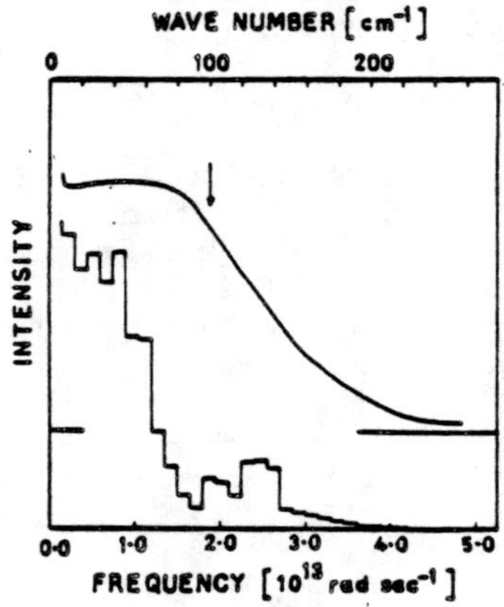

FIG.2. Raman spectrum of KCN at 279K/7/. Histogram represent theoretical results based on one phonon process.

1. J.R. Hardy, Phil. Mag. 6, 27 (1961); 7, 315 (1962).
2. Strauch et al, Solid State Commun. 30, 559 (1979).
3. Rowe et al, J. Chem. Phys. 62, 4551 (1975).
4. Rowe et al, Ferroelectrics 16, 107 (1977).
5. Daubert et al, J. Phys. C9, L389 (1976).
6. W. Dultz, J. Chem. Phys. 65, 2812, (1976).
7. W. Dultz, Solid State Commun. 15, 595 (1974).

RAMAN AND IR STUDIES AND PHASE TRANSITIONS IN SUBSTITUTED AMMONIUM NITRATES AND PERCHLORATES.

M.Mylrajan and T.K.K.Srinivasan
Regional Sophisticated Instrumentation Centre
Indian Institute of Technology, Madras 600 036 (INDIA)

Raman and infrared spectra of substituted ammonium nitrates and perchlorates are recorded through their different structural modifications in the temperature range 77-500 K.

DTA studies of trimethyl ammonium nitrate similar to mono and tetramethyl ammonium nitrate, also exhibit phase transition at 362K. It is monoclinic ($P2_1/C$) with $z = 4$ at 300K. Temperature variation of the FWHH of ν_1', ν_2' and ν_4' modes of the nitrate ion and NC_3 stretching show abrupt changes at the phase transition with hysteresis, characteristic of first order involving reorientation of both cations and anions. Since both the ions are situated in C_1 site, the degenerate modes show splittings in the spectra of phase II at 300K.

Ethylene diammonium dinitrate is triclinic ($P\bar{1}$) at 300K with z = 1. DSC studies show successive phase transitions at 408, 414 and 443K. The lattice mode at 58 cm^{-1} considered to arise from a translation of the nitrate ion coupled to a librational mode, shows a tendency of softening. It is shifted to 67 cm^{-1} at 77K, while above 300K it moves slowly to low frequency and appears at 30 cm^{-1} above 414K. It is therefore inferred that this phase transition has a partial nature of the displacive type transition.

Mono, di and trimethyl ammonium perchlorates (1,2,3 MAP) show two phase transitions above 300K, while DSC studies of tetramethyl ammonium perchlorate (4 MAP) reveals a phase transition below room temperature at 228K. The C-N stretching and internal modes of the perchlorate ion show sudden changes at the phase transition temperature of 1 and 2 MAP. In the high temperature phase I, self diffusion and isotropic motions of cations and anions are achieved. 3 MAP is monoclinic ($P2_1/m$) with z = 2 at 300K. Raman and infrared spectra of thin crystals show changes in position and FWHH of the C-N stretch and internal modes of the perchlorate ion at the phase transitions at 397 and 480K. The vibrational spectra of 4 MAP indicates that freezing out of disorder of the perchlorate ion appears to be responsible for the phase transition at 228K.

BROMATE ION DYNAMICS IN ORTHORHOMBIC, MONOCLINIC AND CUBIC METAL BROMATES.

T.K.K. Srinivasan and T. Devanathan
Structural Chemistry Group
Department of Chemistry
Indian Institute of Technology, Madras 600 036 (INDIA)

The dynamics of bromate ion in a series of hydrated bromates of different crystal structures were probed by Raman spectral studies. $Cd(BrO_3)_2.2H_2O$ belongs to D_2^4 orthorhombic space group, in which the bromate ions and water molecules occupy C_1 sites. Thermal treatment of this compound gives rise to interesting changes in the $\nu_1 - \nu_3$ region of the BrO_3^- revealing unusual dynamics of the bromate ion.

At 298K, while the ν_1 mode at 788 cm^{-1} is quite intense the ν_3 mode at 810 cm^{-1} appears weak. On heating, the intensity of the band at 810 cm^{-1} increases gradually, whereas that at 788 cm^{-1} decreases. The band at 810 cm^{-1} attains high intensity in the anhydrous state at 423K, while the band at 788 cm^{-1} shifts to 795 cm^{-1}, the latter appearing as a weak feature. Loss of water of hydration in $Cd(BrO_3)_2.2H_2O$ can lead to an altered structure. In solids, high temperature form is usually of higher symmetry. The spectral observations in the stretching region along with other internal modes of the bromate ion clearly indicate that the lower site C_1 of the ion at 298K possibly changes to a higher symmetry C_3 at 423 K.

$Sr(BrO_3)_2.H_2O$ conforms to monoclinic space group C_{2h}^6 with the bromate ion situated in C_1 site. The Raman spectra at 298K shows bands at 756, 779, 796, 831 and 841 cm^{-1} in the $\nu_1 - \nu_3$ region. On heating to 356K, where the water molecule is lost, the spectrum is completely changed as shown by the appearance of an intense band at 820 cm^{-1} together with bands of moderate intensities at 806 and 839 cm^{-1}. The number of bands in the $\nu_2 - \nu_4$ bending region are doubled at 356K. Significant modification of the structure as indicated by the disappearance of some of the bands present at 298K in the lattice region together with simultaneous changes in the internal modes of BrO_3^- appears to have taken place at 356K. A comparison between the spectral behaviour on thermal treatment of the hydrates of cadmium and strontium bromates suggest that the dynamics of the bromate ion are different inspite of the fact that BrO_3^- in both these compounds lie in C_1 site. These and similar studies in other bromates will be presented.

LOW FREQUENCY MODES IN INCOMMENSURATE PHASE OF QUARTZ

K.Abe, T.Shionoya and T.Shigenari

Department of Engineering Physics
The University of Electro-Communications
1-5-1,Chofugaoka,Chofu-shi,Tokyo 182,Japan

Recently the existence of an incommensurate (IC) phase between α and β-phase of quartz has been proved from various aspects[1][2][3]. Among many other crystals which show IC phase transitons, quartz seems to be one of the best candidates to study the elementary excitations in the IC phase by means of light scattering spectroscopy, since one can obtain a specimen with excellent optical quality. By a careful polishing of surfaces and avoiding stray scattering, the elastic component (Rayleigh scattering) can be so much reduced that the strong Brillouin components could be clearly seen by human eyes.

Raman and Brillouin spectrum were simultaneously measured along Y and -Y axis, while the Ar ion laser is incident along X axis. All data were taken by slowly cooling a sample from β phase. The temperature where the so-called 'opalescence' takes place was taken as T_c which is about 573°C. Both natural and synthetic quartz with high optical quality gave essentially the same results. Fig.1 show the Raman spectra around T_c in configurations X(ZZ)Y and X(ZX)Y. In the IC phase as temperature approaches T_c, we observed two modes in the low frequency region of X(ZZ)Y spectrum[4]. A mode at 7cm^{-1} appears at $T_i=T_c+2.0$K(T_i is β to IC transition point), while the 12cm^{-1} mode appears at a little lower temperature about $T_c+1.0$K. Their intensity increase as $T\rightarrow T_c$ and disappear below T_c. We notice that the Brillouin doublet,observed in the Raman spectra well above and below T_c, turns into a single peak at about $T_x=T_c+1.5$K, which is between T_i and T_c ($T_c<T_x<T_i$)[5].

In addition to these modes in X(ZZ)Y spectra, a low frequency mode was also found in X(ZX)Y spectrum around 10cm^{-1}.It is observed only at temperature close to T_c and disappears just below T_c. Temperature dependences of these modes,as well as those of Brillouin components are shown in Fig.2 . Only the Rayleigh intensity in X(ZZ)Y spectrum shows a sharp peak at $T_x=T_c+1.5$K, while others show no remarkable change in the IC phase.

Comparing these behavior with the expected Raman activities of the amplitude and phase mode in IC phase,one may assign these modes as follows; 7cm^{-1} and 10cm^{-1} mode are the amplitude mode with A- and E-symmetry, respectively. 12cm^{-1} mode correspondsto a high harmonic mode of the amplitude or phase mode. On the assumption that there exists a '1-q' state between the β- and the '3-q' state[6], the Rayleigh component which increases around T_x might be assigned as the A-symmetry phase mode. Further study of a Goldstone-type phase mode is now in progress.

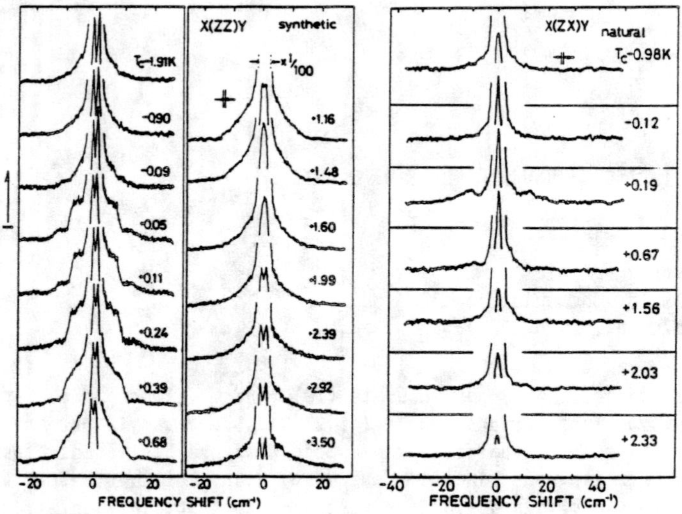

Fig.1

Fig.1(a) Low frequency Raman spectra in X(ZZ)Y. (b) In X(ZX)Y.

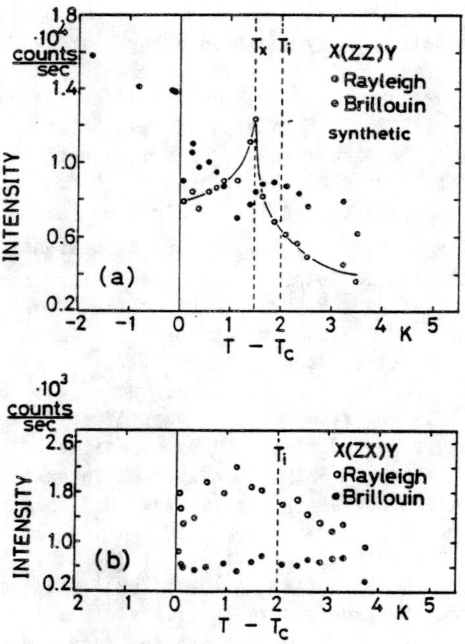

Fig.2(a) Temperature dependence of peak intensities of Rayleigh(○)and Brillouin spectra(●)in X(ZZ)Y (LA). (b) In X(ZX)Y (TA);Rayleigh(○)and Brillouin spectra(●).

References

1) K.Gouhara, Y.H.Li and N.Kato :J.Phys. Soc.Jpn. 52 (1983) 3697.
2) J.P.Bachheimer and G.Dolino : Ferroelectrics 43 (1983) 77.
3) G.van Tendeloo, J.van Landuyt and S.Amelinckx : Phys.Status Solidi(a) 33 (1976) 723.
4) B.Berge, G.Dolino, M.Vallade, M.Boissier and R. Vacher : J.Physique 45 (1984) 715.
5) T.Shionoya, K.Abe and T.Shigenari: J.Phys.C;Solid State Phys. to be published.
6) T.A.Aslanian and A.P.Levanyuk: Solid State Commun. 31 (1979) 547.

RAMAN MICROPROBE (RMP) STUDY OF DIAMOND ANVIL CELL SYNTHETIZED GERMANATE PHASES

H. Boyer *, F. Guyot **

* Instruments S.A., JOBIN YVON
16, rue du Canal, 91163 Longjumeau Cedex, France
** Institut Physique du Globe
4, Place Jussieu, 75230 Paris, Cedex 05, France

Mn GeO_3 pyroxene is considered as a good analogue of natural Mg SiO_3 pyroxene, which is an important constituant of the upper Earth's mantle. At high pressure, it transforms, like Mg SiO_3, to ilmenite and perovskite structures. Vibrational spectra of these phases permit to constrain their thermodynamic and structural properties.

In the RMP spectrum of Mn GeO_3 pyroxene (Fig. 1), the main line at 843 cm^{-1} is attributed to modes characteristic of a tetrahedrally coordinated germanium atom. The corresponding modes of an octahedrally coordinated Ge atom for the ilmenite and perovskite structure of Mn GeO_3 are found at 710 cm^{-1} (Fig. 2 and 3). These germanate spectra are very similar to those of the corresponding silicate phases (pyroxene : White et al., ilmenite : Ross et al.) with peaks shifted to lower frequencies mainly by mass effect.

Until now, no Raman spectrum of the Mg SiO_3 perovskite phase could be obtained probably because of its unstability under normal exciting conditions. Hence, the Mn GeO_3 and Mg SiO_3 perovskite having the same distorted structure, we can assume that the main and higher frequency band in the Raman spectrum of the Mg SiO_3 perovskite should be at 790 cm^{-1}. In this way, it is possible to assign a frequency to the main coordination polyhedron (stronger bands) in a compound and to predict the high frequency cut off of the spectrum necessary for thermodynamic calculations.

REFERENCES

W.B. White; in C. Karr, Ed., Infrared and Raman Spectroscopy of Lunar and Terrestrial Minerals, p. 325 - 358, Academic Press, New York (1975)

N.L. Ross, P. Mc Millan; Amer. Mineralogist, 69, 719 - 721 (1984)

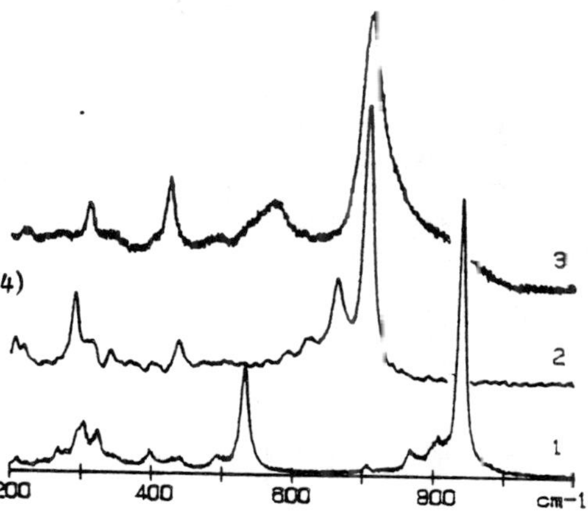

Spectrum 1 - Mn GeO_3 pyroxene,
2 - Mn GeO_3 ilmenite
3 - Mn GeO_3 perovskite

RAMAN MICROPROBE SPECTRA OF NATURALLY SHOCKED QUARTZ

H. Boyer *, B. Velde **

* Instruments S.A., JOBIN YVON
16, rue du Canal, 91163 Longjumeau Cedex, France
** Laboratoire de Géologie, ER 224,
CNRS, Ecole Normale Supérieure, 46, rue d'Ulm,
75320, Paris, France

Raman microprobe spectra were made with the RAMANOR/MOLE U1000 on naturally shocked quartz crystal from the Ries (Germany) meteorite crater. Coexistence of amorphous silica and the high pressure phase coesite, present as submicroscopic crystals, suggest minimum shock pressures of 45 GPa (Stoffler, 1984). Apparent increase in coesite content was accompanied by shifts of 20 cm^{-1} in the major SiO2 glass band envelops.

Comparison of these spectra with normal SiO2 glass and those of densified SiO2 glass, produced under static pressure in the laboratory, (McMillan et al., 1984), indicates that shock and static pressure do not produce the same glass structure. High coesite content shock glass shows an increase in all band frequencies while densified glass (static pressure) shows an increase in the lower band frequency and a decrease in the higher band frequency. Within the same shocked mineral grain about 500 μm diameter one can find zones with spectra of pure SiO2 glass (spectrum 3 in figure), quartz or coesite, as well as mixtures of them. The two latter do not extend some square micrometer.

REFERENCES

D. Stoffler; J. non Cryst. Solids, 47, 465-502, (1984)
P. McMillan, B. Piriou, R. Couty;
J. Chem. Phys., 81, 4234-4236, (1984).

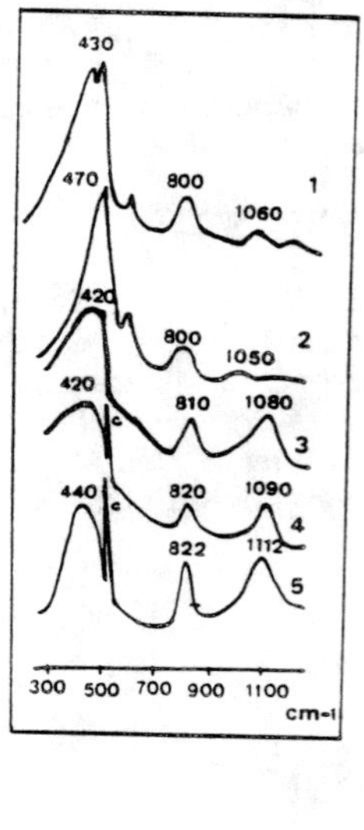

Spectrum 1 - thermal SiO2 glass, from McMillan et al. 1984, 2 - densified glass, idem,
3-4 Raman microprobe spectra of shocked quartz crystal with increasing coesite content, C in figure.

STOKES AND ANTI-STOKES FLUORESCENCE OF Er^{3+} IN THE RAMAN SPECTRA OF ERBIUM OXIDE AND ERBIUM GLASSES

Edgar S. Etz and John C. Travis
Center for Analytical Chemistry
U.S. National Bureau of Standards
Gaithersburg, Maryland 20899, USA

Rare-earth doped materials have been developed which, when excited by infrared radiation, emit in the visible region of the spectrum by what have come to be termed upconversion processes [1]. An example is the trivalent erbium ion, Er^{3+}, incorporated into the host lattice of various phosphors which produce visible fluorescent emissions when stimulated with infrared radiation [1,2]. In microanalytical studies with the NBS Raman microprobe [3] of various materials (crystal hosts, solid solutions, glasses) containing Er^{3+}, we have observed in the micro-Raman spectra excited at 514.5 and 568.2 nm (Ar^+/Kr^+ laser) these reported [1,2] green and red emissions differing in spectral complexities depending on the Er^{3+} host matrix. We ascribe the energetically "forbidden" anti-Stokes fluorescence to a visible-to-visible upconversion, -rarely reported-, but analogous to the infrared-to-visible conversion reported for various Er^{3+} activated host-matrices [1,2]. We are reporting these observations for two of these materials, erbium oxide, Er_2O_3 (<99.99% purity), and one of a series of erbium-bearing glasses, a silicate glass of composition (wt%): SiO_2(48.5), Na_2O(10.9), BaO(9.0), ZnO(4.7), Er_2O_3(26.9). For these samples, the spectra were obtained from single microparticles (size, 10-50 μm) employing power densities (at the sample, in a beam spot of ~10 μm diam.) in the range 5-25 kW/cm^2, recorded with 5 cm^{-1} spectral resolution. In the spectrum of the oxide excited at 514.5 nm (Fig. 1) the fluorescent emissions in the green (~520-~575 nm) and in the red (~640-~700 nm) consist of a complex pattern of sharp bands, relatively so intense that they entirely swamp the Raman spectrum. These green and red emissions reported for Er^{3+} in various host lattices originate from the $^4S_{3/2}$ and $^4F_{9/2}$ electronic manifolds, respectively [1,2,4]. For the glass, this sharp multi-line fluorescence degrades to a pattern of several intense, broad bands. In the spectra excited at 568.2 nm (Fig. 2), the green emissions, --oxide and glass--, appear in the anti-Stokes spectrum at the same wavelengths and in a pattern identical to that observed for 514.5 nm excitation. For both materials, the Stokes side displays the Raman scattering of the sample, free from fluorescence interferences. The Raman spectrum of Er_2O_3 is consistent with literature data from studies where fluorescent emissions were not identified [5,6]. The intense 378 cm^{-1} Raman band (*) of the oxide is the only line detected in the anti-Stokes spectrum. We are investigating possible mechanisms involved in this upconversion such as multiphoton excitation or collisional energy transfer between Er^{3+} ions [1,2,4].

REFERENCES

1. J.C. Wright, "Up-Conversion and Excited State Energy Transfer in Rare-Earth Doped Materials," Chap. 4 in Topics in Applied Physics, Vol 15: Radiationless Processes in Molecules and Condensed Phases, F.K. Fong, Ed., Springer-Verlag, Berlin (1976), p. 239.
2. D.G. Ettinger and T.M. Niemczyk, J. Chem. Phys. 68, 872 (1978).
3. E.S. Etz, Scanning Electron Microscopy/1979/Vol. I, 67 (1979).
4. J.B. Gruber, J.R. Henderson, M. Muramoto, K. Rajnak, and J.G. Conway, J. Chem. Phys. 45, 477 (1966).
5. G. Schaack and J.A. Koningstein, J. Opt. Soc. Amer. 60, 1110 (1970).
6. A.M. Lejus and D. Michel, Phys. Status Solidi (B) 84, K-105 (1977).

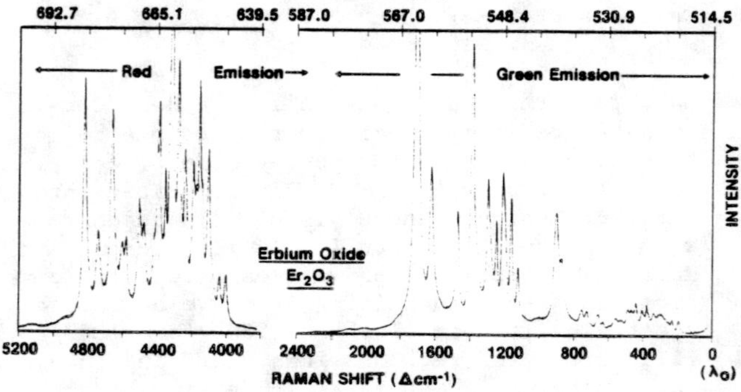

Fig. 1 Raman microprobe spectrum of Er_2O_3 excited at 514.5 nm scanned across the two regions of fluorescent emission.

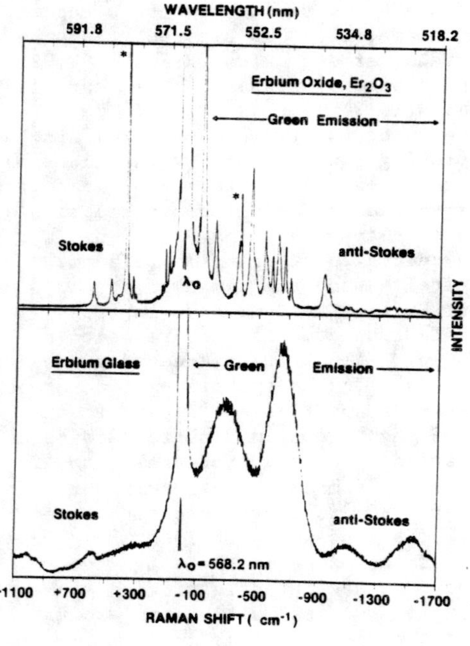

Fig. 2 Micro-Raman spectra of Er_2O_3 and an erbium-silicate glass (~27 wt% Er_2O_3) excited at 568.2 nm showing fluorescence in anti-Stokes region.

RAMAN SPECTRA OF KDP MELTS AND $(KPO_3)_n$ GLASS.

Jong-Jean Kim, Byoung-koo Choi, Ja-Yong Koo and Kwang-Sei Lee

Physics Department, Korea Advanced Institute of Science
P.O. Box 150-Chongyangni, Seoul, Korea

O-H---O Raman bands of KDP melts and KDP single crystals could give informations on the degree of correlation between hydrogen bonds[1] and the off-axis equilibrium sites of hydrogens in O-H---O bonds.[2] We report now the low frequency Raman spectra observed in the various KDP melts including $(KPO_3)_n$ glass.

Melt samples are obtained by melting KDP powders stuffed in the pyrex tube heated in the open air by use of the alcohol burner. Melting temperature was kept as close as possible to the melting point 252°C until complete melting, when the crystal-clear melt was sealed and cooled down slowly to the room temperature (melt 1). Sample melt 2 was given a thermal heating in the 100°C oven for a possible dehydration until the sample became translucent. $(KPO_3)_n$ glass was produced by melting the KDP powder in a quartz tubing directly at 1000°C for 5 hours and then quenching between two copper blocks of room temperature to form a clear glass disk.

We can make the following observations from the spectra :

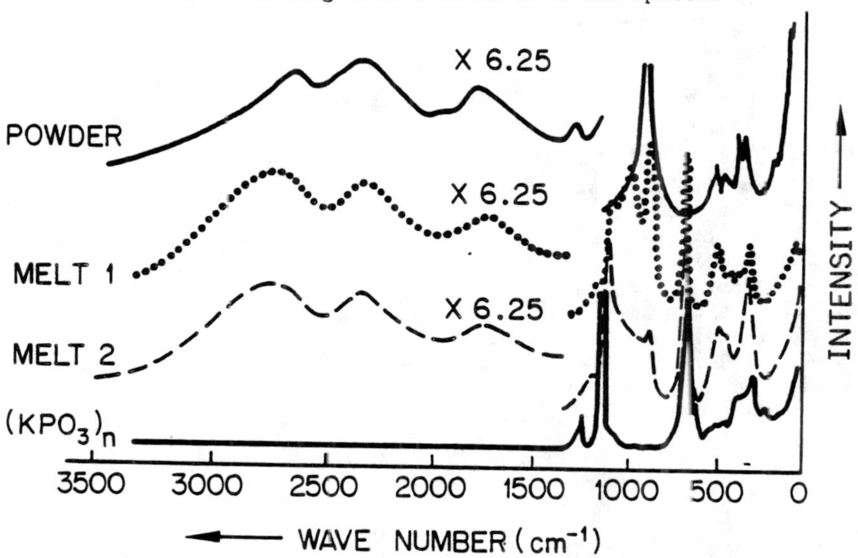

Fig. 1 Raman spectra of KDP powder, KDP melts and $(KPO_3)_n$ glass.

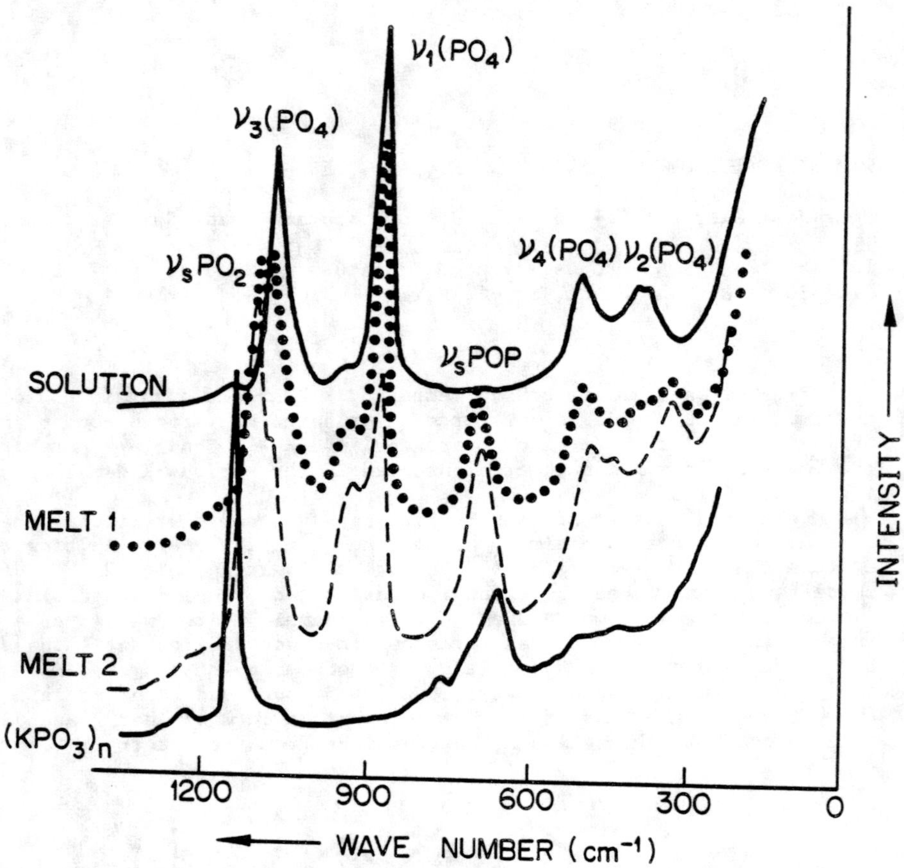

Fig. 2 Raman spectra of KDP solution, KDP melts and $(KPO_3)_n$ glass.

1. No O-H---O bands are observed in $(KPO_3)_n$ sample but O-H---O bonds still exist in melt 2.
2. Gradual changes in frequency and intensity suggests that there are different mixtures of mono, di, triphosphate or higher polymers in the different KDP melts.
3. Melt 1 shows also a broad 40 cm^{-1} peak characteristic of the glassy structure.
4. KDP powder and KDP melts show some difference in the O-H---O bands, which may be understood in terms of the strong coupling theory[3] if we assume that O-H---O bonds are not linear and have a longer bond length in the melts.

References

1) J.J. Kim and B.K. Choi, Sol. St. Commun. 49, 47(1984).
2) B.K. Choi and J.J. Kim, Jpn. J. Appl. Phys. Suppl. 24-2, 914(1985).
3) G.L. Hofacker, Y. Marechal and M.A. Ratner, The Hydrogen Bond, vol. 1 (1976), Chap. 6.

HIGH RESOLUTION RAMAN SCATTERING IN GLASS FORMING LIQUIDS

Dimitrios Samios and Thomas Dorfmüller
Universität Bielefeld, Fakultät für Chemie
Postfach 8640, D-4800 Bielefeld 1, F.R. Germany

Experimental techniques like dielectric relaxation, infrared spectroscopy, nuclear magnetic resonance spectroscopy, time-resolved fluorescence, Raman and Rayleigh scattering are capable of monitoring internal and/or external rotation of molecules. These methods provide us with different information [1] and reflect various aspects of molecular dynamics.

The Rayleigh scattering line and the Raman lines in liquids are shaped by dynamic interactions of molecules with their environment. Depolarized Rayleigh lineshapes are sensitive to collective motions. The Raman process predominates and monitors the single particle molecular motion, however there is some evidence that it can be sensitive to collective molecular motions. In order to obtain reorientational correlation times (RCT) we need the depolarized component of the Rayleigh line. The single particle RCT can be obtained by Raman lineshape analysis [2]. In that case we need both components (polarized and depolarized) of the Raman line under study.

The determination of the width is more difficult in the case of Raman spectra than in Rayleigh spectra where a high-resolution Fabry-Pérot interferometer reduces the instrumental width to 0.002 cm^{-1} which is negligible in comparison to 0.5 cm^{-1} achieved with grating monochromators normally used in a Raman lineshape analysis. In a previous publication [3] we showed that the combination of a grating monochromator and a Fabry-Pérot interferometer (high resolution Raman spectrometer), allows us to determine even very narrow Raman band shapes with high accuracy. It seems that the optimization of this instrumentation is promising and that it will allow us to obtain more accurate information about problems connected to Raman lineshape analysis.

Results

The RCTs (Fig.) of the phenyl group in neat Toluene (▲), neat polystyrene (MW 35000 and 50000) (•), and in 1.4 M/L Biphenyl in Chloroform (×) have been studied by high-resolution Raman lineshape analysis as a function of temperature. A comparison with NMR (o,Δ) and mechanical relaxation (□) data [4,5] in the glassy and solid states indicates that the reorientational motion of the phenyl group continues to persist in the glassy and solid states (γ relaxation) below T_g.

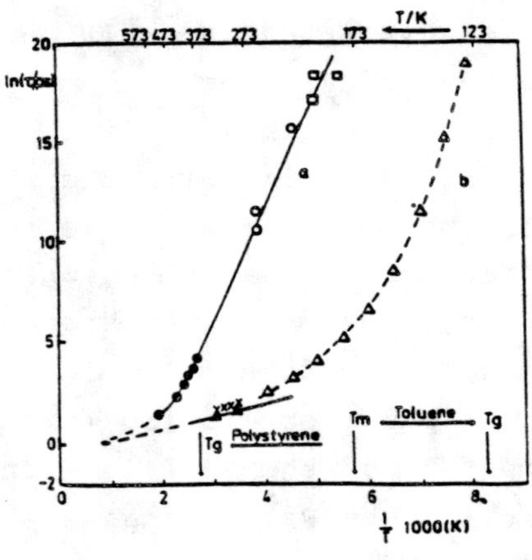

Phenyl group reorientational Activation Energy (Kcal/mol)

States	Toluene	Polystyrene	Biphenyl
Liquid	1.5	2	1.5
Glassy	14	9	-

References

1. B. Berne, R. Pecora, "Dynamic Light Scattering", Wiley-Interscience, N.Y. 1976
2. W.G. Rothschild, "Dynamics of molecular liquids", John Wiley and Sons, N.Y. 1984
3. D. Samios and Th. Dorfmüller, Chem. Phys. Lett. 117, 165, 1985
4. O. Yano and Y. Wada, J. Pol. Sci. A2 9, 669, 1971
5. E. Rössler and H. Sillescu, Chem. Phys. Lett. 112, 94, 1984

OXYGEN-OXYGEN VIBRATIONS IN VITREOUS SILICA.

Frank L. Galeener

Xerox Palo Alto Research Center, Palo Alto, CA 94304, U.S.A.

The structure of vitreous (v-) SiO_2 is still a matter of some controversy, as outlined in several recent publications [1-5]. The most widely accepted model is that the material is chemically ordered, containing only Si-0 bonds, with any wrong bonds (Si-Si, or 0-0) existing only at the concentrations normally associated with defects in solids, viz., less than a few parts per million. A minority of authors have made the interesting suggestion that v-SiO_2 may actually contain wrong bonds at much higher concentrations - greater than or equal to 1 atomic percent. In at least three cases [6-8], the authors have suggested that -0-0- bonds ("peroxl" bridges) may account for one of the two sharp "defect" lines (D_1 or D_2) seen [9] in the Raman spectra of v-SiO_2. In particular, J. C. Phillips [8] has assigned D_1 (seen at 495 cm^{-1}) to the bending mode of a subclass of surface or interfacial peroxyl bridges having a common *minimum* 0-0 distance (and therefore showing a *narrow* vibrational band). He then uses this evidence for peroxyl bridges to provide experimental support for his internal surface model [4] of the structure of v-SiO_2. In this brief report, I shall show that 495 cm^{-1} is actually quite inconsistent with an assignment to ≡Si-0-0-Si≡ vibrations in the glass network.

In a paper quoted by Phillips [8], Khorshev, et al. [10] report infrared spectra of several organometallic peroxides and quote ω(0-0) frequencies for molecules of the form R_3≡Si-0-0-Si≡R_3 where R represents a methyl (Me) or ethyl (Et) group. Note that the atomic mass of a Me group (CH_3) is about 15, and this is quite close to the mass of each of the oxygen atoms that would replace Me groups in the glass. *Thus the methylated molecule is an excellent model for the dynamics of any minimum energy peroxyl group that may exist in the glass.* (Any steric hindrances that act on the Si atoms in the glass can only serve to *raise* the energy of the peroxyl bridge over that of the free methylated peroxide molecule.)

Khorshev, et al. [10], report that the *approximately* 0-0 stretch motion of the above "Me-molecule" occurs at ω(0-0) = 945 cm^{-1}. This changes little when Me is replaced by Et [ω(0-0) = 918 cm^{-1}]. This mode will be one of two dominant Raman active modes of the peroxyl bridge in the molecule.

The other Raman active mode may be pictured *approximately* as in-phase Si-0 stretch motions of both halves of the bridge. This is denoted ω(Si-00) by Khorshev, reminding the reader that the frequency depends *critically* on the fact that each oxygen atom is bonded to one Si and one 0 atom, not two Si atoms, for example. Thus, ω(Si-0-Si) in v-SiO_2 occurs at 420 cm^{-1}, while ω(Si-00) in the Me-molecule occurs at ~833 cm^{-1}.

Both of these Raman active modes, ω(0-0) = 945 cm^{-1} and ω(Si-00) = 833 cm^{-1}, are combinations of symmetric stretch modes [11] and will therefore [11] be highly polarized. No highly polarized Raman line is seen in v-SiO_2 at frequencies above 650 cm^{-1}, so there is in fact no evidence for the peaks to be expected from -0-0- bridges.

Without saying so explicitly, Phillips apparently assumes that the peroxyl bridge frequencies are *much lower* in v-SiO_2, namely ω(0-0) ≡ ω(D_2) = 495 cm^{-1} and ω(Si-00) ≃ ω(Si-0Si) = 420 cm^{-1}, and finds comfort in the observation that ω(D_2)/ω(Si-0Si) = 1.14 is close to the peroxyl ratio 833/945. Since these frequencies in the glass are about half those in the peroxyl bridge *molecule* the force constants must be about *one-fourth* of those in the molecule. This would represent a highly strained and energetic peroxyl bridge, which is quite difficult to imagine in the glass, especially at the high concentrations (~10%) needed in Phillips model for v-SiO_2. Unpublished force constant calculations by the author show that this dilemma is not alleviated even when arbitrary Si-0-0 angles are assumed (from 90° to 180°). [Note that it is quite erroneous to assume that ω(Si-00) ≃ ω(Si-OSi).]

In summary, the known Raman active frequencies of a model Me_3≡Si-0-0-Si≡Me_3 molecule occur at frequencies far above those of any candidate polarized Raman features seen in v-SiO_2. The candidate features actually used by Phillips appear at less than half the expected frequencies, so that the force constants would have to be about one-quarter of those in the real model molecule. This would represent an unreasonably high energy situation. The Phillips interpretation of D_1 = 495 cm^{-1} as a bending mode of a peroxyl bridge is thus found untenable. In addition, neither D_2 nor any other feature of the Raman spectrum of v-SiO_2 is consistent with the known frequencies of peroxyl bridges in the organometallic peroxide model molecules here discussed.

REFERENCES

[1] R. Zallen, *The Physics of Amorphous Solids* (Wiley, New York, 1983).
[2] S. R. Elliott, *Physics of Amorphous Materials* (Longman, London, 1984).
[3] F. L. Galeener, D. L. Griscom and M. J. Weber, *Defects in Glasses* (Materials Research Society, Pittsburgh, 1986).
[4] J. C. Phillips, Solid State Phys. 37, 93 (1982).
[5] F. L. Galeener and A. C. Wright, Solid State Commun. 57, 677 (1986).
[6] E. J. Friebele, D. L. Griscom, M. Stapelbroek and R. A. Weeks, Phys. Rev. Letters 42, 1346 (1979).
[7] A. R. Silin, P. J. Bray and J. C. Mikkelsen, Jr., J. Non-Crystall. Solids 64, 185 (1984).
[8] J. C. Phillips, Phys. Rev. B 32, 5350 (1985); 33, 4443 (1986).
[9] See, e.g., Fig. 2 in Ref. 5.
[10] S. Y. Khorshev, S. E. Skobeleva, A. N. Egorochkin, N. P. Sluchevskaya, N. V. Yablokova and V. A. Yablokov, J. Gen. Chem. (U.S.S.R.) 47, 1247 (1977).
[11] F. L. Galeener, in *Raman Spectroscopy, Linear and Nonlinear*, eds., J. Lascombe and P. V. Huong, (Wiley-Hayden, Chichester, 1982), p. 529.

RAMAN SPECTROSCOPY FOR INDUSTRIAL GLASS MATERIALS AND OPTICAL FIBERS

Fumihiko Fujiki, Seiichiro Higuchi, Shigeyuki Tanaka

Department of Industrial Chemistry, Faculty of Engineering,
The University of Tokyo, Hongo, Bunkyo-ku, Tokyo, Japan

In the present study, we examined the Raman spectroscopy for the characterization of optical fibers and their material glasses. The results obtained suggest that this spectroscopy is one of the most useful method to investigate glass materials and optical fibers.

Optical fibers are made of two kinds of glasses. The core glass has high refractive index, and the cladding is made of low refractive index glass. Both of the glass are made of pure silicate and silicate glass doped some kind of compounds to control the refractive index.

At first, we have measured those material glass. The spectra exhibit clearly characteristic bands of each dopant. When F is doped, a new band appeares at 940 cm-1,when B_2O_3,at 930 cm-1,and GeO_2,at 680 cm-1. The intensity of those charactaristic band increase with the increase of the amount of dopant. Generally, The 800 cm-1 band of quartz glass is used as an internal standard to calculate relative intensities. But as shown in Fig.1, the shapes of 800 cm-1 band changes remarkably when B_2O_3 is added as a dopant, while the shapes of the band remains unchanged when dopant is F, which suggests that the structural interaction between glass and the dopants is different between two dopants. This indicates that in the case of quantitative treatements, the 800 cm-1 band of quartz glass must be treated more carefully, according to the kind of dopants.

Here, the 800 cm-1 band of B_2O_3 doped glass was analized in more detail. It was assumed that this 800 cm-1 band consists of two Gaussian curves.[1] This band shape analysis was made by using of mathematical method and digital computer. The result of this analysis is shown in Fig.2 It is clear that the peak position of these two resolved curves shift with an increase of the amount of dopant. It suggests that the dopant, B_2O_3, induce the change of glass structure.

Next, we have observed spectrum of fiber. Measurement system of optical fiber is different from ordinary Raman scattering system. That is, the laser beam inserted into one end of fiber can be transported to the other end.[2] The Raman scattering light produced in the fiber glass is also transported to the end of fiber. So we can obtain Raman scattering spectra by leading the light which is emitted from the end of fiber to the measurement system. This emitted light include laser and Rayleigh scattering lights. But the monochrometer can separate Raman scattering spectra from other unnecessary lights. Thus, we measured the Raman spectra of fibers whose length is 5-1000 m.

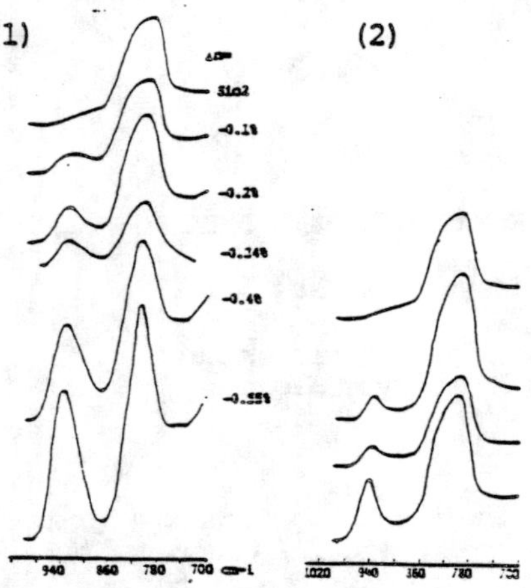

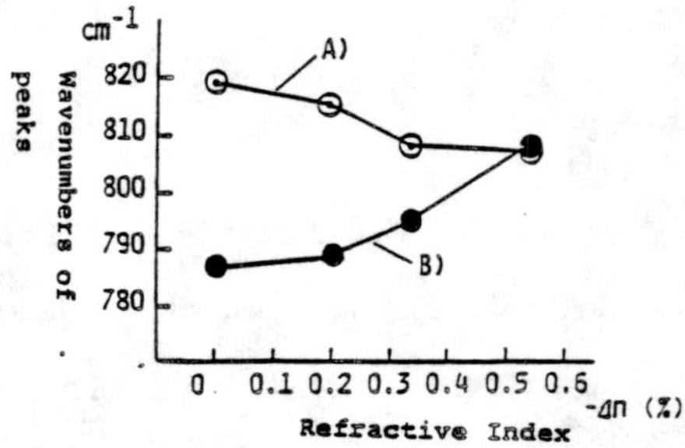

Fig.2 The relations between the refractive indices of B_2O_3 doped glass and wavenumber of component bands

Fig.1 The changes of the shapes of the bands
(1) 800 and 930 cm-1 band by addition of B_2O_3
(2) 800 and 940 cm-1 band by addition of F

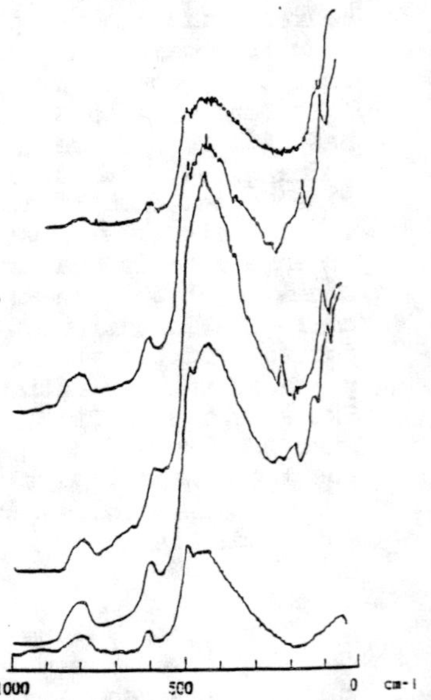

Fig.3 The Raman spectra of the different kind of optical fibers

Fig.3 shows that, as a whole, the spectra of fibers are similar to those of material glass. However, when we see the results in more detail, there are some cases in which the bands which are not observed for material glasses newly appear in 100-500 cm^{-1} region. It indicates that some change of structures occured in the process of manufacturing optical fibers. It proves that the Raman method is very sensitive to the change of glass structures.

Concerning the spectra of optical fibers, some interesting phenomena have been observed. We measured the spectra of two fibers, one is treated with H_2 gas and the other, untreated. These two fibers are the same in other conditions, but their spectra showed clear difference. Another problem is the change of spectral patterns with the measurement time in the case of some optical fibers. We found that some characteristic bands appeared in 100-500 cm^{-1} region, which vanished and the baseline of the spectra became smooth with time. It is not clear from what origin such spectral change occurs

The above-mentioned results suggest that the Raman spectroscopy can be one of the powerful tools for the characterization of industrial glass materials and optical fibers.

1) B. O. Mysen, et al., American Mineralogist, 67, 686(1982).
2) K. Noguchi, et al., Appl. Phys. Lett., 44, 491(1984).

OH STRETCHING BAND SHAPE ANALYSIS AND PROTON DYNAMICS IN OH...O HYDROGEN BONDED CRYSTALS AT LOW TEMPERATURE

Soulard L. and Fillaux F.

Laboratoire de Spectrochimie Infrarouge et Raman
2 rue Henri-Dunant, 94320 THIAIS, France.

OH stretching band-shape analysis in OH...O hydrogen bonded crystals offers an opportunity to get informations concerning the structure, the dynamics and the electrical properties of the hydrogen bond. This communication is focused on two typical examples of rather strong hydrogen bonded systems: potassium hydrogen carbonate ($KHCO_3$) and cesium dihydrogen phosphate (CsH_2PO_4) which show broad and complicated OH stretching Raman band profiles [1,2].

Several band-shaping mechanisms are generally considered to account for the profiles encountered in strong hydrogen bonded systems: mechanical and electrical couplings of the fast OH stretching (n coordinate) and the slow O...O stretching (N coordinate) or O...O bending (Q coordinate) modes of the hydrogen bond, Fermi resonances of the OH stretching with combinations involving internal modes of the complex, Evans interactions, proton tunelling along the hydrogen bond, ...

Recently, a new theoretical approach for infrared and Raman band-shape analysis in hydrogen bonded systems was proposed [3,4] which provides mathematical equations relating the observed profiles with the shape of the Born-Oppenheimer potential, along the N coordinate, in the upper state of the fast n mode. Owing to the strong n,N coupling, the broad features stems from the very fast decrease of the autocorrelation function, the Fourier transform of which determines the spectral profile. The system may be thus approximated by the propagation of a semi-classical wave packet on the upper potential curve. At very low temperature the Raman intensity is:

$$I(\nu) \approx (\nu_e+\nu)^4 \left| \int_{-\infty}^{+\infty} x_{00}(N) M_{01}(N) x_{1\nu}(N)\, dN \right|^2$$

where $x_{00}(N)$ is the wave function in the fundamental state, $M_{01}(N)$ is the transition moment operator integrated over all the internal coordinate of the complex, except M and:

$$x_{1\nu}(N) = \pi^{1/2} \hbar^{-2/3} (2M)^{1/3} |F|^{-1/6} \mathbf{Ai}\left[(N-N_\nu)(2M|F|\hbar^{-2})^{1/3}\right]$$

Ai is an Airy function, F is the slope of the upper potential at frequency ν and M the reduced mass associated with the N coordinate.

Cesium dihydrogen phosphate is a pseudo-one-dimensional ferroelectric which contains two different hydrogen bonds (fig.1). The shorter bond (R =2.48A) links the phosphate groups into chains running along the b axis and the longer one (R =2.54A) crosslinks the chains to form (100) layers. The Raman OH stretching bands due to short and long hydrogen bonds could be identified using different scattering geometries (fig.2): the shortest bond gives the characteristic triplet, usually called ABC bands, centered near 2100cm^{-1} while the longer one shows a strong single band around 2700cm^{-1} and weak narrow components at 2430, 2400 and 1680cm^{-1}. Potential functions for the N mode in the upper state and n,N electrical couplings are determined through band fitting procedure (fig.2).

The same analysis was carried out for the Raman bands of $KHCO_3$(R =2.59A) and the upper potential is quite similar to that obtained previously for the long hydrogen bond of CsH_2PO_4.

References

[1].F.Fillaux, Chem.Phys., **74**, 405, 1983.
[2].F.Fillaux, B.Marchon and A.Novak, Chem.Phys., **86**, 127, 1984.
[3].F.Fillaux, Chem.Phys., **74**, 395, 1983.
[4].L.Soulard and F.Fillaux, unpublished results.

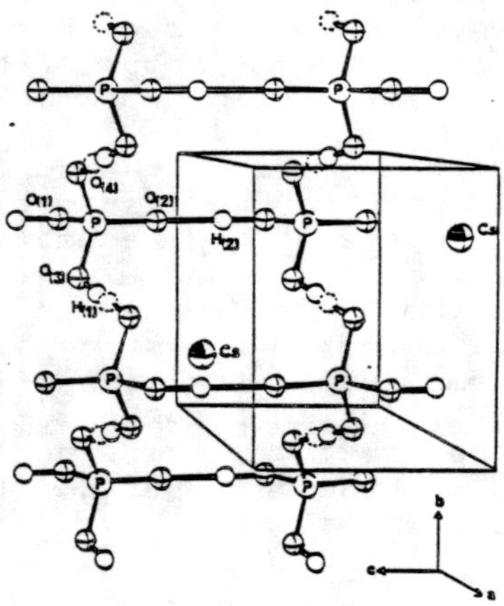

Fig.1.Crystalline structure of CsH_2PO_4.

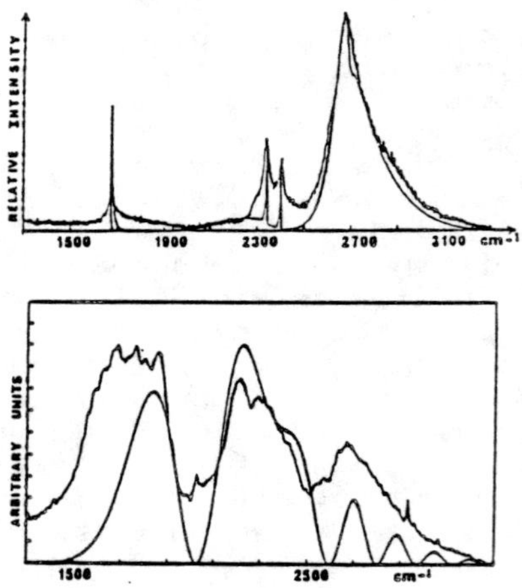

Fig.2.Experimental and calculated Raman band profiles of CsH_2PO_4 at 5K in the OH stretching region corresponding to the short (bottom) and to the long (top) hydrogen bond.

Extended Defect Structures in Bulk As_2S_3 Glass.

M.Malyj and J.E.Griffiths
AT&T Bell Laboratories, Murray Hill NJ 07974

The structure of bulk As_2S_3 glass is widely considered to be an archetypical Continuous Random Network (CRN). Many experimental results have been adequately interpreted in terms of the CRN Molecular Model put forth by Lucovsky and Martin[1]. However, recent experiments, such as Kawamura's study of the low frequency Raman peak as a function of quench conditions[2], have cast some doubt on the sufficiency of the CRN model. In the present study we analyze high frequency Raman bands to determine the effects of differing thermal treatments on the structure of bulk As_2S_3 glass.

A single 280mg sample, sealed in an evacuated quartz ampoule, was subjected to a number of thermal cycles. During cycle 6, the sample was heated to 450°C overnight, and then "air-quenched" by removal from the furnace. In cycle 7, the same sample was slowly cooled within the furnace from 600°C at a rate of 10°/h. In cycle 8, the sample was given a low-temperature anneal at 450°C overnight, then air-quenched, while in cycle 9, the sample was given a high-temperature soak at 600°C overnight, then lowered 50°/h to 450°C before air-quenching. Polarization analyzed Raman spectra were excited by 7525Å laser light at a typical power level of 30mW, and measured at room temperature; photodarkening was not generally evident. Figure 1 depicts the depolarized HV Raman spectra.

Slight changes in the spectra are clearly evident in Figure 1. In comparing cycles 6 and 7, we see that the result of slow cooling is a decrease in the strength of the band at $232cm^{-1}$, commonly assigned to As-As wrong bonds in the structure[3], and simultaneously a slight decrease in the $492cm^{-1}$ band, assigned to the symmetric stretch of S-S wrong bonds[3]. This suggests that the chemical composition of the bulk material remains the same, but that the sample has become more chemically ordered.

Additionally, there is slight growth of the $316cm^{-1}$ band relative to the broad asymmetric band roughly centered at $340cm^{-1}$. This result seems to conflict with the original predictions of the CRN Molecular Model, which assigns these two bands to different vibrations of the same $AsS_{3/2}$ pyramidal unit: $340cm^{-1}$ is designated as the symmetric stretch, and $316cm^{-1}$ the asymmetric stretch[1,4]. Being assigned to the same molecular unit, the two bands ought to move in tandem, but evidently they do not.

In proceeding from cycle 7 to cycle 8, the result of low-temperature annealing is growth of the bands at 232 and $492cm^{-1}$ back to roughly their original amplitudes, suggesting a reinstatement of chemical disorder. But the $316cm^{-1}$ band, instead of shrinking, appears to broaden on the high-frequency side, and perhaps even grows somewhat in intensity. Only after high-temperature soaking (cycle 9), does the Raman spectrum finally revert back to the original form of cycle 6. The results of cycles 6-9 are thus reversible.

Figure 2 is a schematic representation of our Extended Defect Model, which provides a more comprehensive view of the structure of As_2S_3 and other bulk chalcogenide glasses, and explains our experimental results. A fast cooled sample (cycle 6) contains many localized defects, generally structural in character. In the case of As_2S_3, many chemical defects are also frozen into the structure[5]. Naturally, there are spatial fluctuations in the concentration of such defects, resulting in rather indistinct regions of higher and lower strain (regions A and B in Figure 2a).

When a sample is subjected to slower cooling (cycle 7), many localized defects originally present are allowed to coalesce and couple with one another, being replaced by extended defect structures (region B in Figure 2b). These extended defects are one or two (but not zero) dimensional, and are finite in extent; they might be likened to internal surfaces or local line dislocations. The extended defects disrupt the continuity of the bulk glass structure. They have a definitely different character than the adjacent bulk structure, and so are manifested by distinct vibrational modes that are perhaps delocalized in nature[6], appearing as "companion lines" in Raman spectra[7] (the band at $316cm^{-1}$ in As_2S_3). The extended defects allow for a reduction of structural and chemical disorder, and a relaxation of strain.

Under conditions of annealing, localized defects are generated once more. At a low annealing temperature (cycle 8), the extended defect structures still retain their identity, but are placed under stress by the new localized defects (region B in Figure 2c), resulting in a broadening of the associated vibrational bands. Eventually, when soaking the sample at a sufficiently high temperature, the extended defect structures are torn apart, and the sample reverts to its original state (cycle 9).

The Extended Defect Model, as presented here, naturally fits in with paracrystalline or cluster models of glass structure[8]. It can, however, also be considered as a supplement to the Continuous Random Network theory, if one removes one of the major tenets of this theory - that network glass structure is, above all, continuous. Our Extended Defect Model no longer pictures the "ideal glass" structure as continuous in nature, but as containing a significant amount of "scar tissue" in the form of extended defect structures.

References

1. G.Lucovsky and R.M.Martin, J.Non-Cryst.Solids 8-10 185 (1972).
2. H.Kawamura, K.Fukumasu and Y.Hamada, Solid State Commun. 43 229 (1982).
3. P.J.S.Ewen, M.J.Sik and A.E.Owen, in The Structure of Non-Crystalline Materials, ed. P.H.Gaskell, Taylor & Francis, London (1977), p.231.
4. R.J.Kobliska and S.A.Solin, Phys.Rev.B8 756 (1973).
5. F.Kosek, J.Chlebny and Z.Cimpl, Philos.Mag.B47 627 (1983).
6. M.Malyj, G.P.Espinosa and J.E.Griffiths, Phys.Rev.B31 3672 (1985).
7. P.M.Bridenbaugh, G.P.Espinosa, J.E.Griffiths, J.C.Phillips and J.P.Remeika, Phys.Rev.B20 4140 (1979).
8. J.C.Phillips, J.Non-Cryst.Solids 43 37 (1981).

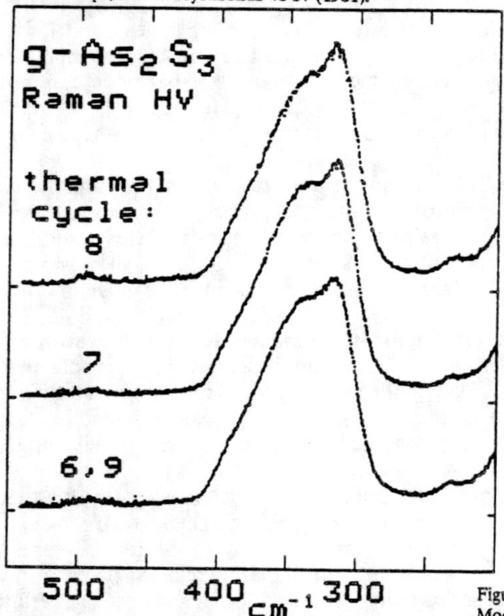

Figure 1. Raman spectra for g-As_2S_3 sample subjected to differing thermal treatments. See text for details of thermal cycles 6-9.

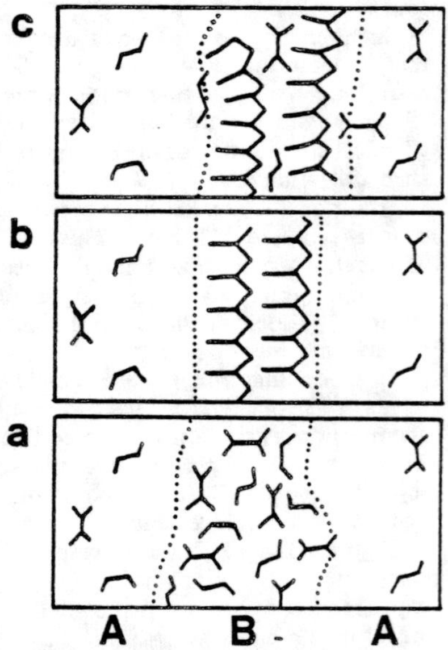

Figure 2. Schematic representation of the Extended Defect Model. Local structures drawn in regions A and B are As-As and S-S chemical defects. Coupled structures in region B represent the extended defects. White space in region A is the pervading bulk structure. See text for further details.

MOLECULAR ORIGIN OF GLASS FORMING TENDENCY IN CHALCOGENIDES

P. Boolchand, R. N. Enzweiler, George H. Lemon, Dave Ruffolo* and J. G. Hernandez+
University of Cincinnati, Cincinnati, OH 45221-0011 USA
*University of Chicago, IL 60637
+Energy Conversion Devices, Troy, MI 48084

Glasses based on the $Ge_{2-2x}Sn_{2x}S_3$ ternary with cation/anion ratio 2/3 rather than the more common ratio of 1/2 or less have attracted basic interest in decoding the origin of glass forming tendency[1] in the chalcogenides. Glasses in this ternary can be prepared in the usual way by water quenching over a wide range of Sn concentrations $0<x<0.62$. Insights into the molecular structure of these glasses has been greatly facilitated by the existence of two crystalline compounds, c-$GeSnS_3$ and c-Sn_2S_3, corresponding to the stoichiometry $x = 1/2$ and $x = 1$ in the present ternary. The molecular structure[2] of c-$GeSnS_3$ (fig. 1) consists of corner sharing chains of tetrahedral $Ge(S_{1/2})_4$ units that are laterally coupled by weak Sn-S bonds. In this structure the covalently bonded zig-zag chains can be characterized as forming a floppy molecular fragment. c-Sn_2S_3[3] on the other hand is made up of infinitely long, thin narrow and planar molecular ribbons (fig. 2) that couple through weaker Van der Waals forces. In the ribbons, two types of ionically bonded (Sn^{4+} and Sn^{2+}) cations occur (fig. 2) and the molecular fragment can be characterized as more rigid in character.

The pivotal role of the floppy c-$GeSnS_3$ fragment in promoting and the rigid c-Sn_2S_3 ribbons in suppressing bulk glass formation in the present ternary has been elucidated in differential scanning calorimetry-isothermal crystallization measurements,[4] Raman scattering and ^{119}Sn Mossbauer spectroscopy experiments. The first clue indicating a close analogy in the molecular structure of c- and g-$GeSnS_3$ derives from the glass transition temperature Tg(x) variation in the present ternary that displays a local maximum at $x = 1/2$ corresponding to c-$SnGeS_3$ stoichiometry.

To understand the existence of a glass forming threshold at $x = 0.62$, the results of Raman scattering and ^{119}Sn Mossbauer spectroscopy experiments are particularly insightful.[1] These experiments performed as a function of composition x in both the bulk alloy glasses and crystalline alloys unequivocally demonstrate that in the compositional window $0.5 \leq x \leq 0.62$, only the zig-zag molecular chain structure occurs with the excess Sn atoms replacing Ge sites in the corner sharing chains. For $x \geq 0.62$, the Sn for Ge replacement in the corner sharing chains disrupts the chain structure on account of strain and leads to nucleation of the Sn-rich ribbons based on c-Sn_2S_3. The drastic loss of glass forming tendency at $x \geq 0.62$ thus correlates exactly with a change in molecular structure of the corresponding crystalline alloys. Specifically the floppy zig-zag molecular chains in which cation and anions bond conforming to the 8-N coordination rule promote glass formation.

Chalcogenide glassy alloy networks largely consist of elastically floppy and rigid regions which are based on specific crystal-like molecular fragments.[5] A bulk glass is obtained when the floppy regions percolate[6] through the solid where as a polycrystalline powder is produced when the rigid regions dominate. These general considerations also permit one to understand[7] glass forming thresholds in the Ge_xS_{1-x} and Ge_xSe_{1-x} binaries at $x \approx 0.42$. Similar behavior but less dramatic because of greater complexity in the phase diagrams has been observed in the glass forming tendency in other chalcogenide glasses.[7]

This work was supported by NSF grants DMR 82-17514 and DMR-85-21005.

1. D. Ruffolo and P. Boolchand, Phys. Rev. Letts 55, 242 (1985).
2. D. Mootz and H. Puhl Acta Cryst. 23, 471 (1967).
3. J. Fenner and D. Mootz Z. Anorg Allg. Chem. 427, 123 (1976).
4. G. H. Lemon, MS Thesis (unpublished) University of Cincinnati 1986.
5. W. Bresser, P. Boolchand and P. Suranyi Phys. Rev. Lett (in press).
6. M. F. Thorpe Jour. Non Cryst. Solids 57, 355 (1983); J. C. Phillips and M. F. Thorpe Sol. State Comm. 53, 699, (1985) J. C. Phillips Jour. Non Cryst. Solids 34, 153 (1979).
7. P. Boolchand, J. Grothaus and J. C. Phillips Solid State Comm. 45, 183 (1983) P. Boolchand, Hyperfine Interactions 27, 3 (1986).

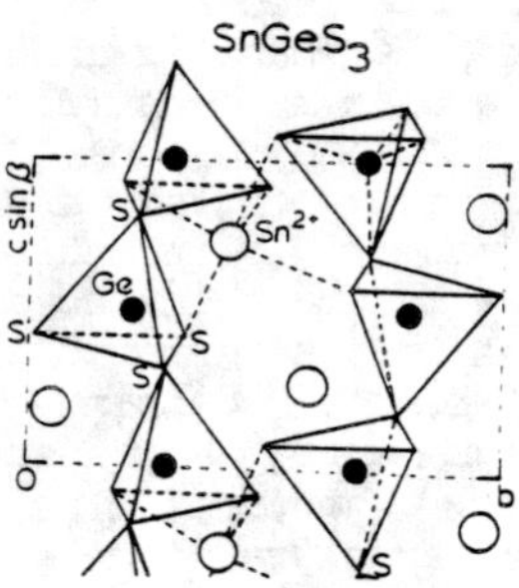

Fig. 1 Crystal structure of $SnGeS_3$ showing zig-zag chains of corner sharing $Ge(S_{1/2})_4$ tetrahedral units running along c-axis.

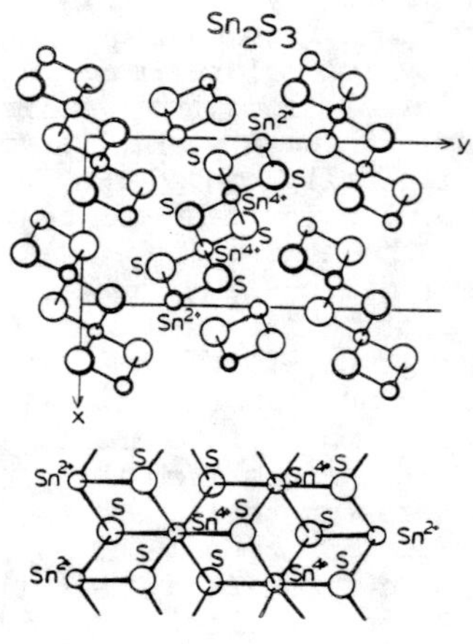

Fig. 2 Crystal structure of Sn_2S_3 showing (top) cross section of the planar ribbons running along z axis. A ribbon has interior (bottom) cation sites (Sn^{4+}) that are octahedral and edge cation sites (Sn^{2+}) that are 3-fold coordinated.

RAMAN STUDY OF LOW TEMPERATURE CURED METAPHOSPHATE COACERVATE- GLASS

G. Palavit*, P. Vast*, A.M. Bera *, G. Vergoten**, P. Lagant**
* Laboratoire de Chimie Inorganique Appliquée Bat C8 Université des Sciences et Techniques de Lille Flandres Artois F 59 655 Villeneuve d'Ascq
**Laboratoire G. B. M. U 279 INSERM Faculté de Pharmacie Lille France

1 - Introduction

The application of colloïdal state of metaphosphoric polyanions (polyelectrolytes) gives us two different kinds of products : coacervates and gels [1]. There are many attractions to develop a process for curing metaphosphoric coacervates, gels at low temperatures (≈100°C). We don't know existence of some metaphosphate glasses below 400°C. The interpretation may be postulated on structural informations of these states.

First, it has been well established that polyphosphates form chain structures in the solid state.

In solution, these chain polyphosphates exist as helixes of PO_4 tetrahedra [2].

Secondly, Umegaki and Kanazawa [3] suggested the presence of tightly coiled highpolyphosphate chains by measurements on dilute solutions of coacervates.

In our case[4], by differential scanning calorimetry, the Cp curve of coacervate shows a second order transition and a peak of dissolution in the non-bounded water in the range of temperature (70 - 90°C).

Raman spectroscopy seems to be a good tool of investigation with an important bibliography on the Raman studies of metaphosphate glasses[5].

2- Experimental

Experiments were performed onto OMARS 89 Multichannel device (DILOR Lille France) using 351 nm excitation line of 2000 Spectra Physic Argon ion laser .

3- Results and discussion

Figure 1 shows the Raman spectra of the coacervate 0.57 Na_2O - 2.4 CaO - 3 P_2O_5 - x H_2O at various temperatures .

At 20°C , the bands observed in the frequency range 500 -1500 cm^{-1} were caused by chain vibrations .(PO_3^-)n.

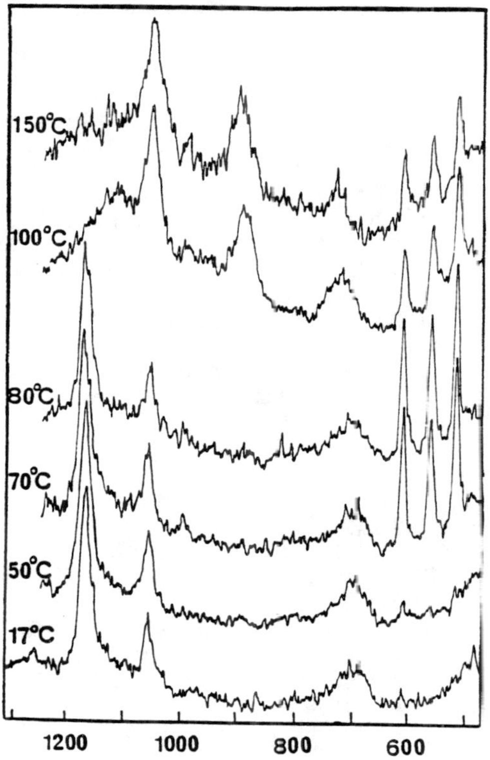

fig. 1

The bands are observed and assigned as as (OPO) at 1272 cm^{-1}, s (OPO) at 1165 cm^{-1}, as (POP) at 1055cm^{-1}, s(POP) at 700 cm^{-1} and skeleton deformations at 613 cm^{-1}, 563 cm^{-1}, 520 cm^{-1} such as bending modes.

The intensity of the 613 cm^{-1}, 563 cm^{-1}, 520 cm^{-1} bands varie considerably during heating (fig. 1), the transition appeared at 70°C.

Changes on configurations seem to depend on the ratio of helixes in coils.

The mechanism can be illustrated by fig.2

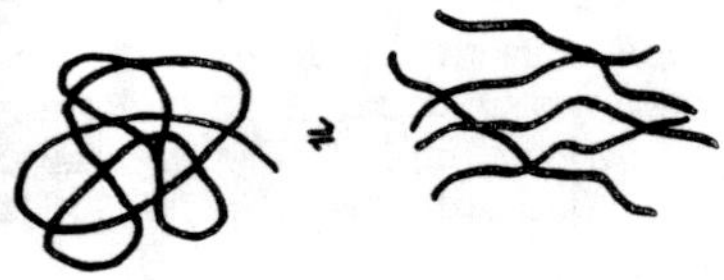

fig. 2

This is confirmed by ^{31}P NMR of the coacervate versus temperature. Those differences don't depend upon the formation of hydrolysis products like short metaphosphate cycles or pyrophosphates below 80°C.

At higher temperatures (100°C), a band appeared at 900 cm^{-1}. We explain this change as Garrigou - Lagrange and coworkers do (6) for $LiPO^3$ by a Davidov splitting. The 1053 cm^{-1} adsorption in the coacervate is caused by the as POP vibration, split because the Ca^{2+} ion introduces a dissymetry in the skeleton.

At these temperatures some hydrolysis occur.

The changes in the spectrum (shiftings from 700cm^{-1} to 732 cm^{-1}, from 1165 cm^{-1} to 1115cm^{-1}, new band at 996cm^{-1}, ...) come from a gradual breakdown to shorter chains in the polymeric structures as it was explained by Bertoluzza and coworkers (7) for oligophosphate glasses.

After being heating at 100°C, the coacervate is quenched to obtain a low temperature glass. Raman spectra of this new compound is given in figure 3.

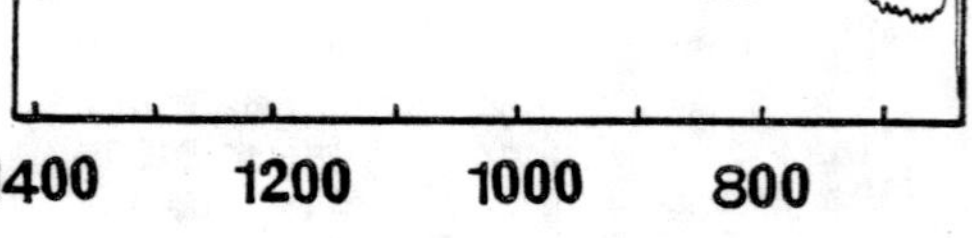

References:

(1) M. DRAOUI, P. VAST, G. PALAVIT, Revue de chimie minérale, 22, 256 (1985)
(2) T. GLONEK, J.R. VAN WAZER, M. MUDGETT, T.C. MYERS, Inorg. Chem., 11, 567 (1972)
(3) T. UMEGAKI, T. KANAZAWA, Bull. Chem. Soc. Jpn, 48(5), 1452 (1975)
(4) G. PALAVIT, P. VAST, A.M. BERA, M. DRAOUI, to be published
(5) P.J. MILLER, J. Chem. Phys., 71(2), 997 (1979)
(6) C. GARRIGOU-LAGRANGE and coworkers, Can. J. Chem., 63, 1436 (1985)
(7) A. BERTOLUZZA, C. FAGNAGNO, J. Ram. Spectr., 11(1), 10 (1981)

RAMAN SCATTERING FROM A NEW CLASS OF DISORDERED SOLIDS.

L. A. Farrow and C. J. Sandroff

Bell Communications Research, Inc.

A debate has raged for over half a century as to whether amorphous solids should be described as a continuous random network or as a collection of randomly oriented microcrystallites[1,2]. Here we describe a different approach to the study of solid state materials composed of microcrystalline subunits. Rather than preparing a non-crystalline substance by melt-quenching or vapor deposition, we synthesize materials by allowing microcrystallites in a colloid suspension to aggregate into a solid. Specifically, we have prepared colloidal suspensions of the layered semiconductor PbI_2, and have shown that it is possible to limit the growth of the crystallites to a single layer sandwich, ≈7Å thick, with in-plane dimensions of 12, 18, and 29Å, depending on the solvent used[3]. If the solvent is gently evaporated, the colloidal particles aggregate to form a finely powdered precipitate, composed of microcrystalline subunits with well-characterized dimensions.

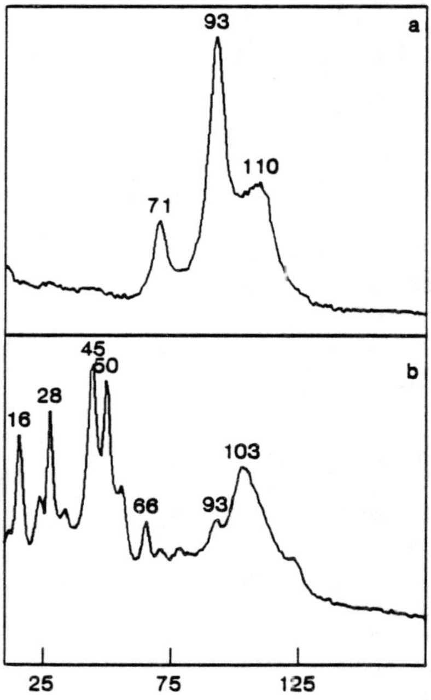

Fig. 1. Raman spectra of PbI_2 (Stokes shift in cm^{-1} vs. intensity in arbitrary units). a. Single crystal. b. Aggregate.

Raman spectroscopy has proven effective in studying the nature of disordered or amorphous materials[4–6], and therefore is used here to investigate the structure of the PbI_2 aggregate. A micro-Raman apparatus[7] concentrated the 647.1nm radiation from a Kr^+ onto the samples. spectral resolution was $1\,cm^{-1}$. Fig. 1a shows the spectrum from a PbI_2 single crystal, whose layered structure consists of two hexagonally close-packed layers of iodine sandwiching a similar layer of lead. The strong band at $93\,cm^{-1}$ is the totally symmetric vibrational stretch, with the lead sheet remaining stationary, and the band at $71\,cm^{-1}$ is the doubly degenerate shear mode. Fig. 1b is the Raman spectrum of the aggregate; it is obvious that these bands have virtually disappeared. Equally startling, there appear a series of strong bands at lower frequency shifts.

Some insight is gained concerning the origin of these intense bands by considering Raman spectra of higher order polytypes[8] of PbI_2. Table I compares vibrational data from bulk studies of PbI_2 crystallized in various polytopic forms with results obtained on the aggregated material. The anomalous strength of these bands in the aggregate is consistent with a structure comprised almost exclusively of microcrystalline domains with lateral extent of ~29Å and a thickness averaging three PbI_2 layers. The virtual disappearance of the bulk bands implies a loss of the 2D symmetry found in the crystalline solid. We propose that the layer symmetry is destroyed by the bonding of the excess iodine to the large number of lead atoms located at the microcrystallite periphery. To our knowledge this is the first direct spectroscopic evidence for distortion in the intrinsic structure of microcrystalline domains due to edge effects, and it could be a general feature in a broad class of disordered materials.

References

1. R. Zallen, *Physics of Amorphous Solids*, Wiley, New York, 1983.
2. J. C. Phillips, *J. Non-Cryst. Solids*, **43**, 37 (1981)
3. C. J. Sandroff, D.M. Hwang and W. M. Chung, *Phys. Rev. B*, in press
4. P. M. Bridenbaugh, G. P. Espinosa, J. E. Griffiths, J. C. Phillips, and J. P. Remeika, *Phys. Rev. B*, **20**, 4140 (1979)
5. J. E. Griffiths, G. P. Espinosa, J. P. Remeika, and J. C. Phillips, *Phys. Rev. B*, **25**, 1272 (1982)
6. J. E. Griffiths, M. Malyj, G. P. Espinosa and J. P. Remeika, *Phys. Rev. B*, **30**, 6978 (1984)
7. J. B. Hopkins and L. A. Farrow, *J. Appl. Physics*, **59**, 1103 (1986)
8. J. I. Hanoka and V. Vand, *J. Appl. Physics*, **39**, 5288 (1968)
9. M. Y. Khilji, W. F. Sherman and G. R. Wilkinson, *J. Raman Spec.* **13**, 127 (1982)

Table 1. Raman frequencies in PbI_2 (cm^{-1})

polytype			aggregate
2H	4H‡	6H‡	
---	14 m	12 m	16 s
---	29 w	29 m	28 s
---	47 w	46 m	45 s
---	---	50 m	50 s
---	---	65 m	66 m
---	72 m	71 s	---
70 s	74 m	73 w	---
93 s	96 s	94 s	93 vw
105 m	106 m	109 m	103 m

‡Reference 9.
vw(very weak), w(weak), m(moderate), s(strong)

ABOUT THE SHARP INCREASING OF LOW FREQUENCY RAMAN SCATTERING SPECTRAL INTENSITY NEAR THE PHASE TRANSITIN POINTS IN CRYSTALS

V.S. Gorelik

P.N. Lebedev Physical Institute, Academy of Sciencies, 117924, Moscow, Leninsky pr., 53, USSR

We report about the effect of the spectral Raman scattering intensity sharp increasing(in 10^2- 10^3 times)near the structural phase transitions in crystals for low frequency region: 1-10 cm^{-1}. Such effect was detected with the help of isofrequency (at fixed frequency and slow temperature changing) procedure and was investigated for a number of crystals: SiO_2, $BaTiO_3$, $KNbO_3$, $LiNbO_3$, $LiTaO_3$.

The sharp intensity increasing takes place for nonelastic scattering of light and so the discussed effect is different from the known static opalescence phenomenon and can be classified as nonelastic(dynamic)opalescence. Such effect have been observed not only for monocrystals, but for crystalline powders and solid mixture systems too. Figure gives us the illustration of spectral intensity increasing for several frequencies(in cm^{-1} at the Figure)near the phase transition points of $KNbO_3$.

The explanation of such phenomenon is given on the base of soft mode theory and classical nonelastic scattering of light description /1/.

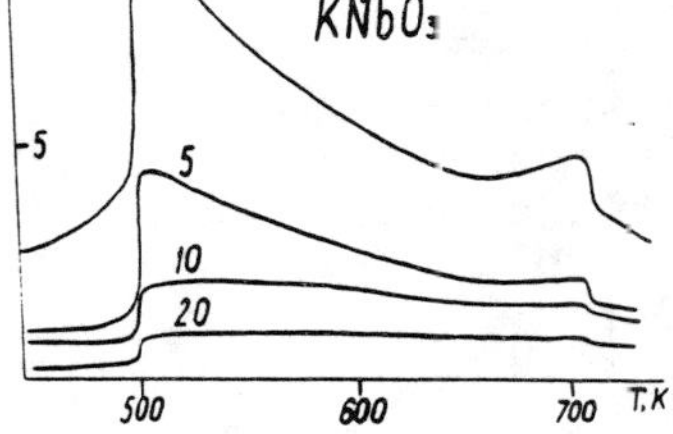

The observed spectral intensity sharp increasing gives us the new sensitive tool for phase transition detection and investigation. Besides, such effect opens the opportunity of stimulated scattering generation at low threshhold near the phase transition in crystals.

Literature.

/1/ Gorelik V.S, Izv. of the Acad. of Sciencies, USSR, ser. phys., 1985, v.49, p.282-286.

RAMAN SCATTERING OF LIGHT IN AMORPHOUS AND CRYSTALLINE FILMS OF SILICON AND CARBON

V.S.Gorelik, S.I.Sagitov, T.P.Phaizullov
P.N. Lebedev Physical Institute, Academy of Sciences, 117924, Moscow, Leninsky pr., 53, USSR

The pecularities of Raman scattering in micron and submicron silicon and carbon films, received at dielectric and metallic substrates at different conditions are reported.

The silicon films have been prepared by diffusing on quartz substrates at high vacuum; the thickness of such films was 6-60 nm. Then we have used the laser annealing of such films with the help of YAG:Nd-laser second harmonic and pulsed copper laser. At the last case the annealing took place in the 10 μm spot. The sample was scanned under laser beam relating to this spot.

As a result of Raman investigations we recorded the sharp enough silicon Raman peak at 517 cm^{-1} and so made the conclusion about the cristallisation of silicon films after annealing. The observed Raman frequency decreasing (on 4 cm^{-1}) was explained as a result of expending of the film. If so the corresponding stress was $10^9 H/m^2$.

The carbon coating have been prepared in discharge at carbonhydrogene atmospher on monocrystall silicon substrates. The shape of Raman spectra in this case is illustrated on Figure. As can be seen, the distinct diamond Raman peak at 1332 cm^{-1} is revealed. Besides, there is the wide band with small intensity at 1550 cm^{-1}; such band is due to amorphous or grafite phases.

So with the help of Raman investigations we have found the conditions for receiving of crystalline silicon and "quasi" diamond films on different substrates.

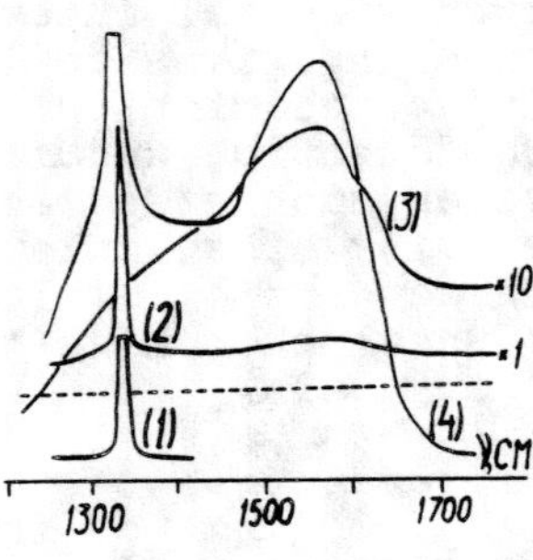

Figure 1.
(1)-diamond monocrystall, (2), (3)-"quasi" diamond film, (4)-amorphous carbon film.

chapter 12

SOLID STATE: ORGANIC CRYSTALS AND POLYMERS

TIME-RESOLVED STUDIES OF PHONON DYNAMICS IN MOLECULAR CRYSTALS

Jeffrey R. Hill, Eric L. Chronister*, Jay C. Postlewaite, and Dana D. Dlott

School of Chemical Sciences
University of Illinois
Urbana, Illinois 61801

* Current Address: Department of Chemistry, University of California, Los Angeles, CA 90024

Research supported by the National Science Foundation, Solid-State Chemistry by grant DMR 84-15070, and a fellowship from the Alfred P. Sloan foundation to Dana D. Dlott.

Stimulated Raman scattering with picosecond pulses has been used for many years to study vibrational dynamics in condensed matter. Recent development of tuneable high power, high repetition rate dual dye laser systems has greatly increased data acquisition rates permitting investigations of the effect of chemical structure, vibrational frequency, temperature, and solvent effects on vibrational relaxation and dephasing.

In this work, we discuss results obtained on aromatic molecular crystals. Coherent Raman scattering has been used to study vibron and phonon decay in naphthalene, anthracene, and durene at low temperature. Lifetimes as long as 100 ps have been observed. Increasing temperature leads to a decrease in lifetime, the onset of "pure" dephasing processes, and ultimately vibrational localization.

Vibrational dynamics in disordered materials are studied using pure crystals of asymmetrically deuterated naphthalene molecules and binary solid solutions of naphthalene and various deuterated naphthalenes. In highly disordered systems, the low temperature vibrational lifetimes may be decreased by an order of magnitude by disorder effects and vibrational energy transfer between the various chemical species.

Energy transfer between a vibrationally excited molecule [naphthalene] and the host crystal matrix [durene] has been studied using picosecond photon echo spectroscopy and stimulated Raman scattering. It is found that nearly isoenergetic vibrations of host and guest molecules can have vastly different vibrational lifetimes. The factors which control these lifetimes will be discussed.

References

Schosser, C.L.; Dlott, D.D. J. Chem. Phys. 80, 1394 (1984).

Chronister, E.L.; Dlott, D.D. J. Chem. Phys. 79, 5286 (1983).

Chronister, E.L.; Hill, J.R.; Dlott, D.D. J. Chimie Phys. 82, 159 (1985).

Dlott, D.D. Ann. Rev. Phys. Chem. 1986 [in press].

Hill, J.R.; Chronister, E.L.; Postlewaite, J.C.; Dlott, D.D. Ultrafast Spectroscopy V [in press].

RAMANSPECTROSCOPIC INVESTIGATIONS ON MOLECULAR CRYSTALS: PRESSURE AND TEMPERATURE DEPENDENCE OF EXTERNAL PHONONS IN NAPHTHALENE-D_8 AND ANTHRACENE-D_{10}

W. Häfner[1] and W. Kiefer[2]

1) Physikalisches Institut, Universität Bayreuth, D-8580 Bayreuth, Federal Republic of Germany
2) Institut für Experimentalphysik, Karl-Franzens-Universität Graz, Universitätsplatz 5, A-8010 Graz, Austria

In the last few years there has been a considerable interest in the study of anharmonicities in molecular crystals. Particularly spectroscopic techniques (Raman scattering, neutron scattering and far-infrared absorption) have been applied in combination with hydrostatic pressure to lattice dynamical studies of molecular solids. Since molecular crystals such as naphthalene or anthracene experience a volume compression of 10% or more when hydrostatic pressure of the order of 1 GPa is applied, this kind of study provides the opportunity of influencing and therefore investigating intermolecular bondings in molecular crystals.

Theoretical "ab-initio" calculations done by Jindal and Kalus /1/ based on superimposed atom-atom-potential of Van-der-Waals type

$$V(r_{ij}) = -A\cdot r_{ij}^{-6} + B\cdot \exp(-\alpha r_{ij}) \qquad (1)$$

(r_{ij} is the distance between two atoms of different molecules; A, B, and α are parameters, which have to be varied for describing the C-C, C-D or D-D interaction between the molecules) require the knowledge of mode Grüneisen parameters γ_i for the lattice modes, which are defined as

$$\gamma_i = -d(\ln\omega_i)/d(\ln V) = (1/\varkappa)\ d(\ln\omega_i)/dp \qquad (2)$$

where ω_i is the frequency of mode i, V the crystal volume (extrapolated from data in the literature /2,3,4/), $\varkappa$ the compressibility and p the pressure. These quantities have been experimentally investigated until now only at room temperature /5,6/. We have undertaken a comprehensive study of naphthalene-d_8 and anthracene-d_{10} by means of linear Raman spectroscopy in the pressure range up to 1 GPa and simultaneously at temperatures from ambient down to 10 K. Fig. 1 shows e.g. Raman-spectra in the external mode region of naphthalene (A) and anthracene (B) measured for various pressures at 65 K and 85 K, respectively.

These data enabled us to derive a linear temperature dependence of the mode-Grüneisen parameters for some of the Raman-active -rotational lattice modes at the Γ-point of the Brillouin-zone (Fig. 2).

Regarding the temperature induced frequency-shift of the lattice modes, one easily recognizes a twofold dependence. First an implicit part, which is due to the change in crystal volume and second, an explicit part, which depends on the thermal population and the change in selfenergy (eqn. 3).

$$\frac{d\omega}{dT} = \left(\frac{\partial\omega}{\partial V}\right)_{T=const.} \frac{dV}{dT} + \left(\frac{\partial\omega}{\partial T}\right)_{V=const.} \quad (3)$$

The latter was calculated Jindal and Kalus /1/. Our data for the mode Grüneisen parameters allow to determine the implicit (volume) part in the whole temperature range (0-300 K), so that it is possible to derive the explicit part as difference of the total and the implicit temperature induced frequency-shift (fig. 3 A:naphthalene B:anthracene). A comparison between theory and experiment shows good agreement for the high frequency modes, whereas for the low frequency modes the explicit shift differs in sign.

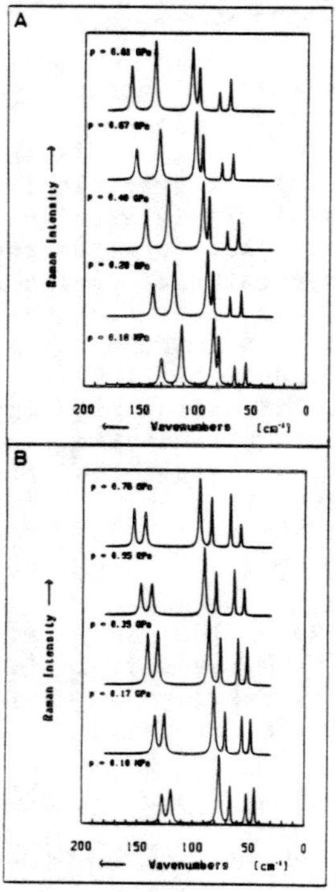

Fig. 1 Raman Spectra of
(A) naphthalene-d_8 at 65K
(B) anthracene-d_{10} at 85K

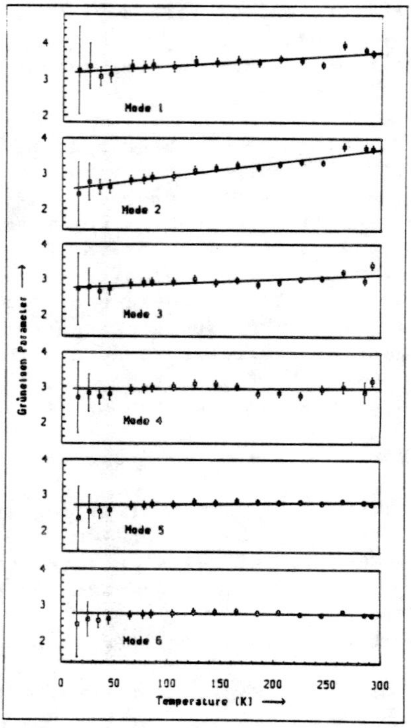

Fig. 2 Temperature dependence of the Grüneisenparameters for the lattice modes of naphthalene-d_8

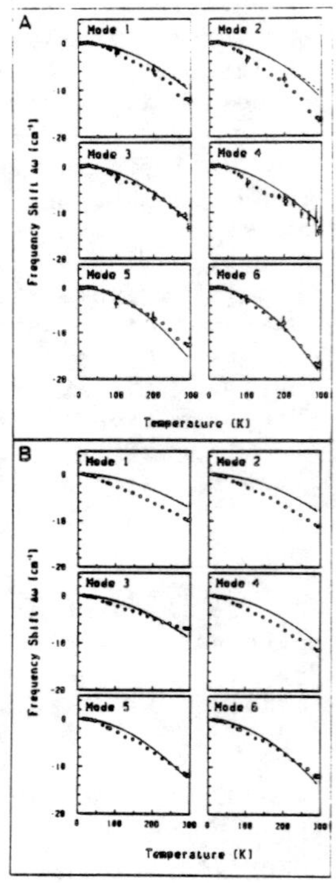

Fig. 3 Comparison of the total (•) temperature induced frequency shift and the implicit part (-)

REFERENCES

/1/ Jindal V. K., Kalus J., J. Phys. Chem. C.: Solid State Physics, 16, 3061 (1983).
/2/ Alt H. C., Dissertation, Universität Bayreuth (1981).
/3/ Vaidya S. N., Kennedy G. C., J. Chem. Phys., 55, 987 (1971).
/4/ Jordan J. F. J., Axmann A., Egger H., Kalus J., Phys. Stat. Sol. (a), 71, 457 (1982).
/5/ Backer M., Häfner W., Kiefer W., J. Raman Spec., 13,247 (1982).
/6/ Nicol M., Vernon M., Tak Woo J., J. Chem. Phys., 63, 1992 (1975).

POLARIZED RAMAN AND INFRARED STUDY OF 5-AMINOTETRAZOLE MONOHYDRATE.

Adriano Bigotto

Department of Chemical Sciences,University of Trieste,Trieste, Italy

Although the crystals of 5-aminotetrazole monohydrate have been exhaustively studied from the structural point of view (1,2),no detailed interpretation of the vibrational spectra has been reported in the literature.Investigations on the infrared and Raman spectra were mainly directed to the elucidation of the problem of the amino-imino tautomerism in this molecule.In particular,some different patterns observed in the Raman spectra of the hydrated and anhydrous materials have been invoked as an evidence of an amino-imino reversible interconversion in the solid state (3).Our study of the polarized Raman spectra of single crystals,combined with deuteriation and low temperature measurements of both Raman and infrared spectra has led to a different interpretation for these patterns in terms of Fermi resonance effects (4).These results stress the importance of a detailed interpretation of the Raman and infrared spectra for a correct understanding of the chemical properties of the molecule. The aim of the present work is to provide this interpretation on the basis of the polarized Raman and infrared measurements on single crystals.

Crystals of 5-aminotetrazole monohydrate are monoclinic, space group $P2_1/c$ (C^5_{2h}),with four formula units in the primitive cell.An extensive three-dimensional network of hydrogen-bonds connects the 5-aminotetrazole molecules to each other and to the water molecules.Each molecular fundamental gives, for k=0,two Raman active components (A_g and B_g) and two infrared active components (A_u and B_u).24 Raman active and 21 infrared active external modes are also predicted.

The analysis of the polarized Raman spectra enabled the unambiguous attribution of seven in-plane ring vibrations at 1636,1453,1301,1158,1098-1058 (F.R.), 996 and 762 cm^{-1}. The assignments of the vibrations associated to the groups affected by hydrogen-bonding was carried-out by combining the polarized Raman data and the results of the polarized absorption and reflection measurements.For example,the ring N-H stretching gives an A_g component of noticeable intensity at 2650 cm^{-1} in the -(cc)- polarization (Fig.1):the corresponding infrared band,which moves at 2040 cm^{-1} upon deuteriation, is almost completely polarized along the c crystal axis both in the absorption and ATR spectra (Fig.2),in agreement with the predictions of the oriented gas model.

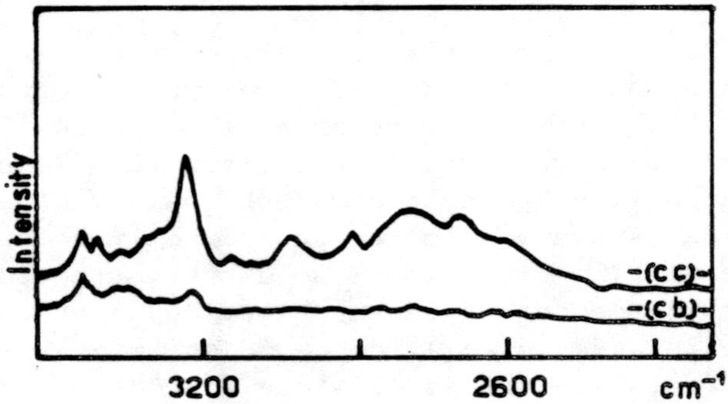

Fig. 1. Single crystal Raman spectra

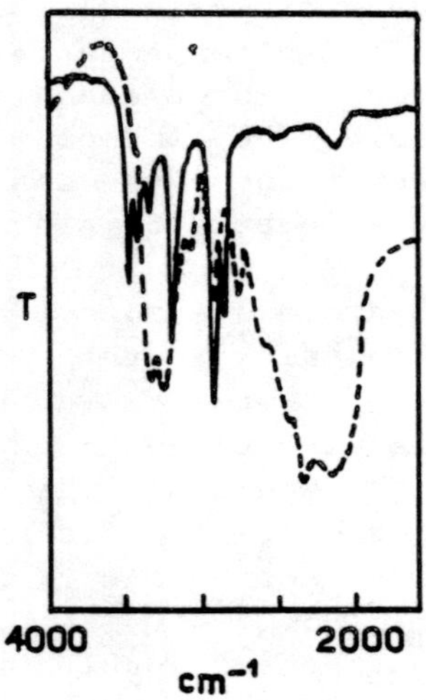

Fig. 2. Single crystal ATR spectra;
solid line: TE,E ∥ b
dashed line: TE,E ∥ c

REFERENCES

1) K.Britts and I.L.Karle, Acta Cryst. 23,308(1967)
2) D.D.Bray and J.G.White, Acta Cryst.B35,3089(1979)
3) J.H.Nelson and F.G.Baglin, Spectrosc.Letters 5,101(1972)
4) A.Bigotto and G.Castellani, J.Raman Spectrosc. 15,232(1984)

RAMAN SPECTRA OF N-H STRETCHING MODES IN UREA CRYSTAL.

[1]M.Y. Khilji
[2]W.F. Sherman and G.R. Wilkinson

[1]Physics Department, University of Karachi, Karachi, Pakistan
[2]Department of Physics, King's College London, London, England.

Urea is a commonly known molecular organic crystal. The vibrational spectra of urea has been extensively studied during the past two decades (1-3), and the assignment of the lattice modes of urea is now quite well established (2,3). However, very little has been published on the high frequency Raman spectra of urea crystals (1). The present work is concerned with the Raman and IR spectra of internal (N-H stretching) modes of urea in the frequency region 3100 - 3500 cm^{-1}.

Urea crystallizes in a tetragonal structure and the primitive unit cell contain two planar molecules with symmetry D_2d. Hence 36, of the 48 degrees of freedom of the unit cell are due to the internal modes of urea. These 36 internal modes are: $7A_1+2A_2+2B_1+7B_2+9E$; where A_1 and B_1 are Raman active, B_2 and E are both Raman and IR active and A_2 is a silent mode. There are eight N-H bonds in the unit cell and these result in eight well-defined N-H stretching degrees of freedom which are represented by $2A_1 + 2B_2 + 2E$. These modes are schematically shown in Figure 1. Out of these, two ($2A_1$) are Raman active only and four ($2B_2 + 2E$) are both Raman and IR active. Thus, we expect six bands in the Raman spectrum (and four bands in the IR spectrum) due to the N-H stretching vibrations.

Polarised Raman spectra of urea single crystals in the frequency region 3100 - 3500 cm^{-1} at room temperature and liquid nitrogen temperature are shown in Figures 2 and 3. There are two bands observed at 3250 and 3420 cm^{-1} in Y(XX)Z, X(YY)Z and X(ZZ)Y polarisations and are assigned to A_1 mode. Two pairs of bands (3350, 3490 and 3260, 3434 cm^{-1}) have been observed. Their corresponding IR frequencies 3347, 3485 and 3259, 3440 cm^{-1} respectively, were observed and were subsequently assigned to the B_2 and E type modes. These assisgnments are summarized in Table I.

INFRARED		RAMAN		
297K	80K	297K	77K	Assignment
	-	3246	3250	A_1^2
3259	3258		3260	E^5
3326	3327	3318.8	3319	$61+1626+1626(A_1)$
3349	3347	3352.7	3349.8	B_2^2
			3420	A_1^2
		3426		
w	3440	-	3434	E^6
-	-	w	3448	$2 \times 1720(A_1)$
w	3485	3480	3490	B_2

Table 1.

Assignment of the vibrational modes of urea crystal.

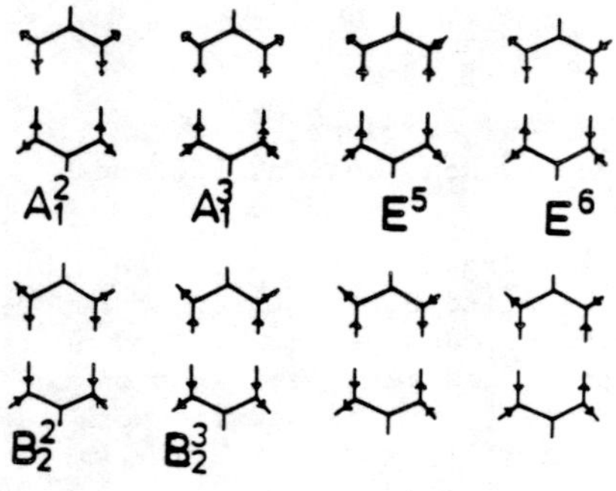

Fig. 1. Sechematic view of the N-H stretching modes of urea molecule in the crystal.

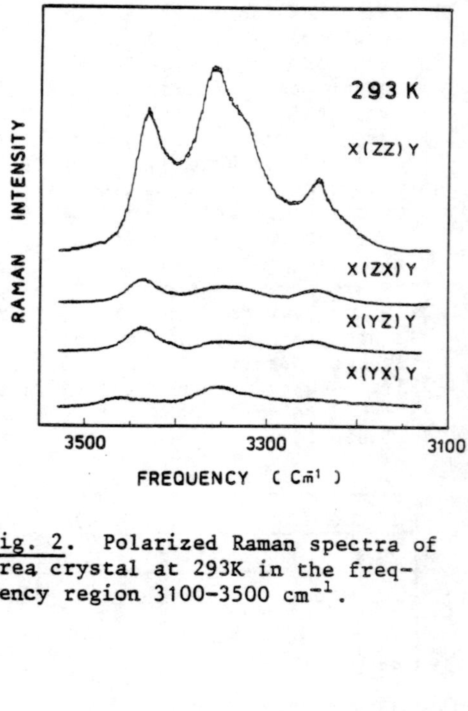

Fig. 2. Polarized Raman spectra of urea crystal at 293K in the frequency region 3100-3500 cm^{-1}.

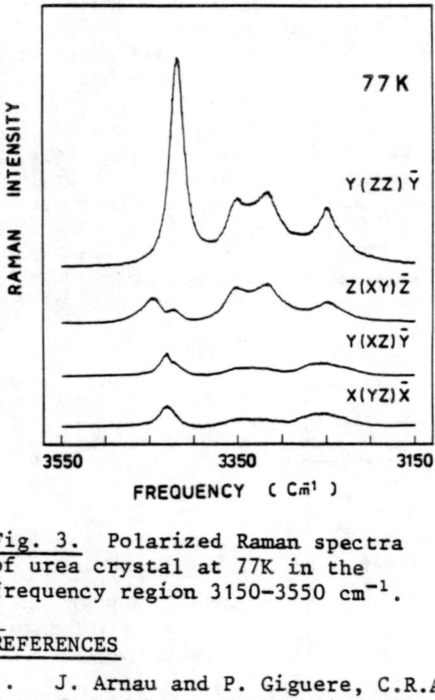

Fig. 3. Polarized Raman spectra of urea crystal at 77K in the frequency region 3150-3550 cm^{-1}.

REFERENCES

1. J. Arnau and P. Giguere, C.R.Acad.Sc.(Paris) 268, 245, (1969).
2. J. Lefebvre, H. Fontaine and R. Fouret, J.Raman.Spectr. 4, 173, (1973).
3. M.Y. Khilji, W.F. Sherman and G.R. Wilkinson, J.Mol.Struc. (1986).

RAMAN STUDY OF THE LATTICE VIBRATIONS OF METHANOL*.

A. Anderson, B. Andrews, E. Meiering and B.H. Torrie

Guelph-Waterloo Program for Graduate Work in Physics, Waterloo Campus, University of Waterloo, Waterloo, Ontario, Canada, N2L 3G1

Methyl alcohol is an interesting example of a simple hydrogen-bonded organic crystal. It is known to exist in at least three different solid modifications(1). There is a low temperature ordered α-phase stable below 160K, a disordered β-phase stable between 160K and the melting point, 175K, and a glassy phase obtained by vapour deposition at low temperatures, which transforms to the α-phase at about 130K(2). There is also some evidence that another phase exists over a small temperature range (3K) near the α-β transition (3). Crystal structures of the α and β phases have been investigated by Tauer and Lipscomb using X-ray diffraction (4).

Only one previous Raman study of solid methanol has been found in the literature (5). On the other hand, a number of infrared studies have been reported, the most comprehensive by Falk and Whalley (6). Later measurements have included the far infrared region (7). In the present study, we report on the Raman spectrum of the lattice region of the α-phase at 20K, a region not examined in detail in earlier work, presumably because of the intrinsically weak scattering signals. Spectra of the deuterated species, CH_3OD and CD_3OD, have also been recorded to aid in the assignment of the observed peaks.

Low frequency Raman spectra of the three crystals at 20K are shown in the Figure and peak frequencies are listed in the table with ratios of frequencies of corresponding peaks. From the known molecular geometry (8), values of the principal moments of inertia for the three isotopic species have been calculated. These and the ratios of their square roots, together with similar ratios for the molecular masses have been used for the assignments given in the right hand column of the table. Deviations from the theoretical ratios may be due to mixing between modes of the same species and similar frequencies or to changes in the hydrogen bond strength on deuteration.

The number of lattice peaks observed is not compatible with the monoclinic unit cell, space group C^2_{2h}. containing 2 molecules on mirror planes, suggested by Tauer and Lipscomb (4). However, these authors found some evidence of a super-lattice, and, of course, their X-ray study did not locate hydrogen positions. Further spectroscopic work is in progress, including Raman spectra of the internal modes and far infrared spectra of the lattice modes, on crystals at lower temperatures than in previous studies. It is hoped that these investigations and a planned neutron diffraction structural determination will lead to a fuller understanding of the vibrational properties of these interesting systems.

*Research supported by grants from NSERC, Canada.

Lattice Frequencies of CH_3OH, CH_3OD and CD_3OD at 20 K

(1) CH_3OH	(2) CH_3OD	(3) CD_3OD	RATIOS (1)/(2)	RATIOS (1)/(3)	MODE* TYPE
54.7	54.6	50.7	1.00	1.08	T
64	64.2	60.1	1.00	1.06	T
80.9	79.0	74.1	1.02	1.09	T or L
85.3	85.3	87.9	1.00	0.97	T
101.2	102.6	80.9	0.99	1.25	L
119.5	119.9	102.0	1.00	1.17	L
133.0	132.0	116.0	1.01	1.15	L
147.1	144.9	129.1	1.02	1.14	L
167.9	168.8	147.1	0.99	1.14	L
180	180.5	161.4	1.00	1.12	L

* T = Translation; L = Libration.

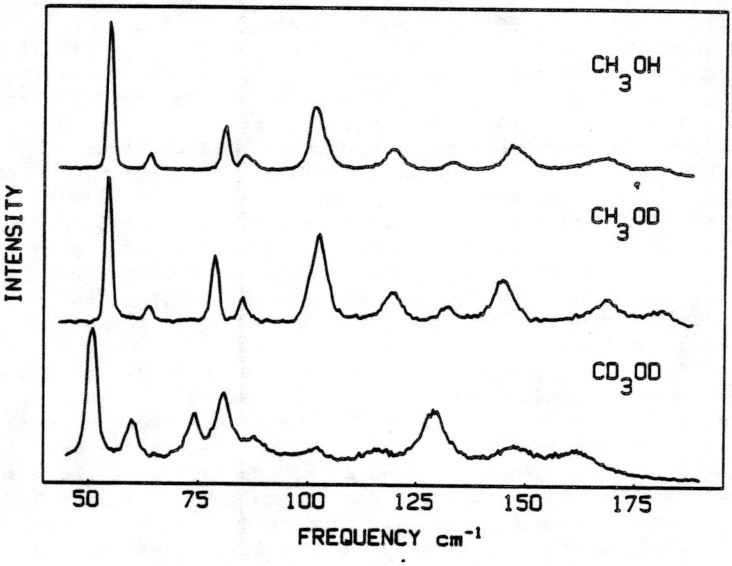

References

1. G.S. Parks, J. Amer. Chem. Soc. 47, 338 (1925).
2. A.B. Dempster and G. Zerbi, J. Chem. Phys. 54, 3600 (1971).
3. L.A.K. Staveley and M.A.P. Hogg, J. Chem. Soc. 1013 (1954).
4. K.J. Tauer and W.N. Lipscomb, Acta Cryst. 5, 606 (1952).
5. J.R. Durig, C.B. Pate, Y.S. Li and D.J. Antion, J. Chem. Phys. 54, 4863 (1971).
6. M. Falk and E. Whalley, J. Chem. Phys. 34, 1554 (1961).
7. W.F. Passchier, E.R. Klompmaker and M. Mandel, Chem. Phys. Lett, 4, 485 (1970).
8. P. Venkateswarlu and W. Gordy, J. Chem. Phys. 23, 1200 (1955).

RAMAN AND FAR INFRARED SPECTRA OF CRYSTALLINE METHYLENE IODIDE.

B.H. Torrie*, A. Anderson, B. Andrews, D.G. Laurin, J.K. White and W.W.E. Zung

Guelph-Waterloo Program for Graduate Work in Physics, Waterloo Campus, University of Waterloo, Waterloo, Ontario, Canada N2L 3G1

The lattice vibrations and internal modes of polycrystalline CH_2I_2 and CD_2I_2 have been studied as a function of temperature using Raman and far infrared spectroscopy. The results show that CH_2I_2 has two stable phases, one of which is metastable in the temperature range of the other, and a second metastable phase. CD_2I_2 has only one solid phase.

Raman spectra of the lattice region of CH_2I_2 in the three phases are shown in Figure 1. In order to clarify the conditions under which the three phases are obtained, a phase diagram is given in Figure 2. Above 258K there is a stable phase (phase I) which transforms reversibly but sluggishly (~½ hour) to a second phase (phase II) if the temperature changes are not too rapid. A sudden drop in temperature causes phase I to be retained and this metastable phase I' transforms reversibly to a second metastable phase (phase III) near 50K. On warming, phase I transforms to phase II near 200K. The spectrum of the deuterated sample is similar to that of phase I of the undeuterated sample. Comparison with the spectra of CH_2Br_2 (1) shows that it is this phase which is isomorphous to the stable phase of CH_2Br_2.

The structure of CH_2I_2 was determined at about -20C (253K) (2) which is close to the transition temperature but on the phase II side. The evidence cited above, however, indicates that the determined structure corresponds to that of phase I, i.e. phase I has the same structure as CH_2Br_2 (also described in reference 2). In this structure there are four molecules on general sites in the centrosymmetric primitive unit cell. The splittings of the internal modes and number of external modes are consistent with this structure.

It is interesting to speculate on the structure of the low temperature stable phase (phase II) of CH_2I_2. The internal modes show no splitting and the Raman lattice spectrum is much simpler with only 3 peaks being observed. For both internal and external modes there is coincidence or near coincidence between Raman and infrared frequencies. These results are consistent with a molecule being on a site of C_{2v} symmetry in a unit cell belonging to the point group C_{2v}. A neutron diffraction study of CH_2I_2 is in progress (3) which will test these predictions.

The structure of phase III must be very closely related to that of phase I. Internal mode frequencies and intensities are different but the splittings of the internal modes are similar. The low temperature of the transition also suggests that there is only a minor rearrangement of the molecules. Further confirmation of the similarity of these phases comes from the neutron diffraction results which showed nearly identical powder patterns for the two phase III and I'.

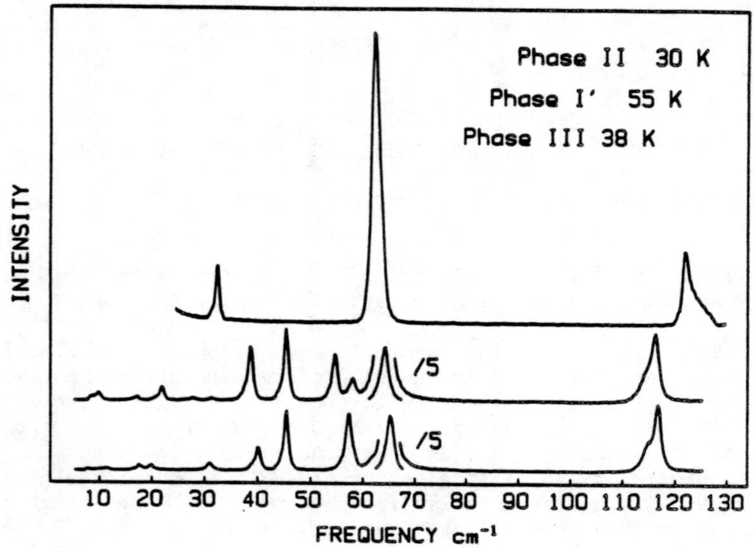

Figure 1 Raman spectra of the lattice region of CH_2I_2

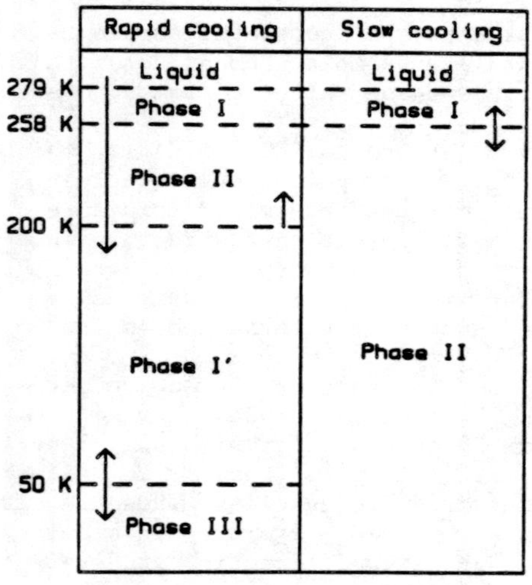

Figure 2 Phase diagram of CH_2I_2

References

1. A. Anderson, B.H. Torrie, D.J. Danagher, D.G. Laurin, J.K. White and W.W.E. Zung, J. Raman Spectrosc., in press.
2. T. Kawaguchi, V. Schettino and S. Califano, J. Chem. Phys. 45, 1400 (1966).
3. B.H. Torrie, P.N. Gerlach and P.M. Powell - work in progress.

METHYL DYNAMICS AND MOLECULAR MECHANISM DETERMINING PHASE TRANSITIONS IN 4-METHYL-PYRIDINE

Le Calvé N., Pasquier B., Soulard L. and Fillaux F.

Laboratoire de Spectrochimie Infrarouge et Raman, 2 rue Henri-Dunant, 94320 Thiais, France.

This communication deals with phase transitions and molecular dynamics in the 4-methyl-pyridine ($CH_3-C_5H_4N$) crystal and its d_7 analog. The purpose is to understand the phonon scattering processes which monitor the thermally activated motions involved in the phase transitions.

This molecule is interesting since beside the low frequency external modes generally involved in phase transitions, the methyl group linked to the aromatic ring has a very low potential barrier for internal rotation which may be strongly perturbed by crystal environment. The large amplitude motions of this group are thus related to the crystal symmetry and dynamics.

Three crystal phases in 4-methyl-pyridine are observed by infrared and Raman between 5-100K (III), 100-254K (II) and 254-276.8K (I). In phase III, where the site symmetry for the molecule is C_1 [1], different potential-shapes for methyl torsion are compared and only a non-periodic function (fig.1) accounts satisfactorily for the five τCH_3 and two τCD_3 bands observed in Raman and by inelastic neutron scattering [2] (the corresponding levels are denoted ▲ and ■, respectively). A molecular model is proposed where the methyl groups are correlated along infinite chains in the crystal (fig.2). The non-periodic potential corresponds to the out-of-phase rotations (a) while the in-phase mode (b) is not observed.

In phases II and I, where the molecular site symmetry are C_2 [3] and C_{2v} [4], respectively, the methyl group rotation is free.

Several transitions ascribed to combinations of a libration mode (R'_1), located near $61cm^{-1}$ at 5K, and methyl torsion levels are observed. It is shown that the hindered non-periodic methyl group in the fundamental state behaves as a free rotator in the first excited state and probably in upper states of the R'_1 mode which is likely to correspond to molecular librations around an axis closely parallel to that of the infinite methyl chain. The thermal excitation of this mode is thus the main source of methyl disordering in phase II and temperature effects observed in Raman are quantitatively represented on this ground.

The II ⟷ I transition is essentially driven by a libration (R'_z), near 116cm^{-1} at 5K, which corresponds to molecular rotations around an axis going through the N atom and the methyl group. Its Raman frequency decreases rapidly at higher temperature and the corresponding potential function is derived from this variation.

References

[1].S.G.Biswas, Ind.J.Phys., **35**, 261, 1961.
[2].B.Alefeld, A.Kollmar and B.A.Dasannacharya, J.Chem.Phys., **63**, 4415, 1975.
[3].U.Ohms, H.Güth, W.Treutmann, H.Dannohl, A.Schweig and G. Heger, J.Chem.Phys., **83**, 273, 1985.
[4].A.Péneau, M.Gourdji and L.Guibé, J.Mol.Struct., **111**, 227, 1983.

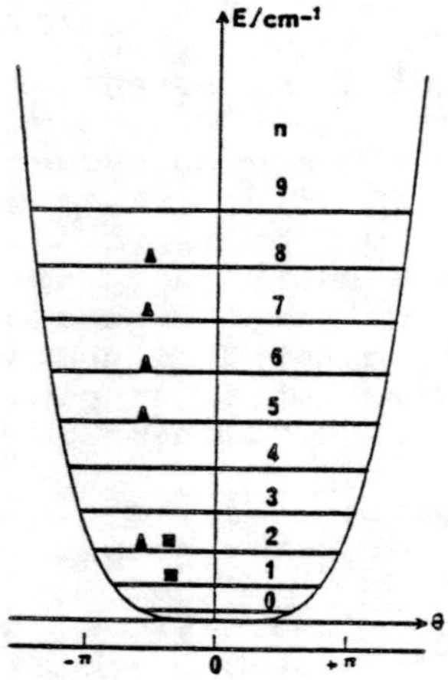

Fig. 1. Potential function and energy levels for the τCH_3 mode at 5K.

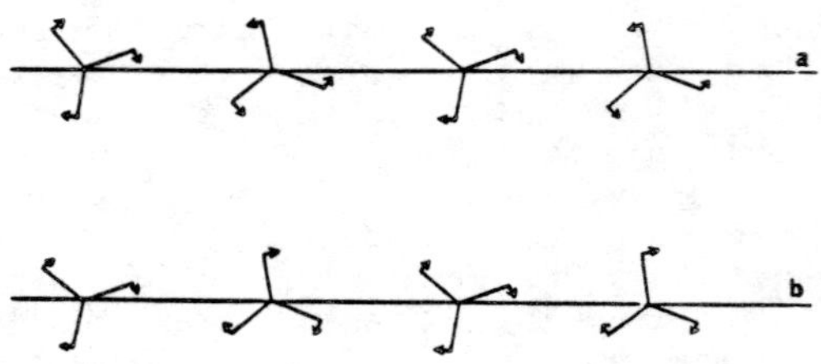

Fig. 2. Tentative structure consistent with methyl torsions correlated along infinite chains.

Raman Linewidths in Alkane-urea Clathrates

Herbert L. Strauss and Robert G. Snyder

Department of Chemistry, University of California, Berkeley, CA 94720

Urea clathrates containing n-alkane chains are excellent systems for studying the vibrational dynamics of n-alkane chains. The clathrate lattice holds the chains in straight, narrow channels, which constrain the chains to an all-trans configuration. It is particularly significant, then, that in the disordered phase of the clathrate some of the n-alkane Raman bands are found to broaden appreciably as the temperature is raised.[1] Particularly illuminating experiments have been carried out on single crystals of the clathrate.[2,3] These experiments determine the line widths using different crystal polarizations. The results have been shown to be qualitatively in agreement with the prediction of a model that is used in the theory of dielectric relaxation.[4,5] In this model a molecule is assumed to move rapidly between sites. In the application to the clathrate, the sites are taken to be the six equivalent orientational positions possible for the n-alkane chain. A group theoretical analysis is used to arrive at results to compare with experiment.

The dielectric model correctly models both the symmetry of the experiment and the relative sensitivity of the different vibration frequencies to the orientational motion. However, the widths of the vibrations bands cannot be simply interpreted as due to the rate of site hopping. For a site hopping model to be completely valid, the motion between sites must be rapid.[6,7] For the n-$C_{20}H_{42}$ molecule investigated, the moment of inertia about the z-axis (long axis) is on the order of 127 amu · $Å^2$. If we assume no potential barrier to rotation and a kinetic energy of kT, the time it takes for the molecule to rotate $2\pi/6$ radians will be $\Delta t = 0.7$ psec. A better lower limit for Δt can be obtained by taking the rotational correlation time of 5 psec measured by neutron scattering in alkane crystals.[8] For the model to apply, Δt must be <u>fast</u> compared to the reciprocal of the linewidth, $\tau = 1/(2\pi c\Delta\nu)$. For a linewidth of $\Delta\nu = 10\ cm^{-1}$, $\tau = 0.5$ psec. Clearly the condition necessary for the site hopping model to be valid is not fulfilled. Instead the widths of the bands are related to the time correlation function of the frequency or the polarizability derivatives of the vibration.[7,9] These time correlation functions are determined by the interaction of the internal vibrations with the torsions and with the overall rotation about the long axis. The torsions and the rotation have relatively small amplitudes ($<<2\pi/6$ radians) and therefore cannot be described by site hopping model.

References

1. R. G. Snyder, J. R. Scherer, and B. P. Gabor, Biochem. Biophys. Acta, 601, 47 (1980).
2. Y. Cho, M. Kobayashi, and H. Tadokoro, J. Chem. Phys. to be published.
3. Y. Cho, M. Kobayashi, and H. Tadokoro, J. Chem. Phys. to be published.
4. L. Onsager and L. K. Runnels, J. Chem. Phys. 50, 1089 (1969).
5. G. Williams and M. Cook, Trans. Faraday Soc. 67, 990 (1971).
6. G. D. Patterson, Macromolecules, 15, 204 (1982).
7. R. A. MacPhail and H. L. Strauss, J. Chem. Phys. 82, 1156 (1985).
8. J. Doucet and A. J. Diamoux, J. Chem. Phys. 81, 5043 (1984).
9. R. A. MacPhail, R. G. Snyder, and H. L. Strauss, J. Chem. Phys. 77, 1118 (1982).

RAMAN SPECTROSCOPIC STUDY OF THE CONFORMATIONAL VARIETY OF n-ALKYLOLIGO(OXYETHYLENE)S IN THE SOLID STATE

Hiroatsu Matsuura and Koichi Fukuhara

Department of Chemistry, Faculty of Science, Hiroshima University, Higashisenda-machi, Naka-ku, Hiroshima 730, Japan

α-n-Alkyl-ω-hydroxyoligo(oxyethylene)s, $CH_3(CH_2)_{n-1}(OCH_2CH_2)_mOH$ (C_nE_m), constitute a typical series of nonionic surfactants, and their physical properties in solution are highly dependent upon their molecular conformations. Conformational studies of these compounds in the solid state are fundamentally important for elucidating the conformational properties in solution, since the solid-state conformation is closely related to the lowest-energy forms in solution. In our previous studies [1], we have investigated the molecular conformation of C_nE_m in the solid state by paying attention to the accordion vibration and have found three conformational forms called the α, β, and γ forms (Fig. 1) for $C_{10}E_m$ (m=4, 6, 8), $C_{12}E_m$ (m=1-8), and $C_{16}E_m$ (m=4, 6, 8).

In the present work, we have extended the conformational study of C_nE_m to shorter members to accomplish a thorough investigation of the conformational morphology. The materials newly treated are C_5E_2, C_6E_1, C_6E_2, C_6E_3, C_8E_2, C_8E_3, C_8E_4, and $C_{10}E_3$, and their Raman spectra were measured in the crystalline solid state at liquid nitrogen temperature. In the analysis of the spectra, we utilized the key bands characteristic of particular local conformations [2], together with the overall-vibration bands, to identify the molecular forms. For C_8E_3 and $C_{12}E_3$, two different spectra giving distinguishing features were obtained (Fig. 2); one obtained by the rapid cooling of the liquid and the other by the much slower cooling. According to the normal coordinate treatment, the Raman band of C_8E_3 with the slow cooling at 253 cm^{-1} is assigned to the accordion vibration of the planar structure consisting of 8 backbone atoms in the alkyl part and the bands at 1231 and 940 cm^{-1} are characteristic of the ($O\overset{T}{-}CH_2\overset{G}{-}CH_2\overset{T}{-}$) helical structure of the oxyethylene chain. These spectral features for C_8E_3

	alkyl chain	oxyethylene chain
	C-C-C-····-C-C-C-C-	O-C-C-O-C-C-O-····-O-C-C-O-C-COH
α form	T T T T T T T	T G T T G T T T T G T T G
$α_T$ form	T T T T T T T	T G T T G T T T T G T T T
β form	T T T T T G T	T G T T G T T T T G T T G
γ form	T T T T T T T	T T T T T T T T T T T T G

Fig. 1. Molecular conformations of C_nE_m (T: trans, G: gauche).

(slow cooling) lead to the molecular conformation of the β form. On the other hand, the molecular conformation of C_8E_3 with the rapid cooling is identified as the γ form on the basis of the observation of the accordion vibration at 152 cm^{-1}, corresponding to the planar structure of 17 backbone atoms, and of the absence of the characteristic bands of the helical oxyethylene chain.

In the spectrum of C_6E_1, two prominent bands are observed at 255 and 239 cm^{-1} with nearly equal intensities; they are assigned to the accordion vibrations of the planar structure of 9 and 10 backbone atoms, respectively. The appearance of the two accordion bands, along with the observation of the 859 cm^{-1} band, associated with the $O^{T}CH_2^{G}CH_2OH$ group, and of the 981 cm^{-1} band, associated with the $CH_2^{T}CH_2^{T}O^{T}CH_2^{T}CH_2OH$ group, proves the coexistence of the α and α_T forms in the C_6E_1 solid. These forms differ from the other in respect of the conformation of the terminal OCH_2-CH_2OH group (Fig. 1).

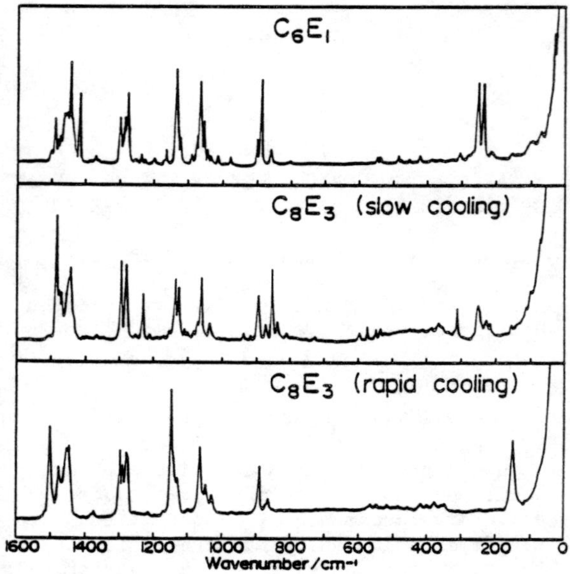

Fig. 2. Raman spectra of C_nE_m in the solid state.

Table 1. Conformational forms of C_nE_m in the solid state

n	m								
	1	2	3[a]	3[b]	4	5	6	7	8
5		γ							
6	$\alpha+\alpha_T$	γ	β	β					
8		γ	γ	β	β				
10			γ	γ	β		β		β
12	α_T	γ	γ	β	β	β	β	α	β
16					β		β		β

[a] Rapid cooling. [b] Slow cooling.

Molecular conformations of the other C_nE_m compounds were determined in a similar way in the light of the accordion vibration and the conformation-sensitive internal vibrations. The conformational forms thus established are summarized in Table 1. These results show that the homologous C_nE_m compounds assume a variety of molecular conformations and that the trend is noted that the conformation changes from the highly extended γ form to the helix-dominative β form as the number of the oxyethylene units (m) increases.

REFERENCES

1. H. Matsuura and K. Fukuhara, IXth International Conference on Raman Spectroscopy, Tokyo, August-September 1984; Chem. Lett., 1984, 933.
2. H. Matsuura and K. Fukuhara, J. Polym. Sci., Polym. Phys. Ed., in press.

RAMAN AND BRILLOUIN SCATTERING OF AMMONIUM HYDROGEN OXALATE, SEMI HYDRATE NH_4 H C_2O_4, 1/2 H_2O AND PHASE DIAGRAM.

J.L. Godet, J. Berger, M. Krauzman, N. Toupry

Département de Recherches Physiques, Université P. et M. Curie
75230 Paris Cedex 05, France.

Raman and Brillouin studies of NH_4 H C_2O_4, 1/2 H_2O have allowed us to determine its phase diagram in the range 80-300 K and 0-8 kbars and to obtain information on the dynamical process of the phase transition at normal pressure.

At room temperature, (phase I), this compound belongs to the space group Pmnb with Z = 8 formulae units in the unit cell. The NH_4^+ ions form two non-equivalent sets, one of which is orientationally disordered.[1] At 146 K, the crystal structure undergoes, a second order equitranslational ferroelastic transition (phase II), which is accompanied by ordering of the NH_4^+ ions[2].

1 - Brillouin scattering. The velocity of an acoustic wave propagating along z, and polarized along x, shows a spectacular softening, as the temperature is lowered and approaches T_c. The thermal evolution of the corresponding elastic constant C_{55} can be explained, using the theory of Landau-Yamada[3] by the coupling of the strain with the B_{2g} Fourier component of a pseudospin associated with the two orientations of the NH_4^+ ions.

The same coupling also explains the behaviour of the sound velocity below T_c, but the existence of a static deformation in the monoclinic low temperature phase requires a more complicated analysis[4].

2 - Raman scattering. Evidence for the dynamical ordering of the pseudo-spins is provided by a central mode observed in the B_{2g} spectrum which exhibits a characteristic critical behaviour (narrower width and larger intensity) in the vicinity of T_c. In fact, the minimum of the line width is obtained slightly below T_c; this unusual result can be explained by postulating that, below T_c, the ferroelastic strain produces an increase in the potential barrier separating the two potential wells of the NH_4^+ ions.

3 - Phase diagram. An internal mode, the C-C stretching of the oxalate ion, observed at 853 cm^{-1}, with no perceptible correlation field splitting at room temperature is found to be sensitive to the transition. It splits into as many lines as there are non-equivalent sets of oxalate ions, in the unit cell.

Using this technique, the phase diagram (fig.1) has been determined: A new phase (III) is observed with a second order I-III transition, while the transition II-III is strongly first order. In phase III, two new Raman lines appear in the external mode region, the frequencies of which, vary by a factor of two, over the pressure range 4 to 8 kb, but are insensitive to temperature. It is suggested that phase III is incommensurate, with a wave vector depending on pressure but not on temperature, with a macroscopic C_{2h} (z axis) symmetry. The central mode is insensitive to the I-III transition.

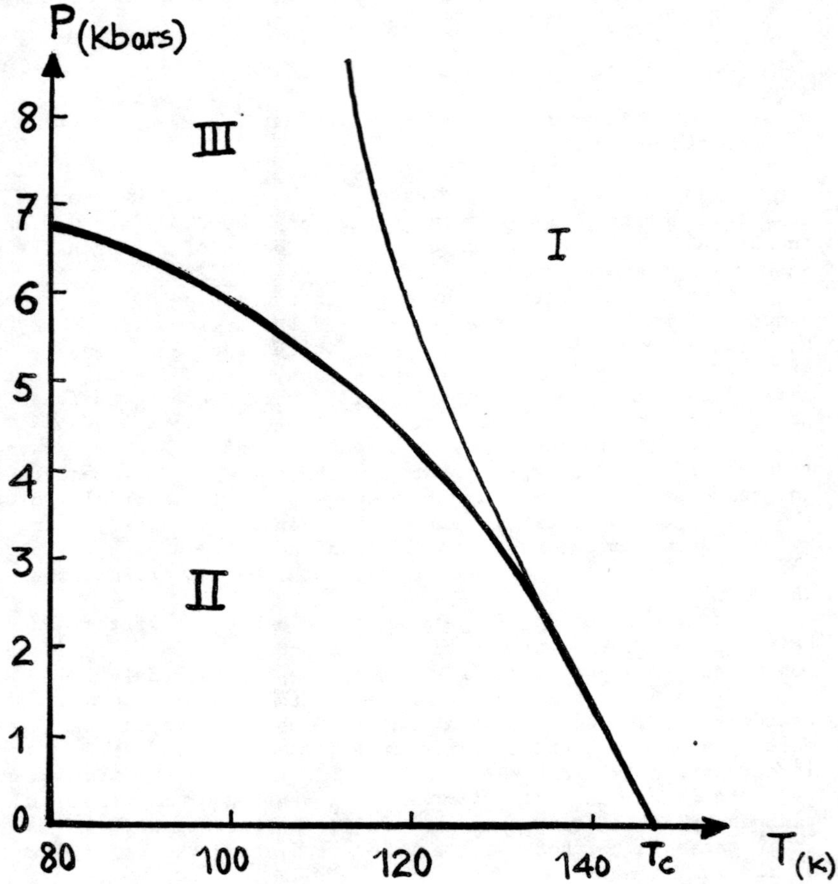

1 H. KUPPERS, Acta Cryst. (1973), B 29, 318
2 H.L. KELLER, D. KUCHARCZYK et al. ZEIT. f. Krist (1982), 158, 221
3 Y. YAMADA et al. J. Phys. Soc. Jap. (1974), 36, 641
4 J.P. BENOIT, J. BERGER, M. KRAUZMAN et J.L. GODET, J. de Physique, to be published (1986)

RAMAN SPECTROSCOPIC STUDY OF CRYSTALLIZATION KINETICS IN POLYESTER FIBERS.

Bernard J. Bulkin, Frank DeBlase, Jong Kim, and Menachem Lewin

Polytechnic Institute of New York, Brooklyn, New York 11201

The annealing of polyester fibers results in a complex set of processes. Among these are changes in crystallinity, conformation, and orientation. In previous work, we have shown that the Raman spectrum is able to distinguish these three changes. There are bands in the Raman spectrum of polyesters that are true crystallinity bands. These correlate with long range order as measured by such techniques as wide angle X-ray scattering. Other bands are sensitive to conformational changes, independent of whether those changes are coupled to an increase in crystallinity. Finally, certain modes are related to orientational order. These may be further divided into amorphous and crystalline orientation. To accomplish all of these determinations from the Raman spectrum requires use of the low frequency region as well as the usual internal modes which have been well studied in the past.

In this paper, we present data on the kinetics of annealing from the viewpoint of the Raman spectrum. Data are obtained using rapid scanning Raman spectroscopy, in which the gratings are oscillated under computer control. This technique suffices to characterize the annealing because the time scale involved is relatively slow (seconds to minutes).

In the case of poly(ethylene terephthalate) (PET) the kinetics observed depend very much on the particular functional group which is being studied. The activation energies and rate constants for loss of orientation induced in fiber spinning are, for example, lower than those for gauche to trans conformer conversion, which are in turn lower than those for development of crystallinity.

For poly(propylene terephthalate) (PPT) the kinetics are quite different from those of PET, and the Raman spectrum shows this. The stable crystalline conformer of the flexible linkage in PPT contains a gauche conformer, while PET is all trans. Thus, in PET, annealing is always the conversion of gauche to trans, while in PPT is may involve trans to gauche conversions. Kinetic data derived from the Raman spectrum are consistent with this difference.

Poly(butylene terephthalate) (PBT) represents the third and most complex system studied. Here the tetramethylene linkage can undergo a separate transition in conformation, under strain, which is independent of crystallization. Again, these processes can be followed by rapid scanning Raman Spectroscopy.

RAMAN MODE ATTRIBUTIONS AND FERRO-PARAELECTRIC TRANSITION IN THE 70/30 VINYLIDENE FLUORIDE-TRIFLUOROETHYLENE COPOLYMER.

M. Abénoza and A. Armengaud.

Laboratoire de Science des Matériaux Vitreux, associé au CNRS(UA 1119), Université des Sciences et Techniques du Languedoc, 34060 MONTPELLIER Cédex, France.

The ferro-paraelectric transition of vinylidene fluoride-trifluoroethylene random copolymers, p(VDF - Tr FE), have been investigated by different analytical technics (1-7) and particularly by Raman spectroscopy (8). Nevertheless a systematic assignment of Raman bands to the modes relatives to the copolymer is not yet carried out. We studied here the evolution of Raman bands frequencies, intensities and polarizations relative to an uniaxially oriented sample (stretched about 3,5 times the original length) of p(VDF - Tr FE) 70/30 copolymer versus temperature when the 25°C→110°C→25°C temperature cycle is carried out. We then propose an attribution of the different modes which exhibit Raman caracteristic changes

Ferro-paraelectric transition.

The transition was found to occur at about 105°C rising temperature and about 80°C when the temperature is lowering. At 105°, the 842, 1300 and 1350 cm^{-1} Raman band intensities strongly decrease, on the contrary the 611, 811 cm^{-1} band intensities increase.

On the same time we also note an important modification of the Raman bands polarization occuring at the transition, this change being not completely reversible.

Copolymer Raman band attributions.

A comparative and systematic study of Raman spectra relative to unoriented and uniaxially stretched samples of α and β p(VDF), p(Tr FE) and co-polymer allows us a coherent attribution of Raman bands 70/30 copolymer. This attribution is consistent with our earlier investigations of observed and calculated Raman mode intensities of p VDF β (9) and p VDF α (10).

Conclusion.

The modes, the Raman intensities of which exhibit a strongly change at the ferro-paraelectric transition of p(VDF - Tr FE) copolymer, are essentially the $\nu_s(CC)$, $\nu_s(CF_2)$ and $\omega(CF_2)$ characteristic modes of the polymer chain configuration. At the transition, the Raman spectra evolution is consistent with a conformational change between the extended "all trans" form and a rich gauche sequences configuration. Moreover polarization changes along the studied temperature cycle shows that the paraelectric phase is highly disordered and that after the cycle is carried out a part of this chain desorientation remains in the ferroelectric phase.

25°C		110°C		Attributions [b]
$\nu(cm^{-1})$	I observed [a]	$\nu(cm^{-1})$	I observed	
612	W	614 m		$\omega(CF_2)$
811	m	805 S		$\nu_s(CF_2)$ gauche
851	V S	850 W		$\nu_s(CF_2)$ trans
889	m	891 W		$\rho(CH_2)$
1083	large band	1083 V W		$\nu_s(CC)$
1105	W			
1250	V W	V W		
1293	m	1290 W		$\nu_s(CC)$ trans
1353	m	abs		$\nu_s(CC)$ trans
1402	W	1402 W		$\omega(CH_2)$
1436	S	1432 S		$\delta(CH_2)$
2977	V S	2978 V S		$\nu_s(CH_2)$
3013	W	3013 W		$\nu_a(CH_2)$

Raman bands for an uniaxially oriented sample of p(VDF - Tr FE) 70/30 in the region 600 - 3050 cm^{-1} at 25°C and 110°C.

[a]W = weak; V W very weak; m = medium ; S = strong; V S = very strong;

[b]ν_s: symmetric stretching; ν_a: antisymmetric stretching; ρ: rocking; δ: bending; ω: wagging.

References

1. T. Yagi, M. Tatemoto and J. Sako. J. Polym. 12, 209 (1980).
2. Y. Tajitsu, A. Chiba, T. Furakawa, M. Date and F. Fukada. Appl. Phys. lett., 36, 286 (1980).
3. T. Furakawa, G.E. Johnson, H.E. Bair, Y. Tajitsu, A. Chiba and E. Fukada. Ferroelectrics, 32, 61 (1981).
4. Y. Higashihata, J. Sako and T. Yagi. Ferroelectrics, 32, 85 (1981).
5. T. Yamada, T. Ueda and T. Kitayama. J. Appl. Phys. 52, 948 (1981).
6. J. Lovinger, G.T. Davis, T. Furakawa and M.G. Broadhurst. Macromolecules, 15, 323 (1982) ; 15, 329 (1982).
7. J . Lovinger, T. Furakawa, G.T. Davis and M.G. Broadhurst. Polymer, 24, 1225 (1983) ; 24, 1233 (1983).
8. K. Tashiro, K. Takano, M. Kobayashi, Y. Chatani and H. Tadokoro, Polymer, 25, 195 (1984).
9. A. Armengaud, J.L. Ramonja and M. Abénoza, J. Raman. Spectrosc (1986).
10. A. Armengaud, M. Abénoza, to be published.

PICOSECOND RAMAN TRANSIENTS IN POLYACETYLENE.*

D. L. Weidman and D. B. Fitchen

LASSP, Cornell University, Ithaca, NY 14853

Polyacetylene [$(CH)_x$] is a quasi-one-dimensional organic semiconductor in which the photogenerated carriers show novel properties. We use fast optical probes to study the picosecond dynamics of these carriers. Here we report measurements of picosecond transient changes in the Raman scattering intensity and lineshape in *trans* $(CH)_x$. These results are compared with our picosecond time-resolved reflectivity data and with our model of quasi-one-dimensional diffusion-limited recombination in the presence of disorder on the polymer chains.

The experiment was done using 3 picosecond pulses from a synchronously pumped dye laser with energies above the optical gap (1.8-2.2eV) and repetition rates from 5 to 94 Mhz. The pulse train was split into a pump beam and a mechanically delayed, orthogonally polarized probe beam which were then recombined and focused onto a free-standing film of thermally isomerized *trans* $(CH)_x$. The backscattered Raman light was dispersed in a double monochromator and detected with an intensified diode array. Difference spectra were taken, subtracting the total Raman signal with a large (Δt>2nsec.) pump-probe time separation from the total Raman signal at a small ($0 \le \Delta t \le 20$psec.) time separation.

An example of a typical difference spectrum is shown in figure 1 for pump-probe separation Δt=0 (solid curve), where we show the 2 percent decrease in Raman scattering intensity for the 1050 cm^{-1} Raman line (single bond stretch mode). We compare the integrated change in Raman intensity under this peak (triangles) to the change in reflectivity (solid curve) in figure 2, where the reflectivity data has been scaled (actual peak change = -4×10^{-3}). The transient change in reflectivity was measured using the same experimental setup as for the Raman experiment. The specularly reflected

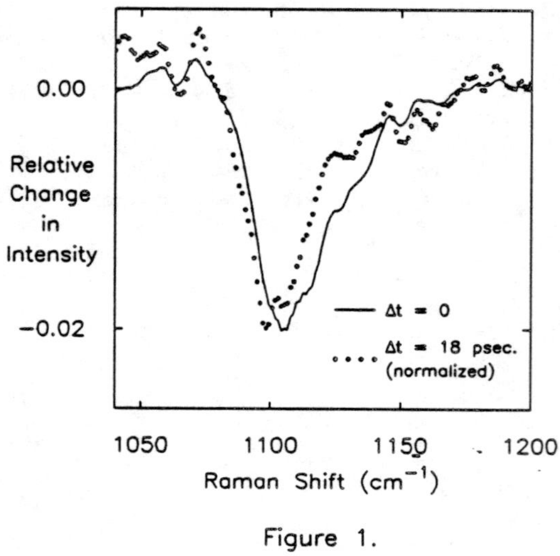

Figure 1.

light was detected with a photodiode. In both cases the effect is due to ground state depopulation, and we are monitoring the recovery of the ground state. We see that the recovery is roughly the same in both measurements.

The interesting new information obtained from the picosecond Raman transients is shown in figure 1, where we show for comparison the result for $\Delta t = 18$ picoseconds normalized to the $\Delta t = 0$ result. As can be seen, there is a shift in the Raman band towards lower wavenumber. A similar shift is seen in the 1450 cm^{-1} (double bond stretch) Raman line.

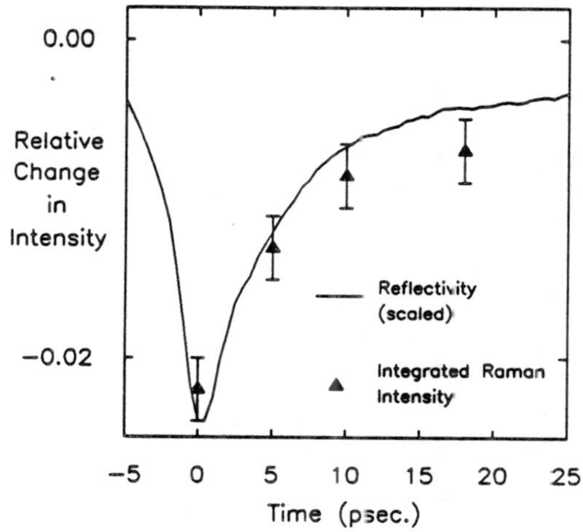

Figure 2.

One might attempt to explain this via electron screening. There is evidence from the transient reflectivity (and transmission) data, however, which indicates that the initially hot carriers thermalize very rapidly (≲100 fsec.) to the band edge. This in turn implies that the maximum carrier density occurs at the shortest times. Since the maximum screening is at short times, one would expect to see mode softening near $\Delta t = 0$, opposite to the experimental data. In any case, the initial carrier population is small, about 5×10^{17}. Instead, we believe that this spectral shift reflects different recovery kinetics across the inhomogeneously broadened Raman line. We will examine a model in which the one dimensional $(CH)_x$ chains consist of segments of random length on which the carrier recombination is modified from simple 1D diffusion by the effects of confinement.

*Work supported by the National Science Foundation through the Materials Science Center.

FULLY ORIENTED TRANS-$(CH)_x$ SYSTEMS : ANALYSIS OF POLARIZED RESONANT RAMAN SCATTERING.*

S. Lefrant[a)], E. Mulazzi[b)], E. Faulques[a)] and G.P. Brivio[b)]

a) Laboratoire de Physique Cristalline (U.A. CNRS n° 802), Université de Nantes, 44072 Nantes Cédex, France
b) Dipartimento di Fisica dell'Università di Milano, Milano, Italy

Polarized Resonant Raman Scattering (RRS) spectra of trans-$(CH)_x$ taken at different excitation wavelengths (λ_L = 647 nm and λ_L = 458 nm) have been recorded for oriented samples obtained with different methods[1,2]. The experimental results given in Fig. 1 are highly polarized and show different properties at the different incident laser wavelengths. In particular, the // // spectrum at λ_L = 458 nm shows two main bands whose peak positions and widths can be explained by considering the contributions of long and short conjugation length segments. On the contrary, the ⊥//, //⊥ and II polarized spectra only display two structureless bands peaked at higher frequencies with respect to the previous ones. In, the red region, however, for λ_L = 647 nm, the peaks of the bands of the //⊥, ⊥// and II spectra show a very small frequency displacement with respect to those of the // // spectrum. These results confirm the theoretical predictions[3] and their comparison with the experimental data as published in Ref. 4. In fact, following the model based on a double length segment distribution, only short segments with N < 10 (N is the number of double bonds) contribute to the RRS spectra in the violet region, since they have electric dipole moment components also perpendicular to the stretching chain axis. In this way, we can explain why the corresponding bands show higher displaced peaks than those in the // // polarized RRS spectra for λ_L = 458 nm (Fig. 2). In the red region instead, the main contribution to the depolarized spectra comes from the distribution of long misaligned segments in resonance with the incident laser wavelength. The dependence of the depolarization ratios on the incident laser wavelengths, sample quality and history is also determined.

* Work partially supported by Progretto Finalizzato Chimica Fine Secondaria del CNR

1. G. Leising, Polym. Bull. 11, 401 (1984)
2. G. Lugli, V. Pedretti and G. Perego, Mol. Cryst. Liq. Cryst. 117, 43 (1985) (Ital. Patent Appl. 22722/A82 to Assoreni).
3. E. Mulazzi, Sol. St. Comm. 55, 807 (1985)
4. E. Faulques, E. Rzepka, S. Lefrant, E. Mulazzi, G.P. Brivio and G. Leising, Phys. Rev. B (1986) in press.

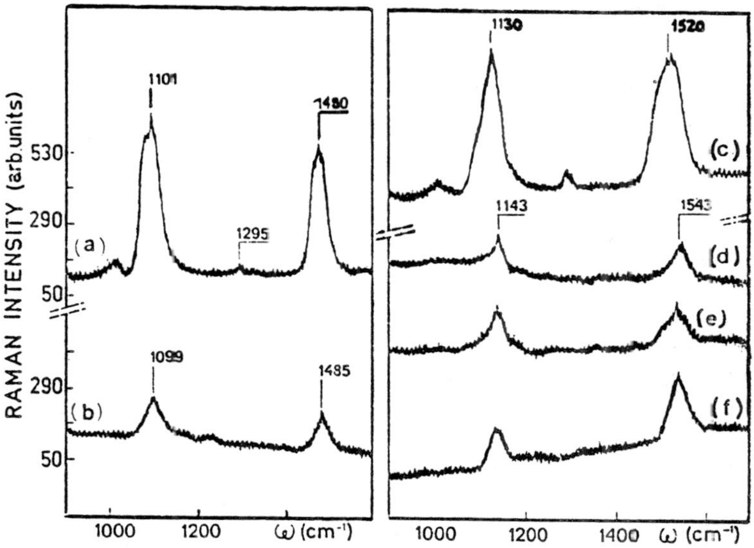

Fig. 1 Experimental RRS spectra of oriented trans-$(CH)_x$ at 77K
λ_L = 676.4 nm a) polarized // // ; b) ⊥ // ;
λ_L = 457.9 nm c) polarized // // ; d) ⊥ // ; e) //⊥ ;
f) ⊥⊥.

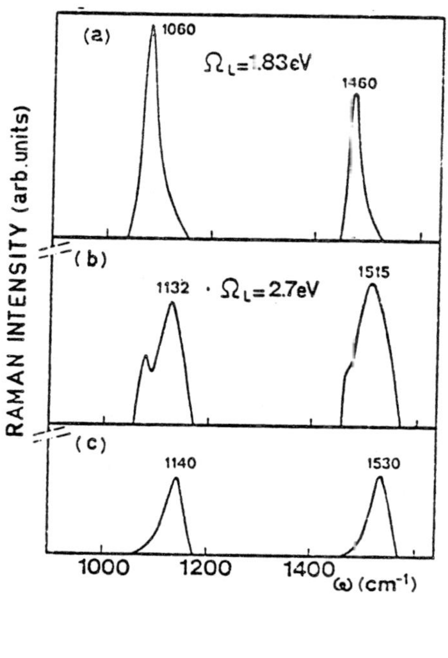

Fig. 2 Calculated polarized RRS spectra of oriented trans-$(CH)_x$ (stretching axis z)
a) I_{zz} spectrum for Ω_L = 1.83 eV
b) I_{zz} spectrum for Ω_L = 2.7 eV
c) I_{zz} spectrum for Ω_L = 2.7 eV.

DISPERSION OF THE DEPOLARIZATION RATIO IN THE RESONANCE RAMAN SPECTRA OF SOLUBLE POLYACETYLENES (+)

S.Fuso, R.S.Cataliotti*, G.Dellepiane, S.Destri**, G.Paliani*, R.Tubino**

Istituto di Chimica Industriale, Università, Genova
*Dipartimento di Chimica, Università, Perugia
**Istituto Chimica Macromolecole, CNR, Milano

Recently accurate Resonant Raman excitation profiles have been measured for a soluble form[1] of polyacetylene, by using the solvent lines as internal standard[2]. These experimental data have provided information on the lifetime of the electronic excited state responsible for the resonant enhancement and on the distribution of the electronic energies in this material. In this note we report preliminary data on the dispersion of the depolarization ratio for the C=C mode and C–C stretching modes. The electronic absorption spectrum of the sample studied is shown in fig.1. The low frequency tail of this spectrum is indicative of the presence of highly conjugated segments as it is also apparent from the Raman bandshapes which exhibit only a sharp and asymmetric peak at 1420 and 1040 cm^{-1}. Small amount of the cis isomer is also detected in the spectra. In fig.2 is reported the variation of the depolarization ratio ρ ($\rho = I_{\perp}/I_{\parallel}$) as a function of the exciting laser frequency. For this sample the extinction coefficient at the absorption maximum (4500 l $mole^{-1}cm^{-1}$) is considerably lower that the value observed for shorter polyenes[3]. Also the depolarization ratio of polyacetylene appears to be anomalous if compared with the values reported for β-carotene and diphenylpolyenes which are roughly 1/3[4,5]. As a matter of fact a polarization of 1/3 is predicted by considering that the $\pi \rightarrow \pi^*$ electronic transition in trans polyenes is strictly polarized along the molecular axis (z) and therefore only the α_{zz} component contributes to the scattering intensity. If a slight misalignement of the transition moment occurs only slightly higher values are calculated. As the measurements have been carried out in the liquid phase, any kind of macroscopic preferential orientation can be ruled out. The observed depolarization ratio appears therefore to be an intrinsic molecular property.

Our tentative explanation is the following. As trans polyacetylene is obtained by thermal isomerization of the cis isomer it is not unlikely that a number of cis linkages, as those depicted in fig.3, hold together

trans segments without interrupting the conjugation. The presence of a cis linkage has two main effects on the electronic absorption spectrum :
i) reduction of a factor 1/2 of the oscillator strength for the optical absorption
ii) the appearance of a new absorption with polarization perpendicular to the molecular axis (see fig.3).
These two perpendicular states can be coupled vibronically by a phonon of the proper symmetry, thus originating unusually high values for the depolarization ratio as well as its variation with the exciting light.

This model is presently being investigated and it will be presented in detail in a forthcoming publication.

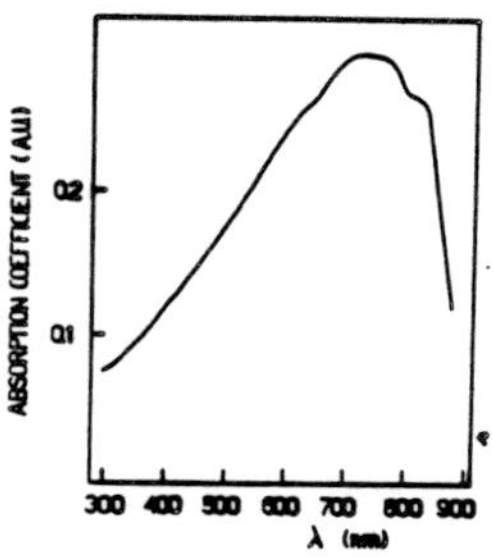

Figure 1. Electronic absorption spectrum of trans-polyacetylene in toluene solution.

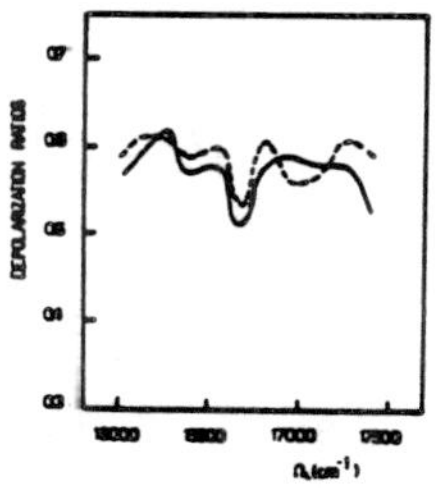

Figure 2. Depolarization ratios as a function of exciting frequency for C=C (—) and C-C (---) vibrations.

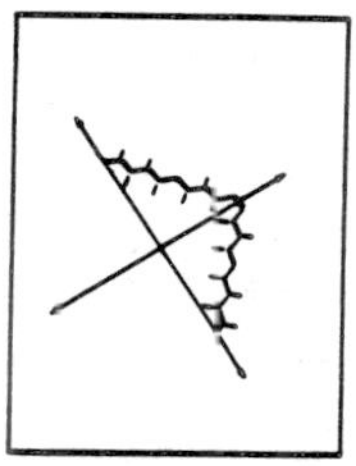

Figure 3. Effect of a "cis linkage" on the polarization properties of electronic transition in trans-polyacetylene.

REFERENCES

1. S.Destri et al., Makromol.Chem., Rapid Commun. 5, 353 (1984)
2. R.S.Cataliotti et al., J.Chem.Phys. 82, 2223 (1985)
3. C.N.R. Rao, "Ultraviolet and visible spectroscopy", Butterworth, London 1975
4. T.A.Moore and P.S.Song, J.Mol.Spectry. 52, 209 (1974)
5. L.Margulies and M.Stockburger, J.Raman Spectry. 8, 26 (1979)

(+) - *Work partially supported by C.N.R. Progetto Strategico Nuovi Materiali.*

RAMAN CHARACTERIZATION OF AN IMPLANTED CONDUCTING POLYMER

A. Bidarian, M. S. Mathur*, P. K. Kadaba and E. B. Bradley
Department of Electrical Engineering
University of Kentucky
Lexington, KY (USA) 40506

Polymers, in general are non conducting, but can be made conducting by the introduction of suitable dopants. A polymer of current interest is PBT, (polyphenyline benzobishiazole), which does not lend itself to chemical doping on account of lack of suitable solvent. However, a change in the conductivity can be brought about my ion implantation. Samples studied here have been implanted with iodine ions and annealed in argon atmosphere. We have used Raman spectroscopy to detect changes in structure and bondings. Preliminary spectra of the iodine implanted PBT polymer show that Raman bands and their depolarization ratios, especially in the region 40-300 cm^{-1} and 1500-2000 cm^{-1}, differ between the raw and implanted samples. We believe that these differences can be attributed to changes in structural and skeletal parameters due to the implanted ions. A correlation is being attempted between structure and the variation of conductivity with ion type.

* On leave from Department of Physics, University of Manitoba, Winnipeg (MB) Canada R3T2N2

RAMAN SPECTRA OF POLYPYRROLE AND ITS 2,5-^{13}C-SUBSTITUTED AND C-DEUTERATED DERIVATIVES AT THEIR NEUTRAL AND DOPED STATES.

Y. Furukawa, S. Tazawa, Y. Fujii and I. Harada

Pharmaceutical Institute, Tohoku University, Aobayama, Sendai 980, Japan

Polypyrrole (PP) is a good electric conductor at acceptor-doped state and an insulator at neutral state [1]. ^{13}C NMR [2,3] and infrared [4] studies have suggested that neutral polypyrrole consists mostly of α,α'-linked all-*s-trans* coplanar structure (Fig. 1a). Results of a combined study of ESR and optical absorption spectra on doped films have been interpreted by the formation of dication (bipolaron) [5]. A quantum chemical calculation has indicated that the formation of bipolaron causes a significant change in the structure, including the inter-ring bond length from 1.474 to 1.357 Å [6] (Fig. 1b).

We studied the vibrational spectra of neutral and doped polypyrrole and its 2,5-^{13}C-substituted and C-deuterated analogs. Assignment of each band was made on the basis of the shifts in wavenumber on isotopic substitutions and normal coordinate calculations on the neutral polymer. The results are discussed in relation to the structures in Fig. 1.

Doped polypyrrole films were obtained by electrochemical polymerization of pyrrole and were converted to neutral films by reduction. C-Deuterated polypyrrole (PPD) films were prepared from pyrrole-2,3,4,5-d_4. 2,5-^{13}C-Substituted polypyrrole (PPC) films were prepared from the mixture of pyrrole-2,5-$^{13}C_2$ (83.7%), pyrrole-2-^{13}C (15.5%) and pyrrole (0.8%). Normal coordinate calculations were performed for the infinitely long all-*s-trans* coplanar structure (Fig. 1a). For such a structure repeating unit consists of two pyrrole rings and its factor group is isomorphous to D_{2h} point group. Force constants obtained for pyrrole and its derivatives [7] were used without modification.

Resonance Raman spectra (λ_{max} = 412 nm, λ_{exc} = 457.9 nm) of neutral polymers are straightforwardly assigned as are shown in Fig. 2. All the a_g modes expected in the region for the D_{2h} structure are observed. The calculated values correspond fairly well with the observed. Moreover, mutual exclusion rule holds between the Raman and infrared [2] spectra. These observations are consistent with the assumed structure. The broad feature of Fig. 2b is due to isotopic inhomogeneity of the parent monomer. The ^{13}C-isotopic shift is most significant for the 1562-cm^{-1} band which is assigned to a hybrid mode of the C=C stretch ($\nu_{C=C}$) and the inter-ring CC stretch ($\nu_{CC'}$).

Raman spectra of doped polypyrrole (λ_{max} = 462 and >800 nm, λ_{exc} =

Fig. 1. Proposed structures of neutral PP (a) and doped PP (b)

632.8 nm) are rather complicated (Fig. 3). However, it is possible to correlate most of the peaks with those of neutral films by studying the spectra of films with intermediate dopant concentration. The $\nu_{CC'}$ contributes considerably to the 1593-cm^{-1} band since it shifts down by 25 cm^{-1} on the ^{13}C-substitution. This band corresponds to the 1562-cm^{-1} band of the neutral film. An increased contribution from intra-ring C_3C_4 stretch (ν_{CC}) is indicated by the increased C-deuteration shift of this band on doping (-3 in PP vs. -18 cm^{-1} in doped PP). If the force constants of $\nu_{CC'}$ and ν_{CC} are increased by the doping at the expense of a decrease of the $\nu_{C \cdot C}$, and the effect of the increase overcomes that of the decrease, the upshift of this band on doing is reasonably understood.

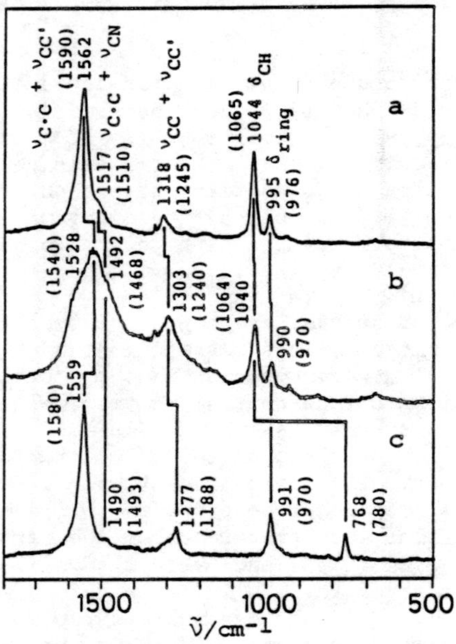

Fig. 2. Raman spectra of neutral PP (a), PPC (b) and PPD (c) excited at 457.9 nm, with ν_{obs} (ν_{calc}).

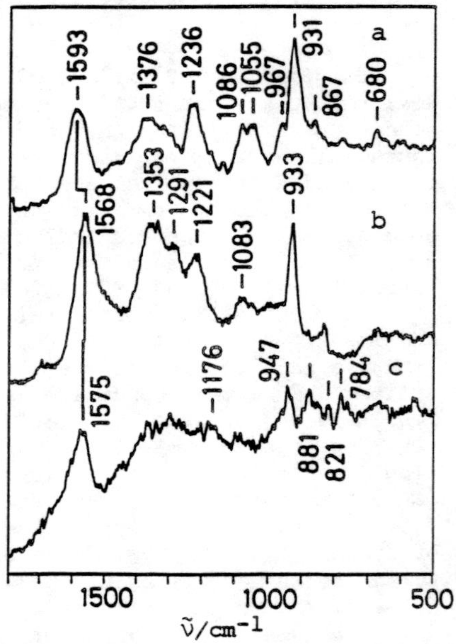

Fig. 3. Raman spectra of BF_4^--doped PP (a), PPC (b) and PPD (c) excited at 632.8 nm.

REFERENCES

1. A. F. Diaz, K. K. Kanazawa and G. P. Gardini, J. Chem. Soc., Chem. Commun., 635 (1979).
2. G. B. Street, T. C. Clarke, M. Krounbi, K. K. Kanazawa, V. Y. Lee, P. Pfluger, J. C. Scott and G. Weiser, Mol. Cryst. Liq. Cryst., 83, 1285 (1982).
3. F. Devreux, G. Bidan, A. A. Syed and C. Tsintavis, J. Phys., 46, 1595 (1985).
4. G. B. Street, S. E. Lindsey, A. I. Nazzal and K. J. Wynne, Mol. Cryst. Liq. Cryst., 118, 137 (1985).
5. J. C. Scott, J. L. Bredas, K. Yakushi, P. Pfluger and G. B. Street, Synth. Met., 9, 165 (1984).
6. J. L. Bredas, B. Themans and J. M. Andre, Phys. Rev. B, 27, 7827 (1983).
7. D. W. Scott, J. Mol. Spectrosc., 37, 77 (1971).

chapter 13

SPECIFIC MOLECULAR SYSTEMS: INORGANIC AND ORGANOMETALLIC

TIME-RESOLVED RESONANCE RAMAN SPECTROSCOPY OF THE $\delta\delta^*$ EXCITED ELECTRONIC STATE OF $Re_2Cl_8^{2-}$.

Richard F. Dallinger

Department of Chemistry, Wabash College, Crawfordsville, Indiana 47933

The excited state properties of molecules which contain multiple metal-metal bonds have been probed in many different ways, yet much debate still exists concerning both the molecular and electronic excited state structures of these species. This paper presents the results of a time-resolved resonance Raman (TR^3) spectroscopic study of the $\delta\delta^*$ excited electronic state of the quadruply-bonded octachlorodirhenate dianion ($Re_2Cl_8^{2-}$), the prototypical molecule of this class of compounds.[1a]

The TR^3 spectrum of the $\delta\delta^*$ excited state of $Re_2Cl_8^{2-}$ was obtained using a two color pump-probe technique. The pump wavelength was at 640 nm (Quanta-Ray PDL-1 pulsed dye laser with DCM dye pumped by the second harmonic of a Quanta-Ray DCR-1 Nd:YAG laser); this wavelength is just to the blue of the allowed $\delta \rightarrow \delta^*$ electronic transition at 685 nm. Pump excitation near this transition populates the photochemically important $\delta\delta^*$ excited state (assigned as the singlet by Gray), which has a lifetime of 75 nsec in CH_2Cl_2. Population of this excited state was confirmed by luminescence saturation measurements, in which the plot of luminescence intensity at 790 nm versus pump laser energy per pulse is non-linear. The probe wavelength was at 416 nm (the first hydrogen Stokes-shifted line from the third harmonic output of the same Nd:YAG laser used to pump the dye laser), near the transient absorption band at 390 nm assigned as $e_u(Cl) \rightarrow \delta$ charge transfer. The probe laser beam arrived at the sample ($\sim$6 mM $Re_2Cl_8^{2-}$ in deoxygenated CH_2Cl_2) approximately 12 nsec after the pump beam.

When the pump laser (640 nm) alone illuminated the sample, the ground state resonance Raman spectrum of $Re_2Cl_8^{2-}$ was observed. It was virtually identical to the spectrum obtained using red cw laser illumination, with an intense Re-Re stretching mode at 272 cm^{-1}, a weaker Re-Cl stretch at 359 cm^{-1}, overtones of the Re-Re stretch and a 272 cm^{-1} + 359 cm^{-1} combination mode. When the probe laser (416 nm) alone illuminated the sample, only the CH_2Cl_2 solvent spectrum was observed. When the pump-probe combination was used, the spectrum of the excited state was obtained. The Figure shows the TR^3 difference spectrum of the 640 nm pump/416 nm probe spectrum minus the 416 nm only (i.e. solvent) spectrum. Thus, the three bands seen in the difference spectrum can only be due to the $\delta\delta^*$ excited state population generated by the 640 nm pump beam. All three excited state bands are polarized; three polarized Raman bands are predicted by group theory for either D_{4h} or D_{4d} symmetry.

The 366 cm^{-1} excited state band is assigned to Re-Cl stretching. This represents a shift of only 7 cm^{-1} going from the ground state to the $\delta\delta^*$ excited state, consistent with the fact that the $\delta \rightarrow \delta^*$ transition is highly localized along the metal-metal axis. The 138 cm^{-1} excited state band is assigned to Re-Re-Cl bending by virtue of its low frequency. The ground state A_{1g} bend has not been observed.

The 204 cm^{-1} excited state band is assigned to Re-Re stretching. It was expected that the excited state Re-Re stretch would be lower than the 272 cm^{-1} ground state Re-Re stretch since the formal metal-metal bond order decreases from 4 in the ground state to 3 in the $\delta\delta^*$ excited state. However, a shift of 68 cm^{-1} is unexpectedly high. The 204 cm^{-1} frequency (room temperature solution) is also much lower than the 248 cm^{-1} excited state Re-Re stretch deduced from low temperature, vibronically resolved single crystal absorption measurements. One possible interpretation is that the 248 cm^{-1} value corresponds to a short timescale eclipsed excited state (Franck-Condon) geometry while the 204 cm^{-1} value corresponds to the fully relaxed, possibly staggered or partially staggered, excited state geometry. Other explanations involve reassignment of the emissive excited state, possibly to the triplet $\delta\delta^*$ state. Further work on the TR^3 spectroscopy of multiply metal-metal bonded dimers is currently in progress to help sort out the excited state structural features of $Re_2Cl_8^{2-}$.

[1] R. F. Dallinger, J.Am.Chem.Soc. 1985, 107, 7202.

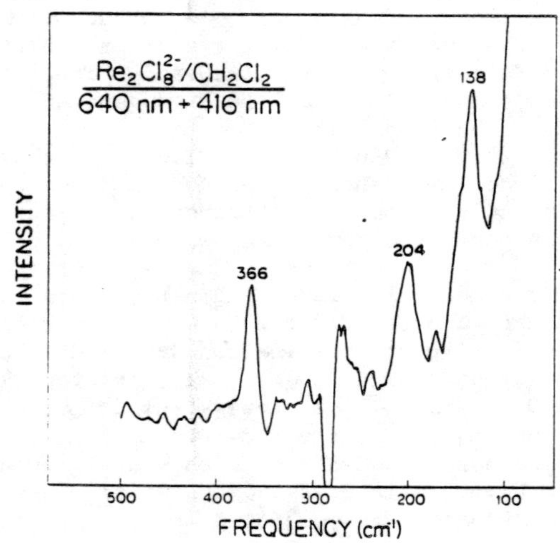

SIMULATION OF RESONANCE RAMAN SPECTRA OF MOLYBDENUM AND RHENIUM COMPLEXES

John M. Jean, Mahfoud Belhadj, and Richard A. Friesner

Department of Chemistry, University of Texas at Austin, Austin, TX 78712

Quantitative simulations of absorption and resonance Raman spectra for the complexes $Mo_2Cl_8^{4-}$ and $Re_2Cl_8^{4-}$ have been performed. Overtones, combination bands and temperature dependence have been accurately reproduced. From this analysis, the Franck-Condon accessible part of the excited state potential surface is deduced through second order. Mode mixing in the excited state can be inferred from the relative intensities of the overtone and combination bands.

RESONANCE RAMAN STUDY OF THE LINEAR-CHAIN MIXED-VALENCE COMPLEX $[Pd(-chxn)_2Br]Br_2$.

P. Waage Jensen and H. Toftlund

Department of Chemistry, Odense University, DK-5230 Odense M, Denmark

The most studied mixed-valence complexes are the halogen bridged derivatives of platinum. One of the oldest and most famous and typical examples is Wolffram's red salt $[Pt(NH_2Et)_4][Pt(NH_2Et)_4Cl_2]Cl_4 \cdot 4H_2O$. This and several analogous complexes are highly dichroitic due to intense absorptions associated with the inter-valence transition (IT) occurring when the electric vector of an incident light beam is parallel to the stacking axis.

Excitation of this type of mixed-valence complexes within the contour of the IT bands generally leads to strong enhancement of the Raman bands associated with the symmetric X-Me(IV)-X stretching vibration ν_1 and the development of a long associated overtone progression (1).

The resonance Raman spectra have been studied of a number of linear-chain complexes of platinum and cyclohexanediamine (chxn) and a few palladium cyclohexanediamine complexes (2), but no spectra have been published of complexes involving cyclohexanediamine, palladium, and bromine.

$[Pd(-chxn)_2Br]Br_2$ was prepared by slow diffusion of bromine gas into an aqueous solution of $[Pd(-chxn)_2]Br_2$. The crystals are dark violet with a pronounced metallic sheen.

The crystal structure of this complex has not yet been determined, but the powder diffraction diagram shows it to be isomorphous with $[Ni(chxn)_2Br]Br_{1.77}$ (3) and $[Pt(chxn)_2Cl]Cl_2$ (4) with a Pd-Pd distance of 5.267Å. Raman spectra have been recorded of the complex in KBr discs at room temperature and at liquid nitrogen temperature using argon and dye lasers for excitation. Strong resonance enhancement was found in the spectral region of the argon laser, whereas excitation with a dye laser in the region 570-630 nm resulted in weak lines on the slope of an extraordinarily broad Rayleigh line.

The Raman spectrum of $[Pd(-chxn)_2Br]Br_2$ at a laser wavelength of 514.5 nm is shown in Figure 1 and listed in Table 1. It differs from the spectra of similar complexes in some respects. There seems to be two progressions which coalesce for $v \geq 3$. Only overtones up to $v = 5$ are found. This corresponds to what is found for other mixed-valence complexes involving Pd and Br (2) and it is less than for the corresponding Pt and Cl compounds. This indicates a relatively small change in the Pd-Br distance in the excited state compared to the ground state.

In the tables in ref. (1) and (2) are listed Pd-Br stretching frequencies in the range 150-160 cm^{-1}. Compared to this the present value of 180.5 or 189 cm^{-1} is extremely high. The listed ν_1 values are however mostly measured at a very long excitation wavelength. In (2) is reported a remarkable variation of ν_1 with the excitation wavelength, the largest

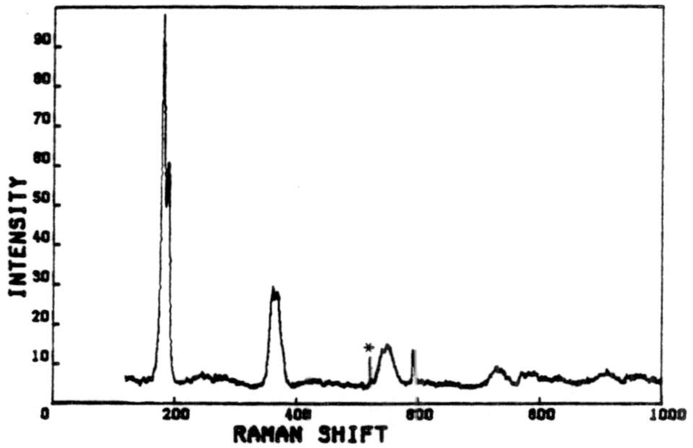

Figure 1. Raman spectrum of $[Pd(-chxn)_2Br]Br_2$ with 514.5 nm excitation.
* = laser emission line.

Wavenumber	Assignment
180.5	ν_1
189	
361	$2\nu_1$
366.5	
548.5	$3\nu_1$
592.5	ν(ring)
730	$4\nu_1$
913	$5\nu_1$
cm^{-1}	

Table 1. Vibrational wave-numbers and assignments.

ν_1 being found for the smallest laser wavelength. Shifts up to 59 cm^{-1} have been measured for complexes of Pd with Cl and cyclohexanediamine. Taking this into account the high ν_1 value is not extraordinary.

A frequency shift of this type is not found in this compound. The Raman spectra for 488.0 and 514.5 nm excitation are exactly alike. Clark et al. (2) explain the wave number shift as a result of the cationic units being a mixture of geometrical isomers. This is apparently not so in the present case. The doublet structure could maybe be due to an overtone of an infrared active vibration becoming enhanced via Fermi resonance.

In contrast to other Wolffram's red type salts the present compound exhibits a rather high AC conductivity ($\sigma_{||}$(35 GHz) = 1 $\Omega^{-1}cm^{-1}$ at 300 K).

The specular reflectance spectrum has a pronounced plasma edge at 16 kK, more typical of highly conducting systems than of the very slightly conducting Wolffram's red salts.

The magnetic susceptibility and the EPR spectrum clearly show this material to be non-stoichiometric although far from as much as the corresponding nickel compound.

The reflectance spectrum of the palladium complex suggests that the band gap is below 10 kK so the Raman spectrum in Figure 1 is far from being in exact resonance.

It is therefore possible that what is seen in the spectrum are the Br-Pd(IV)-Br defects in a Pd(III)-Br extended chain.

References

1. R. J. H. Clark, in "Advances of Infrared and Raman Spectroscopy", Vol. 11, Chapter 3, Wiley, Chichester, 1984.
2. R. J. H. Clark, M. Kurmoo, and D. N. Mountney, J. Chem. Soc. Dalton Trans. 1851 (1982).
3. H. Toftlund and O. Simonsen, Inorg. Chem. 23, 4261 (1984).
4. K. P. Larsen and H. Toftlund, Acta Chem. Scand., Ser. A. 31, 182 (1977).

VIBRATIONAL SPECTRUM OF $MMoO_4$ (M = Mn OR Co)

I. Matsuura, H. Hashiba, H. Kawahara, and I. Kanesaka

Faculty of Science, Toyama University, Gofuku, Toyama 930, Japan

The crystal structure [1] of α-$MnMoO_4$ and β-$CoMoO_4$ belongs to the space group $C2/m-C_{2h}^3$ with the isotypic one, whereas that of α-$CoMoO_4$ [2], which belongs to the same space gruop, is not isotypic with them. The basic difference between the α-$MnMoO_4$ and α-$CoMoO_4$ structures is the coordination of Mo which is tetrahedral in α-$MnMoO_4$ and essentially octahedral in α-$CoMoO_4$. The ir spectrum of α-$CoMoO_4$ has been reported [3,4], whereas no Raman study of the present systems has been done except for our recent study on the ^{18}O-substituted catalysis [5]. We found that α-$CoMoO_4$ transfers into β-$CoMoO_4$ under the laser beam, which will be also reported.

$MMoO_4$ was prepared by means of stirring mixture of ammonium paramolybdate with metal nitrate, followed by filtration, drying and anneal at 540 °C. The purity of α-Mn(Co)MoO_4 was confirmed by the X-rays analysis. The Raman spectrum was observed by use of a JASCO-R800 spectrometer and 514.5 and 488.0 nm lines of an Ar^+ ion laser at room temperature. The ir spectrum except for β-$CoMoO_4$ was observed by use of a DIGILAB-FTS spectrometer. The Raman spectra are given in Fig. 1.

In Fig. 1 the strong band at 701 cm^{-1} in α-$CoMoO_4$ decreases gradually in intensity under the laser power of about 400 mW, resulting in Fig. 1-b, whose spectral feature is similar to that in α-$MnMoO_4$. This indicates clearly that α-$CoMoO_4$ transfers into β-$CoMoO_4$, which is consistent with the fact that β-$CoMoO_4$ is the stable structure at room temperature; α-$CoMoO_4$ is the stable sturcture at higher than 500 °C.

The normal modes are factored in both systems as Γ = 19Ag(R) + 17Bg(R) + 14Au(ir) + 19Bu(ir). The Mo-O stretchings are as: Γ = 5Ag + 3Bg + 3Au + 5Bu in α-$MnMoO_4$ and Γ = 7Ag + 5Bg + 5Au + 7Bu in α-$CoMoO_4$. It should be noted that there are two types of arrangements around Mo in both systems with a wide range of Mo-O distances lying between 0.1724 and 0.1851 nm in α-$MnMoO_4$ or 0.1719 and 0.2328 nm in α-$CoMoO_4$.

The higher bands than 900 cm^{-1} correspond well in the present systems, as shown in Fig. 1. This is true for the ir bands. The band at 701 cm^{-1} in α-$CoMoO_4$ corresponds to that at 737 cm^{-1} in α-$MnMoO_4$ in Fig. 1. The same is found in the ir spectra. On the other hand two bands at 880 and 821 cm^{-1}

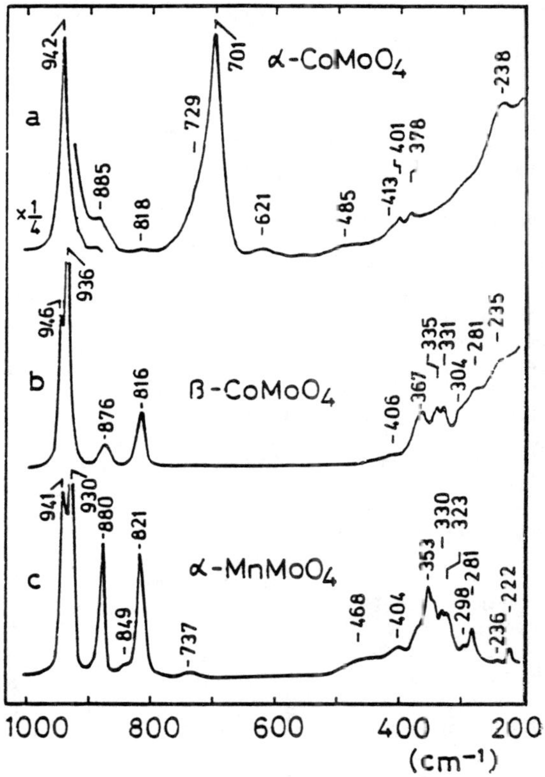

Fig. 1. The Raman spectrum of $MMoO_4$.

in α-$MnMoO_4$ seems to have no counter parts in both the ir and Raman spectra of α-$CoMoO_4$, except for the weak bands, as seen in Fig.1. These are explained in terms of the slight coupling of one or two Mo-O stretchings with the others in the tetrahedron or octahedron, as suggested [4]. Hence the bands at about 940, 840 and 720 cm^{-1} are attributed to the bonds with distances of about 0.172 0.180 and 0.190 nm, respectively. This is consistent with our recent study [5]. These suggest that the Mo-O stretchings due to the bonds with distances of 0.232 nm are quite low; one of them is probably assigned to the band at 238 cm^{-1} in Fig. 1-a.

In Fig. 1-b and -c the Mo-O stretchings coincide almost each other. On the other hand some bands in the region of 400 300 cm^{-1} shift to the higher frequency from α-$MnMoO_4$ to β-$CoMoO_4$, that is, 353 to 367, 330 to 335, 323 to 331 and 298 to 304 cm^{-1}, respectively. That is, these are sensitive to M and may be assigned to translational modes of M. The band at 404 or 406 cm^{-1} corresponds well to ν_2 [6], in the MoO_4^{2-} ion and the band at 281 cm^{-1} in both compounds may result from ν_4, 318 cm^{-1} [6]. That is, these as well as the Mo-O stretchings are insensitive to M. This supports, thus, adequacy of translational modes described above.

References

1. S.C.Abrahams and J.M.Reddy, J. Chem. Phys. 43, 2533 (1965).
2. G.W.Smith and J.A.Ibers, Acta Cryst. 19, 269 (1965).
3. G.M.Clark and W.P.Doyle, Spectrochim. Acta 22, 1441 (1966).
4. C.G.Barraclough, J.Lewis, and R.S.Nyholm, J. Chem. Soc. 3552 (1959).
5. I.Matsuura, I.Kanesaka, and H.Hashiba, Chem. Lett., in press.
6. R.H.Busey and O.L.Keller,Jr., J.Chem.Phys. 41, 215 (1964).

TETRANUCLEAR PLATINUM COMPLEXES STUDIED BY RESONANCE RAMAN SPECTROSCOPY.

Paul Stein and Hans K. Mahtani

Department of Chemistry, Duquesne University, Pittsburgh, PA 15282

Raman spectroscopy provides a probe to Pt-Pt bonding in dimeric [1,2] and tetrameric platinum compounds. Structurally similar tetranuclear Pt complexes with formal Pt oxidation states 2.0[3], 2.25[4,5], and 2.5[6] have been analyzed by X-ray crystallography and shown to consist of two amidate bridged Pt dimers with the exocyclic oxygens coordinated to the inner platinum atoms (Pt_2 and $Pt_{2'}$ atoms in fig. 1). The bidentate ligands are deprotonated α-pyridone, α-pyrrolidone or 1-methyluracil. The tetramers are nearly of C_{2h} symmetry with the Pt atoms arranged in a zigzag fashion.

α-Pyridone blue has been the most extensively investigated and its low frequency Raman spectrum is given in figure 2. Raman bands between 61-84 cm^{-1} are reported (Table I) for other crystallographically analyzed tetramers and assigned to the Pt_2-$Pt_{2'}$ stretch. This mode should be of low frequency since the Pt_2-$Pt_{2'}$ bond is not reinforced with bridging ligands although hydrogen bonding may provide some stabilization. The Pt-Pt stretching frequency in Pt(II)tetrakis(diphosphite) occurs at 116 cm^{-1}[1]. A nice inverse correlation between the Pt_2-$Pt_{2'}$ distance and $\nu(Pt_2-Pt_{2'})$ is observed in Table I. An increase in $\nu(Pt_2-Pt_{2'})$ with an increase in the formal oxidation state provides spectroscopic evidence for a Pt-Pt antibonding HOMO. The Pt_2-$Pt_{2'}$ stretch is not possible for binuclear complexes and therefore this Raman band is diagnostic for the tetranuclear structures[7]. We have not observed this band in any of the structurely characterized binuclear platinum complexes of 1-methyluracil or α-pyridone. The second Raman band at 149 cm^{-1} is assigned to a vibration of the outer Pt atoms with the overtone of Pt_2-$Pt_{2'}$ stretch unresolved. A combination/overtone is assigned to the broad band at 218 cm^{-1} in α-pyridone blue.

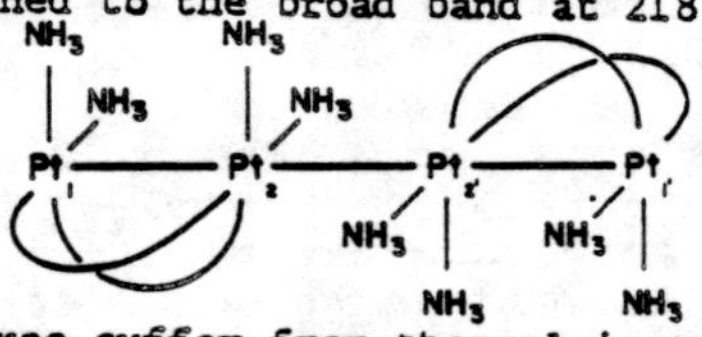

Fig. 1. Structure of the tetranuclear cation $[(Pt_2(NH_3)_4L_2)_2]^{n+}$ n = 4,5,or 6

These complexes suffer from thermal instability and show sensitivity to pH and anions causing the optical absorptions to vary. Interest has been given to the assignments of the optical absorptions and to their variation with solution conditions. The low frequency Raman bands completely dominate the rR spectra of the blue complexes indicating the optical absorptions (680-750 nm) arise from either mixed valent or metal-centered transitions. Solutions of α-pyrrolidone tan do show weak rR bands from carbonyl stretching and Pt-O stretching suggesting some charge transfer character is involved with the optical absorptions (470-490 nm). These observations are in accord with X-α calculations of α-pyridone blue[8].

TABLE I.

COMPOUND[1]	Pt_2-$Pt_{2'}$ dist(Å)	av. Pt oxidn state	Raman $\nu(Pt_2$-$Pt_{2'})$ (cm^{-1})	optical absorption[2] (nm)	ref
$[Pt_2(NH_3)_4(C_4H_6NO)_2]_2^{6+}$	2.702	2.5	84	480[3]	6
$[Pt_2(NH_3)_4(C_5H_4NO)_2]_2^{5+}$	2.877	2.25	69	680	4
$[Pt_2(NH_3)_4(1\text{-MeU})_2]_2^{5+}$	2.865	2.25	66	740	5
$[Pt_2(en)_2(C_5H_4NO)_2]_2\cdot(NO_3)_4$	3.236	2.0	61	—	3

1. Cations were prepared from ref procedures. en = ethylenediamine, Raman spectra of the fourth entry was collected on single crystals.
2. Major bands only
3. Absorptions range from 476–490 nm with different batches of α-pyrrolidone tan as prepared from air oxidation of α-pyrrolidone green solutions. See figure 3.

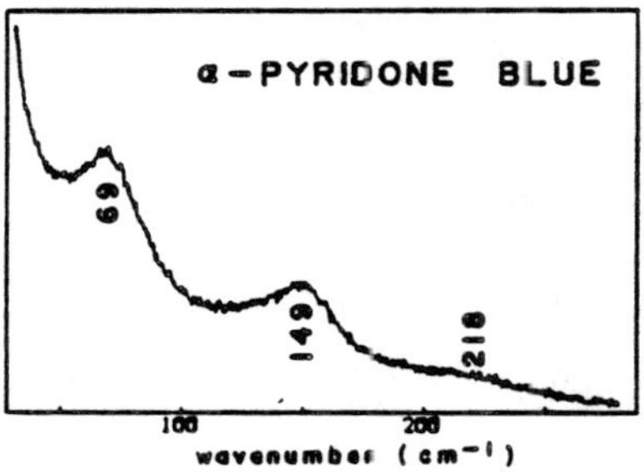

Fig.2. rR spectra of α-pyridone blue (excitat = 6764 Å) (0.012 gr PPB. 0.042 gr NH_4NO_3 in 3 ml of 0.05 M HNO_3).

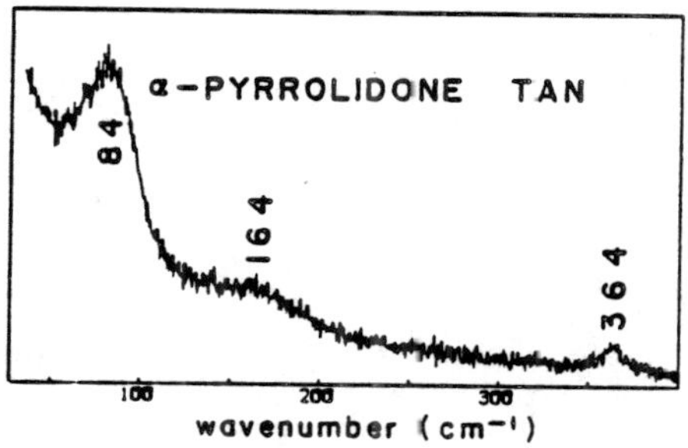

Fig.3. rR spectra (excitat = 4765 Å) of dark brownish red solution with optical absorption at 480 nm.

ACKNOWLEDGEMENT: This research was supported by a grant from Research Corporation and a grant from NSF (CHE-8307651).

REFERENCES:

1. P. Stein, M.K. Dickson, D.M. Roundhill, J. Am. Chem. Soc. 105,3489(1983).
2. P. Stein, Proc. VIIIth Int. Conf. on Raman Spect., 651 (1982) in Raman Spectroscopy: Linear and Nonlinear (wiley).
3. L.S. Hollis, S.J. Lippard, Inorg. Chem. 22, 2600 (1983).
4. L.S. Hollis, S.J. Lippard, J. Am. Chem. Soc. 105, 3494 (1983).
5. P.K. Mascharak, I.D. Williams, S.J. Lippard, ibid. 106, 6428 (1984).
6. K. Matsumoto, H. Takahahi, K. Fuwa, Inorg. Chem. 22, 4086 (1983).
7. P. Stein, H.K. Mahtani, J. Am. Chem. Soc. (submitted for publication).
8. A.P. Ginsberg, T.V. O'Halloran, P.E. Fanwick, L.S. Hollis, S.J. Lippard, J. Am. Chem. Soc. 106, 5430 (1984).

RAMAN AND MULTINUCLEAR NMR SPECTRA OF PEROXO COMPLEXES OF Mo, W AND V IN AQUEOUS SOLUTION.

Nicholas J. Campbell, Andrew C. Dengel and William P. Griffith

Department of Chemistry, Imperial College, London SW7 2AY, Great Britain

Raman spectra (50–1200 cm^{-1}) have been measured of aqueous solutions of $[MoO_4]^{2-}$, $[WO_4]^{2-}$ and $[VO_4]^{3-}$ at 0.1–1 M metal concentrations as a function of pH (> 14 to 0) in the absence and in the presence of H_2O_2 (metal: H_2O_2 ratios from 2:1 to 1:10). Such solutions were also studied by IR (800–1200 cm^{-1}) and NMR (^{95}Mo, ^{51}V). Solute species were identified by their Raman activities, their NMR chemical shifts and also by comparison of their Raman profiles with those of solid peroxo and polyperoxo complexes of known structures. In particular we have used normal and ^{18}O-substituted $K_2[M_2O_3(O_2)_4(H_2O)_2]$ (M = Mo, W) and $K_3H[V_2O_3(O_2)_4]$ as models; X-ray structures on the Mo[1] and W[2] complexes are known, and we have recently reported that of the vanadium complex.[3]

The results, which extend those of our earlier studies, suggest that the complex polymerisation processes observed for molybdates, tungstates and vanadates are much simpler in the presence of H_2O_2 than in its absence, presumably because the side-bonded O_2^{2-} ligand competes successfully for O^{2-} in high-oxidation state early transition metals so that fewer oxo sites are available to form μ-oxo bridges.

A simplified description of the principal equilibria involved (for M metal and excess H_2O_2) involves prior formation of tetra peroxo species $[M(O_2)_4]^{n-}$ (n = 2 for Mo, W; 3 for V) at high pH; near pH 6 triperoxo intermediates $[Mo(O_2)_3]^{n-}$ may be formed and at lower pH the dimers $[M_2O_3(O_2)_4(H_2O)_2]^{n-}$ (n = 2 for Mo, W, 4 for V) predominate. Protonated forms of this may occur at still lower pH for the Mo and V systems, and in acid solution (e.g. HCℓ) the monomeric species $[Mo(O_2)Cℓ_4]^{2-}$ (Mo, W) and $[VO(O_2)Cℓ_4]^{3-}$ are formed.

Acknowledgements. We thank the Science and Engineering Research Council (SERC) and Interox Chemicals Ltd. for support of this work.

1. R. Stomberg, Acta Chem. Scand., 1968, 22, 1076.

2. P.W.B. Einstein and B.R. Penfold, Acta Cryst., 1964, 17, 1177.

3. N.J. Campbell, J. Flanagan, W.P. Griffith and A.C. Skapski, Trans. Met. Chem., 1985, 10, 353.

4. W.P. Griffith and P.J.B. Lesniak, J. Chem. Soc. (A), 1969, 106.

ABSOLUTE RAMAN INTENSITIES AND EOPs OF ClO_n^- ANIONS IN AQUEOUS SOLUTIONS.

Hans H. Eysel

Anorganisch-Chemisches Institut der Universität Heidelberg
Im Neuenheimer Feld 270, D-6900 Heidelberg-1, Germany

A four-channel photon counting technique is devellopped in order to simultaneously assess and separate isotropic and anisotropic Raman scattering from liquid samples and liquid standards. To this purpose a Spex 1403 double monochromator and two Elscint real time single channel photon counters are coupled with an Apple IIe PC. The computer controls monochromator movement, polarization switching and data aquisition. Photon counter gates are controlled by the rotation of the twin-chamber spinning cell.

Spectra are then numerically corrected for the influence of the refractive index (Lorentz local field and solid angle correction) and for the total apparatus+detector throughput. Finally the spectra are set on a molar scale and reduced according to the harmonic oscillator model. Hence our representation of isotropic and anisotropic Raman scattering are immediately proportional to $45\bar{\alpha}^2$ and $3\gamma^2$,respectively.

Accuracy and reproducibility of this procedure were tested with numerous organic standard solvents such as CCl_4, C_6H_6, C_6D_6, C_6H_{12}, CH_3CN, CH_2Cl_2, CS_2, and aqueous solutions of Na_2SO_4 and $NaClO_4$. Agreement with many of the literature data is within 5%. Reproducibility of successive independent measurements is within 2%.

The method was then applied to study the intensities of the various chlorine oxo-acids. Solutions of NaClO, $NaClO_2$, $NaClO_3$, and $NaClO_4$ were studied at moderate concentrations from about 0.1 molar to 2 molar. Within this range of concentrations, the integrated molar Raman intensities decrease generally very little with increasing concentration, so that extrapolation to "zero concentration" might not be essential, but is nevertheless included in our EOPs calculations.
Using force fields from the literature we derive polarizability data of the Cl-O bond of anions in the the oxidation states +1,+3,+5,and +7 of the central chlorine atom. The variation of $(d\alpha/dQ)$ and $(d\alpha/dr)_{Cl-O}$ is discussed with respect to π-bond and lone-pair contributions to the derived bond polarizability.

RECENT APPLICATIONS OF MICRORAMAN SPECTROSCOPY IN INORGANIC CHEMISTRY.

Pierre D. Harvey[a] and Ian S. Butler[b]

[a]Division of Chemistry, California Institute of Technology, Pasadena, CA 91125 and [b]Department of Chemistry, McGill University, Montreal, Quebec, Canada H3A 2K6.

The new microRaman technique is proving extremely useful to chemists because of the microscopic amounts of material needed for analysis and the very low laser powers that can be used for spectral excitation (sometimes < 1 mW) owing to the high collection efficiency of the microscope optics. The common problem of laser-induced, thermal damage to strongly-absorbing samples can almost always be avoided (especially when using a defocused laser beam) because of the rapid heat loss from the microscopic sample, and high-quality Raman data can now be obtained for materials earlier considered totally unsuitable for analysis. As a continuation of our work on the applications of microRaman spectroscopy in inorganic chemistry,[1,2] we have succeeded in recording the spectra of several other types of inorganic complexes that are difficult from the Raman point of view. Examples of the complexes studied are: (1) mixed-metal organometallic complexes such as $Mn(CO)_4(CNBu^t)Re(CO)_5$ and $Mn(CO)_5Re(CO)_4(CNBu^t)$; (2) rhodium cluster complexes such as $Rh_6CO)_7(Ph_3P)_9$; and (3) halocopper(II) anions such as $[CuCl_4]^{2-}$ and $[CuCl_5]^{3-}$. Vibrational assignments have been made for the major features appearing in the Raman spectra. For instance, the ν(MnRe) mode in the metal-metal bonded species appears at ca. 158 cm^{-1}, while a comparison of ν(CN) for the free and complexed $CNBu^t$ ligand indicates non-linear isonitrile groups. The modes of the triply-bridged CO groups in the rhodium clusters are too weak to be detected even at 100 K, but the terminal ν(CO) modes are readily observed. Data for the Cu(II) complexes were also fairly easily obtained under the microRaman conditions. It is clear that microRaman spectroscopy will continue to prove to be another important spectroscopic tool for the modern inorganic chemist.

1. P.D. Harvey and I.S. Butler, Microbeam Analysis 1985, ed. J.T. Armstrong, San Francisco Press Inc., San Francisco, 1985, pp. 49-51.

2. I.S. Butler, P.D. Harvey and G.C. Allen, J. Raman Spectrosc., 1986, in press.

THE PRE-RESONANCE RAMAN EFFECT OF THE SOLVATED IONS $I^-.SO_2$ AND $SCN^-.SO_2$.

P.S.Santos,J.G.N.Santos and D.L.A.Faria
Instituto de Química da Universidade de São Paulo
São Paulo,BRASIL

The species $I^-.SO_2$ and $SCN^-.SO_2$,existing in solution and that can also be isolated in the form of crystaline alkaline salts,have been described in the literature as sulfur dioxide solvates or ,alternatively,as halosulfinate ions[1].Although some previous spectroscopic studies have been reported for the iodide solvate,the nature of the interaction in such solvates was not satisfactorily investigated.

In the present study the Raman spectra of the forementioned species in acetonitrile solutions were obtained using several exciting radiations from 647.1nm to 457.9nm.The excitation profiles,obtained with the acetonitrile Raman band at 920 cm^{-1} as internal standard,show the pre-resonance enhancement of the SO_2 symmetric stretching mode as the exciting radiation approaches the electronic transition frequencies at ca. 380 nm and 360 nm ,for the iodide and thiocyanate,respectively.In the case of the iodide it was possible to detect the overtone of the enhanced mode in the Raman spectrum excited by the 457.9nm radiation.The SO_2 symmetric stretching frequency was observed at 1115 cm^{-1} and 1128 cm^{-1} for the iodide and thiocyanate solvates,respectively,and these values compared with 1151 cm^{-1} for gaseous SO_2 indicate that the interaction between SO_2 and the ions is weaker than the existing in the well studied trimethyl amine.SO_2 molecular complex[2] ,but much stronger than the existing in the SO_2 clathrate of hydroquinone[3] .On the other hand the observed frequency shifts in the SCN^- Raman bands are quite small.All these facts strongly suggest that the species $I^-.SO_2$ and $SCN^-.SO_2$ are better described as charge transfer molecular complexes formed by the electron donation into the lowest lying molecular orbital of sulfur dioxide.The absorption bands in the near UV can be assigned as due to charge transfer transitions,that according to the various existing resonance Raman theories give origin to enhancementof totally symmetric modes and appearance of their overtones by means of the Franck Condon mechanism.

1.D.F.Burrow,Inorg.Chem.,11,573(1972).
2.C.S.Sass and B.S.Ault,J.Phys.Chem.,88,432(1984).
3.P.S.Santos and P.C.Isolani,J.Mol.Struct.,67,487(1979).

RAMAN SPECTRA OF HALOMERCURIOMETHANES $CH_2(HgX)_2$ AND $CH(HgX)_3$ (X = Cl, Br and I) IN VARIOUS STATES*.

D. K. Breitinger and W. Kress

Institute of Inorganic Chemistry, University of Erlangen-Nürnberg, Egerlandstrasse 1, D-8520 Erlangen, F. R. G.

INTRODUCTION

In the course of structural [2] and spectrometric (multinuclear NMR, vibrational, UV) [3,4] investigations of anionomercuriomethanes $CH_{4-n}(HgX)_n$ ($1 \leq n \leq 4$) the vibrational spectra of bis- and tris(anionomercurio)methanes (n = 2 and 3; X = Cl, Br and I; CN, SCN and OAc) have been studied [5]. Raman and IR spectra of the polycrystalline compounds $CH_2(HgX)_2$, and of deuterated derivatives, together with polarization data for solutions, allow reliable assignments of most of the observed fundamentals. Additional support of the assignments and insight into some details of the solid-state spectra are achieved with low-temperature vibrational spectra of polycrystalline samples and with single-crystal Raman spectra [1]. In this contribution the influence of solvation on some fundamentals of the molecules $CH_2(HgX)_2$ and the Raman spectra of representatives of the compounds $CH(HgX)_3$ will be discussed.

RESULTS AND DISCUSSION

The molecular solids $CH_2(HgX)_2$ (X = Cl and Br) are held together by strong intermolecular donor-acceptor interactions [6], which influence the frequencies of mainly the valence vibrations $\nu(CHg_2)$ and $\nu(HgX)$. On dissolution of these compounds in various solvents these interactions are substituted for by weaker solvent-solute interactions, and, as a consequence, the frequencies of the fundamentals observed for solutions generally increase, as compared with the solids (Table 1). On the basis of these frequency shifts a decrease of the solvent-solute interactions in the series py > DMSO > TEMUR

Table 1. Raman bands of the solids $CH_2(HgX)_2$ (X = Cl, Br and I) and of solutions in pyridine (py), dimethylsulfoxide (DMSO) and tetramethylurea (TEMUR) (p = polarized)

		$CH_2(HgCl)_2$	$CH_2(HgBr)_2$	$CH_2(HgI)_2$
$\nu_s(CHg_2)$	solid	512	501	486
	py	512 p	500 p	
	DMSO			503 p
	TEMUR	527 p	518 p	
$\nu_s(HgX)$	solid	300	213	177
	py	320 p	221 p	
	DMSO		223 p	181 p
	TEMUR	331 p	229 p	

* Metallomethanes XV. For XIV see [1].

Table 2. Raman spectra of solid $CH(HgX)_3$ (I) and $CD(HgX)_3$ (II) (X = Cl)

	$\nu(CH/D)$	$\delta(H/DCHg)$	$\nu_d(CHg_3)$	$\nu_s(CHg_3)$	$\nu_d(HgX)$	$\nu_s(HgX)$	$\delta(CHgX)$	$\delta(CHg_3)$
(I)	2948 w	971 m	654 m	482 m	323 sh	309 vst	120 st	73 m
	2940 sh	940 m	625 m	474 m			86 m	58 m
(II)	2191 vw	760 m,br	638 w,br	471 m,br	326 sh	309 vst	121 vst	67 vst
			598 w,br					

is suggested. Possible intermolecular interactions in the molten compounds $CH_2(HgX)_2$ will be studied by Raman spectrometry.

The vibrational spectra of tris(anionomercurio)methanes $CH(HgX)_3$ are largely understood on the basis of polarization data for solutions, of solid-state effects and of frequency shifts on deuteration. Some points of interest will be discussed shortly for the pair $CH(HgCl)_3/CD(HgCl)_3$ as representatives for the compounds $CH(HgX)_3$ (Table 2). The assignments of the vibrations involving the H-atoms are obvious. To some extent, the vibrations of the CHg_3-entity are also affected by deuteration. The large splittings of the originally degenerate modes $\delta(HCHg)$ and $\nu_d(CHg_3)$ are to be attributed to lowering of the symmetry of the molecules in the solid state, whereas the small splittings of the symmetric modes $\nu(CH)$ and $\nu_s(CHg_3)$ can only be caused by correlation coupling. The strong bands at 309 and 120 cm^{-1} are the in-phase valence and deformation modes of the three HgX- and CHgX-groups, respectively. Up to now, the three low-frequency bands of $CH(HgCl)_3$ cannot be assigned unambigously to distinct deformations of the CHg_3-entity.

As with the compounds $CH_2(HgX)_2$ the frequencies of the $CH(HgX)_3$ molecules in solution shift to higher values. E.g. in the case of $CH(HgCl)_3$ in DMSO $\nu_s(CHg_3)$ and $\nu_s(HgCl)$ appear at 484 (p) and 327 cm^{-1}(p), respectively (for the frequencies of the solid see Table 2). From this solution a labile solvate $CH(HgCl)_3 \cdot DMSO$ can be obtained. The vibrations of the $CH(HgCl)_3$ molecules in this solvate are not systematically affected; thus, $\nu_s(CHg_3)$ increases slightly, whereas the $\nu(HgCl)$ decrease by a few cm^{-1}. However, a drop of the $\nu(SO)$ vibration of the DMSO molecule by 35 cm^{-1} indicate interaction of DMSO via oxygen, possibly in a similar way as in $C(HgCl)_4 \cdot DMSO$ [2].

CONCLUSION

The influence of solvation onto the vibrations of bonds involving mercury atoms suggest that the solvent molecules interact as donors with the mercury atoms of the $CH_{4-n}(HgX)_n$ molecules as acceptors.

REFERENCES

1. D.K. Breitinger and W. Kress, J. Mol. Struct. 142 or 143(1986), in press.
2. D.K. Breitinger, G. Petrikowski, G. Liehr and R. Sendelbeck, Z. Naturforsch. 38b(1983)357, and references therein.
3. J. Mink, D.K. Breitinger, W. Kress, W. Morell and R. Sendelbeck, J. Organomet. Chem., in press, and references therein.
4. D.K. Breitinger and W. Kress, J. Organomet. Chem. 256(1983)217.
5. W. Kress, Ph.D. Thesis, Erlangen (1983).
6. K.P. Jensen, D.K. Breitinger and W. Kress, Z. Naturforsch. 36b(1981)188.

RESONANCE RAMAN SPECTRA OF METALLO-PHTALOCYANINS.

P.V.THOAI

S.R.S., B.P.441 - 75161 PARIS Cedex 04 - FRANCE

INTRODUCTION

Metallo-phtalocyanins are recently revealed to have very interesting electrochemical behaviour, as they can be very easily oxidized or reduced. Some of them are good potentials as electrochemichromes for possible use in displays or in wide planar screens. Some others are expected to be photosensible semi-conductors with potential implication in the photo-conversion of solar energy.

If Red-Ox reactions and electric conductivity of some of metallo-phtalocyanins are extensively explored, their other physical properties and structures are not yet well known.

RESULTS AND DISCUSSION

In this paper, we shall present our results concerning Magnesium-Phtalocyanin (Mg-Pc), Zinc-Phtalocyanin (Zn -Pc) and Iron-Phtalocyanin (Fe-Pc).

As the color of these materials changes with their environment and temperature and their practical stability ranges are rather drastically limited, we have also studied their ligand with a strong electron donor, pyridine. Thus the Raman spectra of complexes (Py-Mg-Pc), (Py-Zn-Pc) and (Py-Fe-Pc) are recorded and analyzed in correlation with their possible structure.

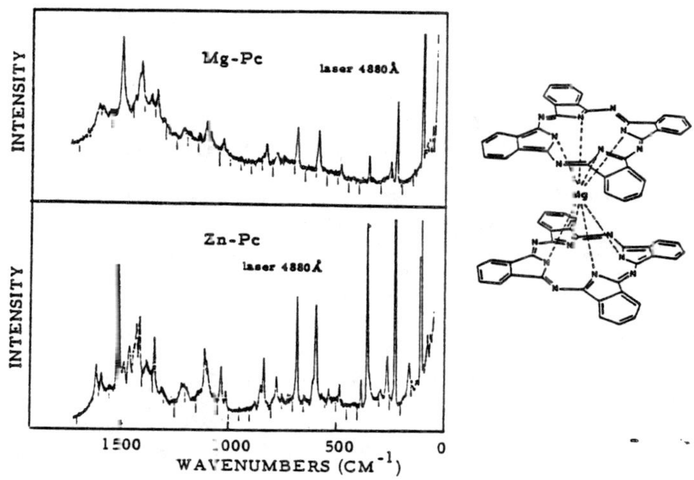

Fig.1 - Raman spectra of metallo-phtalocyanins.

Metallo-phtalocyanins

Resonance Raman spectra of the mentionned metallo-phtalocyanins are recorded with nine lines of a ion-Argon Laser from 454.5 nm to 514.5 nm (fig.1). Big changes are observed when compared with free phtalocyanin recorded in the same conditions. The shifts of bands in the spectral range from 1000 to 1700 cm^{-1} can be assigned to a doming effect, similar to some metalloporphyrins (1-4).

Pyridine-metallo-phtalocyanin complexes

When complexing with Pyridine, the metallo-phtalocyanin exibits new bands due to the new external ligand M-N. Its stretching frequency is very sensitive with the nature of the metal (fig.2). The macro cycle

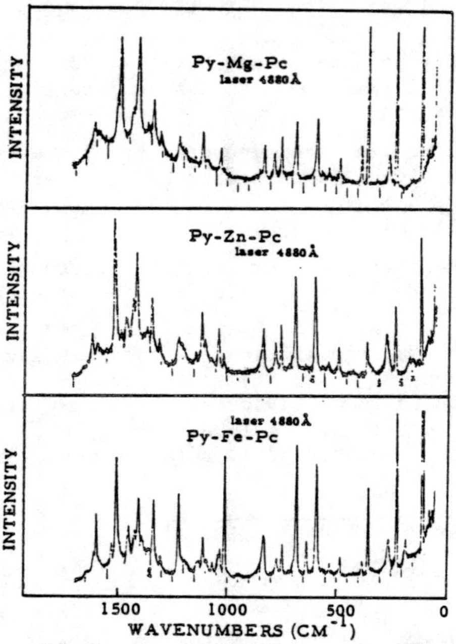

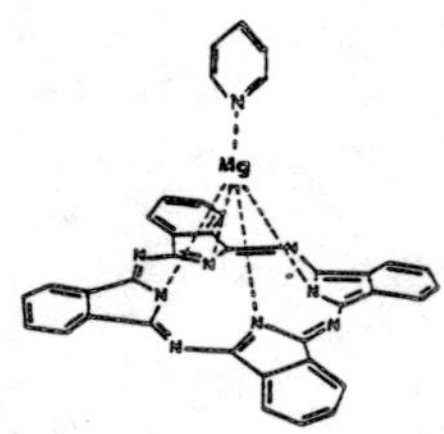

Fig.2 - Raman spectra of Pyridine-metallo-phtalocyanin complexes.

vibrations about 1500 cm^{-1} are noticeably shifted in particular with Mg and Fe derivatives. The sens of these shifts indicates that the doming angle is wider, in other terms, the metal atom is nearer to the average plane of the bounded nitrogen atoms.

REFERENCES

1 - P.V.THOAI , These Doctorat ès Sciences, Paris VI, 1976.

2 - P.V.THOAI,J.P.LEICKNAM, Proc.6th Intern.Conf.Raman Spect., Bangalore, 1978, 152.

3 - P.V.THOAI,J.P.LEICKNAM,M.MOMENTEAU,B.CODDE, Spectr.Lett.,1979, 72,333.

4 - P.V.THOAI, in Raman Spectroscopy, Linear and Non Linear, Wiley, Chichester, 1982, 743.

OUT-OF-PLANE DEFORMATION MODES OF NI(II)-OCTAETHYLPORPHYRIN IN RESONANCE RAMAN SPECTRA.

Xiao-yuan Li, R.S.Czernuszewicz, J.R.Kincaid* and T.G.Spiro

Department of Chemistry, Pinceton University,
Princeton, N.J. 08544, U.S.A.
*Department of Chemistry, Marquette University,
Milwaukee, WI 53233, U.S.A.

Although extensive resonance Raman(RR) studies and structural correlations have been done on metalloporphyrin in-plane modes in the 1000 - 1700cm^{-1} region[1], the modes below 1000cm^{-1} are not well understood. This region is of much interest, however because it contains four different modes; a) in-plane deformations, b) out-of-plane deformations, c) metal-ligand stretches (both in- and out-of-plane), and d) peripheral substituent deformations. The porphyrin deformation modes are expected to be sensitive indicators of the ring distortions (ruffling, doming, etc.), whereas metal-ligand vibrations can directly provide information about the metal-ligand bonds.

We report high resolution RR spectra of octaethylporphinatonickel(II) (Ni-OEP) and its meso-deuterated and pyrrole-^{15}N substituted derivatives in two different crystalline forms; the triclinic form with D_{4h} symmetry of the square planar Ni-OEP molecule and the tetragonal form, in which the molecular symmetry is effectively reduced to D_{2d} by S_4 ruffling of the porphinato ring. The RR spectra of both forms, excited at 514.5nm, have previously been reported by Spaulding et al. [2], but only four high frequency (> 1500 cm^{-1}) structure-sensitive bands were discussed.

In this work pure D_{4h} and D_{2d} Ni-OEP crystal forms were excited within both Soret and Q bands absorption maxima, with several lines of the Ar^+ (488.0 and 514.5nm) and Kr^+ (406.7, 413.1, 530.8, and 568.2nm) lasers. The use of variable wavelength excitation has provided more complete spectra, and several previously unrecognized bands are detected. As expected, the tetragonal form is shown to display more RR bands in the region below 1000 cm^{-1} than the triclinic form. These additional bands are assignable to the porphyrin out-of-plane vibrations that are activated by symmetry lowering of the porphinato core. For example, a medium-intense depolarized band appeared at 842 cm^{-1} in the tetragonal form at both Soret (413.1nm) and Q band (568.2nm) excitations, which shifted to 679 cm^{-1} upon deuteration at the meso carbons (C_m). These frequencies coincide with the solution IR $A_{2u}(D_{4h})$ C_m-H(843cm^{-1}) and C_m-D(680cm^{-1}) out-of-plane deformation modes, reported by Kitagawa et al. [3]. The A_{2u} mode is Raman inactive in D_{4h} point group but it can be activated once the symmetry is lowered to D_{2d}, $A_{2u}(D_{4h}) \rightarrow B_2(D_{2d})$. Thus, we assigned the 842 cm^{-1} RR band to the C_m-H out-of-plane deformation vibration.

Another example is the low frequency, depolarized band at 357 cm^{-1}, which occurs only in the Q band excited RR spectrum. It shifts down

22cm^{-1} upon meso-d_4 substitution. Again, coincident frequency (355 cm^{-1}) and similar meso-d_4 downshift (21 cm^{-1}) can be found in the solution IR spectra of Ni-OEP [3], although the origin of 355 cm^{-1} IR band was not assigned. We believe that the 357 cm^{-1} RR band and its 355 cm^{-1} correspondance arise from a pyrrole out-of-plane tilting vibration, B_2 (D_{2d}) and A_{2u} (D_{4h}), respectively.

In order to ascertain accurate assignments of these modes and all other structure-sensitive RR bands, a great effort was made to develop detailed valence force fields for both in- and out-of-plane vibrations of Ni-OEP [4]. Our calculated values are in excellent agreement with the experimental data and the pyrrole-^{15}N and meso-d_4 isotopic shifts are well reproduced. As a result a more coherent picture of in- and out-of-plane vibrational assignments than has hitherto been available emerges from this work.

The resonance enhancement mechanism for the out-of-plane vibrations will be discussed. Ni-OEP was also examined in CS_2 solution and a direct influence of structure on the RR spectra will be demonstrated.

REFERENCES:

[1] T.G. Spiro in "Iron Porphyrins" Part Two, A.B.P. Lever and H.B. Gray, eds., Addison-Wesley, Reading, Mass. (1982).

[2] L.D. Spaulding, C.C. Chang, Nai-Teng Yu, and R.H. Felton, J. Am. Chem. Soc. 97, 2517 (1975).

[3] T. Kitagawa, M. Abe, H. Ogoshi, J. Chem. Phys. 69, 4516,4526 (1978).

[4] Xiao-yuan Li and T.G. Spiro, Paper Abstracts, INOR. 252, 191st National ACS Meeting, New York City, New York (1986).

RESONANCE RAMAN SPECTRA OF COBALT PHTHALOCYANINE AND LUTETIUM DIPHTHALOCYANINE IN THE SOLID STATE.

C.R. Bartholomew, A.A. McConnell and W.E. Smith

Department of Pure and Applied Chemistry
University of Strathclyde
GLASGOW G1 1XL
SCOTLAND

High quality Raman spectra have been obtained for copper phthalocyanine by presenting the sample as a compacted metal disc [1] made from a mixture of silver powder and finely divided particles of the phthalocyanine. The disc method improves the ratio of scattering to absorption, gives a heat sink and in some cases reduces or shifts fluorescence. Unfortunately the incorporation of a non-resonant standard into the mixture requires such high concentrations of standard that the effect of the metal matrix is reduced and on changing temperature the weak standard signal is often further reduced in intensity by changes in the physical packing of the non-metallic components of the disc. This effect is clearly shown by the resonance Raman spectra at low temperature of cobalt phthalocyanine with the standard mixed into the disc. The standard signal is very weak and therefore a poor comparison for the strong phthalocyanine peaks (Figure 1a) but it can be improved by placing a single crystal of the same standard on the surface. This is illustrated for lutetium diphthalocyanine (Figure 1b).

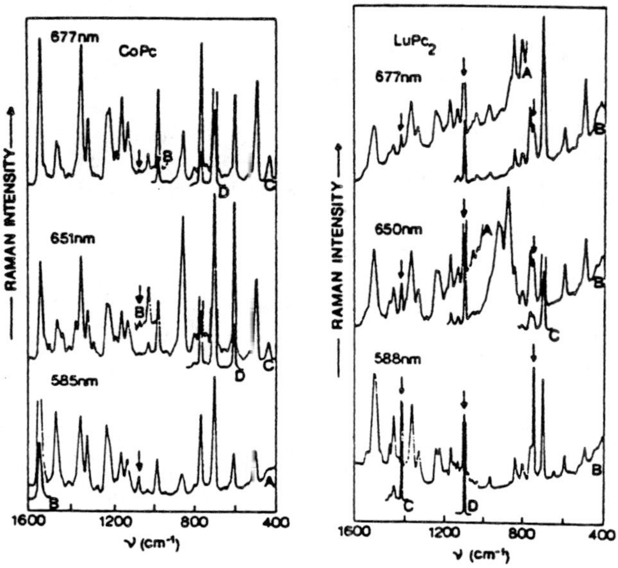

Figure 1

Raman spectra of cobalt phthalocyanine and lutetium diphthalocyanine in the Q band region using three different excitation wavelengths at a temperature of 10 K. The sensitivities are A:- 3000, B:- 10,000 cps, C:- 30,000 D:- 100,000 cps Slits 4 cm^{-1} Standard arrowed.

The single crystal technique is faster in alignment, achieves greater stability due to reduction in interaction between sample and the standard and provides a more accurate method for relative intensity measurements since a sufficiently large signal is achieved with both standard and sample. Thus when combined with the advantages of the silver disc technique, quality spectra are obtained particularly at low temperature. In the Q band region the resonance enhancement is mainly on the inner ring system and the main peaks have been assigned previously using as a basis Abe and Kitegawa's porphyrin nomenclature. The effect of the metal on frequency is illustrated in Table 1 (680 nm excitation) for four bands, two higher frequency modes ν_3 and ν_4 and two lower frequency modes ν_{15} and ν_7 .

Table 1 (650 nm excitation)

ZnPc	NiPc	CuPc	CoPc	$LuPc_2$	H_2Pc	Assignment	Symmetry
1514	1561	1530	1548	1534	1541	ν_3	A_{1g}
1341	1348	1341	1344	1341	1341	ν_4	A_{1g}
756	754	747	740	751	724	ν_{15}	B_{1g}
690	686	682	694	685	682	ν_7	A_{1g}

The frequency of ν_3 is more sensitive to the central metal ion than ν_4 [3]. In the case of ν_3 , C_α and N_m atoms move in phase causing a contraction and expansion of the bridging region.

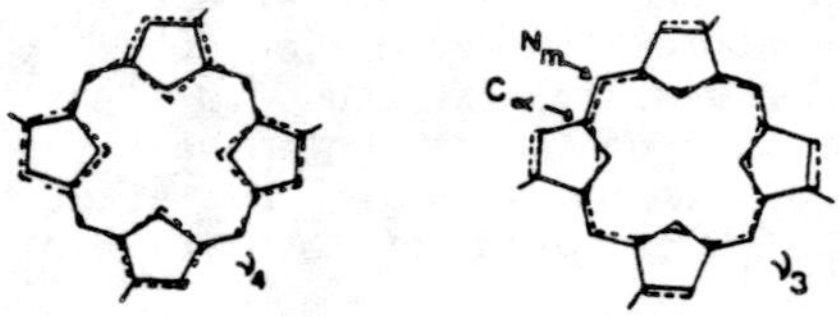

The ordering of the frequency of ν_3 is not simply related to size but requires a consideration of the doming of the ring and the bonding of the metal ion as illustrated by the apparent importance of the bridging region. In lutetium diphthalocyanine further changes such as the breadth of ν_3 and the sharp rise in intensity of the bands at about 800-900 cm^{-1} with 650 nm excitation reflects the fact that the metal does not fit into the ring. This effect in lutetium diphthalocyanine may be due to charge transfer between the rare earth and phthalocyanine or $\pi \rightarrow \pi^*$ transitions within the phthalocyanine.

References

1. J.A. Nimmo, A.J. Bovill, A.A. McConnell and W.E. Smith, J. Raman Spectrosc., 1985, 16, 245-250.

2. In Press

3. R. Kotz, E. Yeager, J. Electroanal. Chem., (1980), 113, 113-125.

RESONANCE RAMAN SPECTRA OF Co(II),Ni(II),Cu(II),Zn(II) PROTOPORPHYRIN IX DIMETHYL ESTER COMPLEXES

L.T.He and T.L.Jiang

Lanzhou Institute of Chemical Physics, Academia Sinica,Lanzhou, China

We report here the resonance Raman studies of Co(II), Ni(II), Cu(II) and Zn(II) protoporphyrin IX dimethyl ester (PP) complexes,using an assembled and computerized Raman spectrometer with different excitation lines of an argon ion laser.

Some resonance Raman bands worthy of attention and their assignments are listed in Table 1. The resonance Raman spectra of Cu(II)PP are shown in Figure 1. The 1638 cm^{-1} and 1581 cm^{-1} bands in Cu(II) are greatly en-

Table 1. Some resonance Raman frequencies (cm^{-1}) of metalloPP and their assignments*

Ni(II)PP	Co(II)PP	Cu(II)PP	Zn(II)PP	Assignment
1662	1650	1638	1618	$\nu_{C_a-C_m} + \nu_{C_a-N}$
1608	1596	1581	1564	$\nu_{C_a-N} + \nu_{C_a-C_m}$
1524	1512	1503	1487	$\nu_{C_a-C_m}$
1379	1375	1374	1369	ring breathing
754	754	753	756	$\delta_{C_a-C_m-C_a}$
676	676	675	676	ring deformation

*data obtained from samples in KBr pellets, using rotating cell technique.

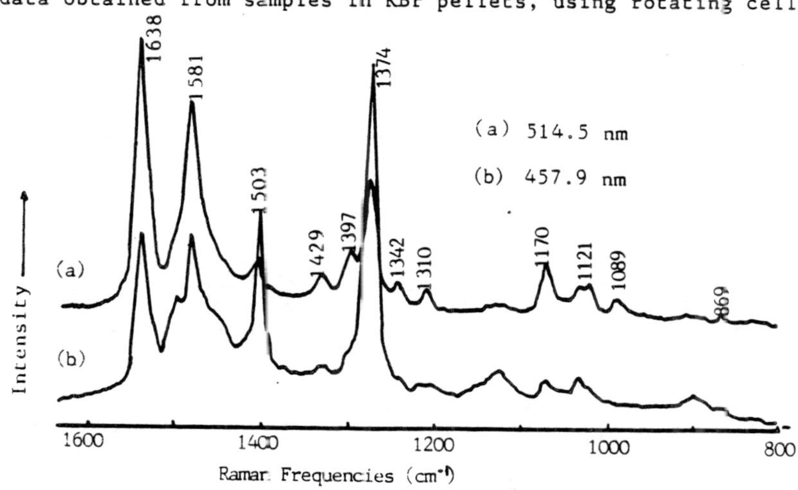

Fig. 1. Resonance Raman spectra of Cu(II)PP in CS_2 solution

hanced by 514.5 nm excitation which is near the β band of its optical absorption spectrum, while 1374 cm^{-1} and 1503 cm^{-1} bands show distinct enhancement by 457.9 nm excitation which is close to the soret band.

Table 1 shows that the Raman frequencies, except for 753 cm^{-1} and 676 cm^{-1} bands, decrease in the order of Ni(II) > Co(II) > Cu(II) > Zn(II). It can be attributed to the effects of sizes of metal ions and their electronic structures. For square planar complexes, the electron distribution of the splitted metal-ion d orbitals in Co(II) and Ni(II) is $3d^0_{x^2-y^2}$, in Cu(II), $3d^1_{x^2-y^2}$ and in Zn(II), $3d^2_{x^2-y^2}$. Since $d_{x^2-y^2}$ orbital lies in a plane, pointing towards the four porphyrin nitrogen atoms, the electrons present in $d_{x^2-y^2}$ orbitals of Cu(II) and Zn(II) would repel the nitrogen atoms and result in "core expansion" of the porphyrin ring, thus lower the vibrational frequencies. In Co(II) and Ni(II)PP, the $3d_{x^2-y^2}$ orbitals are empty, but radius of the former (0.72 Å)(1) is larger than that of the latter (0.69 Å); these two effects, perhaps, determine the lowering of the Raman frequencies in Co(II)PP.

With reference to (2-4) and the fact that the highly polarized band near 1375 cm^{-1} is relatively insensitive to the nature of the central metal ion, ~ 754 cm^{-1} and 676 cm^{-1} bands are almost unaffected by the central metal ion or the peripheral substituents, we assign the ~1375 cm^{-1} band to ring breathing mode and the ~ 754 cm^{-1} and ~676 cm^{-1} bands to $\delta C_a-C_m-C_a$ and ring deformation vibrations, respectively.

The dispersion of depolarization ratios for the four "structure marker" bands of Cu(II)PP is shown in Table 2. The depolarization ratios of the polarized 1375 cm^{-1} and 1503 cm^{-1} bands slightly change when varying the excitation wavelengths, but 1638 cm^{-1} and 1580 cm^{-1} bands vary from 0.59 to 1.17 and 0.31 to 2.80, respectively. For these two bands, rather large dispersion is observed, which denotes the lowering of the symmetry of the molecule as a whole.

Table 2. Depolarization dispersion (ρ_l) of Cu(II)PP in CS_2 solution

Excit.lines	457.9	476.5	488.0	496.5	514.5 (nm)
1638 cm^{-1}	0.59	0.67	0.81	0.92	1.17
1581	0.31	0.53	0.74	1.06	2.80
1503	0.25	0.34	0.35	0.42	0.43
1375	0.15	0.20	0.21	0.23	0.34

References

1. R.C. Weast and H.J. Astle, ed.,"CRC Handbook of Chemistry and Physics" 63rd ed., F-179, CRC Press Inc., Florida, 1982
2. S.Sunder, R. Mendelsohn and H.J. Bernstein, J. Chem. Phys. 63 573(1975)
3. T. Kitagawa, M. Abe and H. Ogoshi, J. Chem. Phys. 69 4516 (1978)
4. J.A. Shelnutt, D.C. O'shea, N.T. Yu, L.D. Cheung and R.H. Felton, J.Chem. Phys. 64 1156 (1976)

LIGAND DEUTERATION AS A MEANS OF DIFFERENTIATING BETWEEN METAL-METAL AND LIGAND-SENSITIVE VIBRATIONS OF DIMETAL-TETRAACETATES.

Robin J H Clark[a], Colin D Flint[b] and Andrew J Hempleman[a,b]

[a]Christopher Ingold Laboratories, University College London, 20 Gordon St., London WC1H 0AJ, U.K.

[b]Department of Chemistry, Birkbeck College, Malet St., London WC1E 7HX, U.K.

The assignment of metal-metal stretching vibrations in some dimetal tetraacetates remains controversial[1-3]. The main difficulty lies in the distinction between the metal-metal vibrations and the metal-oxygen vibrations since both may occur in similar spectral regions[4]. The most satisfactory method of assignment is to make use of the small shifts that occur on isotopic substitution of the metal atoms[5]. This method is, however, not generally applicable since suitable isotopes of some metals are not readily available and, in other cases, the cost may be prohibitive. The alternative procedure of ^{18}O substitution of the acetate grouping is also expensive[4].

In the course of our studies of the Raman, resonance Raman and electronic spectra of dirhodium tetracarboxylates[6,7], we observed that the wavenumber shifts in the metal-carboxylate modes induced by deuteration of the carboxylate ligand are larger than those caused by oxygen isotopic substitution. We were unable to confirm our assignments by metal isotope substitution since there are no stable isotopes of rhodium other than naturally occurring ^{103}Rh. It seemed possible that acetate deuteration could provide a low cost, readily available, and generally applicable method for differentiating between predominantly metal-metal and metal carboxylate stretching vibrations. In order to test this hypothesis we have applied the method to the well-established case of $Mo_2(O_2CCH_3)_4$ where ν(Mo-Mo) has been shown to occur at 404 cm^{-1} by metal isotope substitution[5].

The wavenumbers of the prominent features below 450 cm^{-1} in the 20 K Raman spectra of $Mo_2(O_2CCH_3)_4$ and $Mo_2(O_2CCD_3)_4$ are given in the Table. It is immediately clear that the band at 404 cm^{-1} is essentially unshifted on deuteration whereas the group of three bands in the 323-301 cm^{-1} region shift by about 10 cm^{-1} to low wavenumbers. This is consistent with the assignment of the 404 cm^{-1} band to the metal-metal stretch and the lower group of bands to the three Raman-active Mo-O stretching modes expected on a Mo_2O_8 vibrational model. These Mo-O stretching modes shift by about 2 cm^{-1} to lower wavenumber on ^{92}Mo substitution of natural-abundance molybdenum. The relative insensitivity of the two bands at 203 and 189 cm^{-1} to both ^{92}Mo [5] and deuterium substitutions suggests that they involve bending modes of the Mo_2O_8 skeleton. We also note that the insensitivity of the main progression-forming mode in the lowest energy electronic transition to deuteration confirms that this progression is in the excited state ν(Mo-Mo) mode which has a wavenumber ca. 370 cm^{-1}. The 80 K f.t.i.r.

spectra are consistent with these assignments.

The results of this study indicate that deuteration of the carboxylate group provides a useful, low cost alternative to ^{18}O- and metal-isotope-substitution experiments for the assignment of metal-metal stretching vibrations in dimeric tetracarboxylates. The application of this procedure to the assignment of ν(RhRh) and ν(OsOs) for dirhodium- and diosmium-tetraacetates will also be described.

Table. Vibrational data for isotopomers of $Mo_2(O_2CCH_3)_4$ in the range 450-150 cm^{-1} (a)

Raman				Infrared	
$^{NA}Mo_2$	$^{NA}Mo_2$ (b)	$^{92}Mo_2$ (b)	$^{NA}Mo_2$-d_{12}	$^{NA}Mo_2$	$^{NA}Mo_2$-d_{12}
404	404	413	403		
				373	363
				353	343
				342	321
323	321	322	313,310		
315	311	314	302		
301	298	299	293		
				311	285
				235	224
				230	220
203	200	200	201		
189	186	186	187		
				183	172

(a) NA = natural abundance; (b) Values taken from ref.5. The excitation wavelength was 514.5 nm with the samples held at room temperature.

References

1. San Filippo, J.; Sniadoch, H.J. *Inorg. Chem.* 1973, *12*, 2326.
2. Ketteringham, A.P.; Oldham, C. *J. Chem. Soc., Dalton Trans.* 1973, 1067.
3. Miskowski, V.M.; Schaefer, W.P.; Sadeghi, B.; Santarsiero, B.D.; Gray, H.B. *Inorg. Chem.* 1984, *23*, 1154.
4. Clark, R.J.H.; Ferris, L.T.H. *Inorg. Chem.* 1981, *20*, 2759.
5. Shriver, D.F.; Cooper, C.B. in *Advances in Infrared and Raman Spectroscopy*, ed. R.J.H. Clark and R.E. Hester, Heyden, London, vol.6, 1980, p.127.
6. Clark, R.J.H.; Hempleman, A.J.; Dawes, H.M.; Hursthouse, M.B.; Flint, C.D. *J. Chem. Soc., Dalton Trans.* 1985, 1775.
7. Clark, R.J.H.; Hempleman, A.J.; Flint, C.D. *J. Am. Chem. Soc.* 1986, *108*, 518.
8. Trogler, W.C.; Solomon, E.I.; Trajberg, I.; Ballhausen, C.J.; Gray, H.B. *Inorg. Chem.* 1977, *16*, 828.

THE INFLUENCE OF SELECTED METALS ON THE AROMATIC SYSTEM OF SALICYLIC ACID

H.Barańska and W.Lewandowski*

Industrial Chemistry Research Institute
Rydygiera 8, 01-793 Warsaw

*Agricultural University of Warsaw
Rakowiecka 26/30 , 02-528 Warsaw
Poland

This work is continuation of our previous study on the influence of metals on ligands with aromatic systems [1-3] .We present results on the influence of: Na,K,Mg,Zn,Cd, Fe(III), Al,La (III),Pb (II), Hg (I and II) on the aromatic system of salicylic acid.Salicilic acid is regarded here as a model compound representing a wide group of aromatic ligands which are components of enzymes.Vibrational (Raman and IR) and electronic (UV)spectra were applied in this study. The number,frequency and relative intensities of bands of the aromatic system recorded by vibrational spectra (e.g. La and Pb salicylate,see Table 1) and frequencies of $\pi \rightarrow \pi^*$bands in electronic spectra were analysed.The analysis and assignment of vibrational bands showed that,during complex formation,there are significant change of the frequencies and relative intensities of bands no 1,7a,8a,12,13,14,19a and 19b, whereas in bands no 6a,6b,9a,11,16a,16b,18a and 18b the changes are very small.

From a biochemical point of view,most interesting is the observation that the largest disturbance of the aromatic system is caused by heavy,toxic,metals: Hg (I and II) and Pb.A decrease in the number and frequency of bands originating in the aromatic system vibrational spectra of toxic Hg and Pb metal salicylates occurs simultaneously with a hypsochromic shift of the $\pi \rightarrow \pi^*$bands in the UV spectra.

References :
1. W.Lewandowski, J.Mol.Struct., 101, 93 (1983).
2. W.Lewandowski, Pol.J.Chem., 58, 1199 (1984).
3. W.Lewandowski, H.Barańska,J.Raman Spectrosc.17,17 (1986).

Table 1.

Band frequency of the aromatic ring in R and IR spectra of lanthanum (III) and lead (II) salicylates.**

Lanthanum (III) salicylate		Lead (II) salicylate		Normal vibration of the aromatic ring*
Frequency (cm^{-1})				
Raman	IR	Raman	IR	
3070 w	3066 w	3065 w	-	20 a
-	-	-	3050 vw	20 b
1623 m	1626 m	1620 m	1623 s	8 b
1605 m	1596 m	-	-	8 a
1508 w	1508 s	1480 w	1481 vs	19 b
1478 m	1483 m	-	-	19 a
1327 w	1326 sh	1312 m	-	14
1310 m	1307 w	-	1297 m	7 a
1165 w	1163 vw	-	1156 sh	9 a
1150 m	1148 m	1145 m	1144 m	18 a
1100 vw	1106 w	1090 vw	1091 vw	13
1043 s	1040 sh	-	-	18 b
1037 sh	1031 m	1032 vs	1032 m	
825 s	826 sh	820 vs	813 m	12
-	752 s	-	748 s	11
702 vw	699 m	670 w	664 m	1
573 m	566 m	572 m	576 vw	6 a
536 vw	546 vw	534 w	529 w	16 a
	526 m			
464 w	467 m 462 sh	455 m	-	6 b
420 s	416 vw	421 m	-	16 b

* For symbols see: G.Versanyi "Assignments for Vibrational Spectra of 700 Benzene Derivatives", Akademiai Kiado, Budapest 1973 .

** s= strong ; m= medium ; w= weak ; sh= shoulder ; v=very

RAMAN SPECTROSCOPIC STUDY OF DISPROPORTIONATION OF IODINE IN BASIC SOLUTIONS.

J.C. Wren, S. Sunder and J. Paquette.

Atomic Energy of Canada Limited, Whiteshell Nuclear Research Establishment, Pinawa, Manitoba, Canada R0E 1L0.

In the course of energy production in a nuclear power reactor, radioactive iodine is formed in the nuclear fuel by fission of uranium and by the decay of fission products. The radioactive iodine could, under certain accident scenarios, escape from the fuel and end up in aqueous solutions as $I^-(aq)$, from where, under adverse conditions, it could be released to the environment in the form of undesirable volatile compounds. In this respect, it is important to understand the aqueous chemistry of iodine and, in particular, the disproportionation reaction

$$3I_2 + 3H_2O \rightleftharpoons 5I^- + IO_3^- + 6H^+ \quad (1)$$

which leads from an involatile and unreactive I^- to a volatile and reactive I_2.

Although there have been many studies of the above reaction using several techniques including UV-visible absorption spectroscopy, the transient iodine species have never been positively identified. We have studied this reaction with kinetic Raman spectroscopy and have identified the transient intermediates as well as obtained kinetic data for the reactions of these intermediates.

The kinetic studies were initiated by reacting I_2/I^- solutions with NaOH solutions. A flow system was used to record the spectra of the transient species so formed as a function of time. The Raman spectra were excited with either the 514.5-nm or the 488.0-nm radiation from an argon-ion laser. To avoid interference from bands due to glass, silica, quartz or sapphire, the usual materials for

Raman cells, the Raman spectra were excited in an open-air stream of the reacting solutions in the sample compartment of the Raman spectrometer [1]. This arrangement not only removed the extraneous bands, but also enhanced the observed Raman intensities by a factor of about ten compared with those obtained using a glass capillary.

As expected, the spectrum of the unreacted I_2/I^- solutions showed Raman bands due to I_2 and I_3^- [2]. These bands decreased in intensity as a function of reaction time and two new bands, not seen in the spectra of the reactants, appeared at 430 and 560 cm^{-1} immediately after mixing. After a short time, additional bands at 685 cm^{-1} and 800 cm^{-1} also appeared. At longer reaction times (> 10 min), the spectra showed only one iodine band at 800 cm^{-1}, in addition to the solvent bands. The intensity of the 430- and 560-cm^{-1} bands decreased steadily with time, but the I_{560}/I_{430} ratio remained constant throughout the reaction. The intensity of the 685-cm^{-1} band increased from zero, at short reaction time, went through a maximum and then decayed to zero at long reaction times. The intensity of the 800-cm^{-1} band, which was initially zero, increased with time but levelled off at long reaction times.

The band at 800 cm^{-1} was assigned to IO_3^-, in agreement with reaction (1) and existing spectroscopic data [3]. We have assigned the bands at 430, 560 and 685 cm^{-1} to the intermediates OI^-, HOI_2^- and IO_2^-, respectively. Plots of the inverse of the Raman intensity at 430 and 560 cm^{-1} against reaction time gave linear relationships, indicating that both OI^- and HOI_2^- decayed by a second-order process. A detailed mechanism for the disproportionation of iodine in basic solutions was obtained by measuring the time and concentration dependence of the Raman spectra [1]. We found that, in basic solutions, the major channel for the disproportionation of iodine in the presence of I^- was the reaction between OI^- and HOI_2^-.

References

1. J.C. Wren, J. Paquette, S. Sunder and B.L. Ford, Can. J. Chem. submitted for publication (1986).
2. W. Kiefer and H.J. Bernstein, Chem. Phys. Lett. 16, 5 (1972).
3. J.R. Durig, O.D. Bonner and W.H. Breazeale, J. Phys. Chem. 69, 3886 (1965).

IN SITU MICRO-RAMAN STUDY OF DISCHARGE PRODUCTS OF $Li/SOCl_2$ CELLS

M. C. Dhamelincourt, F. Wallart, M. Delhaye

Laboratoire de Spectrochimie IR et Raman, C.N.R.S. (LP 2641), USTL Bat. C 5, 59655 VILLENEUVE D'ASCQ Dedex, France

If the overall reaction for thionyl chloride cell is well established now there continues to be controversy regarding the mechanism and the chemical nature of intermediates of the reaction. Particularly, the pressure generated in $LiSOCl_2$ cells has been found to be significantly less than that predicted on the basis of the accepted discharge reaction. This is believed to be due to the SO_2 generated being complexed by $LiAlCl_4$ and/or absorbed by the carbon cathode (1,2). Several studies were carried on this problem but no Raman study has been hitherto described in the literature. Nevertheless, Raman spectroscopy and especially micro Raman spectroscopy is a technique particularly well adapted to this kind of problem allowing an in situ study during the discharge.

So, we have constructed a microcell, directly adaptable to the microscope stage of the spectrometer -Microdil 28, Dilor, France- in order to observe the phenomena appearing near the electrodes in the solution and also at the electrodes.

The working electrode was made of lithium ribbon mechanically attached to the upper end of a stainless steel rod protected by teflon coating. The reference and the counter electrodes were identical carbon sticks. The top of the cell, especially made for micro-Raman observations, was an optical plane window. The electrolyte was 1.8 M $LiAlCl_4$ thionyl chloride.

The figure shows the spectra of electrolyte obtained in the spectral range of the vibrational stretching mode S-O, spectrum 2 before the discharge, spectrum 1 during the discharge; spectrum 3 was obtained when the laser beam was focused on the surface of the lithium electrode.

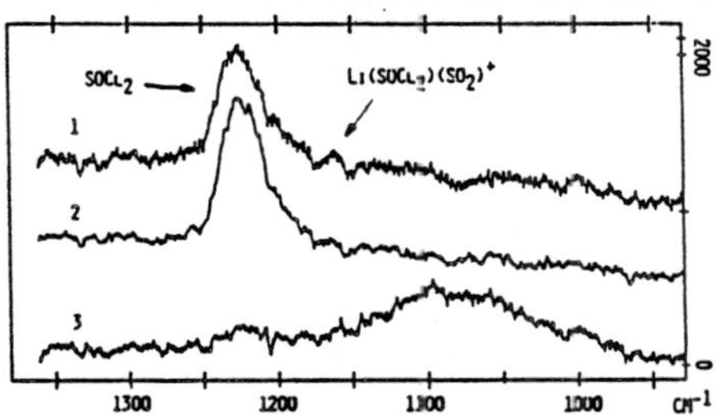

After several hours, we observed the formation of the $Li(SOCl_2)(SO_2)^+$ complex characterized by the band appearing at 1160 cm^{-1} (2). The frequency of the broad band observed on spectrum 3 suggests that a complex specie, likely $Li(SO_2)_3^+$, was observed at the surface of the lithium electrode.

Acknowledgement:

Mr. Lelieur from Polytechnicum of Lille is gratefully acknowledged for his help in designing the cell.

References:

1. K.M. Abraham, L. Pitts, W.P. Kilroy, J. Electroch. Soc. 132, 10, 2301 (1985).

2. M.C. Dhamelincourt et al., Power Sources, 14, 77 (1985).

chapter 14

SPECIFIC MOLECULAR SYSTEMS: ORGANIC

CONFORMATIONAL ANALYSIS OF SOME ORGANIC MOLECULES BY GAS PHASE RAMAN SPECTROSCOPY.

J.R. Durig, J.F. Sullivan, T.S. Little, T.J. Geyer and S.E. Godbey

Department of Chemistry, University of South Carolina, Columbia, SC 29208

Four-membered ring molecules have one low-frequency ring-puckering mode and the nature of the potential governing this vibration determines the conformational stability of the ring. In the earlier studies, far infrared and microwave data were used to determine whether the ring was planar or puckered as well as the form of the potential function governing the ring-puckering mode. In later studies, laser-Raman spectroscopy of molecules in the gas phase has been applied to the study of these anharmonic vibrations. In a number of cases the differences in selection rules between the infrared and Raman transitions have been extremely useful in that they frequently provide a check on the assignments to specific energy levels. Additionally, it has been shown that gas phase Raman spectroscopy can be used to obtain the barriers to internal rotation of both asymmetric and symmetric rotors. More recently both electron diffraction and ab initio studies have been utilized to determine the conformational stability of several substituted four-membered ring molecules. In the ab initio studies it has been predicted that the equatorial position of monosubstituted cyclobutane molecules is more favored when the substituent is more electronegative. Therefore, as a continuation of our earlier studies on four-membered ring molecules, we have investigated the Raman spectra of methylcyclobutane, $c\text{-}C_4H_7CH_3$, cyclobutylsilane, $c\text{-}C_4H_7SiH_3$, and cyclobutylgermane, $c\text{-}C_4H_7GeH_3$, for determining the relative stabilities of the two conformers and the barrier to internal rotation of the methyl, silyl and germyl groups.

For methylcyclobutane, a series of Q-branches was observed beginning at 161 cm^{-1}, with successive transitions falling to lower frequencies and these transitions have been assigned to the ring-puckering vibrations of both the low energy equatorial and high energy axial conformers. The energy difference between the equatorial and axial forms is found to be 225 ± 20 cm^{-1} (643 cal/mol) and a barrier of 630 ± 20 cm^{-1} (1.80 kcal/mol) was found for the interconversion. The difference between the puckering angle for the two conformers was found to be 3.4° with the equatorial conformer having the larger value of 20.6°. Experimental values for the ΔH of both the liquid and gas from relative intensities of different Raman band pairs over 37°C and 85°C temperature ranges, respectively, were found to be consistent with the value obtained from the potential function.

From a comparison of the spectra of cyclobutylsilane in the fluid phases with that of the solid, it is clear that two stable conformers exist in the fluid phases at ambient temperature. It was determined that the equatorial conformer is the predominant form at ambient temperature in the gas phase, and from a temperature study of the Raman spectrum in this phase, the enthalpy difference between the equatorial and axial conformers was determined to be 147 ± 50 cm^{-1} (420 cal/mol). Similar studies of the Raman spectrum of the liquid phase gave an enthalpy difference of 231 ± 25 cm^{-1} (660 cal/mol), again with the equatorial conformer being thermodynamically preferred and the only conformer present in the solid state. A series of Q branches obtained from the low-frequency Raman and far-infrared spectra of the vapor as well as sum and difference bands in the 1600-1200 cm^{-1} region have been assigned to the ring puckering vibration of both the low-energy equatorial and high-energy axial conformers and fitted to an asymmetric potential function of the form $V(cm^{-1}) = (3.13 \pm 0.09) \times 10^5X^4 + (1.36 \pm 0.04) \times 10^4X^3 - (2.30 \pm 0.08) \times 10^4X^2$ with a reduced mass of 198 amu. This potential is consistent with the equatorial conformation being more stable than the axial by 188 cm^{-1} (538 cal/mol) and the equatorial to axial barrier is 531 cm^{-1} (1.52 kcal/mol). From this potential function, puckering angles of 21.3° and 17.8° were obtained for the equatorial and axial conformers, respectively. Combination band spectra arising in the 2300-2000 cm^{-1} region were attributed to the sum and difference bands of the SiH_3 torsion with an SiH stretching mode. Analysis of these data leads to barriers of internal rotation of the SiH_3 group of 655 and 763 cm^{-1} for the equatorial and axial conformers, respectively. Similar studies have also been carried out for cyclobutylgermane.

The Raman and infrared spectra of gaseous and solid 1-fluoro-2-methylpropane, 1-chloro-2-methylpropane and 1-bromo-2-methylpropane, as well as the Raman spectra of the liquids, have been recorded and assigned. The _gauche_ asymmetric torsions of the 1-fluoro-2-methylpropane and 1-chloro-2-methylpropane molecules have been observed at 114 and 110 cm^{-1}, respectively, in the Raman spectra of the gases. For the 1-bromo-2-methylpropane molecule, both the _trans_ and _gauche_ asymmetric torsions have been observed at 106.70 and 103.94 cm^{-1}, respectively, with three additional transitions for the _gauche_ conformer being observed. From these data, the asymmetric potential function for the bromide molecule has been calculated to be V_1 = -493 ± 16, V_2 = 595 ± 18, and V_3 = 2006 ± 6 cm^{-1} with the _trans_ conformer being more stable than the _gauche_ conformer by 44 ± 20 cm^{-1}. The _trans_ form was found experimentally to be more stable in the liquid phase by 30 ± 14 cm^{-1} (83 ± 40 cal/mol). From the relative intensities of the C-Cl stretches in the Raman spectra as a function of temperature, the _gauche_ conformer of the chloride molecule was found to be 167 ± 71 cm^{-1} (479 ± 203 cal/mol) more stable than the _trans_ conformer in the gas phase, and 73 ± 10 cm^{-1} (208 ± 29 cal/mol) in the liquid phase. The methyl torsions for the _gauche_ and _trans_ conformers of all three molecules have been tentatively assigned in the gas phase and the barriers have been calculated. The results of this study will be compared with previous studies on some similar molecules.

DETERMINATION OF VIBRATIONAL POTENTIAL ENERGY SURFACES FROM RAMAN AND INFRARED SPECTRA.

J. Laane
Department of Chemistry, Texas A&M University, College Station, TX 77843

Low-frequency infrared and Raman spectra have been used for more than twenty years to study the potential energy functions governing the conformational changes of small ring molecules. The Raman spectra generally correspond to double quantum jumps between the energy states, whereas the infrared transitions for the same molecules result primarily from single quantum jumps. The infrared and Raman data are thus complementary, although in actual fact, the Raman data, typically recorded at lower resolution, often adds very little useful information. Exceptions to this rule are molecules such as cyclobutane for which the puckering transitions are infrared inactive but observable in the Raman spectra. In recent years, as more complicated molecules have been studied (sometimes with multi-dimensional potential energy surfaces), Raman spectroscopy has played a more vital role in the determination of the potential surfaces associated with conformational changes. Improved methods for the calculation of kinetic energy and potential enrgy functions have also significantly helped to better characterize these systems.

For illustrative purposes, Figures 1 and 2 show respectively the Raman spectrum and and the one-dimensional potential energy function for 1,3-disilacyclobutane (DSC3).[1] The data demonstrate that the ring is non-planar and the barrier to inversion is 87 cm^{-1}. The degree of puckering of the ring can also be deduced if the reduced mass, based on an assumed vibrational model, can be correctly calculated. However studies of DSCB-d_4 showed that the data for both isotopic species of the molecule can not be fitted with the same one-dimensional potential function. This is due to the interaction of the puckering with two other motions (the ring deformation and the SiH_2 rocking). Calculations for a three-dimensional potential surface were required to satisfactorily analyze the spectra.[2] Figure 4 shows a two-dimensional slice of the three-dimensional potential which was derived from additional data such as the Raman combination bands shown in Figure 3.

For small ring systems, for which one-dimensional analyses are adequate, the kinetic energy expansions have been carefully calculated and the corresponding potential functions have been redetermined. For ring molecules with two or more out-of-plane motions (cyclopentane, cyclopentene, cyclohexene, various bicyclic molecules, and analogous compounds), both the far-infrared and Raman spectra have been analyzed using appropriate reduced mass expressions[3] and newly devised computational schemes for accurately calculating the energy levels.[4]

Cyclopentane and five of its deuterated derivatives show infrared combination bands arising from the pseudorotation, which is almost totally unhindered; Raman transitions from the radial mode occur between 215 and 280 cm^{-1}. The overtones of the radial modes can be seen as intense features in the infrared. The barrier to planarity[5] has been determined to be 1808 cm^{-1}. Molecules with hindered psuedorotation, such as silacyclopentane, also have Raman bands corresponding to the radial (twisting) motion in the 200-300 cm^{-1} frequency range. The calculation of meaningful potential energy surfaces for such ring systems can only be achieved if adequate and appropriate basis sets are utilized.

The far-infrared and Raman spectra of cyclohexene and several of its deuterated species have also been analyzed in terms of a two-dimensional potential energy surface. The Raman data near 270 cm^{-1} correspond to the ring-twisting motion. Figure 5 shows a preliminary potential energy surface calculated for the cyclohexene bending and twisting motions.

The Raman and far-infrared data for other molecule, including several bicyclohexanes and bicycloheptenes, have also been investigated to better understand the potential functions associated with the conformational changes of these molecules.

References

1. R. M. Irwin, J. M. Cooke, and J. Laane, J. Amer. Chem. Soc., **99**, 3273 (1977).
2. P. M. Killough, R. M. Irwin, and J. Laane, J. Chem. Phys., **76**, 3890 (1982).
3. J. Laane et al., J. Mol. Spectrosc., **91**, 286 (1982); **91**, 300 (1982).
4. M. A. Harthcock and J. Laane, J. Phys. Chem., **89**, 4231 (1985).
5. L. A. Carreira et al., J. Chem. Phys., **56**, 1440 (1972).

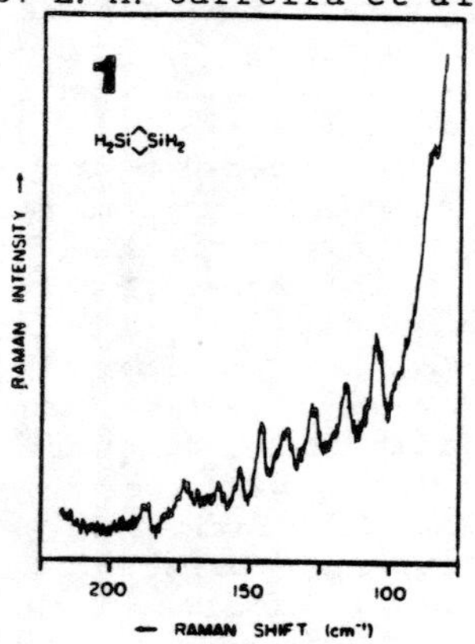

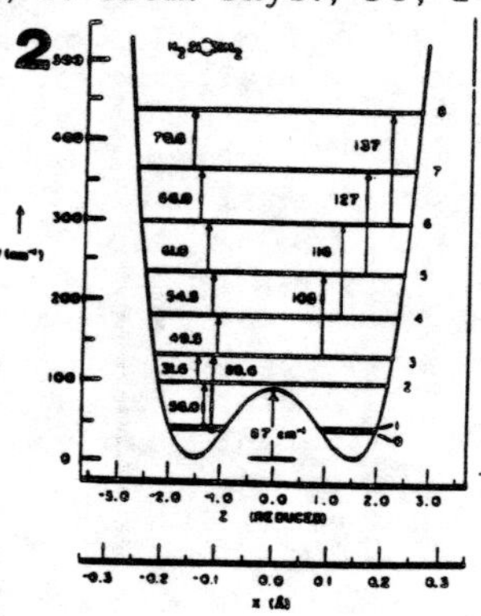

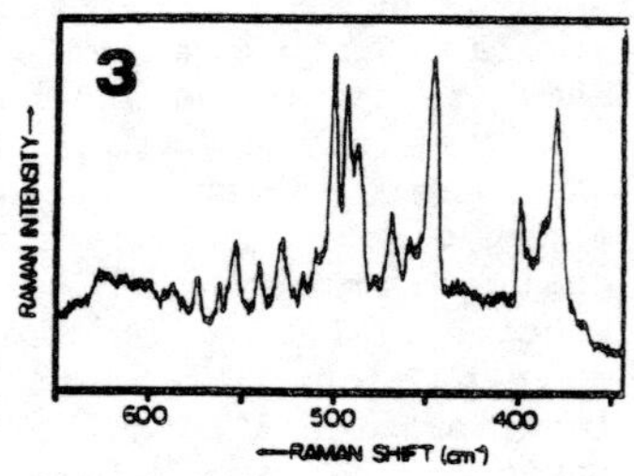

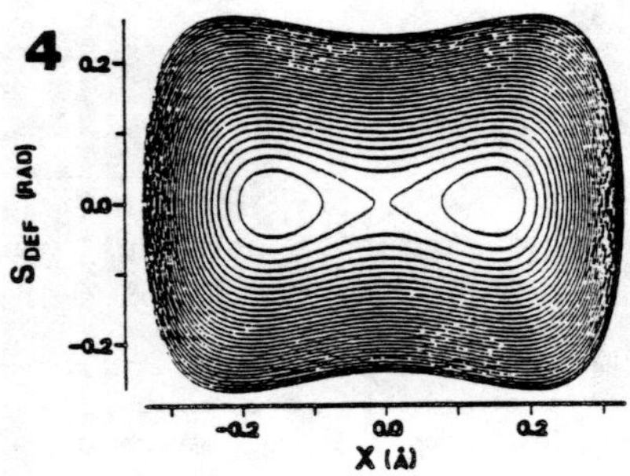

RAMAN SPECTRA AND CONFORMATIONS OF DIBENZO- AND DICYCLOHEXYL-18-CROWN-6.

Hideo Takeuchi, Takaki Arai, and Issei Harada

Pharmaceutical Institute, Tohoku University, Aobayama, Sendai 980, Japan

The polyoxyethylene ring of a crown ether is flexible and shows various conformations depending on the chemical species with which the crown ether forms adducts. In a previous study, we investigated the Raman spectra of 18-crown-6 (18C6), 15-crown-5 (15C5) and their complexes with alkali and alkaline earth metal cations [1]. The analysis of the Raman spectra based on the normal coordinate calculations has revealed that the frequencies of Raman bands due to the C-O stretching and CH_2 rocking vibrations provide key information on the local conformations of crown ring: *trans* (t) or *gauche* (g) about the C-O bonds and *trans* (T) or *gauche* (G) about the C-C bonds. As an extension of the study, we have investigated Raman spectra of dibenzo-18-crown-6 (DB18C6), dicyclohexyl-18-crown-6 (DC18C6), and their complexes with metal cations. Conformations of the crown rings will be discussed.

Fig. 1 shows the Raman spectra of DB18C6 and its complex with NaBr in the solid state. The strong band at 786 cm^{-1} is assigned to a vibration characteristic of ortho-substituted benzene. In the spectrum of the NaBr complex, two bands at 946 and 856 cm^{-1} and a shoulder around 839 cm^{-1} correspond to the 943, 868, and 833 cm^{-1} bands, respectively, of the D_{3d}-symmetric conformer of 18C6 [1], indicating the presence of tGttG't conformation. This is in accord with the conformation $(tCttGttG't)_2$ found by X-ray analysis (C stands for the *cis* configuration at the benzene ring) [2]. On the other hand, uncomplexed DB18C6 gives a completely different spectral pattern in the 800-1000 cm^{-1} region. According to the analysis on the C_i-symmetric conformer of 18C6 [1], the 998 cm^{-1} band is ascribed to a vibration localized at the CH_2-CH_2-O-CH_2-CH_2 segment in the G'gtT form, the 931 and 890 cm^{-1} bands arise from the G'g conformation, and the 822 cm^{-1} one indicates the presence of the T conformation. Therefore, it is most likely that uncomplexed DB18C6 takes the $(tCttTtgG't)_2$ conformation in the solid state. This conformation is different from that found for two uncomplexed DB18C6 rings in a crystal of $(Rb_{0.55},Na_{0.45})NCS\cdot DB18C6$ [3].

There are two isomers of DC18C6, which are different from each other in the relative direction of two cyclohexane rings (*syn* and *anti* in isomers A and B, respectively). The isomers are not interconvertible by conformational change. An X-ray study revealed the crystal structure of the NaBr complex of isomer B [4], where the crown ring has a conformation of three consecutive tGttG't forms similar to the D_{3d} structure of 18C6. The Raman spectrum of the complex shown in Fig. 2b is, however, much more complicated than that of 18C6 and the relationships between Raman frequencies and local conformations described above cannot be simply applied to this complex. The complication must be due to vibrational coupling between the crown and cyclohexane rings. The Raman spectrum of uncomplexed isomer B in the same frequency region (Fig. 2a) seems simpler than that of the NaBr complex, implying weaker vibrational coupling as a result of conformational change in the crown ring. The 982, 948, 845, and 777 cm^{-1} bands are likely due to vibrations mainly involving the cyclohexane ring and these bands were observed to shift slightly in isomer A. The appearance of a band at 823 cm^{-1} is suggestive of the T conformation about, at least, one C-C bond.

Raman spectral measurements in methanol solution have also been made for isomers A and B of DC18C6 and their complexes with Na^+ and K^+. The spectra of both isomers are analogous with each other in all the states studied, though the band shapes are broad in the uncomplexed state and sharp in the complexed state. The close resemblance between the spectra of the complexes in solution and that of crystalline complex of isomer B•NaBr lead to the conclusion that the ring conformation of DC18C6 is $(tGttG't)_3$ in the Na^+ and K^+ complexes in solution. A metastable form, which gives rise to a Raman band at 828 cm^{-1}, was also found for isomer A•Na^+ complex in methanol.

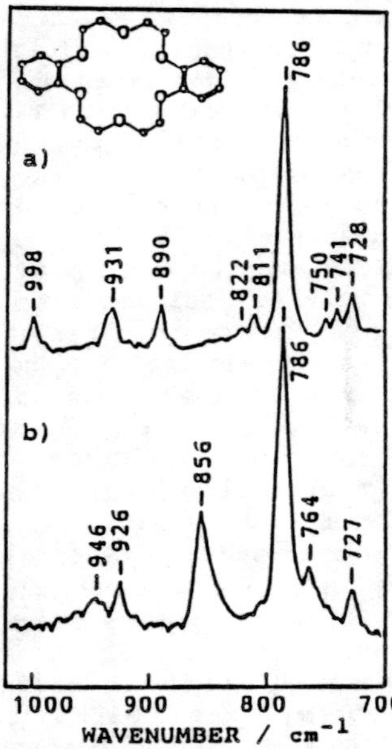

Fig. 1 Raman spectra of a) DB18C6 and b) its NaBr complex in the solid state

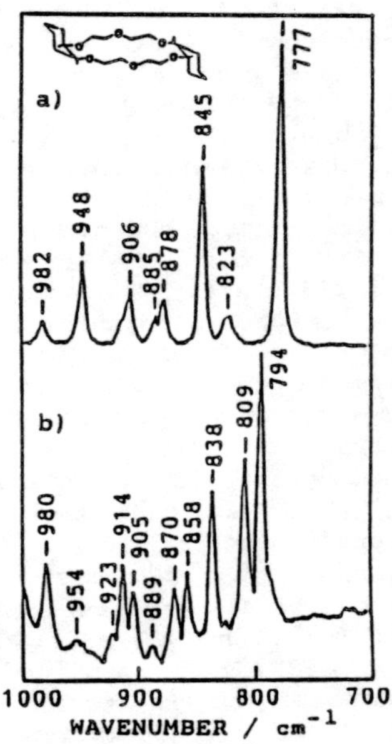

Fig. 2 Raman spectra of a) DC18C6 isomer B and b) its NaBr complex in the solid state

REFERENCES

1. H. Takeuchi, T. Arai, and I. Harada, J. Mol. Structure, in press.
2. M. A. Bush and M. R. Truter, J. Chem. Soc. (B), 1971, 1440.
3. D. Bright and M. R. Truter, J. Chem. Soc. (B), 1970, 1544.
4. M. Mercer and M. R. Truter, J. C. S. Dalton, 1973, 2215.

COMPARATIVE STUDY OF PHENYLACETYLENE, DIPHENYLACETYLENE AND DIPHENYLDIACETYLENE. ASSIGNMENT OF THE ACETYLENIC BRIDGE VIBRATIONS

L. Colombo+, G. Baranović+ and J.R. Durig[x]

+Rudjer Bošković Institute, Zagreb, Yugoslavia
[x]University of South Carolina, Columbia S.C. 29208, USA

Long molecules of the type phenyl ring - bridge - phenyl ring have been subject of extensive spectroscopic investigations with the aim to get information on coupling of the phenyl fundamentals, of the dynamics of the bridge and on possible conformations due to the specific structure of these molecules [1,2]. Due to the complexity of the corresponding spectra, this aim was not easy to be attained.

Therefore, in the present study we try to solve this problem by treating simultaneously three molecules having similar structural parameters: phenylacetylene (PA), diphenylacetylene (DPA) and diphenyldiacetylene (DPDA). The assignment of observed spectra is supported by normal coordinate analysis based on a transferable force field. The experimental basis for this study were a large number of Raman and infrared measurements on molecules and their isotopomers performed in solutions as well as in solids. In this way we could use 253 observed frequencies to fit 31 in-plane force constants, and 87 frequencies to fit 17 out-of-plane force constants. The starting force field for the aromatic ring is after LaLau and Snyder [3], while the force field for the acetylenic bridges had to be determined for the first time.

A few of aromatic ring vibrations are strongly dependent of the supstituent or bridge, the rest being typical for the monosupstituted benzenes. The frequencies characteristic of the vibrations in the acetylenic supstituent (PA) and in the single (DPA) or double (DPDA) acetylenic bridge are given in Table 1.

The conformation of the DPA (tolane) has been also considered in earlier investigations [1,2]. Crystal structure data indicate planar configuration [4], while electronic spectroscopy investigations indicated that there is a quasi-free rotation of the phenyl rings about the -C≡C- group [5]. Depending on dihedral angle ϕ between two phenyl rings, the conformation is $D_{2h}(\phi=0^o)$ or $D_2(0^o<\phi<90^o)$ and consequently the number of polarized Raman bands is different. Depolarization ratio measurements have indicated the presence of twisted molecules in solution.

Table 1. Specific vibrations of the acetylenic bridge

		PA	DPA	DPDA
ν(C≡C)		2120 a_1	2220 a_g	2211 a_g + 2156 b_{1u}
bend Ø-Z ip.		513 b_2	473 b_{2u} + 759 b_{3g}	515 b_{2u}+ 450 b_{3g}
bend Ø-Z op.		530 b_1	535 b_{2g} + 511 b_{3u}	530 b_{2g}+ 530 b_{2u}
linear bend. of the bridges	ip.	152 b_2	(44)* b_{2u}+(157) b_{3g}	218 b_{2u}+ (66) b_{3g} 885 b_{3g}+ (33) b_{2u}
	op.	349 b_1	380 b_{2g} + (60) b_{3u}	(110) b_{2g}+ 450 b_{3u} 608 b_{2g}+ (33) b_{3u}

*Frequencies in brackets are calculated values

References

[1] C. Pecile and B. Lunelli, Can. J. Chem. 47, 235 (1969)

[2] B. Kellerer, H.H. Hacker and J. Brandmüller, Ind. J. Pure Appl. Physics, 9, 903 (1971)

[3] C. LaLau and R.G. Snyder, Spectrochim. Acta 27, 2073 (1973)

[4] A.A. Espiritu and J.G. White, Z. Kristallogr. 147, 177 (1978)

[5] K. Okuyama, T. Hasegawa, M. Ito and N. Mikami, J. Phys. Chem. 88, 1711 (1984).

ON THE VALIDITY OF USING RAMAN INTENSITY MEASUREMENTS FOR DETERMINING CONFORMER POPULATIONS AS A FUNCTION OF PRESSURE AND TEMPERATURE.

Derek J Gardiner, Nicholas A Walker

and Martin P Dare-Edwards*.

School of Chemical and Life Sciences, Newcastle upon Tyne Polytechnic, Newcastle upon Tyne, UK, NE1 8ST.

* Thornton Research Centre, PO Box 1, Chester, UK, CH1 3SH.

The Raman spectra of cis and trans isomers of 1,2, 1,3, and 1,4-dimethylcyclohexane have been recorded in the liquid phase at pressures up to 11 kbar and temperatures ranging from -80°C to 120°C. Large relative integrated intensity changes (up to 30% in some cases) were observed for all six fluids as both temperature and pressure were varied. These changes cannot be explained soley by invoking temperature and pressure induced shifts in conformer populations.

In the optical skeletal regions both increasing pressure and decreasing temperature suppress the intensity arising from symmetrical modes which involve significant volume changes. The C-H stetching regions behave differently in that increasing pressure and increasing temperature suppress the intensity arising from symmmetric modes relative to that arising from antisymmetric modes.

These observations suggest that the relative intrinsic scattering activities of vibrational modes are not independent of environment but are a function of both density and temperature. An explanation is offered which relates the relative intensity change as a function of density to the volume change associated with the vibration.

14-10

DEPHASING AND THE VIBRATIONAL FREQUENCY CORRELATION FUNCTION IN LIQUID CRYSTALS.

Z. Gburski

Department of Physics, Silesian University, Uniwersitecka 4, 40-007, Katowice, Poland

The reorientaional correlation functions of selected Raman bands in ordered liquids have been recently measured [1]. Here we report a study of the vibrational correlation function Gvib(t) of the Raman active C≡C stretching vibration /v_o = 2224 cm^{-4}/ in the aligned nematic phase of PMT/ 4,n-pentyl 4'-methoxytolane/ at the temperature T = 316 K and the three geometries [1]. In case a), the electric vector of the incident light is paralle to the long axis of the homogeneously aligned mesogen molecules. In case b), the molecular configuration is the same but the laser polarization is perpendicular to the director n. In case c), the sample is aligned homeotropically, and the polarization of the incident light is the same as in case a).

The vibrational correlation functions Gvib(t) are different for each of the different geometries involved. The geometry dependence of the vibrational frequency correlation function $G_\omega(t) = \langle(\omega(t)\omega(0))\rangle \langle\omega^2(0)\rangle^{-1} = -d^2/dt^2$ (ln Gvib(t)) [2], which directly reflects the fluctuations of the local structure in liquid, is much more prounanced. G (t) can be described by eq.,

$$\langle \omega(t)\omega(0)\rangle = \omega_h^2 \exp(-t/\tau_h) + \omega_{inh}^2 \qquad (1)$$

Both homogeneous ω_h^2, τ_h an inhomogeneous ω_{inh}^2 contributions to $G_\omega(t)$ were extracted on the basis of experimental data; interesting that ω_{inh}^2 is anisotropic, i.e. varies according to whether the geometry a), b) or c) is applied.

The anisotropy of the local environment seeing by the molecular oscillator could be a consequence of the relatively high degree of spacial order existing in the liquid crystal phase, if compared with the normal (isotropic) liquid.

1 . N. Kirov, I. Dozor, M.P. Fontane, J.Chem.Phys. 83 (1985) 5267

2 . Z. Gburski, T.W. Żerda, Acta Phys.Polon., A57 (1980) 447

CONFORMATIONAL ANALYSIS OF ORGANOPHOSPHORUS COMPOUNDS BY GAS PHASE RAMAN SPECTROSCOPY.

T. J. Hizer[1], R. D. Johnson[2], P. Groner[2] and J. R. Durig[2]

[1]Department of Chemistry, Claflin College, Orangeburg, SC 29115
[2]Department of Chemistry, University of South Carolina, Columbia, SC 29208

Low frequency gas phase Raman spectroscopy has been proven to be invaluable in the determination of the conformational stability of a large number of organophosphorus molecules. Molecules of formula $CH_3XP(Y)Z_1Z_2$ with X = CH_2, O, and S; Y = O, S, BH_3 or simply the lone pair of nonbonded electrons; and Z_1Z_2 = CH_3, F, Cl, or possibly OR, potentially have two or more conformers present at ambient temperature. Vibrational studies can provide sufficient data from which the conformational stability of these molecules may be determined. Until very recently, such studies were carried out utilizing mid-infrared and Raman spectra of the liquids. However, we have found that gas phase Raman and far infrared spectroscopy are much more useful in determining the conformational stabilities of these compounds. With advances in Fourier transform far infrared spectroscopy and laser Raman spectroscopy, it is relatively routine to record the spectra of these molecules in the gaseous state where they are free from the intermolecular forces in the liquid or intramolecular forces in a solvent, which may affect the true relative conformational stabilities. Utilizing the gas phase data, one can study the conformational changes of the molecules as they go from the gaseous state to the liquid and finally to the solid state. Under favorable circumstances, it is possible to calculate from the low frequency Raman and far infrared data the potential function for the asymmetric torsion from which the relative energies of the conformers can be obtained along with the barriers to interconversion. Therefore, the combined systematic studies of the gas phase Raman spectrum, along with the far infrared data, can yield conformational stabilities of the free molecules and, in some cases, potential functions for internal rotation for these organophosphorus molecules. Examples of such studies will be given for ethylphosphine ($CH_3CH_2PH_2$), ethyldifluorophosphine ($CH_3CH_2PF_2$), chloromethylphosphonic difluoride ($ClCH_2P(O)F_2$), chloromethylphosphonothioic difluoride ($ClCH_2P(S)F_2$), ethyldimethylphosphine ($CH_3CH_2P(CH_3)_2$) and ethyldimethylphosphine sulfide ($CH_3CH_2P(S)(CH_3)_2$). For these molecules, the more stable conformer has been identified, and the ΔH value between the conformers has been determined. Additionally, potential functions for the internal asymmetric rotation have been determined for four of these molecules.

CONFORMATIONAL STABILITY AND LOW FREQUENCY VIBRATIONAL ASSIGNMENT OF 1,1,1-TRIFLUOROISOPROPANOL, HEXAFLUOROISOPROPANOL, 1,1,1-TRIFLUOROISOPROPYLMETHYLETHER, AND HEXAFLUOROISOPROPYL METHYLETHER.

R. A. Larsen, F. O. Cox and J. R. Durig

Department of Chemistry, University of South Carolina, Columbia, SC 29208

Since gas-phase Raman spectroscopy directly measures the frequencies of torsional transitions, it offers an immediate advantage over some of the other techniques which depend upon the effect of the torsion on other molecular properties. As in the case of infrared spectroscopy, one has only to measure the frequencies of torsional transitions, and these can very quickly be related to the basic quantum mechanical properties of the molecule -- namely its energy levels. Quite often torsional fundamentals are extremely weak in the infrared owing to an insignificant dipole moment change, or they may give rise to band contours containing no Q branches from which reliable frequencies may be obtained. In the Raman spectrum, both mechanical and electrical anharmonicity allow the torsional overtones to be active and also give them sufficient intensity to be frequently observed. Since the overtones (or in some cases, the fundamentals) are totally symmetric, only isotropic polarizability terms can exist and will lead to stronger Q branches than for anisotropic fundamentals. From these sharp Q branches, frequencies can be easily extracted. However, perhaps the greatest advantage gas-phase Raman spectroscopy offers over infrared spectroscopy is in the study of molecules with more than one conformation present. Depolarization data, readily available in the Raman spectrum, quite often offers firm evidence for assigning torsional peaks observed for one conformer or another.

For 1,1,1-trifluoroisopropanol, three conformers have been identified from the gas-phase Raman spectra. From the asymmetric torsional frequencies, the potential function governing the internal rotation has been determined. The ΔH value between two of the conformers has been experimentally measured from the temperature study of the Raman spectra of the gas, and the values are consistent with those obtained from the potential function. Similar studies on hexafluoroisopropanol have been carried out, where two conformers have been identified. However, for the ethers, only a single conformer has been identified in each case. These results will be compared with those obtained for some similar molecules.

VIBRATIONAL SPECTRA AND NORMAL COORDINATE ANALYSIS OF NITRO SUBSTITUTED BENZENES

Tian Yongchi, Zhao Wenyun and Liang Yingqiu

Institute of Theoretical Chemistry,
Jilin University, Changchun, China

INTRODUCTION

Emperical assignments of vibrational bands of ortho-, meta- and para-dinitrobezenes have already been proposed(1), and a set of force field of 1,3,5-trinitrobenzene based on symmetrical coordinates has been recently reported(2). However, satisfactory forcefield should be transferable and obtained on the basis of the systemetical investigation of a series of molecules.

In the present paper, normal coordinate calculations of ortho-, meta- and para-dinitrobenzenes have been carried out with internal coordinates of these molecules.As a direct result, a set of force field is obtained and vibrational assignments of the molecules are made. The set of force field may be transferred to the calculation of 1,3,5-trinitrobenzene, and has been extended to deal with the analysis of picric acid.

MOLECULAR GEOMETRY AND CALCULATION METHOD

Geometrical parameters used in the calculations are listed in the following table:

parameter	value
C-C	1.397Å
C-H	1.086Å
C-N	1.45Å
N-O	1.207Å
C-C-C, O-N-O	
C-C-N, C-C-H	120°

NRCC Molecular Force Constant Calculation Program is employed and refinement of force field has been made with the observed frequencies.

RESULTS AND DISCUSSION

The vibrational frequencies of the five nitro substituted benzenes mentioned above, where most experimental frequeny values could be determined with high accuracy are well reproduced by the normal coordinate analysis. More detail properties of vibrational bands of the molecules have been known and in agreement to the emperical assignments given by literature(1).

Some main values of force constants in this work are tabulated in the following:

stetch(mdyn·Å⁻¹)		bend(mdyn·Å)	
C-N	5.420		0.35
N-O	9.067	O-N-O	1.565
		C-C-N	1.154
		C-N-O	0.897
		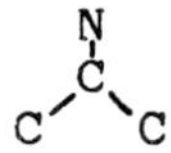	0.505

interaction

C-N/N-O	o.398	C-N/O-N-O	-0.215
C-N/C-C-N	0.909	C-N/C-N-O	0.355
N-O/O-N-O	1.034	N-O/C-N-O	0.367
C-C-N/C-N-O	-0.279		

Information about a possible conjugation between benzene ring and nitro group in these molecules may be obtained from the frequeny values of the vibrations concerning C-N bond and from the force constants belonging to the bond. Raman and FIR spectrum of 1,3,5-trinitrobenzene shows bands at 196$cm^{-1}$and 91cm^{-1}, which can be assigned to the torsional vibration of C-N bonds. Tortional vibration bands are not observed in the spectra of the three dinitrobenzenes, but the present calculation results predict the existence of these bands in the range of 90-140cm^{-1}.

REFERENCE

(1) J. H. S. Green and H. A. Harrison, Spectrochim Acta, 27A, 817(1971)
(2) J. P. Huenne et al, J. Mol. Struct. 118, 177(1984)

14-16

RAMAN SPECTROSCOPIC STUDY ON MOLECULAR CONFORMATIONS OF METHANESULFONATE ESTERS AND DEUTERATED DERIVATIVES

Hiroatsu Matsuura and Haruhiko Tamaoki

Department of Chemistry, Faculty of Science, Hiroshima University, Higashisenda-machi, Naka-ku, Hiroshima 730, Japan

Structual studies of sulfur compounds are important for understanding the nature of the chemical bonds in which the sulfur atom is involved [1]. In our previous Raman spectroscopic work [2], we have studied molecular conformations of sodium 1-alkanesulfonates RSO_3Na and sodium alkylsulfates $ROSO_3Na$. In order to acquire extensive conformational properties of sulfur compounds, we have extended our study to a series of 1-alkanesulfonate esters RSO_2OR' and have established the conformation about the S–C bond as well as that for the alkyl group. However, the conformation about the S–O bond has not been determined unambiguously. In the present work, we have studied the Raman spectra of methanesulfonate esters $MeSO_2OR$ (R=Me, Et, and n-Pr) and various deuterated derivatives of $MeSO_2OMe$ ($CD_3SO_2OCH_3$, $CD_3SO_2OCD_3$, etc.) in order to examine conformational stability of the S–O bond. Some of the observed Raman spectra are shown in Fig. 1. In analyzing the spectra, normal coordinate treatment has been utilized to establish vibrational assignments of the observed bands. Molecular orbital calculations have also been made on $MeSO_2OMe$ by using the 3-21G* basis set and the energy difference between the trans and gauche conformations about the S–O bond has been evaluated.

The most fundamental compound $MeSO_2OMe$ has two rotational isomers, trans and gauche, since, in this molecule, only the S–O bond is responsible for the rotational isomerism. Spectral features of this compound in the liquid and solid states resemble each other and the number of the observed bands is consistent with the presence of a single rotational isomer in these states. This observation indicates that either of the trans and gauche isomers has much higher stability than the other. The normal coordinate analysis on $MeSO_2OMe$ (undeuterated) by using the force constants transferred from RSO_3Na and $ROSO_3Na$ did not clarify the predominant isomer in the liquid and solid states. The analysis was then extended to $MeSO_2OEt$ and $MeSO_2OPr$, and correlations between the length of the ester chain and vibrational wavenumbers were found. For the gauche form about the S–O bond, the vibrational coupling between the SO_2 rocking and wagging modes depends largely on the chain length; the wavenumber of the coupled vibration of lower wavenumber increases as the chain length increases, while that of the higher-wavenumber vibration remains almost constant. For the trans form, on the other hand, the vibrational wavenumbers for the SO_2 group do not vary with the chain length. The Raman spectra of $MeSO_2OMe$ (undeuterated), $MeSO_2OEt$, and $MeSO_2OPr$ exhibit in common the bands of the SO_2 group at 550, 530, 470, and 340 cm^{-1} in both the liquid and solid states, independently of the chain length (Fig. 1). This observation indicates that the predominant conformation about the S-O bond is trans for these compounds. In order to establish the assignments of the observed bands, the deuterated derivatives of $MeSO_2OMe$ were also included in the normal coordinate analysis. The results show that the above-mentioned four bands of the SO_2 group are assigned as follows; the 550 cm^{-1} band to the SO_2 wagging and SOC bending, the 530 cm^{-1} band to the SO_2 scissoring, the 470 cm^{-1} band to the SO_2 twisting, and the 340 cm^{-1} band to the SO_2 rocking. For $CD_3SO_2OCD_3$, the first band is shifted largely to 500 cm^{-1}, in conformity with the normal coordinate analysis.

The molecular orbital calculation on $MeSO_2OMe$ gave the energy difference between the trans and gauche forms to be 2.7 kcal mol^{-1}, the trans form being more stable. This result is consistent with the present experimental finding that the trans form is predominant in the liquid and solid states.

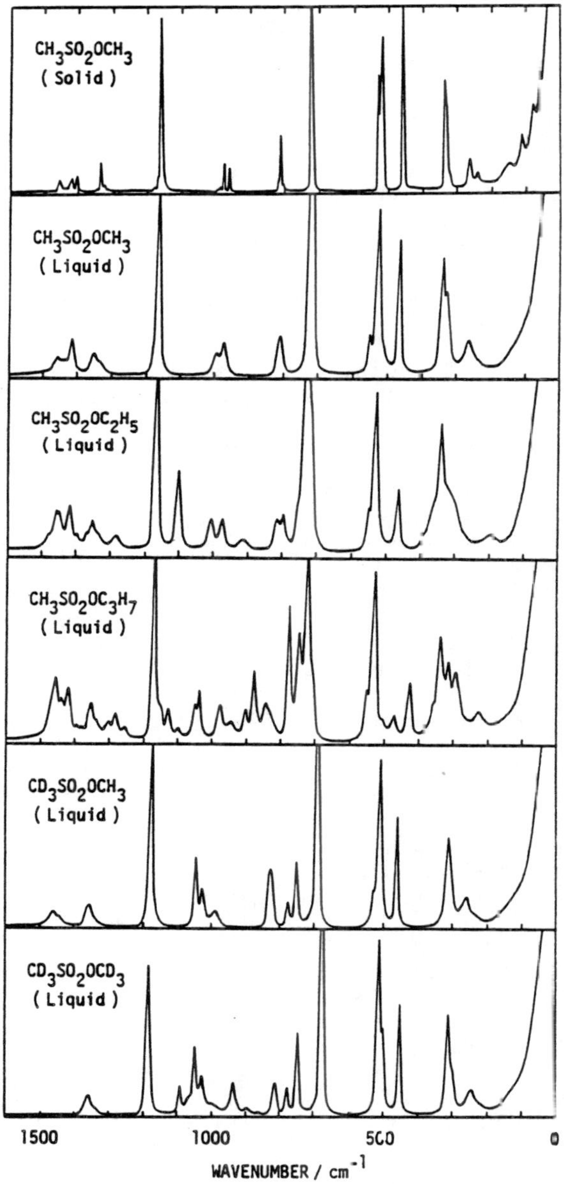

Fig. 1. Raman spectra of $MeSO_2OR$.

REFERENCES

1. I. Hargittai, "The Structure of Volatile Sulphur Compounds," D. Reidel, Dordrecht (1985).
2. H. Matsuura and H. Tamaoki, unpublished work.

Raman Spectra and Structures of 1,5-Dithiacyclooctane and its Monocation Radical and Dication

Koichi Itoh, Masanori Serita, Makiko Tamaoki and Yasushi Shiratori

Department of Chemistry, School of Science and Engineering, Waseda University, Shinjuku-ku, Tokyo 160 Japan

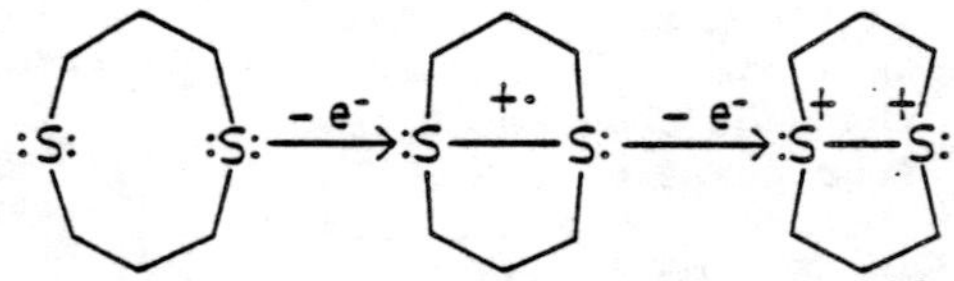

Fig. 1. Oxidation steps of DTCO

1,5-Dithiacyclooctane (DTCO, Fig.1) is known to undergo stepwise one-electron oxidations resulting in the formation of a monocation radical ($DTCO^{+\cdot}$) and a dication($DTCO^{2+}$)(1,2). The oxidation site is the sulfur atom and the electron removed is one of the non-bonded 3p electrons. An interesting aspect of this process is that intra-molecular sulfur-sulfur bonds containing three and two electrons are formed in the monocation radical and the dication, respectively (Fig. 1). In the present paper we measured the Raman and infrared spectra of DTCO and related cyclic dithioethers such as 1,5-dithiacyclononane(DTCN) and 1,6-dithiacyclodecane (DTCD), the resonance Raman spectrum of $DTCO^{+\cdot}$ and the Raman spectrum of $DTCO^{2+}$ under various conditions. On the basis of the spectra and the results of normal coordinate treatments, molecular mechanics calculations and ab initio MO calculations, we discussed possible structures of DTCO, $DTCO^{+\cdot}$ and $DTCO^{+2}$.

$DTCO^{+\cdot}$ was prepared either by a chemical oxidation with $NOBF_4$ or by a controlled potential electrolysis(2). $DTCO^{2+}$ was obtained by a chemical oxidation with $NOBF_4$ as well as by an acidification of 1,5-dithiacyclooctane 1-oxide with H_2SO_4(2). A HITAC M-280H at Tokyo University Computer Center was used to perform normal coordinate treatments, molecular mechanics calculations, and ab initio MO calculations and structural optimizations with the STO-3G basis set (Programs HONDO and GAUSSIAN 80 were used.).

Parts A and B of Fig. 2 show the Raman spectra of DTCO in a liquid and solid states, respectively. From these spectra it is clear that the bands at 781 and 742 cm^{-1} observed for the liquid state disappear in the spectrum of the solid state in which the C-S stretching vibrations are observed at 692, 639 and 612 cm^{-1}. This result indicates that DTCO in the solid state takes only one kind of conformation, whereas the sample in the liquid state takes at least two kinds of conformations. Structural optimizations based on the molecular mechanics calculation as well as ab initio MO calculations with the STO-3G basis set suggest that twisted boat-boat(TBB) and chair-chair(CC) conformations of DTCO are relatively more stable than the other structures. The results of the calculations indicate also the importance of a repulsion between the non-bonded 3p electrons of the sulfur atoms in determining the structure of DTCO. By using a general force field, which satisfactorily reproduces the vibrational spectra of DTCN and DTCD with a known structure, normal coordinate treatments were performed for a series of assumed structures. The calculated frequencis for TBB and CC similarly conform well to the observed ones. In order to determine whicn one is taken by DTCO in the solid state we are now analyzing the vibrational spectra of a partially deuterated DTCO.

Part C of Fig. 2 represents the resonance Raman spectrum of $DTCO^{+\cdot}$ excited by a 457.9-nm laser line , whose feature is quite different from that of DTCO itself. This can be explained by considering that a resonance effect from the 405-nm absorption band associated with a transition from a bonding orbital to an anti-bonding one of the $S-S^{+\cdot}$ bond exclusively enhances the vibrational modes in which the S-S bond length changes appreciably. The Raman spectrum of $DTCO^{2+}$ in Fig. 2(D) again shows a feature quite different from that of DTCO. Especially, the former gives several prominent bands in the 600 - 400 cm^{-1} region, where DTCO shows only weak bands near 467 cm^{-1}. The 644-, 603-, 550- and 502-cm^{-1} bands of the dication can be ascribed to the C-S streching vibration. The frequency lowering of this mode as compared to that of DTCO indicates that the dication formation is accompanied by an appreciable weakening of the C-S bond. Ab initio MO calculations and structural optimizations with the STO-3G basis set were also performed to the dication assuming the TBB and CC conformations . The optimized structures have the S^+-S^+ bond lengths of 2.140 and 2.233 Å for TBB and CC, respectively, and the C-S bond lengths of 1.890 and 1.871 Å for TBB and 1.849 Å for CC, which are appreciably longer than the C-S bond lengths of DTCO itself (average = 1.808 Å). This result is consistent with the weakening of the C-S bonds predicted from the Raman data. In order to assign the other bands characteristic of the dication (including the S^+-S^+ stretching vibration) we are now preparing a series of compounds containing the S^+-S^+ bond and trying to measure their Raman spectra.

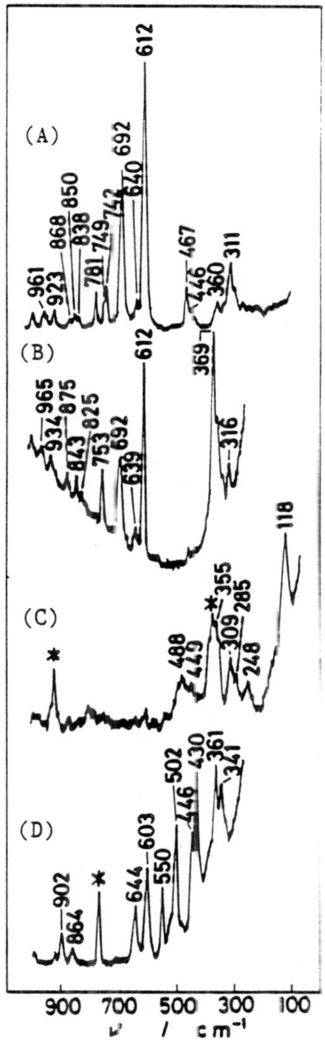

Fig. 2. (A),(B) Raman spectra of DTCO in a liquid(at r.t.) and a solid state(at 80 K), respectively, (C) Resonance Raman spectrum of $DTCO^{+\cdot}$ in CH_3CN, (D) Raman spectrum of $DTCO^{2-}(BF_4^-)_2$. (A),(B),(D):514.5-nm ex. (C):457.9-nm ex.

References

(1) Klaus-Dieter Asmus, Acc.Chem.Res. 12,436(1979).
(2) W.Kenneth Musker, Acc.Chem.Res. 13,200(1980).

RAMAN AND INFRARED SPECTRA OF 2–AMINO–4*H*–PYRAN DERIVATIVES

J. Bellanato
Instituto de Optica, C.S.I.C., Serrano 121, 28006–Madrid

N. Martín and C. Seoane
Departamento de Química Orgánica, Facultad de Química, Universidad Complutense, 28040–Madrid
Spain

The Raman and infrared spectra of eight 2–amino–4*H*–pyran derivatives have been studied in the ν(N–H), ν(C≡N) and double bond regions. For the purposes of comparison 2,6–dimethyl–3,5–dicyano–4–phenyl–4*H*–pyran has also been studied.

The table below shows the 4*H*–pyran derivatives studied.

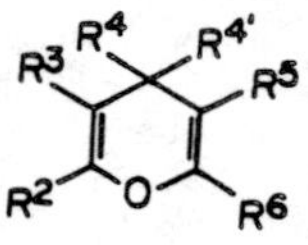

Compound	R^2	R^3	R^4	$R^{4'}$	R^5	R^6
1	NH_2	CN	*i*–C_3H_7	H	CN	CH_3
2	NH_2	CN	*s*–C_4H_9	H	CN	C_6H_5
3	NH_2	CN	CH_3	C_2H_5	CN	C_6H_5
4	NH_2	CN	C_6H_5	H	CN	H
5	NH_2	CN	*i*–C_3H_7	H	CO_2 Et	CH_3
6	NH_2	CN	C_6H_5	H	$COCH_3$	CH_3
7	NH_2	CN	C_6H_5	H	C_6H_5	C_6H_5
8	NH_2	CO_2 Et	C_6H_5	H	CN	C_6H_5
9	CH_3	CN	C_6H_5	H	CN	CH_3

The Raman spectra of (1) in solid and in chloroform solution are characterized by three bands in the double bond region. The very strong band at *ca* 1680 cm^{-1} is strongly polarized in solution, the other two bands at lower frequency (*ca* 1640 and 1590 cm^{-1}) being comparatively weak, specially in solution. The band at higher frequency is assigned to $\nu(C^5=C^6)$ and the frequency value is in the range of that reported in the literature for cyclic ketene acetals and *N*–disubstituted 2-amino γ-pyrans. The bands at *ca* 1640 and

1590 cm^{-1} are considered to arise from skeletal vibrations of the enaminonitrile system $H_2N-C=C-CN$. Both bands disappeared upon *N*–deuteriation and were replaced by another weak band (partially polarized) at 1629 cm^{-1} due mainly to the $\nu(C=C) + \nu(C-N)$ vibration of the *N*–deuteriated species.

The corresponding infrared spectra of (1) were also characterized by three bands, and as in the Raman spectra the band at higher frequency was predominant. However, the bands at 1640 and 1590 were strong in infrared and on disappearing after *N*–deuteriation gave place to a very strong $\nu(C=C)$ band at 1628 cm^{-1}. Similar behaviour for enamine bands has been observed, for example, in 3–aminocrotonic esters and 3–amino–2–cyanoacrylic esters with primary amino–groups.[1,2]

Our results suggest that due to the near planarity of the ketene acetal system[3] some vibrational coupling exists between the two $\nu(C=C)$ vibrations.

The infrared spectrum of (1) showed in the $\nu(N-H)$ region two medium bands at 3415 and 3335 cm^{-1} in the solid state which were shifted at 3510 and 3408 cm^{-1} in chloroform solution due to the disappearance of the weak hydrogen bonds present in the crystal, which has been demonstrated by X–ray measurements.[3]

Two $\nu(C\equiv N)$ bands were observed in the solid state and in solution, both polarized, in the Raman spectrum. In the infrared spectrum the lower frequency band is stronger than the higher frequency band and therefore the former is assigned to the $\nu(C\equiv N)$ of the enaminonitrile system the frequency shift being attributed to the conjugation effect.

The seven compounds studied containing the enaminonitrile structure presented similar spectroscopic characteristics. However, the electronic properties of the substituents at the C^5 and C^6 induce frequency and intensity changes in the $\nu(C^5=C^6)$ band ("γ–pyran" band) and consequently in the degree of coupling with the $\nu(C^2=C^3)$ band (enamine band). Differences in the hydrogen bond strength and in the more or less ring planarity are also clearly reflected in the spectra.

When we consider compound (8) and due to the presence of the mesomeric enamino-ester system $H_2N-C=C-C=O$ and of an intramolecular hydrogen bond, its spectra showed significant differences with those of compounds (1)–(7) containing the enaminonitrile structure. It must be mentioned that in compound (8) both $\nu(C=C)$ vibrations lay further apart than in (1)–(7) and consequently no vibrational coupling takes place.

REFERENCES

1. A. Gómez Sánchez, A.M. Valle and J. Bellanato, J.C.S. Perkin II, 15 (1973).
2. J. Bellanato, A. Gómez Sánchez and P. Borrachero. An. Quím., **72**, 876 (1976); J. Bellanato and A. Gómez Sánchez. Unpublished results.
3. F. Florencio and S. García-Blanco. Private communication.

INTENSITY PARAMETERS FOR THE RAMAN TRACE SCATTERING OF n-ALKANES.

K.M. Gough and W.F. Murphy

Division of Chemistry, National Research Council, 100 Sussex Drive, Ottawa, Canada K1A 0R6

One of our current interests is the characterization of deuterated species of simple gaseous hydrocarbons on the basis of their Raman spectra. The dependence of observed intensities on both the abundance of the scattering species and the molar scattering cross section must be taken into account. We deal with the trace scattering component (also known as the isotropic or mean polarizability component) because of the advantages of working with the sparse spectrum of well defined bands on a flat baseline which results from the restrictive vibrational and rotational selection rules in this case.

The trace scattering cross section may be expressed in terms of the parameters $\partial\bar{\alpha}/\partial S_j$, the derivatives of the mean molecular polarizability with respect to the vibrational symmetry coordinates. (See [1] for the relations in SI units.) Under the assumption of the nuclear mass independence of the molecular electronic properties (Born-Oppenheimer approximation), these parameters are independent of isotopic substitution [2]. A second assumption made at this level of approximation is that the contribution of non-fundamental vibrational transitions to the observed intensity is negligible.

Before these concepts may be used to interpret spectra, it is necessary to determine the limits to the approximations. To do this, we have extended a previous study [1] of ethane.

The absolute trace scattering cross sections of ethane-h_6 and ethane-d_6 have been measured and fit by the three non-vanishing $\partial\bar{\alpha}/\partial S_j$ parameters. The geometry, symmetry coordinates and general harmonic force field were taken from the literature [3]. Absolute trace scattering cross sections were then measured for 1,1,2,2-ethane-d_4. In this case, the spectrum consists of bands due to two rotational conformers, since the two protons may be either trans or gauche with respect to rotation about the CC bond. Thus, in the fit of the observed cross sections, the relative abundance of the two conformers must be included in addition to the $\partial\bar{\alpha}/\partial S_j$ parameters.

$\partial\bar{\alpha}/\partial S_j$ intensity parameters for ethane (10^{-30} CmV^{-1})

	Experimental			Theoretical	
Symmetry Coordinate	CHD_2CHD_2	C_2H_6,C_2D_6	[1]	[4]	[5]
S_1 (sym. CH str.)	3.21 ± 0.09	3.32 ± 0.11	3.003	3.22	3.22
S_2 (sym. CH_3 def.)	-0.160 ± 0.008	-0.181 ± 0.019	-0.245	-0.11	
S_3 (CC str.)	1.31 ± 0.04	1.32 ± 0.04	1.161	1.49	1.01

In the table, the parameters are compared for our two determinations, and the previous experimental and theoretical work. In the CHD_2CHD_2 case, we found the mole fraction of the trans conformer to be 0.33 ± 0.02, as expected on the basis of a random distribution of the proton among the possible sites. We also found that the electrical anharmonicity contributions to the total integrated cross sections are insignificant in terms of the measurement accuracy, although they are necessary for understanding the relative intensities of perturbed bands [6].

This approach is currently being used in the analysis of the trace Raman scattering spectrum of various isotopic species of propane. In this case, the harmonic vibrational force field must first be obtained.

REFERENCES

1. J. Martin and S. Montero, J. Chem. Phys. 80, 4610 (1984).
2. B. Crawford, J. Chem. Phys. 20, 977 (1952).
3. J.L. Duncan, R.A. Kelly, G.D. Nivellini and F. Tullini, J. Mol. Spectrosc. 98, 87 (1983).
4. E.N. Svendsen and T. Stroyer-Hansen, Mol. Phys. 56, 1025 (1985).
5. M.P. Bogaard and R.S. Watts, Chem. Phys. Lett. 97, 494 (1983).
6. D. Bermejo and S. Montero, J. Chem. Phys. 81, 3835 (1984).

THE RAMAN SPECTRUM OF 1,4-NAPHTOQUINONE INTERPRETED BY USING A TRANSFERABLE VALENCE FORCE FIELD.

B. Mannfors and K. Palmö

University of Helsinki, Dept. of Physics, Accelerator Laboratory, Siltavuorenpenger 20 M, SF-00170 Helsinki, Finland

The Raman spectrum of 1,4-naphtoquinone (NQ) has been measured at liquid helium temperature using Kr^{+}-laser excitation. The resolution used was $1cm^{-1}$, which is about the natural Raman line width at the low temperature in question. The lines in the measured spectrum have been resolved by computer fits of Voigt profiles to all the observed band configurations. The assignments were made using force field calculations based on our previous studies (1-5).

Due to the low symmetry of the NQ molecule (C_{2v}) its vibrational spectra are rather complex and therefore difficult to assign, especially when only the spectra of deuteromers and related molecules are compared. The main disadvantage is that the modes may change on deuteration in an unpredictable way which easily leads to misinterpretations. Instead the assignments of complex vibrational spectra can be made most reliably by utilizing physically reasonable valence force fields. Previously we have developed a transferable overlay valence force field for a series of weakly coupled conjugated molecules e.g. p-benzoquinone, benzaldehyde, styrene and ethynylbenzene. However, X-ray diffraction measurements have shown (6) that the structure of the NQ molecule is modified as compared to the structures of benzene and p-benzoquinone. This means that, strictly speaking, NQ does not form a weakly coupled system. Still, the aim of this work has been to study the possibilities of applying the force field of weakly coupled conjugated systems to an intermediate case as in NQ. According to this model the NQ molecule is considered to consist of an ortho-disubstituted benzene ring and a conjugated O=C-CH=CH-C=O system similar to that of p-benzoquinone. Thus, NQ would share most of its force constants with benzaldehyde, ortho-disubstituted benzene and p-benzoquinone, all systems of which have been extensively studied by our group. The differences in the bond lengths of NQ as compared to benzene and p-benzoquinone have, though, been taken into account in the stretching and torsion force constants by interpolating them from our previous results (1-5). The assignments of NQ and NQ-d_6 made by using "weakly coupled model" agree with the Raman depolarization ratios and they also account for all high and medium intensity Raman and IR bands. The rms frequency deviations are 9.71cm^{-1} and 13.0cm^{-1} for the in-plane and out-of-plane vibrations, respectively. In the in-plane force field the diagonal CC-stretching force constants excluding those in the O=C-CH=CH-C=O group have been optimized. The observed frequencies have not been corrected for Fermi resonance. Girlando et al. (7) have previously assigned the in-plane vibrations of NQ and NQ-d_6 by calculating an overlay force field for NQ, 9,10-anthraquinone and p-benzoquinone. Our results are essentially in agreement with theirs. We conclude that, although the NQ molecule has stronger coupling between its subsystems than e.g. benzaldehyde the assignments of its vibrational spectra can still be carried out using the "weakly coupled model."

References:

1. L-O.Pietilä, K.Palmö and B.Mannfors, J.Mol.Spectrosc.112,104(1985)
2. L-O.Pietilä, K.Palmö and B.Mannfors, J.Mol.Spectrosc. (in press)
3. B.Mannfors, L-O.Pietilä and K.Palmö, J.Mol.Struct. (in press)
4. K.Palmö, B.Mannfors and L-O.Pietilä, Spectrochim. Acta (submitted)
5. B.Mannfors, K.Palmö and L-O.Pietilä, in preparation
6. J.Bernstein, M.D.Cohen and L.Leiserowitz, The structural chemistry of quinones, in: The Chemistry of the Quinonoid Compounds (Edited by S.Patai), Chapter 2, Wiley-Interscience, London(1974)
7. A.Girlando, D.Ragazzon and C.Pecile, Spectrochim. Acta 36A,1053(1980)

ROTATIONAL ISOMERISM OF O,O'-DIALKYL ESTERS OF PHOSPHORODITHIOIC ACID.

D.N. Waters and J.L. Paddy

Department of Chemistry, Brunel University, Uxbridge, UB8 3PH, England.

Dialkyl esters $(RO)_2P(:S)SH$ (1) of phosphorodithioic acid (1, R=H), are the starting point for the preparation of metal dialkyldithio phosphates, e.g. $[(RO)_2PS_2]_2Zn$, important for their application as anti-wear additives in petroleum- based lubricants. The ir spectrum of the dimethyl ester has been studied, and it was found [1] that several bands appear as doublets attributable to the presence of rotational isomers. Similar behaviour is exhibited by many other organophosphorus compounds [2], but little information on the quantitative aspects of the equilibria between rotational isomers in these compounds is available.

Figure 1 shows the ν(S-H), 2600-2400 cm^{-1} and ν(P-S), 600-500 cm^{-1} regions of the Raman spectrum of di*iso*propylphosphorodithioic acid (1,R=Pr^i), as a 10% solution in *n*-tetradecane. In the pure compound the ν(S-H) region shows an additional broad feature centred at 2450 cm^{-1}, due to hydrogen bonding, but this disappears in dilute solution in an inert solvent, to leave a well resolved doublet at 2590, 2550 cm^{-1}. The higher frequency component is assigned to structure (a) below, and the lower frequency component to structure (b).

a 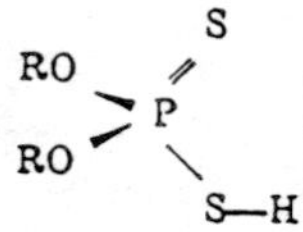

b 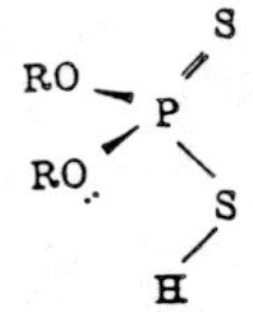

The temperature dependence of the Raman intensities of the ν(S-H) and ν(P-S) doublet components was investigated for the *iso*propyl and *n*-butyl esters, (1, R=Pr^i and Bu^n), in 10% solutions in *n*-tetradecane (at which concentration hydrogen bonding is essentially absent). From the plots of intensity ratios against T^{-1}, ΔH values for the isomer equilibria were deduced. The experimental values are:- for R=Pr^i, ΔH=2.5±0.3 J mol^{-1}; for R=Bu^n, ΔH=3.0±0.5 J mol^{-1}. The ΔH values appear to be independent of whether the S-H or the P-S doublet is used as the basis of the determination.

These observations confirm the occurrence of rotational isomerism about both the P-S and P-O bond axes in the molecules studied. Moreover, the apparent independence of the measured ΔH values with respect to the choice of doublet component used for the determination,

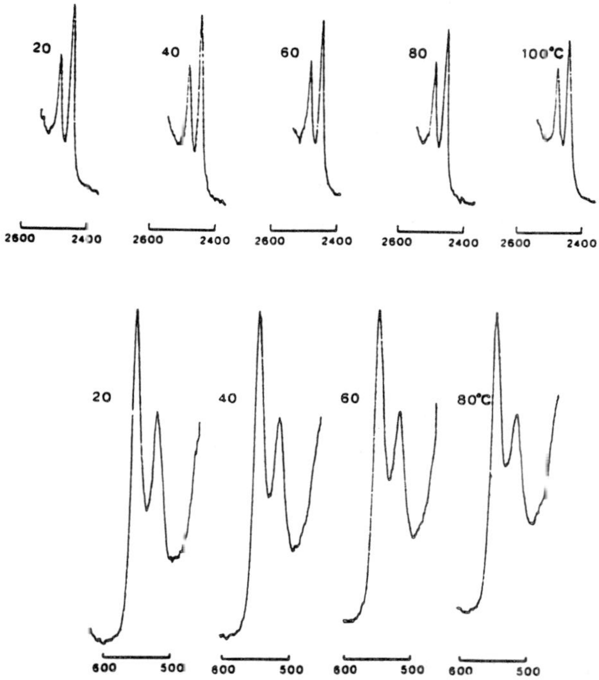

Figure 1. Top, ν(S–H); Bottom, ν(P–S) vibrations of di*iso*propylphosphorodithioic acid; 10% solution in *n*-tetradecane.

lends some support to an earlier suggestion [1], *viz*, that the two internal rotation processes act concertedly, so that each conformation of the P–S–H group is preferentially associated with one particular conformation of the P–O–C group.

ACKNOWLEDGEMENT

We thank DGDQA/TS, Ministry of Defence, for support.

REFERENCES

1. R.A. Nyquist, Spectrochim. Acta A, 24A, 47 (1969).
2. L.C. Thomas, Interpretation of the Infrared Spectra of Organophosphorus Compounds, Heyden, London, 1974.

RAMAN AND INFRARED SPECTROSCOPIC STUDIES OF NOVEL POLY-AZA CHAINS AND RELATED METAL COMPLEXES

R. BARIZ, C. BREMARD, J.C. MERLIN

Laboratoire de Spectrochimie Infrarouge et Raman, CNRS (LP.2641) Université des Sciences et Techniques de Lille Flandres Artois, Bât. C.5 59655 Villeneuve d'Ascq Cedex, France.

The number of linear conjugated macromolecules for which a detailed spectroscopic studies have been done, concern mainly those possesing C=C and C≡C bonds[1] : polydiacetylenes, polyacetylenes, carotenoids. In this work,we reported the preliminary results concerning the structural and spectroscopic studies of a new series of polyaza ligands and the related copper (II) and zinc (II) complexes. The new molecules under study are the following :

1

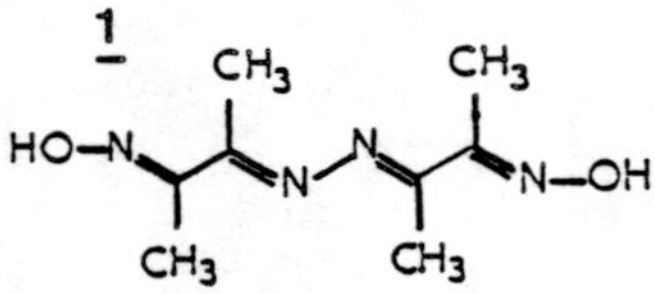

2

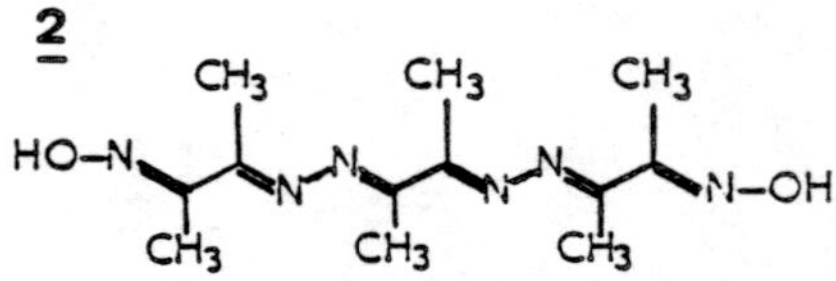

3

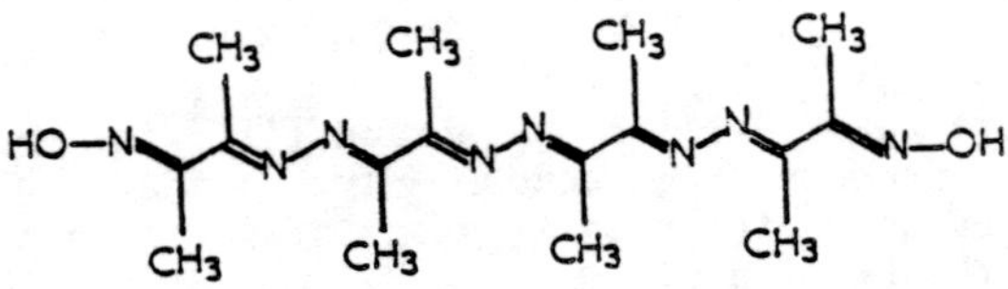

Although all the X-ray structure of the poly-aza ligands are not undertaken yet, it is probable that among the possible conformations, all the poly-aza ligands exhibit a planar backbone with an all-trans geometry in the solid state.

Typical Raman spectrum of a sexta-aza molecule 3 is shown on figure 1. It is very similar to those of di and tetra-aza analogs and look like to those of other polyene compound (carotenoids, polyacetylenes, polydiacetylenes).

We propose the following assigments.The C=N stretching mode appears near 1550 cm^{-1}. The wavenumber decreases as the length of the chains increases. The shift in wavenumber can be related to the change in π-electron delocalization in the poly-aza-molecules.For oxime derivatives an additional band observed at 1632cm^{-1} can be assigned to terminal C=N-OH group stretching. The medium lines observed in the spectral range, 1300-1400 cm^{-1}, can be assigned to methylgroup deformations. The C-C and N-N stretching mode assigments are not straitforward and are supported by ^{14}N-^{15}N isotope spectral shift.

FIGURE 1

Raman spectra of <u>3</u>

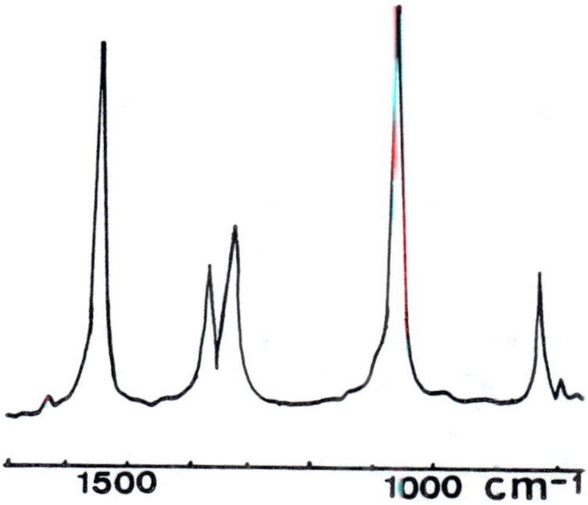

The poly-aza-molecules <u>1</u>, <u>2</u>, <u>3</u> are respectively bi, tri and tetra nucleating ligands and the corresponding zinc (II) and copper (II) complexes have been isolated.

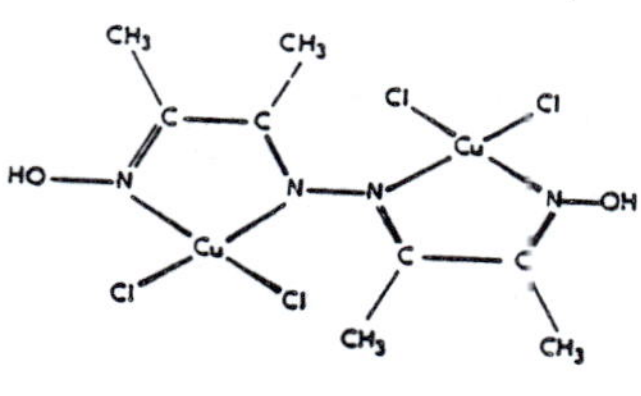

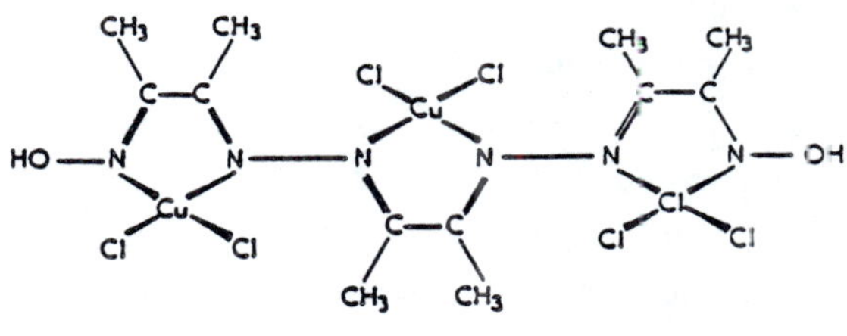

The <u>trans</u> <u>cisoid</u> conformations [2] of the poly-aza ligands are stabilized by the chelation to Cu(II) and Zn(II) ions. The vibrational spectra are interpreted in relation to these conformations.

References :

1. N.M. Batchelder and D.Bloor, in Advances in infrared and Raman spectroscopy, vol. 11, R.J.H. Clark and R.E. Hester Eds 133 (1984).
2. C.J.O'Connor, R.J. Romananch, D.M. Robertson, E.E. Educk, Inorg. Chem. <u>22</u>, 449, (1983).

PREDICTION OF THE VIBRATIONAL FREQUENCIES OF TRANS-STILBENE AND TOLANE BY USING A TRANSFERABLE VALENCE FORCE FIELD.

K.Palmö and B.Mannfors

University of Helsinki, Dept. of Physics, Accelerator Laboratory, Siltavuorenpenger 20M, SF-00170 Helsinki, Finland

Large organic molecules which consist of weakly coupled conjugated systems form an important class of compounds. Vibrational spectra can, in principle, give valuable information about the properties of these molecules. However, the observed spectra are often very complex and almost impossible to interpret without normal coordinate analysis. Reliable force fields are therefore indispensable in order to obtain correct assignments. Thus, the majority of the force constants to be used in the normal coordinate analysis cannot be derived simultaneously with the assignments, but have to be available in advance. The only way to achieve this seems to be through a series of successive overlay calculations where force constants are assumed transferable between geometrically similar subsystems. This means that one must start deriving force constants by using small or symmetric well-assigned molecules and gradually add more complicated molecules to the calculation along with assigning their fundamentals.

In this study we present the calculations of the vibrational frequencies of trans-stilbene (tSB) and tolane (T) and several of their deuteromers. The work has been preceded by overlay force field calculations according to the method described above (1-5). The molecules treated in refs. 1-5 include, among others, many benzene derivatives e.g. styrene and ethynylbenzene and the force constants proved valid in these molecules have been directly transferred to tSB and T, respectively. No refinement of force constants or addition of new interaction force constants have been made.

As expected the results confirm the transferability of the force field used. A comparison between similar modes in tSB, T and other benzene derivatives clearly shows the correctness of the reassignments made as compared to previous studies (6-7). One reason for faulty assignments of the fundamentals of tSB is the surprisingly low frequencies of the two lowest a_u modes, calculated to 44 and 59 cm^{-1}, respectively. Warshel (8) has calculated the normal frequencies of tSB with the CFF method and obtained similar results for these low a_u modes.

As a conclusion it may be said that the systematically constructed force field applied in this work constitute a sound physical basis for the interpretation of vibrational spectra of large conjugated molecules where the transferability of force constants can be fully utilized.

References:

1. L-O.Pietilä, K.Palmö and B.Mannfors, J.Mol.Spectrosc. 112,104-114 (1985)
2. L-O.Pietilä, K.Palmö and B.Mannfors, J.Mol.Spectrosc. (in press)
3. B.Mannfors, L-O.Pietilä and K.Palmö, J.Mol.Struct. (in press)
4. K.Palmö, B.Mannfors and L-O.Pietilä, Spectrochim. Acta (submitted)
5. B.Mannfors, K.Palmö and L-O.Pietilä, in preparation
6. C.Pecile and B.Lunelli, Can.J.Chem. 47,234-250(1969)
7. Z.Meić and H.Güsten, Spectrochim. Acta 34A,101-111(1978)
8. A.Warshel, J.Chem.Phys. 62,214-221(1975)

A PRIORI CALCULATIONS OF THE VIBRATIONAL FREQUENCIES OF METHYLATED POLYENES.

M.M. ABO ALY, J. FAVROT.

Laboratoire de Spectrochimie Infrarouge et Raman, C.N.R.S, 2 rue henri Dunant, 94320 THIAIS, FRANCE.

Recently some valence force fields obtained from ab initio studies have been scaled on very accurate experimental data. Such work have been carried out not only for small molecules as, for example, ethene (1,2), propene (3), 1,3 - butadiene (4) but also for larger molecules with higher symmetry as benzene (5). Many force fields have been calculated from experimental results for linear or branched alkanes.On the other hand there is little accurate force fields of unsaturated methylated molecules except for propene. The influence on the force field of the double bond configuration of 2 - butene or of the central single bond conformation of isoprene is not known with sufficient precision. Furthermore there is still some confusion in the definition of the coordinates and the force constants especially for out of plane vibrations. Finally a simplified geometry, such as regular tetrahedron for methyl group, is often chosen for sake of convenience even if that geometry is not the real or most probable one.

We undertake vibrational calculations to fit valence force fields on accurate experimental data (small molecules) or to determine the normal modes (frequency and potential energy distribution) for large molecules. We choose the most probable geometry from experimental results (X-Rays, electron diffraction, microwaves, CH lengths from isolated vibrators ...) and ab initio calculations (HCH angles). We use a non-redundant set of local symmetry coordinates (group coordinates) such those of the I.U.P.A.C (6) and the japanese team (7). We hope to obtain univocal and more easily transferable sets of force constants.

For the initial force field of small molecules as ethene we use force constants from experimental results (8) and from ab initio calculations (1,2) and we fit this force field on vibrational frequencies (8). Similarly we calculate a force field from vibrational results of isotopic species of propene (9), of (E)-and (Z)-2-butenes (10) and of isobutylene (11). We transfer the force fields of these molecules to confirm or to modify the vibrational assignments for isoprene, (E)- and (Z) - 1,3 - pentadienes or our assignments for (E,E) -, (Z,E) - and (Z,Z) - 2,4 - hexadienes (12), 2 - methyl - and 4 - methyl- 1,3 - pentadienes and 2,5 - dimethyl - 2,4 - hexadiene (13). Analogous studies can be made from vibrational spectra of medium size polyenals (14) or of photopigments chromophores.

We hope to precise by these studies the force field of the methylgroup bound to a sp^2 carbon and the influence of the C = C double bond configuration on the force field. The preliminary results will be submitted to the conference.

References

1. C.E. BLOM and C. ALTONA, Mol. Phys., 34, 177 (1977)
2. Peter PULAY, Géza FOGARASI, Gabor PONGOR, James E. BOGGS and Anna VARGHA, J. Amer. Chem. Soc., 105, 7037 (1983).
3. C.E BLOM and C. ALTONA, Mol. Phys., 33, 875 (1977).
4. Ch. W. BOCK, YU. N.PANCHENKO, S.V. KRASNOSHCHIOKOV, V.I. PUPYSHEV. J. Mol. Struct. , 129,57 (1985)
5. Peter PULAY, Géza FOGARASI and James E. BOGGS, J. Chem. Phys., 74, 3999 (1981).
6. IUPAC Commission on Molecular Structure and spectroscopy, Pure Appl. Chem., 50, 1707 (1978)
7. Hiroatsu MATSUURA and Mitsuo TASUMI, in "Vibrational Spectra and structure", Vol. 12, Ed. J.R. DURIG, Elsevier (1983) ch.2, p.59.
8. J.L. DUNCAN and E.HAMILTON, J. Mol. Struct. Theochem., 76, 65 (1981)
9. B. SILVI, P. LABARBE et J-P PERCHARD, Spectrochim. Acta, 29A, 263 (1973)
10. D.C. Mc KEAN, M.W. MACKENZIE, A.R. MORRISSON, J.C. LAVALLEY, A. JANIN, J. FAWCETT and H.G.M. EDWARDS, Spectrochim. Acta, 41A, 435 (1985)
11. W.C. HARRIS and I.W. LEVIN, J. Mol. Spectrosc., 39, 441 (1971)
12. M.M. ABO ALY, M-H BARON, J. FAVROT, F. ROMAIN et M. REVAULT, Spectrochim. Acta, 40A, 1037 (1984)
13. M.M.ABO ALY, M-H BARON, M-J COULANGE and J. FAVROT, Spectrochim. Acta (in press)
14. M.M. ABO ALY, M-H BARON, J. FAVROT, J. BELLOC and M. REVAULT, Can. J. Chem., 63, 1587 (1985)

INCREASE IN THE C=N STRETCHING FREQUENCY UPON COMPLEXATION OF TRANS-RETINYLIDENE-N-BUTYLAMINE WITH GENERAL LEWIS ACIDS.

J. J. López-Garriga and G. T. Babcock

Department of Chemistry, Michigan State University, East Lansing, MI 48824-1322 USA.

Resonance Raman studies of retinal Schiff's bases, their protonated derivatives and the corresponding isotopically substituted compounds have been used effectively in assigning key modes in the spectra of rhodopsin and bacteriorhodopsin[1] and in showing that a protonated retinal Schiff's base occurs in various intermediates in the photocycles of both proteins.[2] A red-shifted optical absorption spectrum and an increase in the C=N stretching frequency, relative to the unprotonated derivative, are characteristic of the protonated form of the chromophore. The optical absorption shift has been attributed to increased delocalization of the π system while the increase in ν(C=N) has been explained by invoking an interaction between the C=N stretching mode and the C=N-H bending motion. The stretch-bend interaction has been postulated to obscure the expected correlation between increased π-delocalization and decreased C=N stretching force constant.

Resonance Raman studies on trans-retinylidene-n-butylamine, a retinal Schiff's base model for the chromophores in rhodopsin and bacteriorhodopsin, are presented here and show that the C=N stretching frequency increases by 29 cm^{-1}, 29 cm^{-1}, and 31 cm^{-1}, respectively, upon complexation of the free Schiff's base with the general Lewis acids, BCl_3, BBr_3, and BF_3 (Figure 1a). The complexes also show pronounced red shifts in their optical absorption maxima (Figure 1b). This behavior is analogous to that observed upon protonation of retinal Schiff's bases by a variety of protic acids and demonstrates that the absorption red shift and the increases in ν(C=N) are general properties of Lewis acid: retinal Schiff's base complexes. Normal

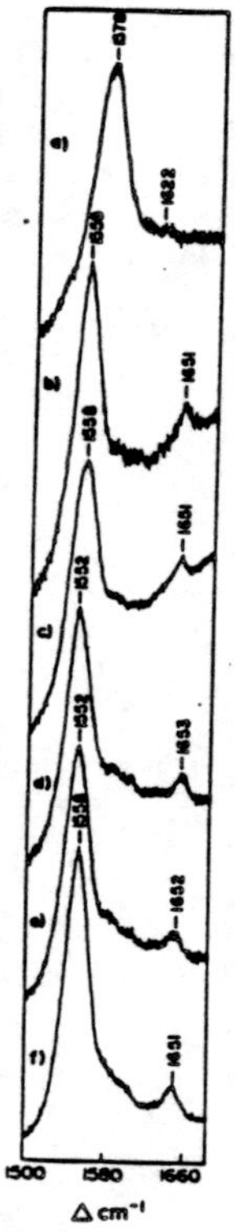

Figure 1a

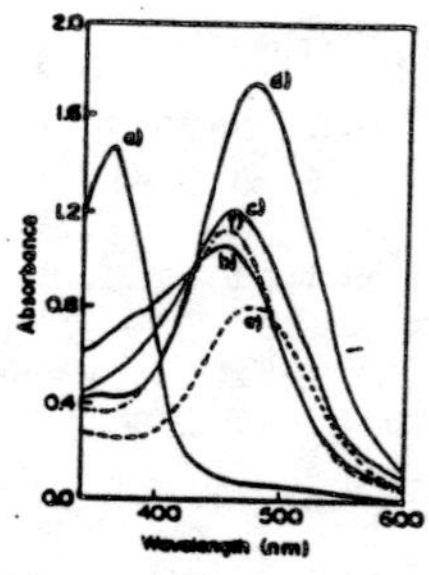

Figure 1b

Resonance Raman (1a) and absorption (1b) spectra of trans-retinal Schiff's base and Lewis acid derivative in CH_2Cl_2. (a) Trans-retinylidene-n-butylamine. (b) BCl_3 complex (c) BBr_3 complex. (d) BF_3 complex. (e) $HClO_4$ complex. (f) HCl complex. RRS of the samples was obtained by spinning the solution (λ_{ex} = 457.9 nm).

coordinate analyses have been carried out for the $CH_3HC{=}NCH_3$ Schiff's base model and its protonated, deuterated and BF_3 derivatives. With reasonable values for the C=N/C=N-H or C=N/C=N-B stretch-bend interaction force constants (0.53 mdyn) and for the C=N/N-B stretch-stretch interaction force constant (0.53 mdyn/Å), our calculations show that we cannot account for the observed increase in C=N upon complex formation, if we assume that the C=N stretching force constant decreases upon reaction.

To explore this discrepancy in more detail we have carried out Generalized Valence Bond (GVB) calculations for methylimine and protonated methylimine.[3] The results show that the C=N bond length decreases upon protonation and concomitantly that the stretching force constant increases by 0.5 mdyn/Å. Incorporation of our calculated C=N stretching force constant into the normal coordinate calculations predicts an increase in the C=N stretching frequency of 30 cm^{-1} upon protonation.

The analysis of these results indicates that reaction with a Lewis acid alters the electronic properties of the Schiff's base nitrogen's lone pair in such a way that there is a transfer of s character from the protonated or complexed nitrogen to the sp^2 hybrid C=N bond and that this effect rather than, for example, changes in mechanical coupling, is primarily responsible for the unusual vibrational properties of the complexed Schiff's base.

REFERENCES

[1](a) Ottolenghi, M. Adv. Photochem. 1980, 12, 97-200; (b) Mathies, R.A. in "Spectroscopy of Biological Molecules", Sandorfy, C.; Theophanides, T., eds., D. Reidel Publishing Company, Dordrech/Boston/Lancaster, 1983, pp. 303-328; (c) Argade, P.V.; Rothschild, K.J. Biochemistry 1983, 22, 3460-3466; (d) Aton, B.; Doukas, A.G.; Narva, D.; Callender, R.H.; Dinur, U.; Honig, B. Biophys. J. 1980, 29, 79-94.

[2](a) Smith, S.O.; Myers, A.B.; Mathies, R.A.; Pardoen, J.A.; Winkel, C.; van den Berg, E.M.M.; Lugtenburg, J. Biophys. J. 1985, 47, 653-664; (b) Kakitani, H.; Kakitani, T.; Rodman, H.; Honig, B.; Callender, R. J. Phys. Chem. 1983, 87, 3620-3628; (c) Sheves, M.; Nakanishi, K. J. Am. Chem. Soc. 1983, 105, 4033-4039; (d) Bagley, K.A.; Valogh-Nair, V.; Croteau, A.A.; Dollinger, G.; Ebrey, T.G.; Eisenstein, L.; Hong, M.K.; Nakanishi, K.; Vittiotow, J. Biochemistry 1985, 24, 6055-6057.

[3]López-Garriga, J.J., Hanton, S., Babcock, G.T. and Harrison, J.F. J. Am. Chem. Soc. 1986, submitted.

EXCITED STATE STRUCTURE AND DYNAMICS OF ISOPRENE FROM ABSOLUTE RESONANCE RAMAN INTENSITIES

Mark O. Trulson and Richard A. Mathies

Chemistry Department, University of California, Berkeley, CA 94720 USA

The lowest optically allowed $1B_u$ state of butadiene and its derivatives has drawn considerable attention as a prototype for electronic excitations in conjugated π-systems. The understanding of the electronic states of butadiene has proven experimentally difficult, because the absorption spectrum displays too little vibronic structure to be amenable to Franck-Condon analysis; only the vibronic progression of the ethylenic stretching mode is evident, even under isolated molecule conditions. This spectral diffuseness is observed for all substituted butadienes yet examined.[1] The relative contributions of unresolved Franck-Condon progressions and single vibronic level linewidths have remained undetermined. Of particular interest is the role of distortions from planarity in generating spectral broadening.[2]

Resonance Raman intensity arises from changes in potential energy for nuclear motion in the resonant electronic state, allowing mode-specific potential surface properties (origin shifts and/or frequency changes) to be extracted from the spectrum.[3-6] In addition, the absolute resonance Raman cross sections may be employed to partition single vibronic linewidths into homogeneous and inhomogeneous components, yielding excited state relaxation rates. In this laboratory, we have applied these techniques to analyze the unresolved absorption spectra of retinal,[7] trans- and cis- stilbenes,[8,9] and bacteriorhodopsin.[10]

The recent introduction of absolute Raman intensity standards in the deep ultraviolet[11] has permitted us to extend these studies to isoprene (2-methylbutadiene) in dilute cyclohexane solutions. The resonance Raman spectrum shows intensity primarily in a_g fundamentals and overtones. Overtone intensity in a_u and b_g out-of-plane methylene wags indicates that the frequencies of these modes decrease to 400±100 cm^{-1} from 622 and 755 cm^{-1}, respectively. We conclude that no inversion occurs along these or any other out-of-plane coordinates. Our analysis reveals that the initial dynamics on the $1B_u$ surface leads to a structure of D_{2h} symmetry, in agreement with a recent MCSCF study.[12] The very low absolute Raman cross sections of all modes result from an unusually large homogeneous linewidth of 700 cm^{-1}. This is a manifestation of very rapid radiationless relaxation (T_1<10 fsec) that competes effectively with geometric relaxation.

(1) D.G. Leopold, R.D. Pendley, J.L. Roebber, R.J. Hemley, & V. Vaida, J. Chem. Phys. 81, 4218 (1984).

(2) U. Dinur, R.J. Hemley, & M. Karplus, J. Phys. Chem. 87, 924 (1983).

(3) P.M. Champion & A.C. Albrecht, Annu. Rev. Phys. Chem. 33, 353 (1982).

(4) A. Warshel, Annu. Rev. Biophys. Bioeng., 6, 273 (1977).

(5) B.B. Johnson & W.L. Peticolas, Annu. Rev. Phys. Chem. 27, 465 (1976).

(6) A.B. Myers & R.A. Mathies, in Biological Applications of Raman Spectroscopy, edited by T.G. Spiro (Wiley-Interscience, New York, in press).

(7) A.B. Myers, M.O. Trulson, J.A. Pardoen, C. Heeremans, J. Lugtenburg & R.A. Mathies, J. Chem. Phys. 84, 633 (1986).

(8) A.B. Myers, M.O. Trulson & R.A. Mathies, J. Chem. Phys. 83, 5000 (1985).

(9) A.B. Myers & R.A. Mathies, J. Chem. Phys. 81, 1552 (1984).

(10) A.B. Myers, R.A. Harris & R.A. Mathies, J. Chem. Phys. 79, 603 (1983).

(11) M.O. Trulson & R.A. Mathies, J. Chem. Phys. 84, 2068 (1986).

(12) M. Aoyagi, Y. Osamura & S. Iwata, J. Chem. Phys. 83, 1140 (1985).

EXCITATION PROFILES OF RESONANCE RAMAN SCATTERING AND LUMINESCENCE FROM β-CAROTENE: THE NON-MOTIONAL NARROWING EFFECT.

S. Kinoshita, J. Watanabe and T. Kushida

Department of Physics, Osaka University, Toyonaka, Osaka 560, Japan

Excitation spectra for narrow Raman lines and a broad fluorescence band have been studied in detail in β-carotene in isopentane over a wide spectral range in the low-energy side of the $^1A_g \rightarrow {}^1B_u$ absorption band [1]. The secondary emission spectra for various detuning wavenumbers are shown in Fig. 1. The three sharp lines designated as ν_1, ν_2 and ν_3, and the broad band, which shows the 0-0, 0-1 and 0-2 vibrational structures are ascribed to the first-order Raman lines and the luminescence bands of β-carotene, respectively. It is clear that the intensity ratio of the luminescence bands to the Raman lines decreases quickly, when the excitation wavenumber is tuned from resonance to off-resonance. This behavior can not be explained by the theories using the phenomenological relaxation parameters such as population and phase relaxation rates, and it is a typical example of the non-motional narrowing effect, in which the finiteness of the correlation time for the system-reservoir interaction plays an essential role.

Using the stochastic theories for the second-order optical processes derived by Takagahara et al. [2] and Mukamel [3], we can readily explain this feature under the assumption that the intermediate state undergoes the energy modulation which obeys the Gaussian-Markovian process. In Fig. 2, the experimentally-determined intensities of the Raman lines and the luminescence band are plotted against the detuning wavenumber together with the theoretical curves calculated by the stochastic theory. Here, the intensity of the Raman lines is the sum of the integrated intensity for the first-order Raman lines of ν_1, ν_2 and ν_3 modes, and that of the luminescence band is the corresponding integrated intensity for the 0-1 luminescence band. The agreement between the theoretical curves and the experimental data is quite good.

The parameter values used in this calculation are $\Delta = 300 \pm 30$ cm^{-1}, $\gamma_m = 100 \pm 20$ cm^{-1}, $\gamma_b = 12 \pm 2$ cm^{-1} and $\Delta_s = 300 \pm 30$ cm^{-1}, where Δ and γ_m are the amplitude and rate of the energy modulation, respectively, $2\gamma_b$, the population relaxation rate of the intermediate state, and Δ_s, the width of the inhomogeneous broadening, which has been introduced to explain the shift of the luminescence band with the exciting wavenumber and is assumed to have a Gaussian shape. It is evident from this result that the system-reservoir interaction in β-carotene solution is not near the motional narrowing limit where $\Delta/\gamma_m \ll 1$, but near the static limit where $\Delta/\gamma_m \gg 1$. Although most of the analyses on the excitation profile of the resonance Raman scattering in the literature have been done using the Kramers-Heisenberg's equation with the damping constant, our experiment clearly indicates the importance of the non-motional narrowing effect in this material.

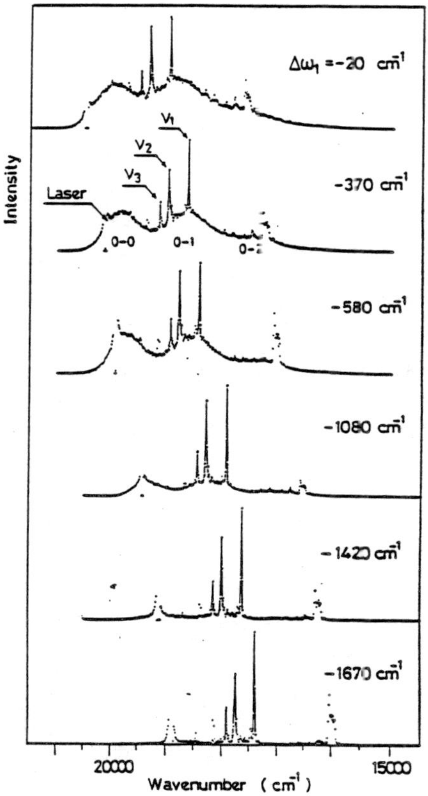

Fig. 1 Secondary emission spectra of β-carotene in isopentane at 177 K for various detuning wavenumbers.

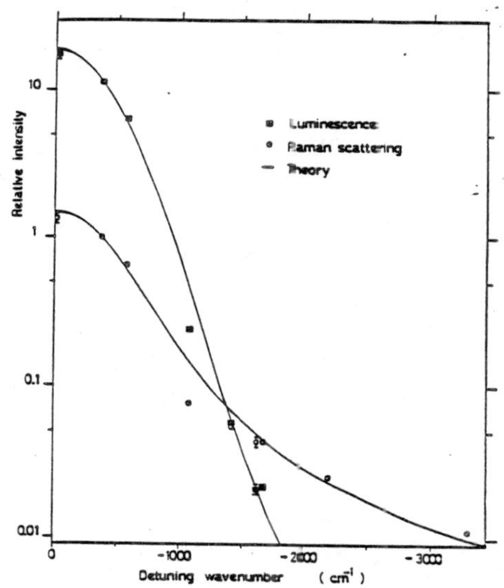

Fig. 2 Intensities of the Raman lines (o) and the luminescence band (■) versus the detuning wavenumber, with the calculated curves using the stochastic theory.

References

1. S. Kinoshita, J. Watanabe and T. Kushida, J. Physique, Colloque C7, 419 (1985). J. Watanabe, S. Kinoshita and T. Kushida, Chem. Phys. Letters, to be published. J. Watanabe, S. Kinoshita and T. Kushida, submitted.
2. T. Takagahara, E. Hanamura and R. Kubo, J. Phys. Soc. Japan 43, 802, 811 (1977).
3. S. Mukamel, Phys. Rep. 93, 1 (1982), J. Chem. Phys. 82, 5398 (1985).

LOW-FREQUENCY RAMAN SPECTRA OF DISORDERED CHAIN MOLECULES.

R. G. Snyder, V. M. Hallmark, and H. L. Strauss

Department of Chemistry, University of California, Berkeley, California 94720, USA

The conformational disorder of simple, flexible chain molecules can be characterized by low-frequency Raman spectroscopy. The isotropic component of the Raman spectra of such chains consists of a continuum of overlapping bands associated with skeletal-bending modes. These modes, which are highly delocalized, can be described in terms of the phase-angle model that is ordinarily used only for ordered chains. The zone-center skeletal-bending mode appears in the isotropic Raman spectrum as a broad band, which we will refer to as the D-LAM-1 band. The frequency and width of D-LAM-1 is directly related to the conformational statistics of the chain (1). In a sense, the D-LAM-1 band is the disordered-chain analogue of the LAM-1 band, which was first observed for a crystalline polymer by Peticolas et al. (2) and which has since been used extensively to study polymer morphology (3). Recently the D-LAM-1 band has been observed for a variety of disordered polymers in their liquid or amorphous state (4) and has been used to study the nature of the amorphous component in linear and branched polyethylenes (5).

To elucidate the nature of the skeletal vibrations of disordered chains, we have generated D-LAM spectra numerically by summing a large number of individually calculated Raman spectra of randomly generated conformers of some given chain molecule. Intensities are obtained on the basis of a bond polarizability model. These spectra are useful for direct comparison with observed spectra and for providing insight into the nature of the frequency-phase relation.

Figure 1 shows the observed D-LAM spectrum of molten polyethylene along with a numerically computed spectrum of a disordered chain of 20 carbons. The main features of the observed spectrum are reproduced in the synthesized spectrum.

From a series of spectra calculated for various chain lengths and trans/gauche ratios, we have demonstrated that the D-LAM-1 frequency dependence on these parameters is near that predicted by the phase-angle model. A Fourier analysis of the normal coordinates of the skeletal modes of the disordered polymethylene chain reveals explicitly how the frequency and phase-angle are related. Similar Raman measurements and numerical calculations are currently being extended to chains with side groups.

1. R. G. Snyder, J. Chem. Phys. 76, 3921-27 (1982).
2. W. L. Peticolas, G. W. Hibler, J. L. Lippert, A. Peterlin, and H. Olf, Appl. Phys. Lett. 18, 87 (1971).
3. J. F. Rabolt, CRC Critical Rev. Solid State and Mat. Sci. 12, 165 (1985).

4. R. G. Snyder and S. L. Wunder, Macromolecules 19, 496 (1986).
5. R. G. Snyder, N. E. Schlotter, R. Alamo, and L. Mandelkern, Macromolecules, in press; L. Mandelkern, R. Alamo, W. L. Mattice, and R. G. Snyder, Macromolecules, submitted.

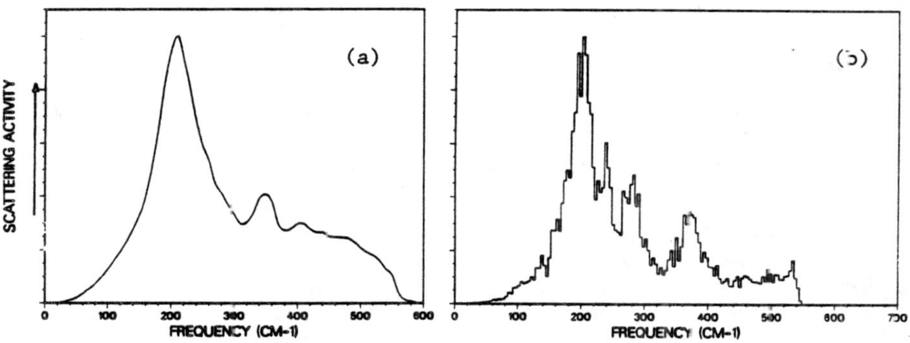

Figure 1. Observed and calculated isotropic Raman spectra of disordered polymethylene chains: (a) Observed spectrum of molten polyethylene; (b) Calculated spectrum of an ensemble of 1000 disordered n-C_{20} chains.

RAMAN SPECTROSCOPY INVESTIGATION OF BRANCHED NONANES

A. Adil[1], G. Turrell[1], G. Vergoten[2], and R. G. Snyder[3]

[1]LASIR, Univ. Sci. Tech. de Lille, 59655 Villeneuve d'Ascq, France

[2]INSERM U279, Faculte de Pharmacie, 59045 Lille, France

[3]Dept. of Chemistry, Univ. of California, Berkeley, CA 94720, USA

The Raman and infrared spectra of a series of nonanes have been measured and analyzed by normal coordinate methods. The molecules studied are 2,2,4,4- and 2,3,3,4-tetramethyl pentane and 2,2,5-, 2,2,4- and 2,3,4-trimethyl hexane. In those cases in which the nonane molecule can exist in more than one conformation, the conformers were identified and their relative concentrations estimated. Due to the relatively high density of vibrational states associated with these molecules and the insensitivity of most of the vibrational frequencies to internal rotation, accurately calculated frequencies and normal coordinates are essential for distinguishing conformers. A valence force field developed earlier for branched saturated hydrocarbons was used in the calculations.[1] The Raman spectrum in the region between 600 and 300 cm^{-1} is especially useful because conformationally sensitive bands appear here and in addition, because it is possible to extimate isotropic intensities for these bands with the use of a bond polarizability model.

Results for two molecules, each having two possible conformations (Fig. 1) are listed in Table I. For each molecule both possible conformers are found in the liquid at room temperature. The dominant conformer is the C_1 form of 2,2,5-trimethyl hexane and the C_2 form of 2,3,3,4-tetramethyl pentane.

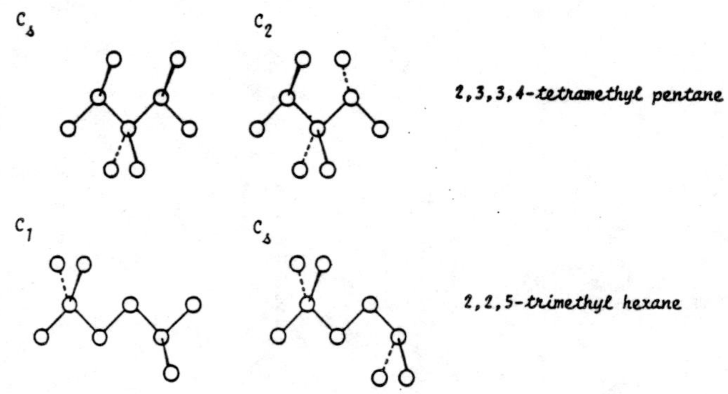

Fig. 1. Conformations identified in some branched nonanes.

TABLE I. Calculated and observed Raman frequencies of some branched nonanes.

2,2,5-trimethyl hexane

ν(*calculated*) C_1	ν(*calculated*) C_s	ν(*obs*)
	541(s)	545(m,p)
495(m)		497(m,p)
468(s)	472(s)	479(s,p)
417(w)		412(m,p)
	408(0)	

2,3,3,4-tetramethyl pentane

ν(*calculated*) C_2	ν(*calculated*) C_s	ν(*obs*)
566(0)		564(w,dp)
	542(0)	
	526(s)	518(w,p)
468(w), 467(0)		466(w)
456(s)	465(s)	475(s,p)

Reference:

1. R. G. Snyder and J. H. Schachtschneider, Spectrochim. Acta 21, 169 (1965).

THE EFFECTS OF ELECTROCHEMICAL PROCESSES ON THE SURFACE STRUCTURAL SYMMETRY OF SINGLE CRYSTAL ELECTRODES AS STUDIED BY OPTICAL SECOND HARMONIC GENERATION

Victoria L. Shannon and Geraldine L. Richmond

Department of Chemistry
University of Oregon
Eugene, Oregon 97403 USA

Because the second-order nonlinear optical process of second harmonic generation is electric dipole forbidden in the bulk of a centrosymmetric medium, it is uniquely sensitive to the structure of the surface, where the inversion symmetry is necessarily broken. The real beauty of second harmonic generation as a probe of surface structure lies in its applicability to the study of interfaces between two dense media, where the use of more conventional surface science probes is prohibited. It has been our interest in this laboratory to study the nature of liquid/solid and solid/solid interfaces [1]. Specifically, in this paper we will present investigations on the structural symmetry of Si(111) and other single crystal electrode surfaces during electrochemical processing.

Second harmonic generation may be used to reveal the structural symmetry of a surface, simply by monitoring the second harmonic intensity as the sample is rotated about its surface normal (or alternately, rotating the polarization of the pump laser and the polarization analyzer) [2]. An example of this may be seen in the Figure for the simple case of a bare crystalline Si(111) surface under ambient conditions.

Some of the interesting questions that we have attempted to address in our work are: What are the effects of electrochemically adsorbing and desorbing ions and molecules on the structural symmetry of a crystalline surface? How do electrodeposited layers of a metal alter the characteristics of the single crystal semiconductor electrode surface? Various aspects of these issues and related topics will be discussed.

1. See for example, G.L. Richmond, Langmuir, 2, 132 (1986).
2. See for example, H.W.K. Tom, T.F. Heinz, Y.R. Shen, Phys. Rev. Lett., 51, 1983 (1983).

This work has been supported by the National Science Foundation (CHE#8451346) and the Petroleum Research Fund of the American Chemical Society (16905-ACS).

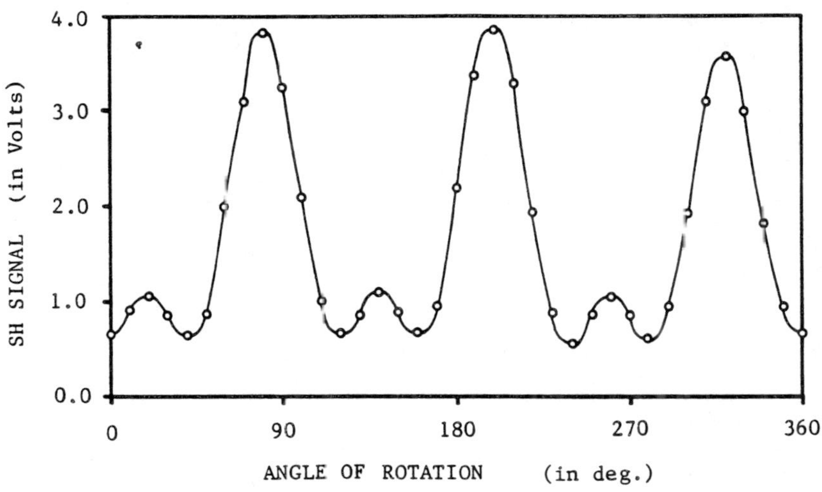

Total reflected second harmonic intensity as a function of rotational angle for the bare crystalline Si(111) surface (with a native oxide layer). Incident pump radiation was p-polarized.

OXYGEN SENSITIVE BACKGROUND IN THE RAMAN SPECTRA OF WOODY TISSUE

U. P. Agarwal and R. H. Atalla

Chemical Sciences Division
The Institute of Paper Chemistry
Appleton, WI 54912

Efforts to record the Raman spectra of woody tissue are generally frustrated by the high levels of laser induced luminescence. In our previous work using the Raman microprobe we found that mounting the samples in water eliminated most of the luminescence.[1,2] Because the water immersion method is not readily adapted to macroscopic heterogeneous samples, we sought alternative approaches. An earlier successful microprobe experiment in which the tissue was immersed in mineral oil suggested that the key factor may be either isolation of the sample from atmospheric oxygen, or thermal quenching of the heating effects of the laser, or perhaps a combination of both effects. A series of experiments designed to test these possibilities was undertaken. This involved recording spectra while flushing the samples with a variety of nonreactive gases of different thermal characteristics. These included He, Ar, N_2 and CO_2. The trials with O_2 followed the unexpected initial observations.

The samples were transverse sections of Picea mariana (Black spruce) approximately 30 micrometers thick, which had been allowed to stand overnight exposed to the ambient atmosphere at room temperature. They were loosely mounted between cover slips and inserted in a specially constructed cell designed to allow acquisition of spectra in a 90 degree scattering geometry while flushing the samples with different gases. The cell was mounted in the macrochamber of the spectrometer system described in References 1 and 2. A cylindrical lens was used to focus the laser beam on the samples with the axis of the illuminated sample region parallel to the entrance slit of the spectrometer.

Results and Discussion

The effects of flushing with O_2 and N_2 are shown in Figure 1. Spectrum A was recorded as the sample was flushed with air, spectrum B as it was flushed with N_2, and spectrum C as the N_2 was replaced with O_2. Figure 2 shows spectrum C of Figure 1 on an expanded scale, together with another spectrum recorded about an hour later. Figure 3 shows the final spectrum in Figure 2, together with one recorded after O_2 flushing was discontinued and N_2 flushing resumed. The effects of flushing with all the other gases were similar to those obtained with N_2. The Raman bands superimposed on the luminescence background are typical for woody samples and have been discussed in our previous reports.

The results in Figure 1 suggest that the quenching effects parallel the changes in the partial pressure of O_2 in the gaseous environment.

Figure 2 indicates that continued exposure to O_2 results in enhancement of the quenching effect, suggesting continued adsorption of oxygen within the woody tissue, or further diffusion within the tissue, or perhaps a combination of both effects. Figure 3 indicates that the effect is substantially reversible, suggesting that an interaction involving a physical adsorption process is the dominant effect.

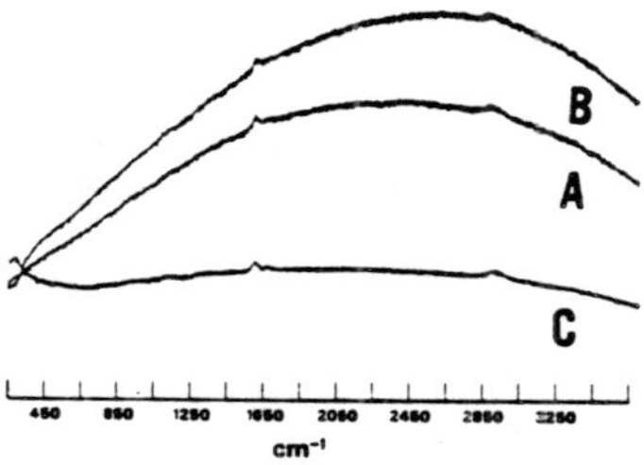

Fig. 1. Woody tissue Raman spectra; (A) Air flushing, (B) N_2 flushing, (C) O_2 flushing; all on the same intensity scale.

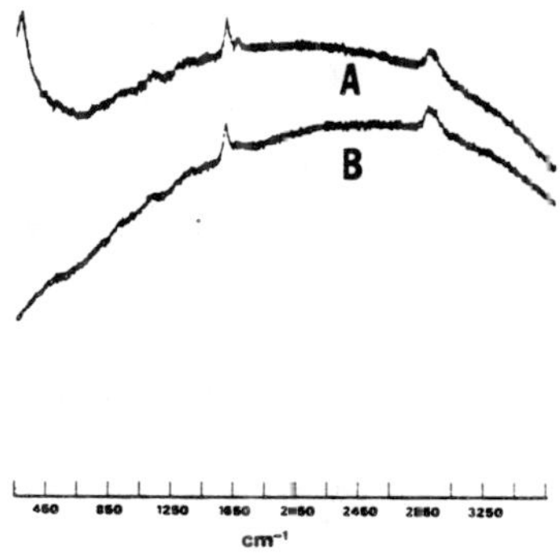

Fig. 2. Raman spectra of the O_2 flushed sample; (A) O_2 flushing just started, (B) one hour later.

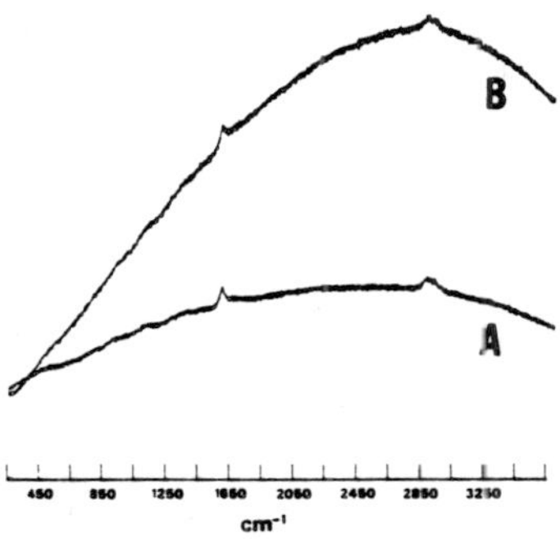

Fig. 3 (Left). Effect of switching over from O_2 flushing (A) to N_2 flushing (B).

Oxygen is known to quench excited states of a number of molecules. However, in most instances the mechanism of its action remains unknown. It appears most likely that the major interaction that we have observed is between O_2 and some of the constituent groups of lignin. Contributions of other minor components of the cell walls cannot be excluded at this time, however.

References

1. R. H. Atalla and U. P. Agarwal, J. Raman Spectr., 17(1986) in press.
2. R. H. Atalla and U. P. Agarwal, Science, 227:636(1985).

MICRORAMAN STUDY OF THE STRUCTURAL ORDER OF THE CARBON NETWORK IN PITCHES

P. Dhamelincourt, D. Cottinet, P. Couderc*

Laboratoire de Spectrochimie IR et Raman, D.N.R.S. (LP 2641), USTL Bat. C 5, 59665 Villeneuve D'Ascq Cedex, France

* Hiles, Goudrons et Derives (Groupe CDF Chimie) BP 189 Lens, France

During the last ten years, Raman spectroscopy has evolved as an effective method for studying the structure of partially ordered or even amorphous carbonacieous materials (1). Microraman spectroscopy, which permits very small areas of a sample to be analyzed nondestructively, has proven to be a very efficient tool for following the evolution of the structural order of the carbon network in pitches according to their origin and thermal treatment.

The materials examined were coal pitches used as hydrocarbon binders in the fabrication of carbon electrodes for the electrolytic production of ferrous or nonferrous metals. Pitches are the concentrate left after the distillation of coal tar. They are very complex mixtures of polyaromatic compounds whose molecular masses may vary over a very broad range. The rheological properties of these materials are very important when mixing petroleum coke and coal tar pitch to prepare prebaked electrodes. Unfortunately pitches having very similar physicochemical characteristics may have different rheological behaviors. We have found that these behaviors are largely dependent on the structural order of the overall carbon network.

Our study of pitches was carried out using a multichannel microraman spectrometer (MICRODIL 28-DILOR France). It is worthy to note that due to the strong absorption of laser radiation by pitches only high sensitivity instruments equipped with multichannel detectors were successful in obtaining results. These results clearly indicate that the ability of a pitch to flow into a granular coke bed, measured by the temperatures at the beginning and the end of the flow (2), is well correlated with the structural order revealed by Raman spectroscopy. More particularly, cracking reactions caused by an overly lengthy heating treatment time which lead to pitches with abnormal flow behaviors are perfectly evidenced by Raman spectroscopy. At last, this technique can give indications on the mode of fabrication of pitches of unknown origin (straight distillation of tars or tar maturation by heating treatment).

Multichannel Raman microspectrometry allows the evolution of the structural order of the overall carbon network of pitches to be followed. So pitches with similar physicochemical properties can easily be distinguished. This technique has proved to be a very useful tool which is complementary to other methods used to characterize industrial pitches.

References:

1. P. Lespade, Doctoral Thesis, University of Bordeaux (France). 1982.

2. P. Couderc and P. Dhamelincourt, Proc. Int. Tar Conf. Biarritz (Fr.) 1983.

NORMAL MODES OF PENTACHLOROBENZENE

R. A. Yadav* and R. Shanker**

*Department of Physics, *Department of Applied Physics, I.T.
Banaras Hindu University, Varanasi 221005, INDIA

The vibrational spectrum of pentachlorobenzene (PCB) was reported earlier[1]. In the present study the vibrational spectrum of PCB is reinvestigated. The accurate measurements of depolarization ratio (ξ) of the Raman lines suggest some assignments proposed earlier [1] to be altered.

PCB of tech. grade (purity 98%), purchased from Aldrich Chemical Co. (USA), was recrystallized twice from CCl_4 solution before use. The Raman spectra in powder phase and CS_2 and $CHCl_3$ solutions were recorded on a Jarrell-Ash 25-300 spectrometer in the region 50-4000 cm^{-1} using the 5145 Å line of an Ar^+ laser. ξ values of Raman lines were measured in CS_2 and $CHCl_3$ solutions. The infrared spectra were recorded, in CsI pellet, nujol mull and CCl_4 solution, in the region 200-4000 cm^{-1} on a Perkin-Elmer-621 spectrometer. The resolution for all the spectra was better than 2 cm^{-1} and the accuracy of the measurements was estimated to be ± 3 cm^{-1}.

PCB possesses C_{2v} point group symmetry with the normal modes distribution as: $11\ a_1 + 10\ b_2 + 3\ a_2 + 6\ b_1$, where yz is the molecular plane and z is the axis of symmetry. The frequencies observed in the Raman and i.r. spectra and assignments of the observed frequencies to different fundamental modes are given in Table 1. The assignments for which we disagree, are briefly discussed here.

Scherer and Evans[1] observed the frequency 1525 cm^{-1} to be depolarized and yet assigned it to the C=C stretching mode under the most symmetric species a_1 which can have only polarized Raman frequencies. However, in the present case this frequency is observed to be polarized and therefore, it appears that the assignment of this frequency made by Scherer and Evans was correct but the depolarization measurement was erroneous. Similarly, Scherer and Evans[1] have correlated the experimental frequencies 199 and 215 cm^{-1} with the calculated frequencies 197 cm^{-1} (a_1) and 209 cm^{-1} (b_2). The Raman frequency 199 cm^{-1}, earlier observed to be depolarized, is observed to be depolarized in the present case also whereas the frequency 215 cm^{-1} was estimated from combination bands by Scherer and Evans[1]. In view of the depolarized nature of the frequency 199 cm^{-1}, we prefer to interchange the assignments of the frequencies 199 and 215 cm^{-1}.

The infrared bands 1210 and 1230 cm^{-1}, the former of which was not mentioned by Scherer and Evans anywhere in their paper[1]

Table-1. Raman and I.R. frequencies and proposed assignments for pentachlorobenzene

	Infrared				Raman				Assignments[b]
	cm^{-1}	Rel. intensity			cm^{-1}	Rel. intensity[a]			
		CsI	Nuj.	CCl_4		Solid	$CHCl_3$	CS_2	
	3065	70	44		3066	11	13 (0.13)	77 (0.36)	ν(C–H)
	1530	20	23		1523	13	15 (0.67)	22 (0.71)	ν(C=C)
	1328	89	100	96					ν(C=C)
	1210	48	57	37	1206	35	77 (0.25)	100 (0.13)	ν(C–C)
	1090	76	77	48	1087	11	17 (0.20)	26 (0.36)	α(C–C–C)
a_1	828	92	85		823	8	10 (0.33)		ν(C–Cl)
	570	64	28		565	39	50 (0.33)	28 (p)	α(C–C–C)
					383	76	100 (0.10)	62 (0.10)	ν(C–Cl)
					345	100	83 (0.34)	48 (0.33)	ν(C–Cl)
					218	21	7 (0.50)	9 (?)	β(C–Cl)
					215[c]				β(C–Cl)
	1550	24	25		1555	23	25 (0.88)	37 (0.88)	ν(C=C)
	1392	84	78	97	1400	5	5 (—)		ν(C=C)
	1230	51	57	37	1236	10	20 (0.76)	18 (0.76	ν(C=C)
b_2	1173	81	83	83	1165	sh	13 (dp ?)	15 (dp ?)	β(C–H)
	865[c]								ν(C–Cl)
	690	100	95	97	687	2			α(C–C–C)
	560	sh	sh						ν(C–Cl)
					319	50	28 (0.88)	17 (0.82)	β(C–Cl)
					225	23	7 (~ 1)	9 (0.22)	β(C–Cl)
					199	87	40 (0.83)	28 (0.78)	β(C–Cl)
	595	sh	sh						φ(CCCC)
a_2					375	sh		sh (-)	γ(C–Cl)
					80[d]				γ(C–Cl)
	872	81	88	62					γ(C–H)
	707	70	57	sh	699	4			φ(CCCC)
	537	32	17		546	sh			φ(CCCC)
b_1					274	20		9 (0.82)	γ(C–Cl)
					152	18	7 (0.80)		γ(C–Cl)
					91	12			γ(C–Cl)

a: Under columns 7 and 8 numbers within brackets represent depolarization ratio; p=polarized, dp=depolarized.
b: ν: stretching, α: angle bending, β: in-plane bending, φ : ring torsion, γ: out-of-plane bending
c: estimated frequencies; d: From Ref.1

and latter was assigned as a combination band, are correlated here respectively with the frequencies 1206 and 1236 cm^{-1}, assigned as fundamentals. The frequency 343 cm^{-1} estimated from combination bands was correlated with the calculated frequency 389 cm^{-1} by Scherer and Evans whereas in the present case, the Raman frequency 375 cm^{-1} observed as a shoulder is assigned to the corresponding mode.

Reference :

1. J.R. Scherer and J.C. Evans, Spectrochim.Acta 19, 1739 (1963).

FUNDAMENTAL FREQUENCIES OF O-ETHYL PHENOL

R. Shanker*, R.A. Yadav**, I.S. Singh** and O.N. Singh*

*Department of Applied Physics, I.T.; **Department of Physics, Banaras Hindu University, Varanasi-221005, INDIA

The i.r. spectrum of o-ethyl phenol (OEP) was reported and out of 51 normal modes only 41 modes were observed [1]. In this study, the polarised Raman and i.r. spectrum of OEP are presented and all the 51 normal modes have been assigned and the nature of hydrogen bonding is studied.

OEP (pure), purchased from Kodak Co., was distilled twice prior to use. The polarised Raman spectrum was recorded (100-3200 cm^{-1}) on a Spex Ramalog spectrometer using the 4880Å line of an Ar^+ laser. The i.r. spectra were recorded (200-4000 cm^{-1}) in liquid and CCl_4 solns. at different conc. on Perkin Elmer 621 and 180 spectrometers. The resolution was better than 2 cm^{-1} and the accuracy of the measurements was estimated to be within ±3 cm^{-1}. From the measurements of depolarization ratio (ρ) all the Raman lines are observed to be polarised suggesting lower symmetry than Cs [2]. Therefore, vibrational assignments are proposed assuming C_1 symmetry for OEP. The observed frequencies along with their relative intensities, ρ values and proposed assignments are presented in Table 1.

For di-substituted benzenes [2], ring-breathing and trigonal C-C-C angle-bending modes interact with other modes resulting in considerably reduced magnitude. On intensity and ρ value considerations, the modes 1029 cm^{-1} and 836 cm^{-1} are assigned to these modes respectively. The frequency at 1038 cm^{-1}, having Raman intensity and ρ value very close to that of 1029 cm^{-1}, is assigned to the C-H in-plane bending mode. Comparable intensities and ρ values for these frequencies suggest that the corresponding modes interact with each other resulting in increased magnitude of the breathing mode as compared to benzene (992 cm^{-1}). The twisting of the CH_2 group in OEP about $C-C_2H_5$ bond is expected to appear at considerably lower magnitude. A strong Raman band at 150 cm^{-1} observed with high ρ value (0.68) is assigned to this mode. The bands at 747 and 725 cm^{-1} are assigned to the $\rho(CH_2)$ and $\nu(C-CH_3)$ modes respectively. The intensities and ρ values for these two modes suggest strong coupling between the corresponding modes.

The nature of hydrogen bonding and its effect on vibrational modes are studied by several workers [3,4]. In the i.r. spectrum of OEP a broad and strong band is observed at 3440 cm^{-1} which shifts to 3620 cm^{-1} in dilute CCl_4 solution (0.05%) and is assigned to the monomeric ν(O-H) mode. The broadening of the band and lowering of the magnitude of the ν(O-H) mode is due to intermolecular hydrogen bonding. The frequencies at 1232 and

1254 cm^{-1} are assigned to α(C-O-H) and ν(C-OH) modes and from intensities and ε values, it appears that the corresponding modes are strongly coupled. The τ(OH) mode appears in the region 500-700 cm^{-1} in bonded state and the band at 528 cm^{-1} is assigned to this mode.

1. R.A.Yadav and I.S.Singh, J.Sci. Res., B.H.U. (India) 32(2), 133 (1982-83).
2. R.A.Yadav and I.S. Singh, Proc. Ind. Acad. Sci. (Chem.Sci.) 95, 471 (1985).
3. J.W.Brasch, R.J.Jacobsen, W.G.Fateley and N.T.McDevitt, Spectrochim. Acta 24A, 203 (1968).
4. A.Hall and J.L.Wood, Spectrochim. Acta 24A, 1109 (1968).

Table 1: Raman and IR Frequencies (cm^{-1})[a] and Vibrational Assignments for O-Ethyl Phenol

Raman	Infrared		Proposed
(Pure-liq.)	Pure liq.	CCl_4 sol.	assignments[b]
	3440(99)	3620(c)	ν(O-H)
	3097(sh)	3100(sh)	ν(C-H)
3080(54,0.65)	3080(64)	3075(25)	ν(C-H)
3055(54,0.49)	3050(63)	3050(35)	ν(C-H)
	3036(sh)	3036(sh)	ν(C-H)
2980(20,0.75)	2980(88)	2975(65)	$\nu_{as}(CH_3)^e$
2952(54,0.18)	2948(72)	2950(54)	$\nu_{as}(CH_2)^d$
2930(54,0.18)			
2890(35,0.24)	2880(66)	2880(43)	$\nu_s(CH_3)$, $\nu_s(CH_2)$
1610(27,0.71)	1606(61)	1608(34)	ν(C=C)
1595(16,0.63)	1588(71)	1587(55)	ν(C=C)
	1498(73)	1497(60)	ν(C=C)
	1485(72)	1488(65)	$\delta_{as}(CH_3)^e$
1454(13,0.69)	1448(100)	1450(85)	ν(C=C), $\delta(CH_2)$
	1370(51)	1375(22)	$\delta_s(CH_3)$
	1342(sh)	1340(sh)	β(C-H)
1330(14,0.36)	1324(53)	1322(75)	$\omega(CH_2)$
	1285(35)	1285(32)	ν(C=C)
1256(41,0.13)	1251(73)	1255(90)	ν(C-OH)
1235(45,0.11)	1228(80)	1215(100)	α(C-O-H)
	1196(61)	1200(76)	β(C-H)
	1168(63)	1166(80)	β(C-H)
1155(18,0.61)	1150(39)	1149(45)	$\nu(C-C_2H_5)$
	1108(69)	1108(86)	$\rho_{\perp}(CH_3)$
	1062(23)	1064(14)	$\rho_{\parallel}(CH_3)$
1046(82,0.08)	1050(35)	1048(23)	β(C-H)
1028(84,0.06)	1030(25)	1030(16)	ν(C-C)
	988(13		γ(C-H)
964(10,0.75)	964(17)	965(20)	γ(C-H)
	930(20)	928(15)	γ(C-H)
936(31,0)	836(72)	836(65)	α(C-C-C)
782(6,0)	786(45)		γ(C-H)
746(90,0.07)	748(98)		$\rho(CH_2)$
725(100,0.06)	725(44)		$\nu(C-CH_3)$
	702(32)		ϕ(C-C-C-C)
590(36,0.31)	588(32)		α(C-C-C)
555(24,0.26)	557(29)		α(C-C-C)
530(13,0.73)	526(32)		τ(C-OH)
482(6,0)	485(31)		ϕ(C-C-C-C)
	475(34)		ϕ(C-C-C-C)
442(4,0)	435(26)		β(C-OH)
	355(16)		$\beta(C-CH_3)$
340(13,0.23)	335(15)		$\beta(C-C_2H_5)$
	300(7)		γ(C-OH)
	274(3)		$\gamma(C-CH_3)$
	250(8)		$\gamma(C-C_2H_5)$
234(16,0.68)			$\tau(C-CH_3)$
150(70,0.68)			$\tau(C-C_2H_5)$

a: IR & Raman Int. & ε are given within bracket, b: where notations have their usual meanings, c: Observed at 0.05% conc., d: Fermi resonance (2940, 2x1485), e: Degenerate mode.

VIBRATIONAL SPECTRA OF CYCLOPROPYLI, 1,-CYANO CARBOXAMIDE (CCC) AND CYCLOPROPYL 1,1, CARBOXAMIDO ACETATE (CCA)

C.R.Gopinath, D.N.Sathyanarayana* and A.Natarajan**

Dept. of Physics,Bangalore University,Bangalore-560001
*Dept.of I.P.C., Indian Institute of Science,Bangalore-12
**Dept of Physics,Bharathidasan University, Trichy-23.

Abstract- Laser Raman and infra red spectra of the two titled compounds have been recorded using Spex spectrophotometer with Argon ion laser (5145A) and Perkin-Elmer model 599 spectrophotometer respectively. The spectra have been interpreted on the basis of the bands arising from Cyclopropyl ring and the substituent moieties. The results are used to discuss the effect of the carboxamide group on cyclopropyl ring.

In Cyclopropane derivatives containing substituent group with multiplebonds the interaction between the cyclopropane ring and the substituent might be expected to modify the electron distribution in the three membered ring. Not unexpectedly Raman and infrared spectroscopy have played key roles in the study of these intriguing molecules (1). There have been no Raman and infrared data on the above titled compounds.

We report the Raman and infrared spectra of CCC and CCA in order to compliment the recent studies on the vibrational spectra of the halogeno and pseudo halogeno(cyano)substituted cyclopropanes(2-6).

Table-I gives the observed infrared and Raman wavenumbers of the compounds in study. Assignments are proposed in terms of group frequencies. The effect of conjugate interaction on those vibrations which usually give rise to true group frequencies in this class of compounds is also examined. The published data clearly demonstrates serious deviations in the interpretation of the spectra of different investigator. The origin of an intense doublet in the cyanide group stretching region and the other results of our study will be discussed in the Conference.

Table I: Raman and Infrared frequencies (cm^{-1}) of CCC and CCA

CCC compound			CCA compound.		
Raman	IR	Assignment	Raman	IR	Assignment
438(m)	440(m)	C-C-N bend	510(vs)	520(s)	CCC bend
685(vs)	-	NH_2 wag	775(s)	750(ms)	CH_2 rock
845(s)	840(w)	CH_2 wag	865(s)	865(w)	CH_2 wag, rock
965(s)	970(s)	C-C wag	972(vs)	975(s)	C-C wag
1038(s)	1040(ms)	sym str CC	1055(s)	1055(w)	CH_2 wag
1275(s)	1275(w)	CH_2 twist	1185(w)	1195(s)	sym str CC
1438(s)	1445(m)	CH_2 sym ip bend	1240(s)	1242(wsh)	CH_2 twist
1685(w)	1690(m)	C=O str	1415(m)	1412(s)	CH_2 asy ip δ
2235(s), 2250(s)	2250(m)	CN str	1420(m)	1440(m)	CH_3 sym ip δ
3028(s)	---	CH_2 asym str	1705(s)	1708(m)	C=O str
3110(s)	---	CH_2 sym str	2960(s)	2900(s)	CH_3 str
3405(mw)	3390(w)	NH_2 str	3430(w)	3400(s)	NH_2 str

Note: s-strong, m-medium, w-weak, ip-in plane, sh-shoulder, str-stretch, sym-symmetric, asym-asymmetric, δ - bend

Acknowledgement: The authors thank Prof. I.D.Brown and Carol Yoshimoto of McMaster University,Canada for the record of Raman spectra of the compounds under our study .One of the authors (C.R.G) is thankful to the U.G.C., New Delhi for financial assistance, to Prof.N.G.Puttaswamy Bangalore University for encouragement, and to Prof.K.Venkatesan Indian Institute of Science, Bangalore for the supply of the cyclopropyl compounds to carry out our investigations.

References:

1. C.J.Wurrey and A.B.Nease, "Vibrational Spectroscopy and Structure of Three-membered Ring Compounds" Chapter 1, Vibrational Spectra and Structure, Vol.7, ed.J.R.Durig, Elsevier Amsterdam,1978.
2. D.D.Keeports and D.F.Eggers,Jr.Spectrochim Acta 40A,7(1984)
3. G.Schrumpt, Spectrochim Acta 39A,487,511 (1983)
4. J.Maillols,V Tabacik and S.Sportouch,J.Raman Spectrosc 11,312(1981)
5. C.J.Wurrey and L.E.Firment Spectrochim Acta,30A,1115(1974)
6. L.H.Daly and S.E.Wiberly,J.Mol.Spectry,28,540,(1968).

LASER RAMAN SPECTRA OF SOME HALOGENO CYCLO PHOSPHAZENES

A.Natarajan, C.R.Gopinath,* C.Raghavendra Rao* B.S.Nagaraj,* and K.C.Kumaraswamy**

Dept. of Physics,Bharathidasan University, Tiruchy-23
* Dept. of Physics,Bangalore University, Bangalore-560001
**Dept.of I.P.C. Indian Institute of Science,Bangalore-12

Abstract- Laser Raman spectra of the two compounds (triphenyl phosphazenyl) cyclotriphosphazenes-$N_3P_3(N= PPh_3)X_5$ (X=Cl or F) have been recorded using spex spectrophotometer with Argon ion laser (5145A). The characteristic vibrational frequencies of the two above cyclic halides have been identified by correlating the assignment with those available for related systems.

INTRODUCTION- The two compounds (Triphenyl phosphozenyl) pentachloro cyclotriphosphazene and fluorine substituted one prepared by Appel and Staudinger reactions can serve as "mechanistic probes" for nucleophillic displacement reactions of halogeno cyclophosphazenes (1-2). It is of our interest to study the Raman spectra since no attention appears to have been paid to the vibrational analysis of these two compounds.

Table-I gives the observed Raman wave numbers of the two compounds proposed in terms of group frequencies.Details of our results will be discussed in the Conference.

Acknowledgements-The authors thank Prof. I.D.Brown and Carol Yoshimoto of McMaster University Canada for the record of Raman spectra. One of the authors (C.R.G) is thankful to U.G.C.New Delhi for financial assistance, Prof. S.S.Krishnamurthy and Prof. D.N.Sathyanarayana of I.I.Sc., Bangalore for useful discussions.

References-

1. K.C.Mumaraswamy, P.Ramabrahmam and S.S.Krishnamurthy, Synth.React, inor.met-org chem 15(8), 1023-35(1985).

2. K.C.Kumaraswamy and S.S.Krishnamurthy,Phosphorous sulfur 18,241, (1983).

Table 1: Raman frequencies (cm^{-1}).

$N_3P_3(N{=}PPh_3)Cl_5$		$N_3P_3(N{=}PPh_3)F_5$	
Raman	Assignment	Raman	Assignment
220	PCl_2 rock	252	ring torsion
260	ring twist	288	PF_2 bend
360	Cl-P-Cl SymStr	465	Phyring O.P def
610	ring deformation	550	ring deformation
660	P-Cl stretch	610	PNP bend
760	PNP ring stretch	674	ring deformation
995	CH bend, ring def.	715	PF sym stretch
1025	CH bend	995	PF Asym stretch
1525	ring stretch phenyl	1024	CH bend
1590	" "	1100	" "
3015	CH Sym. stretch	1585	phenyl ring stretch
3130	" "	3060	CH sym. stretch

Note: O.P. = out of plane, def.= deformation
str. = stretch , Sym. = symmetric, asym.= assymetric

chapter 15

HIGH RESOLUTION STUDIES

CARS DIAGNOSTICS OF DISCHARGES

J.P. Taran

Office National d'Etudes et de Recherches Aérospatiales, BP 72, 92322 Châtillon Cedex, FRANCE.

The study of low pressure gases and discharges is perhaps the most interesting application of CARS. The measurements are free of "dirty" interferences like beam steering by density gradients or complicated phenomena like motional narrowing. Precise measurements of quantum state populations are possible in principle, allowing exciting investigations in molecular physics to be carried out. Several programs are in progress at ONERA. Species N_2, H_2 and O_2 have been studied in three different types of discharges. Scanning CARS was used, and, whenever made necessary by non-homogeneities, the pump beams were crossed in a planar BOXCARS configuration to ensure adequate spatial resolution. Background suppression never was found to be necessary. Spectral resolution was 0.05 cm^{-1}.

The first investigation was that of H_2 in a magnetic multipole. Rotational, vibrational and total populations have been measured at pressures in the range 6 - 55 μbar [1].

Recently, pulsed discharge experiments were conducted to gain insight into the kinetics; rotational and vibrational relaxation could be measured, giving the corresponding lifetimes of the excited rovibrational states. For instance, the vibration decays in about 4 ms at 55 μbar, indicating that the process is dominated by diffusion to the walls and that each wall collision is efficient. Some preliminary results at 6 μbar also seem to indicate that, in collisions wih electrons, ortho H_2 receives less rotational energy and more translational energy than the para form.

Oxygen was studied in a surfatron microwave discharge. The pressure was 2 mbar and the discharge power 180 W. Rotation and vibration were found to be near equilibrium, with temperatures of ≈ 800 and 1200 K, respectively. Excited molecular oxygen ($^1\Delta$) was detected and its lines assigned for the first time. It appears that this species is formed primarily in the post-discharge region, probably as a result of the recombination of O atoms. It was also observed that the rotational and vibrational temperatures decay very rapidly to room temperature in the post discharge region.

Finally, N_2 was observed in a glow discharge, as a preliminary to an investigation of nitridation. High purity N_2 was flowed through a 2 cm dia. tube. Flow rates were about 30 m/s. The pressure was in the range 2.3-4.6 mbar and the discharge current 80 mA. The highest vibrational state detected was v = 13. Low (540 K) static and rotational temperatures were seen, with a very high vibrational excitation of ≈ 4500 K. It is the first time that these high-lying states are seen at such low pressures using Raman spectroscopy. Strong deviations are naturally seen from the Boltzmann law. In the post-discharge region, good agreement is found between the experimental vibrational distribution and the Treanor-like one calculated using Gordiet's model [2]. The agreement is good, however, up to v = 11 only. In the discharge, the distribution is modified by electron collisions. A model by Capitelli and coworkers then gives a better agreement [3].

While discharge studies are attractive in that they can in principle yield precise data and that the results can frequently be compared with models, some difficulties do remain which may spoil the measurements:

- saturation and Stark effects manifest themselves at fairly low pump power densities; these phenomena depend on the vibrational state;
- anharmonicity corrections to the Raman cross-sections must be applied.

These effects and the latest results will be discussed.

[1] - M. Péalat, J.P. Taran, M. Bacal and F. Hillion, J. Chem. Phys., 82, (1985), p. 4943.

[2] - B.F. Gordiets, S.S. Mamedov and L.A. Shelepin Sov. Phys. JETP 40 (1975) p. 640.

[3] - M. Capitelli, M. Dilonardo and C. Gorse, J. Chem. Phys. 56, (1981), p. 29.

HIGHLY TIME-RESOLVED CARS SPECTROSCOPY OF PHOTOFRAGMENTS AND REACTION PRODUCTS.*

James J. Valentini

Department of Chemistry, University of California, Irvine, CA 92717

We describe the use of nanosecond-time-resolved CARS spectroscopy to obtain rotationally and vibrationally state-resolved Raman spectra of photofragments and reaction products in low density gases. The high temporal resolution, sensitivity, and spectral resolution of CARS spectroscopy make it possible to obtain spectra that directly yield the nascent electronic, vibrational, and rotational state distribution of photofragments under collisionless conditions, and reaction products under single collision conditions. The species generality of CARS makes it useful for the state-resolved detection of molecules like O_2, N_2, and H_2, that are otherwise difficult to detect in a quantum state specific way in these chemical dynamics experiments.

Examples of the detection of $O_2(^1\Delta_g)$ and $O_2(^3\Sigma_g^-)$ from the uv and visible photolysis of ozone are presented. Also described is the detection of the H_2 and HD products of the reaction of translationally hot H atoms with hydrogen and deuterium containing species. Problems specific to the acquisition and analysis of CARS spectra in such experiments are discussed. Solutions of these problems are described.

*Research supported by the U.S. Department of Energy, the National Science Foundation, and the American Chemical Society Petroleum Research Fund.

DOUBLE-RESONANCE TECHNIQUES FOR HIGH-RESOLUTION RAMAN-SPECTROSCOPY.*

Peter Esherick and Adelbert Owyoung

Advanced Laser Technology Division 1124
Sandia National Laboratories
Box 5800
Albuquerque, NM 87185, USA

Stimulated Raman spectroscopy (SRS) has established itself in recent years as a viable technique for obtaining rotationally-resolved Raman spectra of gas-phase molecules[1]. Instrumental linewidths on the order of 0.002 cm-1 are routinely achieved using CW probe lasers and near-Fourier-transform-limited-bandwidth pulsed pump lasers. The technique has proved itself especially valuable in resolving the Q-branch rotational structure of the strong symmetric-stretch transitions in spherical top molecules. While SRS has demonstrated remarkably high sensitivity relative to spontaneous Raman techniques for such high-resolution applications, its sensitivity is still relatively poor when compared to laser induced fluorescence or resonant laser ionization techniques.

In an effort to improve the sensitivity of coherent Raman spectroscopies we have been investigating double-resonance techniques that can combine the high-resolution of stimulated Raman spectroscopy with the sensitivity of electronically resonant spectroscopies such as multiphoton ionization. One such technique[2] that has been highly successful in our laboratory is ionization detected stimulated Raman spectroscopy (IDSRS). In IDSRS, resonant laser ionization is used to detect vibrationally excited molecules produced via stimulated Raman pumping. In a two-step process molecules are first excited to v=1 levels via a driven stimulated Raman transition. Following this the rotational/vibrational distribution is probed with resonant laser ionization by a tunable UV source. By using two narrow-bandwidth pulsed lasers for the stimulated Raman step we are able to selectively populate discrete ro-vibrational levels even for a molecule as large as benzene[3]. Raman spectra are then obtained by scanning the frequency of one of the Raman pump sources and monitoring the ionization signal as a function of the frequency difference (Stokes shift) between this laser and the second, fixed-frequency, visible pump laser.

*This work performed at Sandia National Laboratories, supported by the U.S. Department of Energy under Contract Number DE-AC04-76DP00789.

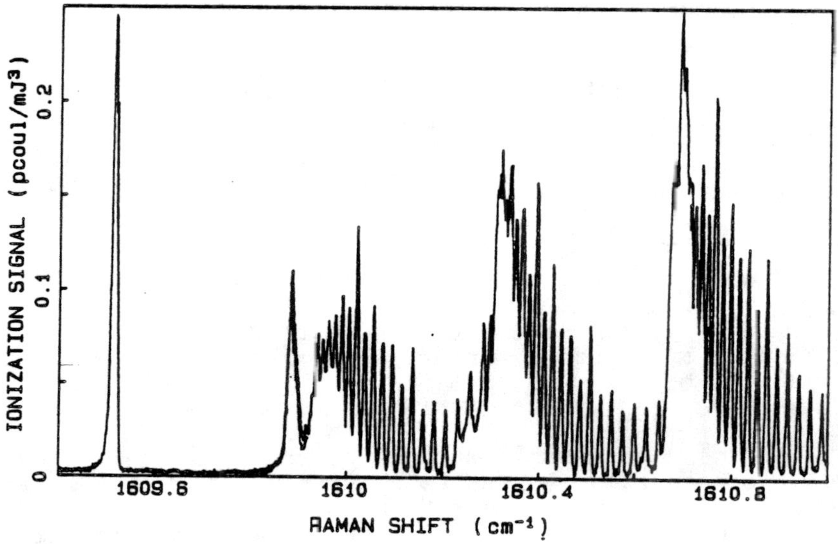

Figure 1. Ionization-detected Raman spectrum of benzene in a free-expansion jet. The ionizing wavelength was 274.12 nm.

A representative ionization-detected spectrum of benzene[3] is shown in Fig. 1 for a region just above the band origin of the upper component of the (ν_{16}, ν_{2+18}) Fermi doublet. The spectrum shows primarily peaks of the ^{S}R branches, while the strong peak at the left is an unresolved portion of the $^{O}Q_1$ branch. Note that the double-resonance aspect of IDSRS has allowed us to cleanly separate this spectrum from the contributions of the lower component of the Fermi doublet.

IDSRS is just one of several closely related optical-optical double-resonance techniques[4] involving folded excitation schemes. While the applicability of any one of these techniques to a particular problem will depend on many factors, we believe vibrational/electronic double-resonance methods can make a significant contribution to our understanding of the complex and congested spectra of polyatomic molecules.

References:

1. P. Esherick and A. Owyoung in Advances in Infrared and Raman Spectroscopy, Vol. 9, Edited by R. J. H. Clark and R. E. Hester, (Heyden & Son, Ltd., London, 1982), pp. 130-187.

2. P. Esherick and A. Owyoung, Chem. Phys. Lett. 103, 235 (1983).

3. P. Esherick, A. Owyoung and J. Pliva, J. Chem. Phys. 83, 3311 (1985).

4. D. A. King, R. Haines, N. R. Isenor and B. J. Orr, Optics Letters 8, 629 (1983).

STUDIES OF THE SELF-BROADENING AND FOREIGN GAS-BROADENING COEFFICIENT DEPENDENCE ON TEMPERATURE FOR DIATOMIC MOLECULES AND MONATOMIC GASES

By: H.G.M. Edwards, D.A. Long and G. Sherwood

Molecular Spectroscopy Unit, School of Chemistry,
University of Bradford, Bradford, England, BD7 1DP.

ABSTRACT

Raman spectroscopic studies of the rotational line broadening in simple gases as a function of gas pressure and temperature have previously been almost wholly confined to ambient temperatures. The J-dependence of the broadening coefficient is well established experimentally (1-4) and theoretically (5-7). In another paper presented at this Conference we gave details of our own studies of the self-broadening of the rotational lines in the pure rotation Raman spectrum of nitrogen in the temperature range 100-375 K and in the pressure range 1-60 bar. The data for this system indicate a linear relationship exists between the self-broadening coefficient, σ, and the temperature, T, over the temperature range 100-250 K with a deviation in linearity around 300 K. Several possible explanations are proposed to explain this behaviour and it is believed that different contributions from quadrupolar and dispersion forces could be involved.

In this presentation, the results of our work on the following systems will be reviewed:

(i) Nitrogen:

- (a) self-broadening;
- (b) foreign gas-broadening by hydrogen chloride;

(ii) Deuterium:

Foreign gas broadening by (a) carbon monoxide;

(b) nitrogen;

(iii) Hydrogen:

Foreign gas-broadening by

(a) carbon monoxide;

(b) nitrogen;

(c) argon;

(d) helium.

These studies extended over the temperature range 100-375 K and pressure range 1-60 bar. For foreign gas-broadening the host gas was maintained at 1 bar pressure and the broadening gas pressures in the range 1-59 bar. The results of these studies will be related to the polar properties of the broadening agents and theories of intermolecular interaction.

The importance of such measurements for the interpretation of CARS spectra will be reviewed.

ACKNOWLEDGEMENT

We acknowledge with gratitude the award of a SERC/CASE studentship to Mr. G. Sherwood.

REFERENCES

1. G.E. St. John, Ph.D. Thesis, University of Toronto (1952).
2. F. Pinter, *Opt. Spectrosc.*, 17, 428 (1964).
3. Yu A Lazarev, *Opt. Spectrosc.*, 13, 373 (1962).
4. K.S. Jammu, G.E. St. John and H.L. Welsh, *Can. J. Phys.*, 44, 797 (1966).
5. J. Van Kranendonk, *Can. J. Phys.*, 41, 433 (1963).
6. C.G. Gray and J. Van Kranendonk, *Can. J. Phys.*, 44, 2411 (1966).
7. R.P. Srivastava and H.R. Zaidi, *Can. J. Phys.*, 55, 542 (1977).

A TEMPERATURE STUDY OF THE PRESSURE-BROADENING COEFFICIENT DEPENDENCE ON ROTATIONAL QUANTUM NUMBER, J, FOR NITROGEN

H.G.M. Edwards, D.A. Long and G. Sherwood

Molecular Spectroscopy Unit, School of Chemistry,
University of Bradford, Bradford, England, BD7 1DP.

ABSTRACT

Self-broadening coefficient data are presented for the Stokes pure rotational Raman lines of nitrogen, J = 2, 6, 10, 14, at temperatures in the range 100-375 K and pressures in the range 1-60 bar. Dependence of the self-broadening coefficient on J and temperature is observed for the first time. The experimental values are compared with previous data at room temperature and with theory.

INTRODUCTION

Raman spectroscopy is an excellent non-invasive technique for probing gaseous environments and for the study of collision processes in the gas phase. We present the first experimental study of the effect of temperature on the self-broadening coefficients of the Stokes pure rotational Raman lines J = 2, 6, 10, 14, of nitrogen. Previous experimental (1-4) and theoretical (5-7) studies for nitrogen have been carried out at room temperature only.

EXPERIMENTAL

The high-pressure gas cell was housed in the vacuum shroud of a liquid-nitrogen cryostat. Temperatures in the range 100-375 K (±1 K) and pressures 1-60 bar (±0.25 bar) were used.

Spectra were recorded with a three-metre Czerny-Turner spectrometer with a rld of 1.4 cm^{-1} mm^{-1} at 488.0 nm in the twelfth order using an argon laser (4.5 W at sample). The detector was an EMI 9789 QA photomultiplier and data acquisition was carried out using photon counting and a Nicolet 1180 system.

Linewidths were measured using a curve-fitting program that deconvoluted overlapping lines.

RESULTS

The self-broadening coefficients, σ, obtained in the present work are shown in the Table.

The J- dependence of the pure rotational Raman linewidth has been observed previously (1-4); Figure 1 compares the present room temperature data with that of St. John (1) and Jammu *et al.* (4). Our results compare favourably with the experimental data of Jammu *et al.* (4).

J LINE	$\sigma \times 10^2$ cm^{-1} bar^{-1}					
	TEMPERATURE/K					
	100	150	200	250	290	375
2	18.3±0.3	16.3±0.3	14.5±0.4	11.3±0.3	10.3±0.5	8.3±0.4
6	17.3±0.3	14.6±0.2	12.7±0.3	11.2±0.5	10.0±0.3	8.0±0.3
10	14.6±0.6	13.7±0.7	11.3±0.4	8.6±0.2	7.3±0.3	6.7±0.1
14	13.6±0.7	12.4±0.5	10.1±0.5	7.5±0.3	6.9±0.2	5.4±0.2

Figure 2 compares our experimental room temperature data with theoretical studies (6,7). Both sets of calculations predict a J-dependence. The initial calculations (6) were based only on quadrupole-quadrupole interactions resulting in a weaker interaction potential than in the second set of calculations (7) where anisotropic dispersion forces were also taken into account.

In Figure 3 the σ values for J = 2, 6, 10 and 14 are plotted against temperature, T. The plots indicate a linear relation between σ and T over the range 100-250 K with a deviation from linearity around 300 K. It is possible that different contributions from dispersion and quadrupolar forces could play a part in the behaviour of σ with T through the intermolecular potential.

FIG. 1 Comparison of Experimental Results

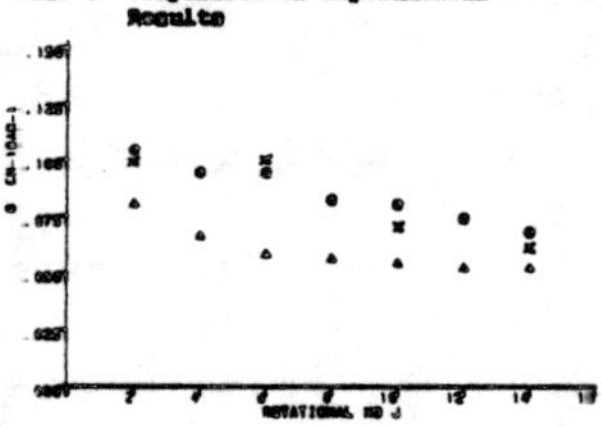

Δ Ref. 1   O Ref. 4   x This work

FIG. 2 Comparison of Our Experimental Results with Theory

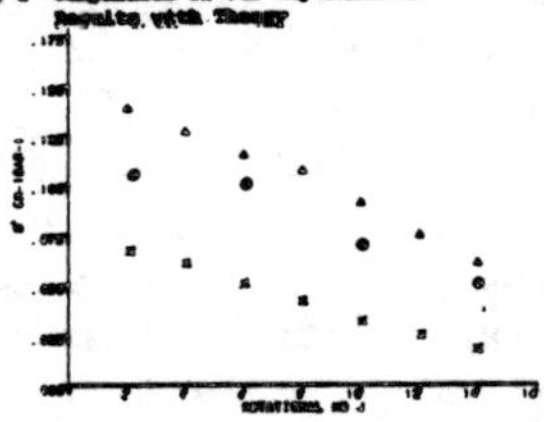

Δ Ref. 7.   x Ref. 6   O This work

FIG. 3 Dependence of σ on T

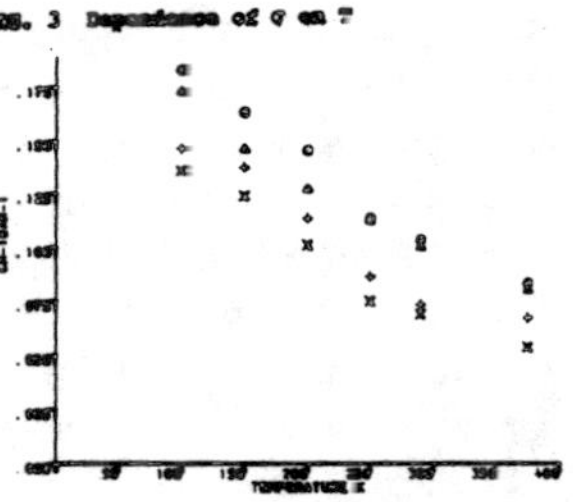

O J = 2; Δ J = 6; ◇ J = 10; x J = 14.

ACKNOWLEDGEMENT

We acknowledge with gratitude the award of a SERC/CASE studentship to Mr. G. Sherwood.

REFERENCES

1. G.E. St. John, Ph.D. Thesis, University of Toronto (1952).
2. F. Pinter, *Opt. Spectrosc.*, 17, 428 (1964).
3. Yu A Lazarev, *Opt. Spectrosc.*, 13, 373 (1962).
4. K.S. Jammu, G.E. St. John and H.L. Welsh, *Can. J. Phys.*, 44, 797 (1966).
5. J. Van Kranendonk, *Can. J. Phys.*, 41, 433 (1963).
6. C.G. Gray and J. Van Kranendonk, *Can. J. Phys.*, 44, 2411 (1966).
7. R.P. Srivastava and H.R. Zaidi, *Can. J. Phys.*, 55, 542 (1977).

DUAL BROADBAND APPROACHES TO VIBRATIONAL AND PURE ROTATIONAL CARS SPECTRA.

Alan C. Eckbreth and Torger J. Anderson

United Technologies Research Center, East Hartford, CT 06108 USA

CARS (coherent anti-Stokes Raman spectroscopy) has become a well-established technique for the remote, spatially and temporally precise probing of instrumentally hostile environments. A major limitation of CARS, when compared to spontaneous Raman scattering, is the inability to interrogate more than one molecular species at a time as normally implemented. A new approach to CARS has been demonstrated which permits CARS to be generated from a multiplicity of species simultaneously.[1] As shown in Fig. 1a, the technique employs two independent broadband Stokes sources in combination with the pump. In addition to the two separate, two-color wave-mixing processes between the pump and Stokes lasers, spectrally-resolved CARS is produced in a three-color process from species whose Raman resonances correspond to the frequency differences of the two broadband Stokes sources. Since the Raman frequencies are well defined by the constituent molecules, spectral smearing does not occur even with dual broadband excitation as long as the pump is narrowband. The narrowband pump scatters off the resultant coherent Raman excitation resulting in CARS with a spectral resolution characteristic of the pump laser or spectrograph employed. Phase matching all of these combinations is readily achieved with planar or folded crossed-beam arrangements or combinations of these. Particularly simple schemes have also been devised which exploit the annular shape of pump beams from diffractively-coupled unstable resonators.[2] Shown in Fig. 2 are CARS spectra from N_2, CO_2 and H_2O simultaneously generated in the postflame zone of a premixed C_2H_4-air flame. Quite serendipitous in combustion is that placement of the two Stokes lasers respectively on the major products of hydrocarbon/air reactions, i.e. CO_2 and H_2O, permits CARS generation from heavy diatomics such as N_2, CO, NO corresponding to the frequency differences of the two broadband sources.

For pure rotational CARS, the two broadband Stokes sources need to be located in such close spectral proximity that they overlap substantially, Fig. 1b. It is clear, as seen in Fig. 1c, that a single broadband source will suffice for pure rotational CARS if properly phase matched and this may be considered a degenerate form of dual broadband CARS. Thus pure rotational CARS and CSRS (coherent Stokes) can be generated from a pump and a spectrally arbitrary Stokes laser.[3] Figure 3 shows the pure rotational CARS/CSRS spectrum of N_2 generated using a 532 nm pump and broadband Stokes source centered at 660 nm. The CARS and CSRS are generated simultaneously and very small Raman shifts are observable since high suppression of the pump is achieved with crossed-beam phase matching. This technique is much simpler than conventional approaches to pure rotational CARS and, thus, may be more broadly applicable. Besides being self-calibrating, the dual integration aspect of this technique may obviate the requirement for spectral referencing during single pulse measurements.

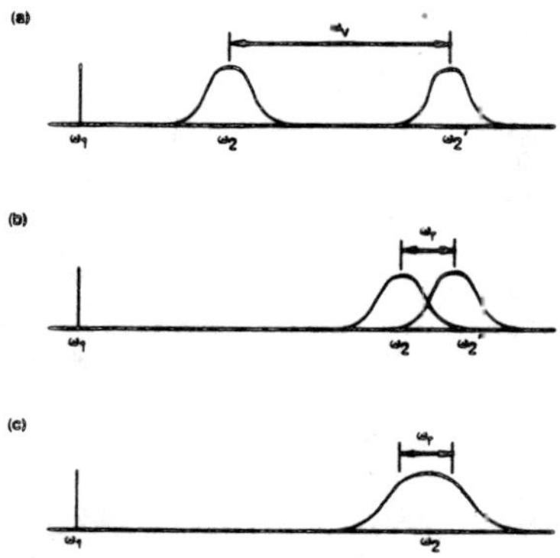

Fig. 1 (above) Dual broadband Stokes placement to excite vibrational (a) and rotational (b) resonances. (c) New approach to pure rotational CARS/CSRS using a spectrally arbitrary Stokes laser.

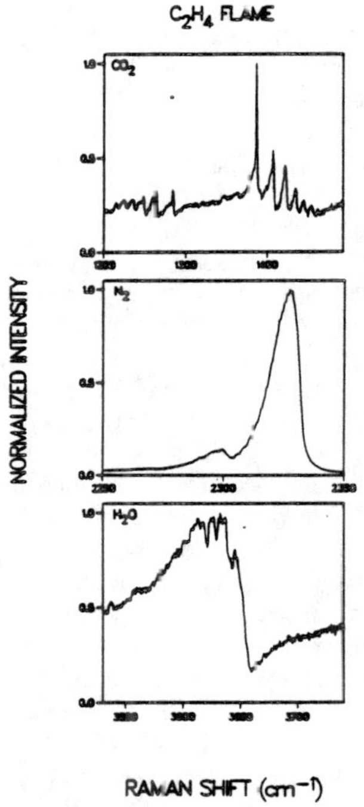

Fig. 2 (above) Simultaneously generated dual broadband CARS signatures from CO_2, N_2 and H_2O in an atmospheric pressure flame.

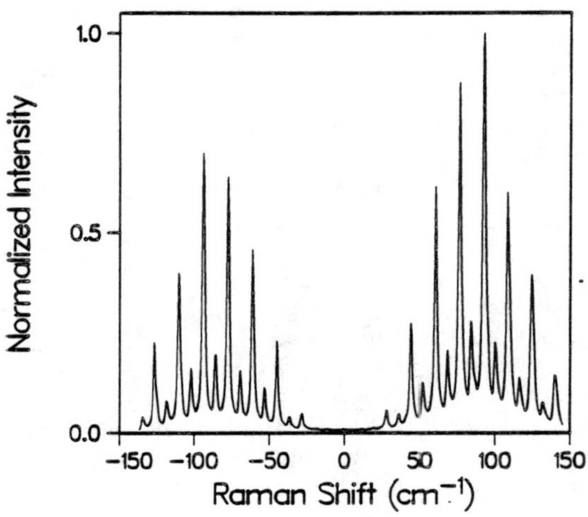

Fig. 3 (left) Pure rotational CARS/CSRS spectrum of N_2.

1. A. C. Eckbreth and T. J. Anderson, Appl. Opt. 24, 2731 (1985)
2. A. C. Eckbreth and T. J. Anderson, Appl. Opt. 25, in press (1986)
3. A. C. Eckbreth and T. J. Anderson, Opt. Lett. 11, submitted (1986)

CARS Measurements of Temperature Profiles and Pressure in a Tungsten Lamp.

Rick Trebino and Roger L. Farrow

Combustion Research Facility
Sandia National Laboratories
Livermore, California 94550

Introduction

The measurement of temperature and pressure inside incandescent lamps is important for understanding heat and mass transport processes and, consequently, lamp efficiency and longevity.[1] Traditional methods, however, are often inadequate. Radiation and optical pyrometers measure the temperature of emitting materials only, thermocouples cannot survive the temperatures of interest, and interferometric techniques measure only integrated quantities. Pressure-measurement devices, such as transducers and barometers, have similar problems. Coherent Anti-Stokes Raman Spectroscopy (CARS), on the other hand, allows temperature and pressure measurement with none of these drawbacks and is a potential alternative technique.

Previous work has established CARS as an accurate temperature probe in a variety of hostile environments, ranging from laboratory flames[2,3] to jet engine exhausts.[4] In addition, Greenhalgh et al[5] have performed CARS temperature measurements in specially constructed incandescent lamps. Very little work has been done, however, to establish CARS as an accurate pressure probe. To our knowledge, only Gustafson and Byer[6] have attempted such measurements, doing so in a 60-atm deuterium fusion pellet. In this paper, we describe the use of CARS to measure the temperature profile and pressure in nitrogen-filled high-intensity lamps. We employ an off-the-shelf General Electric 1000-W FEL lamp, with no special windows or polish, and measure the CARS *Q*-branch spectrum of the nitrogen fill. The resulting temperatures range from below 1000 K beneath the lamp filament to over 2000 K near the top of the filament and have an experimental accuracy of approximately 5%. In addition, we obtain values of the pressure to within 5%, using a specially constructed lamp that permitted independent measurement of the pressure.

Results and Discussion

Using a previously described CARS apparatus[3] (but with a single-axial-mode pump laser), we probed 0.1 mm x 0.1 mm x 1mm regions in a horizontally mounted lamp along a radial direction from just below to just above the filament. Figure 1 shows a typical nitrogen *Q*-branch ($\Delta J=0$) spectrum for a region near the top of the filament of the lamp operating at 207 W of electrical power. Figure 2 shows the measured nitrogen CARS temperature profile in the lamp, obtained from the CARS spectra using a least-squares curve-fitting procedure based on an exponential-gap model for the J-dependent molecular linewidths.[7] Temperatures range from 970 K just below the filament to 2210 K near the top of the filament, consistent with an optical pyrometer reading of the filament temperature of 2260 K. Marked asymmetry with respect to the lamp center exists in the profile, indicating that the effects of convection are significant, as expected for the existing temperature and pressure conditions.[1]

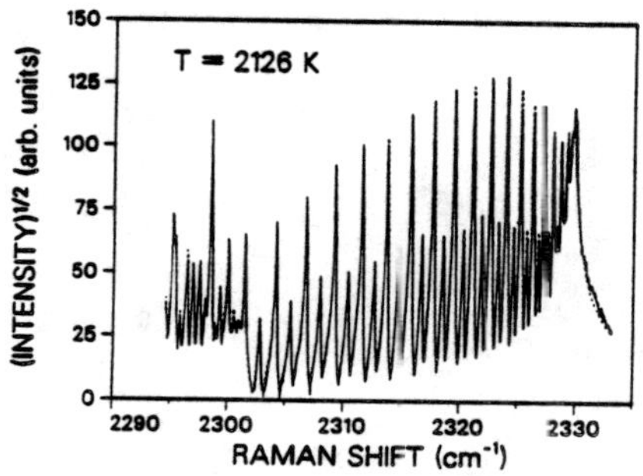

Fig. 1 Experimental (solid line) and theoretical (dashed line) spectra of the nitrogen Q-branch measured near the top of the filament. Least-squares curve-fitting yields a temperature of 2126 K for this region.

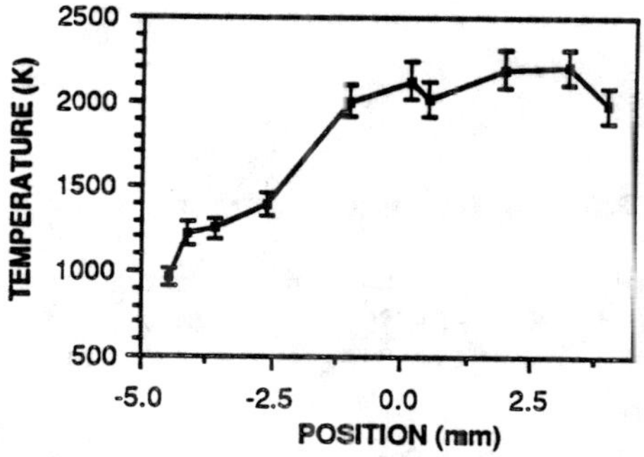

Fig. 2 Temperature profile across lamp, measured using CARS technique. Coiled filament extended from -3.9 mm (low side) to +3.1 mm (high side). Lamp envelope walls occurred at -9.1 mm and +8.9 mm.

To perform pressure measurements, we used a specially constructed lamp with an open fill tube, allowing independent measurement of the lamp pressure with a Baratron gauge with an accuracy of 1%. Because the model of Koszykowski et al[7] underpredicts the CARS lineshapes at high temperatures (>1500 K), we employ a slightly modified model[8] that better predicts high-temperature lineshapes and, hence, better predicts the pressure, too. For three CARS scans with known lamp pressures of 5.46, 5.41, and 5.43 atm, we find fitted CARS values of 5.60, 5.57, and 5.43 atm, respectively, i.e., an average difference of less than 3% and an overall accuracy of about 5%. A high-resolution spectrum of the off-the-shelf lamp yields a pressure of 8.75 atm, a reasonable value, but one that we are unable to corroborate due to the unavailability of alternative techniques.

We anticipate that CARS temperature and pressure measurements may be useful for studies of other hostile environments, including lamps with more extreme internal conditions, such as high-intensity arc lamps. And, conversely, commercial lamps could provide convenient environments for gas-phase nonlinear-optical spectroscopy at high temperature and pressure.

Acknowledgments

The authors would like to thank Michael Drake, General Electric Research and Development Center, for suggesting this series of measurements, and Edward Zubler, also at General Electric, for providing the lamps. This work was supported by the U.S. Department of Energy, Office of Basic Energy Sciences, Chemical Science Division.

References

1. E. Fischer, J. Fitzgerald, and H. Horster, J. Illum. Eng. Soc., p. 271 (1975).
2. A. Eckbreth and R.J. Hall, Combust. Flame **36**, 87 (1979).
3. R.L. Farrow, P.L. Mattern, and L.A. Rahn, Appl. Opt. **21**, 3119 (1982).
4. A.C. Eckbreth, G.M. Dobbs, J.H. Stufflebeam, and P.A. Teller, Appl. Opt. **23**, 1328 (1984).
5. D.A. Greenhalgh, F.M. Porter, W.A. England, D.R. Williams, S.T. Whittley, D.H. Glass, S.A. Barton, R. Devonshire, and I. Dring, Ninth Int. Conf. Raman Spectrosc., p. 342 (1984).
6. E.K. Gustafson and R.L. Byer, Opt. Lett. **9**, 220 (1984).
7. M.L. Koszykowski, R.L. Farrow, and R.E. Palmer, Opt. Lett. **10**, 478 (1985).
8. R.L. Farrow, R. Trebino, and R.E. Palmer, to be published.

RAMAN SPECTROSCOPY ESPECIALLY OF HIGH VIBRATIONAL STATES

W. Knippers, K. van Helvoort, R. Fantoni, J. Reuss and S. Stolte
Physics Department, University of Nijmegen, Toernooiveld,
6525 ED Nijmegen, The Netherlands

In contrast with infrared spectroscopy, overtone transitions are usually not studied by the spontaneous Raman effect because of its inferior sensitivity and limited resolution. However, by employing the very bright excitation light source (λ=488 nm), provided by an intracavity sharply focused ($1/e^2$ waist diameter $\leq$50 μm) cw Ar-ion laser arrangement (see fig. 1) yielding a power density of 10^7 W/cm^2, it turned out to be possible to observe weak Raman transitions possessing a factor of 10^4 less intensity that a typical fundamental (Raman active) vibrational transition. Filling the intracavity volume with one amagat of N_2, O_2 or D_2 gas allowed the observation of both the regular (Δv=1) and the overtone (Δv=2) Q-branch Stokes shifted Raman spectra [1]. For D_2 both spectra are given in Fig. 2. In this case the intensity ratio between corresponding Δv=2 and Δv=1 transitions has been found to be (1.15±0.4)·x10^{-3} and equal for individual rotational lines. This ratio informs directly about the dependence of the molecular polarizability upon the internuclear distance.

The experimental study of combination and overtone transitions of simple polyatomics is especially interesting for molecules with a center of inversion. Here the mutual exclusion principle of Raman and infrared activity for this kind of molecule allows the observation of (higher) vibrational states by Raman effect which cannot be observed by one photon absorption. Information about these levels is important to obtain quantitative insight in the processes of infrared multiphoton excitation and mode specificity of highly vibrationally excited states.
Because of the large number of rotational states occupied in polyatomic molecules, resolution of individual rotational transitions introduces an extra loss of spectral intensity. However, the existence of Q-branches for which rotational lines coincide essentially offers the possibility of observing all rotational contributions of vibrational transitions simultaneously with our limited spectral resolution (2 cm^{-1} typically).

Raman spectra covering transitions up to $\Delta\nu\approx$6000 cm^{-1} have been studied for C_2H_4 [2], C_3H_4 [3] and C_2H_6 [4], and have been shown to possess well defined overtone transitions in this spectral range. The Raman activity of higher excited vibrational states seems to be especially enhanced by the occurrence of a Fermi-anharmonic resonance with a high frequency strong Raman active fundamental transition.

The observation of hot band transitions of Raman overtone excitations turned out to be also very useful for the determination of high lying vibrational states. Practically all vibrational origins observed fitted excellently with a simple Dunham series parametrization.
Vibrational excitation by CO_2 laser pumping has also been studied by

spatially resolved Raman probing in molecular jets. The occurrence of mode selective relaxation and absence of low fluence multiphoton excitation was concluded for C_2H_4 [5].

[1] W. Knippers, K. van Helvoort and S. Stolte, Chem. Phys. Lett. 121 (1985) 279

[2] W. Knippers, K. van Helvoort, S. Stolte and J. Reuss, Chem. Phys. 98 (1985) 1

[3] W. Knippers, K. van Helvoort, M. de Felici, J. Reuss and S. Stolte, Chem. Phys. 105 (1986) 27

[4] W. Knippers, thesis, Nijmegen (1986)

[5] W. Knippers, G. Luijks, S. Stolte and J. Reuss, Chem. Phys. Lett. 109 (1984) 532

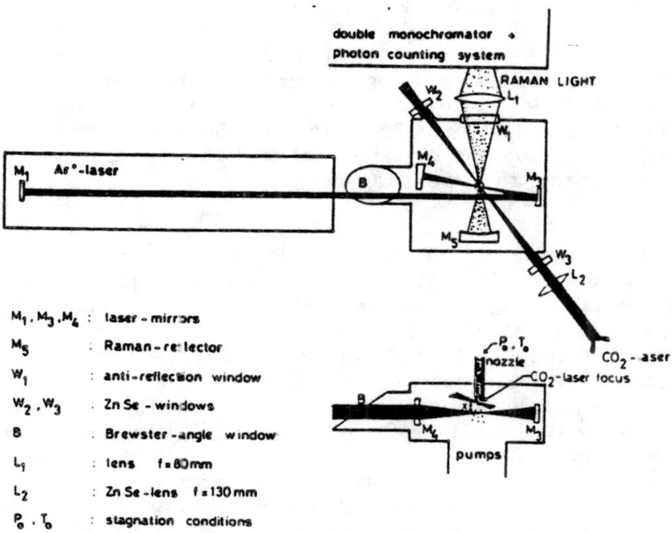

Fig. 1 The intracavity experimental arrangement.

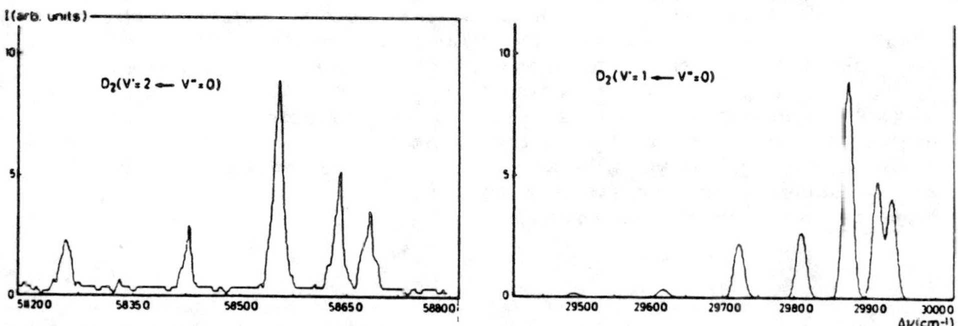

Fig. 2 Totally resolved rotational structure in the Q-branch of the Raman spectrum of the fundamental (left panel) and first overtone (right panel) of D_2.

HIGH RESOLUTION STIMULATED RAMAN SPECTROSCOPY : STUDIES OF MOLECULAR STRUCTURES AND COLLISIONAL EFFECTS ON LINE SHAPES.

B. Lavorel

Laboratoire de Spectronomie Moléculaire et Instrumentation Laser, U.A. CNRS n° 777, 6 Bd Gabriel 21100 DIJON - FRANCE

Introduction

Laser Raman techniques offer the possibility to realize high resolution and high sensitivity molecular spectroscopy. Among these techniques, the stimulated Raman spectroscopy (SRS) can yield Raman spectra without line shape distortion. That explains the ability of SRS to give useful informations on molecular structures and line shapes.

Experiment

An SRS experiment has been developped in Dijon, which uses a pulsed pump laser and a CW probe laser. The pulse pump laser is obtained by amplifying a tunable single mode dye laser in four dye amplifiers pumped by a frequency-doubled Nd-Yag laser. The Raman signal is detected on the CW probe laser, is amplified, and then is averaged and stored in a data acquisition system. In our recent studies, the probe laser (single mode argon ion laser) was actively stabilized with a Fabry-Perot interferometer, and its frequency was locked to a Doppler-free saturated absorption line of I_2. Thus, the absolute frequency of the probe laser is precisely known ($\simeq$1MHz accuracy). A traveling corner cube wavemeter measures the dye laser frequency at the beginning and at the end of the 1 cm^{-1} scan. The Raman shift at each point of the scan is interpolated from transmission fringes of stabilized etalon (300 MHz FSR) with an accuracy of about 10 MHz.

Study of molecular structures

We report in this section two examples of molecular structures studies. The Raman spectrum of the methane like molecule $^{13}CD_4$ has been investigated in several spectral regions corresponding to the pentad ($2\nu_4$, $\nu_2+\nu_4$, ν_1, $2\nu_2$, ν_3) at a pressure of 30 Torr. The observed Raman frequencies have been combined with the infrared data for a simultaneous analysis of the pentad of $^{13}CD_4$. The second molecule recently investigated is C_3H_6. No high resolution spectrum of the ν_1 region (3000-3040 cm^{-1}) was available up to now. We have recorded this spectrum at a 15 Torr pressure and a temperature of 210 K.

Collisional line shapes

As mentioned above, high accuracy linewidth measurements can be performed with SRS. The Dicke narrowing has been demonstrated in a backward scattering configuration on the Q(18) line of N_2 by using a Galatry profile in the fit [1]. If this phenomenon is neglected one gets an error of the order of 10 %. With a view to apply SRS data to laser Raman diagnostic techniques (such as CARS), we have recorded N_2 Q-branch spectra in a forward scattering configuration, over the pressure range 0.25-1.9 atm and from room temperature up to 1310 K [2]. In this study, made in collaboration with J.P. Sala, J.Bonamy and D.Robert of the Besançon University, we report the dependance of the collisional broadening coefficients on the rotational quantum number J and the temperature for Q(J) lines in pure N_2. Semi-classical calculations of these coefficients lead to values consistent with the measured ones, and were extended to higher J values and to higher temperatures. Finally, a polynomial inverse gap law for the rotational energy transfer rates has been also used to predict high temperature line widths.

Recent investigations on the Q-branch of N_2 at higher pressures (5-35 atm) have been performed. In this case, the collisional narrowing plays an important rôle, and a correct description of the observed spectrum requires adequate models [3].

[1]. G. MILLOT, B. LAVOREL, R. SAINT-LOUP et H. BERGER, J. Physique, 46, 1925, (1985).

[2]. B. LAVOREL, G. MILLOT, R. SAINT-LOUP, C. WENGER, H. BERGER, J.P. SALA, J. BONAMY and D. ROBERT, J. Physique, 47, 417, (1986).

[3]. J.P. SALA, J. BONAMY, D. ROBERT, B. LAVOREL, G. MILLOT and H. BERGER, Submitted to J. Chem. Phys.

HIGH RESOLUTION STIMULATED RAMAN SPECTROSCOPY OF $^{13}CD_4$.

G. Millot, B. Lavorel, R. Chaux, R. Saint-Loup, G. Pierre and H. Berger

Laboratoire de Spectronomie Moléculaire et Instrumentation Laser, U. A. CNRS n° 777, 6, Bd Gabriel 21100. DIJON - FRANCE

Coherent Raman techniques offer the possibility of carrying out high resolution Raman spectra ; however, new improvements are yet necessary to compete with infrared spectroscopy in the field of molecular structure determinations. First, the sensitivity must be increased in order to record harmonic or combination bands and secondly, the precision of Raman data must be closed to that obtained in infrared spectroscopy.

We report two improvements in sensitivity and calibration performed on Raman spectra of $^{13}CD_4$. The Raman spectrum has been recorded at a pressure of 30 Torr and room temperature ; in addition to the strong ν_1 Q-branch, the $2\nu_2$ and $2\nu_4$ harmonic bands and the $\nu_2 + \nu_4$ combination band have been obtained with a correct signal to noise ratio. In this spectral region, only a simultaneous treatment of five interacting bands can be successfully envisaged ; so, 300 Raman transitions have been combined in the pentad analysis to infrared transitions recorded on the Fourier transform spectrometer of the "Laboratoire de Spectronomie Moléculaire de PARIS".

Frequency calibration of Raman lines is achieved by using a travelling Michelson wavemeter. Its originality consists in a corner cube travelling vertically in an evacuated cylinder on a distance of 1 meter. The frequency of the probe laser, used as frequency reference, is locked to a Doppler free saturated absorption line of iodine. The precision of the wavemeter is about 3 MHz. Raman shifts are determined with a precision of 10 MHz.

HIGH-RESOLUTION NONLINEAR ROTATION-VIBRATIONAL RAMAN SPECTROSCOPY OF GASES

H.W. Schrötter

Sektion Physik der LMU München, Schellingstrasse 4,
D-8000 München 40, F.R. Germany

In the last 10 years the resolution in rotation-vibrational Raman spectroscopy of gases has been increased by about two orders of magnitude through the application of the nonlinear techniques which are based on the excitation with two lasers, one of which is tunable. Early examples are the resolved Q-branches of the ν_1 and ν_2 bands of acetylene (1,2).

Because the instrumental resolution depends on the convoluted linewidths of the two lasers and the narrowest linewidths can be obtained with stabilized single mode continuous lasers, we have constructed a CARS spectrometer for cw intracavity excitation in Munich (3). It consists of an argon ion ring laser with an etalon, a prism, and a gas cell containing two spherical mirrors in the cavity and a dye ring laser the beam of which is coupled into the gas cell at the prism for collinear excitation.

The CARS spectrum of the Q-branch of nitrogen has been recorded at pressures of 2 kPa and 14.5 kPa (3,4). The resolved structure of the Q-branch of the ν_3 band of methane at pressures of 2.7 kPa and 13.6 kPa was found to be in good agreement with existing theory (5). On the other hand, the analysis of the intensity profile of the ν_1 band of methane recorded at a pressure of 1.4 kPa (4,6) led to an empirical reassignment of transitions with $J = 7$ to $J = 10$ which was only later shown to be in satisfactory agreement with new theoretical calculations by Lolck (7). The pressure dependence of the ν_1 spectrum was also investigated (5) and the lowest pressure at which a satisfactory signal/noise ratio has been obtained was 0.3 kPa (8). The analysis of the Q-branch of the ν_1 band of hydrogen sulfide recorded at a pressure of 1.3 kPa led to excellent agreement with theory (9). The Q-branches of the ν_1 bands of benzene (5) and allene (8) have also been recorded.

In the case of the Q-branch of the ν_1 band of ammonia at a pressure of 1.5 kPa (10) the recorded and calculated spectra could only be brought into satisfactory agreement by introduction of pressure broadened linewidths that depend on the quantum numbers J and K of the symmetric top. This effect has also been found in the infrared spectrum of the ν_1 Q-branch (11) and is being investigated theoretically (12).

In Dijon a CARS spectrometer with flashlamp pumped dye laser amplifiers (13) injection locked to the frequency of stabilized single mode cw krypton or dye lasers has been constructed by Boquillon et al. (see 14). The amplifiers deliver pulses of about 400 ns with peak powers up to 10 kW and a repetition rate of 10 Hz, leading to an instrumental resolution of about 10 MHz. With this spectrometer spectra of the Q-branch of oxygen at pressures between 1.3 kPa and 130 kPa have been recorded and collisional line broadening parameters have been determined for the lines with J = 1 to J = 17.

A spectrometer for inverse Raman spectroscopy has been built in Dijon by Berger et al. (see 14, 15). The Q-branches of the ν_1 bands of silane (15) and germane (16) have been reinvestigated and that of cyclopropane (17) has been observed. The Q-branch of the ν_1 band of the methane isotopomer $^{13}CD_4$ has been recorded and is being analysed together with the infrared spectrum in the C-D stretching region (18). Collisional broadening of the lines in the Q-branch of nitrogen has been studied as function of temperature and density in forward (19) and backward (20) scattering configurations.

References

1 W. Nitsch and W. Kiefer, Opt. Commun. 23, 240 (1977).
2 V.I. Fabelinsky, B.B. Krynetsky, L.A. Kulevsky, V.A. Mishin, A.M. Prokhorov, A.D. Savel'ev, and V.V. Smirnov, Opt.Commun. 20, 389 (1977).
3 H. Frunder, L. Matziol, H. Finsterhölzl, A. Beckmann and H.W. Schrötter, J. Raman Spectrosc. 17, 143 (1986).
4 H. Frunder, D. Illig, H. Finsterhölzl, A. Beckmann, H.W. Schrötter, B. Lavorel and G. Roussel, in H.P. Weber and W. Lüthy (Eds.), Laser Spectroscopy VI, Springer Ser.Opt.Sci. 40, 334 (1983).
5 H. Frunder, Dissertation LMU München, 1983.
6 H. Frunder, D. Illig, H. Finsterhölzl, H.W. Schrötter, B. Lavorel, G. Roussel, J.C. Hilico, J.P. Champion, G. Pierre, G. Poussigue and E. Pascaud, Chem.Phys.Letters 100, 110 (1983).
7 J.E. Lolck, Chem.Phys.Letters 106, 143 (1984).
8 D. Illig (unpublished).
9 H. Frunder, R. Angstl, D. Illig, H.W. Schrötter, L. Lechuga-Fossat, J.M. Flaud, C. Camy-Peyret and W.F. Murphy, Can.J.Phys. 63, 1189 (1985).
10 R. Angstl, H. Finsterhölzl, H. Frunder, D. Illig, D. Papoušek, P. Pracna, K. Narahari Rao, H.W. Schrötter and Š. Urban, J.Mol. Spectrosc. 114, 454 (1985).
11 A.S. Pine (unpublished).
12 G. Buffa (unpublished).
13 J.P. Boquillon and O. Bazemlal, Revue Phys. Appl. 16, 365 (1981).
14 H.W. Schrötter, H. Berger and J.P. Boquillon, in H.D. Bist and J.S. Goela (Eds.), Lasers and Applications, Tata McGraw Hill, New Delhi 1984, p. 321.
15 B. Lavorel, R. Saint-Loup, G. Pierre et H. Berger, J.Physique Lettres 45, L 295 (1984).
16 B. Lavorel, Thèse 3ème Cycle, Université de Dijon, 1984.
17 B. Lavorel, G. Millot, M. Terki, R. Saint-Loup and H. Berger (unpublished).
18 B. Lavorel et al., paper at this conference.
19 B. Lavorel, G. Millot, R. Saint-Loup, C. Wenger, H. Berger, J.P. Sala, J. Bonamy and D. Robert, J. Physique 47, 417 (1986)
20 G. Millot, B. Lavorel, R. Saint-Loup et H. Berger, J. Physique 46, 1925 (1985).

TIME-RESOLVED CARS SPECTROSCOPY: APPLICATION TO PHOTOPHYSICS AND PHOTOCHEMISTRY.

R. Bruce Weisman, Dept. of Chemistry, Rice University, Houston, TX

The detailed study of gas phase molecular photoprocesses requires experimental methods that can selectively probe the kinetics and structures of short-lived species. The most elusive of these are the "dark" product states of photophysical or photochemical decay, for which analysis using spontaneous emission is not fruitful. We describe here the use of transient Coherent Anti-Stokes Raman Spectroscopy (CARS) for measuring vibrational spectra and kinetics of excited polyatomic vapors.

Our apparatus uses two time-sequenced Nd:YAG-pumped dye lasers, one to provide tunable ultraviolet excitation and the other for CARS probing. As shown in Fig. 1, the excitation and probing beams counterpropagate collinearly through the static gas phase sample. A laboratory computer scans CARS frequency at fixed delay (for transient spectra), or delay time at fixed frequency (for transient kinetics). CARS signals are compared at negative and positive probe delays to separate induced from ground state components. The instrument's composite time profile is 7 ns and its spectral resolution is 0.3 cm^{-1} (with an etalon-narrowed YAG beam) or 1 cm^{-1} (without narrowing). The sensitivity is high enough that transient species may generally be detected in samples of about 1 Torr total pressure, thereby allowing observations under pre-collisional conditions.

Styrene, an intriguing photophysical system, was studied using this apparatus to clarify the relaxation processes from its S_1 state. Following excitation at the 34,759 cm^{-1} origin band, transient CARS signals were observed in the C=C and C-H stretching regions. On the basis of their observed decay kinetics, sharp transient bands near 1600 cm^{-1} were assigned to phenyl and ethylenic C=C stretching vibrations on the S_1 surface. Additional transient bands in this region formed an apparent sequence pattern and were found to be quenched by collisions with SF_6. They are thought to be vibrational hot band transitions in S_1 styrene molecules that have absorbed an extra photon and then undergone rapid internal conversion to high vibrational levels of S_1. Other vibrational transitions of the S_1 state were observed at 3017 cm^{-1} and 3026 cm^{-1}. Our findings suggest a significant triplet quantum yield in this system but provide no evidence of a "perpendicular" form of S_1. The rich and detailed vibrational spectra found for this electronically excited polyatomic illustrate the potential of transient CARS probing for studying excited state dynamics and structures.

We have also applied these methods to a long-standing photochemical problem: the question of mechanism in azoalkane photodissociation. In order to determine whether molecules of the form R-N=N-R' dissociate in a stepwise or concerted manner, we have probed for the kinetics of photofragment appearance as well as for the fragment internal state distributions in azomethane, where R and R' are methyl groups. Apart from the data on methyl radical spectroscopy [1] and on the appearance of N_2 and CH_3 [2], new results are now available on the nascent rotational and vibrational excitation of the CH_3 fragments, as well as data from similar studies on an asymmetrical azoalkane (with R'= dimethylallyl). These observations provide a much more detailed view of azoalkane photodissociation than had been available from classical methods.

References

[1] P. L. Holt, K. E. McCurdy, R. B. Weisman, J. S. Adams, and P. S. Engel, J. Chem. Phys. 81, 3349 (1984).

[2] P. L. Holt, K. E. McCurdy, J. S. Adams, K. A. Burton, R. B. Weisman, and P. S. Engel, J. Amer. Chem. Soc. 107, 2180 (1985).

Transient CARS Optical Schematic

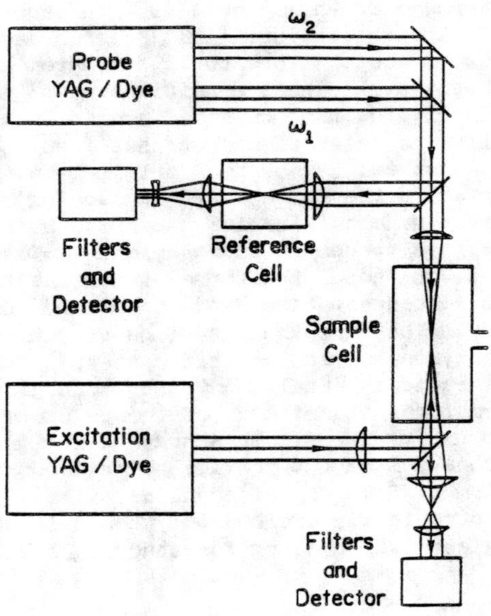

CARS THERMOMETRY AT ULTRA HIGH TEMPERATURES*.

Douglas A. Greenhalgh
Ian Dring, Robin Devonshire•
Fred Boysan‡

Harwell Combustion Centre, AERE Harwell, Oxon U.K.
•Dept of Chemistry, University of Sheffield U.K.
‡Dept of Chemical Engineering, University of Sheffield U.K.

*Work supported by the UKAEA's Underlying Programme

We have performed CARS measurements in approximately 1 atmosphere of nitrogen gas for the temperature range 2,000 to 3,600 Kelvin. In this paper we report a CARS spectrum at 3,467K and its interpretation using a full computer model for the nitrogen molecule, CARP-2 (1). We believe this to be the highest temperature, thermally equilibrated, nitrogen CARS or Raman spectrum yet to be measured. This work is an extension of earlier work on CARS measurements of heat transfer properties around incandescent lamp filaments (2).

Ultra high temperature, stagnant, nitrogen gas is generated inside a specially designed coil of tungsten wire (similar to an incandescent lamp filament) which is heated electrically. Comparison of our CARS nitrogen gas temperatures, with an estimate of the filament surface temperature using a two colour radiation corrected pyrometer, indicates agreement (1σ) of order of 1%. We believe that our high temperature stagnant gas generator forms the basis for a calibration and standardisation system for non-linear and other laser or optically based gas measurement techniques (patent applied for).

The basis for accurate CARS thermometry lies not only in well designed measurement equipment but also in the computer analysis model. Our measurement equipment is similar to that described in reference 3 with the addition of a high voltage pulser to "gate" our multichannel detector on for the 20 nanoseconds corresponding to our CARS signal pulses. Ultra high temperature spectral of nitrogen, of order 3,500 Kelvin, require the inclusion of order 350 transitions for vibrational Q-branches since rotational levels of order J=60 and higher and vibrational levels of order V=6 and higher contain significant population. Also, at pressures of order 1 atmosphere the total molecular line width, Doppler plus pressure broadening, is predicted to be approximately 0.015 cm^{-1} (H.W.). Because of strong line interferences between partially and full overlapping lines the CARS computer model requires very accurate energy level data, even for spectra recorded at our overall resolution of 1.2 cm^{-1}. In our analysis we have used the published data of Gilson et al (4) which we find to be excellent. Our CARS data, the theoretical predictions of our computer model (1), (using the energy level data of reference 4), and agreement of CARS gas temperature with filament surface temperature are all shown in

Figure 1. The principle discrepancy near 2289 cm^{-1} in the difference plot (Experiment - Theory) arises from a probable slight inaccuracy in the energy level data of reference 4 for the levels concerning either or both the transitions with lower states V=0, J=48 or V=1, J=26. We estimate the inaccuracy to be of order .014 cm^{-1} which is of the order of the claimed accuracy of .01 cm^{-1} for the data of reference 4, otherwise we find the energy level data (4) to give excellent agreement with our CARS measurements.

A full account of this work, with a complete discussion of the generation of the ultra high temperature stagnant gas and its relationship to tungsten coil surface temperature is in preparation (5).

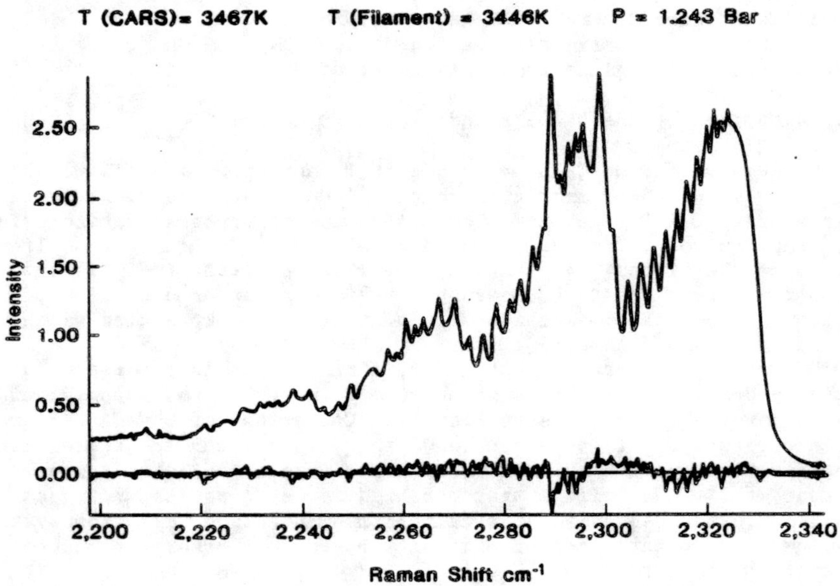

Fig. 1. Experimental CARS spectrum of stagnant nitrogen gas taken inside a coiled tungsten filament. Lower plot is experiment - theory.

References

(1) CARP-2 is a UKAEA copyright computer code for the analysis of CARS spectra; licenses available through Epsilon Research Ltd, Paynes Lane, Rugby, Warwickshire.

(2) R. Devonshire, I. Dring, F.M. Porter, D.R. Williams and D.A. Greenhalgh, to be published.;

(3) D.A. Greenhalgh and S.T. Whittley, Appl. Optics. 24, 907 (1985).

(4) T.R. Gilson, I.R. Beattie, J.D. Black, D.A. Greenhalgh and S.N. Jenny J. Raman Spectrosc. 9, 361 (1980).

(5) D.A. Greenhalgh, I.R. Dring, R. Devonshire and F. Boyson, to be published.

APPLICATION OF HIGH-RESOLUTION COHERENT RAMAN SPECTROSCOPY TO COMBUSTION MEASUREMENTS

L. A. Rahn and R. E. Palmer

Combustion Research Facility
Sandia National Laboratories
Livermore, CA 94550

Precise, high-resolution Raman spectra of combustion-related molecules have been measured using inverse Raman spectroscopy (IRS). The analysis of these measurements is providing broadening coefficients,[1] validation of collisional narrowing models,[2] and, in conjunction with coherent anti-Stokes Raman spectroscopy (CARS), nonresonant susceptibilities.[3] This information allows significant improvement in our ability to extract accurate temperatures and concentrations from CARS measurements in combustion systems.

The "quasi-cw" inverse Raman spectroscopy (IRS) technique used in this study is similar to that first developed by Owyoung.[4] In the IRS technique, a pulsed pump laser induces a small absorption in a cw probe laser via the stimulated Raman effect. The primary advantages of IRS are that the signal intensity has a linear dependence on the imaginary part of the third-order nonlinear susceptibility, and the spectral resolution is limited only by the linewidths of the lasers used. High resolution is attained in the present experiment by using single-frequency argon-ion probe and tunable cw dye pump lasers. The pump laser is amplified to a peak power of ~1 MW by the frequency-doubled output of an injection-seeded Nd:YAG laser operating at 20 Hz. The resulting ~45 MHz (0.0015 cm^{-1}) resolution is complemented by a Raman shift accuracy of ±0.001 cm^{-1}, determined with a traveling corner cube wavemeter.

Molecular nitrogen has been the focus of much of the work, since it is often studied in CARS thermometry and in relative concentration measurements. As part of the development of a comprehensive predictive capability for the nitrogen Raman spectrum in combustion environments, a complete study of N_2 *Q*-branch self-broadening coefficients for temperatures from 500 K to 1500 K and for J = 2 to J = 30 has been performed. Measurements were made in a furnace at pressures from 20 Torr to 760 Torr and were analyzed using the Galatry line profile to account for the Doppler effect.

We have also compare these linewidths to predictions from a modified exponential gap scaling law that predicts transition rates γ_{ij} between rotational states of nitrogen. The rate from state J_i to state J_j $(i < j)$ is given by:[2]

$$\gamma_{ji} = \alpha p \left(\frac{T_0}{T}\right)^n \left(\frac{1 + 1.5E_i/kT\delta}{1 + 1.5E_i/kT}\right)^2 \exp(-\beta\,\Delta E_{ij}/kT) \qquad (1)$$

Here, p is the pressure, T_0 is a reference temperature, E_i is the rotational term energy for state J_i, ΔE_{ij} is the difference in rotational term energy between states J_i and J_j, and α, β, δ, and n are fitting parameters (the factor 1.5 includes N-N internuclear separation distances, characteristic collision interaction lengths, and other fixed parameters). The rate γ_{ij} from J_j to J_i is determined by microscopic reversibility.

The fitting parameters α, β, and δ have been determined[2] from fits of this model to inverse Raman spectra of the N_2 *Q*-branch at 295 K and 1 atm. The resultant values are $\alpha = 0.0235\pm0.002$ cm^{-1} atm^{-1}, $\beta = 1.67\pm0.15$, and $\delta = 1.26\pm0.06$. Using these values, Eqn. 1 can be used to predict collisional broadening coefficients; these coefficients can then be fitted to the experimental values to determine the best value for n. The results of this approach are shown in Fig. 1, where the values from the IRS experiment (symbols) are compared to the scaling law predictions (solid lines) for $n = 1.346\pm0.06$ and $T_0 = 295$ K. As seen in the figure, agreement between the experimental values and scaling law predictions is excellent. Recent experimental results at 1000 K and 20 atm are also well described by this model.

Presently, the scaling law discussed above predicts only collisional self-broadening coefficients for nitrogen. In actual flames, however, other collision partners are present. Two notable examples are H_2O and CO_2. Preliminary inverse Raman measurements have been made[3] of N_2 *Q*-branch transitions in the post-flame gases of an H_2 - O_2 - N_2 flame, where the major (66%) collision partner is H_2O at 1570 K. Other work in progress includes measurements of CO self-broadening and broadening of N_2 by hydrocarbons.

The support of the U.S. Department of Energy, Office of Basic Energy Sciences, Division of Chemical Sciences, is gratefully acknowledged.

Fig. 1. N_2 collisional self-broadening coefficients (FWHM) as a function of temperature and rotational state. The solid curves are predictions from the modified exponential gap scaling law discussed in the text. The 295 K data is from G. J. Rosasco *et al*, Chem. Phys. Lett. 97, 435 (1983).

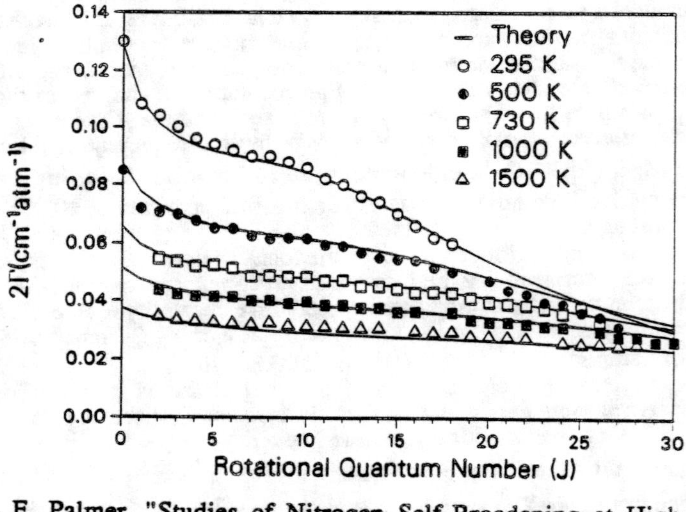

References:

1. L. A. Rahn and R. E. Palmer, "Studies of Nitrogen Self-Broadening at High Temperature," submitted to J. Opt. Soc. B, 1986.
2. M. L. Koszykowski, L. A. Rahn, and R. E. Palmer, "Theoretical and Experimental Studies of High-Resolution Inverse Raman Spectra of N_2 at High Pressure," submitted to J. Phys. Chem., 1986.
3. R. L. Farrow and L. A. Rahn, "Interpreting Coherent Anti-Stokes Raman Spectra Measured with Multimode Nd:YAG Pump Lasers," J. Opt. Soc. B 2, 903 (1985).
4. A. Owyoung, "High Resolution Coherent Raman Spectroscopy of Gases," in *Laser Spectroscopy IV*, ed. by H. Walther and K. W. Rothe (Springer-Verlag, Berlin,1979), pp. 175-187.

HIGH-RESOLUTION CARS FOR COMBUSTION DIAGNOSTICS.

R. L. Farrow and R. P. Lucht

Combustion Research Facility
Sandia National Laboratories
Livermore, CA 94550 USA

The accuracy of analytic measurements obtained with coherent anti-Stokes Raman spectroscopy (CARS) is strongly dependent on the accuracy of spectroscopic parameters used in data analysis. For gas-phase measurements in combustion research, relevant parameters include Raman linewidths, cross-sections, transition frequencies, and nonresonant electronic susceptibilities. While high-resolution inverse Raman spectroscopy (IRS) provides valuable information on Raman linewidths[1,2] and transition frequencies,[3] this technique has limited sensitivity for *in situ* investigations and does not yield nonresonant susceptibilities. High-resolution CARS systems capable of meeting these needs have been used extensively for molecular spectroscopy,[4] but we know of few applications of such systems to combustion studies.

We have constructed a high-resolution CARS system based on single-frequency Nd:YAG and pulse-amplified dye lasers. The experiment has a spectral resolution better than 0.005 cm^{-1}, sufficient to fully resolve transition lineshapes of most species present in atmospheric-pressure combustion environments. We have used the system to measure Raman linewidths of CO and N_2 in a methane/air flame and have found good agreement with the results of other techniques.[5] In this paper, we report measurements of the nonresonant susceptibilities of H_2O, n-C_4H_{10}, and C_3H_8 obtained with the single-frequency Nd:YAG laser and a pulsed dye laser with a linewidth of 0.03 cm^{-1} (FWHM).

Nonresonant electronic susceptibilities(χ_{nr}) were measured with respect to the resonant susceptibility of the N_2 Q-branch. Spectra of gas mixtures containing the species of interest and a smaller, known fraction (5-33 mole %) of N_2 were analyzed by varying the nonresonant susceptibility of the species of interest for best fit to the spectra. This technique is similar in principle to that of Lundeen et al,[6] who used the $Q(0)$ transition of H_2 as a standard by comparing different gas samples.

Spectra of N_2 from binary gas mixtures of 4.8±0.1% N_2 in n-C_4H_{10} and 5.1±0.1% N_2 in C_3H_8 were measured in a gas cell at 296 K. To measure off-resonant contributions from Raman transitions of the buffer gas, we first subtracted all χ_{nr} contributions using polarization techniques (see Fig. 1). The resonant susceptibility of N_2 in the buffer (including J-dependent linewidths and pressure-narrowing effects) was analyzed from this data by using the exponential gap model of Koszykowski et al.[7] The polarization analyzer was then rotated to 50±0.5 degrees to observe χ_{nr} contributions also (see Fig. 2). Varying the total background for best fit gave a value for $\chi_{nr}(C_3H_8)$ of $1.09 \pm 0.12 \times 10^{-16}$ cm^3/erg (STP), after the off-resonant Raman contribution from C_3H_8, $7.9 \pm 2.0 \times 10^{-18}$ cm^3/erg, was subtracted. Similarly, we found $\chi_{nr}(n\text{-}C_4H_{10}) = 1.37 \pm 0.15 \times 10^{-16}$ cm^3/erg, after subtracting an off-resonant Raman contribution of $8.8 \pm 2.2 \times 10^{-18}$ cm^3/erg.

In addition, we measured $\chi_{nr}(H_2O)$ from CARS spectra of N_2 in flame gases containing 67% H_2O and 33% N_2. The gas temperature was determined to be 1570 K from analysis of background-subtracted N_2 Q-branch spectra (we did not detect any off-resonant Raman intensity from H_2O). Since the dye laser linewidth was not narrow enough to resolve the N_2 lineshapes in the flame, we used collisional widths measured with IRS for the same flame by Rahn.[8] Analysis of flame N_2 spectra yielded an STP value for $\chi_{nr}(H_2O)$ of $1.85 \pm 0.18 \times 10^{-17}$ cm^3/erg.

This work supported by the U.S. Department of Energy, Office of Basic Energy Sciences, Chemical Sciences Division.

REFERENCES

1. G. J. Rosasco, W. Lempert, W. S. Hurst, and A. Fein, Chem. Phys. Lett. **97**, 435 (1983).
2. L. A. Rahn, R. L. Farrow, and P. L. Mattern, in "Raman Spectroscopy: Linear and Nonlinear," J. Lascombe and P. V. Huong, Eds., p. 143, Wiley, New York (1982).
3. D. E. Jennings, L. A. Rahn, and A. Owyoung, Astrophys. J. Lett. **291**, L15 (1985).
4. J.-P. Boquillon, "Laser-Induced Processes in Molecules, Vol. 6," K. L. Kompa and S. D. Smith, Eds., p. 43, Springer-Verlag, New York (1978).
5. R. L. Farrow, submitted to the XXIst Symposium (International) on Combustion.
6. T. Lundeen, S.-Y. Hou, J. W. Nibler, J. Chem. Phys. **79**, 6301 (1983).
7. M. L. Koszykowski, L. A. Rahn, and R. E. Palmer, submitted to J. Phys. Chem.
8. Larry Rahn, submitted to the Conference on Lasers and Electro-Optics, 1986.

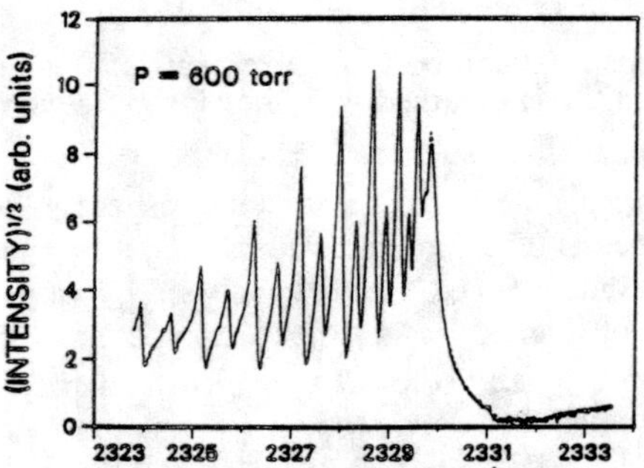

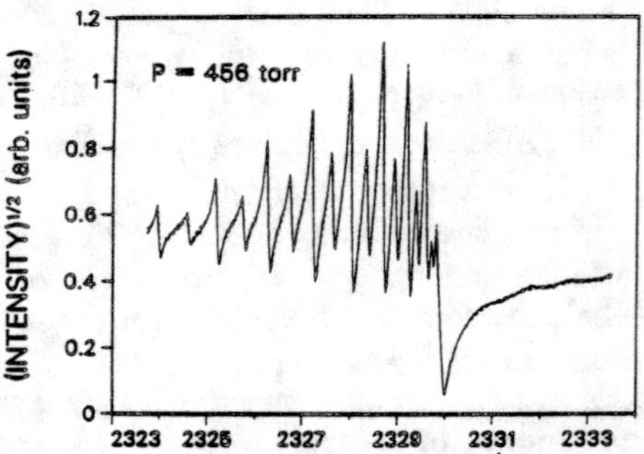

Fig. 1 (left) Spectrum of the N_2 Q-branch (solid line) and fit (dashed line) from a gas mixture of 5.1 ± 0.1 mole % N_2 in C_3H_8 at 296K and a pressure of 600 torr. The resolution was 0.03 cm^{-1}, sufficient to fully resolve the Raman lineshapes. A polarizer angle of 60 deg was used to suppress nonresonant electronic intensities, so that the remaining background is due to off-resonant C_3H_8 Raman transitions.

Fig. 2 (right) Spectrum of the N_2 Q-branch (solid line) and fit (dashed line) at 456 torr from the same gas as Fig. 1. Here, the polarizer angle was 50 deg.

CARS NITROGEN THERMOMETRY AT HIGH PRESSURE FOR I.C. ENGINE APPLICATIONS*

D.R. Williams
C.A. Baker
D.A. Greenhalgh

Harwell Combustion Centre, AERE Harwell, Oxon U.K.

*Work supported by the UKAEA's Underlying Programme and the U.K. Petrol Engine Working Party (Lucas, Ford, Jaguar, B.P. and the U.K. Dept. of Trade and Industry).

We have conducted CARS nitrogen thermometry measurements in an isothermal, isobaric test cell over the pressure range 0-20 bar and the temperature range 300-800K. Preliminary results are reported in this paper. Measurements indicate an accuracy of better than 3% for the full range of temperatures and pressures studied. Analysis of CARS data at high temperatures and pressures requires the use of a sophisticated computer model(1) which takes full account of both finite laser bandwidth effects and motional narrowing(2). Such a model also requires accurate linewidth and rotational relaxation rate data. In our computer model(1) we use a scaling law known as the Differential Energy Gap Law which has been reported previously(3). However, detailed assessment of measurement accuracy has not been previously conducted for the full range of pressures and temperatures of an operating I.C. engine. These preliminary data cover most of the pre-combustion conditions encountered in practical and production I.C. petrol engines.

The analysis method has also been applied to CARS data from a production petrol engine operting under realistic road conditions and excellent agreement between experimental and theoretical CARS spectra have been obtained. Further details concerning the CARS petrol engine experiments have been published elsewhere(4).

In Figure 1 we show a typical CARS spectrum from our test cell and a comparison with our theoretical model. Figure 2 illustrates the accuracy of CARS high pressure nitrogen thermometry using our laser equipment and our computer model.

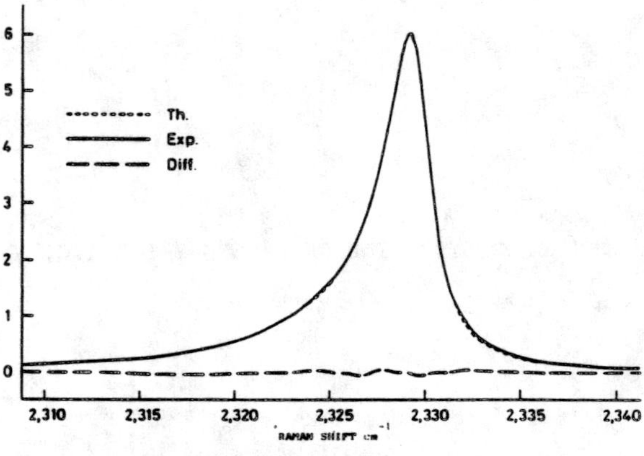

FIGURE 1.

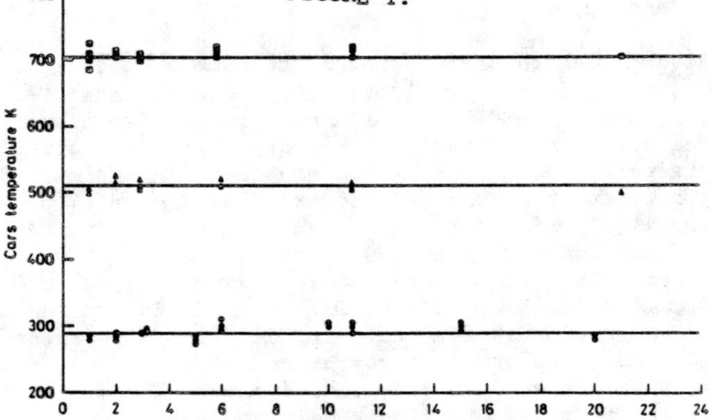

PRELIMINARY CARS MEASUREMENTS
In the high temperature/high pressure cell
(Horizontal lines indicate thermocouple temperatures)

FIGURE 2.

References

(1) CARP-2 is a UKAEA copyright Computer Code for the analysis of CARS spectra; licences are available through Epsilon Research Ltd., Paynes Lane, Rugby, Warwickshire, U.K.

(2) D.A. Greenhalgh and R.J. Hall Opt. Comm. 37, 125 (1986).

(3) D.A. Greenhalgh and F.M. Porter "Cars Applications in Chemical Reactors, Combustion and Heat Transfer" (1985) Proc. 1st Int. Laser Sci. Conf. Dallas, Texas.

(4) D.A. Greenhalgh, D.R. Williams and C.A. Baker "CARS Thermometry in a Firing Production Petrol Engine" (1985) Proc. "Autotech" Conference, I. Mech. E., Birmingham, U.K.

RESONANCE CARS SPECTROSCOPY OF THE OH RADICAL IN A MICROWAVE DISCHARGE

B. Attal-Trétout, D. Houssin and J.P. Taran

Office National d'Etudes et de Recherches Aérospatiales, BP 72, 92322 Châtillon Cedex, FRANCE.

Resonant CARS spectra of the OH radical have been obtained in a low pressure discharge under controlled conditions. A methane-oxygen flame had already been investigated under conditions of double resonance by Hall et al [1]. We have achieved in our experiment a triple resonance condition by using three different laser frequencies $\omega_1, \omega_2, \omega_3$ (fig. 1).

Optimal resonance conditions were selected using the data of Coxon [2]. The frequencies ω_1 and ω_3 have been tuned respectively to the $P_1(0-0)7.5$ and $R_1(1-1)5.5$ lines of the $A^2\Sigma^+-X^2\Pi$ system of OH. The ω_2 frequency was then tuned to scan the fundamental O Raman transition which is located at $\omega_{ba}= 3065.303$ cm^{-1}. A triple resonance is thus achieved for the main resonance CARS line $\left[O_1^{ba}(1-0), P_1^{na}(0-0), P_1^{n'a}(1-0)\right]7.5$ Note that the rules and notations used for the interpretation of resonance CARS spectra have been described in detail previously [3,4]. However some differences occur in the present case:

- the spectral evolution of the CARS profile as a function of ω_1 or ω_3 is different,
- the CARS line assignment is done by following the "N labeling" of Diecke and Crosswhite [5].

The experimental set up is composed of three frequency-doubled dye lasers which produce pulses of spectral width less than 0.05 cm^{-1} with 30-50 mJ energy. Frequency calibrations were carried out first with a commercial Fizeau wavemeter. The absolute calibration was then improved down to 0.01 cm^{-1} by recording excitation spectra of iodine.

The OH signal was searched for by focusing the three laser beams into the afterglow of a microwave discharge. The latter was fed with a flowing mixture of Helium and water vapor at 9 mb total pressure. The Anti-Stokes emission was filtered by a monochromator, detected by a photomultiplier and integrated by an electronic device connected to a PDP 11-23. The OH spectra of figure 2 have been recorded with $\omega_1=\omega_{na}= 32123.492$ cm^{-1}. As the ω_3 frequency is downshifted from $\omega_{n'b} = 32011.598$ cm^{-1}, the electronic CARS line separates from the main Raman line. This phenomenon was already observed in two laser CARS spectroscopy [4]. Spectral positions of the different contributions are marked on the horizontal axis.

The single rotational level which is involved has been unambiguously interpreted. The resonant CARS profile appears strongly above a negligible background caused by the electronic noise. The detectivity ranges between 10^{11} and 10^{12} cm^{-3} depending on the saturation threshold of the one-photon transitions.

[1] - J.F. Verdieck, R.J. Hall and A.C. Eckbreth, AIAA Paper n° 83-1477, (1983).
[2] - J.A. Coxon, Can. J. Phys., 58, 933 (1980).
[3] - S.A.J. Druet, B. Attal, T.K. Gustafson and J.P.E. Taran, Phys. Rev. A18, 1529 (1978).
[4] - B. Attal, D. Debarre, K. Müller-Dethlefs and J.P.E. Taran, Rev. Phys. Appl., 18, 39 (1983).
[5] - G.H. Diecke and A.M. Crosswhite, J.Q.S.R.T., 2, 99 (1962).

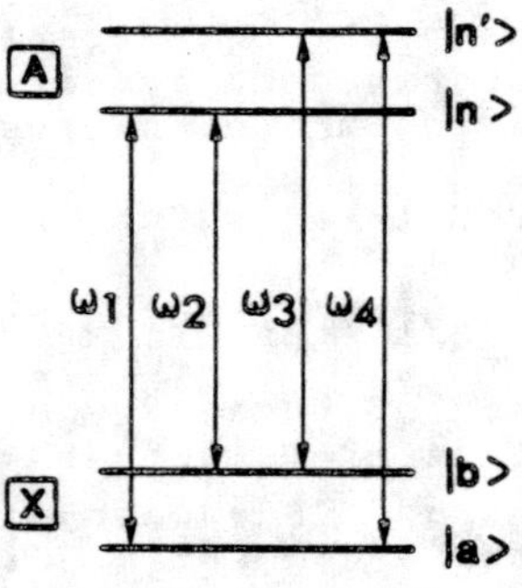

Fig. 1
Energy level diagram of the fully resonant CARS process

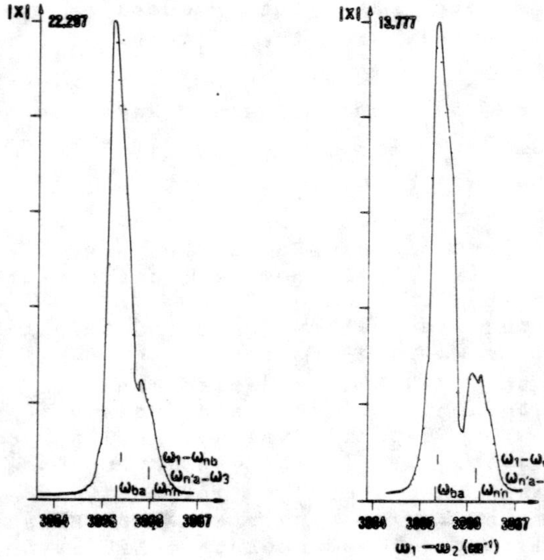

Fig. 2
Resonance CARS spectra of OH in a microwave discharge. The major contributions to the profile are identified following the notation of [4]. Spectral positions are calculated from the data of Coxon [2].

CARS TIME RESOLVED MEASUREMENT OF H_2 ROVIBRATIONAL POPULATION IN A H^- ION SOURCE

M. Lefèbvre, M. Péalat, J.P. Taran
F. Hillion* and M. Bacal*

Office National d'Etudes et de Recherches Aérospatiales
BP 72, 92322 Châtillon Cedex, FRANCE.

*Laboratoire de Physique des Milieux Ionisés
G.R. N° 29 du C.N.R.S.
91128 Palaiseau Cedex, FRANCE

The presence of high fractions of negative hydrogen ions in magnetic multicusp discharges has been attributed to highly vibrationally excited hydrogen molecules which dissociate by electron attachment.

During a previous CARS experiment, vibrational populations for $v \leq 3$ were measured in a steady state plasma of H_2 [1] to verify this hypothesis. These results were compared with numerical models [2]. The calculations predict that transient populations in vibrational levels $v \geq 3$ can exceed steady state populations, giving us an opportunity to detect these levels in a pulsed discharge. Furthermore, after discharge switch off, the experimental conditions are expected to be more favorable because the vibrational populations decay slowly while the rotational temperature is rapidly lowered.

The study of a square-pulsed discharge has therefore been undertaken. The temporal evolution of the rotational temperature and vibrational populations ($v \leq 3$) was measured both during and after the discharge. The plasma duration is 1 ms.

The pressure and the current are identical with those applied under steady state conditions during the previous work (p = 40 m torr, I = 10 A, V = 90 V). The rotational temperature is measured from the relative intensities of Q(J = 3, v = 0) and Q(J = 1, v = 0). A plateau is reached within 500 μs, the temperature being 600 K ± 20 K.

The temperature decreases quickly after the discharge has been turned off. The time scale of the rotational relaxation is 100 µs, two orders of magnitude shorter than the relaxation time of vibrational populations. The pulse duration is too short to obtain a plateau for vibrational temperature. The highest value for the latter, which is measured from the ratio of Q(J = 1, v = 1) and Q(J = 1, v = 0), is attained at the end of the pulse; it remains 200 K below the steady state value. After switch-off, v = 1 relaxes with a time constant of 4 ms. This indicates that the vibrational population decay is controled by diffusion to the walls and that practically each wall collision is efficient.

It was also observed that the v = 2 population depends on how clean the wall is. Metal from the filament is deposited on the copper wall, reducing the v = 2 population by 40% after several hours of operation. These findings confirm the predictions of a the numerical simulation [2].

Finally, some preliminary results point to a possible differential behaviour between ortho and para H_2 with regard to rotational pumping rates and translational temperatures.

[1] - M. Péalat, J.P. Taran, M. Bacal and F. Hillion
J. Chem. Phys., 82, 4943 (1985).

[2] - C. Gorse, J. Bretagne, M. Bacal and M. Capitelli
Chem. Phys., 93, 1 (1985).

HIGH RESOLUTION CARS SPECTROMETER WITH INJECTION LOCKED PULSED DYE LASERS : APPLICATION TO THE Q BRANCH OF OXYGEN.

J. P. Boquillon, Y. Ouazzany and R. Chaux

Laboratoire de Spectronomie Moléculaire et Instrumentation Laser, U. A. CNRS n° 777, 6 Bd Gabriel 21100 DIJON - FRANCE

Non linear experiments in the field of high resolution spectroscopy require tunable lasers of high peak power and very narrow linewidth. They can be obtained by combining the power of pulsed lasers and the spectral qualities of c. w. lasers, with the help of the injection-locking process.

The principle is the following : if single mode radiation is injected into a pulsed cavity and if the optical length of this cavity is adjusted to the injected frequency (longitudinal mode matching), one can force the pulsed laser to deliver its power on a single longitudinal mode at the same frequency as the injected signal. Generally the injected photons come from a c. w. single mode tunable dye laser of high spectral quality and stability which allows to obtain injection locking over wavelength ranges of several hundreds of nanometers.

In our CARS experiment, we use two similar flashlamp pumped dye lasers, each one with a specially designed ring cavity in which we inject tunable laser beams. One mirror of each ring cavity is mounted on a first response Piezoelectric Translator and with the help of an electronic servoloop, we lock the cavity length on the injected frequency.

The main performances of these lasers are : peak power : 5 to 10kW in 0.5 μs pulses. Repetition rate : 10 Hz. Tunability from 572 to 700 nm. Linewidth : < 6 MHz (0.0002 cm^{-1}).

With this new CARS spectrometer, we recorded the spectrum of the Q branch of Oxygen near 1554 cm^{-1} at various pressures between 1000 and 10mbar, for J values from 1 to 17. In order to study collisional line broadening in terms of pressure and J number, we developed a method of calculation on a computer allowing us to extract this Lorentzian width from the line shape and to eliminate the influence of CARS line dissymmetry. Collisional linewidth is obtained by deconvolution of the Lorentzian width and of the Doppler width in the experimental lines. For that, a theoretical profile of the CARS line (Voigt profile is calculated from experimental parameters using Humlicek approximation to estimate the complex probability function. The experimental spectrum is fitted to the calculated profile thanks to a least square method leading to a linearization of the problem through the calculation of partial derivative with respect to selected parameters (Lorentzian linewidth, central frequency of the line, rough intensity factor...).

Examples of line broadening versus pressure for different J value are displayed and pressure broadening coefficients for J values between J = 1 and 17 are compared to experimental and theoretical results obtained previously from infrared spectra by other authors.

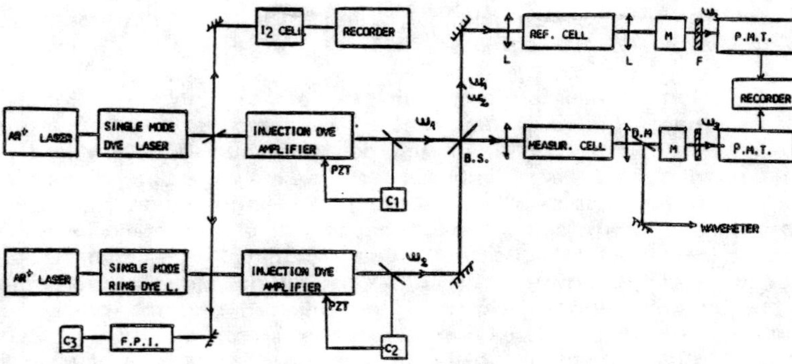

SCHEMATIC ARRANGEMENT OF THE CARS SPECTROMETER WITH FLASHLAMP PUMPED DYE AMPLIFIERS

GENERATION OF SINGLE-PULSE ROTATIONAL CARS SPECTRA USING A MULTIPLE-THREE COLOR SCHEME.

Marcus Aldén, Per-Erik Bengtsson and Hans Edner

Lund Institute of Technology, Department of Physics, P.O. Box 118, S-221 00 Lund, Sweden

Coherent Anti-Stokes Raman Spectroscopy, CARS, has proven to be a very valuable diagnostic tool for measurements of temperatures and species concentration in flames and other gasflows. In this context almost all measurements have been performed using vibrational CARS. Recently it has been shown how two broadband dye-lasers can be used for multiple species detection [1].

We have used a similar approach to generate single pulse rotational CARS spectra [2]. The second harmonic of a Nd:YAG laser was pumping a dye-laser with e.g. Rhodamine 6G as dye. The residual IR beam was frequency doubled in a second crystal and arranged together with two beams from the broadband dye-laser in a folded BOXCARS configuration. The rotational CARS signal is thus produced by several pairs of photons, one from each dye-laser beam, as indicated in Fig. 1. The rotational energy is then coupled to a photon from the green beam producing a CARS photon shifted from 532 nm with the rotational Raman shift. The rotational CARS beam was spatially isolated with an aperture and detected with an optical multichannel analyzer at the exit plane of a spectrograph. The broadband rotational CARS spectrum was easily detected and in Fig. 2 is shown a single shot spectrum from O_2 at room temperature.

An advantage with this technique is that it is also possible to simultaneously generate a rotational and vibrational CARS spectra from e.g. N_2. This could be of interest for measurements in turbulent flows where large temperature fluctuations can occur. Then rotational CARS is used for measurements in cool regions whereas vibrational CARS is used for hot regions, since the population difference factor, involved in the CARS process, decreases the rotational CARS signal intensity more severely than the vibrational CARS signal. Compared with the conventional method of generating rotational CARS spectra, i.e. with a broadband dye laser in the spectral vicinity of the pump beam, there are also other advantages. Rhodamines are much more efficient and stable than coumarin dyes when using a YAG based laser system. The use of several pairs of photons from the dye laser to produce one rotational CARS frequency gives also an averaging effect on the shot-to-shot fluctuations in the dye-laser frequency distribution.

REFERENCES

1. A.C. Eckbreth and T.J. Anderson, Appl. Opt. 24, 2731 (1985).
2. M. Aldén, P.-E. Bengtsson and H. Edner. To be published.

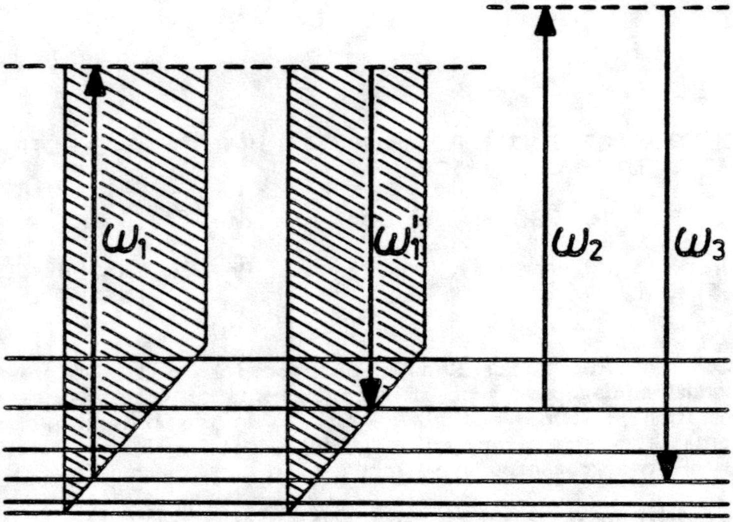

Fig. 1. Rotational CARS generation at frequency ω_3 with a multiple-three color scheme. ω_1, ω_1'= Two frequencies out of many possible pairs from a broadband red dye-laser. ω_2= 532 nm from the Nd:YAG laser.

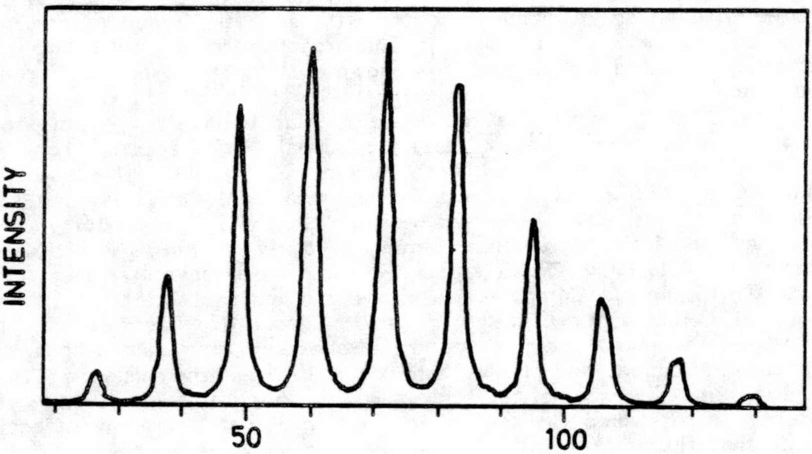

Fig. 2. Single shot rotational CARS spectrum from O_2 at room temperature.

LOW FREQUENCY CARS SPECTROSCOPY.

N.E. TRIGGS, B.J. BOZLEE, G.A. PUBANZ, J.J. YANG, and J.W. NIBLER

Oregon State University, Corvallis, OR 97331

The development of coherent Raman spectroscopy for 0-500 cm^{-1} shifts is desirable for the study of low frequency vibrational and pure rotational transitions. For diagnostic applications, pure rotational measurements are appealing because of the high inherent Raman cross sections (~10 x vibrational cross sections) and the restricted frequency scan range necessary to monitor most molecules. For structural studies, high resolution rotational spectra give directly ground state B_0 constants whereas Raman vibrational Q branch spectra give only ΔB values. The latter are also frequently congested and unresolved due to pressure broadening, so that information on population distributions and corresponding local temperatures is harder to extract than from the widely spaced pure rotational lines. Finally, many molecular modes of special interest such as torsions and vibrations of heavy atom molecules occur in the low frequency shift region.

The major hurdle in achieving low frequency Raman spectra is the separation of the signal (~10^6 photons) from the very intense pump and Stokes beams (~10^{17} photons). We have utilized three dimensional phase matching (folded BOXCARS) to spatially separate the signal from the Stokes and two 532 nm pump beams. The Stokes beam is generated by a Coumarin dye laser pumped by the third harmonic of the primary Nd:YAG source. Residual background stray light and broadband superradiance from the tunable Stokes laser have been greatly reduced by additional spatial filters. For rotational spectra, a final, nearly complete, polarization rejection of pump light from the CARS signal is achieved by using orthogonal pump-Stokes polarizations.

The resultant rotational spectra of gases at a few Torr are clearly resolved with a flat background to ~5 cm^{-1} of zero shift and the sensitivity is such that rotational spectra of molecules in free jet expansions are readily obtained. Such expansions are advantageous for signal enhancement due to rotational cooling and they also favor the production of increased concentrations of interesting weakly-bound complexes. The use of the BOXCARS geometry in jets gives minimal nonresonant contributions and allows for point modeling in jet relaxation studies.

Fig. 1 shows a representative experimental CARS spectrum of neat nitrogen in a free jet expansion and also provides a comparison with a theoretical spectrum calculated assuming a Boltzmann distribution for a rotational temperature of 30 K. Low pump energies (~2 mJ) were used in recording this spectrum to eliminate saturation distortion of the peak intensities which was quite apparent for energies in excess of 3 mJ. Movement of the probe beams along the flow axis yielded the variation of rotational temperature in the early stages of the expansion process as shown in Fig. 2. These data are in good accord with a calculated cooling curve (solid trace of Fig. 2) based on an isentropic expansion model with rotational collision number of 10 at 300 K. Near equilibrium is achieved between translational and rotational degrees of freedom so that the translational temperature variation along the jet axis is also depicted by the solid trace. No evidence of aggregation was seen (or expected) for expansions of neat N_2 or CO_2 but studies are underway of jets under conditions favoring the production of relatively high concentrations of dimeric and trimeric species.

Figure I Neat Nitrogen in Jet P_0=10atm T_0= 300K

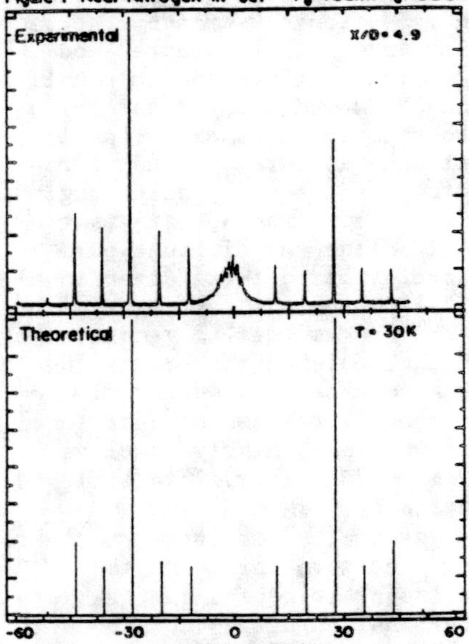

Figure 2 Rotational Temperature as a function of X/D

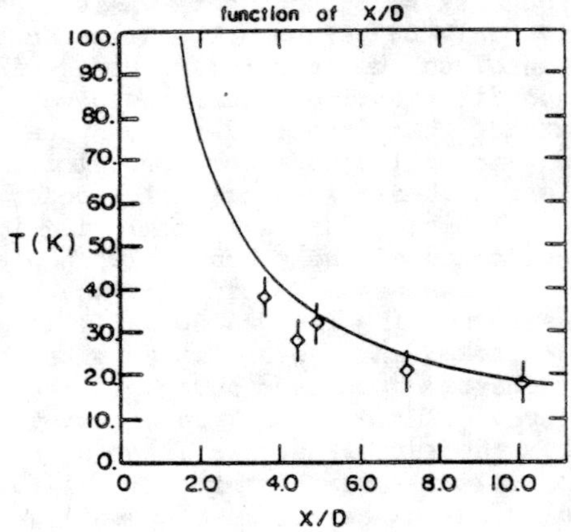

FREE JET SPECTROSCOPY BY COHERENT RAMAN METHODS.

G. A. Pubanz, M. Maroncelli and J. W. Nibler

Oregon State University, Corvallis, OR 97331

Coherent Raman techniques can serve as useful probes of the structural and dynamic properties of monomers and small aggregates formed under the cold, nonequilibrium conditions of free jet expansions. There has been considerable recent interest in the properties of small molecular clusters. As the first products in the nucleation process, dimers and higher polymers provide insight into the fundamental processes of aggregation and, with mixed species, solvation. Structural information such as bond lengths, angles and force constants in these gas phase species is extremely useful in refining models of the attractive part of the potential energy surface. In addition, linewidths can provide information on lifetimes and reveal details of vibrational predissociation pathways.

We have used CARS and photoacoustic Raman spectroscopy (PARS) to probe small hydrogen bonded aggregates. In the case of HCN, PARS and FTIR spectra of dimer and trimer species were obtained for equilibrium samples at 600 Torr. Improved resolution and species discrimination were achieved with jet expansions probed by the CARS technique. The experimental details are outlined elsewhere.[1] The series of spectra in Fig. 1 give evidence for the growth of HCN clusters in cold free jets. Under relatively mild expansion conditions the dimer is the most evident complex. However, as higher driving pressures of argon are

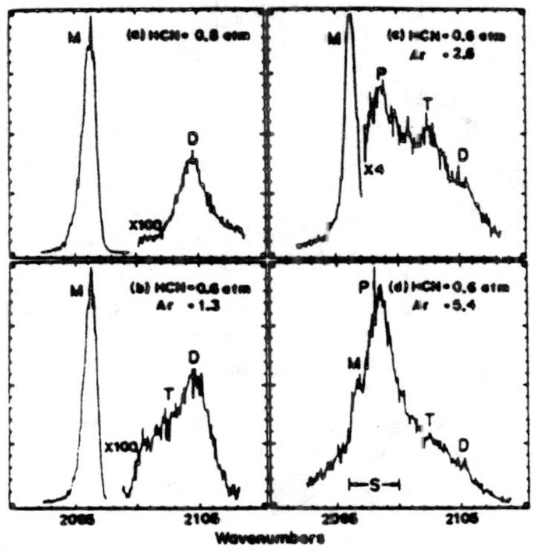

Fig. 1 CARS spectra of the ν_1 region in supersonic expansions of HCN showing bands due to complexes. Peaks are labeled as M = monomer, D = dimer, T = trimer and P = higher polymers. In spectrum (d) the region labeled S denotes the position and width of ν_1 in crystalline HCN at 78 K.

added, the trimer and higher polymers begin to dominate the spectra (Fig. 1 b-d). Further work[1] indicates that both the dimer and the trimer are linear in the gas phase and normal coordinate analysis shows that one hydrogen bond in the trimer is significantly stronger than the other.

The same CARS apparatus has been utilzed for the study of van der Waals complexes such as the CO_2 dimer. The structure of this interesting species has been investigated using a variety of techniques since the IR spectrum was first observed in 1971. Based upon PQR IR structure it was concluded that the CO_2 dimer had a polar, T-shaped conformation (C_{2v}) rather than a non-polar offset parallel geometry (C_{2h}). Subsequently, additional gas phase and matrix infrared studies, electric deflection and infrared vibrational predissociation experiments have lead to conflicting conclusions about the lowest energy form for this simple complex. Empirical and *ab initio* theoretical calculations of the structure also have produced contradictory results.

We have used CARS to study the CO_2 clusters formed in dilute He/CO_2 expansions from a cooled nozzle.[2] Figure 2 compares the CARS spectrum of the ν_1 symmetric stretching band with an infrared spectrum obtained by Kopec.[3] The key observation is that there is no evidence in the Raman spectrum of the dimer Q branch seen in the infrared at 1284.7 cm^{-1} even though, for a T- shaped structure, this mode should have high Raman intensity. There is also no peak in the infrared correlating with the dimer Q branch observed at 1281.3 cm^{-1} in the Raman. Hence the rule of mutual exclusion appears to apply to the CO_2 dimer, a result favoring the offset parallel structure with C_{2h} symmetry. The data do not, however, exclude the possibility of up to 25% of a second, higher energy, polar form. Such a mixture is consistent with some of the theoretical calculations and could account for the electric deflection results.

References

1. M. Maroncelli, G. A. Hopkins, J. W. Nibler and T. R. Dyke, J. Chem. Phys. **83**, 2129 (1985), Chem. Phys. Lett. **114**, 97 (1985).
2. G. A. Pubanz, M. Maroncelli and J. W. Nibler, Chem. Phys. Lett. **120**, 313 (1985).
3. R. L. Kopec, Ph.D. Thesis, Indiana University, (1981).

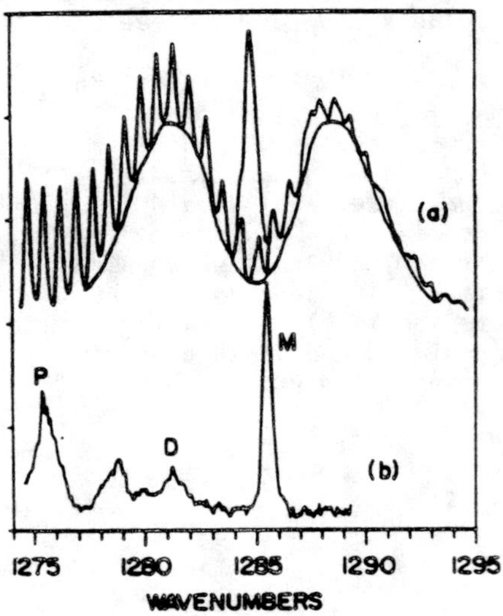

Fig. 2 Comparison of CARS spectrum of CO_2 (b) with a reproduction of the infrared spectrum (a) reported by Kopec.[3] The dimer P-Q-R contours are apparent under the R branch fine structure of 0.2% $CO^{18}O$ monomer. As before M = monomer, D = dimer and P = polymer.

Nonlinear Raman spectroscopy with high-resolution cw-lasers

G.J. Rosasco and W.S. Hurst
Temperature and Pressure Division
U.S. National Bureau of Standards
Gaithersburg, MD 20899

Owyoung and coworkers [1,2] pioneered in the use of cw-lasers for the measurement of Raman gain/loss and OHD-RIKE spectra in liquids and gases. They demonstrated that the primary benefit of these forms of nonlinear Raman spectroscopy was in their ability to provide extremely high resolution spectra (limited only by the laser linewidths). We have implemented cw-Raman gain/loss spectroscopy using high frequency (2-6 MHz) acousto-optic amplitude modulation [3]. The use of high frequency modulation results in approximately a 10-fold sensitivity improvement relative to earlier work because of the reduced probe laser noise at these high frequencies. Recently [4,5], we introduced a phase modulated (modulation frequencies 2-10 MHz) version of stimulated Raman spectroscopy which allows detection <u>at the shot-noise limit</u>. As in OHD-RIKES, this technique yields information on both the real and imaginary parts for the susceptibility [5,6] from signals linear in this susceptibility. We have utilized these two different approaches to cw-stimulated Raman spectroscopy in the study of the pressure dependence of spectral lineshapes, linewidths, and lineshifts in simple molecular systems [7,8,9]. In this paper, we will present our work on the pressure dependence of the Q-branch spectrum of D_2.

Two specific aspects of the pressure dependence are of interest. The first is to accurately determine the proper spectral lineshape function which accounts for the evolution of the spectrum from very low to moderately high densities. In the low pressure limit the lineshape function must be a Gaussian, which is characteristic of Doppler broadening. At higher densities the lineshape function should approach a Lorentzian, which is characteristic of impact-limit, binary collisions in the diffusion regime. The latter regime results from the Dicke narrowing [10] of the Doppler broadened spectrum. We shall discuss the limits of applicability of two well-known models: the Galatry [11] or soft-collision model and the hard-collision [12] model. These models have been tested both as functions of density (0.2-20 Amagat) and collision partner (H_2, He, D_2, CH_4, and Ar). Although neither model completely describes the spectrum [13], the Galatry model gives the most accurate prediction of the spectrum for the light and equal mass collision partners. Neither model accurately predicts (using the mass self-diffusion coefficient) the degree of (Dicke) narrowing for the heavy collision partners. The second area of interest in our studies is the collisional broadening of these lines. In particular, we are interested in the relative importance of pure vibrational dephasing (rotationally elastic) collisions versus collisions in which the J-state of the molecule is changed, i.e.

collisions which involve rotational energy transfer. In order to extract this information, the J-dependence of the line width as a function of collision partner has been studied. These results are analyzed in the context of the many first principle calculations for the hydrogenic systems [14,15]. We also have investigated the use of rate law models [7-9,16-19] to explain these data.

The results presented for D_2 clearly illustrate the importance of narrow bandwidth, cw-laser sources, in that the Q-lines show Dicke-narrowed full widths of only 180 MHz. Further, the use of cw-laser sources, which allows shot-noise-limited sensitivities, also has the advantage of allowing accurate normalization of the signal to the pump and probe power. Therefore, the peak-signal to peak-noise ratios of these D_2 spectra typically approach 300:1. The results of these high resolution, high signal-to-noise ratio measurements are leading to a new understanding of spectral lineshape functions and a more complete picture of the microscopic mechanisms of line broadening and narrowing phenomena.

1. A. Owyoung and E.D. Jones, Opt. Lett. 1, 152 (1977).
2. A. Owyoung, IEEE J. Quantum Electron. QE-14, 192 (1978).
3. G.J. Rosasco, W. Lempert, W.S. Hurst, and A. Fein, in: Spectral Line Shapes, Vol. 2, ed. K. Burnett (de Gruyter, Berlin, 1983) 635.
4. G.J. Rosasco, W.S. Hurst and W. Lempert, Opt. Lett. 9, 19, (1984).
5. G.J. Rosasco and W.S. Hurst, J. Opt. Soc. Am. B 2, 1485 (1985).
6. G.J. Rosasco and W.S. Hurst, Phys. Rev. A 32, 281 (1985).
7. G.J. Rosasco, W. Lempert, W.S. Hurst, and A. Fein, Chem. Phys. Lett., 97, 435 (1983).
8. W. Lempert, G.J. Rosasco and W.S. Hurst, J. Chem. Phys., 4241 (1984).
9. W.S. Hurst, G.J. Rosasco and W. Lempert, SPIE Vol. 482 Applications of Laser Chemistry and Diagnostics (1984), (Society of Photo-Optical Instrumentation Engineers, Bellingham, WA, 1984), pp. 23-30.
10. R.H. Dicke, Phys. Rev. 89, 472 (1953).
11. L. Galatry, Phys. Rev. 122, 1218 (1961).
12. S.G. Rautian and I.I. Sobel'man, Usp. Fiz. Nauk 90, 209 (1966) [Sov. Phys.-Usp. 9, 701 (1967)].
13. G.J. Rosasco, W.S. Hurst, K.C. Smyth, L.B. Petway, and A. D. May, Proc. 8th Int. Conf. on Spectral Line Shapes, 1986(to be published).
14. J. Bonamy, L. Bonamy and D. Robert, J. Chem. Phys. 67,441 (1977).
D. Robert, J. Bonamy, J.P. Sala, G. Levi and F. Marsault-Herail, Chem. Phys. 99, 303 (1985).
D. Robert, J. Bonamy, F. Marsault-Herail, G. Levi and J.P. Marsault, Chem. Phys. Lett. 74, 467 (1980).
15. G. Zarur and H. Rabitz, J. Chem. Phys. 60, 2057 (1974).
16. T.A. Brunner and D. Pritchard, Adv. Chem. Phys. 50, 589 (1982).
17. M.L. Koszykowski, L.A. Rahn, and R. E. Palmer, J. Phys. Chem. (to be published).
18. D.W. Chandler and R.L. Farrow, J. Chem. Phys. (to be published).
19. J.P. Sala, J. Bonamy, D. Robert, B. Lavorel, G. Millot, and H. Berger, Chem. Phys. (to be published).

VIBRATIONAL SPACINGS IN MOLECULAR HYDROGEN.

D. Kirk Veirs and Gerd M. Rosenblatt

Materials and Molecular Research Division, Lawrence Berkeley Laboratory,
University of California, Berkeley, California 94720

Spontaneous Raman spectroscopy is used to determine line positions of the six isotopomers of molecular hydrogen: H_2, HD, HT, D_2, and T_2. State population number densities as low as 1.3×10^8 cm^{-3} are detected with the present experimental apparatus. This sensitivity makes possible measurements of the first overtone Q-branch line positions for H_2 and D_2. The observed line positions are given in Table I.

TABLE I. OBSERVED Q-BRANCH LINE POSITIONS (cm^{-1})[a]

J	H_2 Q_1	H_2 Q_2	HD	HT	D_2 Q_1	D_2 Q_2	DT	T_2
0	4161.200	8087.030	3632.2[b]	3434.9[b]	2993.6[b]	5868.1[b]		
1	4155.281	8075.283	3628.4	3631.6	2991.5	5863.9		
2	4143.493	8051.964	3620.6	3425.1	2987.2	5855.6	2738.4[b]	2461.0[b]
3	4125.903	8017.168	3609.2	3415.6	2981.0	5843.0	2733.7	2457.5
4	4102.592			3402.8	2972.5	5826.3	2727.2	2453.0
5	4073.698		3574.9		2962.2		2719.2	2447.3
6	4039.451				2949.7		2709.8	2440.5
7					2935.2		2698.5	2432.5
8					2918.8		2686.3	2423.5
9					2900.6			2413.2
10								2402.2

[a] estimated uncertainty ± 0.030 cm^{-1}
[b] estimated uncertainty ± 0.1 cm^{-1}

These observations, which include all six isotopomers and a significant number of new transitions, allow one to look for significant trends when comparing experiment with theoretical calculations. Adiabatic ab initio energy levels for molecular hydrogen are constructed from the Born-Oppenheimer potential with adiabatic, relativistic, and radiative corrections. Theoretical ab initio energy levels for all six molecular isotopmers of hydrogen are given in the recent work of Hunt, Poll, Wolniewicz(1). The differences, averaged over the J-levels observed, between the measured line positions and those calculated from the theoretical energy levels and our experimental measurements are presented in the left half of Table II. The differences: 1) are larger than the experimental uncertainty; 2) show an inverse dependence upon the reduced mass and 3) increase approximately linearly with vibrational quantum number. The adiabatic ab initio energy levels are obviously unable to reproduce the experimental measurements.

Nonadiabatic corrections to the energy arise because the electrons do not respond instantaneously to the positions of the moving nuclei, and therefore depend upon the reduced mass. Wolniewicz[2] has reported nonadiabatic correction for H_2, HD, and D_2. Schwartz and LeRoy have extrapolated Wolniewicz's values to the tritium containing molecules.[3] The average of the differences between line positions corrected for nonadiabaticity and these measurements are presented in the right half of Table II. For the low-lying vibrational energy levels (v = 1,2) the nonadiabatic <u>ab initio</u> calculations are in agreement with the present experimental measurements for all isotopomers of molecular hydrogen.

TABLE II. J-AVERAGED DIFFERENCES IN Q-BRANCH LINE POSITIONS BETWEEN EXPERIMENT AND CALCULATED FROM <u>AB INITIO</u> ENERGY LEVELS (cm^{-1})

	ADIABATIC		NONADIABATIC	
	Q_1	Q_2	Q_1	Q_2
H_2	0.854 ± 0.020	1.677 ± 0.022	0.005 ± 0.027	-0.024 ± 0.023
HD	0.63 ± 0.06	----------	0.03 ± 0.06	----------
HT	0.56 ± 0.08	----------	0.00 ± 0.10	----------
D_2	0.37 ± 0.05	0.71 ± 0.07	0.09 ± 0.09	0.06 ± 0.09
DT	0.32 ± 0.10	----------	0.04 ± 0.08	----------
T_2	0.16 ± 0.07	----------	-0.01 ± 0.08	----------

a The standard deviations of the differences are tabulated, not the experimental uncertainty.

References

1) J.L. Hunt, J.D. Poll, and L. Wolniewicz, Can. J. Phys. <u>6</u>, 1719 (1984).
2) L. Wolniewicz, J. Chem. Phys. <u>78</u>, 6173 (1983).
3) C. Schwartz and R.J. LeRoy, private communication.

This work was supported by the Director, Office of Energy Research, U.S. Department of Energy, at Lawrence Berkeley Laboratory under contract number DE-AC03-76SF00098 and at Los Alamos National Laboratory under contract number W-7405-ENG-36.

VIBRATIONAL RAMAN CROSS SECTIONS OF THE OVERTONE TRANSITIONS IN H_2 AND N_2

Winifred M. Huo[1], G. C. Herring[2], and William K. Bischel[2]

[1]Radiation Laboratory, University of Notre Dame
Notre Dame, IN 46556 (mailing address: NASA Ames Research Center, Mail Stop 230-3, Moffett Field, CA 94035)

[2]Chemical Physics Laboratory, SRI International, Menlo Park, CA 94025

Most vibrational Raman studies in diatomic gases have been limited to the fundamental transitions because the overtone transitions are orders of magnitude weaker. However, overtone transitions are related to higher order derivitives of the polarizability and therefore overtones can provide more stringent tests of the quality of theoretical calculations. In this paper, we present new (partially nonadiabatic) *ab initio* cross section calculations for H_2 and experimental results for the first vibrational overtones of H_2 and N_2.

The experimental results were obtained with a standard spontaneous Raman spectrometer that consisted of a cw ion-laser, a double monochrometer, and photon counting electronics. Experimental overtone cross sections were determined by measuring the ratio of the overtone signal to the fundamental signal and using this ratio with previous measurements of the fundamental cross section. In Table 1, our present measurements are compared to theory[1-4] and previous experiments.[5-7]

Past theoretical calculations of vibrational Raman cross sections have used the Born-Oppenheimer approximation (adiabatic). Our new calculations for H_2 are based on a sum-over-state method (nonadiabatic). This calculation explicitely sums over the ro-vibrational levels of the $B^1\Sigma^+_u$ and $C^1\Pi_u$ states (contributing 90% of α_{01}) and then accounts for the remainder of the electronic states (contributing 10% of α_{01}) using the Born-Oppenheimer approximation.

In summary, we present experimental and the first theoretical results for the vibrational overtone cross sections of H_2 and N_2. We find that experiments are in substantially better agreement with nonadiabatic models than with adiabatic models for both the fundamental and first overtone transitions.

G.C.H. and W.K.B. acknowledge support from the Defense Advanced Research Projects Agency under Contract N00014-8884-C-0256 through the Office of Naval Research. W.M.H. gratefully acknowledges the use of unpublished data from S. Langhoff in the calculation of the N_2 overtone polarizability, and acknowledges support from NASA AMES under Cooperative Agreement NCC2-147.

Table I. Vibrational Raman cross sections $\frac{\partial\sigma}{\partial\Omega}$ for the Q-branch transitions in H_2 and N_2 (λ = 488 nm)

	Fundamental ($v=0\to1$) $\times10^{-30}$ cm^2sr^{-1}		1st Overtone ($v=0\to2$) $\times10^{-32}$ cm^2sr^{-1}		2nd Overtone ($v=0\to3$) $\times10^{-34}$ cm^2sr^{-1}	
	H_2	N_2	H_2	N_2	H_2	N_2
Experiment						
Present work			0.51±0.15	0.13±0.05	<3	<1.5
Schwartz et al., Ref. 5			0.56			
Schrötter and Klöckner, Ref. 6	1.3	0.55±0.01				
Bischel and Black, Ref. 7	1.4±0.14	0.57±0.06				
Theory, nonadiabatic						
Present work			0.40		0.32	
Huo (unpublished)	1.63					
Theory, adiabatic						
Rychlewski, Ref. 1	1.57					
Huo and Jaffe, Ref. 2	1.62					
Ford and Browne, Ref. 3	1.53					
Langhoff et al., Ref. 4		0.53		0.064		

References

1. J. Rychlewski, J. Chem. Physics 78, 7252 (1983).
2. W. M. Huo and R. L. Jaffe, Phys. Rev. Lett. 47, 30 (1981).
3. A. L. Ford and J. C. Browne, Phys. Rev. A7, 418 (1973).
4. S. R. Langhoff et al., J. Chem. Phys. 78, 5287 (1983).
5. C. Schwartz et al., Paper 208, 188th ACS National Meeting, Philadelphia, PA (1984).
6. H. W. Schrötter and H. W. Klöckner, in Raman Spectroscopy of Gases and Liquids, edited by A. Weber (Springer, Berlin, 1979).
7. W. K. Bischel and G. Black, in AIP Conference Proceedings, No. 100, Subseries on Optical Science and Engineering, No. 3, Excimer Lasers-1983, edited by C. K. Rhodes et al., (AIP, New York, 1983).

A NEW FEATURE IN THE RAMAN Q BRANCH OF D_2.

A.D.MAY*, G.J.ROSASCO!, W.S.HURST!, K.C.SMYTH! and L.B.PETWAY#

*University of Toronto, Toronto, Can. M5S 1A7. !U.S. National Bureau of Standards, Gaithersburg, MD 20899. #Howard University, Washington D.C. 20059.

At low gas densities, collisions can both narrow spectral lines by modifying the translational motion and broaden spectral lines by changing the internal degrees of freedom. To disentangle these effects one needs a correct form for the self-part of the Van Hove pair correlation function for the translational motion. Traditionally the Galatry or the Nelkin-Ghatak model have been used although theoretical calculations exist for specific intermolecular potentials[1]. We have observed a new feature in the Raman Q branch in D_2 which indicates that these common models or theoretical calculations are incomplete for molecules with internal degrees of freedom. The data were obtained with a cw Raman gain/loss spectrometer employing a multiple pass cell and phase sensitive detection[2]. The high resolution (10 MHz) and high signal-to-noise ratio (500) are important attributes of the spectrometer that have allowed us to detect this new feature.

As an illustration of the effect, Figure 1a shows 10 times the residual between the data points and the computed "best fit" Lorentzian for the Q(1) line of D_2 at 6.1 Amagat units of density (~7 atm, 298K). The fit is not within the noise. However, by slightly adjusting the parameters of the best fit <u>single</u> Lorentzian, a new residual is obtained that is itself a <u>second</u> broader Lorentzian. To illustrate the contribution of this broad component to the experimental profile Figure 1b shows 10 times this new residual and a best fit Lorentzian to it. The fit is excellent being within the noise. Thus the experimental data may be represented as a <u>sum of two Lorentzian profiles</u>, a narrow main component and the new feature, a smaller broad Lorentzian profile. The main narrow component is well described over the entire density regime by a Galatry (soft collision) profile with a width given by the Dicke expression, $k^2D_0/d+\gamma_n d$, with d the density, D_0 the diffusion constant at 1 Amagat, k the momentum transfer, and γ_n the pressure broadening coefficient. This fit to the Dicke expression is improved over that obtained using the widths derived from a single line fit to the data.

For the broad component, the width and integrated intensity (as a % of the total) are shown in Figure 2a and b respectively, as a function of density, for the Q(1) line in pure D_2. The intensity is constant at low densities (~10%) and decreases towards zero near 18 Amagat. Within experimental error the width shows an empirical (A/d)+(Bd) dependence, where A and B are constant. Small errors in the baseline make the uncertainty in the width very large at high and low densities. The value of B is about 1.5 times the value of γ_n found for the narrow main component. Similar results are found for all other observed lines, and D_2 diluted in H_2, He, Ar and CH_4.

A combination of the experimental observations and simple calculations leads us to believe that the effect is associated with the translational motion and in particular with the mean square displacement of the molecule $\langle r^2(t)\rangle$ which appears in all models and calculations of the self-part of the Van Hove pair correlation function. We associate the new feature with the effect of the energetically inelastic collisions on the translational motion. As such the effect should be most prominent, (i) for light molecules colliding with heavy partners, (ii) for those cases where large amounts of internal energy are exchanged with the translational energy, and (iii) where inelastic collisions dominate the broadening process. Existing theories for the translational motion start with the Boltzmann equation and thus explicitly ignore any coupling of the translational and internal degrees of freedom. One must start with the Wang Chang-Uhlenbeck equation where such coupling has been considered for the Rayleigh-Brillouin spectrum[3].

1. R.C.Desai, J. Chem. Phys.44,77(1966)
2. G.J.Rosasco et al, J. Opt. Soc. Am.B2,1485(1985)
3. C.D.Boley et al, Can. J. Phys.50,2158(1972)

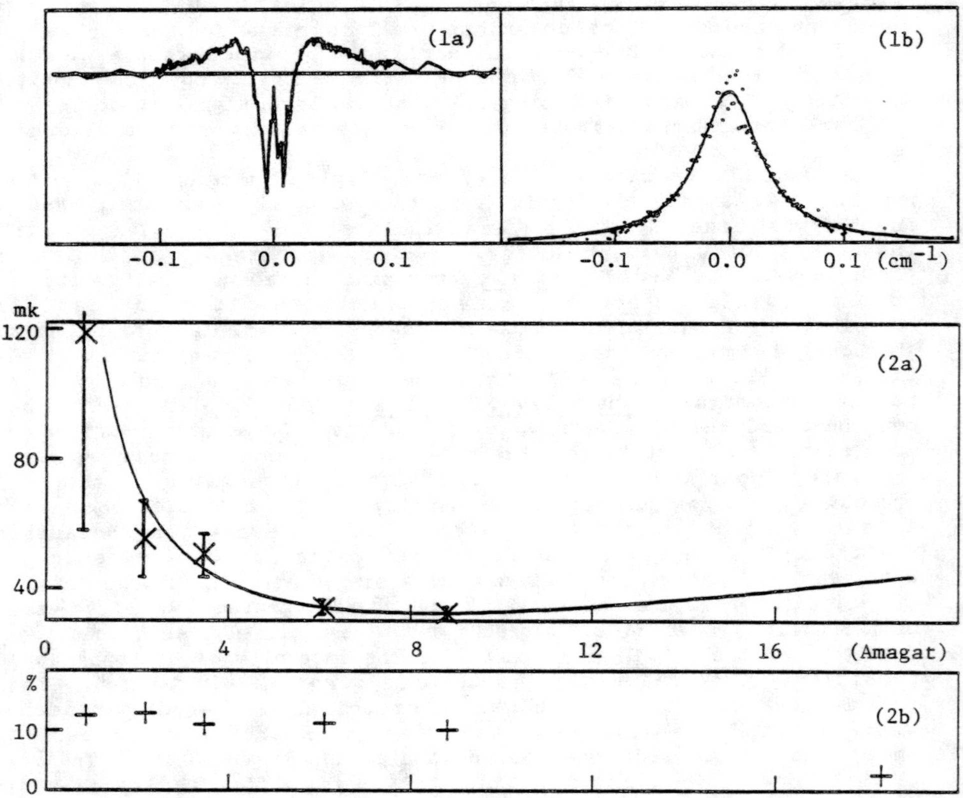

chapter 16

COLLISION INDUCED PHENOMENA and MOLECULAR INTERACTIONS

INDUCED RAMAN SPECTROSCOPY: A GENERAL VIEW.

F. Barocchi

Dipartimento di Fisica, Università di Firenze,
Largo E. Fermi n.2, 50125 Firenze, Italy.

The Depolarized-Interaction-Induced-Light-Scattering (D.I.L.S.) is a very general phenomenon which contributes substantially to the depolarized intensity diffused by any kind of molecular fluid and also in solids. Its spectral region extends between $\sim 1 cm^{-1}$ and $300\ cm^{-1}$ from the exciting line, as well from any allowed transition. Therefore the understanding of the origin of this kind of diffusion is of primary importance for the detailed study of Rayleigh and Raman lineshapes.

The study of this phenomenon, during the last decade, in low density gases has clearified that transient polarizability anisotropies, which develop in the medium because of molecular motion, are the objects responsible of this kind of diffusion of light.

Here, in order to clearify the D.I.L.S. process on a general basis, we will discuss the experiments in simple dilute molecular systems and compare the experimental results with quantum-mechanical ab initio calculations of polarizability anisotropies and lineshapes. The comparison will show that these lineshapes, which can not be reproduced by usual functional forms, can be compared on an absolute scale which extends over 6 decades.

Moreover we will see that, eventhough light is used, this phenomenon permits the determination of new properties which are directly connected with short range spatial properties in the fluid.

INDUCED SPECTRA OF SPHERICAL MOLECULES AT LOW DENSITY

G.C. Tabisz and N. Meinander

Department of Physics, University of Manitoba, Winnipeg, Manitoba, R3T 2N2, Canada

The collision-induced light scattering by compressed inert gases is characterized by a broad depolarized component to the Rayleigh line. The phenomenon is essentially a translational Raman effect, due to a transient incremental polarizability induced in a cluster of atoms through dipole-induced-dipole (DID) and overlap interactions. The width of the spectral feature depends inversely on the interaction time and hence inversely on the square root of the reduced mass of the collision partners for a two-body encounter.

The scattering by isotropic molecules is expected to be similar to that by inert gas atoms and in many ways it is. Much of the early work on collision-induced scattering treated these two species as identical. It is now known however that the molecular spectra contain features not present in the atomic case. In particular these spectra are characterized by a still broader high frequency tail, unexplainable in terms of the models used for the inert gas spectra, and attributable to an induced rotational rotational Raman effect [1].

Tetrahedral molecules, such as CH_4, have an isotropic molecular dipole polarizability α and therefore do not usually exhibit a pure rotational Raman spectrum. Higher order polarizabilities may play a role in the scattering process. For example, a tetrahedral molecule possesses a non-zero, anistropic dipole-quadrupole polarizability A and dipole-octopole polarizability E. The interaction of such a molecule with the field and field gradients, arising from the multipole moments induced in neighbouring molecules by the exciting light beam, yields a fluctuating induced dipole and a rotational spectrum. The intensity of this induced rotational Raman scattering can be significant in relation to the induced translational scattering described above.

For two interacting tetrahedral molecules 1 and 2, the mean square polarizability relative to space-fixed axes is

$$\langle\alpha^2_{xz}\rangle = \frac{48}{35}\left[\alpha_1^2A_2^2+\alpha_2^2A_1^2\right] R^{-8} + \frac{11}{9}\left[\alpha_1^2E_2^2+\alpha_2^2E_1^2\right] R^{-10} + \frac{62912}{4725}\,(A_1A_2)^2R^{-10} + \ldots \qquad (1)$$

where R is the intermolecular distance. The terms in α^2A^2 and α^2E^2 follow selection rules $\Delta J=0, \pm1, \pm2, \pm3$ and $\Delta J = 0, \pm1, \pm2, \pm3, \pm4$ respectively, where J is the angular momentum quantum number of one of the molecules; the term in A_1A_2 results from double transitions.

The theory has had its best success with the spectra of pure CH_4 and CH_4-inert gas mixtures [2]. The shape of the rotational wing has been accounted for by these long-range interactions and an accurate estimate of the magnitude of $\underset{\sim}{A}$ obtained. Moreover, despite the translational broadening of each rotational line, partially resolved rotational structure has been observed.

CH_4 is a particularly propitious case as the molecular mass and large rotational constant conspire to produce a clear separation of the intense purely translational peak and of the rotational branches over a broad frequency range. For most molecules, however, a priori detailed knowledge of the shape of the translational component is necessary in order to specify the rotational contribution. In this regard, an approximate scaling law has been found useful [3].

The pair polarizability anisotropy for translational scattering can be written relative to the intermolecular axis as

$$B(\pi) = 6\alpha^2 R^{-3} + (6\alpha^3 + \gamma C_6/3\alpha)R^{-6} - B\exp(-R/R_0) \qquad (2).$$

The R^{-3} term is first order DID; the R^{-6} term is second order DID corrected for dispersion effects (γ and C_6 are the second hyperpolarizability and the dispersion coefficient); the last term describes the overlap interaction; its range R_0 can be scaled from ab initio calculations of the He-He interaction. There is then only one unspecified parameter, B. Analysis of the inert gas and CH_4 spectra with (2) yields reduced values of B, equal to $BR_m^3/6\alpha^2$, of very similar magnitude for all systems. R_m is the separation at the potential minimum. Accordingly, (2) may be used in lineshape computations to predict translational profiles.

This procedure has been tested successfully for SF_6, where the translational and rotational components greatly overlap [4]. For octahedral symmetry, $\underset{\sim}{A}$ vanishes and consequently a good estimate of the dipole-octopole polarizability was obtained.

It must be emphasized that the theory as developed to date relies on the neglect of translation-rotation coupling.

Among the molecules whose induced scattering has received detailed study are CH_4, CD_4, CF_4, CCl_4, $C(CH_3)_4$, SiH_4, H_2S and H_2 [5].

REFERENCES

1. A.D. Buckingham and G.C. Tabisz, Molec. Phys. 36,583(1978).
2. A.R. Penner, N. Meinander and G.C. Tabisz, Molec. Phys. 54 479,493(1985).
3. N. Meinander, G.C. Tabisz and M. Zoppi, J.Chem. Phys. 84, 3005(1986).
4. S.M. El-Sheikh, N. Meinander and G.C. Tabisz, Chem. Phys. Letter, 118, 151(1985).
5. G.C. Tabisz, N. Meinander and A.R. Penner, Phenomena Induced by Intermolecular Interactions, edited by G. Birnbaum, (Plenum, 1985), p. 345.

LIGHT SCATTERING FROM FLUID CO_2

A. De Santis , R. Frattini , M. Sampoli , R. Vallauri*

University of Venice, DorsoDuro 2137 I-30123 Venezia
*Istituto di Elettronica Quantistica CNR,
via Pianciatichi I-50127 Firenze

Recently, comparisons between real and simulated light scattering experiments have been performed on fluid CO_2 (1,2). The obtained discrepancies suggest that simulated light scattering data could be significantly affected by potential and polarizability models. For we have studied the dependence of the total intensity and second moment from both the interaction potential and polarizability model, in the case of the ν_1 Raman band of fluid CO_2 at ρ=22.37 mol/liter, T=263 K. Induced effects are taken into account within the first order dipole-induced dipole (DID) interactions. The employed potential models are the two center Lennard-Jones (2LJ)(3), the 2LJ plus point quadrupole (2LJQ)(A1 of ref. 4) and the three center LJ potential plus point quadrupole (3LJQ)(4). The polarizability models simulating induced effects are that of point molecular scatterers situate at the center of mass (c-c model) and that of point atomic scatterers situate over the two oxigen sites (s-s model). The last model has been already used by Ladanyi (5). The results, summarized in the table, have been obtained by following the motion of 108 partciles for 12000 time steps of $\Delta t=3.5\times10^{-15}$ sec. The total second moment has been calculated during the MD run by evaluating the correlation function of the first derivatives of the total polarizability of the fluid at equal times. From the table two main effects appear evident: the first one is that, within a fixed polarizability scheme, there is a strong variation of the calculated quantities on the potential model; the second one is that, for a given potential, the variations due to the polarizability model can be also significant. A more detailed analysis (6) has revealed that the different quadrupole strenghts of the potentials are likely responsible for the different results obtained in the c-c scheme. The strong dependence of the induced effects from both potential and polarizability models can explain the above quoted discrepancies (1,2).

TABLE

	2LJ		2LJQ		3LJQ	
	c-c	s-s	c-c	s-s	c-c	s-s
I^{tot}	0.741	0.739	0.873	0.736	0.881	0.790
M_2^{tot}	1.432	1.432	1.260	1.380	1.243	1.238

I^{tot} are normalized to the rotational value and M_2^{tot} are in 6kT/I units.

References
1) A. De Santis and M. Sampoli, Mol. Phys., 51, 97 (1984)
2) C.S. Murthy et al., Ber. Bunsen Phys. Chem., 89, 18 (1985)
3) K. Singer , A. Taylor, J.V.L. Singer, Mol. Phys. 33, 1757 (1977)
4) C.S. Marthy, K. Singer, I.R. Mc Donald, Mol Phys. 44, 135 (1981)
5) B.M. Ladanyi, J. Chem. Phys., 78, 2189 (1983)
6) A. De Santis et al., Europhys. Lett, 00, 00 (1986)

ROTATIONAL RAMAN SCATTERING FROM LIQUID AND GASEOUS H_2S

A. De Lorenzi, A. De Santis, R. Frattini, M. Sampoli

University of Venice, DorsoDuro 2137, I-30123 Venezia

Old data of isotropic light scattering from liquid H_2S (1) are interpreted on the basis of a study performed in the gas phase. The presence of two different spectral features, the low frequency peak and the broad spectrum, suggests the existence of two different scattering mechanisms dominating over different regions. The comparison between H_2S and D_2S spectra (1) has shown that the low frequency peak is due to the translational dynamics whereas the broad one involves orientational dynamics. In the gas phase we have followed the density evolution of the isotropic spectrum and used the integrated intensity of the ν_1 stretching band as internal standard. The relative intensities of the broad spectrum ($\tilde{I}^{iso}_{broad}$) and of the low frequency peak ($\tilde{I}^{iso}_{peak}$) have been obtained after separation of the two spectral shapes. The intensities increase quadratically with density, so the relative ones increase linearly, $\tilde{I}^{iso}_{broad}=2.5\times10^{-29}\rho$ and $\tilde{I}^{iso}_{peak}=3.3\times10^{-29}\rho$ (density is in particles/m^3). To interpret our results we have performed a detailed theoretical analysis by considering the multipolar series expansion and the contributions of polarizabilities of higher orders. It is found that the dipole-quadrupole polarizability A is mainly responsible for the broad spectrum since both the intensity and the spectral shape are satisfactorily reproduced (the estimated value of the tensor components is $A^2_{123,1}+A^2_{123,3}=3$ Å^8) (2). The intensity of the low frequency peak is in agreement with the intensity due to the second order dipole-induced dipole (DID) contributions and to the scalar part of the second hyperpolarizability tensor (2). It is the analogous to isotropic spectra obtained in noble gases (3). So the spectral distribution of isotropic intensity is similar to that of depolarized spectra of globular molecules (4). At the liquid densities, the isotropic spectrum can be interpreted on this same framework: the low frequency peak reflects the DID and the second hyperpolarizability contributions and the broad spectrum the dipole-quadrupole polarizability contribution. Indeed, if we subtract the contribution of the broad spectrum at low frequency, the shape of the low frequency peak results to be exponential with a very good approximation. Such a spectral shape is a further support to our previous conclusions.

References

1) A. De Santis and M. Sampoli, Chem. Phys. Lett. 102, 425 (1983)
2) A. De Lorenzi et al., Phys. Rev. A 00, 00 (1985)
3) M.H. Profitt et al., Can. J. Phys. 59, 1459 (1981)
4) A.R. Penner et al., Mol Phys 54, 479 (1985)

INDUCED EFFECTS IN MOLECULAR LIQUIDS AND LOCAL FIELD PROBLEMS.

Branka M. Ladanyi

Department of Chemistry, Colorado State University,
Fort Collins, Colorado 80523, U.S.A..

Raman spectra of liquids are affected in several important ways by interactions between molecular induced dipoles. These effects can be conveniently divided into intensity and lineshape modifications.

The local field effects on the intensity are due to the fact that intermolecular interactions contribute to the electric field in the liquid. Thus the field a given molecule experiences can differ considerably from the applied field, and the intensity is not simply related to the number of molecules in the scattering volume. This effect depends strongly on the anisotropy in the molecular environment. As a result, it modifies the intensity of isotropic and depolarized spectral components by different amounts.

Due to intermolecular interactions, the collective molecular polarizability is not simply a sum of individual molecule polarizabilities. The interaction dependent part of the polarizability modifies the lineshape of 'allowed' spectra ,i.e. spectra associated with vibrational modes which are Raman active in the gas phase, and gives rise to 'forbidden' Raman spectra. This first phenomenon changes the 'allowed' lineshapes by adding on the components due to orientational-interaction induced cross correlations and to interaction induced polarizability autocorrelations. It can be monitored by measuring the pressure dependence of the second moment of depolarized spectra. The second phenomenon provides a direct measure of intermolecular interactions, since it corresponds to Raman spectra associated with vibrational modes for which the isolated molecule polarizability vanishes.

The static local field effects and the changes in Raman lineshapes due to intermolecular dynamics in several molecular liquids have both been systematically studied by experimental and theoretical methods. These studies have already yielded a great deal of information about intermolecular correlations in liquids. In this talk I will describe some of the key experimental, computer simulation, and theoretical results and discuss our current understanding of interaction induced effects on Raman spectra of liquids. I will also discuss the observed and predicted trends in the dependence of these effects on thermodynamic state and intermolecular interaction parameters.

INDUCED LIGHT SCATTERING IN CRYSTALS AND DISORDERED SOLIDS

G. Signorelli, V. Mazzacurati
Dip.di Fisica University "La Sapienza" of Rome, Italy

G. Ruocco
Dip. di Fisica University of L'Aquila, L'Aquila, Italy

Light scattering in solids has been used since many years as a powerful tool to study vibrational excitations and symmetry properties. If we confine ourselves to "molecular" solids, we can adopt the successful picture which assigns to each unit an effective time dependent polarizability localized in the center of mass. This polarizability is then expressed us the sum of the bare isolated unit polarizability and the modification induced by all surrounding "molecules", so that the volume polarizability of the solid may be written in a microscopic way

$$\bar{\bar{P}}(\vec{r},t) = \sum_\ell {}^{\ell}\bar{\bar{p}}^{eff}(t)\,\delta(\vec{r}-\vec{R}_\ell(t)) \qquad {}^{\ell}\bar{\bar{p}}^{eff}(t) = {}^{\ell}\bar{\bar{p}}^{o}(t) + {}^{\ell}\Delta\bar{\bar{p}}\,(\{\vec{R}_{\ell'}(t)\})$$

The spectrum of the scattered light is the spatial-time Fourier transform of the volume polarizability correlation function which can be expressed as

$$I_{ij}(\vec{q},\omega) = \int dt\, e^{i\omega t} \sum_{\ell\ell'} \langle {}^{\ell}p^{eff}_{ij}(t)\, e^{-i\vec{q}\cdot\vec{R}_\ell(t)}\; {}^{\ell'}p^{eff}_{ij}(0)\, e^{i\vec{q}\cdot\vec{R}_{\ell'}(0)} \rangle$$

This correlation function, apart from the "internal" degree of freedom of the molecular unit itself, derives its time dependence both from the $e^{i\vec{q}\cdot\vec{R}_\ell(t)}$ unit factor and from the relative motions of the units via the induced part ${}^{\ell}\Delta\bar{\bar{p}}\,(\{\vec{R}_{\ell'}(t)\})$. This term, which is quite relevant in dense systems (being also responsible for the renormalization of the dielectric constant in the Clausius-Mossotti relation) can be described using the Buckingham expansion of the induced dipole moment in terms of the local field, obtaining the various induced contributions. This can all be expressed using bare "molecular" quantities (polarizability, and higher order hyperpolarizabilities) which may be eventually modified by electronic overlap (E.O.) and exchange effects due to nearest neighbour environment. In crystals, where the translational invariance implies the selection rule $\vec{q} = \vec{k}_s - \vec{k}_i \simeq 0$ for any excitation contributing to the scattering. one and multi-phonon terms can be obtained expanding both effective polarizabilities and phase factors in series of atomic displacements. The phase factor first order expansion gives only longitudinal acoustic phonon contributions being that of both optical excitations and two phonons about 10^8 times smaller. Therefore the presence of transverse acoustic and optic phonons as well as the two-phonon contributions in monoato mic and ionic crystals can therefore only be attributed to the induced part. We have. for example, recently shown that the Brillouin scattering in rare gas crystals is quantitatively reproduced taking correctly into account these induced terms. (namely dipole induced dipole (DID) effects with a small E.O. correction).

The results of this scheme are even more interesting when dealing with disordered solids. In this case, the disorder induced in the effective polarizability destroyes the $\vec{q}\cong 0$ selection rule and the entire phonon density of states is activated. As an interesting example we can consider the β-AgI crystals. At very low temperature β-AgI is an ordered crystal with a WURTZITE-type structure: its Raman spectrum shows the $\vec{q}\cong 0$ optical phonon peaks, their intensity being well predicted by the D.I.D. mechanism. Increasing the temperature, a small amount of silver atoms become ramdomly distributed. and the density of states starts to appear in the spectrum owing to the C.I.D. (charge induced dipole) mechanism which is rigorously vanishing in the ordered phase. Of course it is not always possible to define exactly the most important terms in the induced polarizability mainly for lack of experimental informations on higher polarizabilities. Nevertheless. thanks to the symmetry properties of the terms in the Buckingham expansion we can separate in a partially disordered solid the different branch contributions. (independently of the exact knowledge of ${}^{\ell}\Delta\bar{\bar{p}}(\{R_{\ell'}(t)\}$)). carefully analyzing the frequency dependence of the polarization ratios. Another example is the Raman spectrum of hexagonal Ice. which is disordered as far as the proton distribution is concerned. We have shown that the entire translational density of states may be reconstructed. separating the single branches contributing to the spectrum, analyzing the polarization ratios between three different scattering geometries.

THE EFFECTS OF DEPHASING COLLISIONS ON RAMAN POLARIZATION.

Robin M. Hochstrasser and A. B. Myers

Univervity of Pennsylvania, Philadelphia, PA 19104

It is well known that a distinction between resonance Raman scattering and fluorescence emission is established in the presence of a pure dephasing interaction. The Raman signal and lineshape are also influenced by collisions which result in a redistribution of the energy of the intermediate state into a new set of levels closer to equilibrium. In the present work we take a closer look at the implications of the damping of the Raman signal to the polarization of the scattered light. One object of this work is to use the Raman intensity as a means of learning about collisional dynamics in liquids. Another is to clarify the relationships between fluorescence and Raman depolarization.

The first part of this study concerns the theory of time resolved emission polarization from an ensemble of free rotors (1). We have also considered nonlinear Raman responses of such ensembles. Time dependent signals were calculated for both parallel and perpendicular transitions at low and high temperatures. At high J both the spontaneous and the nonlinear signals exhibit transients near time zero which level out to essentially constant values within several average rotational periods. The anisotropy for each initial angular momentum state oscillates as a result of the quantum beating of the P, Q and R branch transitions, but at high temperatures the overlapping beat signals average to a constant value except near time zero where the anisotropy is seen to decay on the time scale $(I/kT)^{1/2}$. A typical result is shown in Figure 1 for a molecule having inertial properties like those of trans-stilbene.

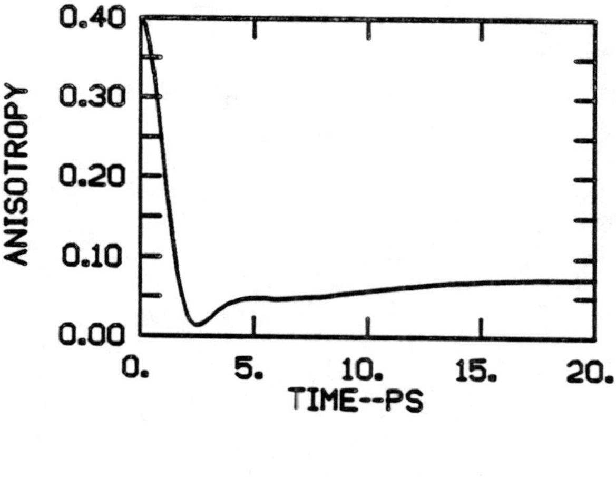

Fig. 1. Time-dependent fluorescence anisotropy calculated for a parallel transition of a prolate symmetric top with rotational constants B_s=0.1 cm^{-1} and B_p=0.01 cm^{-1}, at a temperature of 373 K.

We have recently measured the anisotropy of stilbene emission using a synchronously pumped laser system and nonlinear gating (2). The results clearly show the predicted peak near t=0 corresponding to the in-phase free rotational motion of the molecules.

The free rotation of the molecules is interrupted by collisions which therefore alter the anisotropy measured at each instant. In the high density limit this approach to understanding rotational diffusion was introduced by Gordon (3) who considered that collisions can alter either the orientation (M-diffusion) or magnitude (J-diffusion) of the rotational angular momentum. In considering the time evolution of the linear or nonlinear Raman signal the values of $I_{\parallel}(t)$ and $I_{\perp}(t)$ expected for a rigid rotor must be convoluted with the appropriate Raman decay functions: For example, in the simplest case the Raman signal decays at a rate which is the sum of the pure dephasing, radiative, intramolecular and intermolecular decay rates. Because the Raman emission is damped by these processes, they limit the extent of free rotation that can occur. Hence even the time integrated Raman anisotropy will provide information about the nondiffusive part of the decay.

Recent experimental results on the polarization of resonance Raman transitions will be presented and related to the theoretical ideas summarized above.

References:

1. A. B. Myers and R. M. Hochstrasser, "Time-resolved fluorescence polarization and four-wave mixing from an ensemble of free symmetric rotors," submitted.

2. P. Holt, A. B. Myers, M. Pereira and R. M. Hochstrasser, "Observation of the free rotation regime in the fluorescence of gaseous stilbene."

3. R. G. Gordon, J. Chem. Phys. 44, 1830 (1966).

Acknowledgement: This research was supported by grants from NSF and NIH.

COLLISION INDUCED RAMAN SCATTERING IN HYDROGEN AT LOW DENSITY: TEMPERATURE DEPENDENCE.

Ubaldo BAFILE°, Lorenzo ULIVI°, Marco ZOPPI° and Fabrizio BAROCCHI*

°Istituto di Elettronica Quantistica del C.N.R.
Via Panciatichi 46/30 I-50127 FIRENZE (ITALY)

*Dipartimento di fisica, Università di Firenze
Largo Enrico Fermi 2 I-50124 FIRENZE (ITALY)

Depolarized Interaction-induced Light Scattering has been widely used, in the past few years, to derive unique information on the induced polarizability of the heavier rare gas atoms (1). In this case the interpretation of the spectrum is easier, because the motion of the atoms may be considered classical, ruled by an isotropic interation potential, which, in the majority of cases, is well known.

Experiments on diatomic molecules also present great interest, and hydrogen, which has the advantage of having the first rotational Raman line quite apart from the zero-frequency Rayleigh one, allow a good measurement of the pure traslational pair spectrum. Moreover, hydrogen is an interesting case because the quantum nature associated with its DILS spectrum can be varied as a function of temperature.

The first DILS spectrum of normal hydrogen at room temperature was detected by Lallemand, who also measured the integrated intensity of the pair and triplet specra (2). More recently Le Duff et al. (3) measured the DILS spectrum of para- and thermal equilibrium hydrogen at 123 K and Brown et al. (4) reported a measure of the pair DILS spectrum of normal H2 and D2 at room temperature. We summarize here the result of a measurement of the DILS spectrum at three different temperature, namely 49.5, 103 and 291 K (5).

To be sure to record the pure pair spectrum, a density analysis of the scattered radiation at $\Delta\nu = 20\ cm^{-1}$ and at room teperature has been performed, (fig.1). We have chosen for the experiment the density of 40 amagat, at which three (or more) body effects are considered negligible. Absolute calibration of the spectra recorded was made by mean of a comparison with the intensity of the rotational lines. Fig.2 show a comparision of our room temperature spectrum with the corresponding point of ref. 4.

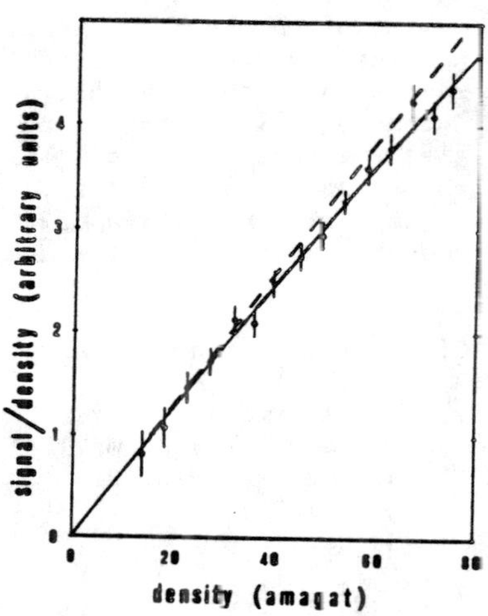

Fig. 1

The experimental moments of the spectra are shown in tab.1, against theoretical ones, calculated classically, (columns 3 and 6) or taking into account the first ($\hbar^2$) quantum correction in a power series of $\hbar$. For the calculation we have assumed DID polarizability and the McConville's isotropic interaction potential (6). No distinction has been made between ortho and para hydrogen in the calculation.

	ZERO MOMENT			SECOND MOMENT		
Temp. K	Exp.	Class.	Q.C.	Exp.	Class.	Q.C.
291	1.20 (±0.05)	1.18	1.17	9.2 (±2.2)	10.5	12.4
103	1.14 (±0.06)	1.10	1.04	4.5 (±1.5)	3.1	4.4
49.5	1.50 (±0.08)	1.28	1.03	2.77 (±0.13)	1.5	3.4

Table 1. Experimental and calculated spectral moments of hydrogen pairs. Unity are 10^{-52} cm^5 for zero moment, and 10^{-49} cm^3 for second moment.

The extrapolation toward zero frequency must be performed with care due to the presence of the unshifted ($\Delta J=0$) rotational Q-branch, which is superimposed at $\Delta\nu=0$ The DILS spectrum and the rotational Q-branch are both fully depolarized and their separation cannot be obtained simply by polarization analysis of the experimental spectra.

We conclude that, within the approximation of our calculation, the DID polarizabilty model and the assumed potential describe fairly well the zeroth moment of the spectrum at room temperature. However, in order to investigate the short-range behaviour of the polarizability anisotropy, low-temperature spectra, which give the possibility of measuring at least the second moment, and a full quantum mechanical calculation of either the moments and the spectrum are needed.

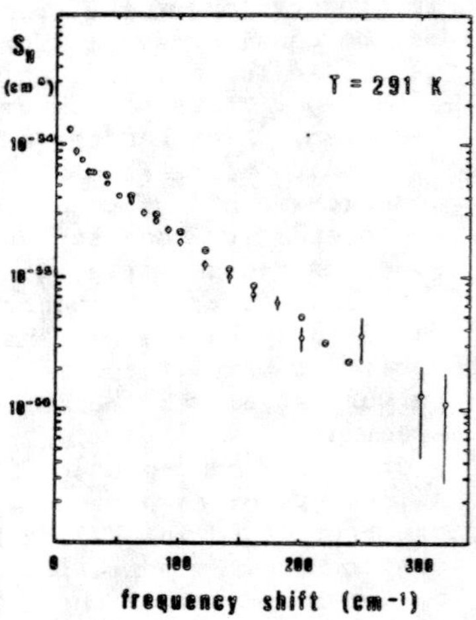

Fig. 2

References.

(1) F. Barocchi and M. Zoppi, in: Phenomena induced by intermolecular interation, ed. G. Birnbaum (Plenum Press, New York, 1985).

(2) P. Lallemand, J.Phys. (Paris) 32, 119 (1971).

(3) Y. LeDuff and R. Ouillon, J.Chem.Phys. 82, 1 (1985).

(4) M.S. Brown, M.H. Proffitt and L. Frommhold, Chem. Phys. Letters 117, 243 (1985).

(5) U. Bafile, L. Ulivi, M. Zoppi and F. Barocchi, Chem. Phys. Letters 117, 247, (1985).

(6) G.T. McConville, J.Chem.Phys. 74, 2201 (1981).

The Relation Between the Band Shape, the Strength of H-Bond and the Solution Viscosity

H. Abramczyk[1], D. Samios[2], A. Marcinek[1], Th. Dorfmüller[2]

1) Technical University, Institute of Applied Radiation Chemistry, PL-93-590 Łódź, Wróblewskiego 15, Poland

2) Universität Bielefeld, Fakultät für Chemie, Postfach 8640, D-4800 Bielefeld 1, F.R. Germany

The peak position, the shape and the band width significantly depend on the strength of the H-bond. The influence of the solvent is believed to be a secondary band shaping mechanism. There are a lot of theories of the band profile [1-3]. The basic idea is that the oscillators $\nu_S(XH)$ and $\nu_\sigma(XH...Y)$ are strongly coupled. The great disadvantage of the theories is the lack of clear link between the strength of the H-bond and the shape profile. The relationship can be disclosed, if the dipole-dipole interactions between the dipole moment μ of molecule under study (proton donor or proton acceptor) and the increment dipole moment $\Delta\mu$ produced by the H-bond are considered. The hamiltonian of the oscillator engaged directly in H-bonding is

$$H = \frac{p^2}{2m} + \frac{1}{2} m\omega^2 q^2 + K_{112} Q_1 q^2 + V_D \qquad (1)$$

where Q_1 is the normal coordinate of the stretching mode of the bridge $\nu_\sigma(XH...Y)$, V_D is the dipolar coupling. Assuming that all isomers involving in H-bond are equally probable and averaging over the angle ϕ, the dipolar coupling can be written as

$$\langle V_D \rangle = -2\, \mu \Delta\mu r^{-3} \cos\Theta(t) \qquad (2)$$

where Θ is the angle between μ and $\Delta\mu$.

The coupling terms (eq. (1)) produce the stochasticity of the hamiltonian. The time modulation of the frequency of the oscillator is

$$\omega(t) = \omega + aQ_1(t) + b\cos\Theta(t). \qquad (3)$$

The two models have been used for describing the dynamics of the bridge XH...Y; (1) the random force model using the generalized Langevin equation, (2) the rotational diffusion model. In both cases the band shape is determined by the coupling constant, which is simply proportional to the increment dipole moment $\Delta\mu$ for weak H-bond. It has been shown, that the two-particle reorientational correlation function $\phi_D = \langle\cos\Theta(0)\cos\Theta(t)\rangle$ is significantly faster than the decrease of the one-particle reorientational correlation function $\langle\vec{u}(0)\vec{u}(t)\rangle$. The ϕ_D correlation function can by no means be neglected in the total vibrational correlation function. It has been found that the time scale of the dynamics of the bridge described on the basis of the random force model and the rotational diffusion model are comparable. The relation between the band width, the enthalpy of formation $-\Delta H$ and the solution viscosity has been obtained. The correlation between the experimental IR band width of proton donors as well as the isotropic Raman band width of proton acceptors and the $\Delta\mu^2\eta$ has been found.

[1] S. Bratos, J. Chem. Phys. 63 (1975) 3499

[2] G.N. Robertson, J. Yarwood, Chem. Phys. 32 (1978) 267

[3] V.P. Sakun, Chem. Phys. 99 (1985) 457

A RAMAN SCATTERING STUDY OF THE LIQUID STATE DYNAMICS OF A HIGH TORQUE MOLECULAR LIQUID.

Donald G. Moses and Frank G. Baglin

Department of Chemistry, University of Nevada, Reno, Nevada 89557

In order to understand liquid state dynamics one would like to formulate a theory in which torques are explicitly considered in a theoretical treatment of the system (1). Recently Lynden-Bell and Steele (LBS) have formulated a gaussian cage model (2) using the generalized Langevin equation of Tokuyami and Mori (3). The formulation of this model has been required because the usual treatments of Gordon (the J diffusion model) and the Fokker-Planck models begin to fail (4,5) when the mean square torque factors (N^2) of a liquid are 30 or greater. Nose and Klein (6) have studied CF_4 near its melting point by molecular dynamic techniques and LBS have applied their model to this system and have found values of N^2 near and above 30.

Data gathered showed that the number density plays a key role in T_2 as expected from Gelbart's work (7). Moreover, the inhomogeneous dephasing correlation functions of MSA showed little variations with viscosity as might be expected. These functions' log plots versus time were all non-linear indicating a non-random distribution of molecules which is consistent with dynamic cage formation.

The most interesting results of our study, however, involved the reorientational correlation functions (RCF) for MSA at 23°C, 51°C and dilute solutions of acetone and CCl_4 (3.6 and 4.9 mole%) MSA respectively. The RCFs for neat MSA at 23°C(I), 51°C(II) and 3.6% in acetone at 23°C(III) are given below.

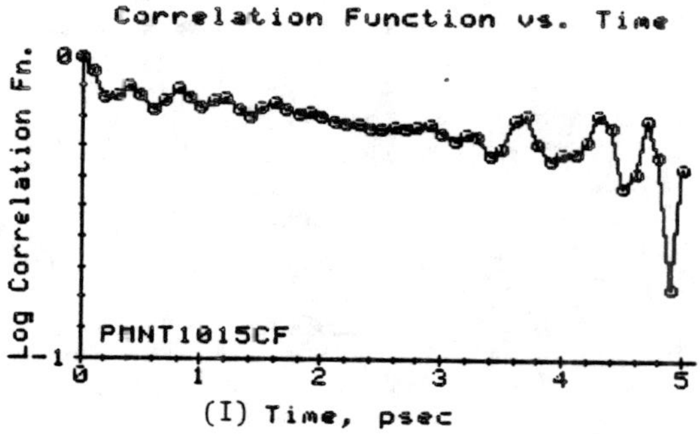

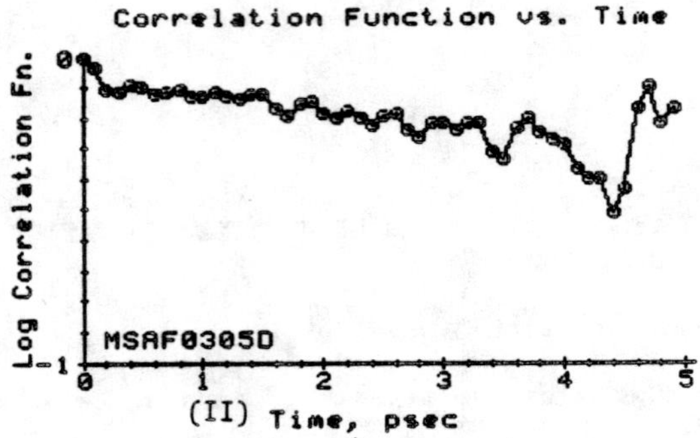

(II)

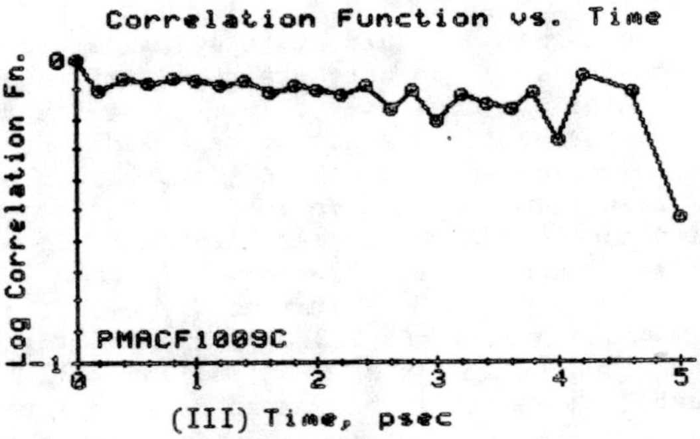

(III)

MSA and CF_4 Results

Neat MSA	KT/I(ps)	(ps^{-1})	$o(ps^{-1})$	o/	N^2	Harmonicity Ratio
T=296°K	0.196	8.83	7.85	0.889	714	$\frac{587}{714} = 0.82$
T=323°K	0.214	7.99	7.85	0.982	587	
Neat CF_4						
T=90°K	0.867	3.14	5.66	1.80	49	$\frac{28}{49} = 0.58$
T=120°K	1.20	3.04	4.83	1.59	28	

One can see the oscillations predicted by LBS (but unseen in CF_4) in MSA(I). Their cooperative nature is indicated by their absences from the acetone dilution studies(III) and their damping out at the higher temperature(II).

References

1. Steele, W.A., in Vibrational Spectroscopy of Molecular Liquids eds. S. Bratos and R.M. Pick (1980), 56, NATO Advanced Study Institute Series, Plenum Press N.Y., N.Y., P. 61.
2. Lynden-Bell, R., Steele, W., J. Phys. Chem. (1984), 88, 6514.
3. Tokuyami, M. and Mori, H., Prog. Theor. Phys. (1976), 55, 411.
4. Breuillard, C., Ouillon, R., Mol. Phys., (1977), 33, 747.
5. Lynden-Bell, R.M. in Molecular Liquids, eds. A.J. Barnes, W. Orville-Thomas and J. Yarwood, Dordrecht, The Netherlands (1984).
6. Nose, S. and Klein, M.L., J. Chem. Phys. (1983), 78, 6928.
7. Gelbart, W.M., in Adv. Chem. Phys. eds. I. Prigogine and S. Rice, (1974), 26, John Wiley and Sons, NY, P. 1.

RAYLEIGH SCATTERING OF DENSE O_2.

Tadeusz W. Zerda

Department of Chemistry, University of Illinois, Urbana, IL 61801

The collision induced effects in light scattering from simple molecular fluids have been investigated both experimentally and theoretically [1-3]. It is well established that the observed spectrum is composed of three components, $I(\omega) = I^R(\omega) + I^{CI}(\omega) + I^{CR}(\omega)$, which occur on the same time scale and therefore cannot be separated. Analysis of the data is complicated by the fact that $I^R(\omega)$, rotational component, includes also cooperative rotational motion and collision induced term, $I^{CI}(\omega)$, is due to two-, three- and many-body collisions. The role of the cross term, $I^{CR}(\omega)$, between rotational motion and collision effects have been studied only recently [1-3]. Analysis of the spectral moments $M(n) = \int I(\omega)\, \omega^n\, d\omega$, combined with the band shape study provides information on relative importance of rotational and non-rotational contributions. It is possible because various components contribute differently to the M(0), M(2) and M(4). In this paper we present the experimental data obtained for O_2 at 300K and pressures varied up to 3000 bar.

A typical experimental high pressure set-up was used with small modifications. To minimalize the stray light 150 fused razor blades were placed in the cell opposite to the observation window, and the windows were antireflection coated. The spectra were recorded in the frequency range 0-600 cm^{-1} with the slit width varied from 20 μm to 200 μm depending on frequency range and pressure applied. At far wings the counting time was extended to several minutes. The data were analyzed after the dark current (4 counts/sec) was subtracted, and the corrections due to the scattering factor were included. Each experiment was repeated at least three times. In order to compare data taken at different days, they were calibrated each day with rotational Raman transition $J = 7 \rightarrow J = 9$ of oxygen at 700 mm Hg.

The experimental values of the spectral moments are listed in Table 1. With growing density, ρ, the ratio $M(0)/\rho$ first increases but only to densities lower than 100 amagats, later the decay is observed. This is a typical observation for the collision induced spectra, and in the dipole-induced dipole approximation it can be explained in part as a cancellation effect due to three-body collisions and in part as a result of increasing role of the negative cross term, as the density is increased.

Similarly to the total intensity the experimental second moment is a sum of three contributions due to orientational motion $M^R(2)$, collision-induced $M^{CI}(2)$ and cross $M^{CR}(2)$ terms. As seen from Table 1, M(2)/M(0) continuously increases with growing density. For other linear molecules, like N_2 or CO, second moment decreased at low pressures and increased only after the density of 150 amagats was exceeded. N_2 and CO have small anisotropic polarizability, $\gamma = \alpha \;\; - \alpha$, and the induced contributions due to the anisotropic part of the polarizability tensor can be neglected. Consequently, the cross term $M^{CR}(2)$ vanishes and the small decay of M(2)/M(0) observed at low pressures is due to the counter-productive contributions of $M^{CI}(2)$. The later increase of M(2)/M(0) can be explained as the effet accompanying normalization procedure. As M(0) decays with density, M(2)/M(0) increases. For oxygen the anisotropic polarizability cannot be neglected leading to a non-zero contribution of $M^{CR}(2)$ to the second moment. This term is responsible for different observation for oxygen and nitrogen Rayleigh spectra. It is in agreement with the computer simulations for O_2 [1] which showed the cross contribution to the spectra moments to be much more important than that due to CI effects.

The differences between measured fourth moment and the theoretical value of $M^{th}(4)$ due to free rotational motion, $M^{th}(4) = 96\ (kT/I)^2$, are large and can be explained also as an effect of the cross term.

Gordon [4] proposed to evaluate intermolecular mean-square torque $\langle(OV)^2\rangle$ from the experimental spectral moments:

$$M(4)/M(2) = 8/3\ M(2) + \langle(OV)^2\rangle/2IkT$$

It is well established now that this relation cannot be applied for dense systems, but most of the theoretical arguments were based on assumption of cooperative motion and collision induced effects contributing to the experimental moments. This study shows that these limitations although correct are less important than those originating from the cross terms. It is interesting to note that the $M^{CR}(4)$ probes the potential derivative, a quantity similar to the torque, but acting on much shorter range than the torque [5]. Since $M^{CR}(4)$ and torque measure the anisotropic part of intermolecular potential, the ratio M(4)/M(2) which is composed mostly of these two terms can be regarded as a probe of the potential anisotropy.

It is expected that the sixth moment is also dominated by the rotational and cross terms. But because of the complicate nature of this quantity, we restrain ourself from further conclusions.

Table 1. Spectral moments of Rayleigh scattering of O_2 at 300^oK. M(0) values are normalized to the 20 bar data.

P [bar]	ρ [amagats]	M(0)	M(2)/M(0) [cm^{-2}]	M(4)/M(0) [$x10^4 cm^{-4}$]	M(6)/M(0) [$x10^8 cm^{-6}$]
20	20.2	1.00	3660	3732	5670
40	40.9	2.22	3784	3830	5766
70	72.6	4.18	3867	4037	6291
100	104.7	4.88	3934	4164	6976
130	130.0	5.92	3956	4286	7079
300	300.6	8.65	4070	4570	8652
600	478	9.31	4121	4859	10004
1000	600	8.89	4205	5302	12200
3000	771	8.90	4348	6064	18670

1. B. M. Ladanyi, J. Chem. Phys. 78, 2189 (1983)
2. A. De Santis and M. Sampoli, Phys. Lett. 100A, 25 (1984), 109A, 28 (1985).
3. T. W. Zerda, X. Song and J. Jonas, J. Phys. Chem. 90, 771 (1986).
4. R. G. Gordon, J. Chem. Phys. 40, 1973 (1964).
5. A. De Santis, R. Frattini and M. Sampoli, Chem. Phys. Lett. 114, 71 (1985), ibid. 114, 297 (1985).

HIGH-RESOLUTION RAMAN AND TIME RESOLVED CARS STUDIES OF THE FERMI DIAD IN CO_2.

R. Ranson and R. Ouillon

Departement de Recherches Physiques -L.A. 71, Universite P. et M. Curie, Tour 22, 4 place Jussieu, 75252 Paris cedex 05, France

Up to now, accurate information concerning vibrational relaxation processes in the solid phase has been mainly obtained by means of coherent Raman spectroscopy. However, we have recently proved that high-resolution spontaneous Raman scattering remains a competitive technique in this field of research (1-3). Particularly, high-resolution easily allows the separation of close lying crystal modes and this can be important for the study of slightly perturbed internal vibrons in molecular crystals. Furthermore, the phonon frequencies can be measured with great accuracy, this being very interesting for the determination of their thermal variations. As typical examples of this selectivity, we present some results about the dynamics of vibrational overtones and biphonon states in molecular crystals of small molecules.

In solid CO_2, we have studied in great details the relaxation of the two hybrid bound states $\Omega_- = 1276\ cm^{-1}$ and $\Omega_+ = 1384\ cm^{-1}$ of the diad $[\omega_1, 2\omega_2]$ which are in strong Fermi resonance. Using a tandem interferometer-spectrometer instrument with a limiting resolution of $0.004\ cm^{-1}$, we succeeded in resolving for the first time the two factor group components, one Ag and one Fg of each of these states (4) and in studying their relaxation separately. The evolution with temperature of the bandwidths has been interpreted in terms of four-phonon decay processes of the bound states into the free two-phonon continuum $\omega_2(k_1) + \omega_2(k_2)$, with annihilation (for Ω_-) or creation (for Ω_+) of a lattice phonon $\omega_\ell(-k_1-k_2)$ (5). Furthermore, our results suggest that the symmetry (Ag or Fg) of the initial state may play a role in the relaxation processes.

At low temperature (T <10K), the lack of thermal lattice phonons ω_ℓ leads to a very narrow residual width $\gamma<0.004\ cm^{-1}$ for the Ω_- components. Thus, we have investigated the vibrational coherence relaxation for the longest living Ω_-(Ag) state at T = 5K by means of time resolved nanosecond CARS spectroscopy. The decay profile was found to be slower than exponential and fitting reasonably well to an $\exp|-(t/\tau)^2|$ time dependence with $\tau = 2.3 \pm .2$ ns. Our results are discussed in terms of weak diagonal structural disorder upon the vibrational relaxation of exciton states located near the band edges.

We also have measured the thermal variation of phonon frequencies in the temperature range 5 K - 80 K. An interpretation is given in terms of explicit and implicit temperature dependent anharmonic contributions.

A similar study was performed for the ω_1 and $2\omega_2$ vibrational modes in solid N_2O. In this crystal the quasi-bound state $2\omega_2$ is degenerate with the free two-phonon continuum $\omega_2+\omega_2$, leading to a quasi temperature independent coherent decay times. Comparisons with recent ps CARS data are given (6).

(1) P. Ranson, R. Ouillon, and S. Califano, Chem. Phys. 86-115 (1984).

(2) R. Ouillon, P. Ranson, and S. Califano, Chem. Phys. 91-119 (1984).

(3) PP. Ranson, R. Ouillon, B. Halac, and S. Califano, J. Chem. Phys. 82-169 (1985).

(4) R. Ouillon, P. Ranson, and S. Califano, J. Chem. Phys. 83-2162 (1985).

(5) P. Ranson, R. Ouillon, and S. Califano, J. Raman Spectroscopy 17-155 (1986).

(6) F. Vallee, G.M. Gale, and Ch. Flytazanis, Chem. Phys. Letters 124-216 (1986).

COMPUTER SIMULATION STUDIES OF RAMAN SCATTERING FROM DENSE FLUIDS.

Branka M. Ladanyi and Lynn Geiger

Department of Chemistry, Colorado State University,
Fort Collins, Colorado 80523, U.S.A..

We present the results of molecular dynamics (MD) computer simulation of rotational and collision induced (CI) contributions to Raman spectra of fluids of diatomic and symmetric linear triatomic molecules. The bond length and potential parameters are chosen to correspond to those of O_2, CO_2 and CS_2. The potential is represented by a set of atom-atom Lennard-Jones interactions. In the case of CO_2, a quadrupole-quadrupole interaction is included as well. We consider two collective polarizability models, one based on dipole-induced dipole (DID) interactions between molecular centers, the other on atom-atom DID interactions. For both models, we study the variations of the effective molecular polarizabilities, integrated intensities, second and fourth spectral moments, and time correlations with density and temperature. We also calculate iteratively the contributions to the collective polarizability from higher order terms in the DID interactions.

16-22

SPECTRUM OF COLLISION-INDUCED SCATTERING : ROLE OF MULTIPOLAR POLARIZABILITIES.

BARREAU A., BERRUE J., CHAVE A., DUMON B., THIBEAU M.

Laboratoire d'Optique des Fluides - Faculté des Sciences - 49045 ANGERS - FRANCE -

The collisional light scattering from fluids induces polarized $I(\nu)$ and depolarized $i(\nu)$ intensities.

The long tails at high stokes and antistokes frequencies observed in Rayleigh spectra of tetrahedral molecules and the increasing variation of the ratio $q(\nu)$ (defined as $q(\nu) = I(\nu)/i(\nu)$) with frequency are attributed to multipolar effects.

The measure value of $q(\nu)$ is not constant and equal to 4/3 - value calculated in the first order DID theory - , so the isotropic and aniso--tropic part of the collision induced polarizability tensor must be computed.

In two different types of theory - CIRRS theory(1) - MIM theory(2) using a conducting spheres model -, the spatial distribution of the local field acting upon a molecule has been taken into account and the DID theory has been extended. The effects of the spatial dependence of the field are related to the multipolar polarizabilities. The computed data $q_{MIM}(\nu)$ (3) and $q_{CIRRS}(\nu)$ - POSCH(4) has computed this ratio for gaseous $C(CH_3)_4$ - increase with frequency : multipolar effects appear to be of importance.

In our measurements (figure I) , we note :

i) a constant value (equal to 4/3) of the ratio $q(\nu)$ in the reference fluid - Nitrogen - whose rotational Raman spectrum depends essentially on the proper anisotropy of the molecule.

ii) an evolution of the ratio with frequency in isotropic molecular fluids such as CH_4 and CF_4.

iii) a same evolution in an atomic gas : Argon.

For atomic gases, only MIM theory can explain the increasing evolution of the ratio. These experimental results show that the different effects - CIRRS, MIM and ...electron overlap - must be taken into account together.

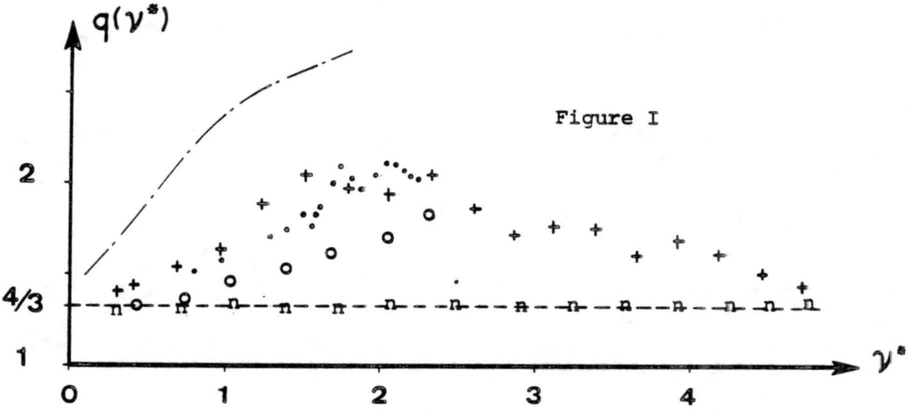

Comparison of the computed ratio $q_{MIM}(\nu)$ with the experimental data. ν^* is the reduced frequency : $\nu^* = \nu.\sigma.(m/kT)^{1/2} / 4$ where σ and m are the diameter and molecular mass.
The experimental data are : + for CH_4 ; . for CF_4 ; o for Argon and n for Nitrogen.

REFERENCES

(1) BUCKINGHAM A.D. and TABISZ G.C. ;Optics Lett 1(1977)220
BUCKINGHAM A.D. and TABISZ G.C. ;Mol. Phys. 36(1978)583

(2) BERRUE J. ; CHAVE A. ; DUMON B. ; THIBEAU M. ;J. Phys. Paris 36(1975)1079

(3) BARREAU A. ; BERRUE J. ; CHAVE A. ; DUMON B. ; THIBEAU M. ; Opt. Comm. 55(1985)99

(4) POSCH H.A. ; Mol. Phys. 46(1982)1213

STRONG COLLISION MODEL FOR THE CALCULATION OF COLLISIONALLY NARROWED ISOTROPIC RAMAN SPECTRA.

J. P. Sala[1], J. Bonamy[1], D. Robert[1], B. Lavorel[2], G. Millot[2], and H. Berger[2].

(1) Laboratoire de Physique Moléculaire, U. A. CNRS n° 772, Route de Gray, 25030 BESANCON - FRANCE
(2) Laboratoire de Spectronomie Moléculaire et Instrumentation Laser, U. A. CNRS n° 777, 6 Bd Gabriel 21100 DIJON - FRANCE

Coherent Raman techniques are now widely used for measuring temperature and concentrations in combustion environments and require a theoretical simulation of the observed spectral profiles.

For most molecules (N_2, CO, CO_2, H_2O, . . .), even at atmosphere pressure, the observed Q-branch exhibits a collisional narrowing which necessitates the study of line coupling effects resulting from rotational population transfers. However at high temperature, the high number of populated levels makes unrealistic the calculation of all the rotational transfer rates. This situation has motivated several authors to modelling the rotational energy transfer rates.

We present a model based on the strong collision approximation which assumes that each collision totally thermalizes the rotational state of the optically active molecule, but, in constrast with previous approaches, the transfer rates satisfy both detailed balance principle and unitarity of the scattering matrix. This model requires only the knowledge of γ_J line broadening coefficients to compute the collisionally narrowing spectra.

Applications to N_2 are reported : an accurate set of γ_J linewidths was established from an experimental study of the isotropic Q-branch performed on the stimulated Raman spectrometer of DIJON and complemented by semi-classical calculations using Bonamy and Robert's model. From this set of γ_J values, collisionally narrowed spectra have been computed and compared with SRS experiments, giving an excellent agreement.

DEPOLARIZED SCATTERING FROM HYDROGEN AT LOW TEMPERATURE

LE DUFF.Y*, CHAVE.A, DUMON.B, IDRISSI.A, THIBEAU.M

Laboratoire d'Optique Des Fluides - Faculté des Sciences-
49045 ANGERS- FRANCE

* on leave from : D.P.R - Université PARIS VI- FRANCE

Recent experiments on hydrogen in the gaseous state at 123 K [1] show that the depolarized intensity scattered with low frequency shifts (ν 40 cm^{-1}) is relatively higher for the thermal equilibrium H_2 than for the para H_2. Different effects may contribute to explain this observation. At 123K most of the para H_2 molecules are in the rotational state J=0 which is a very particular state for a diatomic molecule. Indeed in this case, from a scattering point of view a diatomic molecule behaves like an atom with an anisotropic polarizability equals to zero and no depolarized scattering appears. On the contrary for the thermal equilibrium H_2 at 123 K about 67% of the molecules are in the rotational state J=1 for which the polarizability tensor is anisotropic. In this case, depolarized scattering coming from the permanent anisotropy of the molecule may take place.

In this work, we have studied the contribution from the interference of the permanent anisotropy of the ortho molecule (J=1) and the collisional anisotroy in the dipole-induced dipole approximation. Using an anisotropic potential to take into account the orientational correlation between intermolecular and intramolecular axis of molecules in the low density range, we show that the interference effect [2] contributes very weakly to the depolarized scattering and cannot explain the difference between the para H_2 and the ortho H_2 spectra. Other possible effects such as the contribution from dimers and the collisional broadening of the depolarized band due to permanent anisotropy are also discussed.

References

[1] . LE DUFF.Y and OUILLON.R J.Chem.Phys. 82, 1(1985)

[2] . BERRUE.J, CHAVE.A, DUMON.B and THIBEAU.M Can.J.of Phys.59,1510,(1981)

COLLISION-INDUCED POLARIZED AND DEPOLARIZED RAMAN SCATTERING BY ANISOTROPIC MOLECULES (H_2,D_2).

Michael S. Brown, Wang Shu-Kun[1] and Lothar Frommhold

University of Texas at Austin, Austin, Texas 78712

[1]Peking University, Peking, China

The collision-induced Raman spectra of hydrogen and deuterium are recorded on an absolute intensity scale with the incident beam polarized both perpendicular to and parallel to the direction of observation. Detailed analysis of the hydrogen spectra allows for the separation of the induced polarized and depolarized components. As was seen in the rare gases (Can. J. Phys. 59,1459(1981)), the polarized component is very weak and broad compared to the depolarized component.

In the isotropic potential approximation, quantum lineshapes are calculated using the dipole-induced-dipole (DID) model of the pair polarizability. These profiles agree very well with the observed depolarized spectra at low frequency shifts. At high frequency shifts, overlap corrections are needed for complete agreement with the measurements. Currently an attempt is being made to understand the weak polarized spectrum on the basis of ab initio calculations and semi-empirical models of the trace of the pair polarizability.

In the vicinity of $S_o(0)$ and $S_o(1)$, broad depolarized wings are seen superimposed on the allowed rotational lines due to induction and pressure broadening. Current work includes trying to understand the relative importance of induction and broadening in and near the allowed lines. Independent measurements in deuterium confirm the observations in hydrogen.

INTERMOLECULAR POTENTIALS AND HIGH PRESSURE RAMAN STUDIES.

V. CHANDRASEKHARAN[+] and E. UZAN[++]

+ Laboratoire des interactions moleculaires et des hautes pressions, C.N.R.S., Univ. Paris-Nord, 93430 Villetaneuse.FR.
++ Laboratoire des rayonnements et des structures, C.M.S.R., Ile de Saulcy, 57045 Metz, FR.

Several years ago (1,2) the authors showed that the lattice mode Grüneisen parameters obtained from high pressure Raman studies of solids is important in deciding between different potential models. In the case of solid N_2, the observed Grüneisen parameters of Raman active modes favor the addition of an effective quadrupole moment to the atom-atom potentials in vogue. On the other hand the pressure shifts of the intramolecular vibrational modes are relatively small. Despite this fact, the value of such studies was stressed by the authors (2) since the vibrational frequencies can be calculated by perturbation methods. The crystal field acts as a small perturbation on the high frequencies of the internal modes of the free molecule. At few hundreds Kbar attainable with diamond anvil cells, much larger shifts have been observed. The value of these shifts provides a useful tool for refining the effective potentials.

At room temperature and pressures >40 Kbars a new phase of solid N_2 was discovered with Pm3n structure. There are 8 molecules per unit cell with two types of rotationally disordered molecules (2 spherical and 6 disc type) and only a molecular dynamics simulation is really appropriate for studying this phase change. But the authors (2,3) correctly suggested that the perturbation theory could still be invoked for studying the frequency shifts of the internal modes. This novel method yielded the correct estimates of the relative shifts of the disc and spherical molecules.

Later this idea was pursued (4) to calculate the equilibrium structure and molecular orientations of the ordered crystal with well defined orientations of all the molecules. The total potential energy was minimized with respect to the five parameters: the external rhombohedral angle β for a given volume, Δ the center of mass coordinate (Δ, 1/4, 1/2, $-\Delta$), ϕ the orientation and the 2 bond lengths r_d and r_s of the disc and spherical molecules. Using an ab initio pair potential(5) and intramolecular potential with quadratic and cubic force constants, the p-V curves were calculated for different structures γ-N_2 and $R\bar{3}c$. It showed that a phase transition occurs at 19 Kbars from γ-N_2 to $R\bar{3}c$ calcite structure. This agrees with later Raman scattering measurements. But the curves for the shifts of internal mode of N_2 at both sites versus pressure was consistently steeper than that observed. This clearly indicated the necessity of softening the pair potential at small volumes. This was achieved by fitting a softer pair potential at small intermolecular distances to an earlier potential at large distances with a spline. This gives a good agreement with experiment for the p-V and frequency shifts. Thus the intramolecular frequencies at high pressures provide a useful handle to refine effective intermolecular pair potentials.

The lattice dynamics of this structure at different pressures gives the frequencies of the $4A_{1g}+4A_{2g}+4A_{1u}+4A_{2u}+8E_u+8E_g$ modes of the crystal at k=0.

REFERENCES:

1. V. Chandrasekharan, D. Fabre, M.M. Thiery, E. Uzan, M.C.A. Donkersloot and S.H. Walmsley, Chem. Phys. Letters 26 (1974) 183.
2. M.M. Thiery and V. Chandrasekharan, J.Chem.Phys.67 (1977) 573.
3. V. Chandrasekharan and E. Uzan, 8th Int. Conf. on Raman spectroscopy (Bordeaux 1982).
4. V. Chandrasekharan, R.D. Etters and K. Kobashi, Phys. Rev. B28 (1983) 1095.
5. T. Luty and A. van der Avoird, J. Chem. Phys. 72 (1980) 6107.

VIBRATIONAL AND REORIENTATIONAL RELAXATION OF LIQUID CARBON DISULFIDE AT PRESSURES UP TO 10 KBAR.

Shun-ichi Ikawa* and Edward Whalley

Division of Chemistry, National Research Council of Canada, Ottawa, Ontario, K1A 0R9, Canada.
*Department of Chemistry, Faculty of Science, Hokkaido University, Sapporo, 060, Japan.

Raman band shapes are an important source of information about the dynamics of molecules in liquids, and many theoretical models have been proposed to interpret them. Pressure is a useful variable for testing these models because compression reduces the intermolecular distances and so exposes the molecules to different ranges of the intermolecular potential. In this work, we have measured the Raman line widths of the ν_1 band of liquid carbon disulfide at pressures up to 10 kbar. Details of the spectroscopic measurements have been decribed elsewhere [1]. A 25-kbar hydrostatic Raman cell [2] was operated at room temperature and the sample pressure was measured with a manganin gauge. The depolarized scattering was measured with the geometry Z(YZ)Y, so that the effect of the polarization scrambling by stressed windows of the cell could be made small enough to be neglected [3]. Following the usual procedure [4], the vibrational and reorientational relaxation times, τ_v and τ_r respectively, were obtained from the isotropic and anisotropic line widths, δ_{iso} and δ_{aniso} respectively, as follows:

$$\tau_v = 1/[2\pi c\delta_{iso}], \qquad \tau_r = 1/[2\pi c(\delta_{aniso} - \delta_{iso})].$$

Fig.1 shows a plot of τ_v against pressure. The value at 1 bar, 19±2 ps, is in good agreement with the literature values of 21 ps [4] and 19.4 ps [5]. It decreases to one fourth as the pressure increases to 10 kbar. Since liquid carbon disulfide is non-associating, the effect of the inhomogeneous broadening is neglected and τ_v is attributed to phase relaxation owing to a repulsive interaction. Among several models proposed, a hydrodynamic model by Oxtoby [6] and a kinetic model by Schweizer and Chandler [7] are examined. In the hydrodynamic model, the dephasing time is inversely proportional to the viscosity of the liquid, while in the kinetic model it is related to the Enskog collision time, which is a function of the density of the liquid. Using the viscosity and density of liquid carbon disulfide under pressure [8], the dephasing times were calculated and are compared with the experimental values in Fig.1. The kinetic model (solid line) successfuly reproduces the experimental relaxation times over the pressure range observed, but the hydrodynamic model (broken line) fails to explain the pressure dependence.

As shown in Fig.2, τ_r is linear in the viscosity. The best fit is obtained by the equation

$$\tau_r = [0.555 + 0.780(\eta_p/\eta_1)] \text{ ps}.$$

The finite value of τ_r at $\eta = 0$ can be explained by the effect of the inertial rotation. The reorientational relaxation time τ_f of the classical free rotor, which is given by $(2\pi/9)(I/kT)^{1/2}$ [9], is estimated to be 0.556 ps, which is in good agreement with the experimental value, 0.555 ps. The slope of τ_r vs. η_p can be estimated by a hydrodynamic model with a slip boundary condition [10]. Treating a carbon disulfide molecule as a prolate with semiaxes of 1.85 Å and 3.41 Å, the slope was estimated to be 260 ps/poise. It is in good agreement with the experimental value, 220 ps/poise, estimated by using η_1 = 0.00374 poise.

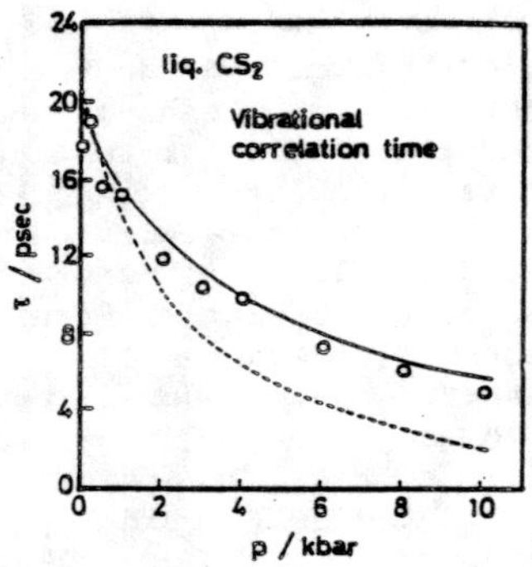

Fig.1 The vibrational relaxation time of liquid CS_2 under pressure at 22°C. o:observed, ——:calculated by a kinetic model [7], ----:calculated by a hydrodynamic model [6].

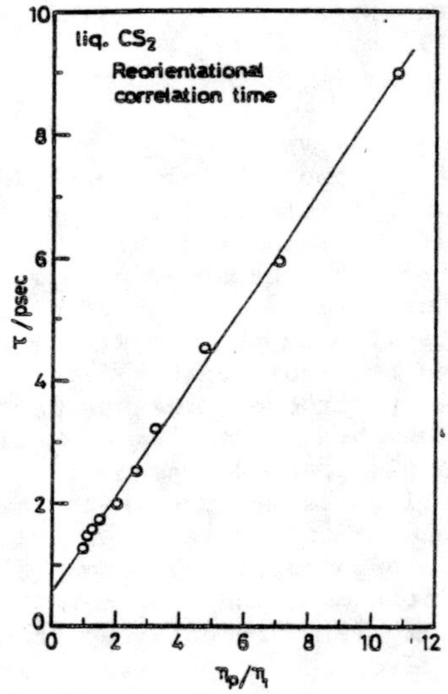

Fig.2 The reorientational relaxation time of liquid CS_2 plotted against the relative viscosity. The straight line is given by τ_r=0.555+0.780(η_p/η_1).

References

1. S. Ikawa and E. Whalley, to be published.
2. E. Whalley, P.T.T. Wong, and A. Lavergne, Rev. Sci. Instrum. 47, 845 (1976).
3. S. Ikawa and E. Whalley, Rev. Sci. Instrum. 55, 1273(1984).
4. F.J. Bartoli and T.A. Litovitz, J. Chem. Phys. 56, 404(1972).
5. T.I. Cox, M.R. Battaglia, and P.A. Madden, Mol. Phys. 38, 1539(1979).
6. D.W. Oxtoby, J. Chem. Phys. 70, 2605(1979).
7. K.S. Schweizer and D. Chandler, J. Chem. Phys. 76, 2296(1982).
8. P.W. Bridgman, The Physics of High Pressure (G. Bell & Sons, 1958).
9. D.R. Bauer, J.I. Brauman, and R. Pecora, J. Am. Chem. Soc. 96, 6840 (1974).
10. C.-M. Hu and R. Zwanzig, J. Chem. Phys. 60, 4354(1974).

MOTIONAL NARROWING IN THE RAMAN SPECTRUM OF IODINE INTERACTING WITH BENZENE OR ITS METHYLATED DERIVATIVES

M. Paatero, S. Eskola, and F. Stenman

University of Helsinki, Department of Physics,
Siltavuorenpenger 20 D, 00170 Helsinki, Finland

The effects of chemical exchange on the vibrational spectra of molecular complexes are similar to those observed in NMR spectra (1,2). We report here the results of a study of the Raman spectrum of complexes of iodine with benzene or its methylated derivatives. The complex formation causes the iodine vibrational frequency to decrease. The decrease is roughly proportional to the number of methyl groups (3).

We have studied the Raman spectrum of iodine dissolved in mixtures of an inert solvent (n-heptane) and benzene, toluene, o-xylene or mesitylene. The ratio of inert solvent to complexing solvent was varied between 0:1 and 1:0. Spectra were recorded at room temperature using Kr laser 647.1 nm excitation.

Our results show that in the mesitylene solution there are two single lines present, one around 210 cm^{-1} corresponding to free iodine and one at 200 cm^{-1} to bound iodine. As the number of methyl groups is decreased, the lines begin to coalesce so that in benzene there is only one line visible. The frequency of this line changes continuously with increasing benzene concentration from 210 cm^{-1} to 204 cm^{-1}.

To explain our findings, we have applied Kubo's theory for the lineshape of an oscillator jumping between two frequencies $\bar{\nu}_1$ and $\bar{\nu}_2$ in a Markov process (4). First we define:

n_C = initial concentration of complexing solvent

n_I = initial concentration of iodine

n_{CI} = concentration of complex formed

It is to be noted that n_C can be changed at will by adding a suitable amount of inert solvent. Using these quantities, the following rate equations are obtained:

$$\dot{n}_{CI} = -k_1 n_{CI} + k_2 (n_C - n_{CI})(n_I - n_{CI}) = -(\dot{n}_I - \dot{n}_{CI}) \quad (1)$$

The equilibrium constant for the reaction comes out as

$$K = k_2/k_1 \tag{2}$$

Defining two new rate constants as

$$c_1 = k_1 \;,\; c_2 = k_2(n_C - n_{CI}) \tag{3}$$

c can be derived from the rate equations as

$$c_2 = \frac{1}{2}Kc_1\left((n_C - n_I - \frac{1}{K}) + \sqrt{(n_C - n_I)^2 + \frac{2}{K}(n_I + n_C) + \frac{1}{K^2}}\right) \tag{4}$$

Kubo's theory gives the following expression for the spectrum (C is a proportionality constant):

$$I(\nu) = C\frac{(\bar{\nu}_2 - \bar{\nu}_1)^2}{(\bar{\nu} - \bar{\nu}_1)^2(\bar{\nu} - \bar{\nu}_2)^2 + (c_1(\bar{\nu}-\bar{\nu}_2) + c_2(\bar{\nu}-\bar{\nu}_1))^2} \tag{5}$$

Thus, the ratio $c_1/(\bar{\nu}_2 - \bar{\nu}_1)$, K and the initial concentrations n_I and n_C determine the spectrum. If K is known, the complex lifetime $\tau = 1/c_1$ can be deduced by fitting (5) to the observed spectrum. Before that, however, the inhomogenous broadening caused by the solvent must be accounted for by deconvolution with the lineshape measured in the inert solvent. Using the values for K given in Ref. (3) we obtain the following values for τ: benzene 6 ± 2 ps, toluene 7 ± 3 ps, o-xylene 8 ± 3 ps and mesitylene 16 ± 5 ps.

REFERENCES:

(1) B. Cohen and S. Weiss, J. Chem. Phys 72, 6804 (1980).
(2) M. Paatero, S. Eskola, and F. Stenman, to be published.
(3) H. Rosen, Y.R. Shen, and F. Stenman, Mol. Phys. 22, 33 (1971).
(4) R. Kubo, in Fluctuations, Relaxation and Resonances in Magnetic Systems, edited by D. ter Haar, Oliver and Boyd, Edinburgh 1963.

RAMAN INTENSITY AND MOLECULAR INTERACTION

E.D.Schmid, E.Brodbek, M.Moschallski and U.E.Hnida

Institut für Physikalische Chemie, Universität Freiburg, F.R.Germany

Our experimental investigations of substituent effects on benzene derivatives (1,2) and solvent effects on the Raman intensities of numerous substrate molecules (3) have shown that the Raman intensity (RI) is very sensitive to molecular interactions. We present two small parts of our extensive and systematic RI studies of intermolecular effects in the liquid phase. In part (A) we consider the case where the solvent effect on RI is solely determined by the preresonance Raman effect and in part (B) we discuss where the resonance is negligible.

(A) With 4-nitroaniline and N,N-diethyl-4-nitroaniline, 4-nitrophenol and 4-nitranisole as an example we will show the dependence of RI on the solvent property. Taft et al. used solvatochromic comparison of uv-visible spectral data to construct a π^*-scale of solvent polarity-polarizabilities and a β-scale ofhydrogen-bond acceptor (HBA) ability. In case of 4-nitroanline - as an example - there is a correlation between the Raman intensity S(NO2) of the symm. NO_2 - stretching vibration and the π^*- and β-solvent parameters:

$$S(NO2) = -162 + 488(\pi^* + .18) + 429\beta$$

Analogical correlations have been found for other nitrobenzene derivatives. These results are consistent with our finding that the solvent dependence of S(NO2) of the nitrobenzene derivatives may quantitatively interpreted in terms of the preresonance Raman effect.

(B) With N,N-dimethylformamide (DMF) and N,N-dimethylacetamide (DMA) we will show an other dependence of RI on the solvent property. In case of DMF, for example, there is a correlation between the RI of the amide I (C=O stretching) vibration S(CO) and the π^*- and α-solvent parameters; Taft's α-scale is used as a measure of hydrogen-bond donor ability:

$$S(C{=}O) = .896 - .229\pi^* - .394\alpha$$

Contrary to S(NO2) of nitrobenzene derivatives the Raman intensity S(C=O) of amide derivatives decreases with increasing solvent parameters. For example S(C=O) of DMA decreases from 4.069×10^{-30} cm^2 in carbon tetrachloride to 0.324×10^{-30} cm^2 in water. This feature is due to hydrogen bonding. Hydrogen bonding between N,N-disubstituded amides (DMF, DMA) and water causes a charge transfer in the amide molecules which can be described with the resonance structures:

The charge transfer contributes to a decrease of C=O bond order and C=O stretching force constant. This indicates a loss of double bond character of the C=O bond and is also reflected in the smaller wavenumber and Raman intensity of the amide I vibration of the DMF- and DMA-water complex, compared to the DMF and DMA molecule. The exciting line is sufficently far off resonance that the denominator of the scattering tensor has the same value for the isolated amides and amide-water complexes.

A more detailed insight into the dependence of RI on hydrogen bond is given by our intensity calculations (4). The results of the Raman intensity calculation of amide-water complexes show that the variational method and the CNDO approximation supply reliable absolute Raman intensities of complex molecular systems. The agreement between measured and calculated Raman intensities makes it possible to correlate the changes in Raman intensities with intermolecular interactions which are considered in a 1:2 amide-water model complex.

1. E.D.SCHMID and R.D.TOPSOM. J.Am.Chem.Soc. 103, 1628 (1981)
2. E.D.SCHMID and R.D.TOPSOM. J.Raman Spectrosc. 14, 191 (1983)
3. M.MOSCHALLSKI. Dissertation, Freiburg. 1984.
4. E.D.SCHMID and E.BRODBEK. J.Chem.Phys. 78, 1117 (1983)
   E.D.SCHMID and E.BRODBEK. Theochem. 108, 17 (1984)
   E.D.SCHMID and E.BRODBEK. Can.J.Chem. 63, 1365 (1985)

RESONANCE RAMAN SPECTRA OF HEXAMETHYLBENZENE-TCNQ ELECTRON DONOR-ACCEPTOR COMPLEX.

M. J. Merriam and J. L. McHale

Department of Chemistry, University of Idaho, Moscow, ID 83843

Raman spectra of CH_2Cl_2 solutions of the electron donor-acceptor complex of hexamethylbenzene and TCNQ are obtained with laser radiation within the charge-transfer absorption band as well as within the locally excited transition of the acceptor (TCNQ). Preresonance enhancement via the locally excited state gives rise to large intensity increases of the totally symmetric vibrations of both the free and complexed acceptor. An additional low-frequency mode (Fig. 1), at ~150 cm^{-1}, is observed in the Raman spectrum of the complex, excited with radiation within the charge-transfer absorption band. Based on comparison to Raman and far-IR data of other EDA complexes, we assign this peak to the first overtone of the totally symmetric intermolecular stretch, ν_{DA}, of the EDA complex. Excitation profiles of the intermolecular mode are measured and compared to those of the totally symmetric TCNQ modes.

Apparent intensity borrowing by the solvent modes from the resonance enhanced TCNQ fundamentals has also been investigated. Using cyclohexane as an internal standard, the excitation profiles of CH_2Cl_2 Raman modes are gathered in the "neat" liquid and in the presence of TCNQ. Intensity borrowing appears to derive from a type of intermolecular Fermi resonance effect, with modes which are close in frequency most likely to share intensity. Theoretical analysis is presented which points out the potential of this effect as a tool for studying solvation of Raman active molecules.

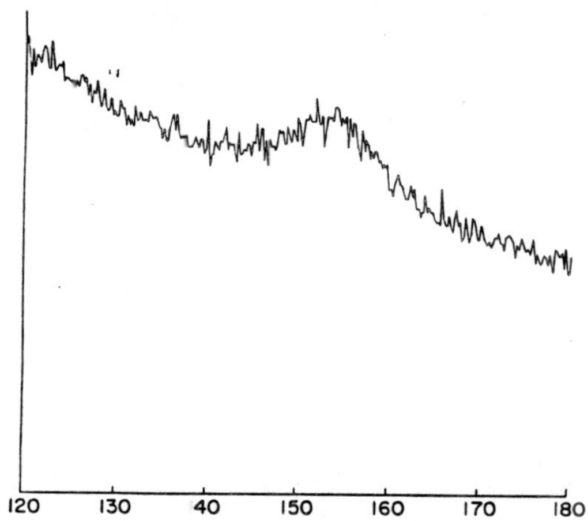

DEPOLARIZED RAYLEIGH WING AND ISOTROPIC/ANISOTROPIC RAMAN SPLITTING IN DIMETHYLFORMAMIDE/CCl_4 SOLUTIONS.

R. Rodriguez and J. L. McHale

Department of Chemistry, University of Idaho, Moscow, ID 83843

Theory and experiment are combined to deduce the role of static and dynamic orientational correlations in the interpretation of three spectroscopic phenomena:

(1) Depolarized Rayleigh wing
(2) Noncoincidence of isotropic and anisotropic peak frequencies of a totally symmetric Raman mode
(3) Far-IR absorption.

Dimethylformamide is a polar liquid known from dielectric studies to have strong orientational correlations. Experimental measurements are made on solutions of dimethylformamide (DMF) and its deuterated derivative (DMF-d_7) in CCl_4, with the goal of elucidating the effects of single particle versus collective orientational correlations on the three types of spectra. Theoretical analysis and comparison of concentration dependent data are employed to deduce the importance of cooperative effects. Comparison of the depolarized Rayleigh scattering and far-IR of DMF to that of DMF-d_7 reveals the contribution of librational motion to the observed spectra. The first moment of the solvent subtracted depolarized Rayleigh Stokes scattering wing is found to vary with the concentration of DMF. The temperature dependence of the first moment of the Rayleigh wing is compared to the predicted temperature dependence of far-IR absorption within collisional and librational models.

Theoretical calculations of the moments of the three kinds of spectra are performed using a model which allows for vibration/rotation coupling and orientation dependent intermolecular interactions. The concentration dependence of the three types of spectra are analyzed and compared to experiment.

ISOTOPIC DILUTION STUDIES OF LIQUID BENZENE BY RAMAN DIFFERENCE SPECTROSCOPY.

N. Meinander and J. Laane

Texas A&M University, Department of Chemistry, College Station, TX 77843

Dilution of benzene in deuterobenzene (and vice versa) can affect its vibrational Raman spectrum in two ways: i) the vibrational relaxation through resonant energy transfer of vibrational energy is hindered, and ii) the small change in molar volume of the liquid causes small shifts in the mean vibrational frequencies. Apart from the small volume effect the average interaction potential between the solute molecules and the solvent should be little affected. The effect of i) is expected to be similar for corresponding vibrational modes of C_6H_6 and C_6D_6, whereas frequency shifts due to ii) should be in opposite directions. Using Raman difference spectroscopy (1), we have studied the small changes that occur in the Raman bands upon mixing of the two isotopic species. The spectra from different samples in a compartmentalized spinning sample cell have been recorded simultaneously by photon counting techniques and stored in digital form. This was done to assure that the different spectra were obtained on a common frequency scale. The difference spectrum was then computed from which the changes in peak position and width of the bands were deduced (Fig. 1) using published methods (2). Observed shifts (3) of some of the Raman bands of benzene determined by this technique are shown in Table I. Also shown is a plausible decomposition of the shifts into a resonance transfer and a volume effect contribution.

In a typical study the four compartments of the spinning sample cell contained benzene and benzene-d_6 in the ratios 1:0, 2:1, 1:2, and 0:1. Comparison of the spectra of the four samples showed clearly that not only are the peak positions and bandwidths affected by isotopic dilution, but also the molar integrated intensities of the bands. In order to confirm this effect quantitatively, carbon tetrachloride was used as an internal standard to obtain a relative intensity calibration of the four spectra. Figure 2 shows the change in molar intensity of the $\nu_2(a_{1g})$ band of C_6H_6 upon dilution.

Wang (4) and Bratos and Tarjus (5) have shown that the distinct terms $\langle a_i(0)\, a_j(t)\rangle$, $i \neq j$, of the vibrational correlation function do not contribute to the integrated intensity of a Raman band, if the macroscopic correlation function is assumed to be the sum of the molecular properties of the isolated molecules. The observed changes of the band intensities must therefore be due to collision induced contributions to the macroscopic correlation function of the neat liquid. The observed intensity changes are in conformity with molecular dynamics results for some diatomic molecular fluids (6).

1) J. Laane and W. Kiefer, Appl. Spectrosc., 35, (1981) 267; 35, (1981) 428.
J. Laane, Vib. Spectra and Structure, 12, (1983) 405.

2) J. Laane and W. Kiefer, J. Chem. Phys., 72, (1980) 5305; J. Laane, J. Chem. Phys., 75, (1981) 2539.

3) A. Johnson M.S. Thesis 1984 Texas A&M University

4) C. H. Wang, Mol. Phys., 33 (1977) 207.

5) S. Bratos and G. Tarjus, Phys. Rev. A 24 (1981) 1591.

6) D. Frenkel and J.P McTague, J. Chem. Phys., 72 (1980) 2801.

Table I Frequency shifts (in cm^{-1}) of the a_{1g} bands of Benzene upon infinite isotopic dilution

(cm^{-1})	solute	solvent	$\Delta\nu_{obs}$	$\Delta\nu_{res}$	$\Delta\nu_{vol}$
992	C_6H_6	C_6D_6	-0.24	-0.42	+0.18
941	C_6D_6	C_6H_6	-0.59	-0.41	-0.18
3060	C_6H_6	C_6D_6	+0.53	+0.43	+0.10
				(0.48)	(0.05)
2292	C_6D_6	C_6H_6	+0.32	+0.42	-0.10
				(0.36)	(-0.04)

Shifts in parentheses proportional to ν.

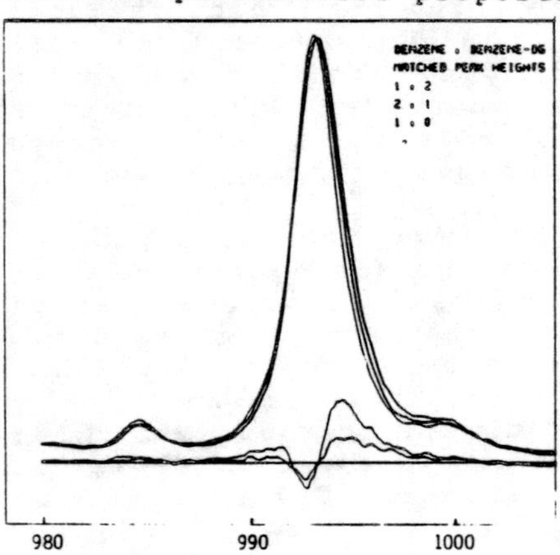

Figure 1.

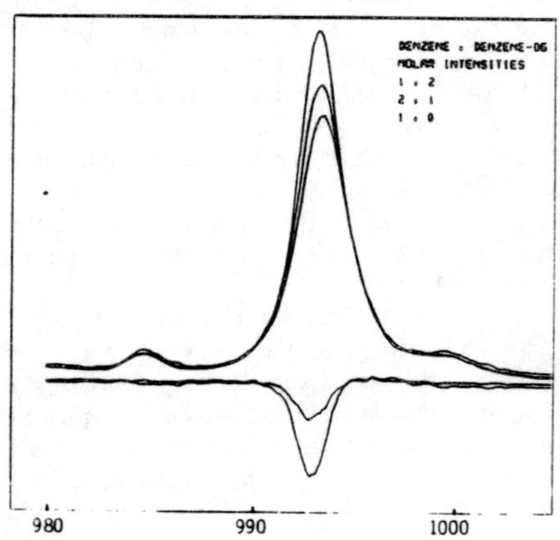

Figure 2.

Lower curves are the difference I(soln) - I(neat liquid)

A MOLECULAR DYNAMICS STUDY OF THE OH STRETCH VIBRATION IN LIQUID WATER.

R. Bansil and T. Berger

Department of Physics, Boston University, Boston MA 02215

In this paper we report a computer molecular dynamics (CMD) study of a model of *flexible* water molecules to directly generate the configurations and trajectories of all the atoms in water. In this approach, the Raman and IR spectra can be calculated directly as Fourier transforms of the time correlation functions of the polarizability and dipole moment , respectively. The calculations are straightforward and simple and avoid many of the complexities of the previous approaches based on *rigid* molecule CMD.[1-2]

The intermolecular potential model employed in this CMD is the simple-point-charge (SPC) model originally proposed by Berendsen et al.[3] The potential consists of a pairwise Lennard-Jones interaction between the oxygen sites, $V_{OO} = -(A/r)^6 + (B/r)^{12}$ with $A = 0.37122$nm $(KJmol^{-1})^{1/6}$ and $B = 0.3428$nm $(KJmol^{-1})^{1/12}$, together with coulomb interaction between suitable point charges situated at the oxygen (-0.82e) and hydrogen (+0.41e) sites. The intramolecular potential, V is chosen to be:

$$V = D\left\{\left(1-e^{-\rho\Delta r_1}\right)^2 + \left(1-e^{-\rho\Delta r_2}\right)^2\right\} + \frac{1}{2}b\Delta r_3^2 + c(\Delta r_1 + \Delta r_2)\Delta r_3 + d\Delta r_1\Delta r_2. \quad (1)$$

Here $\Delta r_1, \Delta r_2$ are the stretch in the OH bond lengths, Δr_3 is the stretch in H-H distance. The parameters are chosen to be the same as those used by Toukan and Rahman[4] with $D = 0.708 mdyn/\AA$, $\rho = 2.37\AA^{-1}$, $b = 2.283 mdyn/\AA$, $c = -1.469 mdyn/\AA$ and $d = 0.776 mdyn/\AA$.

The system we studied consists of 100 molecules in a constant volume simulation cell with periodic boundary conditions. CMD runs at a density of $1g/cm^3$ and temperatures 325 and 250K have been done with a time increment $\Delta t = 6.5 \times 10^{-16}s$. The system was aged for several thousand steps to ensure reaching thermal equilibrium and the pair correlation functions were checked to see that the correct static structure was generated in the simulation.[4]

The Raman spectrum can be calculated as the Fourier transform of the time correlation function of the molecular polarizability.[5] If each bond is assumed to have a cylindrically symmetric polarizability derivative tensor and it is further assumed that non-stretching co-ordinates do not contribute to the polarizability change due to the stretching motion [1] then $A^k_{\alpha\beta}(t)$, the time-varying part of the $\alpha\beta$ components of the polarizability tensor of the kth OH bond can be written as

$$A^k_{\alpha\beta}(t) = \left[\left(\frac{\alpha_\parallel}{\alpha_\perp} - 1\right)\left(\hat{r}_k(t)\cdot\hat{e}_\alpha\right)\left(\hat{r}_k(t)\cdot\hat{e}_\beta\right) + \delta_{\alpha\beta}\right]\delta r_k(t), \quad (2)$$

where $\alpha, \beta = 1, 2$ or 3. Here $\hat{e}_\alpha, \hat{e}_\beta$ denote unit vectors along the 3 space fixed co-ordinate axes $(\hat{e}_1, \hat{e}_2, \hat{e}_3)$, $\hat{r}_k(t)$ a unit vector along the direction of the kth OH bond and $\delta r_k(t)$ is the change in the kth OH bond length from the equilibrium value. The ratio $\alpha_{\|}/\alpha_{\perp}$ is taken as 5.6, the value measured from Raman scattering in single crystal ice.[6] $\delta_{\alpha\beta}$ is the usual Kronecker delta function. The time correlation function for the $\alpha\beta$ component of the polarizability can be written as

$$C_{\alpha\beta}(t) = \left\langle \frac{1}{N} \sum_{k=1}^{N} \left[A^k_{\alpha\beta}(\tau) A^k_{\alpha\beta}(t+\tau) \right] \right\rangle_\tau . \tag{3}$$

The parallel and perpendicular components of the scattered Raman intensity can be written as :

$$I_{\|}(\omega) = K(\omega) \int cos(\omega t) C_{11}(t) dt \ \ and \ \ I_{\perp}(\omega) = K(\omega) \int cos(\omega t) C_{12}(t) dt. \tag{4}$$

Here $K(\omega)$ is the usual prefactor relating the observed Raman intensity to the 'reduced' Raman intensity.[7]

Fig. 1 shows the two components of the 'reduced' Raman intensity calculated at 250K. The calculated spectra are in reasonably good agreement with the experimentally observed band-shape[7] and the ratio of the intensities of the two components agrees with the measured depolarisation ratio for H_2O. A comparison of the calculated spectra at 325 and 250K shows the correct temperature dependence. Clearly better experimental agreement can be obtained by adjusting the potential parameters. In conclusion we note that a first principles simulation of the Raman spectrum of liquid water with the vibrational potential included in the CMD calculation is both feasible and meaningful.

This research was supported by a grant from the Office of Naval Research.

References

1. A. C. Belch and S. A. Rice, J. Chem. Phys. 78, 4817 (1983).
2. D. F. Coker, J. R. Reimers and R. O. Watts, Aust. J. Phys. 35, 623 (1982) and references therein.
3. H. J. C. Berendson, J. P. H. Postma, W. F. van Gunsteren and J. Hermans in Intermolecular Forces, Ed. B. Pullman (Reidel, Dorolzecht, 1981).
4. K. Toukan and A. Rahman, Phys. Rev. B 31, 2643 (1985).
5. D. A. McQuarrie, Statistical Mechanics (Harper and Row, 1976).
6. J. R. Scherer and R. G. Snyder, J. Chem. Phys. 67, 4794 (1977).
7. J. R. Scherer, M. K. Go and S. Kint, J. Phys. Chem. 78, 1304 (1974).

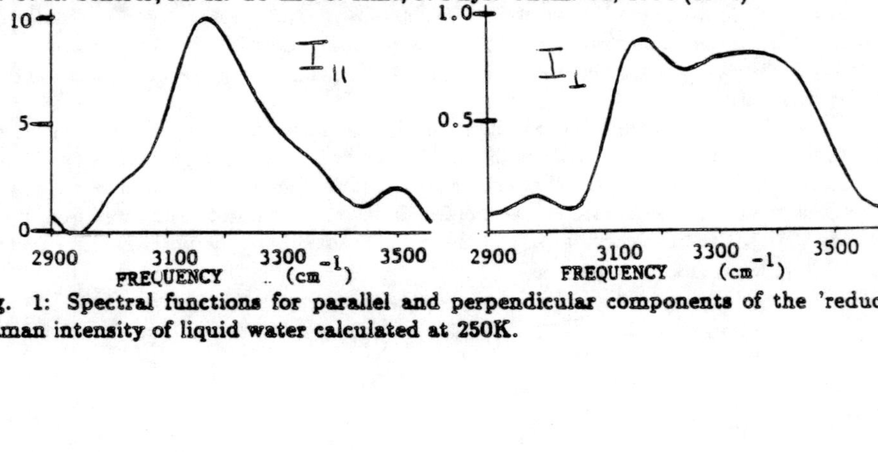

Fig. 1: Spectral functions for parallel and perpendicular components of the 'reduced' Raman intensity of liquid water calculated at 250K.

ROTATIONAL DYNAMICS OF LIQUID WATER: ISOTROPIC INDUCED LIGHT SCATTERING SPECTRA.

G. Briganti[°], M.A. Ricci[°], A. De Santis[+], G. Ruocco[*]

[°] Dipartimento di Fisica, Università di Roma "La Sapienza", P.le Aldo Moro, 2, 00185 Roma, Italy,
[+] Dipartimento di Chimica-Fisica, Università di Venezia, Dorsoduro, Venezia, Italy.
[*] Dipartimento di Fisica, Università dell'Aquila, L'Aquila, Italy.

Raman scattering spectra of the translational-librational region ($0 \div 2000$ cm^{-1}) of water in the normal and supercooled regime ($T = -20 \div +75$°C) have been measured by means of a Jarrel-Ash double monochromator, in two scattering configurations, i.e. VV and HV. At all the investigated temperatures, the depolarization ratio, $\rho = I_{HV}/I_{VV}$, is a function of the frequency shift: it never reaches the value 3/4, even at lower frequencies (i.e. $\omega \simeq 10$ cm^{-1}), and goes down to $\simeq 0.4$ at $\omega \simeq 1600$ cm^{-1} (see fig. 1). This means that an isotropic contribution can be extracted from the measured spectra as:

$$I_{ISO} = I_{VV} - \frac{4}{3} I_{HV}$$

In ref. (1), comparing the isotropic spectrum of H_2O and H_2S at room temperature, we could state that both spectral shapes can be explained in terms of induced light scattering contributions mainly arising from quadrupole-dipole polarizability, as already found in the depolarized light scattering from octahedral molecules[(2)].
Moreover a comparison with the isotropic spectrum of the corresponding deuterated liquids suggests that two different dynamics are related to the low and high frequency part of the spectrum. Indeed either in H_2O or in H_2S the isotropic spectrum scales with the squared root of the inverse mass of the molecule at low frequency, as it is expected for translational dynamics induced contributions. On the contrary, the higher frequency contribution scales with the squared root of the inertial moment, in H_2O and D_2O. Moreover, the temperature behaviour of the spectral shape strongly supports the idea that low and high frequency part of the isotropic spectrum of water are related to translational and librational dynamics respectively.

The integrated intensity behaviour also agrees with the assignement of the two spectral contributions (see fig. 2) to two different terms in the polarizability expansion[3], i.e.

$$\tilde{\Pi}^{(i)}_{\alpha\beta} = \frac{1}{2}\beta^{(i)}_{\alpha\beta\gamma}\, T^{(2)(ij)}_{\gamma\delta}\, \mu^{(j)}_{\delta} + \frac{2}{3}\,\alpha^{(i)}\, T^{(3)(ij)}_{\alpha\gamma\delta}\, A^{(j)}_{\gamma\delta\beta}$$

Developing of a simple librational model for the l=3 rotational correlation functions is in progress, in order to compare the calculated lineshapes with the experimental ones.

References

1) A. De Santis, V. Mazzacurati and M.A. Ricci, submitted to Chem. Phys.
2) A.D. Buckingham and J.C. Tabisz, Molec. Phys. 36, 583 (1978).
3) A.D. Buckingham, Adv. Chem. Phys. 12, 107 (1967).

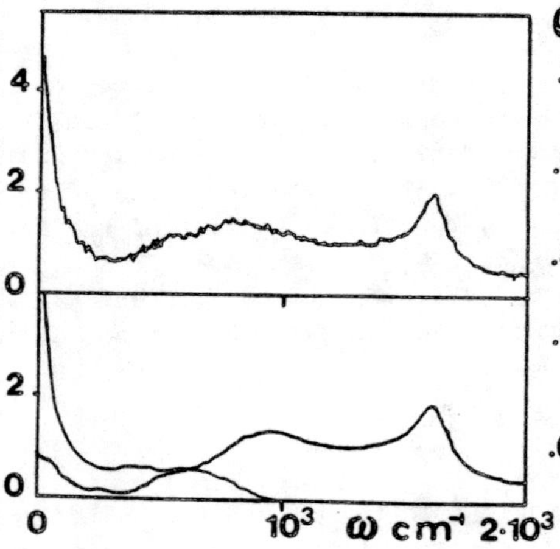

fig. 1 - Depolarization ratio $= I_{HV}/I_{VV}$ at T=-20°C vs

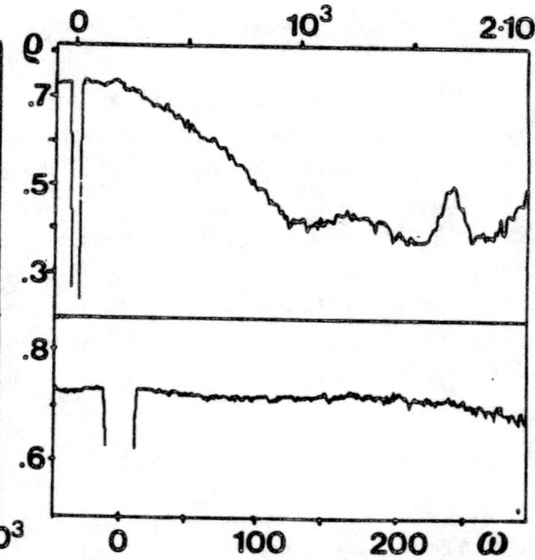

fig. 2 - a) Isotropic spectrum from water at room temperature vs.
b) Low and high frequency components of the isotropic spectrum

VERY LOW FREQUENCY DEPOLARIZED RAMAN SCATTERING IN LIQUID AND SUPERCOOLED WATER

G. Signorelli, V. Mazzacurati, P. Benassi
Dip. di Fisica University "La Sapienza" of Rome, Italy

M. Nardone
Dip. di Matematica e Fisica, University of Camerino, Italy

The depolarized low frequency Raman spectrum of liquid water has been investigated by many authors in the recent years because of its very peculiar temperature dependence, in some way attributed to the dynamics of the hydrogen bond (1). Decreasing the temperature the spectral shape narrows partially merging into the spourious elastic scattering, which represents the low frequency limit of observed spectrum.
Using available multiple grating spectrometers the low frequency limit of reliability for the reported spectra was never reduced below 1 cm^{-1} from the exciting line and an accurate analysis of the very low frequency shape has not yet been performed. In the present work we extend the low frequency limit of the observed H_2O spectrum down to 0.1 cm^{-1} shift, matching the spectra taken both with a double monochromator and with a Fabry-Perot interferometer at different Free Spectral Ranges. The entire spectrum will be than analyzed up to 150 cm^{-1} as a function of temperature. Accurate measurements were also made for the depolarization ratio $\rho = I_{HV}/I_{VV}$ and the results indicate that the low frequency spectrum in liquid water is nearly totally depolarized below 150 cm^{-1}. This suggests to reject the hypothesis that the hydrogen bond breaking dynamics is reflected in this frequency range. General arguments, indeed, may be used to deduce that any change in molecular polarizability due to hydrogen bond relaxation has to be partially isotropic in character yielding $\rho < 0.75$. A big hint for the physical assignement of one of the observed components can be obtained by a comparison between the low frequency relaxation contribution present in hexagonal ice single crystals (2) and the temperature independent lorentzian shape we find in the liquid, the intensity of which progressively decreases above 20°C. This contribution is temptatively attributed to a collective proton mode (3), while all the remaining low frequency spec-

trum may be assigned to collision induced translational scattering, reflecting both shear and self-diffusional dynamics.

(1)
C.J. MONTROSE, J.A. BUCARO et al. J. Chem. Phys. 60, 5025, (1974)

(2)
G. BRIGANTI, V. MAZZACURATI et al., Solid State Comm., 42, 493, (1982)

(3)
K.K. KOBAYASHI J. Phys. Soc. of Japan, 24, 497, (1968).

ON THE REORIENTATION OF α-SUBSTITUTED METHYLBENZENE MOLECULES IN LIQUID PHASE.

P.J.A.Ribeiro-Claro, A.M.Amorim da Costa and J.J.C.Teixeira-Dias
The University Chemical Department, 3000 Coimbra, Portugal

In a previous work, experimental evidence of rotational isomerism was obtained from solvent and temperature variation studies of Raman bands of $C_6H_5CHCl_2$ assigned to the CH stretching vibration (1). In addition, analysis of spin-lattice relaxstion data of α-dichloromethylbenzene in the liquid phase (2) concludes on the anisotropy of the overall molecular motion while confirming the rotational isomerism suggested by the previous Raman studies.

In order to further investigate the reorientational dynamics of α-substituted methylbenzene molecules in liquid phase, we have performed an analysis of the Raman bandshape of the $\nu_{12}(a_1)$ fundamental in these molecules. A comparative study of the Raman reorientational correlation function in C_6H_5-X molecules (X=CCl_3, CCl_2H, $CClH_2$, CH_3, $C{\equiv}CCH_3$) supports the spin-lattice relaxation studies and provides further insight into the overall molecular motion in liquid phase at room temperature.

(1) - P.J.A.Ribeiro-Claro and J.J.C.Teixeira-Dias
J. Raman Spectrosc., 15, 224(1984).

(2) - P.J.A.Ribeiro-Claro, C.F.G.C.Geraldes and J.J.C.Teixeira-Dias, 1986, communication to the annual meeting of Portuguese Chemical Society.

RAMAN SPECTRAL STUDY OF ION-PAIRING AND HYDRATION OF ALKALINE AND EARTH ALKALINE PERCHLORATES

M. Jeremić, A. Jovanović and M. Lalić

Institute of Physilac Chemistry, Faculty of Science, University of Belgrade, P.O. Box 550, 11001 Belgrade, Yugoslavia

The solvated metal ion is the most ubiquitous of all metal complexes and, in consequence of this, the exact composition and structure of its first coordination sphere is of considerable interest. According the model of Frank and Wen a specific number of water molecules, the hydration number, occupy the first coordination sphere and assume more or less defined spacial arrangment. The secondory solvation sheeth is conceived as consisting of region occupied by disordered molecules interposed between the hydrogen bonded structure of bulk water and hydrated ion. Hydration numbers have been estimated by large number of methods, often yielding discordant results, but for many ions there is growing confidence that resonable agreed values have been obtained. These discrepances arise probably due to different experimental methods which can not differentiate between first and second coordination sphere.

Due to weakly interacting character of perchlorate ion in aqueous solution, ion pairing is difficult to detect. Recent application of various instrumental methods including Raman spectroscopy in search of evidence for perchlorate contact association have led to mixed results.

We report here the results of the Raman spectral study of the first coordination sphere of some alkaline and earth alkaline perchlorates in aqueous solution as well as the evidence for ion-pairing. These data are complemented by electrochemical measurments.

The perchlorate anion has nine vibrational degrees of freedom distributed into four normal model of vibration: $\nu_1(A_1)$, $\nu_2(E)$, $\nu_3(F_2)$ and $\nu_4(F_2)$. We measured band positions and half-widths of $\nu_1(A_1)$ mode of the perchlorate ion over wide range of concentrations with various alkaline earth alkaline counterions. The other bands showed no significant changes with concentration variation. The concentrations were varied from very dilute to very concentrate one aproaching pure salt melts. In order to overcome solubility problem, all measurments were made at 99.5°±0.2°. As an example of series of measurments for different cations, the variation of $\nu_1(A_1)$ vibration of magnesium perchlorate solutions versus water to cation molar ratio (R) is shown on Fig. 1. The frequency is almost constant for $R>6$. For molar ratios less than six, it increases linearly with decreasing R. This can be explained in terms of hydration of magnesium by six molecules of water for $R>6$. In this case the polarization of the perchlorate ion by cation is weak because it is mediated through water molecules. Whereas for solutions with $R<6$, it becomes very strong because the perchlorate ion enters the first coordination sphere forming with cation contact-ion-pair. Gradual decrease of the number coordinated water molecules leads to increse of actual electrostatic field of the cation causing stronger interaction with the perchlorate ion, what, at the end, leads to increase of the frequency. It is widely accepted that hydration number for magnesium is six. The studies of sodium perchlorate solutions yielded hydration number 4.2 for sodium. The paralel conductance measurments gave the same value supporting interpretation of Raman data. Hydration numbers of magnesium, in particular, as well as litium, sodium, calcium and strontium are in good agreement with data obtained by other methods (1,2).

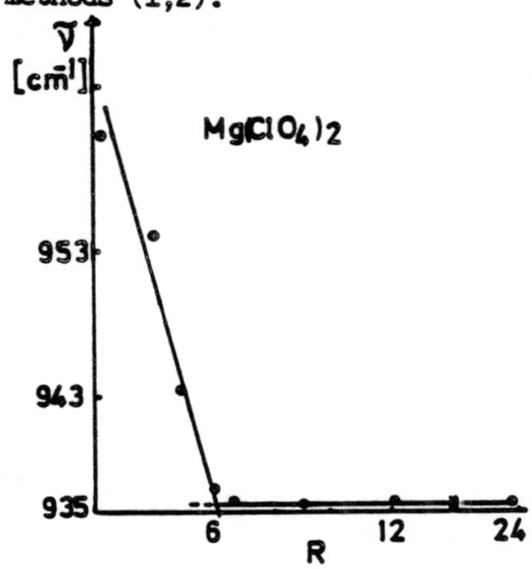

Fig. 1

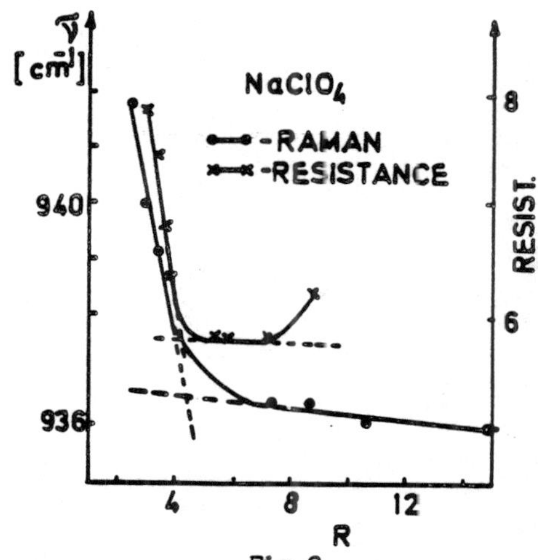

Fig.2

1. J.F. Hinton and E.S. Amis, Chem. Rev., 71, 627, (1971)
2. S.F. Lincoln, Coord. Chem. Rev., 6, 309, (1971).

RAMAN ACTIVE VIBRATIONS OF CATION COMPLEXES IN CONCENTRATED AQUEOUS SOLUTIONS

M.Alves Marques, M.A.Sousa Oliveira e J.Resina Rodrigues

Centro de Física da Matéria Condensada (INIC), Av.Prof.Gama Pinto 2, 1699 LISBOA Codex, PORTUGAL.

Magnesium chloride 3.6M and aluminium chloride 3.1M aqueous solutions were investigated. Heavy water (D_2O) was used, too, as solvent to obtain the isotopic shift of the Raman bands relatively to their spectral situation for solution in ordinary water (H_2O). The modification of the values of the frequency of some bands was interpreted by using models of cation complexes with coordinated anions. A quasi-classical approximation seems to be adequate to calculate intermolecular forces in these idealized agaregates.

RAMAN SPECTRAL STUDY OF SOLUTIONS OF $Mg(ClO_4)_2$ IN SOME ORGANIC SOLVENTS

A. Antić-Jovanović, M. Jeremić and M. Lalić

Institute od Physical Chemistry, Faculty of Science, University of Belgrade, 11000 Belgrade, Yugoslavia

In this work we report the results of Raman spectral study of solutions of $Mg(ClO_4)_2$ and $Mg(ClO_4)_2$-NaSCN system in acetone, acetonitrile and dimethyl-sulfoxide (DMSO) with a view of understanding the ion association and ion-solvent interactions.

For this study, the Raman spectra of solutions and pure liquid solvents were recorded and discussed in terms of positions, relative intensities and half-bandwidths. All measurements were made at room temperature on a Spex 1401 spectrometer employing 514.5 nm line from a Spectra Physics Argon ion laser, model 164.

All solutions of Mg $(ClO_4)_2$ show a strong and sharp band of ClO_4^- stretching mode at ~933 cm^{-1}. In addition to that, a much weaker band at ~910 cm^{-1} is observed for solutions in acetone and DMSO. Simultaneously, significant changes occur on the characteristic Raman bands of pure liquid solvents.

In acetone-perchlorate system, there is a doubling of the bands at ~1710 cm^{-1} and that at ~790 cm^{-1}. The additional components are located at about 1720 cm^{-1} and 810 cm^{-1} while the primary bands, detected for pure liquid acetone, are slightly shifted to ~1706 cm^{-1} and ~795 cm^{-1}, respectively.

In acetonitrile-perchlorate system, there is a splitting of the bands in the C≡N region at 2255 cm^{-1} and 2295 cm^{-1}, as well as the C-C≡N band at ~385 cm^{-1}. As in acetone solutions, the new bands are positioned at higher frequencies, that is at ~2290 cm^{-1}, ~2320 cm^{-1} and ~405 cm^{-1}.

In the case of DMSO-perchlorate system, only perturbation in the structure of S=O stretching band was observed.

It was found that the positions of detected bands and the half-band-widths did not depend on concentration of perchlorate over investigated concentratiom range. However, an addition of NaSCN to solutions caused significant changes on respective bands which were important in explanation of the nature of solveted species and ion-pairs formed in solutions. Namely, it was noticed that the position and relative intersity of weak low frequeny component of the ClO_4^- band at ~910 cm^{-1} remained unchanged in the presence of NaSCN. On the other side, there is a reduction in intensity of the new components of the split bands in acetone and acetonitrile solutions on addition of NaSCN. When the NaSCN/$Mg(ClO_4)_2$ mole ratio $\geqslant$ 2, these components disappear completely.

Considering observed effects in terms of the changes in electronic redistribution of acetone and acetonitrile molecules when bound the metal ion, we concluded that doubling of acetone and acetonitrile modes in perchlorate solutions is due to cation-solvent molecul associations. Such conclusion was supported by variations in band contour in the presence of NaSCN. Its addition to the solution caused the structure breaking of solvates due to formation of ion-pairs Mg^{2+} - $(NCS)_2^-$ and, in a consequence of this, the new components of doubletes disappeared. According to our data, the solvates formed must be $[Mg(CH_3COCH_3)_2]^{2+}$ and $[Mg(CH_3CN)_2]^{2+}$.

The lower frequency of Cl-O stretching band at 910 cm^{-1} was reported earlier[1-3]. However, there is not any proper assignment for this band. Our results suggest that it could arise from the solvated perchlorate ions.

The understanding of interaction of DMSO and $Mg(ClO_4)_2$ is not straightforward and further studies are necessary.

---

1. M.S. Greenbarg and A. I. Popov, J. Solution Chem., 5,653 1976 .
2. D. W. James and R. E. Mayes, Aust. J. Chem. 35, 1775 1982.
3. M. I. S. Sastry and Surjit Singh, Can. J. Chem., 63, 1351 1985.

chapter 17

RESONANCE RAMAN SCATTERING: THEORY AND EXPERIMENT

THEORY OF RAMAN EXCITATION PROFILES OF MULTIMODE SYSTEM WITH DUSHINSKY ROTATION AND JAHN-TELLER EFFECT.

V.Hizhnyakov

Institute of Physics, Estonian SSR Acad. Sci., 202400 Tartu, Riia t. 142, USSR.

The resonant Raman scattering (RRS) of vibronic multimode systems is studied by using the time-correlator formulation [1] which takes exactly into account the contribution of intermediate states to the scattering cross-section. In this formulation, the total Raman cross-section for the given frequency of the excitation Ω_1 and scattering is described by a three-time correlation function. However, a partial RRS, i.e. the Raman scattering of a given order by a given mode, can be described by the one-time correlation function - the Fourier-amplitude of the partial RRS. For example the Fourier-amplitude of the first-order RRS of the mode j in the Condon approximation is

$$F_{1j}(\tau) \sim \Theta(\tau) \langle e^{-i\tau H_1} e^{i\tau H_2} a_{1j}^{+}\rangle,$$

where H_1 and H_2 are the vibrational Hamiltonians of the ground and excited electronic states, a_{1j}^{+} is the generation operator of the phonon with frequency ω_{1j}, $\langle\ldots\rangle$ denotes averaging with the weight factor $\exp(-H_1/\kappa T)$, $\Theta(\tau)$ is the Heavyside step function, $\hbar=1$.

On the basis of this formulation the transform method for Raman excitation profiles (REPs) [2,3] is considered with allowance made for the mixing of vibrational modes on electronic transition (the Dushinsky rotation). It is shown that in the multimode systems with weak and intermediate vibronic coupling, the first order RRS amplitudes in the Condon approximation are described by the following simple formula:

$$A_{1j}(\Omega_1) \sim \phi(\Omega_1) - \sum_{\kappa} c_j \, \phi(\Omega_1 - \omega_{2\kappa}).$$

Here $\phi(\Omega_1)$ is the function of the complex refraction index, $\omega_{2\kappa}$ are the vibrational frequencies in the excited electronic state and $c_{j\kappa}$, parameters of the mode mixing which coincide with the elements of the matrix of Dushinsky rotation within accuracy of a normalization factor.

The REPs of multimode systems with regard to the Jahn-Teller effect in the excited state in the approximation of a weak non-diagonal (in the matrix of vibronic coupling) but arbitrary diagonal vibronic coupling are considered. In such a case, the REPs of the "non-diagonal" modes cannot be found from the ordinary absorption spectrum $\kappa(\Omega_1)$ only, but they can be computed if the change of this spectrum $\kappa^{(1)}(\Omega_1)$ under a low-symmetric external field has also been measured. The corresponding Fourier amplitude of the first-order RRS is

$$F_{1j}(\tau) \sim \Theta(\tau)F(\tau)S(\tau)\phi^{-1}(\tau),$$

where $\phi(\tau)$ will be found as the solution of the equation

$$\phi(\tau) = [F(\tau)/F^{(1)}(\tau)] \int_0^{\tau} ds\phi(s)\phi(\tau-s),$$

$F(\tau)$ and $F^{(1)}(\tau)$ are the Fourier-transforms of the spectra $\kappa(\Omega_1)$ and $\kappa^{(1)}(\Omega_1)$, $S(\tau)$ is the Fourier-transform of the function $\chi(\omega)\chi(\omega-\omega_j)$, $\chi(\omega)$ is the Fourier-transform of the function $\Theta(\tau)\phi(\tau)$. The model calculations are carried out for REPs of the "non-diagonal" non-symmetrical modes.

1. Hizhnyakov V., Tehver I., phys. stat. sol. 21, 755 (1967).
2. Tonks D.L., Page J.B., Chem. Phys. Lett. 66, 449 (1979).
3. Champion P.M., Albrecht A.C., Ann. Rev. Chem. 33, 353 (1982).

THEORY OF TEMPERATURE-DEPENDENT RESONANCE RAMAN SCATTERING BY MULTIMODE SYSTEMS--A MANY BODY APPROACH.*

J. B. Page

Physics Department
Arizona State University
Tempe, AZ 85287

This talk will survey our recent and current research in using many-body techniques within the time-correlator framework [1,2] to develop non-perturbative and tractable theories of the temperature-dependent RR scattering and optical absorption. We focus on multimode systems with intermediate electron-phonon (el.-ph.) coupling, characterized by absorption spectra having well-resolved but overlapping vibrational structure. Representative systems range from complex organic molecules to solid state impurities. Owing to multimode interference, such systems' RR excitation profiles are highly structured, and their study yields detailed information on the vibrational potential energy for the resonant electronic states. However, the traditional theoretical approach of direct summation over the numerous individual intermediate states becomes rapidly intractable as the number of modes increases, especially with thermal effects included.

The time-correlator framework for RR scattering was introduced by Hizhnyakov and Tehver [1] in the course of developing a cumulant expansion scheme which is exact for the important case of linear el.-ph. coupling (displaced oscillators), but which is a poor approximation for other, realistic, cases such as linear plus quadratic el.-ph. coupling (general mode mixing, frequency shifts and equilibrium position shifts with electronic excitation). In our work, we have dropped the cumulant expansion scheme of Ref. 1, but have instead exploited the time-correlator framework as a natural basis for the use of suitably adapted finite-temperature many-body (diagrammatic) techniques. We have thereby achieved non-perturbative solutions for a number of realistic models and have used them to analyze experiments on systems whose vibrationally structured RR profiles and optical absorption spectra provide stringent tests of the theory [3]. Our results allow one to carry out explicit multimode model calculations via one-dimensional fast Fourier transforms. The integrands of the FFTs are convenient analytic functions of time, temperature and the model parameters, and our computing times are orders of magnitude shorter than those for the conventional sum-over-states method. Our results thus offer an analytic framework in which to use model parameters computed from _ab initio_ calculations in order to compute RR profiles and the absorption for complicated molecules which are too large to treat with conventional techniques. In addition to these modeling aspects, our use of the linked cluster theorem brings out clearly the general fundamental relations between the optical absorption and RR profiles. Within some of the models, these relations lead to the development of "transform" techniques for computing profile lineshapes _directly_ from measured temperature-dependent absorption spectra. Such analysis can allow one to extract from _just the data_, well-determined values for the el.-ph. coupling parameters, _a priori_ of explicit model calculations. Transform techniques can thus form a useful prelude to explicit model calculations, and they have been extensively developed by ourselves and others [4]. The many-body approach also leads to useful renormalizations of model parameters [3], and our solutions naturally provide a rigorous and convenient treatment of inhomogeneous and multimode thermal broadening in the absorption and RR profiles [5].

Very recently, we have shown that all of these general features hold for the case of _simultaneous_ linear plus quadratic non-Condon coupling _and_ linear plus quadratic el.-ph. coupling, i.e. arbitrary mode mixing, frequency shifts, equilibrium position shifts and arbitrary linear plus quadratic Herzberg-Teller coupling. This work encompasses many of our earlier results as special cases. New interference effects are found with both types of coupling present, and new absorption-profile transform relations are obtained. These new results will be highlighted here and detailed in another paper at this meeting [6].

Additional recent results, for Jahn-Teller systems and for "hot" and "ordinary" luminescence spectra will also be discussed, as time permits.

[1] V. Hizhnyakov and I. Tehver, Phys. Stat. Sol. _21_, 755 (1967).
[2] J. B. Page and D. L. Tonks, J.C.P. _75_, 5694 (1981).
[3] See, for instance, C. K. Chan, J. B. Page, D. L. Tonks, O. Brafman, B. Khodadoost and C. T. Walker, J.C.P. _82_, 4813 (1985), and earlier refs. cited therein.
[4] See, for instance: C. K. Chan and J. B. Page, J.C.P. _79_, 5234 (1983); P. M. Champion and A. C. Albrecht, Ann. Rev. Phys. Chem. _33_, 353 (1982); and refs. therein.
[5] S. A. Lee, C. K. Chan, J. B. Page and C. T. Walker, J.C.P. _84_, 2497 (1986).
[6] H. M. Lu and J. B. Page, cont. paper at this conference.
* This work was supported by NSF Grant DMR 8204586.

TIME-DEPENDENT APPROACHES TO SPECTROSCOPY INVOLVING NON-ADIABATICALLY COUPLED EXCITED ELECTRONIC STATES.

Rob D. Coalson

Los Alamos National Laboratory
MS J569
Los Alamos, New Mexico 87545

The theory of optical spectroscopy involving excitation to coupled Born-Oppenheimer excited potential surfaces is developed from a time-dependent point of view. Although we stress that the study of many experimentally accessible phenomena (e.g., one and two photon electronic absorption spectra, nonradiative decay rates, etc.) may benefit from this point of view, our main focus here will be on resonance Raman scattering cross sections. In particular, we will discuss the utility of techniques involving localized Gaussian wavepacket propagation and path integration, particularly with regard to processes involving photo-dissociation or molecules characterized by a high vibrational density of states.

INCORPORATING PROTEIN AND SOLVENT ENVIRONMENTS IN RESONANCE RAMAN CALCULATIONS

Arieh Warshel

Department of Chemistry, University of Southern California
Los Angeles, California 90089-0482

Our original method for calculations of RR spectra of conjugated molecules[1,2] did not include the environment of the chromophore in the calculations. Including the environment in an explicit or implicit way might be crucial for relating protein-chromophore interactions to RR spectra. In recent years we developed various strategies for including the environment in RR calculations. These strategies involve several levels of sophistication and include the following approaches:

(a) The static electrostatic field from the protein (or solvent) cage is incorporated into the quantum mechanical calculations of the relevant ground and excited state bond orders. This approach was used in studies of hemoglobin[3,4] and trypsin[5].

(b) The steric strain of the protein is included in the vibrational calculations and its effect on the RR spectrum is evaluated. This approach has been used in studies of rhodopsin[6] and in preliminary studies of hemoglobin.

(c) The fluctuations of the protein electrostatic potential are evaluated by molecular dynamics simulations[7,8] and the corresponding power spectra as used to explore spectral line widths.

The combination of the above strategies appears to provide a powerful tool for interpretation of RR spectra of biological molecules.

REFERENCES:

1. A. Warshel, Ann. Rev. Biophys. Bioeng. 6, 273 (1977).
2. A. Warshel and P. Dauber, J. Chem. Phys. 66, 5477 (1977).
3. A. Warshel and A. Lappicirella, J. Am. Chem. Soc. 103, 4664 (1981).
4. A. Warshel and R.M. Weiss, J. Am. Chem. Soc. 103, 446 (1981).
5. A. Warshel and S. Russell, Quart. Rev. of Biophys. 27, 283 (1984).
6. A. Warshel and N. Barboy, J. Am. Chem. Soc. 104, 1469 (1982).
7. A. Warshel, Proc. Natl. Acad. Sci. 81, 444 (1984).
8. A. Warshel and J-K Hwang, J. Chem. Phys. 84, 4938 (1986).

FEMTOSECOND ROTATION-VIBRATION DYNAMICS AS REVEALED BY RESONANCE ROTATIONAL RAMAN STUDIES OF AMMONIA.

L.D. Ziegler

Department of Chemistry, Northeastern University, Boston, MA 02115

Off-resonance rotational Raman scattering intensities are given by the well-known Placzek-Teller factors weighted by the initial level population. In resonance, the explicit sum over all possible intermediate rovibronic levels must be performed to yield intensity expressions which are dependent on excited state parameters. We have derived expressions for the intensity of rotational scattering transitions of symmetric top molecules using the irreducible tensor representation. Isotropic, anisotropic and antisymmetric scattering contributions are calculated for $\Delta J = 0$ and $\Delta J \neq 0$ transitions. The pattern of observed relative intensities is found to be a sensitive function of laser excitation frequency and resonant excited rovibronic dephasing rate. When the dominant dephasing mechanism is population decay, for example due to photodissociation, then a fit to the observed pattern of relative rotational resonance Raman intensities will determine subpicosecond lifetimes.

This technique is demonstrated by resonance Raman scattering spectra of ammonia. Due to its small moment of inertia, and efficient photodissociation, ammonia is an ideal molecular candidate with which to test this lifetime measurement technique. The lowest lying electronic transition of ammonia extends from $\sim$ 217 - 185 nm. Vibronic absorption bands are well resolved but are rotationally diffuse. Raman excitation frequencies throughout this wavelength region are produced by antistokes Raman shifting Nd:YAG harmonics or Nd:YAG pumped doubled dye.

As the Raman excitation frequency is tuned to different vibronic bands, femtosecond excited state lifetimes may be determined as a function of total energy and vibrational specificity. In addition, the analysis of resonance rotational Raman excitation profiles yield excited state lifetimes with rotational specificity. A fit to three excitation profiles of S transitions are shown in figure 1 for Raman excitation resonant with the $v_2' = 2$ band of NH_3.

Once the excited state vibronic origin and geometry are fixed the damping constant is the only parameter used to fit the rotational scattering excitation profiles. The breadth and height of the profiles are controlled by these damping constants. The best fit excited state damping constants (i.e. photodissociation rates) are plotted as a function of J for a series of S transitions resonant with the $v_2' = 2$ absorption band in figure 2. The rates found by this method span the timescale from $\sim$ 70 to 140 fs and reveal that at least on this portion of the excited potential surface, photodissociation rates are significantly dependent on the particular rotational level excited. Such fast rotational dependent rates could not be measured by direct pump-probe techniques since the energy width of the short pump-probe pulses would measure only an average decay over many nearly resonant rotational levels at these femtosecond lifetimes. The quantum specific lifetimes will be discussed in terms of the mechanism of ammonia photodissociation and show the importance of centrifugal effects even on the timescale of fractions of a rotational period.

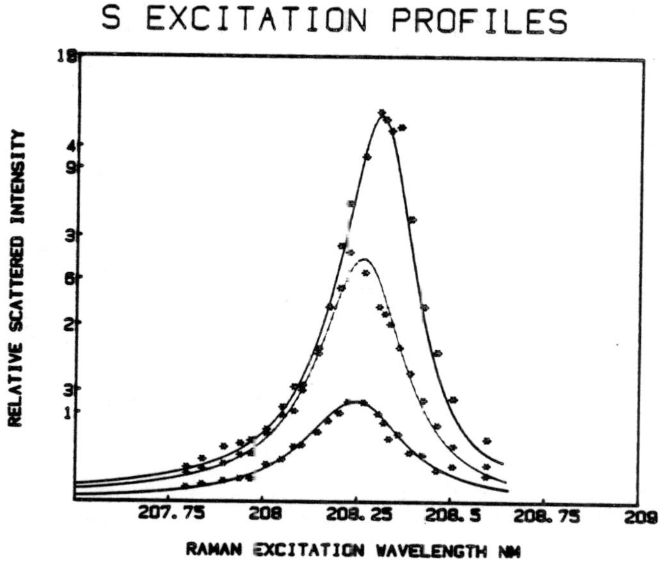

FIGURE 1. Three experimental and theoretical best-fit S rotational excitation profiles are shown for initial J values of 3,5 and 7 theoretical curves determine rovibronic excited state lifetimes of 96, 82 and 76 fs respectively.

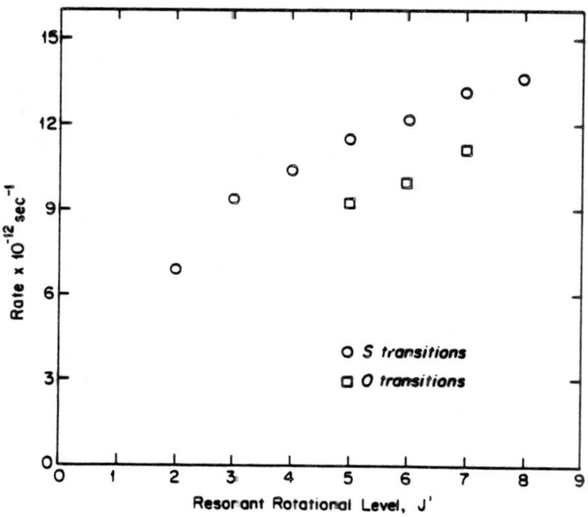

FIGURE 2. Rotational dependence of ammonia photodissociation rate of the $v_2' = 2$ band as determined by rotational excitation profile analysis.

ULTRAVIOLET RESONANCE RAMAN SPECTRA OF NH_3 AND ND_3

W. Hess[*+], L. D. Ziegler[#+] and B. Hudson[+]

[+]Department of Chemistry and Chemical Physics Institute, University of Oregon, Eugene, Oregon
and
[#]Department of Chemistry, Northeastern University, Boston, Massachusetts

Resonance Raman scattering from ammonia has been the subject of several recent studies [L. D. Ziegler and B. Hudson, J. Phys. Chem. 88, 1110 (1984); L. D. Ziegler, P. B. Kelly and B. Hudson, J. Chem. Phys. 81, 6399 (1984)]. Ammonia has a vibrationally resolved electronic transition to its lowest energy excited state (the A state). A long progression is observed in the pyramidalization mode (v_2) due to the planarity of the A state. The individual vibrational bands of the X -> A transition are not rotationally resolved, however, even in a cold jet [V. Vaida, W. Hess, and J. L. Roebber, J. Chem. Phys. 88, 3397 (1984)]. This is due to the rapid dissociation of NH_3 to form NH_2 and H atoms.

One of the interesting features of the resonance Raman spectrum of ammonia obtained with 212.8 nm radiation, resonant with the $v''=0^-$ to $v'=1$ vibronic transition, is that it is dominated by antisymmetric (minus) levels of the ground state. This means that the Raman scattering polarizability expression is dominated by the single vibrational level near resonance. In the present study, the Raman spectrum of NH_3 has been obtained with excitation at 208.8 nm, resonant with the $v''=0^+$ to $v'=2$ transition. In this case, as expected, the vibrational transitions observed are due entirely to symmetric (plus) levels. Because of the dominance of the scattering process by a single vibronic transition, the relative intensities of the Raman lines are given by Franck-Condon factors describing the overlap of the upper $v'=2$ level with the various $v''=1^+,2^+,3^+,4^+$ etc. levels of the ground state. These vibrational overlap factors have been calculated on the basis of a pyramidalization coordinate described by a harmonic potential for the A state and a harmonic plus Gaussian potential for the ground state as used by Harshbarger in an analysis of the absorption. A plot of the observed intensities of the members of the progression in v_2 versus the claulated values shows a good correlation with four out of five members of the series. This includes the very low intensity of the $v''=2^+$ relative to the $v''=1^+$ and 3^+ bands. The weak transition is to a state close to the top of the inversion barrier in the ground state. However, the transition to the $v''=1^+$ level has an intensity that is only 2/3 of the value expected on the basis of this correlation.

The spectra of NH_3 observed with either 212.8 or 208.8 nm radiation show a strong progression in the combinations of n quanta of v_2 appended to one quantum of v_1, the N-H stretching mode. This is of interest in itself as discussed below. In the context of a separable model for the motions along these two coordinates, the relative intensities of the members of this series should also be proportional to the same Franck-Condon factors calculated above. Again, a good correlation is observed with four out of the five bands with the $v''=1^+$ transition deviating from the correlation. In this case, however, this band is larger than expected by a factor of about 1.3. At face value, it appears that the separated mode picture of the motions of this molecule is valid for 8 of 10 transitions but fails badly for the same final level in each series. This behavior is hard to understand.

The intensity of the N-H stretching mode in the 212.8 nm and, especially, the 208.8 nm spectra indicates that there is a significant displacement of this coordinate upon excitation. This is in agreement with the rotational analysis of the few resolved rovibronic lines observed for ND_3. However, no progression in this mode is observed in the absorption spectrum. This behavior is also difficult to explain but, qualitatively, implies that there is significant coupling between the stretching and bending modes in either the ground or excited electronic states. On the other hand, the pattern discussed above implies that the separable mode model is applicable for most of the spectrum. Extensive MCSCF-CI calculations [P. Rosmus, private communication] including two-dimensional overlap integrals give a correlation similar to that discussed above for the relative intensities of the members of either v_2 progression but fail to reproduce the large intensity of the v_1 mode series.

Resonance Raman spectra have also been obtained for ND_3. Excitation at 217.8 nm, red of the 0-0 of the absorption, results exclusively in hot band transitions involving one quantum of $v_2=1$ in the ground state. The rotational side band structure is very strong, often stronger than the Q-branch. This is consistent with dominance of the contribution of particular rovibronic states to the Raman scattering process. This is expected for ND_3 as compared to NH_3 because of the slower dissociation rate and the correspondingly sharper (and resolved) rovibronic transitions. The assignment of several spectra of ND_3 will be discussed.

* Present address: Department of Chemistry and JILA, University of Colorado, Boulder, Colorado

† Address as of September, 1986: Department of Chemistry, Boston University, Boston, Massachusetts

THE INVERSE TRANSFORM IN RESONANCE RAMAN SCATTERING

J. R. Cable and A. C. Albrecht

Department of Chemistry, Cornell University, Ithaca, New York 14853

Transform theory provides a powerful approach for predicting Raman excitation profiles (REPs) in large multimode systems. Except for the mode whose REP is being modeled, the necessary vibronic parameters are implicitly supplied by an experimental absorption spectrum (ABS). Similarly, an inverse transform can be used to allow the resonant ABS responsible for the enhanced Raman scattering to be determined from an experimentally measured REP. Just as in the forward direction, only parameters for the mode being studied are needed explicitly and may be treated parametrically if their values are unknown.

The REP of any vibrational mode can be expressed as the magnitude squared of a difference function involving the complex linear electrical polarizability. The first step in a direct inverse transform is therefore the recovery of the complex difference function from its experimentally measured square magnitude. With the use of an analytic extension function, the magnitude-to-phase Hilbert transform can be directly applied. The complex linear polarizability, whose imaginary component is simply related to ABS, can then be determined by a simple algebraic procedure. By application to a model REP, the validity of this approach can be verified. The inverse transform may also be applied to several recently measured REPs[1] in resonance with the azulene $S_4 \leftarrow S_0$ transition. Here the inverse transform is found to accurately reproduce the experimental ABS only with those modes whose REPs are accurately treated by the forward transform.[2] This behavior is expected since both the forward and inverse transforms are based on the same set of underlying assumptions.

The forward transform can also be linked with the inverse transform to allow the prediction of other REPs from that measured for a single vibrational mode. This capability is clearly demonstrated on a set of three experimental REPs[3] resonant with the azulene $S_1 \leftarrow S_0$ transition. Application of the inverse transform to the 900 cm^{-1} REP allows both the 1260 and 1400 cm^{-1} REPs to be accurately reproduced via a linked forward transform.

The inverse transform bears a strong resemblance to the problem of inverting an experimental reflectance spectrum to recover the conventional ABS. In both cases, the experimentally observable quantity is related to the magnitude squared of a specific function of the linear polarizability. In the reflectance case, however, the form of this function prevents the practical use of an analytic extension function to perform the magnitude-to-phase Hilbert transform.

We gratefully acknowledge support from the materials Science Center of Cornell University and NSF Grant No. CHE-8311401.

References

1. J. R. Cable and A. C. Albrecht, J. Chem. Phys. 84, 1969 (1986).

2. J. R. Cable and A. C. Albrecht, J. Chem. Phys. 84, 0000 (1986).

3. O. Brafman, C. K. Chan, B. Khodadoost, J. B. Page and C. T. Walker, J. Chem. Phys. 80, 5406 (1984).

CONSISTENT NON-PERTURBATIVE THEORY OF TEMPERATURE-DEPENDENT RESONANCE RAMAN SCATTERING FOR SIMULTANEOUS MODE MIXING AND NON-CONDON COUPLING.*

H. M. Lu and J. B. Page

Physics Department
Arizona State University
Tempe, AZ 85287

Vibrational mode mixing (Duschinsky rotation) and non-Condon effects in resonance Raman (RR) scattering have been considered individually in several theoretical and experimental studies, but no general non-perturbative theory has been given which includes both of these effects simultaneously. Nevertheless, it was shown some fifteen years ago [1] that the presence of non-Condon coupling for two or more modes <u>requires</u> the existence of mode mixing. We report here an analytic calculation of the temperature-dependent first-order RR profiles and optical absorption for multimode vibronic systems in the general case of quadratic plus linear non-Condon coupling <u>and</u> quadratic plus linear electron-phonon (el.-ph) coupling (i.e. vibrational mode mixing, frequency shifts and equilibrium positions shifts with electronic excitation). The calculation is carried out within the time correlator theoretical framework [2,3] of RR scattering and optical absorption. We generalize the many-body diagrammatic techniques developed previously by Tonks and Page [4,5] in solving the RR problem for full quadratic plus linear el.-ph. coupling, and we also employ an undetermined coefficients method similar to that used by Chan [6] in treating linear non-Condon coupling for the case of purely linear el.-ph. coupling. The several new infinite series of diagrams occurring in the resulting theory are summed analytically and non-perturbatively, yielding results which are exact for arbitrary strengths of the non-Condon and el.-ph. couplings. Both the temperature-dependent first-order RR profiles and the optical absorption are expressed in terms of <u>one</u>-dimensional Fourier transforms of convenient time-dependent analytic functions of the model parameters. With fast Fourier transform techniques, this provides a very efficient procedure for numerical modeling of complex polyatomic systems.

Illustrative numerical studies are reported for a simple two-mode model with parameters appropriate to typical molecular systems. Of particular interest are new interference effects arising from the simultaneous presence of both types of coupling. Indeed, with both couplings present, we find appreciable new features in the RR profiles and optical absorption. Depending on the values of the model parameters, either constructive or destructive interference is found to occur, and this behavior can be different for the absorption than for the profiles.

For the case of no mode mixing but with frequency shifts and non-Condon coupling present, a new approximate "transform" relation is obtained which allows one to compute directly a mode's RR profile solely from the temperature-dependent optical absorption and the mode's model parameters. It has been shown by us [7] and by others [8] that such "transform" calculations within simpler models constitute a powerful adjunct to explicit model calculations in practical RR studies of multimode vibronic systems ranging from organic molecules to solid state molecular defects.

Extensions, further ramifications and applications of the new theoretical results will be discussed.

REFERENCES

[1] B. Sharf and B. Honig, Chem Phys. Lett. 7, 132 (1970) and G. J. Small, J. Chem. Phys. 54, 3300 (1971).

[2] V. Hizhnyakov and I. Tehver, Phys. Stat. Sol. 21, 755 (1967).

[3] J. B. Page and D. L. Tonks, J. Chem. Phys. 75, 5694 (1981).

[4] D. L. Tonks and J. B. Page, Chem. Phys. Lett. 79, 247 (1981).

[5] C. K. Chan, J. B. Page, D. L. Tonks, O. Brafman, B. Khodadoost and C. T. Walker, J. Chem Phys. 82, 4813 (1985).

[6] C. K. Chan, J. Chem. Phys. 81, 1614 (1984).

[7] D. L. Tonks and J. B Page, Chem. Phys. Lett. 66, 449 (1979); D. L. Tonks and J. B. Page, J. Chem. Phys. 76, 5820 (1982); C. K. Chan and J. B. Page, J. Chem. Phys. 79, 5234 (1983).

[8] See, for instance, P. M. Champion and A. C. Albrecht, Ann. Rev. Phys. Chem. 33, 353 (1982) and references therein; D. Blazej and W. Peticolas, J. Chem. Phys. 72, 3134 (1980).

* This work was supported by NSF Grant DMR 8204586.

VIBRONIC COUPLING IN THE FIRST EXCITED STATE OF ANTHRACENE.

A. BREE

Chemistry Department, University of British Columbia, Vancouver, Canada.

In this work, interest is focussed on vibronic coupling of the first excited electronic state (B_{1u}) of anthracene through the activity of b_{3g} vibrations to other B_{2u} electronic states. There are two main sources of information. The most direct source is the one-photon spectrum, both absorption and fluorescence. We will also use data concerning the excitation profile of the active b_{3g} modes in the pre-resonance region of the Raman spectrum. We first determine the absolute intensity of the one-photon absorption or emission, or of Raman scattering, and then proceed to model this behaviour in order to extract vibronic coupling constants as parameters.

In the one-photon spectrum at low temperature, the pure electronic transition is usually anomalously weak because of the effect of re-absorption or re-emission. Thus, one needs intensities at least into the overtone region, and the many published spectra do not encompass this range. Instead, data were taken from a photographic record of the fluorescence spectrum of anthracene in heptane at 4.2K measured in this laboratory some years ago [1]. Spectra in single crystal matrices were avoided because of the difficulty in correlating intensities in the two polarizations.

Schmid, Derner and Berthold [2] have measured standard intensities (S), based on the intensity of the 459 cm^{-1} line of $CC\ell_4$ as unity, for 10 Raman lines of anthracene in solution. We have repeated these measurements in part and extended them to the single crystal; the agreement is good. Since Schmid et al. used a more extended range of excitation, we use their data, converting it to $d\sigma/d\Omega$ and then to α_{yz} by making use of the value of $d\sigma/d\Omega$ for the 459 cm^{-1} line of $CC\ell_4$ given by Kato and Takuma [3].

We used the simplest model possible. Only the first two excited B_{1u} and the first two excited B_{2u} electronic states were explicitly included; this introduced four vibronic coupling constants as parameters. The energy of the first excited B_{2u} electronic state and its transition strength from the ground state were taken from the recent analysis [4] given the MCD spectrum of anthracene observed by Steiner and Michl [5]. A constant contribution to α_{yz} (ie constant over the excitation range examined) was included as a fifth parameter in an effort to mimic the effect of vibronic coupling between more highly excited electronic states. The model used here makes no attempt to include: (i) Herzberg-Teller coupling involving the ground electronic state, (ii) coupling between adjacent electronic states (such as the first excited B_{1u} and B_{2u} states) through the nuclear kinetic energy operator or (iii) rotation of the normal coordinates for the different electronic states. The results of the modelling will be presented for the three b_{3g} modes (at 1634, 1183 and 523 cm^{-1}) which are active in the fluorescence and Raman spectra.

[1] A. Bree and S. Katagiri, J. Mol. Spectry. 17, 24 (1965).
[2] E.D. Schmid, H. Derner and G. Berthold, J. Raman Spectry. 4, 329 (1976).
[3] Y. Kato and H. Takuma, J. Chem. Phys. 54, 5398 (1971).
[4] M. Zgierski, J. Chem. Phys. 83, 2170 (1985).
[5] R.P. Steiner and J. Michl, J. Am. Chem. Soc. 100, 6861 (1978).

TIME CHARACTERISTICS OF SECONDARY EMISSION FROM RESONANTLY EXCITED I_2.

W.Overlaet, M.Kimura, S.Kinoshita and T.Kushida

Dept. of Physics, Osaka University, Toyonaka, Osaka 560, Japan

The detailed investigation of the time behavior of the secondary emission due to Raman scattering and luminescence has been performed by Rousseau and Williams[1] for a diatomic molecule excited by nearly resonant light pulses. The experimental data have been well explained qualitatively by the theory of Mukamel et al.[2]. However, no quantitative comparison between the theory and experiment has been reported yet. Therefore, we re-examined the time behavior of the secondary emission in I_2 and compared the results with the theory in detail.

The iodine molecules were excited with light pulses from a cavity-dumped single-frequency Ar^+ laser at 514.5 nm, and the time behavior of the secondary emission was measured by means of the time-correlated single-photon counting. Measurements were made at various detunings $\Delta\nu$ of the laser frequency from the center of resonance of the $B^3\Pi(43,12) \leftarrow X^1\Sigma(0,13)$ transition of I_2. The high-resolution studies of the secondary emission spectra and time characteristics in the vibrational overtone regions Δv = 1, 2 and 3 revealed that the contamination of other re-emission lines is negligible for the overtone Δv = 3, $B^3\Pi(43,12) \rightarrow X^1\Sigma(3,11)$. Therefore, we selected this line by a double monochromator with a resolution of at least 0.4 cm^{-1} and measured the time behavior.

A typical example of the obtained result is shown in Fig.1, in which two components are clearly observed. One is the fast Raman-like component which follows the time profile of the exciting laser pulse, and the other is the slow luminescence-like component whose decay time coincides with the lifetime of the excited state. Figure 2 shows the $\Delta\nu$-dependence of the ratio of the integrated intensity of the slow-decay component to the total intensity. The ratio was obtained by decomposing the time behavior of the secondary emission using the least square fitting method.

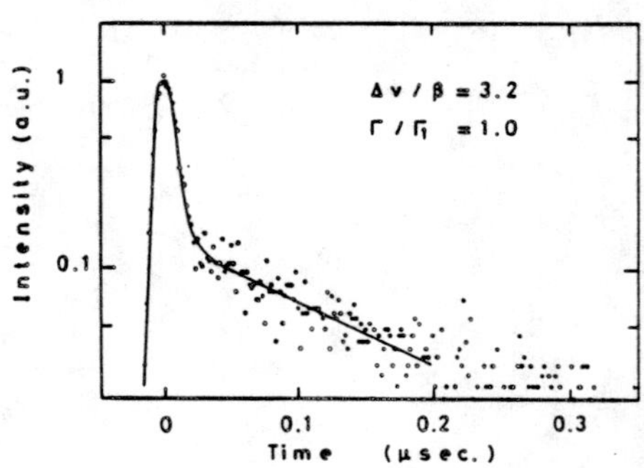

Fig.1: Measured (dots) and calculated (solid line) time behavior of the secondary emission at $\Delta\nu$= 1.4 GHz. The iodine cell temperature was 20°C.

We calculated the time behavior and the relative contribution of the slow-decay component using the theory of Mukamel et al. approximating the exciting pulse-shape as a Gaussian with a FWHM of 15 ns. As shown by solid lines in Figs.1 and 2, our experimental data fit well by putting $\Gamma/\Gamma_1 = 1.0$, where Γ_1 is the decay rate of the luminescence-like component and Γ is the pure dephasing rate. In this calculation, it was necessary to take account of the 21 hyperfine lines besides the Doppler broadening (FWHM β = 447 MHz) in order to obtain a reasonable agreement between the theory and experiment.

Figure 3 shows the excitation spectrum of the total secondary emission. The solid curves are Voigt-profiles calculated for different Lorentzian widths. From the fitting, the Lorentzian FWHM $\Gamma_2/2\pi$ at 20 °C is estimated at 1.8 MHz. Since $\Gamma_2=\Gamma+\Gamma_1/2$ and $\Gamma_1/2\pi$ = 1.3 MHz at 20 °C, we obtain the ratio Γ/Γ_1 as 0.9. This almost coincides with the above-mentioned value estimated from the analysis of the time behavior.

From the above results, the cross sections for the phase-changing collisions and population-changing collisions were obtained as 490 Å^2 and 230 Å^2, respectively. They are close to the values estimated for a different rotational level of I_2 by means of optical multiple-pulse spectroscopy[3].

In summary, it has been confirmed that the time behavior of the secondary emission from I_2 at moderate pressures under pulsed excitation is well explained quantitatively by the theory of Mukamel et al., provided the hyperfine structure of the energy levels as well as the homogeneous and Doppler broadenings are taken into account.

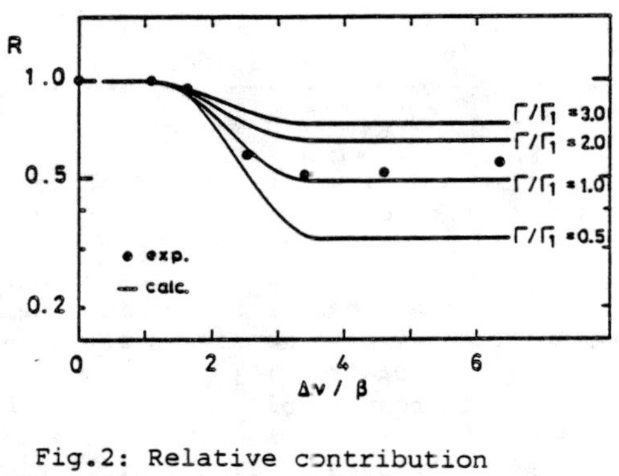

Fig.2: Relative contribution of slow-decay component versus detuning.

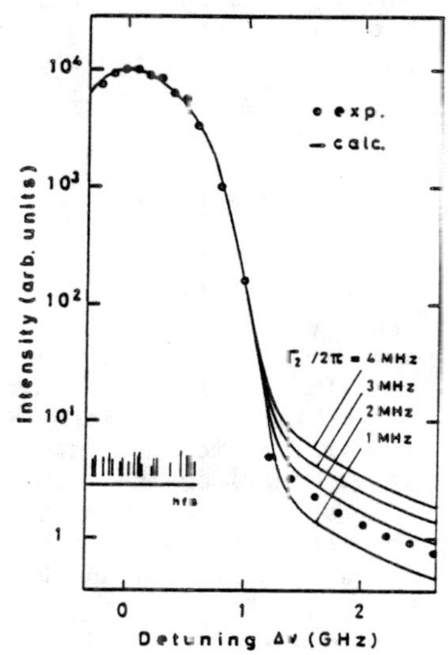

Fig.3: Excitation profile of the 634.5 cm^{-1} line. The positions of the hyperfine lines (hfs) are also indicated.

References:

[1] D.L.Rousseau and P.F.Williams, J.Chem.Phys. 64 (1976) 3519.

[2] S.Mukamel, A.Ben-Reuven and J.Jortner,Phys.Rev.A 12 (1975) 947.

[3] E.T.Sleva and A.H.Zewail,Chem.Phys.Letters 110 (1984) 582.

Calculation and Measurement of Polarization Dispersion Curves for the Resonance Raman Scattering of I_2 in Solution

Takamichi Kobayashi, Roseanne J. Sension, and Herbert L. Strauss

Department of Chemistry, University of California, Berkeley, CA 94720

It is well known that the polarization dispersion curve for resonance Raman scattering, that is the depolarization ratio as a function of excitation frequency, is determined by the nature of the relevant excited electronic states. This has proved to be particularly useful in cases where the Raman polarizability tensor has an anti-symmetric component [1, 2]. However, the polarization dispersion curve also provides important information on the excited electronic states in cases where the polarizability tensor is symmetric.

For a diatomic molecule, such as iodine, the depolarization ratio, ρ, is dependent on the degeneracies of the excited electronic states contributing to the resonance Raman cross section. For a non-degenerate ground state, ρ will be 1/3 if the excited states contributing to the cross section are all non-degenerate and 1/8 if the states are doubly degenerate. Interference effects between resonant states of differing degeneracy will result in polarization dispersion curves strongly dependent on the magnitude of the cross terms between the electronic states. Specifically, the depolarization ratio is given by [3]:

$$\rho(\nu) = \frac{|\alpha_{\gamma\gamma}(\nu)|^2 + |\alpha_{\beta\beta}(\nu)|^2 - (\alpha^*_{\gamma\gamma}(\nu)\alpha_{\beta\beta}(\nu) + \alpha^*_{\beta\beta}(\nu)\alpha_{\gamma\gamma}(\nu))}{3|\alpha_{\gamma\gamma}(\nu)|^2 + 8|\alpha_{\beta\beta}(\nu)|^2 + 2(\alpha^*_{\gamma\gamma}(\nu)\alpha_{\beta\beta}(\nu) + \alpha^*_{\beta\beta}(\nu)\alpha_{\gamma\gamma}(\nu))}$$

where $\alpha_{\gamma\gamma}(\nu)$ is the contribution to the polarizability tensor of non-degenerate excited states and $\alpha_{\beta\beta}(\nu)$ is the contribution due to doubly degenerate excited states.

In the Figure we have plotted the calculated polarization dispersion curves for the fundamental and first two overtones of iodine in n-hexane. The calculation includes the three excited electronic states of iodine which contribute to the visible absorption spectrum [4]. Of these three states the non-degenerate $BO_u^+(^3\Pi)$ state is responsible for ~73% of the absorption and $\geq$98% of the total resonance Raman enhancement. The doubly degenerate $A1u(^3\Pi)$ and $B''1u(^1\Pi)$ states contribute negligibly (A) or only slightly (B") to the total resonance Raman cross sections, and yet they have a large effect on the polarization dispersion. Also, small changes in the locations or slopes of these states have significant effects on the calculated polarization dispersion curves.

Preliminary measurements of I_2 in n-hexane follow the general trend of the calculation, indicating that the gas phase potentials are not perturbed much in n-hexane.

The polarization dispersion curves should prove useful in resolving the contributions of the A, B, and B" states in cases where the solvent is known to perturb the excited state potentials of iodine.

References

1. O. S. Mortensen and S. Hassing, Adv. in Infrared and Raman Spec. 6, 1 (1980).
2. W. Siebrand and M. Z. Zgierski, Excited States, 4, 1 (1979).
3. R. J. Sension and H. L. Strauss, submitted to J. Chem. Phys.
4. J. Tellinghuisen, J. Chem. Phys. 76, 4736 (1982).

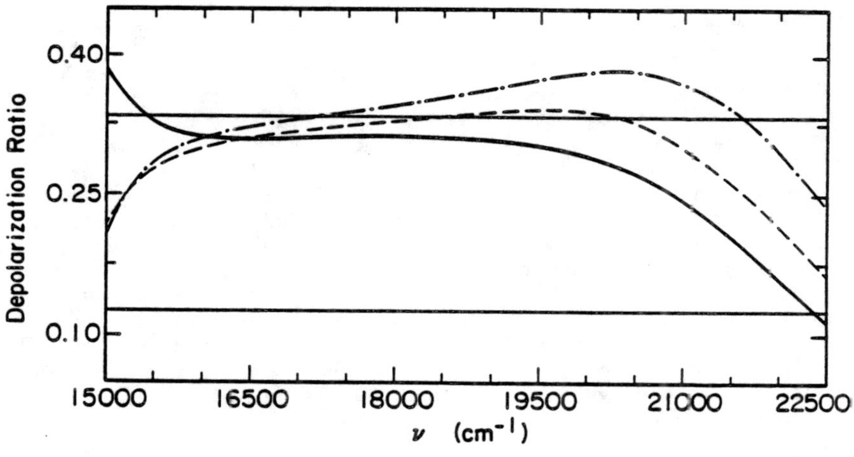

The calculated depolarization ratio, ρ, as a function of incident laser frequency for the iodine fundamental (solid line), first overtone (dashed line), and second overtone (dot-dashed line). The horizontal line at 1/3 represents the depolarization ratio if only non-degenerate excited states, such as the B state, contribute to the Raman cross section, that is if $\alpha_{\beta\beta} \equiv 0$ in the Equation. The horizontal line at 1/8 represents the depolarization ratio if only doubly degenerate states, such as the A and B" states contribute to the Raman cross section. That is, if $\alpha_{\gamma\gamma} \equiv 0$ in the Equation. The calculation was performed using known gas phase parameters for the excited electronic states [4].

Comparison of Theory and Experiment for the Raman Excitation Profile of I_2 in Solution

Roseanne J. Sension and Herbert L. Strauss

Department of Chemistry, University of California, Berkeley, CA 94720

The use of resonance Raman spectroscopy, specifically the resonance Raman excitation profile (REP), as a probe of the excited state properties of molecules has grown rapidly in recent years. The REP is particularly useful in solution where the broadening caused by the solvent obscures much of the vibrational structure in the absorption spectrum. The REP may provide more detailed information on the excited states even in solution, since the REP can resolve more of the vibrational structure.

There has also been much interest lately in the use of the REP with an absolute intensity scale as a probe of the excited state properties [1]. The absolute Raman cross section for excitation in resonance with a bound excited electronic state is directly proportional to the reciprocal line-width, T_2, of the excited electronic state. In addition, the absolute Raman cross section for excitation in resonance with an unbound state is proportional to $|\partial V(Q)/\partial Q|^{-1}$ where $V(Q)$ is the potential energy of the excited state as a function of the normal coordinate of interest.

Much of the previous work however, has concentrated on large organic and biological molecules [1], or on relatively complicated inorganic systems [2]. For these systems, little, if anything, is known quantitatively about the excited electronic states involved. It is useful to attempt an accurate comparison of theory and experiment for a simple molecule in a relatively non-interacting solvent. Consequently, we have obtained the absolute REPs for the fundamental and first two overtones of iodine dissolved in n-hexane with excitation frequencies from 15000 to 22000 cm^{-1}. We have also performed extensive calculations of these REP's as a function of the various parameters.

In the Figure we show the experimental REP of the iodine fundamental and the calculated REP as a function of the excited state linewidth, Γ, where $\Gamma \propto 1/T_2$, for values of Γ ranging from 10 to 200 cm^{-1}. Gas phase potentials for the excited electronic states of iodine were used in the calculation.

Comparison of the calculated and experimental absorption spectra and REP's indicates that the homogeneous linewidth of the bound excited state is $\Gamma \sim 15\text{-}20$ cm^{-1} (HWHM). The spectra also have an inhomogeneous component of 400 cm^{-1} (HWHM).

Calculations including pre-resonance effects of the higher electronic states of iodine, specifically the D state at 53500 cm^{-1}, indicate that these effects may be important, changing the calculated cross sections by as much as 10%.

1. A. B. Meyers and R. A. Mathies in Biological Applications of Raman Spectroscopy edited by T. G. Spiro (Wiley-Interscience, New York, in Press) Vol. 2.
2. R. J. H. Clark and B. Stewart, Structure and Bonding, 36, 1 (1979).
3. D. L. Rousseau, J. M. Friedman, and P. F. Williams, Topics in Current Physics (Springer, Berlin, 1979), Vol. 11, Chap. 6, especially Fig. 6.14.

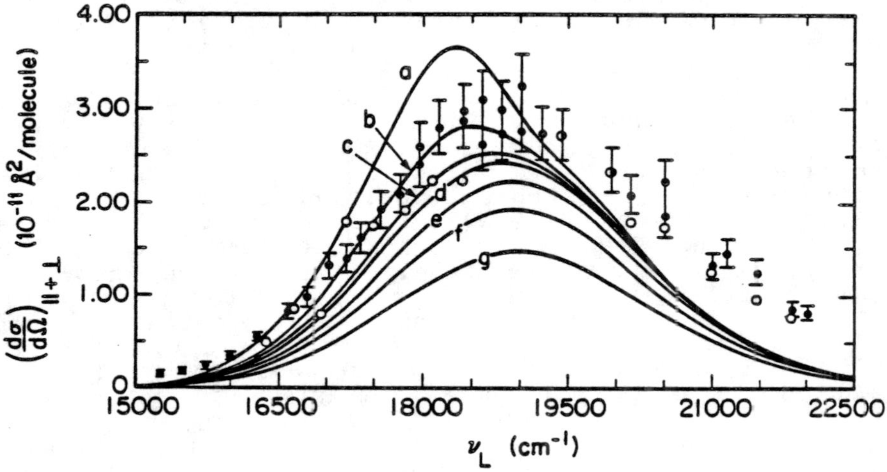

Calculated and experimental REPs for the iodine fundamental in n-hexane plotted on an absolute scale. The data points with error bars are our experimental values (the error bars shown are ±10%). The results of repeated experiments are shown as multiple data points. The open circles represent data points from Rousseau et al. [3]. These points were scaled to our values of the absolute cross section at 5145 Å. The calculated REPs shown all have an inhomogeneous excited state linewidth of 400 cm^{-1}, the solvent effect on the transition strength given by $L^2 = 1.75$ and a Boltzmann distribution of initial states equivalent to T = 300 K. The homogeneous linewidths, or damping factors, are as follows: Γ = (a) 10 cm^{-1}, (b) 15 cm^{-1}, (c) 20 cm^{-1}, (d) 25 cm^{-1}, (e) 50 cm^{-1}, (f) 100 cm^{-1}, and (g) 200 cm^{-1}. The inhomogeneous linewidth and the solvent effect on the transition strength were both determined by fitting the absorption spectrum.

ULTRAVIOLET RESONANCE RAMAN SPECTROSCOPY OF BUTADIENE: DETECTION OF THE LOW ENERGY 2^1A_g STATE

Richard R. Chadwick* and B. Hudson

Department of Chemistry and Chemical Physics Institute, University of Oregon, Eugene, Oregon

Resonance Raman spectroscopy can be used to locate and determine the symmetry of excited electronic states that give rise to forbidden transitions. This ability is based on the appearance of specific two-quantum transitions of the vibronic promoting modes in the Raman spectrum observed under conditions of resonance with a forbidden transition [L. D. Ziegler and B. Hudson, J. Chem. Phys. 74, 982 (1981)]. A long standing problem in the spectroscopy of linear conjugated polyenes is the location of the 2^1A_g excited state of butadiene [cf., B. Hudson, B.E. Kohler and K. Schulten, Excited States 6, 1 (1982)]. This prototypical polyene is completely non-fluorescent making its study by other methods difficult or impossible. Extrapolation of the known energy levels of longer polyenes to the case of butadiene indicates that the 2^1A_g state will lie just slightly below the strongly allowed 1^1B_u state. We have recently described the use of ultraviolet resonance Raman spectroscopy to determine the approximate location of this excitation of butadiene in the vapor phase [R. R. Chadwick, D. P. Gerrity and B. Hudson, Chem. Phys. Lett. 115, 24 (1985)]. We conclude that the $2A_g$ state is indeed just below the $1B_u$ state by about 300 cm^{-1}. This is the first demonstration of the presence of a low-energy state of A_g state for a polyene in the vapor phase. The evidence for this conclusion is the excitation frequency dependence of a band corresponding to the two-quantum overtone of the lowest frequency mode of b_u symmetry and a combination of this transition with the totally symmetric C=C stretch. In the present work the details of the assignment of these vibrations will be discussed. This includes an investigation of 1,1,4,4-tetradeuterobutadiene. Other interesting vibrations involving torsional motions will also be discussed. The variation of the relative intensities of some of the totally symmetric modes in the region of the $1B_u$ state are also compared to a model calculation performed by Hemley, Dawson and Vaida [J. Chem. Phys. 78, 2915 (1983)]. The resonance Raman data indicates that the vibrational lines in this spectral region are homogeneously broadened to a width of 500 cm^{-1}. The origin of this broadening will be discussed.

*Present address: Department of Chemistry, University of California, Riverside, California

ULTRAVIOLET RESONANCE RAMAN SCATTERING STUDIES OF METHYLENE CYCLOPROPANE: THE PROPENSITY OF THE V STATE TO FORM A TRIMETHYLENEMETHANE STRUCTURE FROM THE VERTICALLY EXCITED STRUCTURE

R. R. Chadwick*, D. W. Parker#, J. E. Baldwin+ and B. Hudson

Department of Chemistry and Chemical Physics Institute, University of Oregon, Eugene, Oregon 97403

The nature of the excited electronic surface of methylenecyclopropane (MCP, I) is of interest in several respects. On the one hand, this ethylene-cyclopropane system provides an interesting example of a substituted ethylene. Ethylene itself exhibits extension and twisting of its C=C double bond in the excited V state as revealed in its resonance Raman spectrum [L. D. Ziegler and B. Hudson, J. Chem. Phys. 79, 1197 (1983)]. The geometrical changes of methyl substituted ethylenes upon electronic excitation revealed by ultraviolet resonance Raman spectroscopy are discussed elsewhere in this volume. This study of MCP is thus an extension of these studies. Of more general interest is the fact that MCP can be photochemically and thermally converted to the protypical diradical species trimethylene-methane (TMM, II). The present study provides evidence that the vertically accessed excited electronic surface of MCP has a slope in a direction that would lead it to form TMM.

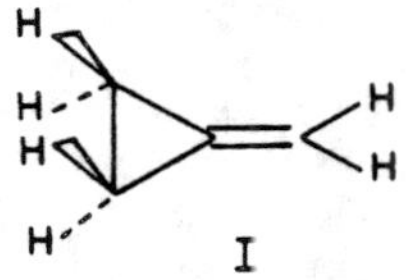

I

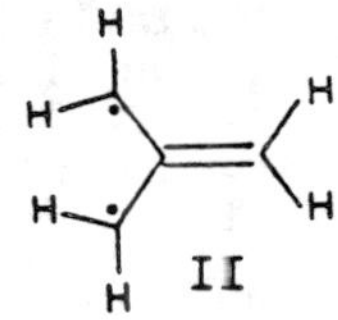

II

Resonance Raman spectra of TMM were obtained with excitation at 213, 200 and 192 nm. This is in the region of the first strong absorption band, presumably dominated by the $\pi \rightarrow \pi^*$ transition. The strongly enhanced vibrations involve stretching of the cyclopropane ring bond opposite to the exocyclic double bond, torsion of the CH_2 groups on the carbons forming this bond and stretching of the C=C bond. No excitation of the C=C torsional mode is observed. Combinations of the ring C-C stretch and the CH_2 twisting motions are very strong. These motions, taken together, are along the reaction coordinate for conversion of MCP to TMM.

* Present address: Department of Chemistry, University of California, Riverside, CA
# Present address: Union Camp Corporation, Lawrenceville, NJ
+ Present address: Department of Chemistry, Syracuse University, Syracuse, NY

ULTRAVIOLET RESONANCE RAMAN STUDIES OF BENZENE DERIVATIVES

S. Li and B. Hudson

Department of Chemistry and Chemical Physics Institute, University of Oregon, Eugene, Oregon 97403

Previously reported Raman studies of methylsubstituted benzenes using 212.8 nm excitation [L. D. Ziegler and B. Hudson, J. Chem. Phys. 79, 1197 (1983)] have been extended to other excitation wavelengths, other substitution patterns and to include fluoro- and chloro-benzene derivatives. Interest in these systems stems from the fact that lowering the symmetry of the D_{6h} benzene system can result in induction of an allowed transition component in the electronic transition from the A_{1g} ground state to the B_{1u} ($1L_a$) transition of the benzene ring. There is, however, still a significant vibronic component to this transition due to mixing with the allowed E_{1u} excited state by e_{2g} vibrations, primarily mode 8. In the absence of substituents or with symmetric 1,3,5 substitution, the only mechanism for resonance enhancement due to proximity of the excitation frequency with the $1L_a$ transition is purely vibronic C-type activity. In the linear vibronic coupling scheme with harmonic vibrations and only small force constant changes, this results in strong activity in the two-quantum transition of e_{2g} vibrations. This, in fact, is what is observed in these systems. In the lower symmetry species, the cross term between the allowed and vibronic transition mechanisms results in activity in the fundamental vibrations of these modes via the B-type mechanism. The relative intensity of the fundamental to overtone transition is, to a first approximation, a measure of the allowed to vibronic intensity in the transition.

The present studies have shown that mono- and para dimethyl and chloro benzene derivatives have significant intensity in the fundamental vibrational transition of the mode corresponding to mode 8 of benzene. This indicates that substituent-induced allowed intensity is comparable to that of the vibronic component in these cases. On the other hand, fluoro substitution results in very weak intensity in the fundamental transition relative to that of the C-type overtone indicating that the electronic transition is still primarily vibronic. These results will be discussed in terms of the differing nature of the perturbations introduced by these substitutents.

Another interesting observation concerns the effect of the geometrical pattern of the substitions. As noted above, symmetric 1,3,5-tri-substitution results in an effectively D_{3h} system so that the $1L_a$ transition is of B' symmetry and is forbidden. This results in no intensity in the fundamental transition of mode 8 as expected. Considerable intensity in the first overtone remains. On the other hand, a study of the asymmetric 1,2,3-trimethylbenzene also shows no activity in the corresponding fundamental band. This result is in agreement with early studies on the effects of such a substitution pattern on the induced absorption intensity of the vibronic states of benzene and will be discussed in terms of simple perturbation arguments.

The dependence of the intensities of these vibronic bands and of the symmetric ring stretching mode on the excitation wavelength will also be discussed. This is of interest because of its relevance to the interpretation of the variation of the spectrum of benzene in terms of an anti-resonance interaction between contributions from the B_{1u} state and the allowed E_{1u} state at higher energy [D. P. Gerrity, L. D. Ziegler, P. B. Kelly, R. A. Desiderio and B. Hudson, J. Chem. Phys. 83, 3209 (1985)]. The studies to be presented include excitation from 250 to 192 nm. This spans the region of the ^1B state as well as the ^1L_a state of the chloro and methyl benzene derivatives.

The spectrum of chloro- and 1,4-dichlorobenzene exhibit a number of binary overtone and combination transitions involving out-of-plane motion of the iodine atoms (b_2 in C_{2v} and b_{3u} in D_{2h}). The activity of these modes clearly demonstrates that the chlorine atom is displaced out of the benzene plane in the excited ^1L_a excited state. This is in agreement with the results of magnetic resonance studies of the corresponding triplet state of 1,3-dichlorobenzene.

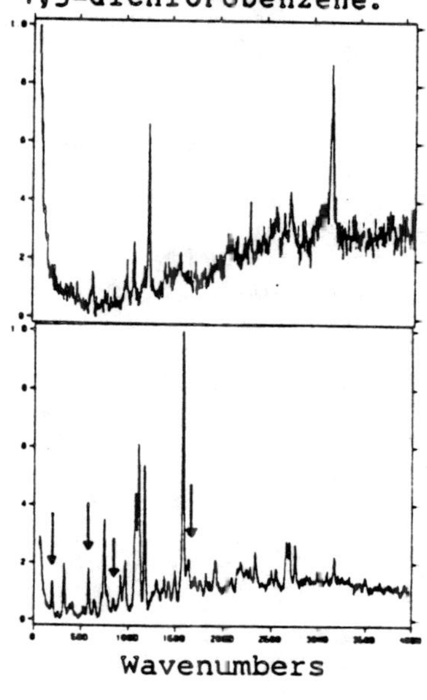

The Raman spectrum of 1,2,3-trimethylbenzene obtained with 218 nm radiation. Note the absence of any intensity at 1600 cm^{-1} and the strong intensity of the overtone at 3200 cm^{-1}.

The Raman spectrum of 1,4-dichlorobenzene obtained with 229 nm excitation. The out-of-plane b_{3u} modes are indicated. Note the strong B-type fundamental of "mode 8a" at 1582 cm^{-1} and the weak overtone at 3165 cm^{-1} due to C-type activity.

RESONANCE RAMAN SPECTROSCOPY OF CHLOROETHYLENES

Peter B. Kelly* and Bruce Hudson

Department of Chemistry and Chemical Physics Institute
University of Oregon

Resonance Raman spectra of chloroethylenes have been obtained with 240nm to 184nm excitations. The N -> V(π-π^*) transition in chloroethylenes is broad and structureless at energies below 50,000 cm^{-1}. The very broad nature of this absorption suggests a significant lengthening of the C=C distance in the excited state, although no structural information can be determined from the absorption spectra. Qualitatively, normal modes that show intensity in a resonance Raman spectrum are either displaced and/or have a force constant change in the excited state. Normal modes with small or no displacement or force constant change are not enhanced. We have used resonance Raman scattering to probe the geometry of the π^* excited state of dichloroethylenes.

The N -> V transitions of dichlroethylenes extend from 235 nm to shorter wavelengths. The resonance Raman spectra show resonance enhancement of the C=C stretch, the overtone of antisymmetric CCl_2 stretching, the overtone of torsion about the C=C axis, symmetric CCl_2 stretch, CCl in-plane bending, and combination bands. This indicates that the excited state geometry has a lengthened C=C bond and is twisted about this bond as in ethylene. There are also changes in the C-Cl bond distance and the Cl-C-Cl bond angle that may be indicative of short time dissociation dynamics of the upper state.

In conclusion, the resonance Raman spectra of the dichloroethylenes show that the π^* states of these molecules are similar to that of ethylene with lengthened olefin C=C bond and displacement in the twisting coordinate. The enhanced activity of the torsional mode in the **cis** compound relative to that of the **trans** compound indicates that there is a significant steric component to the twisting motion but the presence of this distortion in the upper state of 1,1-dichloroethylene indicates that steric effects are not entirely dominating the upper state potential. The significant activity of antisymmetric CCl stretch and symmetric CCl stretch may indicate a dissociation pathway that would contribute to the diffuse nature of the absorption spectrum.

* Address as of September 1986: Department of Chemistry, University of California, Davis

RESONANCE RAMAN SPECTROSCOPY OF METHYLETHYLENES

Peter B. Kelly* and Bruce Hudson

Department of Chemistry and Chemical Physics Institute
University of Oregon

Resonance Raman spectroscopy can provide unique information about the nature of excited electronic states that is often not available from absorption spectra. The features of a Raman spectrum that are strongly enhanced under resonance conditions can determine the electronic symmetry or the geometry of an excited electronic state and photodissociation pathway even when the absorption transitions to that state are so diffuse or so complex that analysis is not possible.

For the isomeric butenes, the 3S Rydberg state lies lower than the π^* state. The absorption spectra of these Rydberg states (240 nm-200 nm) show some sharp features. However, precise analysis of the absorption spectra is difficult without knowledge of the excited state vibrational frequencies. Resonance Raman spectroscopy has the advantage that displacements in the upper state normal modes are made apparent by increased intensity in the Raman spectrum.

The resonance Raman spectra of the butenes show strong intensity in C=C stretch and its overtone. The olefinic C-H in plane bend, CH_3 asymmetric and symmetric deformation, CH_3 asymmetric stretch, and in-plane CH_3 rock all show enhancement as the laser approaches resonance with the 3S Rydberg state. A most interesting observation is the lack of intensity in the overtone of the C=C torsion mode for any of the butenes with excitation in the Rydberg region.

The lengthening of the C=C bond is indicated by appearance of the overtones and combination tones of C=C stretch and by enhancement of the C-H bend similar to the CH_2 in-plane scissors motion observed in ethylene. The most surprising feature of the methyl substituted olefin Raman spectra is the lack of activity in the torsion about C=C. This lack of intensity indicates that the molecule remains planar in the 3S Rydberg state for the isomeric butenes. Recent theoretical calculations have shown that the 3S Rydberg state is planar in ethylene.

Enhancement of the symmetric and asymmetric CH_3 deformation and asymmetric stretch of the methyl C-H bonds is consistent with dissociation of a methyl C-H bond on the Rydberg excited state potential surface for the butenes studied.

* Address as of September 1986: Department of Chemistry, University of California, Davis

PRERESONANCE RAMAN SCATTERING IN QUINOXALINE AND SOME VITAMIN A-TYPE MOLECULES

P.K. Mallick

Department of Physics, Burdwan University
Golapbag, Burdwan - 713104, West Bengal, India.

When the exciting radiation is in resonance with an allowed electronic transition of a molecular system, certain Raman vibrations get enhanced. This resonance enhancement effect helps a lot in studying the properties of the excited electronic state. A few years ago Zgierski and Pawlikowski[1] have shown that under certain conditions resonance with a forbidden electronic transition can give rise to enhancement of a Jahn-Teller active degenerate fundamental. Otherwise forbidden electronic states, in general, do not play any significant role on Raman scattering.

Informations about the presence of hidden excited states [2,3] can also be obtained from Raman excitation profiles (REP). Rimai and others [3] have found peaks in REPs where electronic absorption peaks were absent in Napthalene and Vitamin A-type molecules, Retinal and Retinol. They assigned such peaks to some $\pi\pi^*$ - triplet states of the corresponding molecules.

If the exciting radiation is tuned through appropriate frequency range, vibronic and spin-orbit couplings between the state of interest and nearby allowed electronic states will produce interference effects in the ground state scattering intensity. Here we have considered both vibronic and spin-orbit couplings in relation to vibrational Raman intensity and calculated preresonance REPs of some totally symmetric vibrations of Quinoxaline and Vitamin A-type molecules, Retinal and Retinol. The theory has been set forth in several ways ; we have utilized expressions similar to those of Clark and Dines [4] for calculating REPs. Here the contributions from the lowest triplet and some low lying allowed singlet states have been taken into consideration.

The agreement between the calculated and observed [3-6] profiles is good. For Quinoxaline we have calculated REPs of some totally symmetric vibrations corresponding to skeletal distortion, CH bending and ring stretching modes. The calculation shows that the off-resonance triplet state contribution and the interference between the contributions from the lowest $\pi\pi^*$ - triplet and $n\pi^*$ - singlet states are negligible. But all other $\pi\pi^*$ - singlet states not only contribute but also interfere constructively very significantly. For the other molecules, Retinal and Retinol, we have calculated the profiles of the ethylenic mode of frequency near 1580^{-1} cm. Wonderful agreement between the observed and calculated profiles has been found for some selected values of the relevant parameters such as spin-orbit matrix, triplet state vibronic linewidths etc.

REFERENCES

1. M.Z. Zgierski and M. Pawlilowski, Chem.Phys.Letts. 63, 221 (1979).

2. C.M. Phillips, W. Brown and G.E. Leroi, in proceedings of Eighth International Conference on Raman Spectroscopy, Ed. J. Lascombe and P.V. Huong, P-135, John Wiley and Sons (1982).

3. L. Rimai, M.E. Heyde, H.C. Heller and D. Gill, Chem.Phys.Letts. 10, 207 (1971).

4. R.J.H. Clark and T.J. Dines, Mol-Phys. 42, 193 (1981).

5. N. Ohta and M. Ito, J. Mol. Spectrosc. 59, 396 (1976).

6. A. Aminzadeh, V. Fawcett and D.A. Long, J. Raman Spectrosc. 9, 214 (1980).

RAMAN EXCITATION PROFILES OF 1.8-DIHYDROXYANTHRAQUINONE AT 8 K.

G. Smulevich and M.P. Marzocchi

Dipartimento di Chimica, Università. V. G. Capponi, 9, 50121 Firenze, Italy.

Dihydroxyanthraquinones provide model chromophores for antitumor anthracyclines /1/. In addition 1.5 and 1.8 dihydroxyanthraquinone show dual fluorescence associated to intramolecular excited state proton transfer /2,3/. In order to obtain information on the vibronic states and proton dynamics we measured the Raman spectra and excitation profiles together with the absorption and polarized fluorescence spectra of solid 1.8 dihydroxyanthraquinone at low temperature. It has been possible to measure a well resolved Raman spectrum, of this highly fluorescent compound, by a mixture with CsI which acts as a fluorescence quencher. Most totally symmetric fundamentals, several overtones and combinations were observed. By the combined analysis of the resonance Raman and single crystal fluorescence spectra together with the matrix site selected fluorescence /3/ and the polarized infrared spectra /4/, the ground state vibrational assignment was obtained.

The relative intensities of the bands change by varying the exciting frequency. In particular the excitation profile of ν_5 at 484 cm^{-1} shows two maxima in correspondence with the absorption and fluorescence origins respectively. All the other bands show the profile maximum correspondent to the absorption, while the maximum at lower frequencies is less pronounced or absent (Fig. 1).

We have calculated the excitation profiles of the five prominent bands in terms of a multimode, two-state approach taking as a starting point the single-mode calculations. The good agreement between the calculated and experimental results allowed us to determine the shift parameters between the ground and the two excited states.

The occurrence of negative interference was established. In fact the shift parameters of at least some modes show opposite sign for the two transitions. The same parameters were also used to calculate the vibrational structures of the absorption and fluorescence spectra.

The results allowed us to evidentiate the existence of an intermediate state involved in the excited state proton transfer. This state is a consequence of proton tunneling in a asymmetric double minimum potential function along the O-H···O normal coordinate.

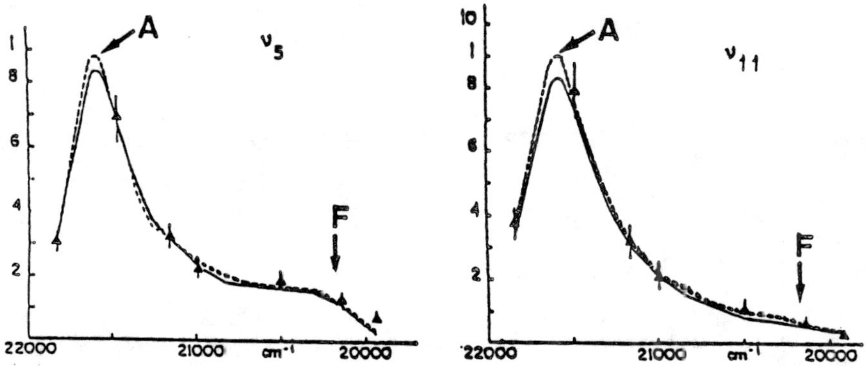

Fig. 1: Multimode calculations of the Raman excitation profiles for ν_5 and ν_{11} (▲). Full and dotted lines indicate two different sets of parameters. Arrows indicate the absorption (A) and fluorescence (F) origins.

References

1. M.J. Waring, Ann. Rev. Biochem. 50 (1981) 159.
2. M.H. Van Benthem and G.D. Gillespie, J. Phys. Chem. 88 (1984) 2945.
3. G. Smulevich, J. Chem. Phys. 82 (1985) 14.
4. G. Smulevich and M.P. Marzocchi, Chem. Phys. 94 (1985) 99.

RAMAN EXCITATION PROFILES OF POLYATOMIC MOLECULES IN CONDENSED PHASES - A STOCHASTIC THEORY.

S. Mukamel[†], J. Sue, Y. Yan and Z. Deng

Department of Chemistry, University of Rochester, Rochester, NY 14627

The absorption and the Raman excitation profiles of polyatomic molecules in condensed phases are analyzed using a stochastic model of line broadening. The electronic energy gap is assumed to be stochastically modulated by the random force resulting from the interaction with the solvent. The model interpolates all the way from homogeneous broadening (fast modulation) to inhomogeneous broadening (slow modulation). We analyze the conditions whereby the Raman excitation profiles can be expressed as the square of an ensemble-averaged amplitude. An efficient procedure for modelling both absorption and excitation profiles and extracting the relevant structural and dynamical information for large polyatomic molecules is proposed. Application is made to the Raman profiles of azulene in CS_2 and in methanol, and the magnitude and the inverse correlation time of the random force are extracted from the experimental data in both cases. The present model can be used to calculate nonlinear susceptibilities and coherent Raman (CARS) lineshapes as well. A detailed comparison of CARS and ordinary Raman lineshapes will be made.

The support of the National Science Foundation, the Office of Naval Research, the U.S. Army Research Office and the Petroleum Research Fund administered by the American Chemical Society is gratefully acknowledged.

References:

1. S. Mukamel, J. Chem. Phys. 82, 5398 (1985)
2. J. Sue, Y. Yan and S. Mukamel, J. Chem. Phys. (in press)
3. Z. Deng and S. Mukamel, J. Chem. Phys. (submitted)
4. S. Mukamel, Phys. Rep. 93, 1 (1982); Phys. Rev. A 28, 3480 (1983)

† Camille and Henry Dreyfus Teacher-Scholar

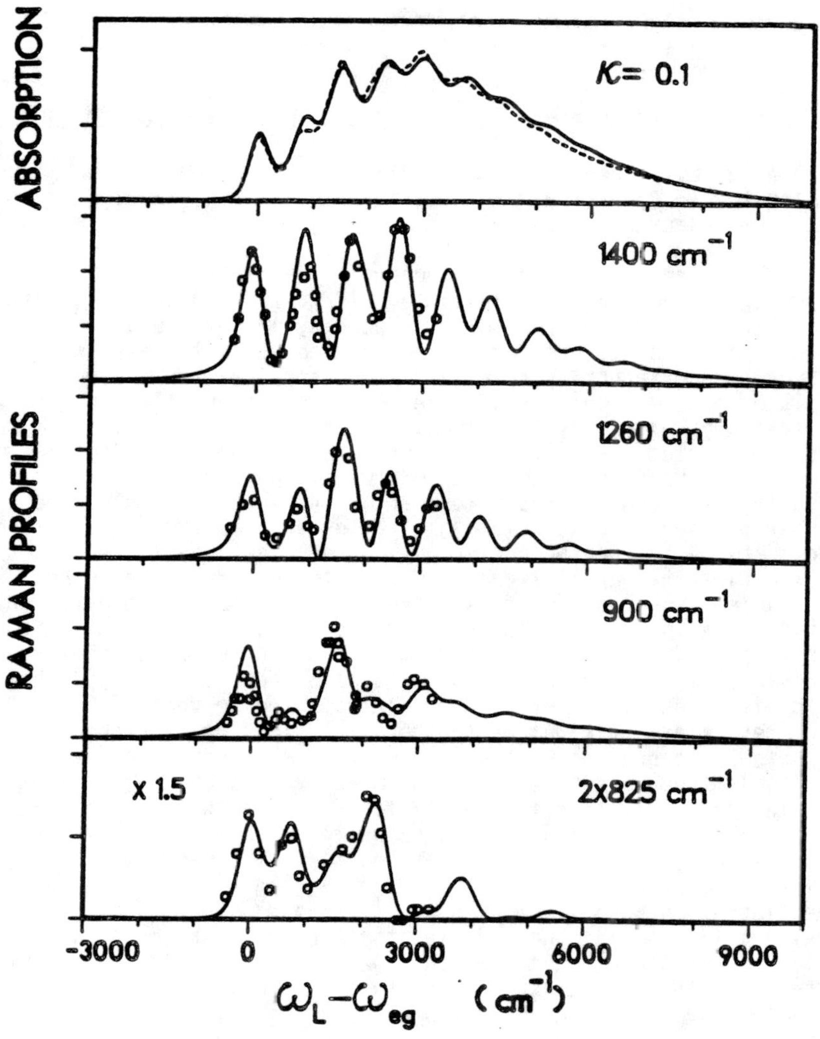

Fig. 1. Absorption and resonance Raman profiles for azulene in CS_2 at 300° K. The solid lines are theoretical curves computed using our seven-mode stochastic harmonic model with Dushinsky rotation. The dashed absorption curve (upper panel) is experimental. The Raman profiles are in the lower four panels. Shown are the experimental data (circles) and the calculated profiles for four different Raman transitions as indicated in each panel. All experimental data are from C.K. Chan, J.B. Page, D.L. Tonks, O. Brafman, B. Khodadoost, and C.T. Walker, J. Chem. Phys. 82, 4813 (1985).

INTERFERENCES IN AN RAMAN EXCITATION PROFILE OF AGGREGATED CHLOROPHYLL.

N. Binnie, L.V. Haley, T.A. Mattioli, D.L. Thibodeau, W. Wang and J.A. Koningstein*. Chemistry Dept., Carleton University, Ottawa, Ont. Canada
* Killam Research Fellow 1985-87

The Raman spectrum of Chlorophyll a insolution has been studied using CW and pulsed laser radiation [1-2]. Due to the fact that aggregation of this large molecule occurs, in particular in such solvents as hexane, carbontetrachloride and benzene etc., the assignment of the frequency and intensity spectrum is not straight forward. In addition, it is by now well known that even for dry Chlorophyll in dry solvents, at least one water molecule remains ligated to Chl a and it is thought that this molecular unit is also present in some of the aggregates. The simplest aggregates is composed of two of such units linked via the C=O keto group of ring V of one monomer to the Mg atom of the other. The former monomer is a donor and the latter an acceptor. X-ray studies have shown that - for Bacteriochlorophyll - a cofacial dimer exists where the macrocycles are parallel and shifted with regard to each other [3]. The energy level diagram for these simple dimeric forms are different because the interaction between the two constituent monomers is quite different. In the weakly interacting donor acceptor complex the individual molecules are expected to have their own energy level diagram which is expected to be different if compared to that of the monomer, the energy level diagram of the cofacial dimer on the other hand is expected to contain series of close lying electronic levels because of excitonic type interactions of the monomers. In Fig. 1 is shown the excitation profile for the intensity of the Raman shift at 1560 cm^{-1} of Chl a in hexane. Solution A is known to contain primarily the donor acceptor complex, solution B contains more of other aggregates [4]. The relative value of the concentration of aggregates can be deducede from the fluorescence spectrum of these solutions. The maxima of curve A are assigned to the prescence of electronic origins of the donor and acceptor molecules in the blue part of the spectrum. The profile of curve B resembles that [5] for the intensity of the bulk LO mode of Ga:P. This profile, in particular the sharp holes are thought to be due to interference effects between leading terms in the scattering tensor due to the prescence of three or four close lying electronic excited states. The wavelength of the laser is resonant with these electronic states and this results in the denominator - of one of the above mentioned leading terms - becoming negative. Although the Raman intensities are proportional to the square of the value of the tensor, the prescence of mixed products of the leading terms can result in lower values of the tensor at specific frequencies. Sharp decreases in ground state population can also give profiles exhibiting interference effects. We have looked for the occurence of stimulated effects under conditions of illumination of our samples which are required to record the light scattering spectrum but the search was negative. This lead us to the conclusion that the sharp holes in the excitation profile of curve B is indeed due to the prescence of close lying electronic levels which may well belong to that of the cofacial dimer of Chlorophyll a in dry hexane.

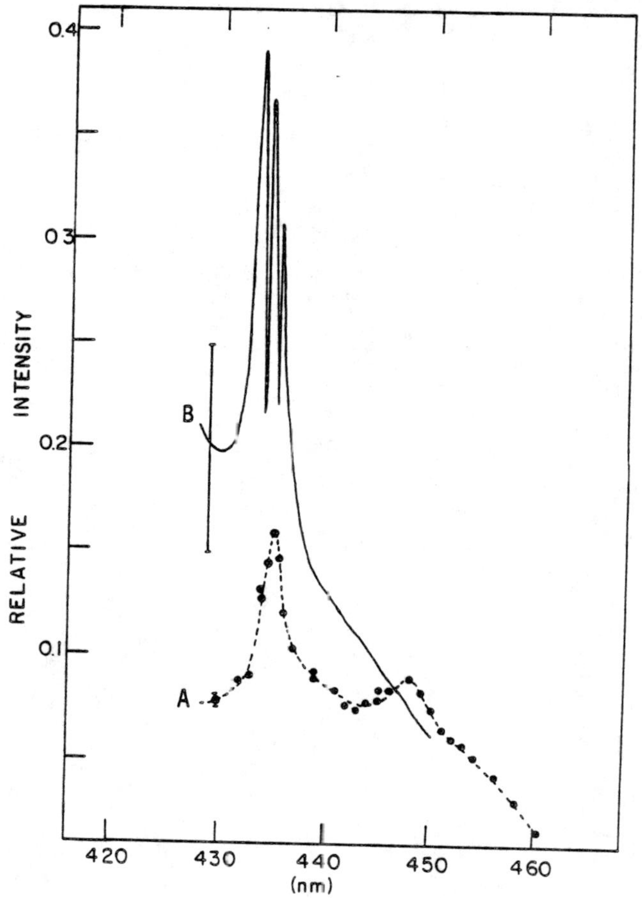

Fig. 1

References:

1) M. Lutz, J. Raman Spectrosc. 2, 497-517 (1974).
2) A.C. deWilton, L.V. Haley and J.A. Koningstein, J. Raman Spectrosc. 14, 288-290 (1983).
3) J. Deisenhofer, D. Epp, K. Miki, R. Huber and H. Michel, J. Mol. Biol., 180, 385-398 (1984).
4) A.C. deWilton, L.V. Haley and J.A. Koningstein, Chem. Phys. Letters 97, 538-541 (1983).
5) C. Hirlimann, M.P. dos Santos and M. Balkanski, J. Phys. C.: Solid State Phys. 18, 6297-6302 (1985).

chapter 18

TIME RESOLVED RAMAN STUDIES

PICOSECOND TIME-RESOLVED RESONANCE RAMAN SPECTROSCOPY.

G. H. Atkinson, G. Rumbles, T. L. Brack, and D. Blanchard

Department of Chemistry and Optical Sciences Center, University of Arizona, Tucson, Arizona

The measurement of vibrational Raman spectra for reaction intermediates formed within a few picoseconds of optical excitation has been pursued through several different instrumental approaches. Those efforts aimed at recording spontaneous resonance Raman scattering have used either (1) dye laser amplifiers to increase the pulse energy of picosecond pulses or (2) cavity dumpers to provide high repetition rate, picosecond pulses from mode-locked dye lasers. Both instrumental designs have been successful. The latter approach, however, encompasses several advantages which are described here together with results from the picosecond time-resolved resonance Raman (PTR^3) spectroscopy of retinal chromophores.

The two experimental techniques can be characterized in general by their repetition rates and pulse energies. Amplified laser systems are limited in repetition rate by the laser used to pump each amplifier. Until the recent utilization of copper vapor lasers operating at several kilohertz, this rate was about 10 Hz. These systems perform, therefore, at low repetition rates and high peak powers (several millijoules per pulse). The large photon flux in each picosecond pulse can significantly complicate the resonance Raman data obtained because of either multiple excitation or secondary photochemistry. Examples of these phenomena are evident in the PTR^3 spectra of BR [1-2]. The alternative experimental approach designed to use high repetition rate, low peak power excitation permits quantitative, independent control over the experimental parameters associated with PTR^3 spectroscopy because of the larger average powers involved. As a result, the optical conditions used to initiate and probe the dynamic by PTR^3 spectroscopy can be optimized with respect to the chemistry under study.

The instrumentation used in this study is shown in Figure 1 [3]. A cw Nd:YAG laser is mode-locked using an actively stabilized resonant mode-locker. The focused output of the Nd:YAG is frequency-doubled in a KTP crystal to produce a train of 70 ps pulses at 532 nm. The 532 nm radiation is used to synchronously pump two jet stream dye lasers (Coherent 701-3). A three-plate birefringent filter is used to select the dye laser wavelength and produce 7 ps pulses (autocorrelation pulsewidth). Acousto-optic cavity dumpers are used to vary the pulse repetition rate of the dye lasers from 150 kHz to 8 MHz. An optical delay line controls the time period between laser pulses used for excitation and probing. The pump and probe laser pulses are focused into a flowing sample from which Raman scattering is collected. Raman spectra are analyzed by a triple monochromator and an intensified reticon array detector.

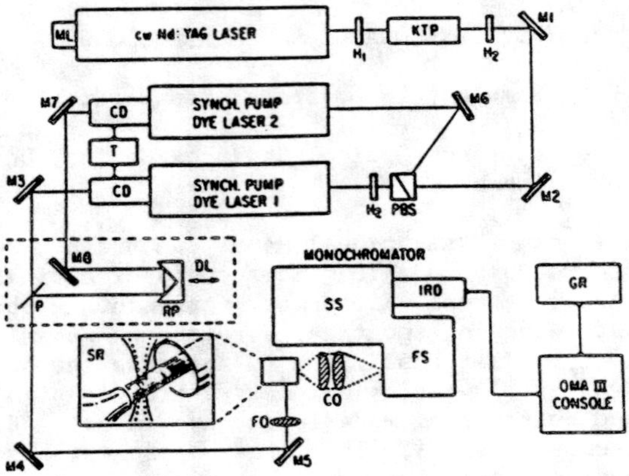

Two laser PTR^3 instrumentation. ML: mode locker; H_1: halfwave plate (1064 nm); H_2: halfwave plate (532 nm); M1-M8: front surface dielectrically coated mirrors; PBS: polarizing beam splitter; CD: cavity dumper; T: timing synchronization; DL: delay line; RP: retroprism; P: dielectrically coated pellicle beam splitter; FO: focusing optics; SR: sample region showing colinear laser beams; CO: collecting optics; FS: filter stage; SS: spectrometer stage; IRD: intensified reticon detector; GR: graphics.

PTR^3 spectra of intermediates in the bacteriorhodopsin photocycle recorded with this instrumentation reveal conformational and structural changes in the retinal chromophore during the 40 ps interval immediately following excitation [4-5]. The quality (e.g. signal-to-noise ratio and time resolution) of these PTR^3 data are comparable to those obtained for stable samples with cw laser excitation.

References

1. I. Grieger and G. H. Atkinson, Biochemistry, 24, 5660, (1985)
2. G. H. Atkinson, I. Grieger, and G. Rumbles, in Time-Resolved Vibrational Spectroscopy, (A. Laubereau and M. Stockburger eds.), Springer-Verlag, Berlin, W. Germany, (1985)
3. G. Rumbles, T. L. Brack, D. Blanchard, and G. H. Atkinson, Opt. Com., (submitted)
4. G. H. Atkinson, T. L. Brack, D. Blanchard, I. Grieger, G. Rumbles, and L. Siemenkowski, in Time-Resolved Vibrational Spectroscopy, (G. H. Atkinson ed.), Gordon and Breach, New York, N.Y., (in press)
5. T. L. Brack, D. Blanchard, L. Siemenkowski, and G. H. Atkinson, Biophys. J., (submitted)

PICOSECOND RESONANT LIGHT SCATTERING IN SILVER HALIDES.

H. Stolz and W. von der Osten

Universität-GH Paderborn, Fachbereich Physik
D-4790 Paderborn, W. Germany

The relaxation behaviour of excitons with excess kinetic energy is determined by two processes: interaction with phonons that provide the necessary energy-loss mechanisms and trapping at defects and impurities that reduce the number of free excitons. In indirect exciton systems like the silver halides AgCl and AgBr, these processes are reflected in the secondary emission spectrum. If excited resonantly within the exciton absorption at varying photon energies, the spectra at low temperatures roughly correspond to the distribution function of the hot exciton system. On the basis of energetic positions, shapes and intensities of the scattering lines the spectra can be analyzed in terms of intra- and intervalley scattering of excitons by various phonons as well as exciton trapping and considered as in principle understood.

The possibility to employ tunable picosecond laser pulses for optical excitation together with fast time-resolving detection methods recently allowed to study the dynamics of the hot excitons in a more direct way. By measuring the temporal evolution of the light scattering spectra after pulsed excitation absolute scattering and trapping rates as function of excitation photon energy can be obtained. From these, deformation potentials for the various exciton-phonon interactions and trapping cross-sections are determined. Taking advantage of the polarization properties of the scattered light, in oriented samples the time behaviour of the spectra is found to be strongly affected by the internal structure of the exciton. From measuring and analyzing the decay of the degree of polarization it is found that in AgBr, besides energy relaxation, an elastic scattering process has to be taken into account. It is strongly energy-dependent and ascribed to cross relaxation of excitons between the degenerate pair of singlet-triplet states resulting in depolarization of the scattered light.

In addition to the processes described above, most recently for both AgCl and AgBr we observed strong quasielastic scattering for excitation below the free indirect exciton. It sets in at the absorption threshold and, in AgCl, increases in strength by a factor 200 reaching its maximum at much lower energies. Although weaker, quasielastic scattering also occurs in AgBr, where we observed a characteristic and polarization-dependent time decay of the quasielastic signal after excitation with picosecond pulses. Preliminary measurements at energies close to the exciton absorption show that decay times of 100 to 500 psec are involved depending strongly on energy. The delayed scattering is suggested to originate from localized exciton states below the exciton gap. A first analysis shows that the strength of the quasielastic signal is mainly due to the lifetime of the states involved, while their absorption probability is too small to be detected by conventional absorption spectroscopy. More detailed measurements directed towards the exploration of the observed scattering are under way and expected to give new insight into exciton relaxation and localization.

TRANSIENT INTERMEDIATES IN PHOTOREACTIONS OF DERIVATIZED ANTHRAQUINONE AND DUROQUINONE.

R.E. Hester[1], J.N. Moore[2], I. McCubbin[2] and D. Phillips[2].

[1]Chemistry Dept., University of York, York YO1 5DD, England.
[2]The Royal Institution, 21 Albemarle St., London W1X 4BS, England.

Time-resolved resonance Raman spectroscopy (TR3S) has been used to characterize vibrational modes in a number of transient species in the solution-phase photoreactions of some sulphonated anthraquinones and duroquinone. Anthraquinone-2,6-disulphonate in water has been most extensively characterized [1,2]. From our further studies of this system, which include band depolarization measurements, we have arrived at the set of assignments shown in Table 1 [3]. The pattern of wavenumber shifts demonstrates clearly that the structural and bonding perturbation in the excited triplet state and its photosolvates B and C, as well as in the free radical reduction products, is heavily concentrated in the carbonyl groups. The data also support the identification of transients B and C as strong charge-transfer complexes.

Comparable data will be presented from TR3S studies of the photoreactions of anthraquinone-2-sulphonate and from anthraquinone-2-methylsulphonate in water and from duroquinone in mixed water/ethanol solutions.

Table 1. Resonance Raman bands (cm^{-1}) from anthraquinone-2,6-disulphonate (AQ26DS) and derived photoreaction intermediates

Parent AQ26DS	Radical AQ26DSH	Anion AQ26DS	Triplet ^{3}AQ26DS	Transient AQ26DS (B)	photosolvates AQ26DS (C)	C_{2h} assignments
1668	-	1496	-	-	-	b_u (C=O) asym
1669	1606	1347	1210	1347	1345	a_g (C=O) sym
1625	-	1598	1597	1598	1595	b_g (C-C) str
1596	1576	1498	1556	1555	1513	a_g (C-C) str
-	-	-	1490	1521	1485	(C-C) str
-	-	-	-	-	1286	(C-C) str
1177	1177*	1182*	1182*	1180*	-	a_g (C-H) ip bend
-	-	1101	1126	-	-	(C-C) str/ (C-C) def

Key: *bands also observed in ground state spectra

REFERENCES

1. J.N. Moore, G.H. Atkinson, D.Phillips, P.M. Killough and R.E. Hester, Chem. Phys. Lett. 107, 381 (1984).
2. J.N. Moore, D. Phillips, P.M. Killough and R.E. Hester, in Time Resolved Vibrational Spectroscopy (A. Laubereau and M. Stockburger, eds.), Springer, 1985.
3. J.N. Moore, Ph.D. Thesis, London University, 1986.

THE GROUND AND EXCITED STATE RAMAN SPECTRA OF α, α'-^{13}C-*trans*-STILBENE.

T.L. Gustafson, D.A. Chernoff, M.A. Hazle, J.F. Palmer, and D.M. Roberts

Standard Oil Research & Development, 4440 Warrensville Center Road, Warrensville Heights, OH 44128

The importance of *cis-trans* photoisomerization in the diphenylpolyenes as a model system for isomerization in visual transduction and photosynthesis is well established and has been reviewed. The structure and ordering of the excited electronic states in these systems remain important areas of study. We are among several groups that are using transient resonance Raman spectroscopy to study diphenylpolyenes.[1-4] The direct structural and dynamical information obtained by transient Raman spectroscopy is a powerful adjunct to fluorescence and absorption measurements. The observed vibrational frequencies of the excited state provide considerable information about its potential energy surface. This is particularly important in the diphenylpolyenes when considering mechanisms for energy transfer from the optical modes to reaction coordinates.

For *trans*-stilbene (tS) we have concentrated on determining the structure of the first excited singlet state, S_1. Hamaguchi *et al.* use nanosecond lasers with spontaneous Raman detection to obtain the excited state spectra of tS and several of its deuterated analogues.[2] They assign bands in the excited state at 1567, 1241, and 1148 cm^{-1} to the olefinic C-C double bond stretch, olefinic C-H in-plane bend, and C-phenyl single bond stretch, respectively. We agree with the assignment of the 1567 cm^{-1} as the olefinic C-C double bond stretch. But we suggest that the bands at 1241 and 1148 cm^{-1} can be assigned as C-phenyl single bond stretch and C-C-H bend, respectively.[1] We proposed that the interpretation of the excited state deuterated spectra is not clear because the normal modes change both form and frequency.[1]

In this work we present the ground state Raman spectra of crystalline α,α'-^{13}C-*trans*-stilbene (^{13}C-tS) and the excited state spectrum of ^{13}C-tS in fluid solution. Our intent was to minimize the degree to which the normal modes changed their forms upon isotopic substitution; we anticipated that ^{13}C substitution would introduce small but perceptible changes in the band positions with no change in the forms of the normal modes.

We have previously described the use of the high repetition rate, amplified, synchronously pumped dye laser for picosecond transient Raman spectroscopy.[1] Briefly, we used a cavity dumped argon ion laser to pump an amplifier for a synchronously pumped cavity dumped dye laser. A temperature tuned ADA crystal was used for second harmonic generation. We separated the ultraviolet and visible beams; the visible probe beam traversed a variable path length and was recombined collinearly with the ultraviolet pump beam. We used a single spectrograph with an intensified reticon array detector for observing the Raman signal. We obtained *trans*-stilbene from Eastman (scintillation grade) and α,α'-^{13}C-*trans*-stilbene from MSD Isotopes (99.6 atom % ^{13}C). We used these materials without further purification to prepare solutions of 25 mM concentration in *n*-hexane.

In Figure 1 we compare the S_1 spectra of ^{13}C-tS and tS over the region from 900 to 1700 cm^{-1}. We have scaled the data so that the intensity of the 979 cm^{-1} band in each spectrum is approximately the same. We observe that most of the band positions and intensities are perturbed with ^{13}C substitution, suggesting that the $C_o{=}C_o$ stretch and C_o-Φ stretch are coupled strongly to many normal modes in the excited state. Based on the bands at 1522 cm^{-1} and 1546 cm^{-1} in α,α'-^{13}C-*trans*-stilbene, we suggest that the broad band at 1567 cm^{-1} in *trans*-stilbene is composed of two bands: a broad band at 1546 cm^{-1} and a band at 1567 cm^{-1}. We suggest that conformations contribute to the broadening of the vibrational bands in the excited states of α,α'-^{13}C-*trans*-stilbene and *trans*-stilbene.

References

1. T.L. Gustafson, D.M. Roberts, and D.A. Chernoff, J. Chem. Phys. 79, 1559(1983); T.L. Gustafson, D.M. Roberts, and D.A. Chernoff, J. Chem. Phys. 81, 3438(1984).
2. H. Hamaguchi, C. Kato, and M. Tasumi, Chem. Phys. Lett. 100, 3(1983); H. Hamaguchi, T. Urano, and M. Tasumi, Chem. Phys. Lett. 106, 153(1984).
3. R. Wilbrandt, N.-H. Jensen, and F.W. Langkilde, Chem. Phys. Lett. 111, 123(1984).
4. A. Kasama, M. Taya, T. Kamisuki, Y. Adachi, and S. Maeda, in: Time-Resolved Vibrational Spectrosc., A. Laubereau and M. Stockburger, Eds.(Springer-Verlag, Berlin, 1985), p. 166.

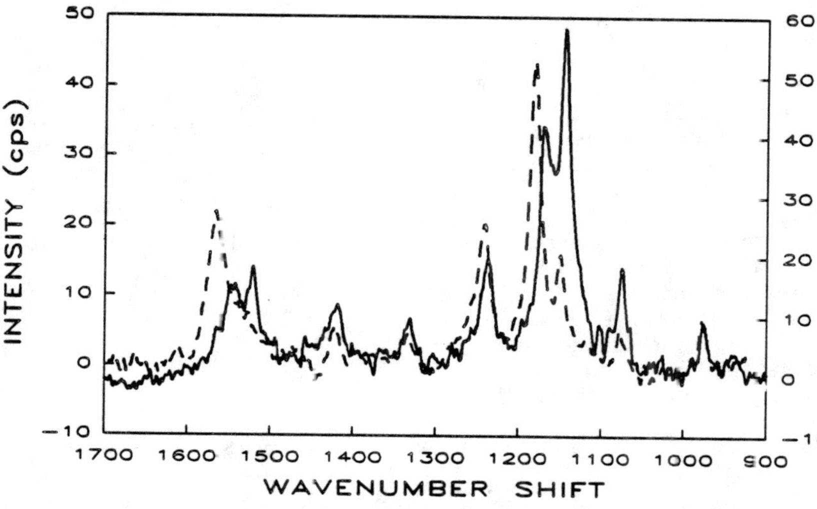

Figure 1. Excited state Raman spectra of α,α'-^{13}C-*trans*-stilbene (solid) and *trans*-stilbene (dash, right ordinate) in *n*-hexane: 296 nm pump; 593 nm probe

TIME-RESOLVED RESONANCE RAMAN SPECTRUM OF CHRYSENE IN THE S_1 AND T_1 STATES.

Shin-ya Koshihara and Takayoshi Kobayashi

Department of Physics, Faculty of Science, University of Tokyo,
Hongo 7-3-1, Bunkyo, Tokyo 113 Japan

Time-resolved S_n-S_1 and T_n-T_1 resonance Raman spectra of chrysene were observed at room temperature with the use of the third harmonic of a Q-switched Nd:YAG laser as an electronic excitation pulse and dye laser pumped by the second harmonic of another Nd:YAG laser as a Raman probe pulse.[1,2] Time-resolved resonance Raman spectra of the highly purified chrysene[3] in tetrahydrofuran in 1600 --- 1300 cm^{-1}, 1300 --- 950 cm^{-1}, and 900 --- 100 cm^{-1} regions are shown in Fig. 1, 2, and 3, respectively. We observed the Raman lines attributed to the S_1 state at 1549, 1369, 1344, 1110, and 990 cm^{-1}, and to the T_1 state at 1528, 1494, 1353, 1199, 1110, 990, 678, 554, 472, and 286 cm^{-1}.

The decay time of the 1369 cm^{-1} (S_1) line, the formation time of 1528 cm^{-1} (T_1) line, that of 1199 cm^{-1} (T_1) line, and the fluorescence lifetime were obtained to be 35^{+6}_{-5} ns, 51^{+19}_{-12} ns, 41^{+10}_{-7} ns, and 41 ± 1 ns, respectively.

The observed S_n-S_1 and T_n-T_1 resonance Raman spectra of highly purified chrysene were compared with the previously reported fluorescence and phosphorescence spectra in Shpolskii matrices, S_n-S_1 and T_n-T_1 resonance CARS spectra, and T_n-T_1 spontaneous resonance Raman spectra. Discussion was made on the geometrical changes of chrysene molecule in the excited states on the assumption that the changes were not very large for the rigid molecular structure. We came to the following conclusions. (1) The resonance-coupled excited singlet state $(S_6)^{2,3}$ is different from the ground state in geometry mainly along the normal coordinates of the vibrational modes which have the S_n-S_1 resonance Raman

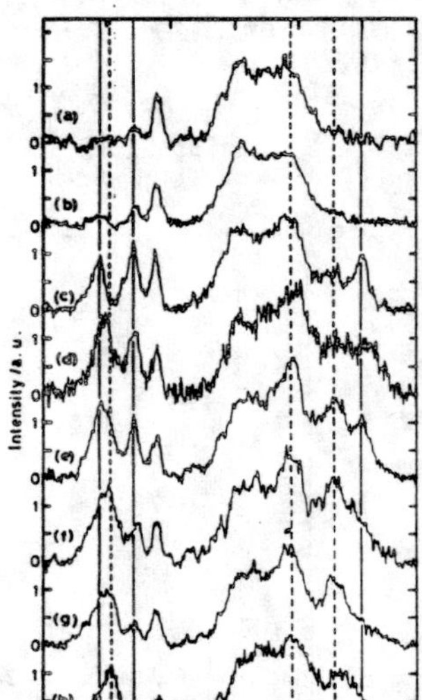

Figure 1: Time-resolved resonance Raman spectra of the electronically excited chrysene in 1600 --- 1300 cm^{-1} region at (a)t_d=-46, (b)-16, (c)-6, (d)4, (e)24, (f)44, (g)64, (h)84 ns. The solid and dashed lines indicate the Raman lines attributed to S_1 and T_1, respectively.

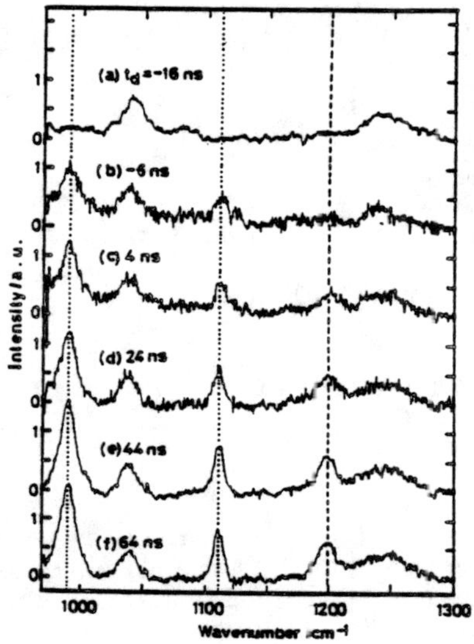

Figure 2: Time-resolved resonance Raman spectra of chrysene in 1300 --- 950 cm^{-1} region. The dashed and dotted lines indicate the wavenumber positions of Raman lines attributed to T_1 and both S_1 and T_1, respectively.

Figure 3: Time-resolved resonance Raman spectra of chrysene in 100 --- 900 cm^{-1} region. The dashed lines indicate the Raman lines attributed to T_1.

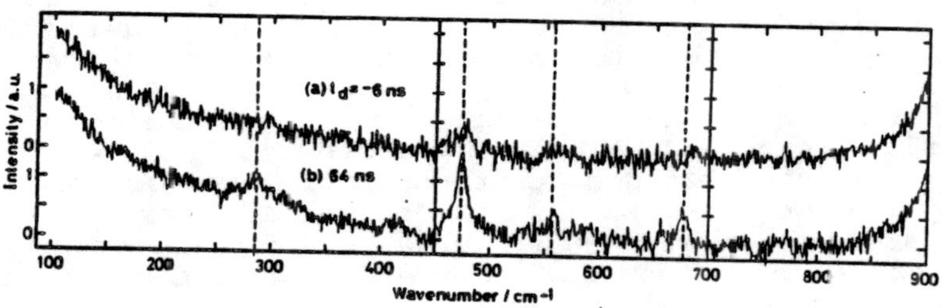

peaks at 1549 and 1344 cm^{-1}. (2) The geometry of the resonance-coupled excited triplet state ($T_3(A_g)$ or $T_4(A_g)$)[2] differs from that of the ground state mainly along five coordinates. Four of them correspond to the vibrational bands observed in the T_n-T_1 resonance Raman spectrum at 1528, 1353, 1110, and 990 cm^{-1}. The other corresponds to the vibrational bands observed in the Shpolskii spectrum of phosphorescence at 1385 cm^{-1} in literature.[4] (3) The excited singlet and triplet states (S_1, S_{rc}, T_1, and T_{rc}) differ along the normal coordinates of the modes which have the T_n-T_1 resonance Raman bands at 678, 554, 472, and 286 cm^{-1}.

REFERENCES

1. T. Kobayashi and S. Koshihara, Chem. Phys. Lett. 104,174 (1984).
2. S. Koshihara and T. Kobayashi, J. Chem. Phys. in press.
3. S. Koshihara, T. Kobayashi, and S. Iwashima, Chem. Phys. Lett. in press.
4. Ch. Jung, A.Lau, H.-J. Weigmann, W. Werncke, and M. Pfeiffer, Chem. Phys. 72, 327 (1982).

TIME-RESOLVED RESONANCE RAMAN STUDY OF THE PHOTOIONIZATION OF N,N,N',N'-TETRAMETHYL-P-PHENYLENEDIAMINE IN SOLUTION.

Hiroaki Takahashi, Hiroaki Isaka, Tomoyuki Ohzeki and Shinichi Suzuki

Department of Chemistry, School of Science and Engineering, Waseda University, Tokyo, Japan

Because of its relatively low ionization potential N,N,N',N'-tetramethyl-p-phenylenediamine (TMPD) readily undergoes photoionization in condensed media by irradiation with UV light to produce the radical cation ($TMPD^+$), Wurster's blue. A marked effect of solvent has been reported on the nature of the photolytically produced transient species and the photoionization mechanism of TMPD.

It now appears to be well established that in polar solvents the ionization takes place monophotonically from the lowest excited singlet state via semi-ionized species, while in non-polar solvents monophotonic ionization does not occur by UV light of the wavelength longer than 300 nm. A biphotonic ionization mechanism in hydrocarbon solutions with the lowest triplet state (^{3}TMPD) as the photoactive intermediate, has been suggested based on photoconductivity measurements.

However, recent time-resolved microwave conductivity measurement revealed that single-photon ionization of TMPD occured even in nonpolar CCl_4 solution, and more recent laser photolysis measurement showed that photoionization of TMPD did not occur in polar acetone solution.

The purpose of this paper is to obtain more information on the photoionization mechanism and the transient species of TMPD in various solvents by time-resolved resonance Raman spectroscopy.

Figure 1 shows the time-resolved resonance Raman spectra of TMPD in deaerated acetonitrile solution. At 200 ns after UV irradiation a transient species characterized by the bands at 1629, 1506, 1423, 1226 and 1171 cm^{-1} and having considerably long lifetime was observed. The spectra were not affected by the presence of oxygen. Since the frequencies are in good agreement with those of Wurster's blue, the transient species is identified with $TMPD^+$. The triplet state was not detected. Since the decay time of ^{3}TMPD in acetonitrile is about 500 ns, this may be attributed to the very low quantum yield of ^{3}TMPD in this solvent.

Figure 2 shows the time-resolved resonance Raman spectra of TMPD in deaerated hexane solution. At 200 ns after UV irradiation, two bands at 1530 and 1158 cm^{-1} were observed, which rapidly decreased in intensity and at 1 μs they almost disappeared. When the solution was aerated, these bands were not detectable. Thus, they are attributable to ^{3}TMPD.

We have tried to observe biphotonically produced $TMPD^+$ in hexane solution by increasing the intensity of the irradiating UV light using an excimer laser (Lambda Phisik EMG 201E). However, no band of $TMPD^+$ could be detected.

In deaerated 2-propanol solution both ^{3}TMPD and $TMPD^+$ were observed, of which the former almost disappeared at 5 μs after UV irradiation. In deaerated acetone solution no band due to ^{3}TMPD was detected. Although a very weak band was observed at 1629 cm^{-1}, this band is undoubtedly assigned to the most prominent Raman band of $TMPD^+$. It is interesting to note that despite the electron affinic property of acetone and its effective quenching of the fluorescence of TMPD, the quantum yield of $TMPD^+$ in this solvent is extremely small.

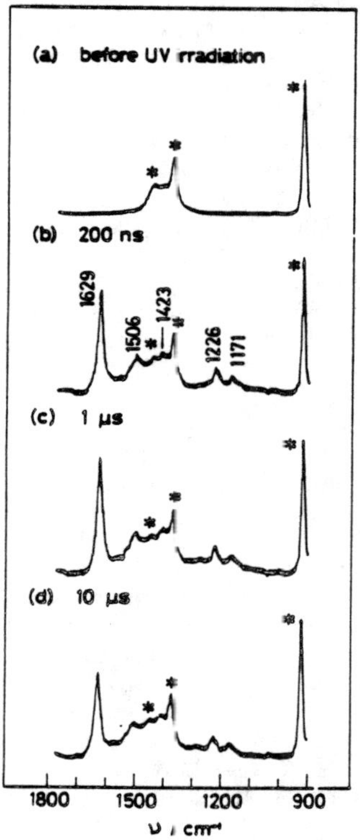

FIGURE 1. Time-resolved resonance Raman spectra of TMPD in deaerated acetonitrile. 580 nm excitation.
*: Solvent band.

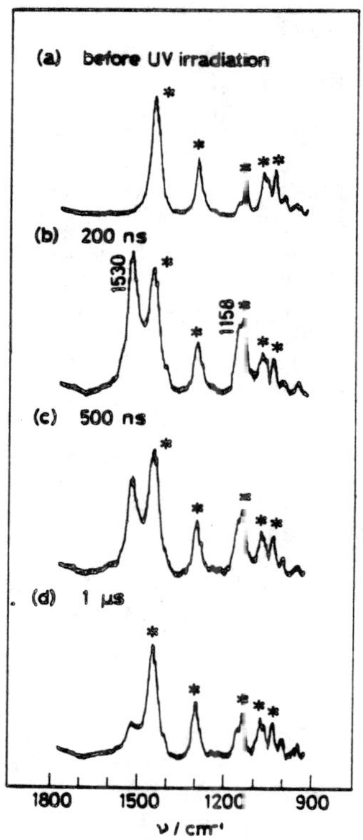

FIGURE 2. Time-resolved resonance Raman spectra of TMPD in deaerated hexane. 580 nm excitation.
*: Solvent band.

CHARACTERIZATION OF THE PHOTODYNAMIC BEHAVIOR OF NICKEL PORPHYRINS IN NONCOORDINATING SOLVENTS USING TRANSIENT RAMAN SPECTROSCOPY.

E.W. Findsen,*[1] J.A. Shelnutt,[2] and M.R. Ondrias[1]

[1]Department of Chemistry, University of New Mexico, Albuquerque, New Mexico 87131

[2]Sandia National Laboratories, Albuquerque, New Mexico

Recent work has detailed the transient resonance Raman behavior of several nickel porphyrins in coordinating solvents and nickel protoporphyrin reconsituted myoglobin and hemoglobin (1,2,3). In this work, these studies have been expanded to explore the excited state dynamics of model nickel porphyrins in noncoordinating solvent systems, as well as the behavior of a tethered imidazole nickel porphyrin.

A number of transient absorption experiments have characterized the decay mechanisms and lifetimes of photoinduced excited state ligation and deligation of nickel porphyrins. The most detailed report of these processes was provided in the work of Kim et al. (4,5).

Using transient resonance Raman spectroscopy we have been able to probe the excited state species formed upon photoexcitation of nickel protoporphyin and to make qualitative assessments as to ligation state, and core size behavior of these species (6). The photoexcitation of nickel porphyrins in strongly coordinating solvents (such as nickel protoporphyrin in pyrrolidine) causes the metal to undergo a net d-d transition from the $^3B_{1g}\bullet L_2(d^1_{z^2}, d^1_{x^2-y^2})$to the $A_{1g}(d^2_{z^2})$ configuration which is an antibonding state for sigma donating ligands (such as pyrrolidine). This transition causes the Ni(PP) to assume a transient 4-coordinate state. Subsequently, the ground state equilibrium is re-established within ~10-20 nsec (4). The results of this study indicate that a state with significant B_{1g} character is formed upon photoexcitation of nickel porphyrins in noncoordinating solvents.

The experimental details have been reported elsewhere (2). Briefly, a nitrogen pumped dye laser, or a Nd:YAG laser (532 nm) was used to both excite and probe the sample within one, 10 nsec wide pulse. The scattered radiation was focused into a Spex scanning double monochromator, and detected using a PMT in a cooled housing. A boxcar integrator collected the signal from the detector and its output was sent into a Datamate controller. The data was transfered to a Apple II computer for permanent storage and graphics manipulaton. The photon density (power) of the laser beam at the sample was controlled in a number of ways. The three most common ways to control power were; use of a spherical or cylindrical lens to image the laser beam on the sample, the actual focus of the above optic, and use of neutral density filters to modulate photon density. The nickel porphyrins were purchased from Porphyrin Products and used without purification. Organic solvents were spectral grade. Cetyltrimethyl ammonium bromide (CTAB) and sodium dodecyl sulfate (SDS) were used without further purification. Sample integrity was monitored using UV/visible absorption spectroscopy. All samples were between 20 and 180 μM in porphyrin and all experiments were performed at ambient temperature (20-25°C). No degradiation of samples due to dissolved oxygen was observed in earlier studies so no sample degassing was performed (1,2).

Data has been obtained from several nickel porphyrins (such as Ni(PPIX), Ni(PPIXDME), Ni(OEP)) in noncoordinating solvents at high and low incident laser powers. The low power spectrum is typical of 4-coordinate nickel porphyrin species. At high power several new peaks are observed, at positions intermediate between those of 4-coordinate and 6-coordinate ground state complexes. We conclude that these arise from a transient B_{1g} unligated state. The Raman spectrum of this B1g state can be used to determine its molecular characteristics. The positions of the core-size sensitive mode, ($\nu 3$, $\nu 2$ and $\nu 10$) in the excited state indicate 60-70% of the shift observed in these modes between 4-coordinate to 6-coordinate ground state complexes results from porphyrin core expansion (from ~1.95Å to ~2.02Å) incurred by the $^2A_{1g} \rightarrow B_{1g}$ transition alone ($d^2_{z^2} \rightarrow d^1_{z^2}, d^1_{x^2-y^2}$). A shift of 6-8 cm^{-1} lower frequency observed for the oxidation state marker line, ν_4, in the excited state is apparently coupled to the transient d-d transition. The excited state properties of Ni(PPIXDME) and Ni(PPIX) in the same solvents (DMSO, and 10% H_2O/acetone) are very similar indicating there is little dependence of excited state properties on perpherial substituents. This behavior is also observed for Ni(OEP) in noncoordinating solvents such as toluene and CH_2Cl_2 and is totally reversible.

In an attempt to model the behavior of the 4-coordinate sites in NiHbA the photodynamic behavior of a tethered imidazole nickel porphyrin was examined. This system is predominantly 4-coordinate in the ground state. Upon photoexcitation new bands are observed in the transient Raman spectra which are attributed (by analogy to previous 5 and 6-coordinate results) to the formation of a 5-coordinate species with the tethered imidazole as the fifth ligand. This behavior is generally similar to that observed in the equilibrium 4-coordinate sites of NiHbA.

1) Findsen, E.W., Shelnutt, J.A., Friedman, J.M. and Ondrias, M.R. (in press), Chem. Phys. Lett.
2) Findsen, E.W. Shelnutt, J.A., Alston, K. and Ondrias, M.R. (in press), J. Am. Chem. Soc.
3) Shelnutt, J.A., Alston, K., Findsen, E.W., Ondrias, M.R. and Rifkind, J.M. (see abstract this conference).
4) Kim, D. and Holten, D. (1983) Chem. Phys. Lett., 98, 584.
5) Kim, D., Kirmaier, C. and Holten, D. (1983) Chem. Phys., 75, 305.
6) Findsen, E.W., Shelnutt, J.A. and Ondrias, M.R. (submitted), J. Phys. Chem.

This work supported by the NIH (63330).

18-14

A TRANSIENT RESONANCE RAMAN STUDY ON THE LOWEST EXCITED TRIPLET-STATE OF β-CAROTENE ISOMERS.

Hideki Hashimoto and Yasushi Koyama

Faculty of Science, Kwansei Gakuin University, Uegahara, Nishinomiya 662, Japan

We reported the resonance Raman spectra (RRS) of the excited triplet-state of β-carotene, which was produced from the all-trans, 9-cis, 13-cis and 15-cis isomers by means of flash-photolysis using a sensitizer, anthracene (1). Recently we have applied time-resolved resonance Raman spectroscopy to the 7-cis isomer in addition to the above isomers and succeeded in recording a set of triplet spectra with a better S/N ratio and with much higher triplet population in the 1500 - 900 cm^{-1} region.

Fig. 1 compares the RRS at the highest triplet population (1.8 μs after excitation) for the (A) all-trans, (B) 7-cis, (C) 9-cis, (D) 13-cis and (E) 15-cis isomers. The RRS of the triplet-state produced from the all-trans and the 15-cis isomers are regarded as identical (within the limit of experimental errors) which is consistent with the result of Wilbrandt et al. (2), who used naphthalene-sensitized pulse-radiolysis. However, the 7-cis, 9-cis and 13-cis isomers give definitely different kinds of triplet spectra from those of the all-trans and 15-cis isomers in the 1240 - 1150 cm^{-1} region, although overall spectral similarity is seen. The all-trans and 15-cis isomers show a peak at 1197 cm^{-1}, while the 7-cis and 9-cis show a peak at 1194 cm^{-1} and 1180 cm^{-1}, respectively. The 13-cis isomer does not give any peak around this region.

HPLC analysis of anthracene-sensitized photo-isomerization through the triplet-state revealed (i) a one-way photo-isomerization from the 15-cis into the all-trans isomer; the all-trans isomer is produced with high efficiency from the 15-cis isomer (Fig. 2-e), while the all-trans isomer has lower efficiency of isomerization and does not produce the 15-cis isomer (Fig. 2-a). (ii) However, the efficiencies of isomerization for the 7-cis and 9-cis isomers are an order of

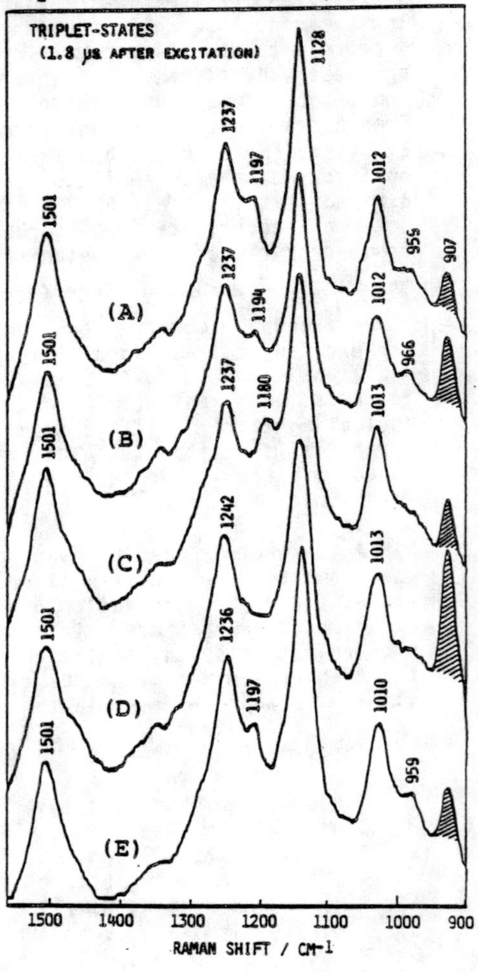

Fig. 1 RRS of the triplet-state of β-carotene isomers

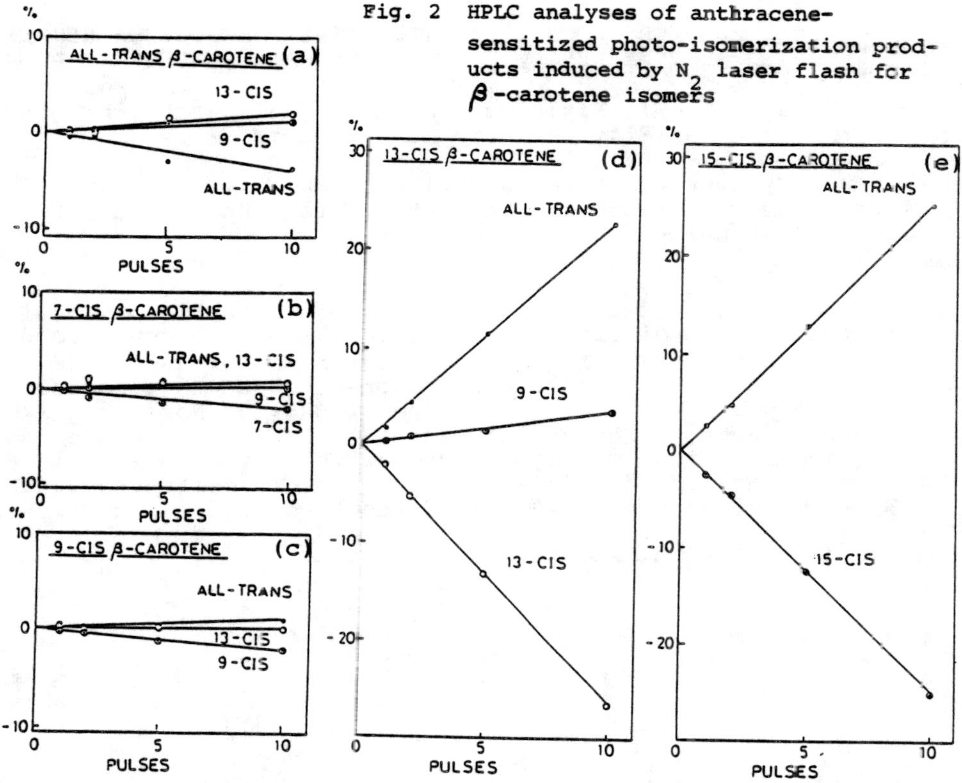

Fig. 2 HPLC analyses of anthracene-sensitized photo-isomerization products induced by N_2 laser flash for β-carotene isomers

magnitude lower than that of the 15-cis isomer (Fig. 2-b,c). This means that these isomers predominantly go back to the original isomers through the triplet-state. (iii) The 13-cis isomer produces the all-trans isomer as the major product, however it produces the 9-cis isomer as the minor product (Fig. 2-d). Assuming vertical relaxation, the above results (i), (ii) and (iii) predict the presence of three different kinds of triplet configurations, which relaxes into the all-trans, 7-cis and 9-cis ground state. We tentively call them "all-trans-like", "7-cis-like" and "9-cis-like" triplet configurations.

The results of the HPLC analyses of the isomerization products lead us to a possible interpretation of the Raman spectra: (I) The "all-trans-like" triplet configuration is responsible for the spectra (A) and (E). (II) The "7-cis-like" and "9-cis-like" triplet configurations are responsible for the spectra (B) and (C), respectively. (III) The spectrum (D) may be due to both the "all-trans-like" and "9-cis-like" triplet configurations.

(REFERENCES)

(1) J. Teraoka, H. Hashimoto, S. Matsudaira and Y. Koyama, Chem. Lett., 1985, 311.

(2) R. Wilbrandt and N.-H. Jensen, "Time-resolved vibrational spectroscopy ", ed. by G. H. Atkinson, Academic Press (1983), p. 273.

THE STRUCTURE OF THE LOWEST EXCITED TRIPLET STATES OF HEXATRIENES STUDIED BY TIME-RESOLVED RESONANCE RAMAN SPECTROSCOPY.

Frans W.Langkilde[a)], Niels-Henrik Jensen[a)],Robert Wilbrandt[a)], Albert M. Brouwer [b)], and Harry J. C. Jacobs [b)]

[a)]Chemistry Department, RISØ National Laboratory, DK-4000 Roskilde, Denmark. [b)]Gorlaeus Laboratory,State University at Leiden, 2300 RA Leiden, The Netherlands

The singlet state photochemistry of hexatrienes involves interconversions of the E and Z isomers and the corresponding 1,3-cyclohexadienes. The importance of the ground state conformation in controlling these photoreactions was expressed in the principle of Non Equilibration of Excited Rotamers (NEER) [1]. The short lifetimes of excited singlet states are among the arguments given in support of the NEER principle. This argument is less valid in the case of triplet excited states.For 1,3,5-hexatriene the predominant conformations in the ground

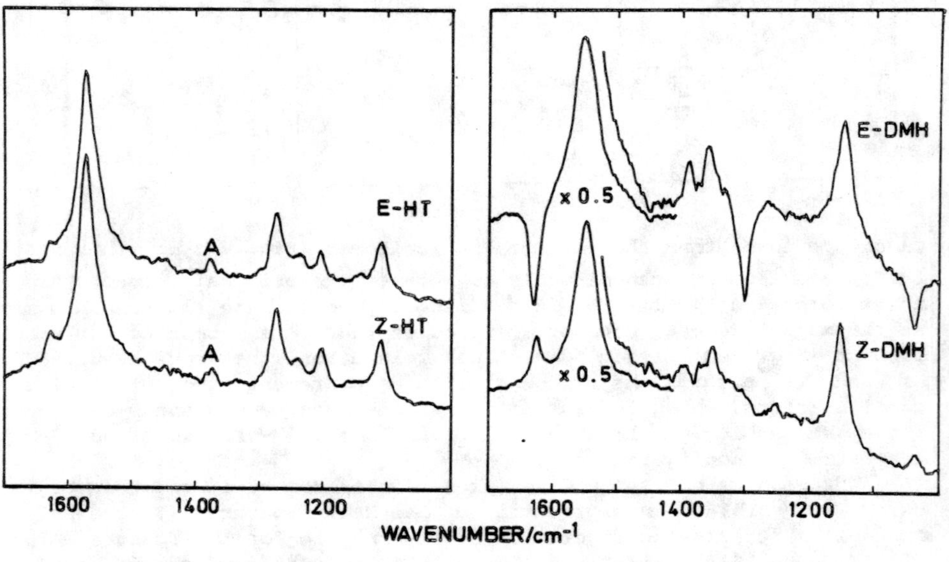

Figure 1. Time-resolved resonance Raman spectra of the lowest excited triplet states of the E (E-HT) and Z (Z-HT) isomers of 1,3,5-hexatriene and the E (E-DMH) and Z (Z-DMH) isomers of 2,5-dimethyl-1,3,5-hexatriene at time delays of 60 (E-HT and Z-HT) , 60-80 (E-DMH) , or 50-130 ns (Z-DMH) between pump and probe pulses. Bands from solvent, sensitizer, and ground state triene have been subtracted. A: Artifact

state are regarded to be tEt for the E isomer (E-HT) and tZt for the Z isomer (Z-HT) [1-3], whereas for 2,5-dimethyl-1,3,5-hexatriene they are regarded to be tEt for the E isomer (E-DMH) and cZc for the Z isomer (Z-DMH) . In order to obtain information on the validity of the NEER principle in the lowest excited triplet state we report the time-resolved resonance Raman spectra of the lowest triplet states of the E and Z isomers of hexatriene and 2,5-dimethyl-hexatriene. The triplet states were created by sensitized laser flash photolysis [4-6].

The time-resolved resonance Raman spectra of the lowest triplet states of E-HT and Z-HT shown in Fig. 1 are identical within limits of experimental error. The corresponding spectra of E-DMH and Z-DMH shown in Fig. 1 are different in the exact position of the band around 1150 cm^{-1} (and probably of the band around 1550 cm^{-1}), and in the bands slightly below 1400 cm^{-1}.

The simplest explanation to these experimental findings is that the differences in conformation around C-C single bonds in the ground state are preserved upon excitation to the lowest triplet state. These results support an extension of the NEER principle from singlet to triplet excited states with much longer lifetimes.

Furthermore, the photochemical interconversion between isomers could be followed as the growth or decrease of Raman bands from ground state species, and was in agreement with the interconversion found from analysis by gas chromatography.

References

1. H.J.C. Jacobs and E.Havinga, Advan. Photochem., 11, Interscience, 1979, pp. 305-373.
2. W.G. Dauben, E. L. McInnis, and D.M.Michno, Rearrangements in Ground and Excited States, Vol. 3, Ed. P. de Mayo, Academic Press, 1980, pp. 91-129.
3. M. Trætteberg, Acta Chem. Scand., 22, 628, 2294 (1968).
4. F.W. Langkilde, R. Wilbrandt, and N.-H.Jensen, Chem. Phys. Lett., 111, 372 (1984).
5. F.W. Langkilde, N.-H. Jensen, and R.Wilbrandt, Chem. Phys. Lett., 118, 486 (1985).
6. F. W. Langkilde, N.-H. Jensen, and R. Wilbrandt, Springer Proc. Phys., 4, 175 (1985).

This work was partially supported by the Danish Natural Science Research Council.

$O_2(^1\Delta_g)$ PHOTOFRAGMENT CARS SPECTROSCOPY AND THE DYNAMICS OF THE UV PHOTOLYSIS OF OZONE.*

Daniel P. Gerrity, David L. Phillips, Jong-Chen Nieh, Kevin T. Tabor, and James J. Valentini

Department of Chemistry, University of California, Irvine, CA 92717

We report the use of CARS spectroscopy to obtain rotationally and vibrationally resolved CARS spectra of $O_2(^1\Delta_g)$ formed in the uv photolysis of gaseous O_3. These experiments are carried out under conditions such that the fragments undergo no collisions prior to their spectroscopic observation. Hence, the CARS spectra directly reveal the nascent quantum state distributions of the $O_2(^1\Delta_g)$. Spectra have been obtained for photolysis of ozone at 20 wavelengths between 230 nm and 312 nm. The rotational and vibrational state distributions derived from these spectra are used to elucidate the photodissociation dynamics. The quantum state distributions show that at every photolysis wavelength all energetically possible $O_2(^1\Delta_g)$ vibrational states are produced, but within each vibrational state only a limited number of the possible rotational states are populated. The rotational states populated usually are high J states. These distributions are interpreted in terms of dissociation from an excited electronic state that is very repulsive along the anti-symmetric stretch coordinate, and that has an equilibrium O–O bond length longer than that in the ground electronic state.

The photofragment spectra reveal a propensity for even rotational states (J = 2, 4, 6, ...) of the $O_2(^1\Delta_g)$ photofragment. We show that this effect is due to selective depletion of $O_2(^1\Delta_g)$ odd rotational states (J = 1, 3, 5, ...) by a curve crossing to $O_2(^3\Sigma_g)$ in the photolysis exit channel. Our experiments indicate that this curve crossing effect leads to isotopic fractionation in the $O_2(^1\Delta_g)$ and $O_2(^3\Sigma_g^-)$ products, the former being relatively depleted of ^{18}O and ^{17}O and the latter relatively enriched in these isotopes. The implications of this isotopic fractionation for atmospheric chemistry and cosmochemistry are discussed.

The photodissociation produces $O_2(^1\Delta_g)$ in high vibrational ($v \leq 6$) and rotational ($J \leq 48$) states, that have not been observed previously. The observation of these high v,J states allows us to determine accurate spectroscopic constants and the first experimental potential energy curve for $O_2(^1\Delta_g)$. The constants and potential are reported.

*Research supported by the National Science Foundation and the American Chemical Society Petroleum Research Fund.

Mechanism of time-resolved ionization detected stimulated Raman scattering from molecules

Y. Nomura, Y. Fujimura, H. Kono,* and S. H. Lin**

Department of Chemistry, Faculty of Science, Tohoku University, Sendai 980, Japan
*Department of Basic Technology, Faculty of Engineering, Yamagata University, Yonezawa 992, Japan
**Department of Chemistry, Arizona State University, Tempe, Arizona 85287 U.S.A.

Time-resolved Raman spectroscopy is one of the powerful methods for studying dynamical processes taking place in the rovibronic states of molecules [1,2]. In this paper we present a theoretical study on application of the ionization detected stimutated Raman scattering method developed by Esheric and Owyoung [3] to investigation of the dynamics in time domain. For this purpose an expression for the transition probability of the time-resolved ionization detected stimulated Raman scattering from the rovibronic states of molecules is derived using the perturbative density mnatrix formalism. This new Raman spectroscopy, which basically consists of three-photon processes, is based on a resonant laser ionization detection of the rovibronically excited molecules produced via the stimulated Raman pumping as a function of time delay τ between the pumping and detection processes.

We consider a time-resolved ionization detected Raman scattering processes from an initial state a to the final state f via states {m}. The stimulated Raman scattering process is assumed to take place via states {s}. The interaction Hamiltonian between molecule and the radiation field, H'_{MR} is given by

$$H'_{MR} = - \sum_{s} M_{sa} T_{sa} E_{I} - \sum_{m} \sum_{s} M_{ms} T_{ms} E'_{I} - \sum_{m} M_{fm} T_{fm} E_{II} + \text{H.C.} ,$$

where M_{sa} denote the matrix elements of the electronic dipole moment, T_{sa} the transition operator from a to s states, and E denote the electric field operators of the lasers. The probability of ionization $P_f(\tau)$ can be expressed as

$$P_f(\tau) = 2 \,\text{Re}[A_1(\tau) + A_2(\tau) + B_1(\tau) + B_2(\tau) + C_1(\tau) + C_2(\tau)],$$

in which A denote the three-photon processes involving the simultaneous stimulated Raman scattering, while B and C represent those involving the sequential stimulated Raman scattering. Each term can be written in terms of the photon field correlation functions and the matrix elements of the time evolution operator of the molecular density matrix [4]

Model calculations of the probability of ionization as a function of τ are performed to discuss the mechanism of time-resolved ionization detected stimulated Raman scattering from the rovibronic states of molecules, and to show dephasing effects in the resonant states.

REFERENCES

1. G. H. Atkinson, "Time-resolved Vibrational Spectroscopy" (Academic Press, 1983).
2. Y. Fujimura, Y. Ohtsuki, M. Arai, and S. H. Lin, J. Chem. Phys. 82, 1246 (1984),
3. P. Esheric and A. Owyoung, Chem. Phys. Lett., 103, 235 (1983).
4. Y. Fujimura, Y. Nomura, and T. Nakajima, Chem. Phys. Lett., 124, 283 (1986).

TIME-RESOLVED RESONANCE RAMAN STUDY OF THE PULSE RADIOLYTIC OXIDATION OF AQUEOUS TRIHYDROXYBENZENES.

L. Qin, G.N.R. Tripathi and R.H. Schuler

Radiation Laboratory and Department of Chemistry
University of Notre Dame, Notre Dame, IN 46556

The vibrational spectra, structure and reaction kinetics of transient radical species produced in the pulse radiolytic oxidation of aqueous trihydroxybenzenes have been examined by time-resolved resonance Raman spectroscopy. The experimental arrangement used for such studies has been described previously.[1,2] We briefly report here some kinetic measurements on oxy radicals related to 1,2,4 trihydroxybenzene where advantage is taken of the resolution made possible by Raman methods.

Aqueous solutions containing 2mM 1,2,4 trihydroxybenzene, with and without 0.1M NaN_3 and bubbled with N_2O, were irradiated by 100 ns electron pulses of 2MeV energy. The reaction was examined in the pH range 3 to 11. In solution without NaN_3, the OH radicals produced initially react with 1,2,4 trihydroxybenzene to form cyclohexadienyl type radicals which at pH < 4, after H_2O elimination, form o-hydroxy-p-benzosemiquinone. This radical absorbs at 400 nm. Its kinetic behavior was monitored by optical absorption as well as kinetic Raman spectroscopy. The second order decay rate constant (2k) has been measured as ~ 3×10^8 $M^{-1}s^{-1}$. The $N_3\cdot$ oxidation is more direct and proceeds via electron transfer. In pH range 6 to 11, two transient anionic radicals, identified as o-hydroxy-p-benzosemiquinone anion (species A) and o-oxo-p-benzosemiquinone anion (species B) radicals are evident in the system. Both of these radicals have similar absorption features, with peaks at ~ 430 nm. Therefore, it is difficult to follow their reactions by optical absorption method. The Raman spectra of the two species are, however, quite distinct. We report here the study of the acid-base equilibria, and reaction of A with OH^- to form B, by kinetic Raman spectroscopy.

Figure 1 shows the pH dependence of the Raman spectra (1480-1680 m^{-1}) excited at 430 nm. The spectra were obtained 30 μs after the electron pulse to allow sufficient time for A and B to equilibrate. The signal intensity of B increases with pH, with corresponding decrease in the signals due to A. To compensate for fast decay of A [2k ~ 1.3×10^8 $M^{-1}s^{-1}$; for B, 2k ~ 6×10^6 $M^{-1}s^{-1}$], the ratio of

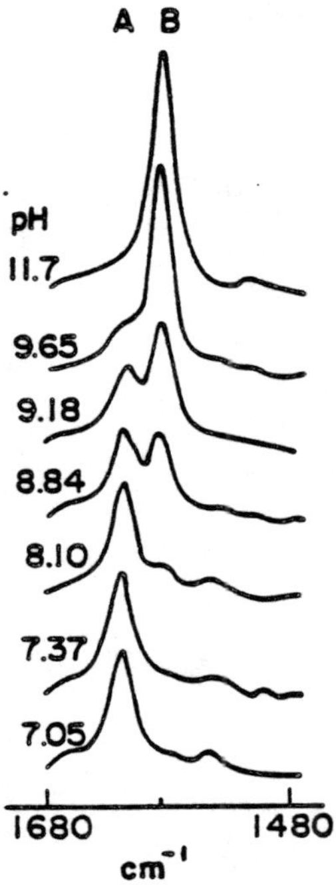

Figure 1

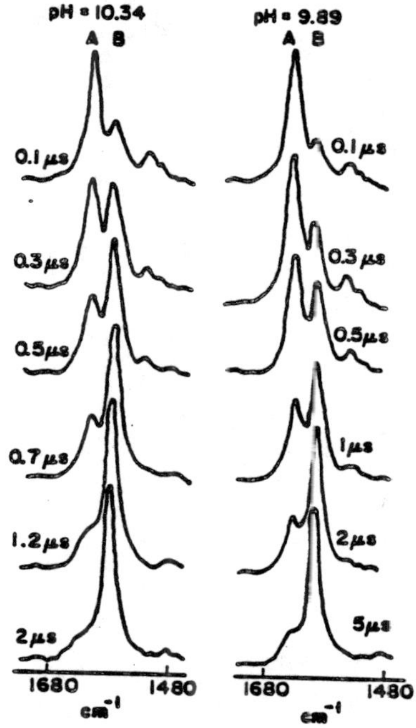

Figure 2

A or B concentration and total radical concentration vs pH was plotted, and the pKa of radical A was determined as 8.85.

Figure 2 exhibits the Raman signals obtained at pH 9.89 and 10.34, 100 ns to 5 μs after the electron pulse. The initial radical concentration was 1.5×10^{-4} M. The decay of the radical A into B is pH dependent and reaction with base is obvious. From the observed half periods of the reaction and OH^- concentrations in solution, a rate constant of $9 \times 10^9\ M^{-1}s^{-1}$ is measured. This may be somewhat over estimation since the transient OH^- concentration, produced by the reaction $N_3^- + \dot{O}H \longrightarrow N_3^{\bullet} + OH^-$ has not been taken into consideration. The rate constant for back reaction of B with water to form A is determined as ~ $2 \times 10^3\ M^{-1}s^{-1}$.

This system provides an interesting example where, because of spectral resolution, the advantage of time-resolved resonance Raman spectroscopy for kinetic measurements over optical absorption method is quite evident.

References

1. G.N.R. Tripathi and R.H. Schuler, J. Chem. Phys. 81 113 (1984); J. Phys. Chem. 88 1706 (1984).
2. G.N.R. Tripathi, Multichannel Image Detectors, II.,(ed. Y. Talmi), ACS Symposium Series, 236 171 (1983).

chapter 19

CARS and OTHER NON-LINEAR OPTICAL STUDIES

TIME-RESOLVED COHERENT RAMAN SCATTERING OF LIQUIDS AND GASES ON THE SUB-PICOSECOND TO NANOSECOND TIME SCALE

A. Laubereau

University of Bayreuth, D-8580 Bayreuth, Fed. Rep. Germany

Coherent Raman techniques have found wide application in the study of fast dephasing phenomena in gases and condensed phases. In these experiments synchronized pump and Stokes pulses excite molecular transitions in the medium. A delayed probe beam is used to produce coherent Raman scattering, the intensity of which develops in time according to reversible and irreversible phase effects. Two topics will be discussed in this paper:(i) time-resolved CARS is studied in liquids on the subpicosecond time-scale with carefully chosen polarization conditions demonstrating a coherence peak and a technique to measure rotational relaxation times; (ii) time-domain coherent Raman scattering is applied for a Fourier Transform Raman Spectroscopy achieving high frequency resolution.

We have performed a theoretical and experimental study of time-resolved coherent Raman scattering under various polarization conditions /1/. Three mechanisms are shown to contribute to the scattering signal in the general case:
(i) scattering off the isotropic part of the resonant material excitation with vibrational dephasing time T_2;
(ii) scattering from a second, oriented component of the resonant material excitation with "anisotropic" time constant τ_{an};
(iii) four wave mixing via the non-resonant part χ_{NR} of the third order susceptibility.
The observed signal transients depend on the different roles the three mechanisms play for different polarization geometries. New effects occur under properly chosen conditions.

For parallel polarization of the four light beams involved in the experiment (input laser and Stokes pulses, probe pulse and scattered light) and equal frequencies of the exciting laser and probe pulses, a "coherence peak" is predicted. The effect occurs for short delay times and is related to the coherence artifact known for the induced grating technique (degenerate four wave mixing). Experimental data are reported for liquid N_2 and benzene. The effect is important for a proper determination of experimental decay times.

For perpendicular polarization planes of the excitation fields (laser ⊥ Stokes) and of the probing process (probe ⊥ coherent scattering) a drastic change of the signal transient occurs compared with the parallel case. The signal decays first rapidly close to the instrumental response function and reaches an asymptotic slope with time constant τ_{an}; the time constant characterices the oriented component of the resonant material excitation. For liquids with rapid molecular reorientation τ_{an} is notably shorter than T_2.

Experimental data of this effect are reported for the first time for the ν_3-mode of methyl iodide /2/. For example, at 247 K our measurements directly yield $\tau_{an}/2=0.72$ ps, while the similar experiment for parallel polarization gives T_2 = 2.1 ps. Comparing the two results a new technique is demonstrated for the measurement of rotational relaxation times of liquids /2/.

The broad spectrum of picosecond pulses of several cm^{-1} allows to excite several neighbouring molecular transitions simultaneously with well-defined phases and amplitudes. Coherent superposition of the excited components leads to a corresponding beating pattern in the subsequent probe scattering process. Analysis by numerical Fourier transformation allows to determine the frequency differences of the excited Raman transitions with high precision depending on the observed delay time interval.

As a demonstration of this novel technique an experimental investigation of methane in the supersonic expansion is presented, using 20 ps pulses derived from a mode-locked Nd: YAG laser /3/. Besides data on the rotational and translational temperature, the frequency spacings of various rotation-vibration transitions are measured with an accuracy $<10^{-3}cm^{-1}$. Our experimental Fourier Transform Raman spectra resolve for the first time the doublet structure of the ν_1,J=2 line of CH_4. The potential of this kind of spectroscopy will be elucidated.

/1/ N. Kohles and A. Laubereau, Appl. Phys. B 39, 141 (1986)

/2/ N. Kohles and A. Laubereau, to be published

/3/ H. Graener and A. Laubereau, Opt. Commun. 54, 141 (1985)

CARS STUDIES OF THE LOWEST EXCITED STATE OF α,ω-DIPHENYLPOLYENES.

Shiro Maeda

Tokyo Institute of Technology, Midoriku, Yokohama 227, Japan

There has been considerable interest in the resonance Raman investigation of short-lived intermediates and electronic excited states. Such a transient vibrational spectroscopy is useful to promote our understanding of molecular photophysical and photochemical behaviors, in cooperation with the electronic spectroscopy, by making up the detailed structural information of the passing states.

CARS and related coherent Raman methods are well known to be very useful to overcome the fluorescence interference in ordinary Raman, which may occur more often in the measurement of those transient species. Besides, the nonlinear signal generation in CARS is advantageous in itself for the investigation of short-lived states, where the use of high peak power pulsed lasers is almost routine. We have observed the vibrations of various conjugated molecules in the lowest excited singlet(S_1) and triplet(T_1) states by using CARS(or CSRS), exploiting the resonance enhancement with the transient S_N-S_1 or T_N-T_1 absorption. In this paper, the resonance CARS and/or CSRS spectra of α,ω-diphenylpolyenes in the S_1 and T_1 states are presented, and the results are discussed, together with some other observations, for the purpose of characterizing those excited states more definitely and investigating the potential surface more closely.

It is a very interesting and significant observation that the S_1 state of many polyenes is of forbidden A_g(in C_{2h}) character with an important contribution from the doubly excited MO configuration, in place of the allowed B_u state of typical LUMO-HOMO single excitation. It is then of much interest to investigate the potential surface of this complex S_1 state by observing the vibrational spectrum. Besides, there still remain some indefinite problems concerning the character of the S_1 state of diphenylbutadiene.

Typical resonance CARS records are shown in Fig.1 for diphenyloctatetraene(DPO). Most striking is the exceptionally intense Raman signal at 1755 cm^{-1} in the S_1 spectrum. Similar vibrational frequency has already been known in the vibronic structures of the excitation spectra of unsubstituted octatetraene(1754 cm^{-1}) and some other longer polyenes. The frequency is unusually high as the polyene chain stretching mode, which never exceeds 1630 cm^{-1} in the ground state and is expected to be lowered in the excited states. However, the assignment other than ν_{CC} fundamental is quite improbable from every evidence. Corresponding signal was also observed in the S_1 CARS of diphenyldecapentaene(DPD) even with higher frequency(1775 cm^{-1}), but not in the S_1 spectra of diphenylhexatriene(DPH) and diphenylbutadiene(DPB) and in all the T_1 spectra. Thus, it seems rather likely that the unusual ν_{CC} reveals the characteristic feature of the potential surface of polyene S_1 state except in some shorter systems, and then, the observation in both of Raman and vibronic spectra may indicate that there is no remarkable difference between the ground and S_1 state conformations.

It is hardly expected that the frequency exaltation is explained by the change of diagonal stretching force constants due to electronic excitation, but the responsible factor may be included in the off-diagonal terms in the conventional quadratic potential function. Thus, we carried out a perturbation calculation of the off-diagonal force constants based on a simple valence bond formalism. The last formalism was employed because the polyene 2^1A_g state is known to be predominantly covalent, and in such a case, the vibronic perturbation treatment can be carried out more straightforwardly than in the MO approach with extensive CI.

The calculation showed that considerable degree of vibronic interaction can take place between the ground(1^1A_g) and 2^1A_g states through the highest symmetric stretching mode of the chain, resulting exceptionally high frequency as observed. Such an interpretation applies to every polyene chain of moderate and longer size unless the electronic and vibrational conditions are greatly changed. The nonappearance of the corresponding high frequency in DPB and DPH may imply such a change.

Raman non-resonant part of the CARS susceptibility includes two-photon resonant contribution, and the study of the interference effect between the Raman and two-photon resonances in the CARS signal provides the information on the two-photon absorption, especially the absolute absorption cross section. Since the cross section is an important information for characterizing the dipole forbidden ^1A_g state of polyenes, we measured the values using CARS method. The result will be used to discuss the nature of the 2^1A_g state.

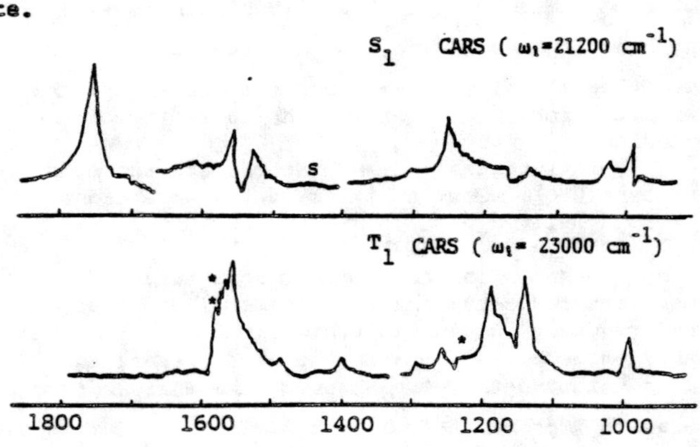

Fig.1 Resonance CARS of the S_1 and T_1 states of DPO. (s: solvent, *: ground state signals)

THE APPLICATION OF NONDEGENERATE CARS TO INVESTIGATE SOLUTIONS

P.A.Apanasevich, V.V.Kvach, V.P.Kozich, V.A.Orlovich

Institute of Physics, BSSR Academy of Sciences,Minsk,USSR

In the present paper, the results are reported of the development of methods of nondegenerate resonance coherent anti-Stokes Raman scattering of light (RCARS) and their application to investigate vibrational modes of the ground and excited electron states of a variety of metalloporphyrins, an important class of complex organic compounds.

We studied CARS spectra of Cu-, Ni-. and Zn-octaethylporphyrins in various solvents with varying intensities and frequencies of exciting fluxes within the absorption Q-band lying in the visible spectral region. Cu-OEP and Ni-OEP belong to nonluminescent and Zn-OEP to luminescent organic compounds. In the latter, luminescence makes it impossible to obtain a vibrational spectrum by spontaneous RS at excitation in the Q-band.

In our experiments, exciting flux frequencies were chosen so that the highest frequency of biharmonic pumping fell within the Q-band. The probing beam was chosen so that the anti-Stokes signal lay in the absorption Q -band. Under such conditions most efficient coherent scattering may be realized.

Measurements were taken by a CARS-spectrometer incorporating three narrow-band dye lasers with controlled radiation frequencies and polarizations. The lasers were pumped by a second harmonic of a pulsed nanosecond YAG laser. The spectrometer comprised optical delay lines, systems of beam convergence and scattering signal recording. The recording of CARS-spectra and the information processing were automatic.

The use of nondegenerate CARS makes it possible to determine three invariants of the light RS tensor in an isotropic medium by measuring the scattering spectra with independent mutual orientations of exciting wave polarizations. It should be noted that the antisymmetric part of the scattering tensor may be separated and measured.

We have developed a method for such measurements and determined symmetry of vibrations observed in the spectrum. It has been found that at excitation to the Q-absorption band, in RCARS spectra incompletely symmetric or antisymmetric vibrations appear, and the characteristics of their corresponding scattering lines have been measured.

With resonance CARS when the exciting flux frequencies are in the electronic absorption band an intense pump makes some of the molecules go to excited electron states. If there is a resonance enhancement of scattering it seems possible to investigate by the CARS method the characteristics of molecular vibrations in these electron states.

Our studies have shown that at exciting flux intensities $\leqslant 10^6 W/cm^2$ in CARS-spectra of Cu-, Ni-, Zn-OEP solutions in tetrahydrofuran (THF) and toluene, only lines of scattering by molecular vibrations in the ground electron state are present. As the intensity of the fluxes in Cu--OEP and Ni-OEP solutions in THF increases, additional lines corresponding to scattering from the excited electron state appear. The measurements of the symmetry, frequencies and efficiency of scattering to these lines, depending on the position of exciting fluxes in the absorption band, as well as the comparison of the obtained information with the data of picosecond absorption spectroscopy, have made it possible to identify the excited electron states from which scattering occurs.

In other solutions, an increase in the flux intensity led to the disappearance of CARS lines. This may be due to the depopulation of the ground electron state and the absence of resonance enhancement of scattering in the excited state into which the molecules go.

The application of RCARS to investigate luminescent solution of Zn-OEP has made it possible to obtain for the first time the vibrational spectrum of this substance at excitation in the Q-absorption band. A comparison of this spectrum with the RRS spectrum obtained at excitation in the Soret band reveals significant differences between them. These differences are due to the fact that in the Soret band, unlike excitation in the Q-band, the resonance enhancement of completely symmetric vibrations predominates. As a result, at excitation in different bands the scattering spectra complement each other nicely.

Our studies have shown that due to the directivity and anti-Stokes shift of the scattering signal CARS makes it possible to study with equal success both luminescent and nonluminescent solutions of metalloporphyrins.

FINDING REAL AND IMAGINARY SUSCEPTIBILITIES GOING THROUGH HILBERT TRANSFORMS, FREIGHT-CARS, A NEW METHOD FOR THE ANALYSIS OF CARS LINESHAPES.

J. M. Bostick and L. A. Carreira

Department of Chemistry, University of Georgia, Athens, Georgia 30602

Coherent anti-Stokes Raman spectroscopy is probably best known as a vibrational (and rotational) probe for highly fluorescent species. In the condensed phase, however, the experimental requirements are quite demanding and the analysis of the resulting spectra is often quite complex if one attempts to extract all of the information present. One of the major drawbacks of condensed phase CARS spectra has been that the spectral lineshapes are often too complex for routine analysis. This complexity is due to the presence of both dispersive and Lorentzian components stemming from the complex solution to the third order susceptibility equation from which the CARS signal intensity is derived. Even the normally simple task of determining Raman vibrational frequencies becomes non trivial since peak maxima do not always correspond to the line center.

In order to extract spectral information, complicated lineshape analysis programs are used to analyze the spectra. These methods use a least squares regression technique to fit the experimental data to known theoretical lineshape functions. These techniques require the user to input original guesses as to the frequency and halfwidth of the transition to be fit. In the hands of the expert user, these methods work reasonably well, however, for congested spectra the initial guesses are often difficult if not impossible to make without other information. This difficulty has led researchers to search for other techniques that reject fluorescence but still retain the easily interpretable Lorentzian lineshape. Unfortunately, these techniques often provide no signal or distorted lineshapes when pumped near an electronic resonance.

A numerical method has been developed which converts the raw CARS spectrum to a corrected spectrum which appears as the appropriately scaled Lorentzian lines. This correction is performed essentially in the Fourier domain and involves the use of a corrected indirect Hilbert transform as an intermediate step in the generation of the phase corrected spectrum. Subsequent generation of a power spectrum from the original and Hilbert transformed data yields a spectrum which is Lorentzian in character. Further treatment of the corrected spectrum through cross correlation of these data sets allows the determination of the relative contributions of Lorentzian and dispersive shapes to the overall peak intensity. With this technique, spectral parameters can be determined quickly and accurately without the necessity of guessing at frequency and halfwidth information.

CSRS WITH INCOHERENT LIGHT FOR VIBRATIONAL DEPHASING TIME MEASUREMENTS

Toshiaki Hattori and Takayoshi Kobayashi

Department of Physics, Faculty of Science, University of Tokyo,
Hongo 7-3-1, Bunkyo, Tokyo 113 Japan

Dephasing dynamics of vibrational states in polyatomic molecules in liquid phase or in solution can be investigated by picosecond coherent Raman experiments [1]. In these experiments, coherent vibrations were excited and probed by picosecond laser pulses and the time resolution was considered to be limited by the pulse durations of them. Since the dephasing time of vibrations in liquid molecules are usually shorter than 10 ps at room temperature, available light sources of short and intense enough pulses are limited, and they are not handy.

However, it was suggested [2], and experimentally verified later [3] that the time resolution is determined not by the pulse durations but by the correlation time of the optical field for dephasing measurements by degenerate four-wave mixing.

We have applied this principle to time-resolved coherent Raman measurements (see Fig. 1), where two beams are originated from a single incoherent light source, with a delay time τ, and the other beam is of different wavelength and assumed to be coherent. A pair of one of the incoherent light and the coherent light excite the vibration and the other incoherent light probes it. When the delay time τ is negative the roles of two incoherent light are exchanged. The correlation time of the incoherent light field should be shorter than the dephasing time while pulse durations of all the beams may be much longer than the dephasing time.

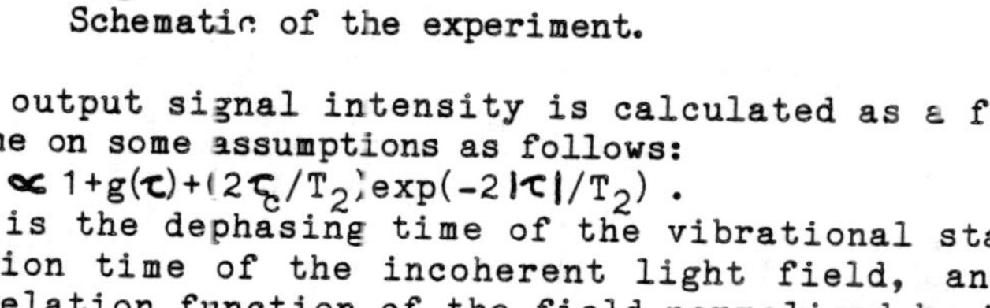

Fig. 1
Schematic of the experiment.

The output signal intensity is calculated as a function of the delay time on some assumptions as follows:

$$I(\tau) \propto 1+g(\tau)+(2\tau_c/T_2)\exp(-2|\tau|/T_2) .$$

Here T_2 is the dephasing time of the vibrational state, τ_c is the correlation time of the incoherent light field, and $g(\tau)$ is the autocorrelation function of the field normalized by its peak value. Since the time resolution of this method is limited only by τ_c, picosecond or subpicosecond dephasing times can be measured with the use of nanosecond laser pulses.

We have performed experiments using the second harmonic (532 nm) of a Q-switched Nd:YAG laser as coherent light and a broad-band dye laser pumped by it as an incoherent light source. Since the correlation time of the second harmonic of Nd:YAG laser is about 30 ps, it may be regarded as coherent light in the time scale of observation. The dye laser was constructed with no tuning elements and tuned to appropriate wavelengths (around 630 nm) by varying the concentration of the laser dye (Rhodamine 640). The correlation time of the optical field was about 100 fs, which was obtained by the autocorrelation measurement. The CSRS signal was detected varying the delay time by a stepping motor.

Figure 2 shows the result for a C-H stretching mode (2914.5 cm^{-1}) in dimethylsulfoxide. The obtained value of T_2 is in agreement with the value 1.4 ps obtained using short pulses [4], and the intensity ratio of the peak to the tail is seven, which is also in good agreement with the theoretical value.

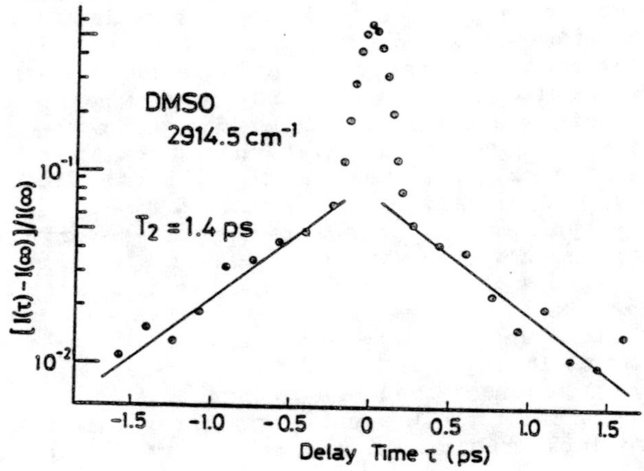

Fig. 2
CSRS signal intensities which are normalized by the background intensity I(∞) after it is subtracted.

REFERENCES

1. A. Laubereau and W. Kaiser, Rev. Mod. Phys. 50, 607 (1978).
2. N. Morita and T. Yajima, Phys. Rev. A 30, 2525 (1984).
3. S. Asaka, H. Nakatsuka, M. Fujiwara, and M. Matsuoka, Phys. Rev. A 29, 2286 (1984).
4. S. M. George, H. Auwester, and C. B. Harris, J. Chem. Phys. 73, 5573 (1980).

HIGHER ORDER SUSCEPTIBILITY IN NON-LINEAR OPTICAL MIXING A COMPARATIVE ANALYSIS

A.N. Weiszmann, N. Chencinski, A.M. Levine, and W.M. Schreiber
CSI of CUNY, Dept. of Applied Sciences, N.Y. 10301

Since the farsighted prediction[1] that coherent processes can be caused by incoherent perturbations, the understanding of the physics of damping in nonlinear optical mixing processes became imperative. The pragmatic introduction of one or more decay terms as effective damping in the iterative solution of the quantum-mechanical Liouville equation for the desnity matrix[2] and the extraction of the 48 terms of the third-order susceptibility, to describe the nonlinear response of matter was a formidable result and experimentally satisfactory for awhile.

The increase in resolution in current experiments[5,6] and the fact that a field theoretical treatment of the incoherent vacuum fluctuations can create coherent processes has suggested the need for more elaborate theories.

The present paper intends to make one step in this direction.

It is a comparative analysis of three different approaches. The classical perturbative density matrix based theory[1,2,3,4] the semi-classical one[7,8] based on an eigenvector description of the physical states and thus the exact solution of the time-dependent Schrodinger equation, and a quantum-theoretical one[9,10] where the states are dressed ones and the fields are fully quantized.

For the sake of simplicity, only three dominant terms: the 31, 32, and 33 will be selected from the classical work of Bloembergen and co-workers[1,2] No additional damping is introduced.

In the semi-classical model, a four-level quantized system interacting with four e-m field will be considered. The time-sequence follows from the diagramatic equivalents[4] of the above terms. The resulting Hamiltonian is time-dependent and can be transformed into a time-independent one using a transformation matrix $T_{ij}(t)$[7,8]. Then the eigenvectors a_{ij} and the eigenvalues of the Hamiltonian can be found. The expectation value of the polarization vector for the time dependent problem is found, and the exact expression for the component oscillating at the frequency ω_p is extracted.

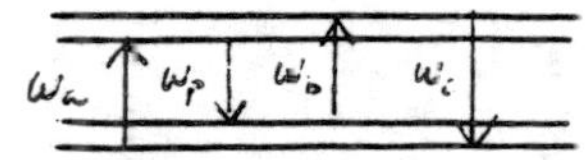

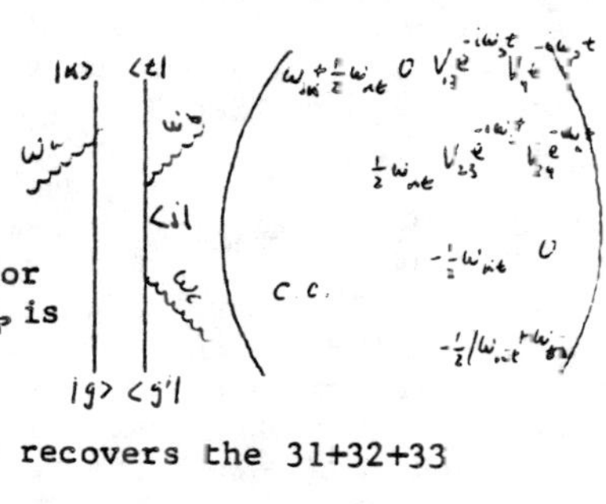

$$P(\omega_p) = P_{41} \sum_{k=1}^{4} a_{1k} |a_{44}|^2 \bar{a}_{4k}$$

The lowest order of this expression exactly recovers the 31+32+33 Bloembergen term.

A five-level system interacting with four coherent signals is considered in the quantum model[9]. The states are dressed atomic ones with the fields in photon number representation, and all one-photon exchanges (repeated as many times as they wish in any order) are allowed. The Hamiltonian is fully quantized with the interaction term[9]

$$H' = \prod_{K=s,c,b,p}' \{\hbar\kappa(J_- a_K^+ + J_- a_K) + \hbar\kappa^*(J_+ a_K^+ + J_+ a_K)\}$$

Thus the matrix form of the Hamiltonian is different for each (31, 32, 33) term-following the correct time sequence-and an exact solution of the Schrodinger equation has been obtained. The expectation value for the time dependent polarization vector has been calculated and the exact component for the term oscillating at the frequence ω_p was extracted. The lowest order again is reduced to the above mentioned Bloembergen term.

$$\langle P(t) \rangle = \sum_{K,j,i=1}^{5} A_{Kj}^*(t)\, A_{ij}(t)\, P_{Ki}(0)$$

Computer simulated perliminary plots for the lineshapes - in all three approaches - for the polarization component oscillating at ω_p shows more than one extra resonances on the tail section of the normal peaks.

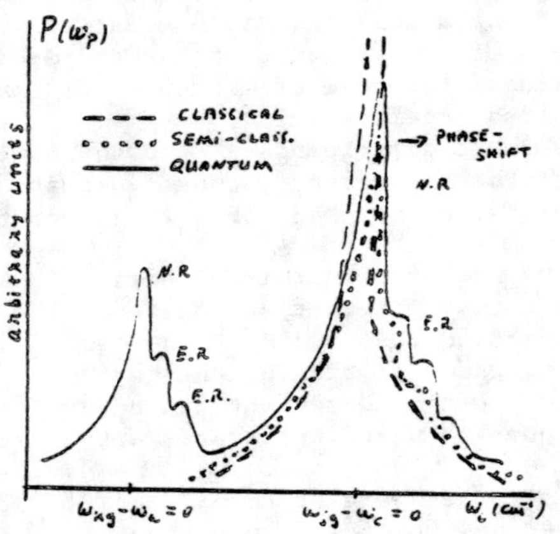

These can be interpreted only (there is no damping in the picture) as a result of higher order terms due to pure quantum effects. Incoherent vacuum fluctuations generating population transfer and acting as proper dephasing, thus creating coherences between dressed states, eliminating the destructive interference between the probability amplitudes. Moreover, there is a phase-shift between the peaks for semi-classical and quantum treatment. Actual values and plots are still under review, and at this time one can only infer the existence of a new QUIER4 (Quantum induced extra resonances) effect.

References

1. N. Bloembergen and Y.R. Shen: Phys Rev. 133 A37, 1964
2. N. Bloembergen, H. Lotem and R.T. Linch, Jr.: Ind.J.Pure and Appl. Phys. 16, 151 (1978)
3. T.U. Yee and T.V. Gustafson: Phys. Rev. A 18, 1597 (1978)
4. Y. Prior: IEEE J-QE 20, 37 (1984)
5. Y. Prior et al: Phys. Rev. Letters 46, 111, (1981)
6. Y. Prior, I. Schek and J. Jortner: Phys. Rev. A 31, 3775 (1985)
7. A.M. Levine, W.M. Schreiber and A.N. Weiszmann: Phys. Rev. A25, 625 (1982)
8. W.M. Schreiber, A.M. Levine and A.N. Weiszmann: Bull. Am. Phys. Soc. 28, 1319 (1983)
9. Y. Prior, and A.N. Weiszmann: Phys. Rev. A 29, 2700 (1984)
10. W.M. Schreiber, N. Chencinski, A.M. Levine and A.N. Weiszmann: Proc. Int. Conf. on Laser Spectr. Rehovot, Israel Dec. 16, 1985

COHERENT ANTI-STOKES RAMAN SCATTERING OF NEMATIC LIQUID CRYSTAL

L.S.Aslanyan, N.K.Badalyan, A.A.Petrosyan, Yu.S.Chilingaryan
Yerevan State University, Yerevan, Mravyan 1, 375049, USSR

The application of laser spectroscopy methods to liquid crystals (LC) mesophase study led to interesting results enabling to extract much information on LC structure, the character of molecular interactions and the behaviour of LC near the phase transitions.

We report the results of determination of order parameter S_4 and α_1 and α_2 components of Raman tensor for vibrations with frequency $\Delta\nu$=1164 and 1576cm^{-1} in MBBA by the method of coherent anti-Stokes Raman scattering (CARS).

CARS application to LC investigations seems very important because of some advantages of this technique in comparison with traditional Raman scattering spectroscopy i.e. high sensibility and spectral resolution, the possibility of local measurements by sharply focused beams. It is important to stress the larger possibilities of polarization measurements by CARS method as well.

The investigations were carried out on standard CARS spectrometer. Biharmonic pumping was provided by second harmonic radiation of YAG:Nd^{3+} laser and by radiation of tunable dye laser with transverse pumping. The anti-Stokes signal was distinguished on spectrometer, registered in digital-code and analised by microcomputer.

The results of measurements are given in Table 1 (see details in [1]).

Table 1

$\Delta\nu$, cm^{-1}	$\chi^{(3)}_{1221}/\chi^{(3)}_{1111}$	$\chi^{(3)}_{1111}/\chi^{(3)}_{3333}$	R_{iso}
1164	0,17±0,04	1,7±0,4	0,28±0,06
1576	0,32±0,07	0,13±0,04	0,29±0,06

The nematic liquid crystal MBBA examined in this work belongs to the group of symmetry $D_{\infty h}$. Tensor $\chi^{(3)}_{ijkl}(\omega_a;\omega_1,\omega_1,-\omega_2)$ has in mesophase only seven independent nontrivial Cartesian components with account of additional symmetry relative to the permutation of two middle indices. These components of nonlinear susceptibility can be connected with the liquid crystal order parameters and molecular parameters in to average the hyperpolarizability over molecular distribution [2]. The expressions obtained this way were used for determination of mentioned orientational and molecular parameters from experimental measurements. The calculations were carried out on a computer by special programme. The results are presented in Table 2.

Table 2

$\Delta\nu$, cm^{-1}	α_1	α_2	S_4
1164	−0,12±0,04	0,14±0,03	0,16±0,04
1576	−0,12±0,02	0,22±0,04	0,16±0,04

The data for Raman resonance with $\Delta\nu$=1576cm^{-1} are in good agreement with those in [3], and for vibrations with $\Delta\nu$=1164cm^{-1} the mentioned parameters were determined for the first time. So, the possibility of using CARS for the study of liquid crystals is demonstrated.

R E F E R E N C E S

[1] Aslanyan L.S., Badalyan N.N., Petrosyan A.A., Khurshudyan M.A., Chilingaryan Yu.S. Opt. i spectr., 1982, 53, p.94.

[2] Aslanyan L.S., Badalyan N.N., Papyan O.S., Petrosyan A. A., Chilingaryan Yu.S., Izv. AN SSSR, ser. Fisicheskaya, 1983, 47, p.1596.

[3] Jen S., Ciark N.A., Preistley E.B., Pershan P.S., J. Chem. Phys., 1977, 66, p.4635.

chapter 20

NEW RAMAN TECHNIQUES

IMPULSIVE STIMULATED RAMAN SCATTERING: AN ENTIRELY TIME-DOMAIN VIBRATIONAL SPECTROSCOPY.

Keith A. Nelson, Margaret R. Farrar and Leah R. Williams

Dept. of Chemistry, Massachusetts Institute of Technology, Cambridge, MA 02139

We have recently developed an entirely time-domain form of vibrational spectroscopy which permits direct observation of coherently vibrating species at various stages of vibrational distortion. This offers unique possibilities for measurement of vibrational dynamics, characterization of vibrationally distorted species and observation of vibrationally-induced rearrangements.

In "impulsive" stimulated scattering (ISS), ultrashort laser pulses are used first to ("impulsively") excite coherent vibrational motion and then to monitor vibrational oscillations and decay in real time. The experimental setup is shown in Figure 1. Figures 2-3 show time-dependent vibrational oscillations and dephasing of acoustic and optic phonons, respectively.

Impulsive stimulated Brillouin scattering (ISBS) has been used to characterize anomalous acoustic behavior near structural phase transitions,[1-2] liquid-glass transitions,[3] and in liquid crystals. Acoustically-induced ordering has been observed directly in ISBS experiments on "soft" acoustic phonons in crystals near structural phase transitions. A wide range of acoustic frequencies (~3 MHz-30 GHz) can be explored, and even heavily damped or overdamped modes can be characterized (See Figure 2).

Impulsive stimulated Raman scattering (ISRS) has been used for femtosecond time-resolved measurements of optic-phonon frequencies and dephasing rates in organic molecular crystals[4] (See Figure 3). The inherently high time-resolution of ISRS permitted measurements of dephasing even at high temperatures. ISRS experiments have recently been carried out in inorganic crystals as well.

Several additional features of ISRS have been predicted and are the subjects of current experimental work. First, ISRS extension to low-frequency <u>molecular</u> vibrations should be straightforward. Second, time resolved probing by <u>absorption spectroscopy</u> (rather than coherent scattering) will yield information about the <u>electronic structures</u> of vibrationally distorted crystals and molecules, as well as electron-phonon interaction strengths. This will be especially interesting in chemically reactive species. Third, direct measurement of vibrational <u>lifetimes</u> (T_1) as well as dephasing rates (T_2) should be possible since ISRS excitation produces a vibrational standing wave whose energy density is spatially periodic. Even after dephasing occurs, the incoherent vibrational population will continue to diffract light with a decay rate given by T_1. Fourth and particularly promising, ISRS excitation occurs not only when two femtosecond pulses are crossed spatially and temporally inside a sample (as shown in Figure 1), but whenever even a <u>single</u> ultrashort pulse passes through a Raman-active medium.[5] In other words, <u>ISRS is a general aspect of ultrashort-pulse intersections with matter.</u> A single pulse excites a spatially uniform (k=0), rather than periodic, coherent excitation. This will greatly facilitate time-resolved absorption spectroscopy of coherently vibrating species.

1. M. M. Robinson *et al.*, Chem. Phys. Lett. **112**, 491 (1984). 2. M. M. Farrar *et al.*, IEEE J. Quant. Electron., in press. 3. Y.-X. Yan and K. A. Nelson, to be published. 4. S. de Silvestri *et al.*, Chem. Phys. Lett. **116**, 146 (1985). 5. Y.-X. Yan *et al.*, J. Chem. Phys. **83**, 5391 (1985).

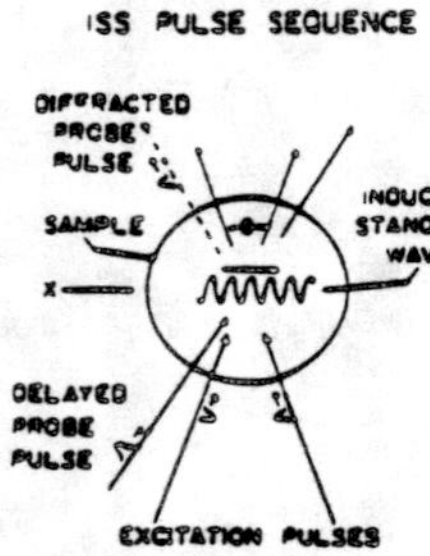

Figure 1. Schematic illustration of the impulsive stimulated scattering experiment. The crossed excitation pulses, with wave vector difference k, excite coherent phonons which form a standing wave of wave vector ±k. The phonon frequency is derived from mixing of frequencies contained within the excitation pulse bandwidth. Time-dependent vibrational oscillations are monitored by coherent scattering of variably delayed probe pulses.

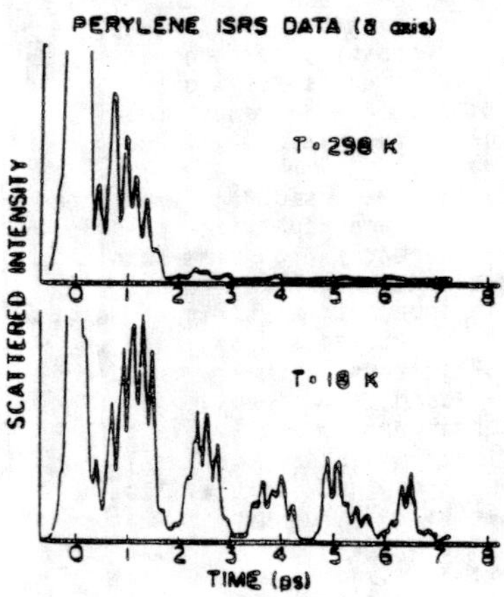

Figure 3. Impulsive stimulated Raman scattering data from α-perylene with the phonon wave vector along the $\hat{a}$ crystallographic axis. The 70-fs excitation pulses excite two optic phonon modes of 80 and 104 cm^{-1}. The data contain sums and differences of these fundamental frequencies to produce the observed "beating" pattern. The phonon decay rates increase from about 10^{11} s^{-1} at 18K to 10^{12} s^{-1} at 298K.

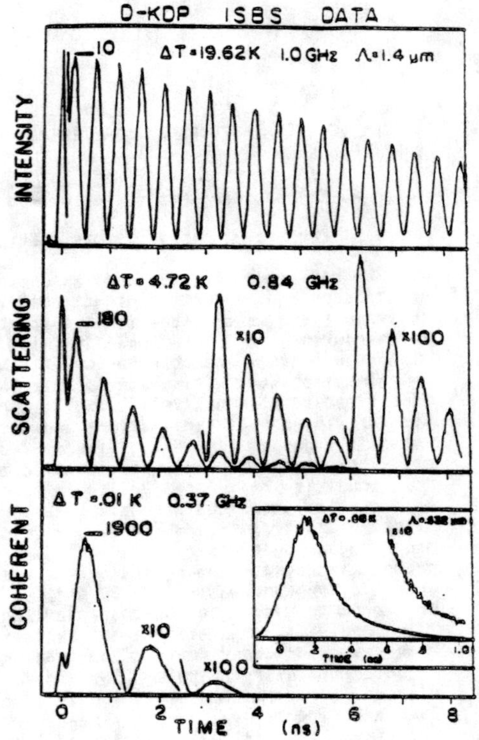

Figure 2. Time-resolved ISBS data showing shear acoustic waves in D-KDP. As the ferro-electric transition temperature, T_c, is approached, the acoustic frequency decreases, attenuation increases, and acoustically-induced ferroelectric ordering causes the scattered light intensity to increase by 1900x over its room-temperature value. $\Delta T = T - T_c$ and Λ is the acoustic wavelength. The inset shows that the acoustic mode becomes <u>overdamped</u> near T_c at higher wave vectors.

INSTRUMENTAL ADVANCES IN RAMAN OPTICAL ACTIVITY.

Max Diem

Dept. of Chem., City University of New York, Hunter College, 695 Park Avenue, New York, NY 10021 USA

Raman Optical Activity (ROA), the scattering of circularly polarized incident radiation by chiral media, was first observed by Barron, et al.[1] and Hug, et al. [2]. It may be observed by modulating the exciting laser beam between left and right circular polarization and monitoring the difference in scattered intensities. Two major problems accompany the observation of ROA: first, the small magnitude of the intensity difference ($<$ .005 of Raman intensities) suggests multichannel detection, which requires the synchronization of modulation and detection electronics. Second, the exciting beam will be, in general, elliptically polarized, which can superimpose artifacts on the ROA signal. Hug [3] reported the design of a ROA instrument which produced excellent data in short acquisition times and free of artefacts. This instrument, which was accidentally destroyed, used custom built optical and electronic components. We have constructed a ROA unit [4], of similar design, using mostly commercial components (Fig.1). In this instrument, we have taken efforts to monitor, by computer, the quality of the circular polarization, and the alignment of the optics, and to display, in real time, the Raman and ROA intensities to permit an immediate assessment of the quality of the alignment. We have succeeded to observe polarized, depolarized and total ROA scattered components on our instrument.

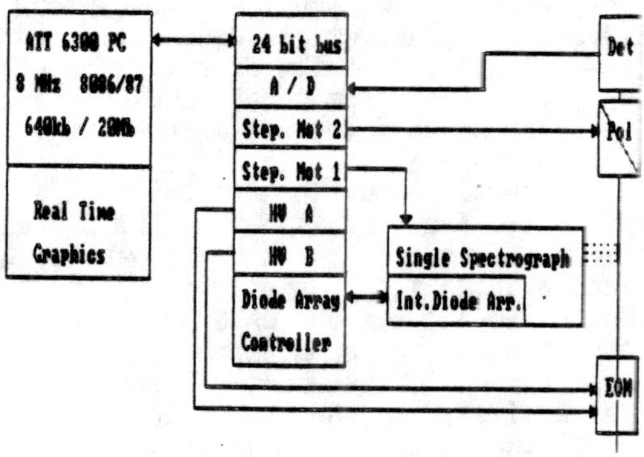

Fig. 1: Block Diagram of ROA Instrument

[1] L.D.Barron, Accts.Chem.Res., 13,90 (1980)
[2] W.Hug,S.Kint,G.Bailey & J.Scherer,J.Amer.Chem.Soc,97,5589 (1975)
[3] W.Hug, Appl. Spectrosc., 35, 115 (1981)
[4] R.Oboodi,M.Davies,U.Gunnia,M.Blackburn & M.Diem, J.Raman Spectrosc.,16, 366 (1985)

RAMAN OPTICAL ACTIVITY IN CARS.

F. W. Schneider and R. Brakel

Institute of Physical Chemistry, University of Würzburg, Marcusstr.9/11
D-87oo Würzburg, FRG

The measurement of Raman optical activity (ROA) by spontaneous Raman scattering (1) has provided valuable stereochemical information. However, the technique may be susceptible to artefacts (f.e. due to incomplete circular polarization) and it is therefore not yet standard. The question arises whether the advantages of CARS (Coherent Anti-Stokes Raman Scattering) spectroscopy, e.g. high signal intensity and collection efficiency, can be utilized for the determination of ROA. The theory of the nonlinear ROA-effect has been formulated recently (2,3), and a detection scheme has been proposed (3). In this work we investigate the limits of detection of the third order susceptibility χ_{2111} arising from magnetic dipole and electric quadrupole interactions and we propose a method how to circumvent the problem of optical rotatory dispersion (ORD).

The wave equation for the CARS amplitude is solved for linearly polarized laser beams. When the planes of polarization of the incident fields (pump and Stokes beams) are parallel, the ROA is predicted to induce a small ellipticity in the CARS beam for strong Raman modes, while ORD rotates the CARS polarization ellipse with respect to the amplitude of the pumplaser. If one of the components of the CARS amplitude which are parallel and perpendicular to the pump amplitude, respectively, is phase shifted by an amount of $\pi/2$, the rotation of the resulting polarization ellipse is only due to ROA. We expect a rotation by an amount of $\mathrm{Im}(\chi_{2111}\chi^*_{1111})/|\chi_{1111}|^2$ where χ_{1111} is the electric dipole susceptibility.

Our experimental investigations are concerned with the strong phenyl breathing-mode of (+)- and (-)-(1)-phenylethylamine (PEA) and toluene. The set-up is similar to that in Ref.(4), except that both laser beams pass through the same polarizer. An additional reference detection scheme is employed in order to reduce the signal fluctuations. The rotation of the CARS polarization ellipse is determined from the interpolated minimum of the CARS intensity transmitted by the analyzing polarizer. A reproducible rotation of the order of $4\cdot 10^{-4}$ rad per mm path length has been measured with opposite sign for (+)- and (-)-PEA, respectively. Since no significant effect ($1\cdot 10^{-4}$) is found for toluene this rotation must be regarded as genuine. The increase of rotation with path length confirms that this effect originates from ORD rather than ROA.

For the ROA measurement a quarter wave plate is placed in front of the analyzer with the fast axis alligned parallel (or perpendicular) to the pump amplitude. No significant rotation of the polarization ellipse has been observed for either substance; i.e. any residual rotation arising from ROA must therefore be smaller than the detection limit of our experimental set-up which is $2\cdot10^{-4}$ including the quarter wave plate. Thus the phenyl breathing mode in PEA shows no appreciable Raman optical activity. This is not surprising, since the asymmetric center is not involved in the Raman mode under investigation. The results indicate that ROA-effects greater than $2\cdot10^{-4}$ should be detectable for strong Raman modes with our present set-up. Unfortunately, Raman bands of molecules which exhibit an appreciable circular intensity differential in spontaneous Raman scattering happen to be weak CARS scatterers and the CARS measurements are affected by the nonresonant background, which prevents complete cancellation of ORD for weak CARS bands. At this writing experiments are in progress to determine the CARS optical activity of (+)- - and (-)- -pinene. They are complicated by the relative weakness of the CARS bands which are interferred with by a strong non-resonant background. Thus the unequivocal measurement of ROA by CARS has not yet been achieved experimentally.

1. L.D. Barron and J. Vrbancich in "Structural Chemistry" ed. by F.L. Boschke, Springer Berlin, 151 (1984).

2. J.O. Bjarnson, H.C. Andersen, and B.S. Hudson, J. Chem. Phys. 72, 4132 (1980).

3. J.L. Oudar, C. Minot, and B.A. Garetz, J. Chem. Phys. 76, 2227 (1982).

4. R. Brakel, V. Mudogo, and V.W. Schneider, J. Chem. Phys. 84, 2451 (1986).

RAMAN OPTICAL ACTIVITY SPECTRA OF (+)-(3R)-METHYLCYCLOHEXANONE OBTAINED WITH A SOLEIL-BABINET COMPENSATOR.

C.G. Zimba[+], T.B. Freedman, K.M. Spencer, X.M. Hu and L.A. Nafie

Department of Chemistry, Syracuse University, Syracuse, NY 13244 USA

Raman optical activity (ROA)[1] spectra of (+)-3-methylcyclohexanone have been obtained by using a Soleil-Babinet compensator to generate left and right circularly polarized laser excitation. The compensator is composed of two stationary crystals, one a plate and the other a wedge, and a moveable crystal wedge, all of crystalline quartz. The optic axes of the wedges are crossed with respect to the plate. Variable retardation is obtained by translating the wedge across the laser beam, thus altering the thickness of the quartz through which the laser beam passes. The compensator is considerably easier to align and control than the device used in previous measurements, an electrooptic modulator. In addition, several measures have been taken to increase the throughput and enhance the S/N ratio including the use of a cylindrical lens immediately in front of the diode array detector to focus more of the dispersed scattered light on the detector.

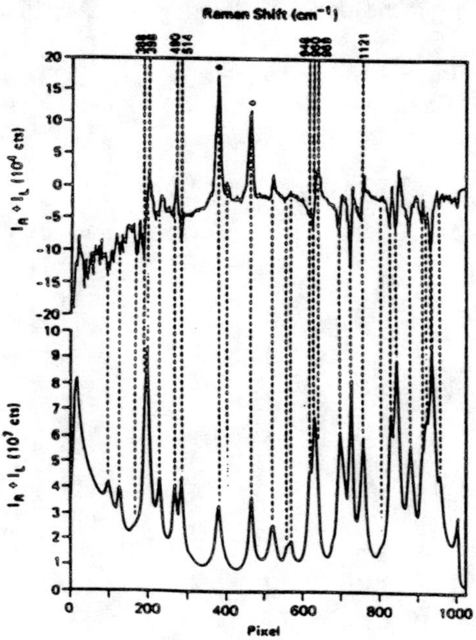

Figure 1. ROA spectrum of (+)-3-meythlcyclohexanone, 150-1400 cm^{-1}. A * identifies polarized bands with contributions from ROA artifacts.

Shown in Fig. 1, the unsmoothed ROA spectrum over the 150-1400 cm^{-1} region was obtained in 2 1/2 hrs. using an 1800 gr/mm holographic grating in an HR-320 spectrograph and a 1024 element intensified diode array and a PAR 1460 optical multichannel analyzer (OMA III). Other experimental conditions were 400 mW excitation at 488 nm, f/2 collection, 45 μm slitwidth (2.1 cm^{-1} bandpass), and 6 cm^{-1} resolution.

Our previous interpretation of ROA couplets in the skeletal region below 650 cm^{-1} in terms of chirally perturbed inherently achiral vibrations[2] can be extended to the bisignate ROA couplet at 960 (-) and 969 (+) cm^{-1}, which has also been observed in dispersive ROA spectra,[1] and the couplet at 1110 (-) and 1120 (+) cm^{-1}, observed for the first time in this study. Each couplet arises from a pair of vibrational modes in the chiral molecule which can be described as in- and out-of-phase combinations of two modes having different symmetries in the point group of a parent achiral molecule. For example, analysis of the nuclear displacements (Fig. 2) demonstrates that the modes at 960 and 969 cm^{-1} in chiral (+)-3-methylcyclohexanone are primarily nonorthogonal methyl rocks which arise from in- and out-of-phase combinations of two orthogonal rocking motions under achiral C_s symmetry.

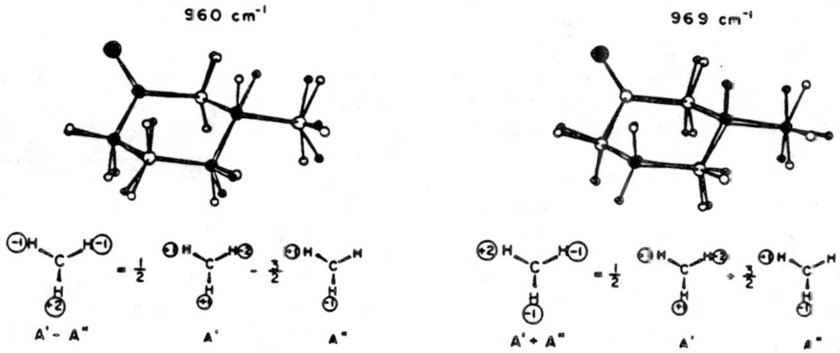

Figure 2. Stereoprojections showing equilibrium (white circles and black bonds) and vibrationally displaced (shaded circles and white bonds) geometries for the methyl rocking modes. The linear combination of achiral motions generating each vibration is shown schematically below each mode.

REFERENCES

1. Barron, L.D. In "Advances in Infrared and Raman Spectroscopy"; Clark, R.J.H., Hester, R.E., Eds.; Heyden:London, 1978; Vol. 4, p. 271.

2. Freedman, T.B., Kallmerten, J., Zimba, C.G.; Zuk, W.M.; Nafie, L.A. J. Am. Chem. Soc. 1984, 106, 1244.

+ Present address: IBM, Almaden Research Center, 650 Harry Road, San Jose, California 95120.

RAMAN OPTICAL ACTIVITY OF SMALL MOLECULES.

Max Diem, Jim Calienni, Mark Davies, Uma Gunnia and Ou Lee

Dept. of Chem., City University of New York, Hunter College, 695 Park Avenue, New York NY 10021 USA

We report the Raman Optical Activity (ROA) spectra of a number of small molecules as neat liquids and in solution, in both depolarized and total scattered components. A comparison between the total and the depolarized components (Fig. 1) reveals that the sign of the ROA signals is maintained for most bands. However, we find a slight bias of the total ROA as compared to the depolarized component. The C-H stretching ROA is generally small, but appears unbiased in the molecules we have investigated. Furthermore, we find no correlation between infrared Vibrational Circular Dichroism and ROA data in molecules were both data are available.

Several small species, which do not contain highly polarizable groups, and exhibit low optical rotations in the visible, have shown exceedingly small or no ROA. In alanine, for example, signals are less than 0.0005 of the Raman intensities, and thus, we have not yet seen ROA data with acceptable S/N ratios, even on our instrument.

Vibrational Optical Activty and detailed vibrational data on some novel, very small chiral species have been obtained and will be reported.

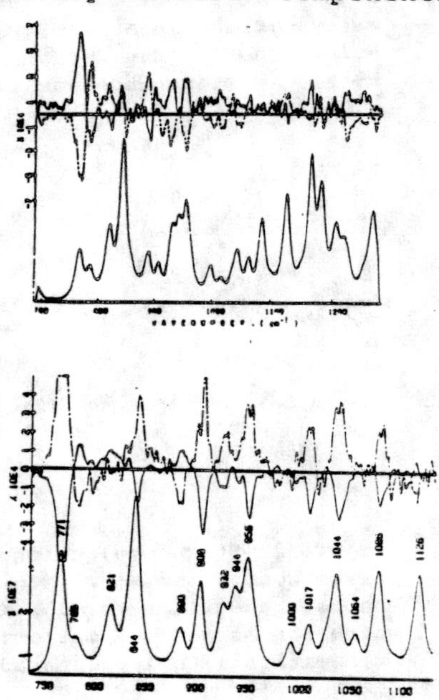

Fig. 1: Depolarized (top) and total (bottom) ROA of α - pinene

INFLUENCE OF ELECTRIC QUADRUPOLE v.s. MAGNETIC DIPOLE IN RAMAN OPTICAL ACTIVITY

Juan R. Escribano* and Laurence D. Barron**

*Chemistry Department, Syracuse University, Syracuse New York 13244,USA.
**Chemistry Department, The University, Glasgow G12 8QQ, UK.

Recent measurements of polarized Raman optical activity (ROA) provide evidence about the relative contributions of the magnetic dipole and the electric quadrupole to this phenomenon [1]. The comparison of natural ROA acquired through an analyzer perpendicular (polarized ROA) and parallel (depolarized ROA) to the scattering plane gives additional information on the relative magnitudes of the optical activity tensors: $G'_{\alpha\beta}$,electric dipole - magnetic dipole and $A_{\alpha\beta\gamma}$, electric dipole - electric quadrupole. For incident light circularly polarized right (R) and left (L) along the z-axis, and scattered along y, the polarized (x) and depolarized (z) circular intensity difference (CID) is [2]

$$I^R_x - I^L_x = K\left(7\alpha_{\alpha\beta}G'_{\alpha\beta} + \alpha_{\alpha\alpha}G'_{\beta\beta} + \frac{\omega}{3}\alpha_{\alpha\beta}\varepsilon_{\alpha\gamma\delta}A_{\gamma\delta\beta}\right), \tag{1}$$

$$I^R_z - I^L_z = 2K\left(3\alpha_{\alpha\beta}G'_{\alpha\beta} - \alpha_{\alpha\alpha}G'_{\beta\beta} - \frac{\omega}{3}\alpha_{\alpha\beta}\varepsilon_{\alpha\gamma\delta}A_{\gamma\delta\beta}\right), \tag{2}$$

If the bond polarizability theory and the two-groups model are combined to obtain phenomenological equations of (1) and (2) in terms of electro-optic parameters, we can obtain very interesting predictions. For chiral molecules with all their bonds axially symmetric,

$$G'_{\alpha\alpha} = 0, \tag{3}$$

$$\alpha_{\alpha\beta}G'_{\alpha\beta} = \frac{\omega}{3}\alpha_{\alpha\beta}\varepsilon_{\alpha\gamma\delta}A_{\gamma\delta\beta}. \tag{4}$$

This leads to

$$(I^R_x - I^L_x)/(I^R_z - I^L_z) = 2. \tag{5}$$

Therefore within the bond polarizability context, for chiral molecules with just axially symmetric bonds, the contribution from the optical activity tensors are predicted to be equal, and polarized spectra will have double the intensity of depolarized ones in all the vibrational bands.

For non-axially symmetric bonds but with internal C_{2v} symmetry (carbonyl or an olefinic group) intrinsic contributions from optical activity tensors must be considered and Eq. (4) no longer holds; for a bond symmetry C_{2v}, Eq. (3) is still valid. It is possible to deduce the relative contribution of $G'_{\alpha\beta}$ and $A_{\alpha\beta\gamma}$ comparing polarized and depolarized ROA spectra [3]. Any deviation from 2 in Eq. (5) provides insight in their relative values.

Polarized and depolarized spectra of (-)-β-pinene are presented. The most interesting feature is the large, equal and opposite, polarized ROA at 716 and 765 cm^{-1} that virtually disappears in the depolarized spectrum. If we assume no contribution from the $A_{\alpha\beta\gamma}$ tensor the ratio Eq. (5) would be 7/6. In the case in which the electric quadrupole contribution increases with respect to the magnetic dipole one, we could expect a growth in the polarized spectrum and a reduction in the depolarized one. Therefore the ratio Eq. (5) could be larger than 2. This is what it is observed in the whole spectra of (-)-β-pinene and it is particularly impressive in the above mentioned bands. Now the larger contribution of $A_{\alpha\beta\gamma}$ over $G'_{\alpha\beta}$ will depend on a particular normal mode, and these conclusions do not have the general character of Eq. (5). We have observed the same results in other chiral molecules with carbonyl and olefinic groups.

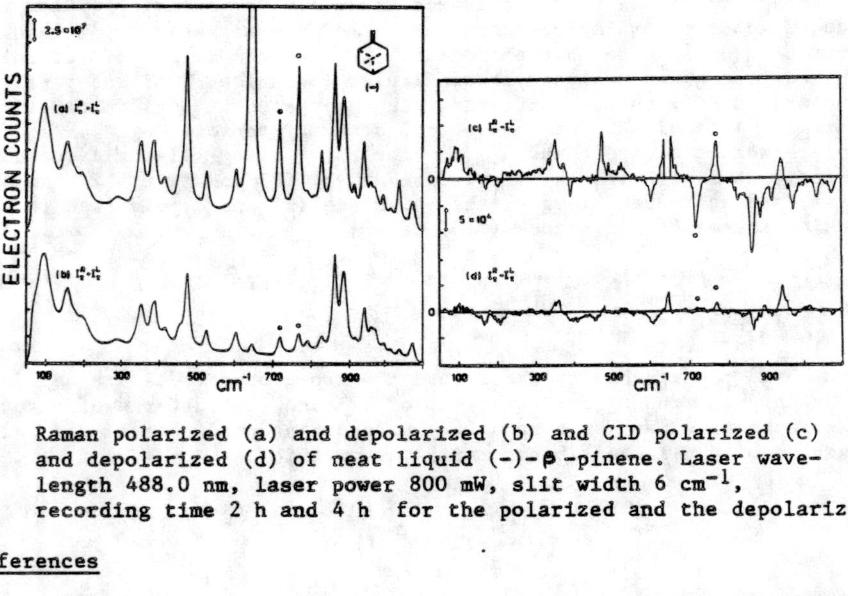

Raman polarized (a) and depolarized (b) and CID polarized (c) and depolarized (d) of neat liquid (-)-β-pinene. Laser wavelength 488.0 nm, laser power 800 mW, slit width 6 cm^{-1}, recording time 2 h and 4 h for the polarized and the depolarized.

References

[1] L.D. Barron, J.R. Escribano and J.F. Torrance, Mol. Phys. 57 (1986) 653.
[2] L.D. Barron, "Molecular light scattering and optical activity", Cambridge Univ. Press, Cambridge, 1982.
[3] L.D. Barron and J.R. Escribano, Chem. Phys. Letters, in press.

TECHNIQUES FOR FAR UV RESONANCE RAMAN SPECTROSCOPY

P. B. Kelly*, A. Ruggiero, S. Li, G. Harhay, G. D. Strahan, and B. Hudson

Department of Chemistry and Chemical Physics Institute, University of Oregon.

Unique information about the symmetry and geometry of excited electronic states maybe obtained by the use of resonance Raman spectroscopy. This is due to the naturally high resolution of Raman spectra and the fact that the vibrational pattern is analyzed with respect to the ground state potential surface where the molecular symmetry and normal modes are well known. This technique is particularly useful when resolved absorption spectra cannot be obtained. Recent advances in laser technology have resulted in the ability to generate ultraviolet radiation at a great variety of wavelengths up to, and beyond, the air transmission cut-off near 195 nm.

The apparatus used for these experiments is based on a Q-switched Nd:YAG laser of unstable resonator design. The fundamental output of this laser is 200 mJ per pulse at 1064 nm with a repetition rate of 30 Hz. Conversion of the IR radiation to higher harmonics in conventional angle tuned KD*P crystals produces radiation at 532 nm with a pulse energy of 100 mJ, at 355 nm with a pulse energy of 40 mJ and at 266 nm with a pulse energy of 20 mJ. The repetition rate of 30 Hz results in an average power at 266 nm of 600 mW. Our initial ultraviolet Raman experiments may be performed with the "fifth" harmonic radiation of 212.8 nm. This is generated by summation of the 1064 nm and 266 nm beams in a KDP (non-deuterated) crystal held at -40° C where it phase matches at roughly 90° to produce a pulse energy at 212.8 nm of about 0.5 mJ (15 mW average power at 30 Hz).

Experiments have been performed using stimulated Raman shifting in hydrogen gas. For Raman shifting in diatomic hydrogen with a vibrational frequency of 4155 cm^{-1} the following wavelengths (in nm) are generated from the initial radiation indicated at the left:

532:	320	282	253	229		209	192	178	166			
355:		309	273	246	223		204	188	175	163		
266:			299	266	240	218		200	184	171	160	
213:						213						

The use of diatomic deuterium with a vibrational frequency of 2991 cm^{-1} results in many more lines with output energy centered near the pump wavelengths. HD with a frequency of 3630 cm^{-1} can also be used. Both D_2 and HD appear to have higher thresholds for Raman shifting than H_2. In some cases, stimulated rotational Raman scattering is observed in the ouput of the shifting cell. The origin of this unwanted radiation will be discussed.

The resulting radiation is then directed to the sample. Either backscattering or 90° geometry may be used. Backscattering has the distinct advantage that the Raman scattered intensity does not critically depend on the optical density of the sample. The samples that we have used so far in our studies have been gases or dilute solutions. Several versions of liquid and vapor sampling systems have been devised and will be described. These include the use of a guided flow system providing a flat liquid film, and shielded gas flow devices for studies of low vapor pressure samples.

The Raman scattered radiation is collected by large S1-UV quartz lenses and directed into a monochromator. Both a half-meter double monochromator and a 1-meter single monochromator have been used. Ultraviolet Raman studies require higher wavelength dispersion monochromators than are commonly used for visible Raman studies because of the reciprocal nature of the relationship between wavelength and wavenumber.

The dispersed radiation at the monochromator exit is detected with an S-5 solar blind photomultiplier. A boxcar amplifier then processes the resulting signal. Absolute wavelength calibration is achieved by simultaneous recording of Raman spectra and mercury spectra. The output of the boxcar is sent to the ADC of a minicomputer for collection. The minicomputer enables multiscan averaging of several such experiments. Also, computer scaling and plotting greatly simplifies data analysis, as this is often performed by comparison of spectral intensities of spectra obtained at different excitation wavelengths.

In a new system which we are developing, the S1-UV quartz optics will be replaced by CaF_2 and MgF_2 optics which have lower wavelength cutoffs (S1-UV is transparent above 180 nm, whereas CaF_2 and MgF_2 are transparent above 130 nm). Purging the surrounding atmosphere with N_2 or Ar will enable the observation of transitions which occur in the VUV region. Furthermore, this system will utilize reflective rather than refractive collection optics, this has the advantage of increasing collection efficiency.

The comparative performance of two commercially available Q-switched Nd:YAG lasers will be presented.

This technique of far UV resonance Raman spectroscopy is described in Spectroscopy 1, 22 (1986) and references therin.

*Address after September, 1986: Department of Chemistry, University of California, Davis.

A NEW METHOD OF STUDYING ENZYME-SUBSTRATE INTERMEDIATES USING ENZYME MICROCRYSTALS AND AN INTRACAVITY LASER RAMAN MICROSCOPE TECHNIQUE.

T.W. Patapoff, G.A. Thomas, and W.L. Peticolas

Department of Chemistry University of Oregon Eugene, Oregon 97403 USA

The study of enzyme-substrate intermediates by Raman spectroscopy is an area of increasing scientific importance. Such intermediates cannot usually be studied by means of X-ray diffraction techniques because of their short life-time. Using time-dependent resonance enhanced Raman techniques it is sometimes possible to obtain a Raman spectrum of such an intermediate. From a comparison of the resonance Raman spectrum of the intermediate with that of model compounds in solutions or in crystals it is often possible to draw important conclusions concerning the structure of the intermediate and the nature of the enzyne induced catalysis (1-3). However, in all of the investigations so far the technique used is that pioneered by Carey and coworkers (1). In this technique the components (enzyme substrate) are dissolved in separate solutions which are mixed in a high-speed mixing device. The stream of liquid from the mixing device is then made to flow through the laser beam. The Raman light scattered from the stream is collected and analyzed. This technique has the disadvantage that it requires the use of large amounts of enzyme which may not be available. To overcome this difficulty we have devised a laser Raman microscope apparatus in which a microcrystal of the enzyme is mounted in a special capillary cell under the microscope and a solution of the chromophoric substrate is made to flow over the crystal resulting in the build-up of an intermediate whose Raman spectrum can be taken. Since only one crystal of the enzyme which is typically about 50 microns on the edge is needed Raman spectra of a transient enzyme-substrate complex can be obtained on less than a microgram of enzyme. The laser beam itself is focused to a spot of 5 cubic microns or less.

The design of the laser Raman microscope follows closely that of Delhaye, Bridoux, and their coworkers (4) but it differs in two important respects. We use an EMI image intensifier which has four amplification stages and is magnetically focused. This is followed by a Tracor Northern reticon diode array which records the spectrum in a multichannel analyzer. This combination gives a factor of 10-100 more signal than the use of the Tracor-Northern reticon system alone. Another feature of our apparatus is that the laser cavity is extended to the reflection from glass capillary cell so that the enzyme crystal is actually in the cavity of the extended Argon laser system.

This technique has been used to study the interactions of crystals of chymotrypsin and a variety of substrates. In order to keep the crystal from dissolving it must be surrounded by a suitable mother liquor. It is found that although the mother liquor may have a high (4M) concentration of sulfate the bands from the sulfate are weak in the Raman spectrum because the microscope is focused into the interior of the crystal where the sulfate concentration is low. Only Raman scattering from the interior of the crystal is obtained. Good Raman spectra of both the crystal of the pure enzyme and that of the enzyme-intermediate are obtained and the intermediate spectrum is obtained by difference.

The bottom spectrum of Figure 1 shows the Raman spectrum taken of a 100 micron crystal of chymotrypsin that was obtained with the intracavity Raman microscope. Upon addition of a chromophoric substate, indoleacryloyl imidazole (IAI), a new band at 1618 cm^{-1} appears as shown in the middle spectrum. Subtraction of the protein spectrum from the enzyme-substrate complex spectrum makes evident many new bands. These bands are believed to be due to the formation and stabilision at high pH of the acyl enzyme. There are many advantages to single crystal Raman studies. In the case of IAI, both the substrate and the end product are highly fluorescent whereas the acyl enzyme fluoresces only slightly. This new technique allows the study of the intermediate within the crystal even while the crystal is being bathed in a solution of fluorescent substrate. In cases where the intermediate has a shorter lifetime, flowing the substrate over the crystal rapidly allows the substrate to diffuse into the crystal (which is 40% water) and the buildup of the intermediate can be followed as a function of the flow rate.

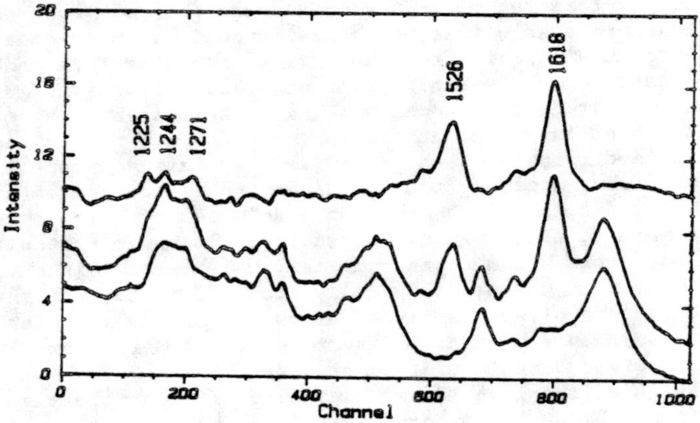

1. P.R. Carey and A.C. Storer, Ann. Rev. Biophys. Bioeng., 13, 25 (1984)
2. P.W. Jagodzinski, G.F. Funk, & W.L. Peticolas, Biochemistry 21, 2193 (1982).
3. J.P. Weber, P.Y.Turpin, S.A.Bernhard,& W.P.Peticolas, Biochemistry, (1986) In Press.
4. M.Delhaye,in Spectroscopy of Biological Molecules, NATOASI series Sandorfy and Theophanides,ed. D. Reidel Publ.Co. Boston (1984),587-607.

MIXING TECHNIQUES FOR TIME-RESOLVED RAMAN SPECTROSCOPY.

James R. Kincaid and Ki-Jung Paeng

Chemistry Department, Marquette University, Milwaukee, WI 53233

In this presentation we will discuss our recent work on the development and applications of rapid mixing devices for use with Raman spectroscopy. These devices can be conveniently coupled with Raman spectrometric detection to provide a capability for acquisition of detailed structural information from vibrational spectra of short lived intermediates.

In the case of one of these devices, schematically diagrammed in Figure 1, two high velocity streams of regularly spaced, uniformly sized droplets (~100 μm diameter) are oriented such that individual droplets from one stream (reactant A) collide and coalesce with droplets from the second stream (reactant B) to produce a product stream of uniformly sized droplets. The product stream is directed through a focussed laser beam and the scattered light is collected and analyzed with a conventional Raman spectrometer.

In our apparatus, the liquid jet is formed by forcing the reagent solution through a specially fabricated glass capillary with compressed gas. The barrel of the capillary is cemented (Duco, Dupont) through a hold drilled in a rectangular piezoelectric strip (Vernitron, Bedord, OH) as shown in Figure 2. The strip is clamped in a cantilever position to a small delrin mount where connections are made to a variable frequency oscillator. The oscillator is of in house design and based upon the circuit of Seymour and Boss. The

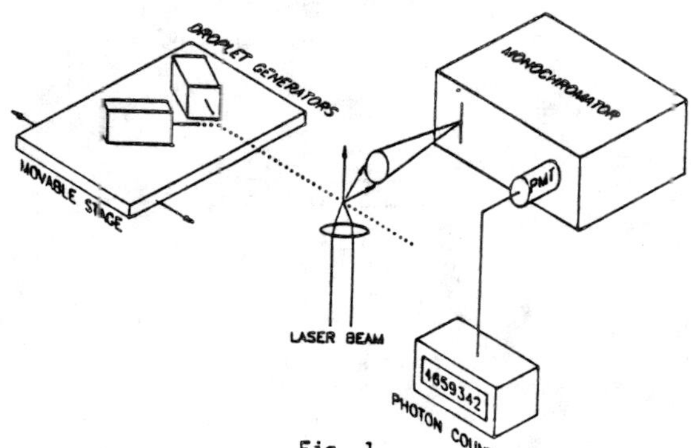

Fig. 1

application of a square wave of sufficient amplitude and frequency to the bimorph results in a regular axial perturbation along the jet stream, thus establishing the necessary conditions for uniform droplet formation. The capillary is connected to a 22 gauge stainless steel syringe needle (Glenco, Houston, TX) using Teflon heat-shrinkable tubing. The needle is housed in a retangular delrin block and provides the interface between the flow system and the capillary.

In order to successfully collide droplets, three conditions must be satisfied: (1) The streams must be made to intersect, (2) the droplets of each stream must be equally spaced, and (3) they must be in phase. The first and third requirements are satisfied by mounting each generator assembly on a precision linear translator (1 μm resolution) as shown in Figure 2. One translator-generator assembly provides vertical movement and is used to cause the two streams to intersect. The second assembly provides axial movement and allows relative phase adjustment of the two streams. The piezoelectric strips are driven by the same osciallator signal to ensure equal droplet spacing in each stream.

Both the generator-translator and generator-tripod mount assemblies are aligned at the desired collision angle and fastened to an aluminum platform. The platform can be quickly mounted on a three dimentional stage (Ardel Kinematic models T-50 translators and PA-32 angle plate) which rests on the optical rail of the Raman spectrometer. In this way, the droplet generators and mounts can be prepositioned and the entire apparatus moved without disturbing the stream alignment.

Applications of this and anther mixing device to the study of the reactions of heme proteins and model compounds will be discussed.

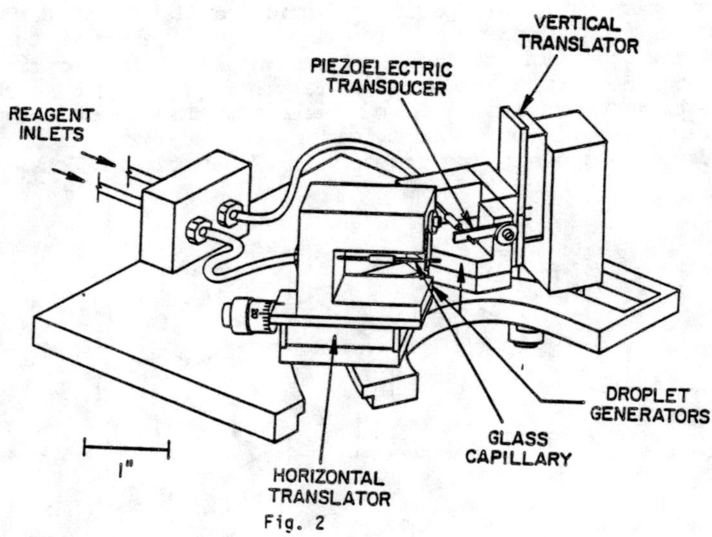

Fig. 2

STUDY OF MIXING PROCESSES IN LIQUIDS BY TIME RESOLVED MULTICHANNEL RAMAN SPECTROSCOPY

A. Lorriaux-Rubbens, N. Castaignos, M. Bridoux, M. Delhaye and F. Wallart

Laboratoire de Spectrochimie Infrarouge et Raman, CNRS (LP. 2641)
Université des Sciences et Techniques de Lille Flandres Artois, Bât. C.5
59655 Villeneuve d'Ascq Cedex, France.

Time Resolved Raman Spectroscopy (TRRS) is presented as a non invasive technique for measuring mixing/concentration profiles in a fully baffled disc trubine agitated tank. In the present experiments, the non reactive liquids chosen were aqueous solutions of sodium nitrate and sulfate whose concentrations could be easily determined by means of theirs νSO and νNO frequencies located at 980 and 1050 cm^{-1} respectively. The evolution of the mixing composition as a function of time during an injection of one of the solutions into the tank was followed for diverse points of the mixing-reactor. The mixing studies were performed in a Rushton type reacting batch (designed at the I.C.I. Society of Runcorn (U.K)) mounted on a translation stage allowing Z. Y and radial movements relative to the entrance slit of the spectrometer. The single stage Raman spectrograph used, was equipped with a concave holographic grating of 2000 lines/mm, a focal length of 1 metre and an f/10 aperture. The multichannel detector was a three stage image intensifier optically coupled by two large aperture objectives to a 512 photodiodes linear array (Reticon serie S). The photodiode linear array was cooled to -20°C in order to reduce the dark noise. The detector was interfaced to an Apple II E microcomputer via a buffer memory which allows the storage of a batch of consecutive acquisitions of 15 Raman spectra. Each spectrum readout of the photodiode array could be the result of a minimum integration time of 20 ms. The spectrum batches were transfered via an Apple II E disc to the mini computer (MINI 6 Bull) for treatments such as band decomposition and measurements of band area. The ensemble of instrumental conditions assures that the element of sample volume analysed by the spectrometer is always approximatively a parallelepiped of 0,09 x 0,07 x 6 mm.

Different points of analysis located in the mixing tank were studied and in the next figure results obtained for the element of volume analysed at 1 cm from the solution surface are presented. The resulting curves show quite cleraly the evolution as a function of time of the mixing composition of elements of analysed volumes and prove that Raman scattering spectroscopy constitutes an interesting probe for the study of mixing dynamics. The results indicate the valuable contribution that TRRS studies can make to the understanding of the mixing processes in an agitated tank, and therefore modeling techniques can be used to solve the practical problem of mixer and reactor scale up.

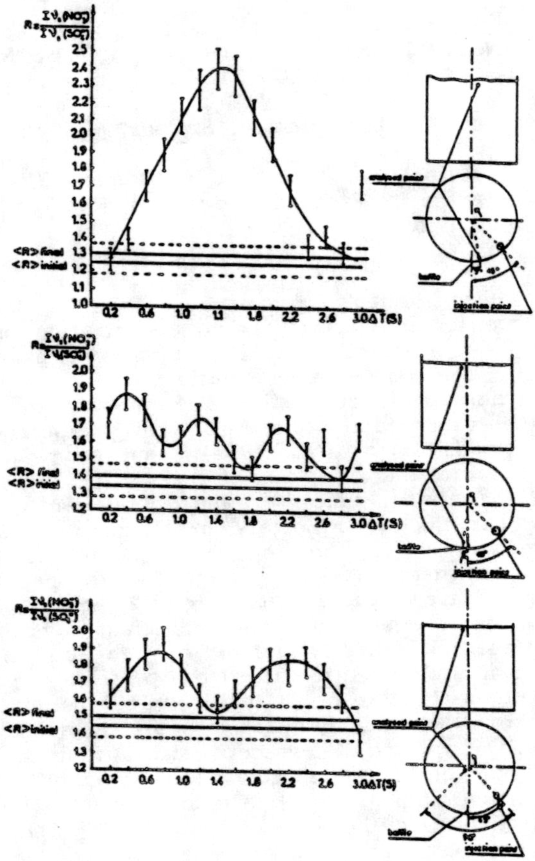

References :

- M. Delhaye, M. Bridoux, F. Wallart, J. Mol. Struct. 72, 51-56 (1982).
- M. Bridoux, B. Cook, M. Delhaye, A. Lorriaux-Rubbens and F. Wallart private communications (1979-80-81).
- A. Deffontaine, M. Bridoux, M. Delhaye, E. Da Silva, W. Hug, Rev. Phys. Appl. 19, 415-421 (1984).
- B. Cook, J.D. Louden, A. Lorriaux-Rubbens, L. Nouël de Kérangué, M. Bridoux and F. Wallart, submitted to Journal of Raman Spectroscopy.

Fluorescence Free Raman Spectroscopy

Bruce Chase
Central Research Department
Du Pont Experimental Station
Wilmington, DE 19898

INTRODUCTION

The general lack of acceptance of Raman spectroscopy in an industrial environment is due to the high degree of experimental failure arising from 1) thermal decomposition and 2) high fluorescence background. Various approaches have been developed to minimize these problems, but none exhibit a truly universal solution to these two phenomenon. Clearly, the most reasonable route to follow is to irradiate in a spectral region where sample (and impurity) absorption is negligible. In the near infrared (>9000 Å) the sole absorption mechanism for most systems of interest is excitations of overtones and combinations of fundamental vibrations. These absorptions are weak and energy redistribution is rapid. In addition the energy deposited per photon in the molecule is half that found with excitation at 5145 Å.

Recently, the feasibility of interferometric detection of Raman scattering has been demonstrated (1,2). The loss in sensitivity due to decreased cross-section and degraded detector performance can be compensated for with a Michelson interferometer and Raman spectra comparable to those obtained in the visible can be obtained.

EXPERIMENTAL

The source for the near infrared Raman experiment is an SL50 CW Nd/YAG laser manufactured by Spectron, operated in a TEM_{00} mode with a power output of 6.0 watts (1.0 W at the sample). Laser line filtering is done using an Anaspec C1000 filter monochrometer, with a transmittance of 25% and a half band width of 150 cm^{-1}. A Nicolet 7199 interferometer equipped with a quartz beamsplitter and a Judson J-16 1 mm Germanium detector was used to analyze the scattered radiation. Collection was done in a 180 degree back-scattering configuration using a 75 mm diameter f/1.5 paraboloid.

The most critical part of an FT Raman experiment involves the filtering of the Rayleigh line. An excitation of 10^8 is required while retaining high transmission to longer wavelength. Various filters were employed, including long pass dielectrics, long wavelength selective absorbers and a Raman notch filter.

RESULTS

Figure 1 shows a Raman spectrum of anthracene taken with 1.064 micron incident laser light. No sample degradation or sublimation was observed in spite of the high power density employed. There is no observable background for this system which exhibits moderate fluorescence in the visible. A more severe test of the fluorescence free nature of the near IR Raman experiment is shown in the next two figures. Rhodamine B gave the spectrum shown in figure two and Fluorescein is shown in figure three. These two materials exhibit immense backgrounds and thermally decompose quickly when excited by visible radiation. The total measurement time for each spectrum was twenty minutes.

CONCLUSIONS

The use of near infrared lasers coupled with interferometric detection can provide Raman data comparable to that obtained with conventional techniques, but avoids the excitation of fluorescence and thermal decomposition. These results represent a feasibility study and further improvements to the sensitivity can be expected upon optimization of collection optics, detectors and filtering schemes.

1. T. Hirschfeld and B. Chase, Applied Spectroscopy, 40(2), 133 (1986).

2. D. E. Jennings, A. Weber and J. W. Brault, Applied Optics 25(2), 284 (1986).

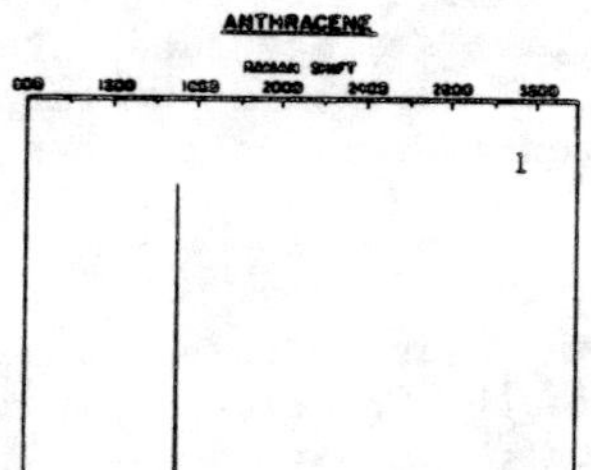

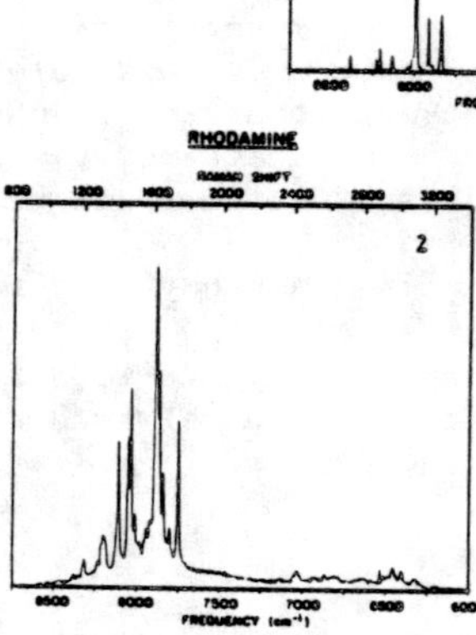

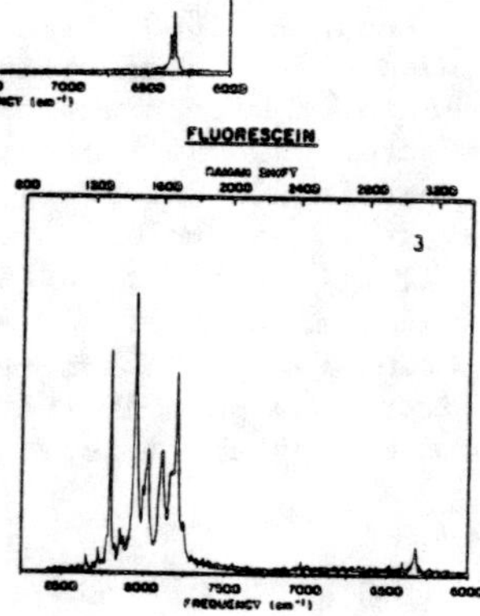

FLUORESCENCE FREE FT-RAMAN SPECTROSCOPY IN THE NEAR INFRARED.

Dr. Henry Buijs

BOMEM INC.

Fourier Transform Raman Spectroscopy has provided access to excitation in the near infrared which has been shown to eliminate fluorescence.

The multiplex and throughput advantages of the Fourier Transform Spectrometer insure a large detected signal which in turn allows the use of relatively noisy Silicon or Germanium photodiodes without compromising signal to noise ratio. These photodiodes have high detection efficiency in the spectral regions from 400 to 1100 mm and 800 to 1700 mm respectively.

Aspects of implementation of Raman Spectroscopy using a BOMEM DA3 FTIR and a Nd:Yag laser for excitation will be presented. The features, advantages and limitations of this Raman/IR spectrophotometer system will be presented along with examples.

DEVELOPMENT OF A NEW RAYLEIGH LINE REJECTION FILTER: THE BASIS FOR A NEW GENERATION OF RAMAN INSTRUMENTATION.

Sanford A. Asher, Perry Flaugh, Sair Hagopian, Paul Rundquist and Guy R. Washinger[1]

Department of Chemistry, University of Pittsburgh, Pittsburgh, Pa. 15260, [1]EG&G Princeton Applied Research, Princeton N.J. 08543

We have developed a new optical rejection filter which rejects a narrow wavelength interval while passing with little attenuation light at wavelengths above and below the rejection wavelength. The rejection efficiency is greater than 10^6 and the bandwidth is less than 50 A.

The filter is based on a crystalline colloidal array of polystyrene spheres which Bragg diffracts the elastically scattered Rayleigh light.

We utilize this new optical filter to construct new high throughput Raman spectrometers which utilize only single gratings. These spectrometers have stray light levels less than conventional double spectrometers and have stray light levels better than the triple spectrometers typically used with optical multichannel detectors. The small number of optical elements result in a very high efficiency, relatively inexpensive instrument optmized for OMA detection. Detailed comparisons of performance will be made between this new spectrometer and the conventional Raman instrumentation.

An example of the Rayleigh line rejection efficiencies is shown in the figure below which shows the Stokes and anti-Stokes region of the Raman spectra of CH_3Cl. This single spectrum was accumulated on the face of an intensified diode array and illustrates attenuation of the "Rayleigh", elastically scattered light.

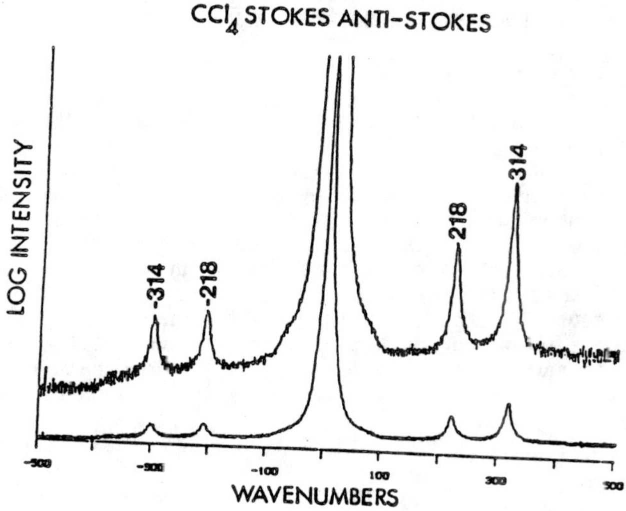
CCl4 STOKES ANTI-STOKES
LOG INTENSITY
-314
-218
218
314
-500
-300
-100
100
300
500
WAVENUMBERS

MEASUREMENTS OF VERY SMALL FREQUENCY SHIFTS IN RAMAN SPECTROSCOPY: DIFFERENCE AND FOURIER TRANSFORM METHODS.

R. Savoie, D. Lévesque, and M. Langlais

Département de chimie, Université Laval, Québec, QC, Canada G1K 7P4.

Minute vibrational frequency shifts, occuring as a result of molecular interactions or changes in conformation, can provide useful clues on the structure and behavior of physical and biological systems. Such shifts can be measured by Raman difference spectroscopy, with an accuracy of the order of $\pm 0.03\ cm^{-1}$ (1-3).

A unique illuminator built in our laboratory permits the simultaneous recording of Raman spectra of two different samples placed in separate samples holders (Fig. 1). The temperature of each sample support is thermoelectrically set by computer control (4).

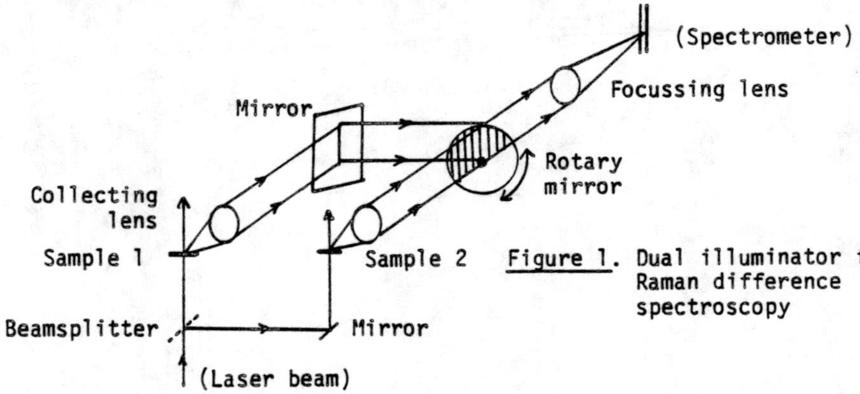

Figure 1. Dual illuminator for Raman difference spectroscopy

An example of application of this system to the study of dipalmitoylphosphatidylglycerol (DPPG) is shown below. In Fig. 2, the spectral changes induced by the melting of this phospholipid are clearly seen in the difference spectrum, even though the difference in the temperature of the two samples was less that 1°C. Fig. 3 shows the frequency shift of the 2845 cm^{-1} band as a function of temperature. The degree of accuracy of these measurements is comparable to that obtained in Fourier transform infrared experiments (5).

Another technique which could permit the measurement of very small frequency shifts in Raman spectroscopy is the use of Michelson interferometers to record the data. This method has recently been shown to provide data of comparable quality to those obtained using conventional dispersive instruments (6). The interferometric method should provide

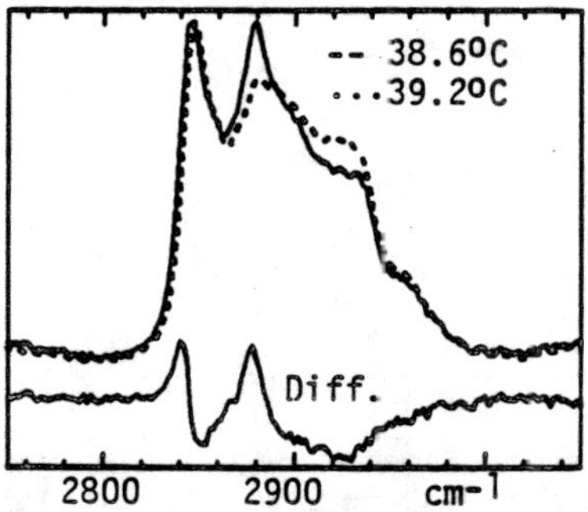

Figure 2. Raman difference spectrum of DPPG.

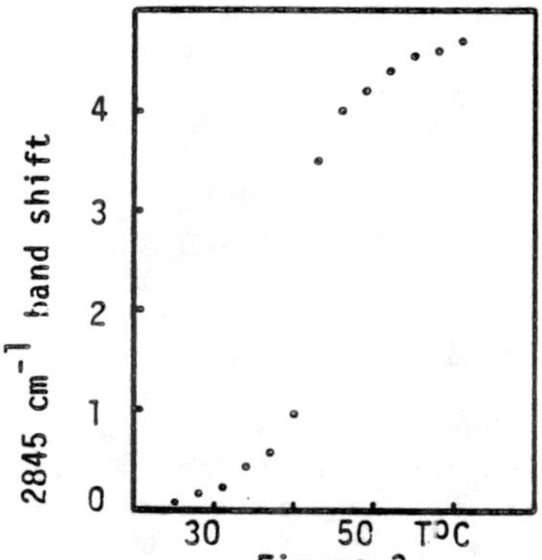

Figure 3

an absolute frequency accuracy which is far better than that of dispersive instruments. An example of Raman spectrum obtained with a commercial interferometer is shown in Fig. 4. This spectrum of cyclohexane at 4 cm^{-1} resolution was obtained on a Bomem-DA3 interferometer with quartz beamsplitter, using a cooled photomultiplier for detection. The spectrum, excited by the 488 nm line of an argon laser (200 mW at the sample), was obtained in approximately 10 minutes.

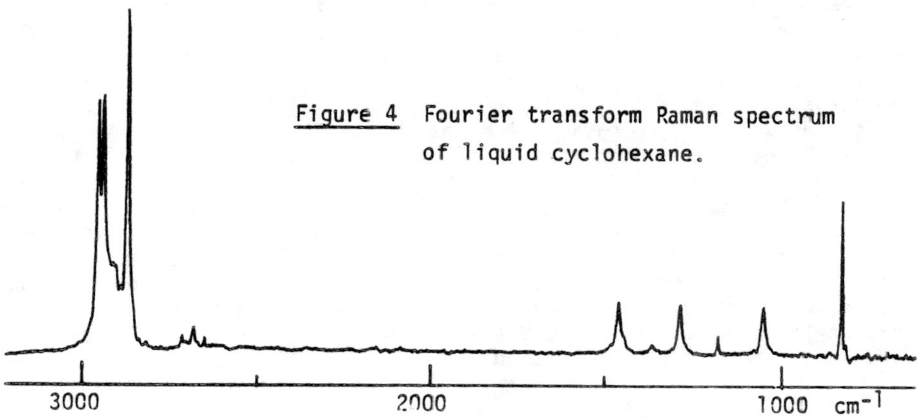

Figure 4 Fourier transform Raman spectrum of liquid cyclohexane.

References:

1. J. Laane, in Vibrational Spectra and Structure, Vol. 12, J.R. Durig, Ed., (North Holland, Amsterdam, 1983), Chap. 6, p. 405 - 467.
2. W. Kiefer, in Advances in Infrared and Raman Spectroscopy, Vol 3, R.J.H. Clark and R.E. Hester, Eds., (Heyden, London, 1979), Chap. 1, p. 1 - 42.
3. J.S. Bodenheimer, B.J. Berenblut, and G.R. Wilkinson. Chem. Phys. Lett. 14, 533 (1972).
4. M. Pézolet, B. Boulé, and D. Bourque. Rev. Sci. Instrum. 54, 1364 (1983).
5. H.L. Casal and H.H. Mantsch. Biochim. Biophys. Acta 779, 381 (1984).
6. T. Hirschfeld and B. Chase. Appl. Spectrosc. 40, 133 (1986).

HIGH-PRESSURE MICRORAMAN SPECTROSCOPY.

Victor Benham, Ian S. Butler, Gabriel Lord, and Denis F.R. Gilson,
Department of Chemistry, McGill University, Montreal, Que. Canada H3A 2K6

One of the difficulties in high-pressure Raman work with a diamond-anvil cell (DAC) is the alignment of the cell in the laser beam. In an effort to overcome this problem, we have now mounted a DAC on the XY-translation stage of the microscope of a microRaman spectrometer. Using a 4X microscope objective, it is possible to focus the laser beam precisely into the 0.3mm hole of the stainless steel gasket that contains the sample and the 10-30μm ruby chip used for pressure calibration. This gasket is located between the parallel faces of the two diamonds in the DAC. The laser beam (slightly defocussed) is then directed simultaneously onto both the sample and the edge of the ruby chip by simply adjusting the position of the DAC with the aid of the XY-translation state and the observation screen of the microscope attachment. Using the new sample arrangement, we have investigated the spectra of mercuric cyanide [$Hg(CN)_2$] and anhydrous calcium nitrate [$Ca(NO_3)_2$] up to pressures approaching 75 Kbar using argon-ion laser excitation. The changes in the symmetric ν(CN) mode of $Hg(CN)_2$ with increasing pressure (using n-octane as the hydrostatic pressure fluid) show the expected phase transition. In the case of $Ca(NO_3)_2$, no pressure transmitting fluid was used because its ν(C-C) modes would interfere with the symmetric ν(NO) mode. This mode does not split under high pressure but gradually shifts to higher energy and broadens appreciably. Plots of the peak position vs. pressure reveal a first-order phase transition at ∿22 kbar. Coupling a DAC directly to a microRaman spectrometer does greatly facilitate high-pressure Raman measurements especially since we have found that realignment of the cell is necessary during the experiments in order to maintain the signal intensity because the point of focus shifts slightly from its original position due to a pressure gradient across the cell.

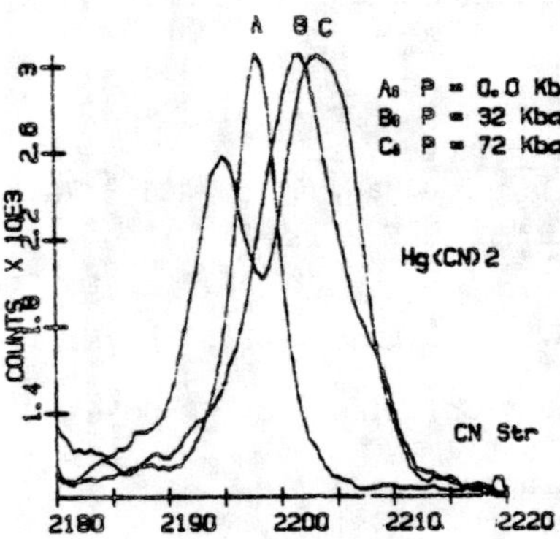

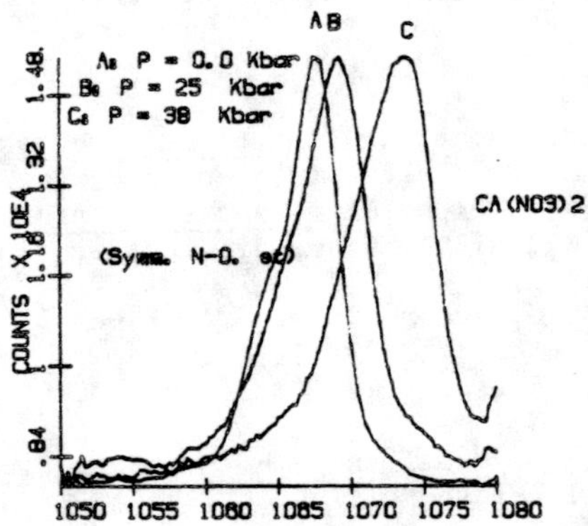

ISOTROPIC AND ANISOTROPIC SAMPLES IN POLARIZED RAMAN MICRO-SPECTROSCOPY

C. Brémard, J. Laureyns and G. Turrell

Laboratoire de Spectrochimie Infrarouge et Raman, C.N.R.S. (LP. 2641) Université des Sciences et Techniques de Lille Flandres Artois, Bât. C.5 59655 Villeneuve d'Ascq Cedex, France.

The possibility of quantitative polarization measurements with Raman microprobe instruments opens up a new area for the investigation of samples of a few micrometers in size, e.g., living cells. However, Raman microspectrometers employ two optical components which distinguish them from the more conventional instruments, viz., a semitransparent mirror ("beamsplitter") and a microscope objective. As these two elements perturb polarization measurements, their effects must be evaluated if quantitative results are to be obtained.

An additional problem arises from the high light intensity produced in the focal region, which often results in destruction of the sample. This problem has recently been at least partially alleviated with the development of multichannel detection systems.[1]

The effects of the beamsplitter can be corrected using the method reported previously,[2] while the depolarization introduced by the microscope objective has been calculated in preliminary publications.[2,3] However, the influence of the index of refraction of the sample has not, as yet, been considered. This aspect of the problem for the case of isotropic samples will now be investigated quantitatively following the earlier analysis.[3] In the present work the interpretation of measurements made on anisotropic samples is only qualitative.

The scattering intensity has been evaluated for the case of a linearly polarized source focused by a large numerical aperture objective and collected by the same objective. The result can be written in the form

$$I = (\alpha^2_{XX}A + \alpha^2_{XY}A + \alpha^2_{XZ}B)(2C_O + C_2) + (\alpha^2_{YX}A + \alpha^2_{YY}A + \alpha^2_{YZ}B)C_2 + (\alpha^2_{ZX}A + \alpha^2_{ZY}A + \alpha^2_{ZZ}B)4C_1, \quad (1)$$

where $C_j = 2\pi \int_0^\infty\int_0^\infty [I_j(u,v,n)]^2 \, v\,dv\,du$, for j = O, 1 and 2. (2)

The integrals $I_j(u,v,n)$ are functions of the optical coordinates $u = kZ\sin^2\theta_m$ and $v = k\sqrt{X^2+Y^2}\sin\theta_m$, where $k = 2\pi/\lambda$, and θ_m is the angular semi-aperture of the objective. They have been evaluated over a range of values of the refractive index, , and a number of different microscope objectives. The parameters related to the scattering cone are defined by

$$A = \pi^2(4/3 - \cos\theta'_m - 1/3\cos^3\theta'_m) \quad (3)$$

and

$$B = \pi(2/3 - \cos\theta'_m + 1/3\cos^3\theta'_m), \quad (4)$$

where $\theta'_m = \sin^{-1}(1/n \sin\theta_m)$ is the effective angular semi-aperture within the sample.

For randomly oriented molecules the scattering intensities $I_\perp$ and $I_\parallel$ have been evaluated from Eq.(1) as functions of the invariants Σ^0, Σ^1 and Σ^2 of the polarizability tensor.[4] The depolarization ratio then follows from $\rho_{cal} = I_\perp/I_\parallel$.

Polarized Raman and resonance-Raman spectra of a variety of isotropic samples, including carbon tetrachloride, β-carotene and ferrocytochrome c, have been recorded with different numerical-aperture objectives. The experimental depolarization ratios, ρ_{exp}, of some Raman bands are compared with the calculated values in the following table. The calculated values are,

	CCl_4 $\nu_4(F_2)$, 314 cm^{-1}		β-carotene $\nu_1(A_g)$, 1525 cm^{-1}		ferrocytochrome c (A_{2g}), 1585 cm^{-1}	
θ_m	ρ_{cal}	ρ_{exp}	ρ_{cal}	ρ_{exp}	ρ_{cal}	ρ_{exp}
71.8°	0.770	0.77	0.35	0.36		
64.8°	0.767	0.77	0.34	0.34		
58.2°	0.765	0.76	0.33	0.32		
39.1°	0.758	0.76	0.33	0.32	22	25
20.5°	0.753	0.75	0.33	0.34	58	> 30
10.4°	0.750	0.75	0.33	0.32	227	> 40
5.2°	0.750	0.75	0.33	0.33	848	> 50
~0°	0.750	0.75	0.33	0.33	∞	> 100

within the experimental error range, in good agreement with the experimental values.

For anisotropic samples such as crystals of the cubic system, e.g., diamond, the calculated values obtained from the theoretical expressions based on Eq.(1) are in accordance with the integrated intensities obtained from the polarized Raman spectra. However, for crystals of the tetragonal system, such as α-quartz, the most important depolarization is attributed to the effect of birefringence. This depolarization is especially important when the incident and scattered light propagate in a direction close to the optical axis of the crystal, but can be minimized by a reduction of the optical path within the sample.

References

1. J. Barbillat and M. Delhaye, in Microbeam Analysis, edited by R. Cooley (San Francisco Press, 1983), p. 280.
2. C. Brémard, P. Dhamelincourt, J. Laureyns and G. Turrell, Appl. Spectrosc. 39, 1036 (1985).
3. G. Turrell, J. Raman Spectrosc. 15, 103 (1985).
4. O. Sonnich Mortensen and S. Hassing, in Advances in Infrared and Raman Spectroscopy, vol. 6, edited by R.J.H. Clark and R.E. Hester (Heyden, London 1980), p.1.

MICROPROBES WITH MULTICHANNEL DETECTIONAND LASER EXCITATION FOR RAMAN AND FLUORESCENCE MICROSPECTROMETRY.

P. Dhamelincourt, M. Delhaye, E. Da Silva, J.P. Cornard, B. Lenain.

G.I.P. "Instrumentation et Spectrométrie"
- *Cité Scientifique, Bât. C.5. 59655 Villeneuve d'Ascq, France.*
- *DILOR, 244 ter rue des Bois Blancs, 59000 Lille, France.*
- *LASIR, CNRS, USTL Flandres Artois, Bât. C.5, 59655 Villeneuve d'Ascq Cedex, France.*

With Lasers as sources of excitation of Raman scattering and more recently fluorescence emission and the ongoing developments in instrumentation for optical spectroscopy, microprobing techniques have matured to the point at which non destructive microanalysis has now become routinely practicable (1,2). The Laser sources present unquestionable advantages in terms of excitation efficiency and spatial resolution. Indeed Laser radiations are intense perfectly monochromatic and can be focused onto a spot whose size is only limited by diffraction. This size can be as small as 0.5 micrometer when high numerical aperture objectives are used.

Moreover precise fluorescence and Raman polarization measurements are possible at this size level because Laser emissions are perfectly polarized (3).

However, these intense sources must be used very carefully when analyzing fragile materials because high irradiances induce generally rapid modification of the probed area of the sample. So, most of the time, studies of fragile samples require recording times as short as possible with low Laser power at sample.

These requirements preclude the use of scanning spectrometers with monochannel detector.

That is why microprobes which employs spectrographic dispersion with a very sensitive multichannel detector are now developping and are especially designed to allow both Raman and fluorescence experiments on the same sample (4).

These instruments will be described and their analytical potential will be illustrated by examples of analysis performed on various materials.

References :

1. P. Dhamelincourt, J. Barbillat, and M. Delhaye
   Laser Raman Microprobing Techniques, Journal de Physique, Colloque C.2, t. 45, 1984.
2. E.S. Etz and J.J. Blaha
   Scope and limitations of single particle analysis by Raman Microprobe analysis, NBS Special publication 533, 1982.
3. C. Brémard, P. Dhamelincourt, J. Laureyns, B. Lenain
   Polarization measurements in microRaman and microfluorescence spectrométries. Applications to biological systems in Spectroscopy of Biological molecules, Alix, Bernard and Manfait eds, J. Wiley Press, 1985.
4. M. Manfait and P. Jeanneson
   In vivo spectroscopic studies of anticancer drug-target interactions in Spectroscopy of Biological molecules, Alix, Bernard and Manfait Eds, J. Wiley Press, 1985.

PULSED MULTICHANNEL RAMAN SPECTROSCOPIC INVESTIGATION OF THE SPATIAL DISTRIBUTION OF MOLECULAR SPECIES IN AN ENGINE.

D. Desenne, M. Crunelle-Cras, F. Grase, M. Bridoux,
J.P. Sawerysyn, L.R. Sochet**

Laboratoire de Spectrochimie Infrarouge et Raman, CNRS (LP. 2641)
Université des Sciences et Techniques de Lille Flandres Artois, Bât. C.5
59655 Villeneuve d'Ascq Cedex, France.
** Laboratoire de Cinétique et Chimie de la Combustion*
Université des Sciences et Techniques de Lille Flandres Artois, Bât. C.2
59655 Villeneuve d'Ascq Cedex, France.

For several years, different optical laser techniques have been used for "in situ" combustion diagnostics, especially, spontaneous and coherent anti-Stokes Raman scatterings (1-3). The application of multichannel pulsed Raman spectroscopy (4) for the determination of the spatial distribution of molecular species in a four-stroke commercial engine has been investigated (Fig. 1). The Raman spectral images obtained with single laser-shot excitation for various crank angles enable the spatial distribution of N_2, O_2, fuel, CO_2, CO and H_2O species to be visualized. The observations are made along 15 mm of the scattering path of the laser beam within the combustion chamber with a spatial resolution of about one mm. Strong heterogeneity of the analyzed volume, in relation to the turbulent propagation of the flame in the engine, was observed at crank angles between 381° and 392°.

A particular study of a Raman spectral image has been achieved (Fig. 2) by means of an electronic gate in order to probe the species located along the analyzed scattering path for a 385° crank angle. Four spatial regions have been analyzed. Regions 1 and 3 show the absence of the fresh mixture (C-H, O_2) whereas in regions 2 and 4, the vibrationnal bands of C-H and O_2 appear simultaneously. The CO, CO_2 and H_2O bands are always detected at a relatively low intensity. For CO and CO_2, this low intensity may be explained by the relatively low concentration of these species as demonstrated by the analysis of exhaust gases which are 8.8 % in CO_2 and 5 % in CO. Althought for H_2O the concentration (12 %) and the Raman cross section of the corresponding line are relatively important, the weak intensity of the spectral line is due to its location near the boarder of the field of view of the detector. The presence of CO, CO_2 and H_2O in all points of the analyzed volume is connected with the low compression ratio of the engine (= 0.8) which leads to a large residual gas fraction from the previous engine cycle. An additionnal line at 1105 cm^{-1} can also be noted which is attributed to an induced fluorescence of C_2 by the laser beam.

However, the weak intensities of the spectral lines obtained by means of a single laser excitation, even using an intensifier tube coupled to a television camera tube as a detection system do not enable accurate quantitative measurements of the concentration of the locally analyzed molecular species. Quantitative measurements may be obtained using an intensified photodiode array as a detector (1,4) if cycle-averaged measurements are acceptable.

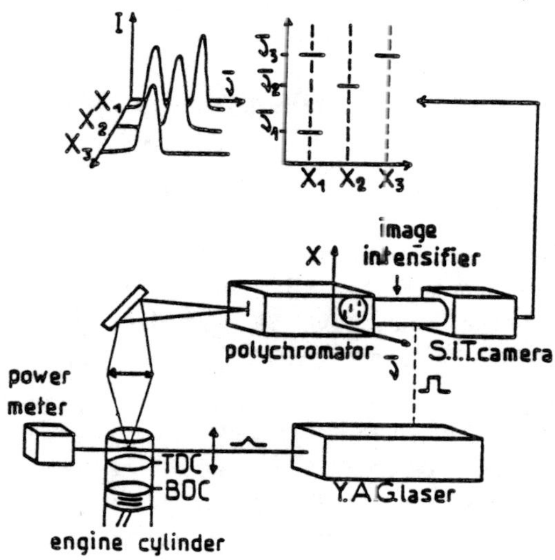

Fig. 1 : Schematic diagram of pulsed Raman multichannel spectroscopy arrangement used to probe heterogenous distribution of molecular species in an internal combustion engine.

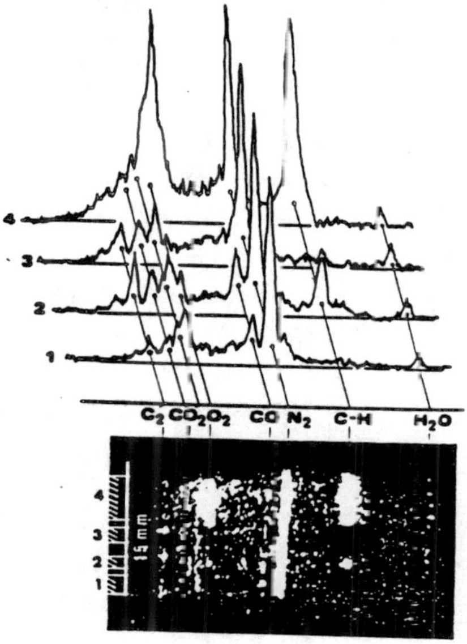

Fig. 2 : Analysis of a Raman spectral image obtained for a 385° crank angle

References :

1. Smith, J.R., Green, R.M., Westbrook, C.K, Pitz, W.J., Twenthieth Symposium (International) on combustion, 1984.
2. Eckbreth, A.C., Dobbs, G.M., Stufflebeam, J.H., and Tellex, P.A., Appl. Opt. 23, 1328, 1984.
3. Klick, D., Marko, K.A, and Rimai, L. Appl. Opt. 23, 1347, 1984.
4. Michael-Saade, R., Sawerysyn, J.P., Sochet, L.R., Buntinx, G., Crunelle-Cras, M., Grase, F., and Bridoux, M. Dynamics of flames and reactive systems (J.R. Bowen, N. Manson, A.K. Oppenheim and R.I. Soloukin, Ed). Progress in Astronautics and Aeronautics, vol. 95, p. 658, 1984.

REMOTE AND *IN SITU* RAMAN SPECTROSCOPY USING OPTICAL FIBRES.

NGUYEN QUY DAO, Michel JOUAN and Pascal PLAZA.

Laboratoire de Chimie et Physico Chimie Minérales
UA 441 CNRS, Ecole Centrale des Arts et Manufactures
Grande Voie des Vignes, F-92290 Chatenay-Malabry, France.

Remote and *in-situ* Raman spectroscopy has been realised using optical fibres to transmit laser and Raman diffusion light. With the equipement devised in our laboratory, we have demonstrated that very simple Raman sensors can be used to carry out qualitative and quantitative analyses of gaseous, liquid and solid samples (1-2). Figure 1 gives the example of the results obtained for a solution of glucose in water. Figure 2 shows the diagramme of a "mobile Raman sensor" for solids. The advantage of such a device is to enable the study of substances that cannot be moved or transported and even when their size is very great. For mixtures of solids, we also have developed a device to examine remote micro-Raman. The diagramme of a remote Raman image terminal is shown in figure 2 and an example is shown in figure 3 (3).

The application of such a method are numerous. Here are some precise examples :

- Sub-marine anti corrosion survey when cathodic protection is applied to the metallic structure (4).
- *In-situ* analyses of the composition of chemical, biological or siderurgical reactors. This allows information of the progress of the reaction to be obtained in real time and can be used as a method of analysis for piloting the process.
- Remote measurements of protected sites (nuclear plant, radioactive reactions, corrosive or hostile sites...)
- Continuous and *in situ* analysis in the pharmaceutical and food science industries.
- *In situ* examination of living cells and applications in bio-engineering.

- Expertise measurements of painting by great masters or examinations of archeological sites.

At the present time, studies are being carried out on the problems of miniaturisation and other specific Raman sensors.

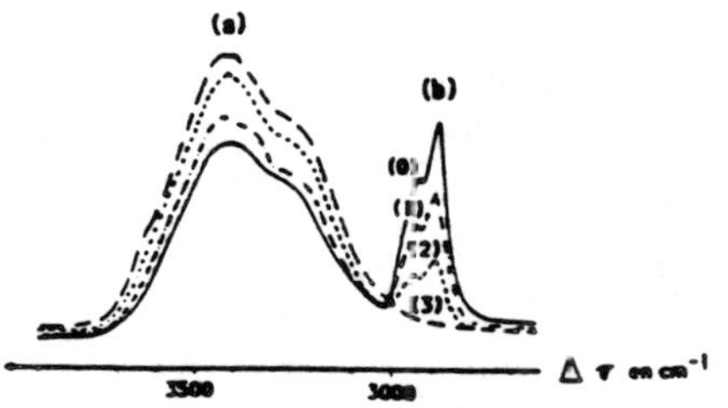

FIG. 1. — *Remote and* in situ *measurements of glucose in water* :
a) Characteristic band of water
b) Characteristic band of glucose
(0) (1) (2) (3) correspond to 0,1,2,3 mol·l^{-1} of glucose in water respectively.
Measurement characteristics :
Distance : 50 m. Duration : 3 min. Excitation : 5145 Å. Power : 600 mW. Fibre carrying excitation light, diameter : 600 µm. Fibre carrying Raman light, diamter : 600 µm. Resolution : 6 cm^{-1}. Scanning speed : 128 cm^{-1}/min.

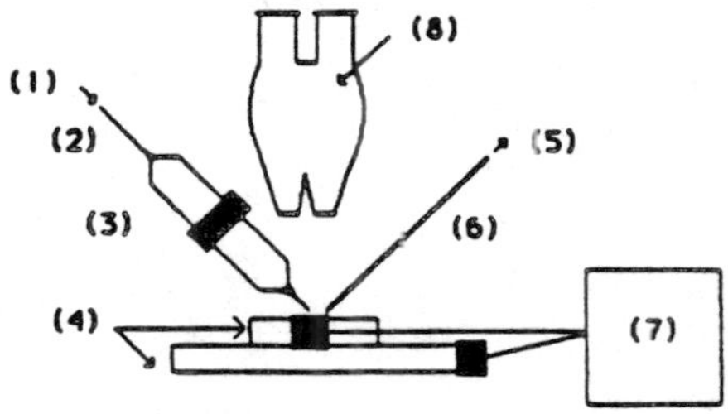

FIG. 2. — *Diagramme of a remote Raman image terminal* :
(1) Laser arrival
(2) "excitation" fibre, diameter 50 µm
(3) Microscope objectives
(4) Stepwise motors
(5) Towards the spectrometer
(6) "Raman" fibre, diameter 600 µm
(7) Control motors
(8) Optical microscope.

FIG. 3 — *Raman image of a* KNO_3 *and* $CaCO_3$ *mixture* :

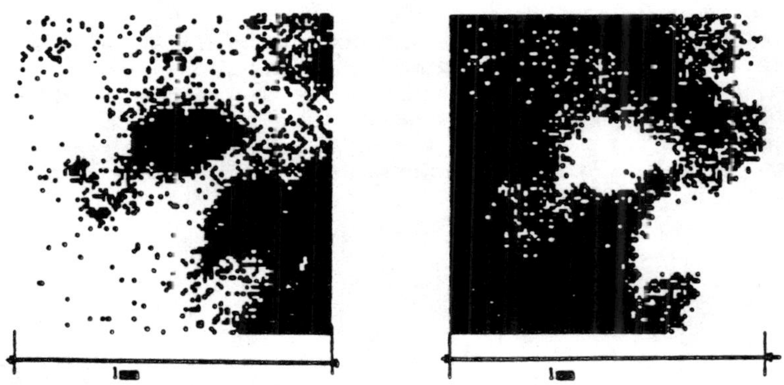

(a) Raman image obtained by setting the spectrometer at the frequency of KNO_3 at 1050.3 cm^{-1}. The compound KNO_3 appears in black on the diagramme.

(b) Raman image obtained by setting the spectrometer at the frequency of $CaCO_3$ at 1086.3 cm^{-1}. The compound $CaCO_3$ appears in black on the diagramme.

References

1. NGUYEN QUY DAO, P. PLAZA, European Conference on industrial line spectrographic analysis, Rouen (France) 19-21 June (1985).
2. NGUYEN QUY DAO, P. PLAZA, Analusis, 14 (1986).
3. NGUYEN QUY DAO, P. PLAZA, M. JOYEUX, Analusis, in press (1986).
4. NGUYEN QUY DAO, M. PROD'HOMME, P. PLAZA, M. JOYEUX, C.R. Acad. SC. Paris, 302, 313 (1986).

THE ASTM E 13.08 COMMITTEE ON RAMAN SPECTROSCOPY AND THE IUPAC COMMISSION 1.5 ON MOLECULAR SPECTROSCOPY - OVERVIEW OF ACTIVITIES.

Bernard J. Bulkin, Laura Powell, Jeanette G. Grasselli
for the members of ASTM E 13.08 and IUPAC

The usefulness of the Raman spectroscopic literature depends on the quality of data which are published. ASTM E 13.08 is a subcommittee of ASTM Committee E 13 (Molecular Spectroscopy) that has spent several years working on such questions as the accuracy of depolarization ratio measurements, definitions of terms for Raman spectroscopy, and, more recently, standard procedures for operation of a scanning Raman spectrometer. The draft format of this procedure will be available for review and comment at the ICORS.

IUPAC has concerned itself with standards for presentation of Raman spectroscopic data. Ongoing activities of IUPAC include recommendations for creating computerized data banks on molecular spectroscopy, recommendations for symbols and nomenclature in Raman spectroscopy, recommendations for the presentation of molecular parameter values for infrared and Raman intensity measurements, and recommendations on nomenclature and symbols for non-linear Raman spectroscopies. This paper reviews the previous work of the committees, discusses work in progress, as well as future directions.

RAMAN SPECTRA OF LIQUEFIED NOBLE GAS SOLUTIONS

G. Davidson and C.L. Davies

Department of Chemistry, University of Nottingham, University Park, Nottingham NG7 2RD, U.K.

Liquefied noble gases were proposed as suitable solvents for infrared spectroscopy by Maier *et al* [1], and have been used extensively by Turner *et al* (e.g. [2]). The advantages of such solvents, including extreme inertness and lack of any interfering solvent spectrum, prompted us to extend their application to Raman spectroscopy.

The experimental arrangements are very simple. The solution (liquid xenon was used as solvent in all the experiments to be reported) was held in sealed, thick-walled glass (only for work below -20°C) or fused silica tubes. These were surrounded by a glass dewar, through which as passed a controlled flow of cold N_2 gas. Temperatures in the range -150°C to 0°C could be maintained with a temperature variation of ± 10°C.

Results on two different systems are reported, to show the possibilities of the technique.

TRIFLUOROHALOMETHANES

Despite their widespread use as refrigerants and propellants, and in infrared photochemistry, published vibrational data on CF_3X, where X = Cl, Br or I, are seriously incomplete. We have therefore recorded their Raman spectra in solution in liquid xenon, which have given complete assignments of all their vibrational modes, together with polarisation measurements.

One problem in previous work has been that a number of bands were overlapping. This made accurate measurement of their wavenumbers very difficult. Figure 1 shows a part of the Raman spectra of pure liquid CF_3I and of a 4M solution in liquid Xe. This shows clearly the improved resolution achieved in the latter case, with ν_6 at 267 cm^{-1}, ν_3 at 286 cm^{-1}.

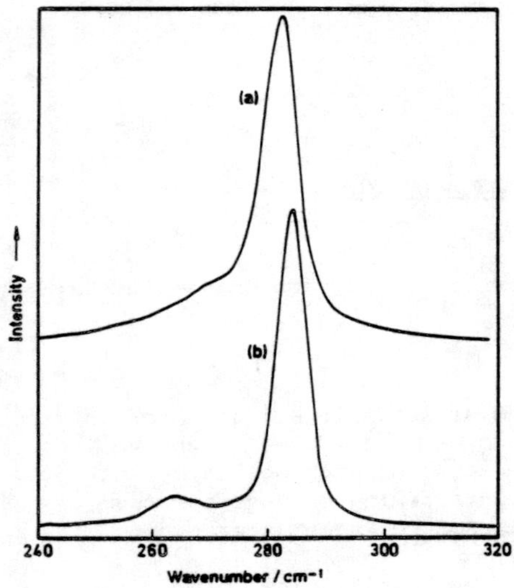

Fig. 1 Raman spectra of (a) pure liquid CF_3I, (b) 4M CF_3I in liquid Xe.

With a complete vibrational assignment for each molecule, including $^{35}Cl/^{37}Cl$ shifts for CF_3Cl, it has also been possible to carry out a reliable normal coordinate analysis for these molecules.

(BUTADIENE)IRON TRICARBONYL

The trifluorohalomethanes do not provide a very severe test of practical problems of using liquid xenon as a solvent for Raman spectroscopy, as they are extremely soluble. In order to test the system more rigorously, we attempted to obtain the Raman spectrum of (butadiene)iron tricarbonyl, $(C_4H_6)Fe(CO)_3$, for which very detailed (pure liquid) and infrared (liquid xenon solution) data are available [3].

Despite the low solubility of $(C_4H_6)Fe(CO)_3$ in liquid xenon, very clear spectra were obtained, and these confirmed earlier assignments. Indeed, the better resolution achieved in the present work made a more detailed assignment possible, e.g. the a' and a" CH stretches of the $CH_2CHCHCH_2$ unit were resolved for the first time (3009, 3017 cm^{-1}, cf. a broad band at 3014 cm^{-1} in the pure liquid).

CONCLUSIONS

We have shown that liquid noble gases can be very useful solvents for Raman spectroscopy, providing an inert and symmetrical environment, with no interfering spectra. A wide range of solutes can be studied, and very good spectra obtained.

REFERENCES

1. W.H. Beattie, W.B. Maier, R.F. Holland, S.M. Freund and B. Stewart, Proc. S.P.I.E. (Laser Spectroscopy), 158 (1978), 113.
2. R.K. Upmacis, G.E. Gadd, M. Poliakoff, M.B. Simpson, J.J. Turner, R. Whyman and A.F. Simpson, J.Chem.Soc., Chem.Comm., (1985), 27.
3. G.E. Gadd, PhD Thesis, University of Nottingham, (1985).

AUTHORS INDEX

AUTHORS INDEX

A

Abe, K. 11-63
Abenoza, M. 12-22
Abo Aly, M.M. 14-32
Abramczyk, H. 16-13
Adams, M.W. 1-50
Adil, A. 14-42
Agarwal, U.P. 14-46
Ahern, A.M. 5-22
Albrecht, A.C. P- 3, 17-12
Alden, M. 15-37
Alix, A.J.P. 1-60, 1-64
Alston, K. 1-15
Alves Marques, M. 16-48
Amorim da Costa, A.M. 11-53, 16-45
Anderson, A. 12- 9, 12-11
Anderson, T.J. 15-10
Andersson, L.A. 1-30
Andrews, B. 12- 9, 12-11
Anraku, Y. 1-17
Antic-Jovanovic, A. 16-49
Apanasevich, P.A. 19- 5
Arai, T. 14- 5
Arakawa, E.T. 10- 1
Armengaud, A. 12-22
Armstrong, R.S. 1-19
Aroca, R. 6-10
Arora, A.K. 7-15
Asher, S.A. 1- 2, 1-82, 20-22
Aslanyan, L.S. 19-13
Atalla, R.H. 14-46
Atkinson, G.F. 5-43
Atkinson, G.H. 1-42, 18-1
Attal-Tretout. B. 15-31

B

Babcock, G.T. 14-34
Bacal, M. 15-33
Badalyan, N.N. 19-13
Bafile, U. 16-11
Baglin, F.G. 16-15
Bajdor, K. 1-52, 2-11
Baker, C.A. 15-29
Baldwin, J.E. 17-24
Balkanski, M. 11- 3
Bangcharoenpaurpong, O. 1-23
Bansal, M.L. 7-23
Bansil, R. 16-39
Baranovic, G. 14- 7
Baranska, H. 13-27
Barbier, P. 11-11
Bariz, R. 14-28
Barocchi, F. 16- 1, 16-11
Barreau, A. 16-22
Barron, L.D. 20- 9
Bartholomew, C.R. 13-21
Bartholomew, D.U. 7-15
Becher, P.F. 11-48
Beer, K.D. 5-22
Begun, G.M. 11-48
Belhadj, M. 13- 3
Bellanato, J. 14-20
Benassi, P. 16-43
Benedek, G. 11- 3
Bengtsson, P.-E. 15-37
Benham, V. 20-26
Bera, A.M. 11-83
Berger, H. 15-18, 16-24
Berger, J. 12-19
Berger, T. 16-39
Berjot, M. 1-60, 1-64
Berquier, J.-M. 1-70
Berrue, J. 16-22
Bhattacharyya, B.D. 1-36
Bhoopathy, T.J. 1-87
Bidarian, A. 12-30
Bigio, I. 1-70
Bigotto, A. 12- 5
Binnie, N. 17-36
Bischel, W.K. 15-47
Bist, H.D. 11-40
Blanchard, D. 1-42, 18- 1
Bloembergen, N. P- 1
Blumenroder, S. 11- 1
Bolton, B.A. 1-66, 3- 5

Bonamy, J. 16-24
Boolchand, P. 11-81
Boquillon, J.P. 15-35
Bostick, J.M. 19- 7
Bouchaour, T. 4- 1
Boyer, H. 11-65
11-66
Boysan, F. 15-23
Bozler, B.J. 15-39
Brack, T.L. 1-42
18- 1
Bradley, E.B. 12-30
Brakel, R. 20- 4
Bree, A. 17-16
Breitinger, D.K. 13-15
Bremard, C. 14-28
20-27
Bridoux, M. 20-17
20-30
Briganti, G. 16-41
Brivio, G.P. 12-26
Brodbek, E. 16-33
Bron, W.E. 7- 7
Brotman, A.D. 5- 7
Brouwer, A.M. 18-16
Brown, M.S. 16-26
Buijs, H. 20-21
Bulkin, B.J. 12-21
20-34
Bulou, A. 11-21
Burstein, E. 5-10
Bush, S.F. 3- 1
Bussian, B.M. 1-62
Butler, I.S. 13-13
20-26
Byler, D.M. 1-58

C

Cable, J.R. 17-12
Cai, M.-Z. 4- 6
Calienni, J. 20- 8
Campbell, N.J. 13-10
Campion, A. 4- 5
5- 6
Cardona, M. 7-17
9- 5
Carmona, P. 2-13
2-15
4- 3
Carreira, L.A. 19- 7
Castaignos, N. 20-17
Cataliotti, R.S. 12-28
Cazzanelli, E. 11-34
Chadwick, R.R. 17-24
17-25
Champarnaud, J. 11-11
Champion, P.M. 1- 8
1-23
1-32
Chandrasekharan, V. 16-27
Chang, C.K. 1-30
Chang, R.K. 8- 7
Chase, B. 20-19
Chaux, R. 15-18
15-35
Chave, A. 16-22
16-25
Chen, S.-C. 1-21
Chen, Y.-L. 4- 6
Chencinski, N. 19-11
Chernoff, D.A. 18- 6
Chilingaryan, Yu.S. 19-13
Chinsky, L. 2- 5
Choi, B.-K. 11-69
Chou, Y.C. 5-12
Chronister, E.L. 12- 1
Chu, P. 5-64
Ciepielewski, P. 11-30
Clark, R.J.H. 13-25
Coalson, R.D. 17- 5
Colombo, L. 14- 7
Comins, J.D. 11-25
11-27
Cooke, R.M. 1-19
Cornard, J.P. 20-29
Corset, J. 5-50
Cottinet, D. 14-48
Cotton, T.M. 5-56
Couderc, P. 14-48
Couzi, M. 10-11
Cox, F.O. 14-13
Crunelle-Cras, M. 20-30
Czernuszewicz, R.S. 1-50
13-19

D

Da Silva, E. 20-29
Dallinger, R.F. 13- 1
Dao, N.Q. 20-32
Dasgupta, S. 1-11
Dasseux, J.-L. 3-11
Davidson, G. 20-35
Davies, C.L. 20-35
Davies, M. 20- 8
de Groot, J. 5-38
De Lorenzi, A. 16- 5

De Santis, A. 16- 4, 16-41, 16- 5
DeBlase, F. 12-21
DeMarco, C. 2-17
Debnath, R. 1-76
Delhaye, M. 13-31, 20-17, 20-29
Delichere, P. 5-45, 6-15
Dellepiane, G. 12-28
Demers, R.T. 11-15
Dendramis, A.L. 5-48
Deng, Z. 17-34
Dengel, A.C. 13-10
Desenne, D. 20-30
Destri, S. 12-28
Devanathan. T. 11-62
Devonshire, R. 15-23
Dhamlincourt, M.-C. 11-11, 11-12, 13-31
Dhamelincourt, P. 14-48, 20-29
Diem, M 20- 8, 20- 3
Dierker, S.B. 6- 8
Diller, R. 1-40
Dines, T.J. 5-26
Dlott, D.D. 12- 1
Dorfmuller, T. 11-71, 16-13
Dozov, I. 3-21
Dring, I. 15-23
Dumon, B. 16-22, 16-25
Dupeyrat, R. 6- 3
Durig, J.R. 11-40, 14- 1, 14- 7, 14-12, 14-13
Dutta, P.K. 2-20

E

Eckbreth, A.C. 15-10
Edner, H. 15-37
Edwards, H.G.M. 15-6, 15-8
Eickmans, J.H. 8- 7
Eklund, P.C. 10- 1
ElJoundi, A. 1-60
Enzweiler, R.N. 11-81
Escribano, J.R. 20- 9
Esherick, P. 15- 4
Eskola, S. 16-31
Etz, E.S. 11-67
Eysel, H.H. 1-62, 13-12

F

Falaras, P. 6-15
Fantoni, R. 15-14
Faria, D.L.A. 13-14
Farrar, M.R. 20- 1
Farrow, L.A. 11-85
Farrow, R.L. 15-12, 15-27
Faulques, E. 12-26
Favrot, J. 14-32
Feng, Q. 5-56
Fillaux, F. 11-77, 12-13
Findsen, E.W. 1-15, 1-34, 18-12
Fitchen, D.B. 12-24
Flaugh, P. 20-22
Flint, C.D. 13-25
Flytzanis, C. 11- 7
Fontana, A. 11-34
Fontana, M. 3-21
Fornari, B. 5-41
Frattini, R. 16- 4, 16- 5
Freedman, T.B. 20- 6
Friesner, R.A. 13- 3
Frommhold, L. 16-26
Fujii, Y. 12-31
Fujiki, F. 11-75
Fujimura, Y. 18-20
Fujiwara, M. 1-44
Fukuda, H. 5-34
Fukuhara, K. 12-17
Funkenbusch, E.F. 11-48
Furukawa, Y. 12-31
Fuso, S. 12-28

G

Gale, G.M. 11- 7
Galeener, F.L. 11-73
Gant, T.A. 9- 2, 9- 4
Gao, Y. 5-18

Gardiner, D.J. 11-52, 14- 9
Garrell, R.L. 5-22
Gburski, Z. 14-10
Geiger, L. 16-21
Gerolymatou, A. 5-16
Gerrity, D.P. 18-18
Gersonde, K. 1-13
Gertsen, J.I. 5- 9
Geyer, T.J. 14- 1
Ghomi, M. 2-21
Gilson, D.F.R. 20-26
Ginsberg, D.M. 11-15
Godbey, S.E. 14- 1
Godet, J.L. 12-19
Gopinath, C.R. 14-53, 14-55
Gorelik, V.S. 11-87, 11-88
Gough. K.M. 14-22
Grase, F. 20-30
Grasselli, J.G. 20-34
Graves, P.R. 5-58
Gredilla, E. 2-15
Greenhalgh, D.A. 15-23, 15-29
Gregora, I. 11-30
Griffith, W.P. 13-10
Griffiths, J.E. 11-79
Grimsditch, M. 9- 9
Groner, P. 14-12
Gu, R.-A. 5-52, 5-68
Gualtieri, G.J. 7-21
Guilment, J. 5-50
Gunnia, U. 20- 8
Guntherodt, G. 11- 1
Gustafson, T.L. 18- 6
Gustafson, W.G. 1-46, 1-47
Guyot, F. 11-65
Guzonas, D. 5-43

H

Hafner, W. 12- 3
Hafsi, S. 11-11
Hagopian, S. 20-22
Haley, L.V. 17-36
Hall, W.H. Jr. 1-48
Haller, K.E. 11-17
Hallmark, V.M. 14-40
Han, S. 1-50
Harada, I. 1-72, 12-31, 14- 5
Harhay, G. 20-11
Harrand, M. 6- 3
Harvey, P.D. 13-13
Hasegawa, H. 11-46
Hashiba, H. 13- 6
Hashimoto, H. 18-14
Hashimoto, S. 1- 6, 1-28
Hattori, T. 19- 9
Hayashi, H. 1-44
Hazle, M.A. 18- 6
He, L.T. 13-23
Hempleman, A.J. 13-25
Henderson, T. 9- 2
Heremans, K. 1-88
Heremans, L. 1-88
Hernandez, J.G. 11-81
Hernandez, L. 2-19
Herring, G.C. 15-47
Hess, W. 17-10
Hester, R.E. 1-52, 5-38, 18- 5
Heyns, A.M. 11-13
Higuchi, S. 11-75
Hildebrandt, P. 1-86, 5-28
Hill, J.R. 12- 1
Hillion, F. 15-33
Hiroshi, H. 1-24
Hiwatashi, A. 1-24
Hizer, T.J. 14-12
Hizhnyakov, V. 17- 1
Hnida, U.E. 16-33
Hochstrasser, R.M. 16- 9
Hoffman, D.M. 10- 1
Holmgren, D.J. 11-15
Hor, A.-M. 6-10
Hotta, T. 1-24
Houssin, D. 15-31
Hu, X.M. 20- 6
Huang, Y. 7- 9
Hudson, B. 1-84, 17-10, 17-24, 17-25, 17-26, 17-28, 17-29, 20-11

Hugot-Le Goff A. 5-45, 6-14, 6-15, 6-16
Huo, W.M. 15-47
Huong, P.V. 2-15, 4- 3
Hurst, W.S. 15-43, 15-49
Hutt, J. 2-20

I

Ichikawa, Y. 1-24
Idrissi, A. 16-25
Iizuka, T. 1-17
Ikawa, S.-I. 16-29
Ikezawa, Y. 5-54
Inoue, K. 7-19
Inoue, M. 8- 5
Irish, D.E. 5-43
Irwin, J.C. 10- 5
Isaka, H. 18-10
Ishida, H. 5-34, 11-50
Ishimura, Y. 1-17
Ishitani, A. 5-34, 11-50
Isono, N. 3-13
Itoh, K. 5-36, 14-18
Iwata, K. 1-44

J

Jacobs, H.J.C. 18-16
Jacquot, J. 1-64
Jagodzinski, P.W. 1-48
Jean, J.M. 13- 3
Jennings, C. 6-10
Jensen, N.-H. 18-16
Jensen, P.W. 13- 4
Jeremic, M. 16-46, 16-49
Jeziorowski, H. 5-46
Jha, S.S. 7-11
Jiang, J.D. 5-10
Jiang, T.L. 13-23
Johnson, M.K. 1-50
Johnson, R.D. 14-12
Jollie, D. 1-23
Jouan, M. 20-32
Jouanne, M. 11- 3
Jovanovic, A. 16-46

K

Kadaba, P.K. 12-30
Kalinski, Z. 11-30
Kanesaka, I. 13- 6
Kash, J.A. 7- 5
Katagiri, G. 5-34, 11-50
Kauschke, W. 7-17
Kawahara, H. 13- 6
Kawamoto, K. 7-19
Kelly, P.B. 17-28, 17-29, 20-11
Keough, K.M.W. 3- 7
Khilji, M.Y. 12- 7
Kiefer, W. 8- 3, 12- 3
Kim, J. 12-21
Kim, J.-J. 11-69
Kim, M. 5-36
Kimura, M. 17-18
Kincaid, J.R. 13-19, 20-15
Kinoshita, S. 14-38, 17-18
Kint, S. 3- 5
Kirov, N. 3-21
Kita, K. 1-17
Kitagawa, T. 1- 6, 1-28
Klein, M.V. 9- 1, 9- 2, 9- 4, 11-15
Klem, J. 9- 4
Knippers, W. 15-14
Kobayashi, H. 5-10, 17-20, 18- 8, 19- 9
Koglin, E. 5-20
Koningstein, J.A. 17-36
Kono, H. 18-20
Koo, J.-Y. 11-69
Koshihara, S.-Y. 18- 8
Koyama, Y. 18-14
Kozich, V.P. 19- 5
Krauzman, M. 12-19
Kress, W. 13-15
Krimm, S. 1-68, 3- 9
Kubasek, W.L. 2- 7
Kudo, S. 11-46
Kumaraswamy, K.C. 14-55

Kushida, T. 14-38, 17-18
Kvach, V.V. 19- 5

L

Laane, J. 14- 3, 16-37
Ladanyi, B.M. 16- 6, 16-21
Lafaut, J.P. 1-56
Lafleur, M. 3-11
Lagant, P. 1-78, 11-83
Laigle, A. 2- 5
Lalic, M. 16-46, 16-49
Lambert, B. 1-64
Lan, G. 11-36
Langkilde, F.W. 18-16
Langlais, M. 20-24
Larsen, R.A. 14-13
Laubereau, A. 19- 1
Laureyns, J. 20-27
Laurin, D.G. 12-11
Lavorel, B. 15-16, 15-18, 16-24
Le Calve, N. 12-13
Le Duff, Y. 16-25
LeFrant, S. 12-26
Leahey, J. 5-39
Lee, B.S. 2-20
Lee, K.-S. 11-69
Lee, O. 20- 8
Lee, S. 10- 3
Lee, S.A. 2-17
Lee, Y.G. 5-39
Lefebvre, M. 15-33
Lemon, G.H. 11-81
Lenain, B. 20-29
Leng, Y.-Z. 5-52, 5-68
Leonard, W. 10- 3
Lesieur, I. 4- 1
Letellier, R. 2-21
Levesque, D. 20-24
Levi, D.H. 9- 4
Levin, I.W. 3- 1, 3- 3
Levine, A.M. 19-11
Lewandowski, W. 13-27
Lewin, M. 12-21
Ley, L. 9- 8
Li, B. 11-36
Li, H.-L. 1-21
Li, S. 17-26, 20-11
Li, X.-Y. 13-19
Liang, N.T. 5-12
Liang, Y. 14-14
Lin, S.H. 18-20
Lindsay, S.M. 2-17
Ling, D. -H. 5-52, 5-68
Little, T.S. 11-40, 14- 1
Littleton, C.J. 11-52
Lively, C.R. 1-47
Loehr, T.M. 1-30
Long, D.A. 15-6, 15-8
Loo, B.H. 5-39
Lopez-Garriga, J.J. 14-34
Lopez-Rios, T. 5-18
Lord, G. 20-26
Lorosch, J. 4- 3
Lorriaux-Rubbens, A. 20-17
Lottici, P.P. 11-38
Loutfy, R.O. 6-10
Lu, H.M. 17-14
Luccas, M.H. 3-15
Lucht, R.P. 15-27
Ludwig, M. 1-82
Lugtenburg, J. 1-38
Lutz, M. 1- 4

M

Macor, K.A. 1-50
Maeda, S. 19- 3
Mahan, G.D. 10- 1, 11-44
Mahtani, H.K. 13- 8
Makino, R. 1-17
Mallick, P.K. 17-30
Malyj, M. 11-79
Mannfors, B. 14-24, 14-30
Marcinek, A. 16-13
Maroncelli, M. 15-41
Martin, N. 14-20
Marx, J. 1-60, 1-64
Marzocchi, M.P. 17-32
Marzouk, H.A. 11-29
Masaki, T. 11-50
Masson, M. 6- 3

Mathies, R.A. 1-38, 14-36
Mathur, M.S. 11-29, 12-30
Matsuura, E. 10-13
Matsuura, H. 12-17, 14-16
Matsuura, I. 13- 6
Mattei, G. 5-41
Mattioli, T.A. 17-36
May, A.D. 15-49
Mayne, L. 1-84
Mazzacurati, V. 16- 7, 16-43
McConnell, A.A. 13-21
McCubbin, I. 18- 5
McFarland, J.T. 1-46, 1-47
McHale, J.L. 16-35, 16-36
McKee, J.S.C. 11-29
McMahon, J.J. 6-12
McMullan, W.G. 10- 5
Meiering, E. 12- 9
Meinander, N. 16- 2, 16-37
Melendres, C.A. 6-12
Melinger, J.S. P- 3
Merlin, J.C. 14-28
Merriam, M.J. 16-35
Merten, L. 8- 9
Mestres, N. 7-17
Michaelian, K.H. 5-30
Millot, G. 15-18, 16-24
Miura, T. 1-72
Mizukami, H. 1-13
Mizuno, J. 11-46
Mohan, S. 1-87
Molina, M. 2-13
Moore, J.N. 18- 5
Morkoc, H. 9- 2, 9- 4
Moschallski, M. 16-33
Moser, B. 5-46
Moses, D.G. 16-15
Mukamel, S. 17-34
Mulazzi, E. 12-26
Murase, K. 7-19
Murphy, W.F. 14-22
Myers, A.B. 16- 9
Mylrajan,M. 11-61

N

Nafie, L.A. 20- 6
Nagaraj, B.S. 14-55
Nagata, H. 5-32
Naik, V.M. 3- 9
Nakashizu, T. 10-13
Natarajan, A. 14-53, 14-55
Nather, H. 7- 3
Nelson, K.A. 20- 1
Ngoepe, P.E. 11-25, 11-27
Ni, F. 5-56
Nibler, J.W. 15-39, 15-41
Nieh, J.-C. 18-18
Nielsen, O.F. 1-70
Nishihara, C. 5-24
Nishimura, Y. 1-17, 2- 3
Nitzan, A. 5- 9
Noel, H. 11-23
Noguchi, T. 1-44
Nomura, Y. 18-20
Nouet, J. 11-21
Nouvel, G. 11-23

O

Ogura, T. 1- 6
Ohzeki, T. 18-10
Okabayashi, H. 1-76
Okamoto, H. 1-44
Okuoka, T. 1-76
Olsen, I. 1-70
Ondrias, M.R. 1-15, 1-34, 18-12
Orlovich, V.A. 19- 5
Otto, A. 5- 1
Otto, K. 2- 5
Ouazzany, Y. 15-35
Ouillon, R. 16-19
Overlaet, W. 17-18
Owyoung, A. 15-4

P

Paatero, M. 16-31
Pacynko, W.F. 3-19
Paddy, J.L. 14-26
Paeng, K.-J. 20-15
Pagannone, M. 5-41

Page, J.B. 17- 3, 17-14
Palavit, G. 11-83
Paliani, G. 12-28
Palings, I. 1-38
Pallotta, C. 6-16
Palmer, J.F. 18- 6
Palmer, R.E. 15-25
Palmo, K. 14-24, 14-30
Paquette, J. 13-29
Parisini, A. 11-38
Parker, D.W. 17-24
Parry, D.B. 5-48
Parthasarathi, N. 1-26
Pascher, H. 9-13
Pasquier, B. 12-13
Patapoff, T.W. 2- 7, 20-13
Payami, F. 1-87
Peacock, R.D. 5-26
Pealat, M. 15-33
Person, W.B. 2-19
Peterson, D.L. 7-15
Peticolas. W.L. 1-52, 1-54, 1-78, 2- 7, 2- 9, 2-11, 20-13
Petroff, B. 3-21
Petrosyan, A.A. 19-13
Petrosyan, K.B. 11-32
Pettenkofer, C. 5- 1
Pettinger, B. 5-16, 5-17
Petway, L.B. 15-49
Pezolet, M. 3-11
Phaizullov, T.P. 11-88
Phillips, D. 18- 5
Phillips, D.L. 18-18
Pierre, G. 15-18
Pinczuk, A. 9-11
Pinnavaia, T.J. 10- 3
Plaza, P. 20-32
Pofahl, G. 11- 1
Pokhsraryan, K.M. 11-32
Postlewaite, J.C. 12- 1
Powell, J.W. 2-17
Powell, L.P. 4- 5, 20-34
Price, R.A. 1-48
Pubanz, G.A. 15-39, 15-41

Q

Qin, L. 18-22

R

Rabolt, J.F. 6- 7
Rafaeloff, R. 11-19
Rahn, L.A. 15-25
Ramdas, A.K. 7- 1, 7-15
Ramnarine, R. 1-74
Ranson, P. 16-19
Rao, C.R. 14-55
Rao, N.R. 11-55
Razzetti, C. 11-38
Reczek, C.M. 1- 9
Reedy, B.J. 1-19
Regis, A. 5-50
Reinisch, L. 1-32
Renucci, M.A. 11-23
Resina Rodrigues, J. 16-48
Reuss, J. 15-14
Rey-Lafon, M. 3-17
Ribeiro-Claro, P.J.A. 16-45
Ricci, M.A. 16-41
Rich, N.H. 3- 7
Richmond, G.L. 5-60, 5-62, 5-64, 5-66
Rifkind, J.M. 1-15
Riveccie, M. 5-50
Rizos, A.K. 1-23
Robert, B. 1- 4
Robert, D. 16-24
Roberts, D.M. 18- 6
Robinson, J.M. 5-66
Rocca, F. 11-34
Rodriguez, R. 16-36
Rodriguez, S. 7- 1
Rojhantalab, H.M. 5-62
Rosasco, G.J. 15-43, 15-49
Rosenblatt, G.M. 15-45
Rousseau, D.L. 1-11
Rousseau, M. 11-21
Roy, A.P. 7-23
Ruffolo, D. 11-81
Ruggiero, A. 20-11
Rumbles, G. 18- 1
Rundquist, P. 20-22
Ruocco, G. 16-41
Rupprecht, A. 2-17
Ruther, W. 6-12

S

Sagitov, S.I. 11-88
Saint-Loup, R. 15-18
Saito, H. 5-54
Sala, J.P. 16-24
Samios, D. 11-71, 16-13
Sampoli, M. 16- 4, 16- 5
Sandroff, C.J. 11-85
Santos, J.G.N. 13-14
Santos, P. 9- 8
Santos, P.S. 13-14
Sassaroli, M. 1-11
Sathyanarayana, D.N. 14-53
Savoie, R. 20-24
Sawerysyn, J.P. 20-30
Scagliotti, M. 11- 3
Schenk, K.J. 3-17
Scherer, J.R. 1-66, 3- 5
Schlotter, N.E. 6- 5
Schmid, E.D. 16-33
Schneider, F.W. 20- 4
Schomacker, K.T. 1- 8, 1-32
Schreiber, W.M. 19-11
Schrotter, H.W. 15-19
Schuler, R.H. 18-22
Schulz, V. 8- 9
Schwartz, G.P. 7-21
Seki, H. 5-14
Sekine, T. 10- 9, 10-13
Sension, R.J. 17-20, 17-22
Seoane, C. 14-20
Sequaris, J.-M. 5-20
Serita, M. 14-18
Shamir, J. 11-19
Shanker, R. 14-49, 14-51
Shannon, V.L. 5-60, 14-44
Sharma, R.M. 11-57, 11-59
Sharma, T.P. 11-57, 11-59
Shelnutt, J.A. 1-15, 18-12
Sherman, W.F. 1-74, 11-42, 12- 7
Sherwood, G. 15-6, 15-8
Shigenari, T. 11-63
Shih, L.B. 3-15
Shindo, H. 5-24
Shionoya, T. 11-63
Shiratori, Y. 14-18
Siemankowski, L. 1-42
Signorelli, G. 16- 7, 16-43
Simons, P. 1-34
Singh, I.S. 14-51
Singh, O.N. 14-51
Sitter, A.J. 1- 9
Sligar, S.G. 1-23
Smith, K.M. 1-30
Smith, S.O. 1-38
Smith, W.E. 13-21
Smulevich, G. 17-32
Smyth, K.C. 15-49
Snyder, R.G. 12-15, 14-40, 14-42, 14-44
Sochet, L.R. 20-30
Solin, S.A. 10- 3
Sood, A.K. 9- 5
Soulard, L. 11-77, 12-13
Sousa Oliveira, M.A. 16-48
Spencer, K.M. 20- 6
Spiro, T.G. 1- 1, 1-26, 1-50, 1-86, 13-19
Spolar, J.G. 10- 1
Srajer, V. 1- 8
Srinivasan, T.K.K. 11-61, 11-62
Stein, P. 13- 8
Stenman, F. 16-31
Stevenson, C. 3- 7
Stockburger, M. 1-40, 5-28
Stoicheff, B.P. P- 2
Stolen, R.H. 8- 1
Stolte, S. 15-14
Stolz, H. 18- 3
Strafford, K.N. 11-52
Strahan, G.D. 20-11
Strauss, H.L. 12-15, 14-40, 14-44, 17-20, 17-22

Subbaswamy, K.R. 11-44
Sue, J. 17-34
Suh, J.S. 5-30
Sullivan, J.F. 14- 1
Sunder, S. 13-29
Sunder, W.A. 7-21
Susi, H. 1-58
Susnow, R.G. 7-21
Suzuki, S. 18-10
Swalen, J.D. 6- 1
Szczepaniak, K. 2-19
Szczesniak, M. 2-19

T

Tabisz, G.C. 16- 2
Tabor, K.T. 18-18
Tadayyoni, M.-A. 5-22
Taga, K. 1-76
Taillandier, E. 2-21
Takahashi, H. 18-10
Takenaka, T. 3-13
Takeuchi, H. 1-72
14- 5
Talwar, D.N. 7-13
Tamaoki, H. 14-16
Tamaoki, M. 14-18
Tanaka, S. 11-75
Tao, M.J. 2-17
Taran, J.P. 15-1
15-31
15-33
Tasumi, M. 1-44
Tatsuno, Y. 1-28
Tazawa, S. 12-31
Teixeira-Dias, J.J.C. 16-45
Teramura, M. 11-50
Terner, J. 1- 9
Thamann, T.J. 1-80
Thibeau, M. 16-22
16-25
Thibodeau, D.L. 17-36
Thoai, P.V. 13-17
Thomas, G.A. 1-78
2- 7
20-13
Thomas, G.J., Jr. 2- 1
Thomas, K.M. 11-52
Thompson, H.M. 1-13
Thurn, R. 8- 3
Tian, Y. 14-14
Tiddy, G.J.T. 3-19
Toba, K. 5-32
Toda, G. 5-54
Toftlund, H. 13- 4
Tomita, S. 5-34
Torrie, B.H. 12- 9
12-11
Toupry, N. 12-19
Travis, J.C. 11-67
Trebino, R. 15-12
Triggs, N.E. 15-39
Tripathi, G.N.R. 18-22
Trulson, M.O. 14-36
Tsang, J.C. 7- 5
Tse, W.S. 5-12
11- 9
Tsubaki, M. 1-24
Tsuboi, M. 1-17
Tubino, R. 12-28
Turpin, P.Y. 2- 5
Turrell, G. 11-12
14-42
20-27
Turrell, S. 4- 1
11-11
11-12

U

Uchinokura, K. 10-13
Ulivi, L. 16-11
Umemura, J. 3-13
Uno, T. 1-17
Uzan, E. 16-27

V

Valentini, J.J. 15- 3
18-18
Vallauri, R. 16- 4
Vallee, F. 11- 7
Van Dael, H. 1-56
van Helvoort, K. 15-14
Vandevyver, M. 7-13
Vast, P. 11-83
Veirs, D.K. 15-45
Velde, B. 11-66
Vergoten, G. 1-78
11-83
14-42
Vincent, J.S. 3- 3
Vogt, H. 11- 5
von der Osten, W. 18- 3
Vorlicek, V. 11-30

W

Wada, N. 10- 7
Wahegoankar, L.D. 11-55
Walker, N.A. 14- 9
Wallart, F. 13-31
20-17
Walsh, M.J. 1-19
Wang, H. 11-36
Wang, S.-K. 16-26
Wang, W. 17-36
Wang, Y. 2- 7
2- 9
2-11
Warner, R.J. 3-15
Warshel, A. 17- 7
Washinger, G.T. 20-22
Watanabe, J. 14-38
Watanabe, N. 1-72
Waters, D.N. 14-26
Weidman, D.L. 12-24
Weismann, R.B. 15-21
Weiszmann, A.N. 19-11
Whalley, E. 16-29
Wharton, C.W. 1-52
White, J.K. 12-11
Wilbrandt, R. 18-16
Wilkinson, G.R. 11-42
12- 7
Williams, D.R. 15-29
Williams, L.R. 20- 1
Wilson, J. 1-54
Wren, J.C. 13-29
Wright, P.E. 1-19

X

Xu, Y. 11-42

Y

Yadav, R.A. 14-49
14-51
Yahia, Z. 11-12
Yamada, H. 5-32
Yan, Y. 17-34
Yang, J.J. 15-39
Yang, X. 11-36
Yarwood, J. 3-19
Yoshida, K. 5-34
Yu, N. 5-45
Yu, N.-T. 1-13
1-24
Yu, P.Y. 7- 9
Yur, G. 11- 9

Z

Zerda, T.W. 16-17
Zeyher, R. 5- 3
Zhang, J.M. 10- 1
Zhang, S.-L. 9- 4
Zhao, W. 14-14
Zhou, Q. 1- 4
Zhu, H.-R. 5-52
5-68
Ziegler, L.D. 17- 8
17-10
Zimba, C.G. 20- 6
Zirngiebl, E. 11- 1
Zoppi, M. 16-11
Zucker, J.E. 9- 7
Zung, W.W.E. 12-11
Zwick, A. 11-23